Dynamical Gauge Symmetry Breaking

A Collection of Reprints

Edited by

E. Farhi and R. Jackiw

World Scientific

World Scientific Publishing Co Pte Ltd
P O Box 128
Farrer Road
Singapore 9128

ISBN 9971-950-24-3
9971-950-25-1 pbk

 This volume consists of selected papers that have appeared in *JETP Letters, Nuclear Physics B, Physics Letters, Physical Review* and *Physical Review Letters*. The editors and publisher are indebted to the original authors, journals and other publishers for their assistance and permission to reproduce these papers.

Cover design Jee Cheng
Printed by Singapore National Printers (Pte) Ltd

PREFACE

Spontaneous breaking of symmetries allows symmetric theories to describe asymmetric reality. This idea has therefore attracted physicists from the time of its discovery a quarter century ago. In our reprint collection, we trace research which explores the possibility that the gauge symmetries describing fundamental interactions of elementary particles are spontaneously broken by dynamical mechanisms.

In the Introduction, we provide a guide, review and critique of the subject. The first articles in our book deal with the original suggestions that dynamical gauge symmetry breaking, familiar from many-body physics, may be relevant to particle physics. Next follow papers on field theoretical aspects of the subject. While no phenomenological applications are attempted in these investigations from the early 1970's, the theoretical possibilities are established. In the late 1970's, physicists tried to use these ideas for realistic model building, and we include articles on such attempts and on their phenomenological/experimental implications. The final reprints concern the many unresolved problems, discuss criteria for symmetry breaking, and examine possibilities that unified electro-weak models need not involve the breakdown of any symmetry.

It cannot be said that the program utilizing dynamical symmetry breaking for model building has succeeded. Not only has no realistic model been constructed, but also it appears that any physically acceptable theory would be inordinately complicated, involving many as yet unobserved particles and interactions. At the same time, research has slowed, and there is evident need for new ideas. We hope that our partial collection of what has been done will aid in finding what should next be done.

E. Farhi
R. Jackiw

Cambridge, 1982

CONTENTS

INTRODUCTION

Two reasons explain why spontaneously broken symmetries have been used by theoretical physicists for building models of fundamental particle interactions. First, it is satisfying to suppose that the ultimate laws of Nature enjoy a high degree of symmetry, and that observed asymmetries arise because symmetric quantum field theory equations can have solutions that do not respect the symmetries. In this way the occurrence and also, one hopes, the amount of symmetry breaking can be predicted. Second, as a practical matter, it is important that theories with symmetry are less afflicted with ultraviolet divergences, and this is not lost when the symmetry is spontaneously broken. As a consequence, spontaneous symmetry breaking is an essential component of current models for strong, weak and electromagnetic interactions, as well as of speculative extensions of Einstein's gravity theory.

That a ground state of a quantal system need not possess the Hamiltonian's symmetries, and is therefore degenerate, was first appreciated in non-relativistic many-body physics and is realized in many condensed matter situations. Heisenberg,[1] Goldstone[2] and Nambu[3] made the seminal suggestion that this may also be true for the vacuum state of a relativistic quantum field theory, and that massless excitations — "Goldstone" particles — will necessarily be present. The suggestion came to fruition in the modern theory of chiral symmetry breaking, where the pion is identified as the [approximate] Goldstone Boson.

However, it was soon realized by many people,[4] that there are exceptions to Goldstone's theorem,[2] which is applicable to spontaneous symmetry breaking only when it is unaffected by gauge invariance, *i.e.* when a global symmetry is broken. In the presence of gauge invariant long-range forces, the "Higgs phenomenon"[4] replaces that of Goldstone; massless vector mesons acquire a mass, and the Goldstone particle disappears from the spectrum by combining with the transverse meson to give the third helicity state of a massive spin-1 particle. In many-body physics, this was appreciated earlier in connection with BCS theory,[5] and suggestions were made that such ideas are applicable to relativistic quantum field theory, as is seen from papers [1, 2, 5]. Whether one should describe the Higgs effect as spontaneously violating a gauge symmetry is still debated, but there is no doubt that the association of a massless vector meson with a gauge invariant theory is not essential.

The Higgs mechanism became phenomenologically important after Weinberg and Salam incorporated it in their unified model of electro-weak interactions,[6] where the massive vector mesons are identified with carriers of the weak force. Spontaneous symmetry breaking is achieved by postulating the existence of a "Higgs sector" in the theory. This sector is populated by scalar fields which acquire symmetry violating vacuum expectation values by virtue of the shape of a postulated scalar-field potential, either in the classical[4] or semi-classical[7] approximation. The Goldstone mechanism gives rise to a massless field, which then is shown to decouple from physical processes, since a gauge transformation unites it with the vector meson field.

While the Weinberg-Salam model is recognized to be a theoretical and experimental success, it is frequently believed that the Higgs mechanism, as described above, is an unsatisfactory feature of the theory. There is no experimental evidence for fundamental scalar fields, which are introduced in an *ad hoc* manner with *ad hoc* interactions solely to effect the symmetry breakdown. There is no other compelling theoretical reason for scalar fields [save supersymmetry, for which there is no experimental evidence]; indeed there are theoretical obstacles: asymptotic freedom may be easily spoiled,[8] there may be difficulties with quantum gravity,[9] and there is the "fine tuning" problem about which more will be said later. Fortunately, we may take an alternate route to massive, yet gauge invariant vector mesons.

Independent of the above development and even before discussions of the Higgs mechanism, Schwinger observed that gauge invariance need not preclude a mass for the gauge field, if the vacuum polarization tensor, $\Pi^{\mu\nu}(p)$, possesses a pole at light-like momenta, $p^2 = 0$. [3] The argument is straightforward. Let j^μ be the current which is the source for the gauge field $F^{\mu\nu}$.

$$\partial_\mu F^{\mu\nu} = j^\nu \tag{1}$$

$$F_{\mu\nu} = \partial_\mu A_\nu - \partial_\nu A_\mu\,, \qquad \partial_\mu j^\mu = 0$$

The vacuum polarization tensor $\Pi^{\mu\nu}$ is the irreducible contribution to the current correlation function, which is necessarily transverse.

$$\Pi^{\mu\nu}(p) = (g^{\mu\nu}p^2 - p^\mu p^\nu)\Pi(p^2) \tag{2}$$

$$\int dx\, e^{ip\cdot x}\langle 0|Tj^\mu(x)j^\nu(0)|0\rangle = -i(g^{\mu\nu}p^2 - p^\mu p^\nu)\frac{\Pi(p^2)}{1-\Pi(p^2)}$$

It is clear that if $\Pi(p^2)$ has a pole at $p^2 = 0$ with positive residue μ^2, then the two-current correlation function and the gauge field propagator have a pole at $p^2 = \mu^2$. Since in most physical systems bound states are expected to exist and to produce poles at time-like momenta in $\Pi^{\mu\nu}$, one may suppose that for sufficiently strong binding, the mass of such a bound state will be reduced to zero, thus generating a mass for the vector meson without upsetting gauge invariance. Schwinger demonstrated his general ideas in 2-dimensional, massless spinor electrodynamics, which is explicitly solvable. $\Pi^{\mu\nu}$ does indeed have a pole at light-like momenta and the "photon" acquires a mass. [4]

Schwinger's approach has the advantage of making no reference to scalar fields, though the reason for a pole in the vacuum polarization tensor is unspecified. In the 2-dimensional example, the pole occurs "dynamically"' without scalar fields; although in the end, after solving the model, it is found that the dynamics are trivial — there are no interactions between the massive vector mesons. The Higgs mechanism is another particular example of the more general Schwinger mechanism. While it is not usually phrased in this way, the Higgs phenomenon can also be seen producing light-like poles in the vacuum polarization tensor, as a consequence of the scalar fields' vacuum expectation value;[10] *i.e.* the poles are due to the Goldstone particles, which in the end decouple from physical processes.

Gauge invariant but massive vector fields can be described in yet another way, which is closer to ideas of many-body theory. An immediate consequence of Eq. (1) [in any number of dimensions] is a second order equation,

$$\Box F_{\mu\nu} = \partial_\mu j_\nu - \partial_\nu j_\mu\,. \tag{3}$$

If the gauge field $F_{\mu\nu}$ is massive, it must satisfy an equation where the d'Alembertian is supplemented by a mass term:

$$(\Box + \mu^2)F_{\mu\nu} = \text{non-linear and non-local terms.} \tag{4}$$

A comparison with (3) shows that the curl of the current must be proportional to the field strength, in order that the gauge field acquires a mass.

$$\partial_\mu j_\nu - \partial_\nu j_\mu = -\mu^2 F_{\mu\nu} + \cdots\,. \tag{5}$$

The dots stand for non-linear and non-local terms. Equivalently, it must be true that the current j_μ satisfies a London *Ansatz*:

$$j_\mu = -\mu^2 A_\mu + \partial_\mu \Theta + \cdots \tag{6}$$

Here Θ is a gauge function which balances the gauge non-invariance of A^μ to produce a gauge invariant combination.

The question now arises whether the current-gauge potential proportionality of Eq. (6) is an operator identity, or whether it is true only for the solutions of the theory. In the Higgs mechanism it is the former. The current is constructed from scalar fields; in an Abelian example we have

$$j_\mu = ie\phi^*(\partial_\mu + ieA_\mu)\phi - ie\phi(\partial_\mu - ieA_\mu)\phi^*. \tag{7a}$$

When ϕ has a vacuum expectation value λ, Eq. (7a) also reads

$$j_\mu = -2e^2|\lambda|^2 A_\mu + \cdots, \tag{7b}$$

which verifies (6) with $\mu = \sqrt{2}|e\lambda|$.

For dynamical mass generation, the London *Ansatz* is established only after solving the theory. For example, in the 2-dimensional Schwinger model there is no *a priori* connection between the photon field and the current, which is bilinear in the massless Dirac fields ψ;

$$j_\mu = -e\bar{\psi}\gamma_\mu\psi. \tag{8}$$

However, by virtue of special properties of 2-dimensional Dirac algebra,

$$\begin{aligned} \varepsilon^{\mu\nu}\gamma_\nu &= i\gamma^\mu\gamma_5, \\ \gamma_5 &= i\gamma^0\gamma^1, \end{aligned} \tag{9}$$

the curl of the vector current is also the divergence of the axial vector current,

$$\begin{aligned} \partial_\mu j_\nu - \partial_\nu j_\mu &= -\varepsilon_{\mu\nu}\varepsilon^{\alpha\beta}\,\partial_\alpha j_\beta = \varepsilon_{\mu\nu}\,\partial_\alpha j_5^\alpha, \\ j_5^\alpha &= e\bar{\psi}i\gamma^\alpha\gamma_5\psi. \end{aligned} \tag{10a}$$

Moreover, just as in four dimensions,[11] the axial vector current is not conserved, even though the Fermions are massless; rather, there is an anomaly.[12]

$$\partial_\alpha j_5^\alpha = \frac{e^2}{2\pi}\,\varepsilon^{\mu\nu}F_{\mu\nu}. \tag{10b}$$

Thus, the London *Ansatz* is exact here,

$$\partial_\mu j_\nu - \partial_\nu j_\mu = -\frac{e^2}{\pi}\,F_{\mu\nu}, \tag{10c}$$

but the field-current identity emerges only after the model has been solved.[13]

Another example, in three dimensions, demonstrates the emergence of the London *Ansatz* dynamically, even though the theory is linear. In 3-dimensional space-time it is possible to introduce a gauge invariant mass for the gauge field.[14] The Lagrangian in an Abelian theory is

$$\mathcal{L} = -\frac{1}{4}\,F^{\mu\nu}F_{\mu\nu} + \frac{\mu}{4}\,\varepsilon^{\mu\nu\alpha}F_{\mu\nu}A_\alpha. \tag{11}$$

The gauge invariant field equation

$$\partial_\mu F^{\mu\nu} + \frac{\mu}{2}\,\varepsilon^{\nu\alpha\beta}F_{\alpha\beta} = 0, \tag{12}$$

may be presented in the form (1) by identifying the second term in (12) with a "current":

$$j^\mu = -\frac{\mu}{2}\,\varepsilon^{\mu\alpha\beta}F_{\alpha\beta}. \tag{13a}$$

The curl of the above is evaluated with the help of (12).

$$\begin{aligned}\partial_\mu j_\nu - \partial_\nu j_\mu &= \varepsilon_{\mu\nu\alpha}\varepsilon^{\alpha\beta\gamma}\,\partial_\beta\left(-\frac{\mu}{2}\,\varepsilon_{\gamma\delta\varepsilon}F^{\delta\varepsilon}\right)\\ &= \mu\varepsilon_{\mu\nu\alpha}\,\partial_\beta F^{\beta\alpha} = -\mu^2 F_{\mu\nu}.\end{aligned} \tag{13b}$$

The London *Ansatz* is again observed, but only for solutions to the equations, and not as an identity. [This model can be extended to a non-Abelian theory.][14]

For a truly general dynamical model in four dimensions, Schwinger's viewpoint and the London *Ansatz* may be reconciled in the following way. [5] To effect the derivation, we introduce an external gauge potential A_e^μ, coupled to the dynamical current j_μ, and compute the latter's vacuum expectation value. In momentum space, the formula for the induced current is

$$\langle j_\mu(p)\rangle = -(g_{\mu\nu}p^2 - p_\mu p_\nu)\frac{\Pi(p^2)}{1-\Pi(p^2)}A_e^\nu(p). \tag{14a}$$

We adopt the Landau gauge, and conclude that (14a) corresponds to an induced potential.

$$\langle A^\mu(p)\rangle = \frac{\Pi(p^2)}{1-\Pi(p^2)}A_e^\nu(p). \tag{14b}$$

The total potential

$$A_T^\mu = \langle A^\mu\rangle + A_e^\mu \tag{15a}$$

is given by

$$A_T^\mu = \frac{1}{1-\Pi(p^2)}A_e^\mu. \tag{15b}$$

Consequently, upon comparing (14a) with (15b), we see that

$$\langle j^\mu(p)\rangle = -p^2\Pi(p^2)A_T^\mu(p). \tag{16}$$

When II (p^2) has a pole at $p^2 = 0$, the current is indeed proportional to the total potential, for light-like momenta.

To be sure, in a theory with interactions, the above general argument is considerably less sharp than those in the dynamically trivial 2- and 3-dimensional examples, whose simplicity derives presumably from the absence of interactions. Nevertheless, it is interesting to identify a property of the two lower dimensional models, which has thus far not been associated with dynamical mass generation for 4-dimensional gauge fields. In both cases, the emergence of massive vector fields is due to mathematical/topological structures that can be constructed in gauge theories: in two dimensions, the mass arises because the axial vector current has an anomalous divergence proportional to $(1/2\pi)\varepsilon^{\mu\nu}F_{\mu\nu}$, which is the 2-dimensional Pontryagin density; in three dimensions, the action for the mass term is proportional to

$\int d^3x \varepsilon^{\mu\nu\alpha} F_{\mu\nu} A_\alpha$, which is the 3-dimensional Chern-Simons secondary characteristic class.[15]

Schwinger's [or London's] idea, that a gauge invariant mass is equivalent to a light-like pole in the momentum space vacuum polarization, can be useful for model building only if assurance can be given that such a pole is indeed present. [Realistic models cannot be solved!] As already mentioned, the Higgs mechanism with scalar fields guarantees this in the classical or semi-classical approximation. But one wants to dispense with scalar fields, and to this end, the suggestion was made in papers [6–8] that the *dynamical* Goldstone mechanism be combined with a gauge theory to give *dynamical* gauge invariant masses to vector mesons. The Goldstone phenomenon would create the pole in $\Pi^{\mu\nu}$ and gauge invariance would ensure that no physical massless particles are present.

Although, most frequently, Goldstone's theorem is exemplified by a scalar field acquiring a symmetry breaking vacuum expectation, Nambu in his initial papers on the subject[3] did not use scalar fields. Rather, the symmetry breaking order parameter is a vacuum expectation value of a Fermion bilinear — analogous to the pair condensates of many-body theory — and the Goldstone Boson is a Fermion anti-Fermion bound state. Moreover, the pion — the only physical candidate for a Goldstone Boson — is not an elementary excitation of a fundamental scalar field; rather, it is formed by quark anti-quark binding. Thus, only a dynamical Goldstone Boson is relevant to chiral symmetry breaking, and it is plausible that a similar dynamical effect can replace the elementary scalar fields of the conventional Higgs method.

It remained to be shown that the massless bound state decouples, as in the Higgs model. This was readily done for various Abelian theories in [6–8] and for a non-Abelian theory in [9]. These investigations establish that 4-dimensional gauge fields can acquire masses without violating gauge invariance, and without introducing scalar fields; apparently dynamical gauge symmetry breaking works exactly the same way as in the Higgs model with scalar fields.

Yet it is curious that there remains in lacuna in our understanding of the dynamical mechanism. In the Higgs model, one useful description of symmetry breaking employs the "unitary gauge."[16] Conceptually, this may be the most satisfactory approach since the Goldstone Boson never appears, and no symmetry is broken. Indeed it is also true that the "unitary gauge" is not even a choice of gauge; rather, for any gauge choice, one may [functionally] integrate the gauge degrees of freedom, leaving a unitary Lagrangian whose fields correspond to physical degrees of freedom.[17] Such a description has not been given in the dynamical framework. Indeed much effort is expended on attempts to establish that massless bound states arise from the dynamics; yet in the end these states are not in the physical spectrum. One should expect that questions about binding of decoupled Goldstone particles must be irrelevant to gauge meson mass generation.

With the exception of an important result about the vector meson mass, to be mentioned below, phenomenological applications are not attempted in the early papers; the models studied there are too simple to be physically relevant. Rather, discussion concerns whether a zero mass bound state does indeed occur. A formalism for studying the vacuum expectation value of composite operators is developed in [10], which is based on similar approaches to many-body theory.[18] However, no definite conclusions could be reached, owing to well-known difficulties of arriving at non-perturbative results in quantum field theory. Nevertheless, some useful qualitative observations can be made.[19]

The existence of any bound state, and in particular of a zero-mass bound state, can be examined by a homogeneous integral equation [*e.g.* a Bethe-Salpeter equation],

$$P(p) = g^2 \int dk\, K(p,k)P(k) \tag{17}$$

where P is an amplitude describing the bound state, K the [Bethe-Salpeter] kernel, and g the coupling constant. For (17) to have a non-trivial solution, it is necessary that in some heuristic sense $g^2 \int K = 1$. In general, if the kernel is Fredholm, this implies a quantization condition on the coupling constant; no bound state will exist for arbitrarily weak coupling. However, in field theory the kernel is not likely to be Fredholm; the integral $\int K$ probably diverges, even though $\int KP$ converges. The growth of the kernel may be in the ultraviolet region or in the infrared; *viz*. the bound state may exist because of ultraviolet or infrared instabilities of the theory. [Equivalently, one can absorb the coupling constant in the kernel, and describe the coupling "constant" as growing in the ultraviolet region, or in the infrared.] In this circumstance, (17) will always have a non-trivial solution, and no eigenvalue condition is required.

In the early investigations [6–8, 10], ultraviolet unstable [asymptotically non-free] theories were considered, and some evidence for a bound state, arising from the ultraviolet instability was found. Today, with dynamics provided by non-Abelian gauge theories, one expects that a bound state will arise from the strong infrared forces.

The distinction between symmetry breaking arising from the ultraviolet or from the infrared has important bearing on symmetry restoring phase transitions at high temperatures, as in the early universe. It is now widely appreciated that the Higgs mechanism is undone at some finite temperature.[20] Finite temperature modifies a field theory in the infrared region, hence a dynamically broken symmetry can be restored only if the forces that bring about the breaking are due to infrared instabilities. On the other hand, ultraviolet dominated dynamical breaking should not be influenced by temperature. This is exemplified in the Schwinger model: the axial vector anomaly — an ultraviolet effect—is here responsible for mass generation. Correspondingly, the massive photon persists in the Schwinger model at all temperatures.[21] Since symmetry breaking in realistic physical models is expected to be an infrared-dominated phenomenon, we may expect transitions to phases with different symmetries as the temperature is varied.

The application of dynamical gauge symmetry breaking to the electro-weak interaction begins with the observation that the relationship between the vector meson M, and the symmetry breaking parameters is the same regardless whether scalar fields or Fermion condensates are responsible for symmetry breaking. [6–8]

$$M = |gF| \tag{18}$$

Here g is the gauge coupling constant and F describes the coupling of the massless Goldstone Boson to the broken gauge current. [In theories with scalar fields, F also equals the vacuum expectation value of the scalar field.] For these vector mesons to mediate the weak interactions, g^2/M^2 must equal the Fermi constant $G_F \approx 10^{-5}$ $(\mathrm{GeV})^{-2}$, so $|F| \approx 300\,\mathrm{GeV}$, which is about 3000 times larger than $f_\pi \approx 100$ Mev—the chiral symmetry breaking parameter of ordinary hadronic physics. Therefore, the characteristic scale of interactions responsible for a possible dynamical breaking of electro-weak symmetry should be about 3000 times the quantum chromodynamical (QCD) scale.

It is very instructive to see how QCD acting alone would break the weak interaction gauge symmetries.[22] Consider QCD with two quarks, u and d, and imagine that in the Lagrangian the quark mass terms are zero so that the Lagrangian has an exact $SU(2)_{L(\mathrm{EFT})} \times SU(2)_{R(\mathrm{IGHT})}$ symmetry. [There are additional U(1) symmetries which do not concern us here.] It is commonly supposed that this theory spontaneously breaks its chiral symmetries and $SU(2)_L \times SU(2)_R$ is reduced to SU(2), resulting in three massless Goldstone Bosons — the pions. [In the real world the pions are not strictly massless because there are small quark mass terms in the Lagrangian, so the symmetries are not exact.] Now consider the effect of turning on the standard $SU(2)_L \times U(1)$ electro-weak gauge interactions and ignore all other sources of symmetry breaking. The electro-weak gauge Bosons couple to generators of the $SU(2)_L \times SU(2)_R$ symmetry group which are spontaneously broken. The broken generators correspond to currents which couple to the pions with strength f_π. Therefore, the electro-weak gauge Bosons get masses gf_π and the pions are now in the spectrum only as longitudinal components of massive vector Bosons.

Obviously, this cannot be a picture of the real world. As discussed before, the scale is off by a factor of 3000 and physical pions are after all observed. [However, regardless of the source for weak interaction symmetry breaking, the Goldstone Boson which is the longitudinal component of the weak interaction vector meson has a small $\approx f_\pi/300$ GeV admixture of the QCD pion, and the physical pion has a small admixture of the weak-interaction Goldstone Boson.]

Weinberg [11] speculated that a new strong interaction, here called "hypercolor", is in fact responsible for the electro-weak symmetry breaking. The hypercolor force is assumed to have many properties in common with QCD: it acts on a set of hypercolor Fermions and induces chiral symmetry breakdown. A set of Goldstone Bosons is produced and those which couple to the electro-weak gauge fields disappear from the spectrum. Weinberg has a general discussion of the fate of all the Goldstone Bosons, including those which do not disappear, and he also gives a formalism for deciding into which direction in group space will the symmetry breaking parameters point. (The application of these ideas about "vacuum alignment" to a variety of semi-realistic models is carried out in papers [21, 22].[23])

The standard electro-weak model[6] with fundamental scalar fields produces charged vector Bosons $W^\pm$, and a neutral vector Boson Z; their mass ratio is determined by the electro-weak gauge coupling constants.

$$\frac{M_W}{M_Z} = \cos\theta = \frac{g_2}{\sqrt{g_1^2 + g_2^2}} \tag{19}$$

Here g_2 is the $SU(2)_L$ gauge coupling and g_1 is the U(1) hypercharge coupling. The exchange of these vector mesons leads to a low energy effective four-Fermion interaction characterized by the scale factor G_F and one dimensionless parameter, $\cos\theta$. This low energy effective interaction has been very successful in fitting all weak interaction data and within the context of the standard model this success can be seen as verification of the mass ratio (19).

Dynamical symmetry breaking can be a serious alternative to the standard model's Higgs fields only if the mass ratio (19) is guaranteed. Therefore, it was crucial to show that Eq. (19) is true when the hypercolor Fermion sector possesses certain symmetries. [12, 13][24] In fact, this symmetry is found in QCD with two massless quarks; it is the residual SU(2) symmetry previously mentioned. Again, QCD alone cannot account for the weak interaction symmetry breaking, but it would produce a W and Z with the correct mass ratio. But hypercolor, which by assumption parallels QCD at a higher mass scale, would also lead to (19), and this realization [12, 13] ushered intense research and model building. The conclusion of this activity is that a hypercolor interaction, at a scale of approximately 300 GeV, along with a set of hypercolor Fermions having the right symmetry properties, can successfully replace the scalar sector in the standard electro-weak theory.[25] The usual weak interaction phenomenology of quarks, leptons and intermediate vector mesons is reproduced at energies below 300 GeV, with the exception, which we shall soon discuss, that a new source for Fermion masses must be introduced. Above 300 GeV, hypercolor dynamics would have obvious experimental implications[26] and physics would be quite different from what is expected in the standard scalar theory.

The replacement of the scalar sector by gauge interactions and Fermions can also solve an important problem faced by "grand" unified models of strong and electro-weak forces.[27] In these theories there is a fundamental scale,[28] M_{GUT}, which is extraordinarily large, close or equal to the Planck scale of 10^{19} GeV, at which the strong force becomes unified with the electro-weak forces. The Higgs scalar field used for electro-weak symmetry breaking has couplings to particles interacting at this scale and one expects the dimensionful ϕ^2 coupling to be on the order of M_{GUT}^2. The vacuum expectation value of the scalar field will then be near M_{GUT} and so will the masses of the W and Z. One way of avoiding this problem — the so-called "hierarchy problem"[29]—is to adjust carefully the parameters of the scalar sector so that the relatively small numbers characterizing electro-weak symmetry breaking emerge due to cancellations between certain large numbers of the grand unification scenario. However, this "fine tuning" appears unnatural and is unstable against tiny changes in values of the parameters at the unification scale.

A more acceptable solution is achieved by eliminating the scalar sector and replacing it by bound Fermion states. The scale of symmetry breaking is now the scale at which the dynamics become strong, *i.e.* the scale at which the dimensionless gauge coupling becomes big. There are no dimensionful couplings, as there are in scalar theories, to set the scale of the dynamics. The dimensionless coupling constant of a non-Abelian gauge theory varies with the logarithm of the energy and it may take a huge change of energy before the coupling becomes large. For example, the QCD coupling constant becomes strong near 1 GeV — 15 orders of magnitude below a typical M_{GUT} of 10^{15} GeV — yet this seems natural. For these reasons, dynamical symmetry breaking is much more attractive in grand unified theories than the usual symmetry breaking by scalar fields.

The scalar sector of the standard model has a function beyond generating massive vector mesons. By introducing Yukawa couplings between scalar fields and Fermion bilinears, the Fermions are provided with masses, coming from non-vanishing vacuum expectation values of scalar fields. The dimensionless Yukawa couplings can have arbitrary magnitudes, and it is possible to arrive at mass terms for all observed Fermions. These Yukawa couplings explicitly break Fermion chiral symmetries; we know that *explicit* chiral symmetry breaking is required because the pion is not actually massless, *i.e.* chiral symmetry is not an exact symmetry of the Lagrangian.

In hypercolor theories, there are no Yukawa couplings [no scalars], so we must find a new source for explicit chiral symmetry breaking. The ordinary gauge interactions of SU(3) × SU(2) × U(1), together with hypercolor will not do, since gauge interactions [at the Lagrangian level] preserve chirality.

The most popular method for producing chiral symmetry breaking interactions in dynamical models is to enlarge the hypercolor gauge group into a larger group, called "extended" hypercolor, which at an energy scale $\Lambda > 300$ GeV breaks down to hypercolor. [14, 15] The Fermion representations of the extended hypercolor group are chosen so that ordinary and hypercolored Fermions belong to the same irreducible representations of extended hypercolor. When extended hypercolor breaks down, the massive gauge Bosons corresponding to broken generators mediate transitions between ordinary and hypercolored Fermions. Total chirality is preserved but not the separate chiralities of ordinary and hypercolored Fermions. These new interactions can lead to ordinary Fermion masses on the order of $(300\ \mathrm{GeV})^3/\Lambda^2$.[30] For $\Lambda \approx 10$ TeV reasonable Fermion masses are obtained. The price paid is yet another interaction above the weak interaction scale.

No satisfactory model of extended hypercolor has been found. One natural and minimal idea is to let the extended hypercolor group unify the hypercolor group with the SU(3) × SU(2) × U(1) gauge group. The hypercolor group breaks off from the other subgroups at the scale Λ and produces the required massive extended hypercolor Bosons which give mass to the ordinary Fermions. These ideas are illustrated in a toy model based on SU(7). [16] The difficulties encountered in this model are typical of the difficulties of the whole approach.[31] The cause of symmetry breaking at the scale Λ is unknown, and the scale Λ itself seems arbitrary. Also, the Fermion masses which are predicted do not agree with experiment.

Even though a realistic model has not been constructed, it is still important to look for low energy [well below 300 GeV] experimental signatures of hypercolor. Many of the models considered have a large number of hypercolor Fermions which typically come in "families." [A "family" is a set of Fermions with the gauge quantum numbers of the up and down quarks, the electron and the neutrino.] Chiral symmetry breaking in the hypercolor sector will lead to many Goldstone Bosons while only three are required to give masses to the $W^{\pm}$, and Z mesons. Usually the remaining Goldstone Bosons are only approximate or "pseudo"-Goldstone Bosons, *i.e.* like the pion their masses are exactly zero only when some parameters are exactly zero. These pseudo-Goldstone Bosons can have masses well below 300 GeV, and might be observable at present accelerators or in the near future. Much effort has gone into estimating the masses, production mechanisms and decay properties of these particles. [15, 17–22][32] Typically the lightest of these states is electrically and color neutral and mass estimates range from a few GeV to tens of GeV's.

Of course, the standard scalar theory also predicts a spin-0 particle whose mass is thought to be somewhere below 300 GeV. If a neutral spin-0 particle is discovered, it will be very important to study its decay characteristics carefully to see whether it is the Higgs scalar, a hypercolor pseudo-Goldstone Boson or something more exotic.

The extended hypercolor scale is roughly 10 TeV. However, any interactions whose scale is less than 100 TeV may lead to tiny but unwanted observable effects in the $K_L - K_S$ mass difference or in rare decay modes of kaons and other particles.[33] Extended hypercolor theories which are complex enough to produce non-trivial Cabibbo angles generally also predict these undesired effects. [15, 23] This is one of the most troublesome difficulties facing hypercolor model builders who try to use extended hypercolor to explain the Fermion masses. In addition, the extended hypercolor theories involve many new particles and interactions with complicated postulated dynamics. This imposing structure must be contrasted with the scalar interactions it is attempting to replace. In models with scalar fields the Fermion mass matrix is simply determined by arbitrary Yukawa couplings. No problems arise but no understanding is gained. Perhaps the origin of Fermion masses is in fact beyond the reach of any weak interaction theory.

Four-dimensional strongly interacting theories have not been solved exactly. Much of our insight comes from the *observed* behavior of hadrons, which we attribute to QCD. Our assumptions about symmetry breakdown due to hypercolor are plausible extensions of what we believe happens in QCD, but all this is still only assumption. For general gauge theories with arbitrary Fermion representations, the interesting dynamical conjecture has been made that the energetics of Fermion condensation is determined by one gluon exchange. This postulate implies that the Fermion condensate which forms is the one with the most negative associated group theory factor — an idea that can be applied to condensates which break global symmetries as well as to condensates which are not gauge singlets. For the latter, the gauge group can be said to "break itself" through its Fermion condensates.[34]

A set of algebraic [rather than dynamical] conditions has been established which must be satisfied by any theory of strongly interacting Fermions not undergoing spontaneous symmetry breakdown. [24] These conditions involve evaluating triangle anomaly diagrams,[11] first using the Fermion fields in the Lagrangian and then using the bound, gauge singlet states which are the particles in the physical spectrum. If there is no symmetry breakdown, some of these bound states will be massless composite Fermions and their contribution to the triangle anomaly must equal that coming from the fundamental Fermions. If the appropriate massless composite Fermions cannot be constructed, the symmetry must be spontaneously broken. These conditions are remarkable because they do not require a solution of the whole theory and they provide a simple algebraic criterion for deciding a dynamical issue.[35]

One example of a theory which does not necessarily break its chiral symmetries and can produce massless composites is QCD with two massless quarks. [24] [Actually the gauge group can be SU(N) with any odd N, not just $N = 3$.] With more than two quark flavors, and with additional assumptions about the decoupling of heavy quarks, the chiral symmetries must be spontaneously broken and this may help explain why chiral symmetry is broken in the real world.[36]

Quarks and leptons may themselves be composites of other particles bound at some scale B. We know that B is probably greater than 500 GeV from the precise agreement of the experimental and theoretical values for the muon's magnetic moment. In terms of the large binding scale B, ordinary quarks and leptons are effectively massless. If we suppose that indeed they are approximately massless bound states of constituents interacting through a force whose characteristic scale is B, then they should satisfy the anomaly conditions. Using Fermions and gauge Bosons as constituents, people have searched for and constructed other solutions to the anomaly conditions which result in massless composite Fermions. [25, 26][37] Unfortunately none of these solutions has a spectrum which resembles that of the real world.

A realistic solution can be constructed, which surprisingly has electro-weak interactions without symmetry breakdown. [27] The constituents are Fermions, gauge Bosons and scalars, although the scalars can be replaced by hypercolor Fermion anti-Fermion bound states.[38] This model uses the same Lagrangian as the standard Weinberg-Salam electro-weak theory but assumes that there is no vacuum expectation value for the scalar field. [If the scalar field is replaced by a bound state the assumption is that no symmetry breakdown is induced by the hypercolor force.] The $SU(2)_L$ interaction is unbroken and becomes strong, binding non-singlet fields into $SU(2)_L$ singlet, massless composite Fermions. The scale of the binding is assumed to be the weak interaction scale $G_F^{-1/2} \approx 300$ GeV. The massless composite Fermions are all left-handed and there is exactly one left-handed bound state for each left-handed quark and lepton. The right-handed fields do not feel the $SU(2)_L$ force and do not participate in the new strong dynamics.

In this model, the weak interactions arise as a consequence of the bound state structure of the left-handed Fermions. By making certain dynamical assumptions, it is possible to show that the low energy effective four-Fermion interaction produced by the new dynamics exactly matches the low energy effective four-Fermion interaction of the standard model and therefore can account for all weak interaction data. This raises the interesting question of what really is the experimental evidence for spontaneously broken gauge theories. The low energy effective interaction of the confining weak interaction theory is mediated by vector Bosons whose masses are not governed by Eqs. (18) and (19). The observation of a W and Z obeying (18) and (19), where the gauge couplings and F have been independently determined, would be the ultimate confirmation of the Weinberg-Salam weak interaction model, based on a spontaneously broken gauge theory.

The anticipated discovery of the $W^{\pm}$ and Z mesons will still not answer the question whether they are associated with dynamical symmetry breakdown or with fundamental scalar fields. Experimentalists must search for evidence of possible new dynamics near or above the weak interaction scale. Experiments are now being done at energies of 500 GeV in the center of mass, and higher energies will soon be reached. We shall then know much more about the nature of weak interactions, and we hope that this experimental information will also advance theoretical ideas about gauge symmetry breaking.

REFERENCES

Numbers in brackets refer to reprinted papers.

1. H. Dürr, W. Heisenberg, H. Mitter, S. Schlieder and K. Yamazaki, "Zur theorie der elementarteilchen," Zeit. Naturforsch. **14a,** 441 (1959).
2. J. Goldstone, "Field theories with superconductor solutions," Nuovo Cimento **19,** 154 (1961); J. Goldstone, A. Salam and S. Weinberg, "Broken Symmetries," Phys. Rev. **127,** 965 (1962).
3. Y. Nambu and G. Jona-Lasinio, "Dynamical model of elementary particles based on an analogy with superconductivity," Phys. Rev. **122,** 345 (1961).
4. F. Englert and R. Brout, "Broken symmetry and the mass of gauge vector bosons," Phys. Rev. Lett. **13,** 321 (1964); P. Higgs, "Broken symmetries, massless particles and gauge fields," Phys. Lett. **12,** 132 (1964); and "Broken symmetries and the masses of gauge bosons," Phys. Rev. Lett. **13,** 508 (1964); G. Guralnik, C. Hagen and T. Kibble, "Global conservation laws and massless particles," Phys. Rev. Lett. **13,** 585 (1964); P. Higgs, "Spontaneous symmetry breakdown without massless bosons," Phys. Rev. **145,** 1156 (1966); T. Kibble, "Symmetry breaking in non-Abelian gauge theories," Phys. Rev. **155,** 1554 (1967).
5. A historical account is given by P. Anderson, "Uses of solid state analogies in elementary particle theory," in *Gauge Theories and Modern Field Theory*, edited by R. Arnowitt and P. Nath (MIT Press, Cambridge, MA, 1976).
6. S. Weinberg, "A model of leptons," Phys. Rev. Lett. **19,** 1264 (1967); A. Salam, "Weak and electromagnetic interactions," in *Elementary Particle Theory*, edited by N. Svartholm (Almqvist and Wiksell, Stockholm, 1968).
7. S. Coleman and E. Weinberg, "Radiative corrections as the origin of spontaneous symmetry breaking," Phys. Rev. **D 7,** 1883 (1973).
8. Increasing the number of Higgs fields decreases the strength of asymptotic freedom. Also, if the Higgs-gauge field coupling is sufficiently strong, the ultraviolet unstable Higgs self-couplings can destabilize the high-energy behavior of the Yang-Mills sector.
9. S. Hawking, D. Page and C. Pope, "Quantum gravitational bubbles," Nucl. Phys. **B170** [**FS1**], 283 (1980).
10. An analysis of the Higgs effect from Schwinger's point of view is given by R. Jackiw, "Dynamical symmetry breaking," in *Laws of Hadronic Matter*, edited by A. Zichichi (Academic Press, New York, NY, 1975).
11. J. Bell and R. Jackiw, "A PCAC puzzle: $\pi^0 \to 2\gamma$ in the σ model," Nuovo Cimento **60,** 47 (1969); S. Adler, "Axial-vector vertex in spinor electrodynamics," Phys. Rev. **177,** 2426 (1969). For a review, see R. Jackiw, "Field

theoretic investigations in current algebra," in *Lectures on Current Algebra and Its Applications*, by S. Treiman, R. Jackiw and D. Gross (Princeton, Princeton, NJ, 1972).
12. K. Johnson, "γ_5 invariance," Phys. Lett. **5,** 253 (1963).
13. The analysis of the Schwinger model in terms of the axial vector anomaly is given in Ref. 10.
14. R. Jackiw and S. Templeton, "How super-renormalizable interactions cure their infrared divergences," Phys. Rev. **D 23,** 2291 (1981); J. Schonfeld, "A mass term for three-dimensional gauge fields," Nucl. Phys. **B185,** 157 (1981); S. Deser, R. Jackiw and S. Templeton, "Three-dimensional massive gauge theories," Phys. Rev. Lett. **48,** 975 (1982) and "Topologically massive gauge theories," Ann. Phys. (NY) **140,** 372 (1982).
15. Topological mechanisms for mass generation are discussed by R. Jackiw, "Gauge invariance and mass, III," in *Asymptotic Realms of Physics,* edited by A. Guth, K. Huang and R. Jaffe (MIT Press, Cambridge, MA, 1983).
16. The "unitary gauge" was originally used to show that spontaneous symmetry breaking in gauge theories gives rise to massive vector mesons, instead of Goldstone Bosons; see Ref. 4. A general exposition is in S. Weinberg, "General theory of broken symmetries," Phys. Rev. **D 7,** 1068 (1973).
17. L. Dolan and R. Jackiw, "Gauge invariant signal for gauge symmetry breaking," Phys. Rev. **D 9,** 2904 (1974).
18. J. Luttinger and J. Ward, "Ground-state energy of a many fermion system, II," Phys. Rev. **118,** 1417 (1960); C. deDominicis and P. Martin, "Stationary entropy principle and renormalization in normal and superfluid systems, I and II," J. Math. Phys. **5,** 14 and 31 (1964).
19. For reviews of early investigations, together with exhortations that dynamical mechanisms be used for gauge symmetry breaking in electro-weak models, see M. Bég and A. Sirlin, "Gauge theories of weak interactions," Ann. Rev. of Nucl. Sci. **24,** 379 (1974), and Jackiw in Ref. 10.
20. D. Kirzhnits and A. Linde, "Macroscopic consequences of the Weinberg model," Phys. Lett. **42B,** 471 (1972); L. Dolan and R. Jackiw, "Symmetry behavior at finite temperature," Phys. Rev. **D 9,** 3320 (1974); S. Weinberg, "Gauge and global symmetries at high temperature," Phys. Rev. **D 9,** 3357 (1974). For a review see A. Linde, "Phase transitions in gauge theories," Rep. Prog. Phys. **42,** 389 (1979).
21. Dolan and Jackiw, Ref. 20.
22. T. Hagiwara and B. Lee, "Proton neutron mass difference in a unified gauge model of leptons and hadrons," Phys. Rev. **D 7,** 459 (1973); M. Weinstein, "Can all hadronic symmetry breaking be due to weak interactions," Phys. Rev. **D 7,** 1854 (1973) and "Conserved currents, their commutators and the symmetry structure of renormalizable theories of electromagnetic, weak and strong interactions," Phys. Rev. **D 8,** 2511 (1973); also L. Susskind in [13].
23. For earlier work on vacuum alignment see R. Dashen, "Some features of chiral symmetry breaking," Phys. Rev. **D 3,** 1879 (1971); F. Wilczek and A. Zee, "Orientation of the weak interaction with respect to the strong interaction," Phys. Rev. **D 15,** 3701 (1977).
24. In Nature, these symmetries are only approximate. Even before the symmetry was identified, and its relevance to maintaining Eq. (19) was realized, the effect of Fermion mass differences which can modify Eq. (19) was studied by M. Veltman, "Limit on mass differences in the Weinberg model," Nucl. Phys. **B123,** 89 (1977); see also P. Sikivie, L. Susskind, M. Voloshin and V. Zakharov, "Isospin Breaking in technicolor models," Nucl. Phys. **B173,** 189 (1980).

25. Recent reviews of hypercolor theories include:
K. Lane and M. Peskin, "An introduction to weak interaction theories with dynamical symmetry breaking," in *XV Rencontre de Moriond, Vol II — Electroweak Interactions and Unified Theories*, edited by T. Thanh Van (Edition Frontières, Gif sur Yvette, France, 1980); P. Sikivie, "An introduction to technicolor," CERN preprint TH 2951 (1980), Proceedings of Varenna Summer School (1980); E. Farhi and L. Susskind, "Technicolor," Phys. Rep. **74,** No.3, 277 (1981).
26. Susskind in [13]; F. Hayot and O. Napoly, "Detecting a heavy colored object at the FNAL tevatron," Zeit. Phys. **C7,** 299, (1981); S. Raby, "The TeV Picture," in *The Second Workshop on Grand Unification*, edited by J. Leveille, L. Sulak and D. Unger (Birkhäuser, Boston, MA, 1981); G. Girardi, P. Mery, P. Sorba, "Heavy quark jets as technicolor signatures in pp and p anti-p collisions," Nucl. Phys. **B195,** 410 (1982).
27. J. Pati and A. Salam, "Unified lepton-hadron symmetry and a gauge theory of the basic interactions," Phys. Rev. **D8,** 1240 (1973); H. Georgi and S. Glashow, "Unity of all elementary particle forces," Phys. Rev. Lett. **32,** 328 (1974).
28. H. Georgi, H. Quinn and S. Weinberg, "Hierarchy of interactions in unified gauge theories," Phys. Rev. Lett. **33,** 451 (1974).
29. For clear expositions see Susskind in [13] and 't Hooft in [24].
30. K. Lane, "Asymptotic freedom and Goldstone realization of chiral symmetry," Phys. Rev. **D 10,** 2605 (1974); H. Politzer, "Effective quark masses in the chiral limit," Nucl. Phys. **B117,** 397 (1976).
31. For a sampling of other models see:
S. Dimopoulos, S. Raby and P. Sikivie, "Problems and virtues of scalarless theories of electroweak interactions," Nucl. Phys. **B176,** 449 (1980); B. Holdom, "A realistic model with dynamically broken symmetries," Phys. Rev. **D 23,** 1637 (1981); M. Dine, W. Fischler and M. Srednicki, "Super symmetric technicolor," Nucl. Phys. **B189,** 575 (1981); S. Dimopoulos and S. Raby, "Super color," Nucl. Phys. **B192,** 353 (1981);
32. A partial list other papers includes:
G. Gounaris and A. Nicolaidis, "Technicolor versus Higgs signatures in e^+e^- collisions," Phys. Lett. **102B,** 144 (1981); and "Discriminating technicolor from Higgs breaking," Phys. Lett. **109B,** 221 (1982); S. Dimopoulos, S. Raby and G. Kane, "Experimental predictions from technicolor theories," Nucl. Phys. **B182,** 77 (1981); S. Chadha and M. Peskin, "Implications of chiral dynamics in theories of technicolor 1. elementary couplings, and 2. the mass of P^+," Nucl. Phys. **B185,** 61 and **B187,** 541 (1981); B. Holdom, "A phenomenological lagrangian from hypercolor," Phys. Rev. **D 24,** 157 (1981); J. Grifols, "Technicolor in gamma gamma processes." Phys. Lett. **102B,** 277 (1981); F. Hayot, "Technicolor pseudo-Goldstone boson couplings to gluon fields," Nucl. Phys. **B191,** 82 (1981); A. Ali, H. Newman and R. Zhu, "Production of charged hyperpions in e^+e^- annihilation," Nucl. Phys. **B191,** 93 (1981); K. Lane, "Hyperpions at the Z," Workshop on Z Physics (Ithaca, NY, 1981).
33. Eichten and Lane in [15]; R. Cahn and H. Harari, "Bounds on the masses of neutral generation-changing gauge bosons," Nucl. Phys. **B176,** 135 (1981).
34. J. Cornwall, "Spontaneous symmetry breaking without scalar mesons. II," Phys. Rev. **D 10,** 500 (1974); Eichten and Feinberg in [9]; H. Georgi, "Towards a Grand Unified Theory of Flavor," Nucl. Phys. **B156,** 126 (1979); S. Raby, S. Dimopoulos and L. Susskind, "Tumbling Gauge Theories," Nucl. Phys. **B169,** 373 (1980). See also some relevant numerical studies by J. Kogut, M. Stone, H.

Wyld, J. Shigemitsu, S. Shenker and D. Sinclair, "Scales of chiral symmetry breaking in quantum chromodynamics," University of Illinois preprint, ILL-(TH)-82–5.

35. A. Zee, "From baryons to quarks through π^0 decay," Phys. Lett. **95B,** 290 (1980); Y. Frishman, A. Schwimmer, T. Banks and S. Yankielowicz, "The axial anomaly and the bound state spectrum in confining theories," Nucl. Phys. **B177,** 157 (1981); S. Coleman and B. Grossman, "'t Hooft consistency condition as a consequence of analyticity and unitarity," Harvard Preprint, HUTP-82/A009.
36. The additional assumptions, which go beyond the strictly algebraic constraints imposed by the axial vector anomaly, have been critically surveyed by J. Preskill and S. Weinberg, "Decoupling constraints on massless composite particles," Phys. Rev. **D 24,** 1059 (1981).
37. Some further examples can be found in:
R. Barbieri, L. Miani and R. Petronzio, "Quarks and leptons as composite states of confined $O(n)$ preons," Phys. Lett. **96B,** 63 (1980); I. Bars and S. Yankielowicz, "Composite quarks and leptons as solutions of anomaly constraints," Phys. Lett. **101B,** 159 (1981); R. Casalbouni and R. Gatto, "Composite description for quarks and leptons from a geometrical picture of complementarity," Phys. Lett. **103B,** 113 (1981); I. Bars, "Spontaneous chiral symmetry breaking in quantum chromodynamics," Phys. Lett. **109B,** 73 (1982).
38. L. Abbott, E. Farhi and A. Schwimmer, "A confining model of the weak interactions in technicolor," MIT Preprint, CTP #962, 1981; Nucl. Phys. (in press).

I. PRE-HISTORY

PHYSICAL REVIEW VOLUME 110, NUMBER 4 MAY 15, 1958

Coherent Excited States in the Theory of Superconductivity: Gauge Invariance and the Meissner Effect

P. W. ANDERSON
Bell Telephone Laboratories, Murray Hill, New Jersey

(Received January 27, 1958)

We discuss the coherent states generated in the Bardeen, Cooper, and Schrieffer theory of superconductivity by the momentum displacement operator $\rho_Q = \sum_n \exp(iQ\cdot r_n)$. Without taking into account plasma effects, these states are like bound Cooper pairs with momentum $\hbar Q$ and energies lying in the gap, and they play a central role in the explanation of the gauge invariance of the Meissner effect. Long-range Coulomb forces recombine them into plasmons with equations of motion unaffected by the gap. Central to the argument is the proof that the non-gauge-invariant terms in the Hamiltonian of Bardeen, Cooper, and Schrieffer have an effect on these states which vanishes in the weak-coupling limit.

I. INTRODUCTION

BUCKINGHAM[1] has questioned whether an energy-gap model of superconductivity, such as that of Bardeen, Cooper, and Schrieffer,[2] can explain the Meissner effect without violating a certain identity derived by Schafroth[3] on the basis of gauge invariance, and by Buckingham using essentially an f-sum rule. This identity is what causes the insulator, which also has an energy gap, to fail to show a Meissner effect; thus, Buckingham and Schafroth[4] argue, a proof of gauge invariance lies at the core of the problem of superconductivity, especially since the Hamiltonian used in B.C.S. is not gauge-invariant.

Bardeen[5] argues that the matrix elements and energy states involved in the gauge problem bring in coherent excitations which will be strongly coupled to the plasma modes, a high-frequency phenomenon presumably unaltered by superconductivity. Unfortunately, while we find that this is indeed exactly the situation, the insulator also often has normal plasma modes. Thus, while the B.C.S. calculation in the London gauge is probably entirely correct, and justifiable on physical grounds, it throws little light on the basic differences between the three cases—insulator, metal, and superconductor.

We also noticed that the operator which is central in the gauge problem as well as the plasma theory,

$$\rho_Q = \sum_n \exp(iQ\cdot r_n) = \sum_{k,\sigma} c_{k+Q,\sigma}^* c_{k,\sigma}, \tag{1}$$

has another interesting property: its separate components $c_{k+Q,\sigma}^* c_{k,\sigma}$, when applied to the B.C.S. ground-state wave function Ψ_g, create excited pairs of electrons k_1, k_2 with momentum pairing

$$k_1 + k_2 = Q \tag{2}$$

instead of zero. The total operator applied to Ψ_g leads to a linear combination of such states, which can be thought of as equivalent to a Cooper bound state[6] of a pair of electrons with nonzero momentum, superimposed on the B.C.S. ground state.

Our discussion of these problems is based on the following physical picture: any transverse excitation involves breaking up the phase coherence over the whole Fermi surface of at least one pair in the superconducting ground state, and so involves a loss of attractive binding energy. This causes the Meissner effect. Longitudinal excitations, however, such as those generated by ρ_Q, do not break up phase coherence in the superconducting state, and so their energies involve only kinetic energy, or electromagnetic energy when plasma effects are included. Thus longitudinal and transverse excitations are different in the superconductor, in a sense in which they are not in either the metal or the insulator, and it turns out to be this difference which allows a gauge-invariant explanation of the Meissner effect.

We proceed further in two stages. First, we discuss the fictitious problem in which the only plasma effect is the screening of the long-range repulsion. In this stage gauge invariance requires, and we indeed find, that the states

$$\Psi_Q = \rho_Q \Psi_g \tag{3}$$

have energies in the energy gap and proportional to Q^2. In a perfectly gauge-invariant theory, their energy would be just the kinetic energy

$$E_Q = (\hbar^2 Q^2/2m)(2\epsilon_f/3\pi\epsilon_0), \tag{4}$$

but we find a small correction going to zero in the weak-coupling limit. Equation (4) follows from the same f-sum rule which leads to gauge invariance.

There is a fundamental difference, which we demonstrate, between the ways in which superconducting and normal substances satisfy this sum rule. Both normal metals and insulators (leaving out the rather confusing effects of long-range Coulomb forces which can be studied later) satisfy this rule with ordinary excitations in such a way that the more familiar optical sum rule—

---

[1] M. J. Buckingham, Nuovo cimento 5, 1763 (1957).
[2] Bardeen, Cooper, and Schrieffer, Phys. Rev. 106, 162 (1957); 108, 1175 (1957). The latter we call B.C.S., and we shall follow its notation as far as possible.
[3] M. R. Schafroth, Helv. Phys. Acta 24, 645 (1951).
[4] M. R. Schafroth (private communication). I am indebted to G. Wentzel for an elegant presentation of these questions in a series of discussions, to which M. Lax and C. Herring also contributed.
[5] J. Bardeen, Nuovo cimento 5, 1765 (1957).
[6] L. N. Cooper, Phys. Rev. 104, 1189 (1956).

which in turn leads to normal diamagnetism—also follows. In these normal cases the states (3) can be shown not to exist separately from the ordinary excitations. In the superconductor, however, the rule is satisfied by matrix elements involving only the longitudinal excitations (3), while the optical sum rule and Meissner effect involve transverse excitations with entirely different behavior from the longitudinal ones. Basically, these transverse excitations involve combinations of pair states which have a finite angular as well as linear momentum and cannot be bound into quasi-"Cooper pairs" like the longitudinal excitations. This difference is the basis for the electromagnetic properties of superconductivity.

At this stage any number of $\mathbf{Q}\neq 0$ pairs could be incorporated into the state with little loss of energy. This alleviates the rigidity of the B.C.S. ground state to some extent, making it easier to fit physical boundary conditions. On the other hand, we will find that when plasma effects are included the $\mathbf{Q}=0$ pairing condition is again enforced, except possibly in particular geometrical situations for very long wavelengths.

In the second stage of the calculation we discuss these plasma effects. Our procedure corresponds to that of Nozières and Pines,[7] who find that the f-sum rules, and particularly the relationship of the optical and longitudinal forms, are best discussed prior to including the long-range Coulomb effects. Appealing to the random-phase approximation, which allows separation of the different momenta $\mathbf{Q}$, we show that the longitudinal f-sum rule becomes the commutation rule of plasma coordinates and momenta, while the transverse behavior is unaffected, whether normal or superconducting, when the long-range Coulomb interactions are included. The states $\Psi_{\mathbf{Q}}$ are recombined to make the plasma excitations, which now have large excitation energies and do not affect the energy gap.

II. COMMUTATORS AND ENERGIES OF COHERENT STATES

In this section we carry out the first stage of the program in the introduction. We naïvely ignore plasma effects and subsidiary conditions and study the energies of coherent excited states, and the f-sum rules, using the B.C.S. ground state derived from phonon and screened Coulomb interactions.

The initial Hamiltonian used by B.C.S. contains the kinetic energy and the second-order phonon interaction between electrons, but ignores the self-energy as well as higher-order terms. It is

$$H=\sum_{\mathbf{k},\sigma}\epsilon_{\mathbf{k}}n_{\mathbf{k},\sigma}+\tfrac{1}{2}\sum_{\mathbf{k},\mathbf{k}',\sigma,\sigma',\boldsymbol{\kappa}}2\hbar\omega_{\kappa}|M_{\kappa}|^2 \times[(\epsilon_{\mathbf{k}}-\epsilon_{\mathbf{k}+\boldsymbol{\kappa}})^2-(\hbar\omega_{\kappa})^2]^{-1}c_{\mathbf{k}'-\boldsymbol{\kappa},\sigma'}{}^{*}c_{\mathbf{k}',\sigma'} \times c_{\mathbf{k}+\boldsymbol{\kappa},\sigma}{}^{*}c_{\mathbf{k},\sigma}+H_{\text{Coul}}. \quad (5)$$

[7] P. Nozières and D. Pines, Phys. Rev. **109**, 741 (1958).

Almost immediately B.C.S. drop, except for later perturbation calculations, all terms except $\mathbf{k}'=-\mathbf{k}$, $\sigma'=-\sigma$; and they replace the coefficient by a constant unless either $|\epsilon_{\mathbf{k}}|$ or $|\epsilon_{\mathbf{k}+\boldsymbol{\kappa}}|>\hbar\omega$, in which case it is zero. The resulting Hamiltonian is their H_{red}.

The Hamiltonian (5) is already not gauge-invariant, if it is to be used with the usual expressions for the current and for the perturbation of an electromagnetic field, because it depends on $\mathbf{k}$ as well as $\boldsymbol{\kappa}$. Fortunately, however, we shall show that the important difficulties are not connected with the momentum dependence of (5) but with the simplification to H_{red}. The difficulty is that in calculating the properties of coherent excited states of the B.C.S. theory one must take into account more than just the $\mathbf{k}'-\mathbf{k}$ terms of the interaction.

When this is done we shall show that the corrections caused by the $\mathbf{k}$ dependence of (5) in the important sum rules and energies become negligible in a well-defined limit which is that applying to most superconductors.

The entire argument is based on a commutation relation used by Buckingham[1] in this connection and by Nozières and Pines[7] for other reasons:

$$[H_K,\rho_{\mathbf{Q}}]=(-\hbar^2/2m)\, i\mathbf{Q}\cdot\sum_n[\nabla_n\exp(i\mathbf{Q}\cdot\mathbf{r}_n)+\exp(i\mathbf{Q}\cdot\mathbf{r}_n)\nabla_n] =(\hbar^2/2m)\sum_{\mathbf{k},\sigma}\mathbf{Q}\cdot(2\mathbf{k}+\mathbf{Q})c_{\mathbf{k}+\mathbf{Q},\sigma}{}^{*}c_{\mathbf{k},\sigma}, \quad (6)$$

where H_K is the kinetic energy and $\rho_{\mathbf{Q}}$ is defined in (1).

In the case of most simple kinds of forces, which are functions only of coordinates or at most of relative momenta, $\rho_{\mathbf{Q}}$ commutes with the potential energy. Then (6) may be written, using

$$[H,\rho_{\mathbf{Q}}]=[H_K,\rho_{\mathbf{Q}}],$$

as

$$(E_m-E_{m'})(m|\rho_{\mathbf{Q}}|m')=(\hbar^2/2m)\,\mathbf{Q}\cdot\sum_j(m|\nabla_j\exp(i\mathbf{Q}\cdot\mathbf{r}_j)+\exp(i\mathbf{Q}\cdot\mathbf{r}_j)\nabla_j|m'). \quad (7)$$

We may also deduce from (7) the basic f-sum rule:

$$[[H,\rho_{\mathbf{Q}}],\rho_{-\mathbf{Q}}]=N\hbar^2Q^2/m;\qquad \sum_{m'}(E_m-E_{m'})|(m|\rho_{\mathbf{Q}}|m')|^2=\tfrac{1}{2}N\hbar^2Q^2/m, \quad (8)$$

or

$$\sum_{m'}|(m|\mathbf{Q}\cdot\sum_j[\nabla_j\exp(i\mathbf{Q}\cdot\mathbf{r}_j)+\exp(i\mathbf{Q}\cdot\mathbf{r}_j)\nabla_j]|m')|^2\times(E_{m'}-E_m)^{-1}=2mNQ^2/\hbar^2. \quad (9)$$

Now in the case of (5) the potential energy does not commute with $\rho_{\mathbf{Q}}$. However, what we can show is that the commutator can be made arbitrarily small compared to (7).

In doing this we shall not use the explicit form (5) but shall carry out the first stage—the cutoff—of the simplification to H_{red}, in full confidence that the results would be practically the same if we used (5) itself, because the new potential is no less k-dependent than the old in the region of interest. One can then show that the full interaction energy, corresponding to the $\mathbf{k}'=-\mathbf{k}$ terms of the B.C.S. H_{red}, is

$$H_V=-\tfrac{1}{2}V\sum_{\mathbf{q}}\Big[\sum_{\mathbf{k}'\neq\mathbf{k}(\epsilon_{\mathbf{k}'}<\hbar\omega)}\ \sum_{\mathbf{k}(\epsilon_{\mathbf{k}}<\hbar\omega)}+\sum_{\mathbf{k}'\neq\mathbf{k}(\epsilon_{\mathbf{k}'-\mathbf{q}}<\hbar\omega)}\ \sum_{\mathbf{k}(\epsilon_{\mathbf{k}-\mathbf{q}}<\hbar\omega)}\Big](c_{\mathbf{k}'\uparrow}{}^{*}c_{-\mathbf{k}'+\mathbf{q}\downarrow}{}^{*}c_{-\mathbf{k}+\mathbf{q}\downarrow}c_{\mathbf{k}\uparrow}). \quad (10)$$

(All notations are as used in B.C.S.) Then first of all we have to calculate the commutator of this with $\rho_{\mathbf{Q}}$. This is done in the Appendix, and the result is:

$$[H_V,\rho_{\mathbf{Q}}]=\tfrac{1}{2}V\Big[\sum_{\mathbf{k}'\neq\mathbf{k}}\ \sum_{\mathbf{k}(\epsilon_{\mathbf{k}}<\hbar\omega,\ \epsilon_{\mathbf{k}+\mathbf{Q}}>\hbar\omega)}-\sum_{\mathbf{k}'\neq\mathbf{k}}\ \sum_{\mathbf{k}(\epsilon_{\mathbf{k}}>\hbar\omega,\ \epsilon_{\mathbf{k}+\mathbf{Q}}<\hbar\omega)}+(\text{same with }\mathbf{k}\to\mathbf{k}',\ \mathbf{k}'\to\mathbf{k})\Big]$$
$$\times(c_{\mathbf{k}'+\mathbf{Q}\uparrow}{}^{*}c_{-\mathbf{k}'+\mathbf{q}\downarrow}{}^{*}c_{-\mathbf{k}+\mathbf{q}\downarrow}c_{\mathbf{k}\uparrow}-c_{\mathbf{k}'-\mathbf{q}\uparrow}{}^{*}c_{-\mathbf{k}'\downarrow}{}^{*}c_{-\mathbf{k}-\mathbf{Q}\downarrow}c_{\mathbf{k}-\mathbf{q}\uparrow}). \quad (11)$$

These terms already refer only to $\mathbf{k}$ states within Q of the cutoff surface, and contain subtractions which make them depend on only the derivatives of the wave function near this surface. Thus, (11) will always be small acting on states close to the ground state, at least for small Q.

We shall see that the most important measure of this smallness is the scalar product.

$$(\rho_{\mathbf{Q}}\Psi_g,[H_V,\rho_{\mathbf{Q}}]\Psi_g), \quad (12)$$

where Ψ_g is the B.C.S. ground state. This will be computed in the Appendix also, but first we will need to compute the wave function

$$\Psi_{\mathbf{Q}}=\rho_{\mathbf{Q}}\Psi_g, \quad (13)$$

and its normalization. Consider a particular term of $\rho_{\mathbf{Q}}$,

$$(\rho_{\mathbf{Q}}{}^{\kappa})_{+}=c_{\kappa+\mathbf{Q}\uparrow}{}^{*}c_{\kappa\uparrow}, \quad (14a)$$

or

$$(\rho_{\mathbf{Q}}{}^{\kappa})_{-}=c_{-\kappa\downarrow}{}^{*}c_{-\kappa-\mathbf{Q}\downarrow}. \quad (14b)$$

Then

$$(\rho_{\mathbf{Q}}{}^{\kappa})_{+}\Psi_g=\prod_{\mathbf{k}\neq\kappa,\kappa+\mathbf{Q}}[(1-h_{\mathbf{k}})^{\frac{1}{2}}+h_{\mathbf{k}}{}^{\frac{1}{2}}b_{\mathbf{k}}{}^{*}]$$
$$\times(1-h_{\kappa+\mathbf{Q}})^{\frac{1}{2}}h_{\kappa}{}^{\frac{1}{2}}c_{\kappa+\mathbf{Q}\downarrow}{}^{*}c_{-\kappa\uparrow}\Psi_v,$$

Ψ_v being the vacuum state. Thus

$$(\rho_{\mathbf{Q}}{}^{\kappa})_{+}\Psi_g=h_{\kappa}{}^{\frac{1}{2}}(1-h_{\kappa+\mathbf{Q}})^{\frac{1}{2}}\Psi_{-\kappa,\kappa+\mathbf{Q}}, \quad (15)$$

where the last notation refers to the presence of a pair of real "single" excitations in the B.C.S. sense in states $-\kappa\downarrow$ and $\kappa+\mathbf{Q}\uparrow$. Similarly,

$$(\rho_{\mathbf{Q}}{}^{\kappa})_{-}\Psi_g=h_{\kappa+\mathbf{Q}}{}^{\frac{1}{2}}(1-h_{\kappa})^{\frac{1}{2}}\Psi_{-\kappa,\kappa+\mathbf{Q}}, \quad (16)$$

and

$$\Psi_{\mathbf{Q}}=\rho_{\mathbf{Q}}\Psi_g\cong2\sum_{\kappa}h_{\kappa}{}^{\frac{1}{2}}(1-h_{\kappa})^{\frac{1}{2}}\Psi_{-\kappa,\kappa+\mathbf{Q}}. \quad (17)$$

Thus $\Psi_{\mathbf{Q}}$ is indeed a certain linear superposition of pair excitations with momentum $\mathbf{Q}$. Its normalization factor is

$$(\Psi_{\mathbf{Q}},\Psi_{\mathbf{Q}})=4\sum_{\mathbf{k}}h_{\mathbf{k}}(1-h_{\mathbf{k}})$$
$$\cong2\pi\epsilon_0N(0), \quad (18)$$

which is finite—a vital point, since the corresponding quantity in metal or insulator approaches zero with $\mathbf{Q}$. Equation (18) already implies the presence of long-range order in the ground state, as we shall see.

Now we can compute the quantity (12) in the Appendix. The result is

$$(\rho_{\mathbf{Q}}\Psi_g,[H_V,\rho_{\mathbf{Q}}]\Psi_g)=\tfrac{2}{3}\epsilon_0{}^2N(0)\hbar^2k_F{}^2Q^2/m^2\omega^2. \quad (19)$$

In order to get a measure of the magnitude of (19), and for later work, let us also calculate the kinetic energy commutator (6). This is also done in Appendix I, and the result is

$$(\rho_{\mathbf{Q}}\Psi_g,[H_K,\rho_{\mathbf{Q}}]\Psi_g)$$
$$=\frac{\hbar^2}{2m}(\rho_{\mathbf{Q}}\Psi_g,\sum_{\kappa,\sigma}\mathbf{Q}\cdot(2\kappa+\mathbf{Q})c_{\kappa+\mathbf{Q},\sigma}{}^{*}c_{\kappa,\sigma}\Psi_g)$$
$$=\tfrac{4}{3}(\hbar^2Q^2/2m)N(0)\epsilon_F. \quad (20)$$

This value is shown there to be essentially independent of the exact structure of the wave function if it resembles a Fermi sea at all.

We can now, first of all, get a numerical measure of the relative magnitude of the ordinary terms (6) and the non-gauge-invariant terms (11) by taking the ratio of (19) to (20). The result is

$$(19)/(20)=2\epsilon_0{}^2/\hbar^2\omega^2=2\exp[-2/N(0)V], \quad (21)$$

which is completely negligible for most superconductors, and zero in the weak-coupling limit (when, of course, in principle superconductivity still exists). Thus we have proved our first contention: that, although exact gauge invariance is still not present except in the weak-coupling limit, the inclusion of the $\mathbf{Q}\neq0$ terms of H_V restores gauge invariance well enough so that no difficulties of principle are encountered by ignoring the terms (11).

We now can use (19), (20), and (18) to show that in this stage, before the introduction of specific plasma effects, states closely related to $\rho_{\mathbf{Q}}\Psi_g$ must lie in the energy gap. To show this we use the identity

$$E_{\mathbf{Q}}(\rho_{\mathbf{Q}}\Psi_g,\rho_{\mathbf{Q}}\Psi_g)=(\rho_{\mathbf{Q}}\Psi_g,H\rho_{\mathbf{Q}}\Psi_g)$$
$$=(\rho_{\mathbf{Q}}\Psi_g,[H\rho_{\mathbf{Q}}-\rho_{\mathbf{Q}}H]\Psi_g)$$
$$+E_0(\rho_{\mathbf{Q}}\Psi_g,\rho_{\mathbf{Q}}\Psi_g), \quad (22).$$

using the fact that the ground state is an eigenstate. (The first equality defines $E_{\mathbf{Q}}$ as the energy of $\Psi_{\mathbf{Q}}$.)

Thus

$$(E_Q-E_0)=(\rho_Q\Psi_g,\rho_Q\Psi_g)^{-1}(\rho_Q\Psi_g,[H,\rho_Q]\Psi_g)$$

$$\simeq\frac{2}{3\pi}\left(\frac{\epsilon_F}{\epsilon_0}\right)\frac{\hbar^2Q^2}{2m}. \quad (23)$$

Since ρ_Q is a state of momentum $\hbar\mathbf{Q}$, it is orthogonal to the ground state; thus some eigenstate must lie below E_Q. One guesses that Ψ_Q is really a very good approximation to this state.[8]

These states, if they existed, would have interesting properties. Their effective mass is about 10^{-4} electron mass. Therefore the specific heat would be tiny and probably below the accuracy of present measurements. The states would exist distinct from the individual particles only up to the energy gap, so that the maximum Q would be about $m\epsilon_0/\hbar^2k_F\simeq 10^4$/cm. Another fact about them is that they cannot be correctly calculated directly from the excitation energies of the pair states and the appropriate matrix elements of H_V. This is because the B.C.S. ground state, while nearly exact, does not diagonalize H well enough for this purpose; we shall discuss this point later.

It is the uniqueness of the states Ψ_Q which will be the central feature in the explanation of the Meissner effect. Ψ_Q is a longitudinal excitation; the corresponding transverse excitations are of a completely different character. Study of the steps which lead to Eq. (11), or a little thought about the properties of Cooper pairs, quickly convinces us that no energy advantage is gained by making an excitation which does not have the same amplitude at all points of the Fermi surface. Any transverse excitation is of this form, and thus necessarily has a finite excitation energy as its wave number approaches zero. An equivalent statement is that no excitation possessing a finite angular momentum exists in the energy gap, since Cooper pairs have no bound state with nonzero angular momentum.

To understand the effect of these statements let us return to the fundamental sum rule (8). [From here on, we consider the question of the gauge invariance of the Hamiltonian itself closed. All statements are exactly true only in the weak coupling limit in which (11)≡0.] Equations (8) and (9) are general statements, and indeed are used in the Nozières-Pines plasmon theory,[7] since they involve only longitudinal excitations. Actually, in the solid as in the atom the more familiar form of (9) is the optical (i.e., transverse) sum rule, to derive which one takes the limit as $Q\to 0$ and assumes

$$\frac{2mN}{\hbar^2}=\lim_{Q\to 0}Q^{-2}\sum_{m'}(E_{m'}-E_m)^{-1}$$
$$\times|(m|\mathbf{Q}\cdot\sum[\nabla_j\exp(i\mathbf{Q}\cdot\mathbf{r}_j)+\exp(i\mathbf{Q}\cdot\mathbf{r}_j)\nabla_j]|m')|^2$$
$$=\tfrac{1}{3}\sum_{m'}(E_{m'}-E_m)^{-1}|(m|2\sum_j\nabla_j|m')|^2. \quad (24)$$

The condition for the validity of this limit process has not been previously discussed; we shall find that it involves the absence of a certain type of long-range order, which is, however, present in a superconductor, in that there is a qualitative difference between transverse and longitudinal excitations even as $Q\to 0$ and in the absence of Coulomb forces.

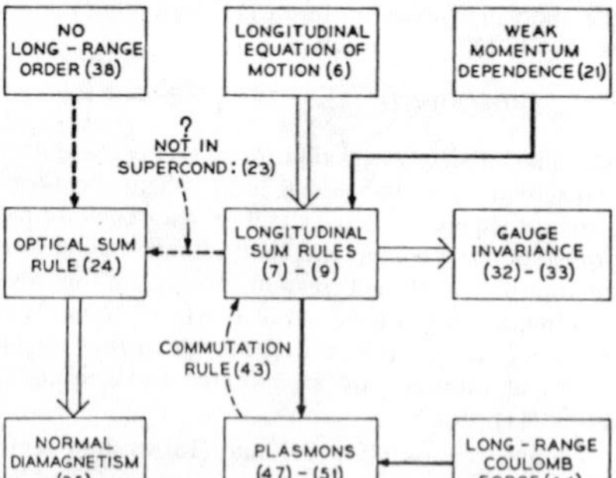

FIG. 1. Logical relations among diamagnetism, sum rules, gauge invariance, and plasmons. Wide arrows are general theorems, solid arrows mean "valid for superconductors at least," and heavy dashed arrows are theorems for normal but not superconducting cases. The numbers in parentheses refer to equations in the text.

III. RELATIONSHIP OF SUM RULES AND ELECTROMAGNETIC PROPERTIES; COMPARISON OF NORMAL AND SUPERCONDUCTING CASES

In this section we shall show how the mathematical questions of the sum rules (6)–(9) and (24) are related to the physical questions of gauge invariance and the Meissner effect. Figure 1 is a schematic diagram of the rather complex relationships involved. The core of the argument is that it is the longitudinal sum rule (9) which implies gauge invariance. As we have pointed out (9) is always true. On the other hand, normal diamagnetism is a consequence of the optical sum rule (24). We shall then show that the normal cases obey the optical sum rule, while the B.C.S. superconductor need not because of the distinction between the states Ψ_Q and the transverse excitations. B.C.S. have in fact shown, in deriving the Meissner effect, that it does not, and there is no need to repeat their calculation here.

To do this we must write down the two quantities involved in the calculation of electromagnetic effects, the current operator:

$$\mathbf{J}(r)=(ie\hbar/2m)(\Psi^*\nabla\Psi-\Psi\nabla\Psi^*)-(e^2/mc)\Psi^*\mathbf{A}\Psi$$
$$=\mathbf{J}_p+\mathbf{J}_d, \quad (25)$$

[8] J. Bardeen has pointed out to me that (23) could also be derived by the method of R. P. Feynman [Phys. Rev. 94, 262 (1954)] using the B.C.S. calculation of the correlation function. Feynman also presents arguments that $\rho_Q\Psi_g$ is nearly the eigenstate.

and the perturbation to the Hamiltonian caused by the vector potential **A**:

$$H_1=(ie\hbar/2m)\int d\mathbf{r}\Psi^*(\mathbf{A}\cdot\nabla+\nabla\cdot\mathbf{A})\Psi. \quad (26)$$

The vector potential may be expanded in Fourier series:

$$\mathbf{A}=\sum_Q \mathbf{a}_Q \exp(i\mathbf{Q}\cdot\mathbf{r}),$$

and one can calculate the appropriate Fourier components of the current $\mathbf{J}_Q$. If we set

$$\mathbf{J}_Q=-K_Q\mathbf{a}_Q, \quad (27)$$

then it is well known that the Meissner effect follows when in the London gauge:

$$\nabla\cdot\mathbf{A}=0,\qquad \lim_{Q\to 0}K_Q=\text{finite}. \quad (28)$$

On the other hand, in a longitudinal gauge there can be no physical current:

$$\nabla\times\mathbf{A}=0,\qquad K_Q\equiv 0. \quad (29)$$

The problem is to reconcile these two [(28) and (29)].

In an energy-gap model, (28) is true because the $\mathbf{J}_p$ term of (25) either is small or vanishes, as a result of small matrix elements and finite energy denominators in perturbation theory. But the finite energy denominators must not cause $\mathbf{J}_p$ to vanish in the longitudinal case (29) (see reference 4).

Let us take up the longitudinal case first. Then a typical component is

$$\mathbf{A}=\nabla\exp(i\mathbf{Q}\cdot\mathbf{r}). \quad (30)$$

The Qth Fourier component of the resulting paramagnetic current is, in perturbation theory,

$$\mathbf{J}_p=\sum_{m'}\left(\frac{e^2\hbar^2}{4m^2c}\right)\times(0|\sum_n[\exp(i\mathbf{Q}\cdot\mathbf{r}_n)\nabla_n+\nabla_n\exp(i\mathbf{Q}\cdot\mathbf{r}_n)]|m')\times(m'|i\mathbf{Q}\cdot\sum_n[\exp(i\mathbf{Q}\cdot\mathbf{r}_n)\nabla_n+\nabla_n\exp(i\mathbf{Q}\cdot\mathbf{r}_n)]|0)\times(E_{m'}-E_0)^{-1}+\text{complex conj.} \quad (31)$$

From the sum rule (9),

$$\mathbf{Q}\cdot\mathbf{J}_p=iQ^2Ne^2/mc, \quad (32)$$

which is exactly the negative of

$$\mathbf{Q}\cdot\mathbf{J}_d=-(Ne^2/mc)(\mathbf{Q}\cdot\mathbf{A}). \quad (33)$$

Thus (9) ensures gauge invariance, and our proof that to a certain order of accuracy (7) is maintained in B.C.S. is a proof of gauge invariance to that order.

Now consider the real diamagnetism problem in which

$$\mathbf{A}=\mathbf{a}_Q\exp(i\mathbf{Q}\cdot\mathbf{r});\quad \mathbf{a}_Q\cdot\mathbf{Q}=0. \quad (34)$$

Now,

$$\mathbf{J}_p=\sum_{m'}(e^2\hbar^2/4m^2c)\times(0|\sum_n[\exp(i\mathbf{Q}\cdot\mathbf{r}_n)\nabla_n+\nabla_n\exp(i\mathbf{Q}\cdot\mathbf{r}_n)]|m')\times(m'|\sum_n[\exp(i\mathbf{Q}\cdot\mathbf{r}_n)\mathbf{a}_Q\cdot\nabla_n+\nabla_n\cdot\mathbf{a}_Q\exp(i\mathbf{Q}\cdot\mathbf{r}_n)]|0)\times(E_{m'}-E_0)^{-1}+\text{complex conj.} \quad (35)$$

As far as possible constant terms are concerned, if the optical sum rule (24) is valid, then (35) is also equal to the corresponding $\mathbf{J}_d$ and normal diamagnetism follows.

Having thus shown how the gauge invariance and sum rule questions are the same, let us compare how the three cases—superconductor, insulator, and normal metal—obey the sum rules (7) and (9). In the normal metal, the ground state may be approximated by a Fermi sea. The appropriate excited states are single-particle excitations from **k** to **k**+**Q**, whose excitation energies are proportional to Q. Equation (7) is satisfied by finite matrix elements of ρ_Q and small excitation energies. However, we can see from (8) that the sum

$$\sum_{m'}|(0|\rho_Q|m')|^2=(\Psi_Q,\Psi_Q) \quad (36)$$

vanishes as Q when $Q\to 0$, as a result of the small number of possible excitations as $Q\to 0$. This means that the identity (22) does not cause any anomalous excited states to appear as $Q\to 0$.

The insulator behaves in an even more regular way. In the insulator there is an automatic energy gap between the ground state and all excitations, but (7) can be satisfied because $(0|\rho_Q|m)\to 0$. This in turn is simply a consequence of the fact that excited states are orthogonal to the ground state. The sum (36) in the insulator vanishes as Q^2, and again (22) does not cause any special excited states to appear.

We see, then, that the failure of (36) to vanish in the superconductor is the major difference from the normal state. We now show that this is related to the presence of a kind of long-range order in the electrons' wave function. A unified way to prove that (36) vanishes in the normal cases is to write the wave function as a determinant of localized one-electron functions:

$$\Psi_0=\sum_P(-1)^P\prod_{j=1}^{N}\Psi(\mathbf{r}_j-\mathbf{R}_{P(j)}). \quad (37)$$

In the insulator case, the Ψ's are the usual Wannier functions. Equation (37) is not in the literature for the case of the Fermi sea, but may be derived as a limit from the insulator by increasing the lattice spacing and combining more and more Brillouin zones to approximate a sphere. Then, as we approach $Q=0$,

$$\rho_Q\Psi_0=\sum_j\sum_P(-1)^P\exp(i\mathbf{Q}\cdot\mathbf{r}_j)\prod_{j=1}^{N}\Psi(\mathbf{r}_j-\mathbf{R}_{P(j)})$$

$$=[\sum_j\exp(i\mathbf{Q}\cdot\mathbf{R}_j)]\Psi_0+\sum_{P,j}(-1)^P$$

$$\times\exp(i\mathbf{Q}\cdot\mathbf{R}_{P(j)})\mathbf{Q}\cdot(\mathbf{r}_j-\mathbf{R}_{P(j)})$$

$$\times\prod_j\Psi(\mathbf{r}_j-\mathbf{R}_{P(j)})\cdots=0+o(Q). \quad (38)$$

In the real physical case, one-electron determinants are only an approximation. Nonetheless, if we retain the definition of "no long-range order" as meaning that the wave function is really a properly symmetrized product of factors, each referring to a volume small in comparison to the whole specimen, then the above proof still holds.

Now we come to the question of showing that (24) holds in the metal and insulator but not in the superconductor. The proof for one-electron function representations of metal and insulator is trivial, but we shall give it here in a form using (37) which indicates how it might easily be generalized to any case without long-range order.

The sum rule we start from may also be written in the mixed form

$$\sum_m(0|\mathbf{Q}\cdot\sum_j[\nabla_j\exp(-i\mathbf{Q}\cdot\mathbf{r}_j)+\exp(-i\mathbf{Q}\cdot\mathbf{r}_j)\nabla_j]|m)(m|\exp(i\mathbf{Q}\cdot\mathbf{r}_j)|0)+\text{c.c.}=2NQ^2. \quad (39)$$

The above is simply the scalar product

$$(\sum_{P,j}(-1)^P\exp(i\mathbf{Q}\cdot\mathbf{R}_{P(j)})2\mathbf{Q}\cdot\nabla_j\prod\Psi(\mathbf{r}_j-\mathbf{R}_{P(j)}),\sum_{P,j}(-1)^P\exp(i\mathbf{Q}\cdot\mathbf{R}_{P(j)})\mathbf{Q}\cdot(\mathbf{r}_j-\mathbf{R}_{P(j)})\prod\Psi(\mathbf{r}_j-\mathbf{R}_{P(j)})).$$

Here we have already used (38). In any case in which the Wannier functions may be taken even or odd, this is the same as

$$\sum_j(\sum_P(-1)^P\exp(i\mathbf{Q}\cdot\mathbf{R}_{P(j)})2\mathbf{Q}\cdot\nabla_j\Psi(\mathbf{r}_j-\mathbf{R}_{P(j)}),\sum_{P'}(-1)^{P'}\exp(i\mathbf{Q}\cdot\mathbf{R}_{P'(j)})\mathbf{Q}\cdot(\mathbf{r}_j-\mathbf{R}_{P'(j)})\Psi(\mathbf{r}_j-\mathbf{R}_{P'(j)})). \quad (40)$$

Since the Ψ are localized, the scalar product will vanish for all but nearby $\mathbf{R}_{P(j)}$'s, which means that at small $\mathbf{Q}$ the exponentials all approach unity. Clearly the only property of the wave function which has been used is the absence of long-range order. The above proof fails completely for the B.C.S. ground state, simply because (37) and (38) are not true. The matrix elements which enter in the current in the B.C.S. case can be shown to lead to states with finite angular momentum, which are orthogonal to the states Ψ_Q which satisfy the sum rule (9). For details of the actual calculation see the B.C.S. paper.

IV. PLASMA CONSIDERATIONS

So far all our work has ignored the long-range order and correlations caused by the plasma effect. The reason is now obvious: that a certain special type of long-range order is required to explain the Meissner effect, which is not at all similar to the order enforced by long-range Coulomb forces. That latter order would only serve to obscure the relationships. Prior to plasma effects, the derivation of (24) from (9) is trivial, as in the Nozières-Pines paper[7]; afterwards, it is not at all simple.

In fact, we shall simply repeat the Nozières-Pines work here, but must first briefly explain the philosophy of what we do. The interaction used in B.C.S. includes both the phonon and repulsive Coulomb interactions, the latter appropriately screened, so that our procedure does require justification, in that this screening implies that in the ground-state wave function the plasma effects have already been included.

The basis of our justification is provided by the observation of Sawada, Brueckner, Fukuda, and Brout[9] that the plasma properties of the free-electron gas follow from a Hamiltonian in which the different Q's are completely decoupled. This observation provides a deeper justification for the "random-phase approximation" of Bohm and Pines.[10] What we shall do is to retain the random-phase approximation in the superconductor.

As Sawada *et al.* show, the random-phase approximation is equivalent to assuming it more probable that Coulomb interactions return an excited particle into the Fermi sea than that they excite still another. Such terms, as shown by Brueckner and Gell-Mann,[11] always lead to the most singular parts of the interactions. Our rather physical argument is that we can show that the part of the interaction retained by these authors is practically unchanged by the energy gap. It is then very hard to see why the less singular parts of the Coulomb interaction would take on a new importance and overwhelm the more singular terms in the superconducting, but not the normal case.

Our procedure is in principle the following: the original B.C.S. Hamiltonian is assumed to contain all the Coulomb interaction except that part involving ρ_Q itself. Thus most of the screening is present, while all momenta except $\mathbf{Q}$ have corresponding subsidiary conditions and plasma terms. For $\mathbf{Q}$, however, we can still prove (7) and (9) and study the transition from (9) to (24) as we did in the last section; the random-phase approximation tells us there is no strong coupling of $\mathbf{Q}$ and other momenta (we shall discuss the explicit point where this is introduced later). We then show that the plasma properties follow from the sum rules, practically

[9] Sawada, Brueckner, Fukuda, and Brout, Phys. Rev. 108, 507 (1957).

[10] D. Bohm and D. Pines, Phys. Rev. 85, 338 (1952).

[11] K. A. Brueckner and M. Gell-Mann, Phys. Rev. 106, 367 (1957).

unaffected by the gap. On the other hand, the states Ψ_Q disappear and are replaced by the plasma states at $E=\hbar\omega_p$. This means that, in a sense, the pairing criterion $\mathbf{Q}=0$ of B.C.S. is enforced only by the subsidiary condition associated with the plasma.

To derive the plasma effects, we shall use the method of the appendix of reference 9. Define a quantity

$$\Pi_Q=(NQ^2)^{-1}$$

$$2i\mathbf{Q}\cdot\sum_j[\nabla_j \exp(i\mathbf{Q}\cdot\mathbf{r}_j)+\exp(i\mathbf{Q}\cdot\mathbf{r}_j)\nabla_j]. \quad (41)$$

The commutation rule (7) gives us

$$[H,\rho_Q]=(N\hbar^2Q^2/mi)\Pi_Q, \quad (42)$$

while (9) gives

$$[\Pi_Q,\rho_{-Q}]=i, \quad (43)$$

so that $\Pi_Q{}^*$ plays the role of a momentum conjugate to the coordinate ρ_Q, and the sum rule becomes their commutation relation. Now we must introduce the interaction

$$H_c=\frac{2\pi e^2}{Q^2}(\rho_Q\rho_{-Q}+\rho_{-Q}\rho_Q), \quad (44)$$

of which we write only the $\mathbf{Q}$ terms. We also assume the existence of a ground state of energy E_0, which we expect to be perturbed from the Ψ_g so far discussed in such a way as to give the part of the long-range correlation energy associated with ρ_Q, and the subsidiary condition. Finally, we observe that if ρ_Q and Π_Q are to play the role of plasma oscillation coordinates, the first excited state Ψ_1 is obtainable by

$$\Psi_1=(\alpha\rho_Q+\beta\Pi_Q)\Psi_0, \quad (45)$$

while the energy condition

$$E_1\Psi_1=(H_0+H_c)\Psi_1 \quad (46)$$

is just

$$[H,(\alpha\rho_Q+\beta\Pi_Q)]\Psi_0=E_1(\alpha\rho_Q+\beta\Pi_Q)\Psi_0, \quad (47)$$

and, as Sawada *et al.* show,[9] the condition that Ψ_0 be the ground state is the subsidiary condition:

$$(\alpha\rho_Q+\beta\Pi_Q)^\dagger\Psi_0\equiv 0. \quad (48)$$

The commutator in (47) with Π_Q is made up of

$$[H_c,\Pi_Q]=+4\pi e^2Q^{-2}i\rho_Q, \quad (49)$$

and of

$$[H_0,\Pi_Q]=mi(N\hbar^2Q^2)^{-1}[H_0,[H,\rho_Q]],$$
$$[H_0,\Pi_Q]_{mm'}=mi(N\hbar^2Q^2)^{-1}(E_m-E_{m'})^2(m|\rho_Q|m'), \quad (50)$$

in the representation in which H_0 is diagonal. The sum rules (7) and (8) show us that this is of lower order in Q than (49) except in the insulator (for which a complete discussion has been given by Nozières and Pines[7]).

The random-phase approximation consists in assuming that (49) is the only important commutator of Π_Q with any of the Coulomb terms. A discussion of this was given earlier.

We can now solve (47).

$$\alpha\left(\frac{N\hbar^2Q^2}{mi}\right)\Pi_Q+\beta\left(\frac{4\Pi e^2}{Q^2}i\right)\rho_Q=E_1(\alpha\rho_Q+\beta\Pi_Q).$$

Since ρ_Q and Π_Q are independent, this can be solved to give

$$E_1{}^2=4\pi ne^2\hbar^2/m=(\hbar\omega_p)^2, \quad (51)$$

$$\alpha/\beta=im\omega_p/N\hbar Q^2, \quad (52)$$

and the subsidiary condition, from (48),

$$(\alpha\rho_{-Q}-\beta\Pi_{-Q})\Psi_0=0, \quad (53)$$

which is the same as that of Bohm and Pines[10] as $Q\to 0$, since $\alpha\gg\beta$, and the no-plasmon states have almost no $Q\neq 0$ pairs present. The old states Ψ_g and Ψ_Q have disappeared; they are related to Ψ_0 and Ψ_1 more or less as eigenfunctions of momentum are to the harmonic oscillator eigenfunctions. The derivation of (51) and (53) completes our program of showing that only (7) and (44) are necessary to normal plasmon behavior.

V. CONCLUSION

The above is by no means a rigorous and complete answer to the original question of whether the B.C.S. theory satisfies gauge invariance and the sum rules and still shows a Meissner effect. Although most of it is fairly rigorous, a few parts have to be considered as a map for how things might be, rather than a proof of how they are. A few points which would bear further discussion follow.

The first one is the question of rigorous rather than approximate gauge invariance. To show rigorous gauge invariance we would have to show that the corrections to $\mathbf{J}$ and H_1 always canceled the momentum dependence of H_0. Many arguments indicate that this is not a basic difficulty. For instance, one can always make a superconductor of arbitrarily large transition temperature by letting $\hbar\omega\to\infty$ and $\exp[-1/N(0)V]\to 0$ simultaneously, thus satisfying gauge invariance arbitrarily well.

Second is the random-phase approximation in the superconductor. As this is an incompletely solved question for normal metals, we have little hope of making much more headway here.

The most serious question revolves around the correctness of calculating E_Q by the commutator argument rather than directly from B.C.S. excited-state energies and matrix elements. These latter are all calculated ignoring the higher order corrections of $\mathbf{Q}\neq 0$ interactions as well as the fact that the B.C.S. ground state is not the exact solution of the reduced Hamiltonian. We believe that the argument is as sound as that of any such intermediate coupling method, in that what has to be assumed is that certain properties of the ground state—the energy commutator, which is very insensitive, and the normalization of Ψ_Q, equivalent to the correlation function of the electrons[8]—are more stable against perturbations than the actual matrix elements themselves. In calculating frequencies of other coherent elementary excitations, such as sound waves or spin waves in metals, the same situation often arises—the energy calculated directly from what appears to be the correct wave function is not the same as that calculated from the equations of motion.

ACKNOWLEDGMENTS

I should like to acknowledge extensive help and advice from J. Bardeen, who independently arrived at a qualitative understanding of much of the above, and to thank him for prepublication use of the B.C.S. paper. I should also acknowledge helpful conversations with C. Herring and G. Wentzel and discussions with E. Abrahams, M. Lax, and G. H. Wannier.

APPENDIX. COMPUTATIONS OF COMMUTATORS

The interaction H_V is given in Eq. (10). We divide $\rho_{\mathbf{Q}}$ into two parts:

$$\rho_{\mathbf{Q}}=\rho_{\mathbf{Q}}^{+}+\rho_{\mathbf{Q}}^{-}=\sum_{\mathbf{K}}(c_{\mathbf{K}+\mathbf{Q}\uparrow}{}^{*}c_{\mathbf{K}\uparrow}+c_{-\mathbf{K}\downarrow}{}^{*}c_{-\mathbf{K}-\mathbf{Q}\downarrow}), \tag{A1}$$

and by direct computation:

$$H_V\rho_{\mathbf{Q}}{}^{+}-\rho_{\mathbf{Q}}{}^{+}H_V=V\sum_{\mathbf{q}}\Big(\sum_{\mathbf{k}(\epsilon_{\mathbf{k}}<\hbar\omega)}\ \sum_{\mathbf{k}'\neq\mathbf{k}(\epsilon_{\mathbf{k}'}<\hbar\omega)}+\sum_{\mathbf{k}(\epsilon_{\mathbf{k}-\mathbf{q}}<\hbar\omega)}\ \sum_{\mathbf{k}'\neq\mathbf{k}(\epsilon_{\mathbf{k}'-\mathbf{q}}<\hbar\omega)}\Big)$$
$$\times(c_{\mathbf{k}'+\mathbf{Q}\uparrow}{}^{*}c_{-\mathbf{k}'+\mathbf{q}\downarrow}{}^{*}c_{-\mathbf{k}+\mathbf{q}\downarrow}c_{\mathbf{k}\uparrow}-c_{\mathbf{k}'\uparrow}{}^{*}c_{-\mathbf{k}'+\mathbf{q}\downarrow}{}^{*}c_{-\mathbf{k}+\mathbf{q}\downarrow}c_{\mathbf{k}-\mathbf{Q}\uparrow}). \tag{A2}$$

In the case of $\rho_{\mathbf{Q}}{}^{-}$, the summations are the same but the last parentheses read:

$$(c_{\mathbf{k}'\uparrow}{}^{*}c_{-\mathbf{k}'+\mathbf{q}+\mathbf{Q}\downarrow}{}^{*}c_{-\mathbf{k}+\mathbf{q}\downarrow}c_{\mathbf{k}\uparrow}-c_{\mathbf{k}'\uparrow}{}^{*}c_{-\mathbf{k}'+\mathbf{q}\downarrow}{}^{*}c_{-\mathbf{k}+\mathbf{q}-\mathbf{Q}\downarrow}c_{\mathbf{k}\uparrow}). \tag{A3}$$

In both (A2) and (A3) the second term can be made to correspond with the first by the appropriate substitutions: in (A2) by

$$\mathbf{k}'-\mathbf{Q}\rightarrow k',\quad \mathbf{k}-\mathbf{Q}\rightarrow\mathbf{k},\quad \mathbf{q}-\mathbf{Q}\rightarrow\mathbf{q}, \tag{A4}$$

and in (A3) simply by

$$\mathbf{q}-\mathbf{Q}\rightarrow\mathbf{q}. \tag{A5}$$

If Q is small, (A4) and (A5) have no effect on the large majority of terms, so that the appropriate parts of the second term cancel the first. Near the cutoff surface, the substitutions affect the presence or absence of the terms: (A4) when the cutoff depends on $\mathbf{k}$ and $\mathbf{k}'$, (A5) when it depends on $\mathbf{k}-\mathbf{q}$, $\mathbf{k}'-\mathbf{q}$. The resulting terms near the cutoff surface are given in (11) of the text.

It is also necessary to use Eq. (11) to compute $(\rho_{\mathbf{Q}}\Psi_g,[H_V,\rho_{\mathbf{Q}}]\Psi_g)$. Equation (17) of the text shows us that only terms of the commutator which break up only two pairs can contribute to the scalar product. Examination of (11) reveals that two types of term can appear. These are the terms involving $\mathbf{q}=0$ or $\mathbf{q}=-\mathbf{Q}$. Upon taking this into account, the important part of (11) becomes

$$\tfrac{1}{2}V\Big\{\sum_{\mathbf{k}'\neq\mathbf{k}}\Big[\sum_{\mathbf{k}(\epsilon_{\mathbf{k}}<\hbar\omega,\ \epsilon_{\mathbf{k}+\mathbf{Q}}>\hbar\omega)}-\sum_{\mathbf{k}(\epsilon_{\mathbf{k}}>\hbar\omega,\ \epsilon_{\mathbf{k}+\mathbf{Q}}<\hbar\omega)}\Big]+\text{same with } \mathbf{k}'\rightarrow\mathbf{k},\ \mathbf{k}\rightarrow\mathbf{k}'\Big\}$$
$$\times(c_{\mathbf{k}'+\mathbf{Q}\uparrow}{}^{*}c_{-\mathbf{k}'\downarrow}{}^{*}c_{-\mathbf{k}\downarrow}c_{\mathbf{k}\uparrow}+c_{\mathbf{k}'+\mathbf{Q}\uparrow}{}^{*}c_{-\mathbf{k}'-\mathbf{Q}\downarrow}{}^{*}c_{-\mathbf{k}-\mathbf{Q}\downarrow}c_{\mathbf{k}\uparrow}-c_{\mathbf{k}'\uparrow}{}^{*}c_{-\mathbf{k}'\downarrow}{}^{*}c_{-\mathbf{k}-\mathbf{Q}\downarrow}c_{\mathbf{k}\uparrow}-c_{\mathbf{k}'+\mathbf{Q}\uparrow}{}^{*}c_{-\mathbf{k}'\downarrow}{}^{*}c_{-\mathbf{k}-\mathbf{Q}\downarrow}c_{\mathbf{k}+\mathbf{Q}\uparrow}). \tag{A6}$$

The effect of the terms in parentheses in this equation on the B.C.S. Ψ_g may now be computed:

$$(\quad)\Psi_g=\Big\{\prod_{\mathbf{K}\neq\mathbf{k}',\mathbf{k}'+\mathbf{Q},\mathbf{k}}[(1-h_{\mathbf{K}})^{1/2}+h_{\mathbf{K}}{}^{1/2}b_{\mathbf{K}}{}^{*}]h_{\mathbf{k}}{}^{1/2}(1-h_{\mathbf{k}'})^{1/2}(1-h_{\mathbf{k}'+\mathbf{Q}})^{1/2}c_{\mathbf{k}'+\mathbf{Q}\uparrow}{}^{*}c_{-\mathbf{k}'\downarrow}{}^{*}$$
$$-\prod_{\mathbf{K}\neq\mathbf{k},\mathbf{k}+\mathbf{Q},\mathbf{k}'+\mathbf{Q}}[(1-h_{\mathbf{K}})^{1/2}+h_{\mathbf{K}}{}^{1/2}b_{\mathbf{K}}{}^{*}](1-h_{\mathbf{k}'+\mathbf{Q}})^{1/2}h_{\mathbf{k}}{}^{1/2}h_{\mathbf{k}+\mathbf{Q}}{}^{1/2}b_{\mathbf{k}'+\mathbf{Q}}{}^{*}c_{\mathbf{k}+\mathbf{Q}\uparrow}{}^{*}c_{-\mathbf{k}\downarrow}{}^{*}$$
$$+\prod_{\mathbf{K}\neq\mathbf{k},\mathbf{k}+\mathbf{Q},\mathbf{k}'}[(1-h_{\mathbf{K}})^{1/2}+h_{\mathbf{K}}{}^{1/2}b_{\mathbf{K}}{}^{*}]h_{\mathbf{k}}{}^{1/2}h_{\mathbf{k}+\mathbf{Q}}{}^{1/2}(1-h_{\mathbf{k}'})^{1/2}b_{\mathbf{k}'}{}^{*}c_{\mathbf{k}+\mathbf{Q}\uparrow}{}^{*}c_{-\mathbf{k}\downarrow}{}^{*}$$
$$-\prod_{\mathbf{K}\neq\mathbf{k}',\mathbf{k}+\mathbf{Q},\mathbf{k}'+\mathbf{Q}}[(1-h_{\mathbf{K}})^{1/2}+h_{\mathbf{K}}{}^{1/2}b_{\mathbf{K}}{}^{*}]h_{\mathbf{k}+\mathbf{Q}}{}^{1/2}(1-h_{\mathbf{k}'})^{1/2}(1-h_{\mathbf{k}'+\mathbf{Q}})^{1/2}c_{\mathbf{k}'+\mathbf{Q}\uparrow}{}^{*}c_{-\mathbf{k}'\downarrow}{}^{*}\Big\}\Psi_v.$$

Now we combine this with (23) and (24) to compute the scalar product:

$$(\rho_{\mathbf{Q}}\Psi_g,[H_V,\rho_{\mathbf{Q}}]\Psi_g)=\tfrac{1}{2}V\Big[\sum_{\mathbf{k}(\epsilon_{\mathbf{k}}<\hbar\omega,\,\epsilon_{\mathbf{k}+\mathbf{Q}}>\hbar\omega)}\ \sum_{\mathbf{k}'\neq\mathbf{k}}-\sum_{\mathbf{k}(\epsilon_{\mathbf{k}}>\hbar\omega,\,\epsilon_{\mathbf{k}+\mathbf{Q}}<\hbar\omega)}\ \sum_{\mathbf{k}'\neq\mathbf{k}}+\sum_{\mathbf{k}'(\epsilon_{\mathbf{k}'}<\hbar\omega,\,\epsilon_{\mathbf{k}'+\mathbf{Q}}>\hbar\omega)}\ \sum_{\mathbf{k}\neq\mathbf{k}'}-\cdots\Big]$$

$$\times\{[h_{\mathbf{k}'}^{\frac{1}{2}}(1-h_{\mathbf{k}'+\mathbf{Q}})^{\frac{1}{2}}+h_{\mathbf{k}'+\mathbf{Q}}^{\frac{1}{2}}(1-h_{\mathbf{k}'})^{\frac{1}{2}}](1-h_{\mathbf{k}'})^{\frac{1}{2}}(1-h_{\mathbf{k}'+\mathbf{Q}})^{\frac{1}{2}}$$

$$\times[h_{\mathbf{k}}^{\frac{1}{2}}(1-h_{\mathbf{k}})^{\frac{1}{2}}-h_{\mathbf{k}+\mathbf{Q}}^{\frac{1}{2}}(1-h_{\mathbf{k}+\mathbf{Q}})^{\frac{1}{2}}]+[h_{\mathbf{k}}^{\frac{1}{2}}(1-h_{\mathbf{k}+\mathbf{Q}})^{\frac{1}{2}}+h_{\mathbf{k}+\mathbf{Q}}^{\frac{1}{2}}(1-h_{\mathbf{k}})^{\frac{1}{2}}]$$

$$\times h_{\mathbf{k}}^{\frac{1}{2}}h_{\mathbf{k}+\mathbf{Q}}^{\frac{1}{2}}[h_{\mathbf{k}'}^{\frac{1}{2}}(1-h_{\mathbf{k}'})^{\frac{1}{2}}-h_{\mathbf{k}'+\mathbf{Q}}^{\frac{1}{2}}(1-h_{\mathbf{k}'+\mathbf{Q}}^{\frac{1}{2}})]\}.$$

If in the first term we interchange the labels **k** and **k**′, it becomes very similar to the second, and the whole sum is symmetric in **k** and **k**+**Q**, and antisymmetric in **k**′ and **k**′+**Q**. We therefore have left only twice the limited sum over **k**′:

$$(\rho_{\mathbf{Q}}\Psi_g,[H_V,\rho_{\mathbf{Q}}]\Psi_g)=V[h_{\mathbf{k}}^{\frac{1}{2}}(1-h_{\mathbf{k}+\mathbf{Q}})^{\frac{1}{2}}+h_{\mathbf{k}+\mathbf{Q}}^{\frac{1}{2}}(1-h_{\mathbf{k}})^{\frac{1}{2}}]$$
$$\times[h_{\mathbf{k}}^{\frac{1}{2}}h_{\mathbf{k}+\mathbf{Q}}^{\frac{1}{2}}+(1-h_{\mathbf{k}})^{\frac{1}{2}}(1-h_{\mathbf{k}+\mathbf{Q}})^{\frac{1}{2}}][h_{\mathbf{k}'}^{\frac{1}{2}}(1-h_{\mathbf{k}'})^{\frac{1}{2}}-h_{\mathbf{k}'+\mathbf{Q}}^{\frac{1}{2}}(1-h_{\mathbf{k}'+\mathbf{Q}})^{\frac{1}{2}}].\qquad\text{(A7)}$$

In view of the symmetry in **k** and **k**+**Q** and the fact that Q is very small, the **k** sum may be simplified to $2V\sum_{\mathbf{k}}[h_{\mathbf{k}}(1-h_{\mathbf{k}})]^{\frac{1}{2}}=2\epsilon_0$. The **k**′ sum is an integral over the cutoff surface $|\epsilon|=\hbar\omega$. When **k**′ points at an angle θ to **Q**, the energy difference is

$$\delta\epsilon=\epsilon_{\mathbf{k}'+\mathbf{Q}}-\epsilon_{\mathbf{k}'}=(\hbar^2k_FQ\cos\theta)/m,\qquad\text{(A8)}$$

and the number of states in the surface element at this angle is $\frac{1}{2}\sin\theta d\theta N(0)\delta\epsilon$. Thus

$$\sum_{\mathbf{k}}\Delta[h_{\mathbf{k}}^{\frac{1}{2}}(1-h_{\mathbf{k}})^{\frac{1}{2}}]=4N(0)\tfrac{1}{2}\int_0^{\pi/2}\sin\theta d\theta\left[\left(\frac{\hbar^2}{m}\right)k_FQ\cos\theta\right]^2\frac{d}{d\epsilon}[\tfrac{1}{2}\epsilon_0(\epsilon^2+\epsilon_0^2)^{-\frac{1}{2}}]_{\epsilon=\hbar\omega}=\tfrac{1}{3}\frac{N(0)\hbar^2k_F^2Q^2\epsilon_0}{(m^2\omega^2)},\qquad\text{(A9)}$$

which gives the value (19) of the text.

Finally, we compute the commutator involved in the kinetic energy:

$$\sum_{\mathbf{K},\sigma}\mathbf{Q}\cdot(2\mathbf{K}+\mathbf{Q})c_{\mathbf{K}+\mathbf{Q},\sigma}{}^*c_{\mathbf{K},\sigma}\Psi_g=\mathbf{Q}\cdot\sum_{\mathbf{K}}(2\mathbf{K}+\mathbf{Q})(c_{\mathbf{K}+\mathbf{Q}\uparrow}{}^*c_{\mathbf{K}\uparrow}-c_{-\mathbf{K}\downarrow}{}^*c_{-\mathbf{K}-\mathbf{Q}\downarrow})\Psi_g$$
$$=\mathbf{Q}\cdot\sum_{\mathbf{K}}(2\mathbf{K}+\mathbf{Q})[h_{\mathbf{K}}^{\frac{1}{2}}(1-h_{\mathbf{K}+\mathbf{Q}})^{\frac{1}{2}}-h_{\mathbf{K}+\mathbf{Q}}^{\frac{1}{2}}(1-h_{\mathbf{K}})^{\frac{1}{2}}]\Psi_{-\mathbf{K},\mathbf{K}+\mathbf{Q}},\qquad\text{(A10)}$$

using (14)–(16). Again using (15) and (16), we get

$$(\rho_{\mathbf{Q}}\Psi_g,[H_K,\rho_{\mathbf{Q}}]\Psi_g)=(\hbar^2/2m)\sum_{\mathbf{K}}\mathbf{Q}\cdot(2\mathbf{K}+\mathbf{Q})(h_{\mathbf{K}}-h_{\mathbf{K}+\mathbf{Q}}).\qquad\text{(A11)}$$

For small Q, this is almost exactly

$$-(\hbar^2/m)\sum_{\mathbf{k}}(\mathbf{Q}\cdot\mathbf{k})\mathbf{Q}\cdot\nabla_{\mathbf{k}}(h_{\mathbf{k}})=(\hbar^2/m)^2\sum_{\mathbf{k}}(\mathbf{Q}\cdot\mathbf{k})^2(dh/d\epsilon).\qquad\text{(A12)}$$

In this form it is clearly independent of the exact form of the distribution, and thus is practically the same as the identical quantity for a Fermi sea. Evaluating (A12) (the assumption must be made that ϵ_0 is small), we get the result quoted as (20) of the text.

Reprinted from THE PHYSICAL REVIEW, Vol. 117, No. 3, 648–663, February 1, 1960
Printed in U. S. A.

Quasi-Particles and Gauge Invariance in the Theory of Superconductivity*

YOICHIRO NAMBU
The Enrico Fermi Institute for Nuclear Studies and the Department of Physics, The University of Chicago, Chicago, Illinois
(Received July 23, 1959)

Ideas and techniques known in quantum electrodynamics have been applied to the Bardeen-Cooper-Schrieffer theory of superconductivity. In an approximation which corresponds to a generalization of the Hartree-Fock fields, one can write down an integral equation defining the self-energy of an electron in an electron gas with phonon and Coulomb interaction. The form of the equation implies the existence of a particular solution which does not follow from perturbation theory, and which leads to the energy gap equation and the quasi-particle picture analogous to Bogoliubov's.

The gauge invariance, to the first order in the external electromagnetic field, can be maintained in the quasi-particle picture by taking into account a certain class of corrections to the charge-current operator due to the phonon and Coulomb interaction. In fact, generalized forms of the Ward identity are obtained between certain vertex parts and the self-energy. The Meissner effect calculation is thus rendered strictly gauge invariant, but essentially keeping the BCS result unaltered for transverse fields.

It is shown also that the integral equation for vertex parts allows homogeneous solutions which describe collective excitations of quasi-particle pairs, and the nature and effects of such collective states are discussed.

1. INTRODUCTION

A NUMBER of papers have appeared on various aspects of the Bardeen-Cooper-Schrieffer[1] theory of superconductivity. On the whole, the BCS theory, which leads to the existence of an energy gap, presents us with a remarkably good understanding of the general features of superconducivity. A mathematical formulation based on the BCS theory has been developed in a very elegant way by Bogoliubov,[2] who introduced coherent mixtures of particles and holes to describe a superconductor. Such "quasi-particles" are not eigenstates of charge and particle number, and reveal a very bold departure, inherent in the BCS theory, from the conventional approach to many-fermion problems. This, however, creates at the same time certain theoretical difficulties which are matters of principle. Thus the derivation of the Meissner effect in the original BCS theory is not gauge-invariant, as is obvious from the viewpoint of the quasi-particle picture, and poses a serious problem as to the correctness of the results obtained in such a theory.

This question of gauge invariance has been taken up by many people.[3] In the Meissner effect one deals with a linear relation between the Fourier components of the external vector potential A and the induced current J, which is given by the expression

$$J_i(q)=\sum_{j=1}^{3} K_{ij}(q)A_j(q),$$

with

$$K_{ij}(q)=-\frac{e^2}{m}\langle 0|\rho|0\rangle\delta_{ij}+\sum_n\left(\frac{\langle 0|j_i(q)|n\rangle\langle n|j_j(-q)|0\rangle}{E_n}+\frac{\langle 0|j_j(-q)|n\rangle\langle n|j_i(q)|0\rangle}{E_n}\right). \quad (1.1)$$

ρ and j are the charge-current density, and $|0\rangle$ refers to the superconductive ground state. In the BCS model, the second term vanishes in the limit $q\to 0$, leaving the first term alone to give a nongauge invariant result. It has been pointed out, however, that there is a significant difference between the transversal and longitudinal current operators in their matrix elements. Namely, there exist collective excited states of quasiparticle pairs, as was first derived by Bogoliubov,[2] which can be excited only by the longitudinal current.

As a result, the second term does not vanish for a longitudinal current, but cancels the first term (the longitudinal sum rule) to produce no physical effect; whereas for a transversal field, the original result will remain essentially correct.

If such collective states are essential to the gauge-invariant character of the theory, then one might argue that the former is a necessary consequence of the latter. But this point has not been clear so far.

Another way to understand the BCS theory and its problems is to recognize it as a generalized Hartree-Fock approximation.[4] We will develop this point a little further here since it is the starting point of what follows later as the main part of the paper.

* This work was supported by the U. S. Atomic Energy Commission.

[1] Bardeen, Cooper, and Schrieffer, Phys. Rev. **106**, 162 (1957); **108**, 1175 (1957).

[2] N. N. Bogoliubov, J. Exptl. Theoret. Phys. U.S.S.R. **34**, 58, 73 (1958) [translation: Soviet Phys. **34**, 41, 51 (1958)]; Bogoliubov, Tolmachev, and Shirkov, *A New Method in the Theory of Superconductivity* (Academy of Sciences of U.S.S.R., Moscow, 1958). See also J. G. Valatin, Nuovo cimento **7**, 843 (1958).

[3] M. J. Buckingam, Nuovo cimento **5**, 1763 (1957). J. Bardeen, Nuovo cimento **5**, 1765 (1957). M. R. Schafroth, Phys. Rev. **111**, 72 (1958). P. W. Anderson, Phys. Rev. **110**, 827 (1958); **112**, 1900 (1958). G. Rickayzen, Phys. Rev. **111**, 817 (1958); Phys. Rev. Letters **2**, 91 (1959). D. Pines and R. Schrieffer, Nuovo cimento **10**, 496 (1958); Phys. Rev. Letters **2**, 407 (1958). G. Wentzel, Phys. Rev. **111**, 1488 (1958); Phys. Rev. Letters **2**, 33 (1959). J. M. Blatt and T. Matsubara, Progr. Theoret. Phys. (Kyoto) **20**, 781 (1958). Blatt, Matsubara, and May, Progr. Theoret. Phys. (Kyoto) **21**, 745 (1959). K. Yosida, *ibid.* **731**.

[4] Recently N. N. Bogoliubov, Uspekhi Fiz. Nauk **67**, 549 (1959) [translation: Soviet Phys.—Uspekhi **67**, 236 (1959)], has also reformulated his theory as a Hartree-Fock approximation, and discussed the gauge invariance collective excitations from this viewpoint. The author is indebted to Prof. Bogoliubov for sending him a preprint.

Take the Hamiltonian in the second quantization form for electrons interacting through a potential V:

$$H=\int\sum_{i=1}^{2}\psi_i^+(x)K_i\psi_i(x)d^3x+\tfrac{1}{2}\int\int\sum_{i,k}\psi_i^+(x)$$
$$\times\psi_k^+(y)V(x,y)\psi_k(y)\psi_i(x)d^3xd^3y$$
$$\equiv H_0+H_{\text{int}}. \tag{1.2}$$

K is the kinetic energy plus any external field. $i=1$, 2 refers to the two spin states (e.g., spin up and down along the z axis).

The Hartree-Fock method is equivalent to linearizing the interaction H_{int} by replacing bilinear products like $\psi_i^+(x)\psi_k(y)$ with their expectation values with respect to an approximate wave function which, in turn, is determined by the linearized Hamiltonian. We may consider also expectation values $\langle\psi_i(x)\psi_k(y)\rangle$ and $\langle\psi_i^+(x)\psi_k^+(y)\rangle$ although they would certainly be zero if the trial wave function were to represent an eigenstate of the number of particles, as is the case for the true wave function.

We write thus a linearized Hamiltonian

$$H_0'=\int\sum_i\psi_i^+K_i\psi_id^3x+\int\int\sum_{i,k}[\psi_i^+(x)\chi_{ik}(xy)\psi_k(y)$$
$$+\psi_i^+(x)\phi_{ik}(xy)\psi_k^+(y)$$
$$+\psi_k(x)\phi_{ki}^+(xy)\psi_i(y)]d^3xd^3y$$
$$\equiv H_0+H_s, \tag{1.3}$$

where

$$\chi_{ik}(xy)=\delta_{ik}\delta^3(x-y)\int V(xz)\sum_j\langle\psi_j^+(z)\psi_j(z)\rangle d^3z$$
$$-V(xy)\langle\psi_k^+(y)\psi_i(x)\rangle, \tag{1.4}$$
$$\phi_{ik}(xy)=\tfrac{1}{2}V(xy)\langle\psi_k(y)\psi_i(x)\rangle,$$
$$\phi_{ik}^+(xy)=\tfrac{1}{2}V(xy)\langle\psi_k^+(y)\psi_i^+(x)\rangle.$$

We diagonalize H_0' and take, for example, the ground-state eigenfunction which will be a Slater-Fock product of individual particle eigenfunctions. The defining equations (1.4) then represent just generalized forms of Hartree-Fock equations to be solved for the self-consistent fields χ and ϕ.

The justification of such a procedure may be given by writing the original Hamiltonian as

$$H=(H_0+H_s)+(H_{\text{int}}-H_s)\equiv H_0'+H_{\text{int}}',$$

and demanding that H_{int}' shall have no matrix elements which would cause single-particle transitions; i.e., no matrix elements which would effectively modify the starting H_0': to put it more precisely, we demand our approximate eigenstates to be such that

$$\langle n|H_{\text{int}}'|0\rangle=\langle n|H|0\rangle=0, \tag{1.5}$$

if in $|n\rangle$ more than one particle change their states from those in $|0\rangle$. This condition is contained in Eq. (1.4).[5] Since in many-body problems, as in relativistic field theory, we often take a picture in which particles and holes can be created and annihilated, the condition (1.5) should also be interpreted to include the case where $|n\rangle$ and $|0\rangle$ differ only by such pairs. The significance of the BCS theory lies in the recognition that with an essentially attractive interaction V, a nonvanishing ϕ is indeed a possible solution, and the corresponding ground state has a lower energy than the normal state. It is also separated from the excited states by an energy gap $\sim 2\phi$.

The condition (1.5) was first invoked by Bogoliubov[2] in order to determine the transformation from the ordinary electron to the quasi-particle representation. He derived this requirement from the observation that H_{int}' contains matrix elements which spontaneously create virtual pairs of particles with opposite momenta, and cause the breakdown of the perturbation theory as the energy denominators can become arbitrarily small. Equation (1.5), as applied to such pair creation processes, determines only the nondiagonal part (in quasi-particle energy) of H_s in the representation in which H_0+H_s is diagonal. The diagonal part of H_s is still arbitrary. We can fix it by requiring that

$$\langle 1'|H_{\text{int}}'|1\rangle=0, \tag{1.6}$$

namely, the vanishing of the diagonal part of H_{int} for the states where one more particle (or hole) having a Hamiltonian H_0' is added to the ground state. In this way we can interpret H_0' as describing single particles (or excitations) moving in the "vacuum," and the diagonal part of H_s represents the self-energy (or the Hartree potential) for such particles arising from its interaction with the vacuum.

The distinction between Eqs. (1.5) and (1.6) is not so clear when applied to normal states. On the one hand, particles and holes (negative energy particles) are not separated by an energy gap; on the other hand, there is little difference when one particle is added just above the ground state.

In the above formulation of the generalized Hartree fields, χ and ϕ will in general depend on the external field as well as the interaction between particles. There is a complication due to the fact that they are gauge dependent. This is because a phase transformation $\psi_i(x)\rightarrow e^{i\lambda(x)}\psi_i(x)$ applied on Eq. (1.3) will change χ and ϕ according to

$$\chi(xy)\rightarrow e^{-i\lambda(x)+i\lambda(y)}\chi(xy),$$
$$\phi(xy)\rightarrow e^{i\lambda(x)+i\lambda(y)}\phi(xy), \tag{1.6}$$
$$\phi^+(xy)\rightarrow e^{-i\lambda(x)-i\lambda(y)}\phi^+(xy).$$

[5] Equation (1.5) refers only to the transitions from occupied states to unoccupied states. Transitions between occupied states or unoccupied states are given by Eq. (1.6). These two together then are equivalent to Eq. (1.4). For the analysis of the Hartree approximation in terms of diagrams, see J. Goldstone, Proc.

It is especially serious for ϕ (and ϕ^+) since, even if $\phi(xy)=\delta^3(x-y)$ times a constant in some gauge, it is not so in other gauges. Therefore, unless we can show explicitly that physical quantities do not depend on the gauge, any calculation based on a particular ϕ is open to question. It would not be enough to say that a longitudinal electromagnetic potential produces no effect because it can be transformed away before making the Hartree approximation. A natural way to reconcile the existence of ϕ, which we want to keep, with gauge invariance would be to find the dependence of ϕ on the external field explicitly. If the gauge invariance can be maintained, the dependence must be such that for a longitudinal potential $A=-\text{grad}\lambda$, it reduces to Eq. (1.6). This should not be done in an arbitrary manner, but by studying the actual influence of H_{int} on the primary electromagnetic interaction when ϕ is first determined without the external field.

After these preliminaries, we are going to study the points raised here by means of the techniques developed in quantum electrodynamics. We will first develop the Feynman-Dyson formulation adapted to our problem, and write down an integral eauation for the self-energy part which corresponds to the Hartree approximation. It is observed that it can possess a nonperturbational solution, and the existence of an energy gap is immediately recognized.

Next we will introduce external fields. Guided by the well-known theorems about gauge invariance, we are led to consider the so-called vertex parts, which include the "radiative corrections" to the primary charge-current operator. When an integral equation for the general vertex part is written down, certain exact solutions are obtained in terms of the assumed self-energy part, leading to analogs of the Ward identity.[6] They are intimately related to inherent invariance properties of the theory. Among other things, the gauge invariance is thus strictly established insofar as effects linear in the external field are concerned, including the Meissner effect.

Later we look into the collective excitations. A very nteresting result emerges when we observe that one of the exact solutions to the vertex part equations becomes a homogeneous solution if the external energy-momentum is zero, and expresses a bound state of a pair with zero energy-momentum. Then by perturbation, other bound states with nonzero energy-momentum are obtained, and their dispersion law determined. Thus the existence of the bound state is a logical consequence of the existence of the special self-energy ϕ and the gauge invariance, which are seemingly contradictory to each other.

When the Coulomb interaction is taken into account, the bound pair states are drastically modified, turning into the plasma modes due to the same mechanism as in the normal case. This situation will also be studied.

Roy. Soc. (London) A239, 267 (1957). Compare also T. Kinoshita and Y. Nambu, Phys. Rev. 94, 598 (1953).

[6] J. C. Ward, Phys. Rev. 78, 182 (1950).

2. FEYNMAN-DYSON FORMULATION

We start from the Lagrangian for the electron-phonon system, which is supposed to be uniform and isotropic.[7]

$$\mathfrak{L}=\sum_p \sum_i [i\psi_i^+(p)\dot{\psi}_i(p)-\psi_i^+(p)\epsilon_p\psi_i(p)] +\sum_k \tfrac{1}{2}[\dot{\varphi}(k)\dot{\varphi}(-k)-c^2\varphi(k)\varphi(-k)] -g\frac{1}{\sqrt{\mathcal{V}}}\sum_{p,k}\psi_i^+(p+k)\psi_i(p)h(k)\varphi(k). \quad (2.1)$$

φ is the phonon field, with the momentum k (energy $\omega_k=ck$) running up to a cutoff value $k_m(\omega_m)$; c is the phonon velocity. ϵ_p is the electron kinetic energy relative to the Fermi energy; $gh(k)$ represents the strength of coupling.[8] ($\mathcal{V}$ is the volume of the system.)

The Coulomb interaction between the electrons is not included for the moment in order to avoid complication. Later we will make remarks whenever necessary about the modifications when the Coulomb interaction is taken into account.

It will turn out to be convenient to introduce a two-component notation[9] for the electrons

$$\Psi(x)=\begin{pmatrix}\psi_1(x)\\ \psi_2^+(x)\end{pmatrix} \quad\text{or}\quad \Psi(p)=\begin{pmatrix}\psi_1(p)\\ \psi_2^+(-p)\end{pmatrix}, \quad (2.2)$$

and the corresponding 2×2 Pauli matrices

$$\tau_1=\begin{pmatrix}0&1\\1&0\end{pmatrix},\quad \tau_2=\begin{pmatrix}0&-i\\i&0\end{pmatrix},\quad \tau_3=\begin{pmatrix}1&0\\0&-1\end{pmatrix}. \quad (2.3)$$

The Lagrangian then becomes:

$$\mathfrak{L}=\sum_p \Psi^+(p)\left(i\frac{\partial}{\partial t}-\epsilon_p\tau_3\right)\Psi(p) +\sum_k \tfrac{1}{2}[\dot{\varphi}(k)\dot{\varphi}(-k)-c^2\varphi(k)\varphi(-k)] -g\frac{1}{\sqrt{\mathcal{V}}}\sum_{p,k}\Psi^+(p+k)\tau_3\Psi(p)h(k)\varphi(k)+\sum_p \epsilon_p$$
$$=\mathfrak{L}_0+\mathfrak{L}_{\text{int}}+\text{const}.$$

The last infinite c-number term comes from the rearrangement of the kinetic energy term. This is certainly uncomfortable, but will not be important except for the calculation of the total energy.

The fields obey the standard commutation relations.

[7] We use the units $\hbar=1$.

[8] For convenience, we have included in $h(k)$ the frequency factor: $h(k)=h_1(k)k_0$.

[9] P. W. Anderson [Phys. Rev. 112, 1900 (1958)], has also introduced this two-component wave function.

Especially for Ψ, we have

$$\{\Psi_i(x),\Psi_j^+(y)\}\equiv\Psi_i(x)\Psi_j^+(y)+\Psi_j^+(y)\Psi_i(x) =\delta_{ij}\delta^3(x-y), \qquad (2.5)$$
$$\{\Psi_i(p),\Psi_j^+(p')\}=\delta_{ij}\delta_{pp'}.$$

We may now formally treat H_{int} as perturbation, using the formulation of Feynman and Dyson.[10] The unperturbed ground state (vacuum) is then the state where all individual electron states $\epsilon_p<0(>0)$ are occupied (unoccupied) in the representation where $\psi_i^+(p)\psi_i(p)$ is the occupation number.

Having defined the vacuum, the time-ordered Green's functions for free electrons and phonons

$$\langle T(\Psi_i(xt),\Psi_j^+(x't'))\rangle=[G_0(x-x',\,t-t')]_{ij},$$
$$\langle T(\varphi(xt),\varphi(x't'))\rangle=\Delta_0(x-x',\,t-t') \qquad (2.6)$$

are easily determined. We get for their Fourier representation (in the limit $\mho\rightarrow\infty$)[10a]

$$G_0(xt)=(1/(2\pi)^4)\int G_0(pp_0)e^{ip\cdot x-ip_0t}d^3pdp_0,$$
$$\Delta_0(xt)=\frac{1}{(2\pi)^4}\int_{|k|<k_m}\Delta_0(kk_0)e^{ik\cdot x-ik_0t}d^3kdk_0,$$
$$G_0(pp_0)=i\left[P\frac{1}{p_0-\epsilon_p\tau_3}-i\pi\,\mathrm{sgn}(\tau_3\epsilon_p)\delta(p_0-\tau_3\epsilon_p)\right] \qquad (2.7)$$
$$=i(p_0+\epsilon_p\tau_3)/(p_0^2-\epsilon_p^2+i\epsilon),$$
$$\Delta_0(kk_0)=i\left[P\frac{1}{k_0^2-c^2k^2}-i\pi\delta(k_0^2-c^2k^2)\right]$$
$$=i/(k_0^2-c^2k^2+i\epsilon).$$

With the aid of these Green's functions, we are able to calculate the S matrix and other quantities according to a well-defined set of rules in perturbation theory.

We will analyze in particular the self-energies of the electron and the phonon. In the many-particle system, these energies express (apart from the self-interaction of the electron) the average interaction of a single particle or phonon placed in the medium. Because the phonon spectrum is limited, there will be no ultraviolet divergences, unlike the case of quantum electrodynamics.

These self-energies may be obtained in a perturbation expansion with respect to H_{int}. We are, however, interested in the Hartree method which proposes to take account of them in an approximate but nonperturbational way. It is true that the self-energies are in general complex due to the instability of single par-

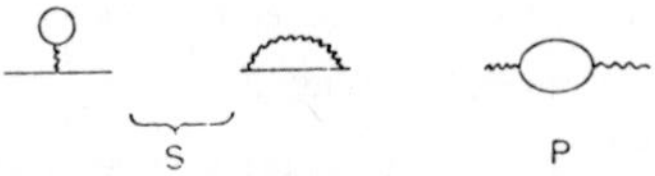

FIG. 1. Second order self-energy diagrams. Solid and curly lines represent electron and phonons, respectively, themselves being under the influence of the self-energies Σ and Π. All diagrams are to be interpreted in the sense of Feynman, lumping together all topologically equivalent processes.

ticles. But to the extent that the single-particle picture makes physical sense, we will ignore the small imaginary part of the self-energies in the following considerations.

Let us thus introduce the approximate self-energy Lagrangian $\mathfrak{L}_s$, and write

$$\mathfrak{L}=(\mathfrak{L}_0+\mathfrak{L}_s)+(\mathfrak{L}_{\mathrm{int}}-\mathfrak{L}_s)$$
$$\equiv\mathfrak{L}_0'+\mathfrak{L}_{\mathrm{int}}',$$
$$\mathfrak{L}_0=\sum_p\Psi_p^+L_0\Psi_p+\sum_k\tfrac{1}{2}\varphi_kM_0\varphi_{-k}, \qquad (2.8)$$
$$\mathfrak{L}_s=-\sum_p\Psi_p^+\Sigma\Psi_p-\sum_k\tfrac{1}{2}\varphi_k\Pi\varphi_{-k},$$
$$L_0-\Sigma\equiv L,\quad M_0-\Pi\equiv M.$$

The free electrons with "spin" functions u and phonons obey the dispersion law

$$L_0(\mathbf{p},\,p_0=\epsilon_p)u_p=0,\quad M_0(\mathbf{k},\,k_0=\omega_k)=0, \qquad (2.9)$$

whereas they obey in the medium

$$L(\mathbf{p},\,p_0=E_p)u_p=0,\quad M(\mathbf{k},\,k_0=\Omega_k)=0. \qquad (2.9')$$

Σ will be a function of momentum p and "spin." Π will consist of two parts: $\Pi(k_0k)=\Pi_1(k)k_0^2+\Pi_2(k)$ in conformity with the second order character (in time) of the phonon wave equation.[11]

The propagators corresponding to these modified electrons and phonons are

$$G(pp_0)=i/(L(pp_0)+i\,\mathrm{sgn}(p_0)\epsilon),$$
$$\Delta(kk_0)=i/(M(kk_0)+i\epsilon). \qquad (2.10)$$

We now determine Σ and Π self-consistently to the second order in the coupling g. Namely the second order self-energies coming from the phonon-electron interaction have to be cancelled by the first order effect of $\mathfrak{L}_{\mathrm{int}}$.

These second order self-energies are represented by the nonlocal operators[12] (Fig. 1)

$$\mathrm{S}((t+t')/2)=\int\int\int\Psi^+(xt)S(x-x',\,t-t')\times\Psi(x't')d^3xd^3x'd(t-t'),$$
$$\mathcal{P}((t+t')/2)=\tfrac{1}{2}\int\int\int\varphi(xt)P(x-x',\,t-t')\times\varphi(x't')d^3xd^3x'd(t-t'), \qquad (2.11)$$

[10] F. J. Dyson, Phys. Rev. 75, 486, 1736 (1949); R. P. Feynman, Phys. Rev. 76, 769 (1949); J. Schwinger, Phys. Rev. 74, 1439 (1948). Although we followed here the perturbation theory of Dyson, there is no doubt that the relations obtained in this paper can be derived by a nonperturbational formulation such as J. Schwinger's: Proc. Natl. Acad. Sci. U. S. 37, 452, 455 (1951).

[10a] P stands for the principal value; $i\epsilon$ in the denominator is a small positive imaginary quantity.

[11] In the same spirit Σ should actually be in the form $\Sigma_1(p)p_0+\Sigma_2(p)$. Here we neglect the renormalization term Σ_1 since the two conditions (2.13) can be met without it.

[12] We use the word nonlocal here for nonlocality in time.

where S and P have the Fourier representation

$$S(pp_0)=-ig^2\tau_3\delta^3(p)\delta(p_0)h^2(0)\Delta(0)\times\int\mathrm{Tr}[\tau_3G(p'p_0)]d^3pdp_0$$
$$-ig^2\int\tau_3G(p-k,\,p_0-k_0)\tau_3h^2(kk_0)\times\Delta(kk_0)d^3kdk_0,\quad(2.12)$$
$$P(kk_0)=ig^2h(kk_0)^2\int\mathrm{Tr}[\tau_3G(pp_0)\times G(p+k,\,p_0+k_0)]d^3pdp_0.$$

In Eq. (2.11) we have chosen more or less arbitrarily $(t+t')/2$ as the fixed time to which we refer the nonlocal operators $\mathcal{S}$ and $\mathcal{P}$. The self-consistency requirements (1.5) and (1.6) mean in the present case that Σ, Π must be identical with S, P (*a*): for the diagonal elements [on the energy shell, Eq. (2.9)], and (*b*): for the nondiagonal matrix elements for creating a pair out of the vacuum.

The pair creation of electrons is possible because Ψ, being a two-component wave function, can have in general two eigenfunctions u_{ps} $(s=1,2)$ with different energies E_{ps} for a fixed momentum, p, only one of which is occupied in the ground state.

Thus taking particular plane waves $u_{ps}{}^*e^{-ip\cdot x+ip_0t}$, $u_{p's'}e^{ip'\cdot x'-ip_0't'}$ for Ψ^+ and Ψ in (2.11), we easily find that the diagonal matrix element of Σ corresponds to $u_{ps}{}^*S(p,E_{ps})u_{ps}$, while the nondiagonal part corresponds to $u_{ps}{}^*S(p,0)u_{ps}{}'$, $s\neq s'$ $(p_0'=-p_0)$.

A similar situation holds also for the photon self-energy Π. Since Π consists of two parts, the diagonal and off-diagonal conditions will fix these.

With this understanding, the self-consistency relations may be written

$$\begin{aligned}\Sigma(\mathbf{p}E_p)_{\mathrm{D}}=S(\mathbf{p},E_p)_{\mathrm{D}},\quad \Sigma(\mathbf{p}0)_{\mathrm{ND}}=S(\mathbf{p}0)_{\mathrm{ND}},\\ \Pi(\mathbf{k}\Omega_k)=P(\mathbf{k}\Omega_k),\quad \Pi(\mathbf{k}0)=P(\mathbf{k}0),\end{aligned}\quad(2.13)$$

where D, ND signify the diagonal and nondiagonal parts in the "spin" space. As stated before, we have agreed to omit possible imaginary parts in S and P. (The nondiagonal components, however, will turn out to be real.)

Before discussing the general solutions, let us consider the meaning of Eq. (2.13) in terms of perturbation theory. Suppose we expand G occurring in Eq. (2.12), with respect to Σ:

$$G=G_0-iG_0\Sigma G_0-G_0\Sigma G_0\Sigma G_0+\cdots,$$

and expand Σ itself with respect to g^2, then we easily realize that Eq. (2.13) defines an infinite sum of a particular class of diagrams, which are illustrated in Fig. 2. The first term in S of Eq. (2.12) corresponds to the ordinary Hartree potential which is just a constant,

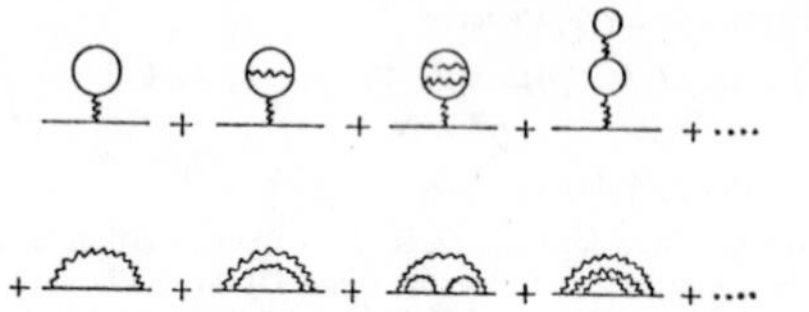

FIG. 2. Expansion of the self-consistent self-energy $\Sigma\sim S$ in terms of bare electron diagrams.

whereas the second term gives an exchange effect. In the latter, the approximation is characterized by the fact that no phonon lines cross each other.

It must be said that the Hartree approximation does not really sum the series of Fig. 2 completely since we equate in Eq. (2.13) only special matrix elements of both sides. For in the perturbation series the Σ obtained to any order is a function of p_0, whereas in Eq. (2.13) it is replaced by a p_0-independent quantity. Hence there will be a correction left out in each order (analogous to the radiative correction after mass renormalization in quantum electrodynamics).

In this perturbation expansion, S in Eq. (2.13) is always proportional to τ_3 on the energy shell since $H_0\propto\tau_3$. Accordingly Σ will be $\propto\tau_3$ and commute with H_0, so that no off-diagonal part exists.[11]

It is important, however, to note the possibility of a nonperturbational solution by assuming that Σ contains also a term proportional to τ_1 or τ_2. Thus, take

$$\begin{aligned}\Sigma(p)&=\chi(p)\tau_3+\phi(p)\tau_1,\\ H_0'&=(\epsilon+\chi)\tau_3+\phi\tau_1\\ &\equiv\bar{\epsilon}\tau_3+\phi\tau_1.\end{aligned}\quad(2.14)$$

This form bears a resemblance to the Dirac equation. Its eigenvalues are

$$E=\pm E_p\equiv\pm(\bar{\epsilon}_p{}^2+\phi_p{}^2)^{\frac{1}{2}}.\quad(2.15)$$

Since H_0' describes by definition excited states, we have to adopt the hole picture and conclude that the ground state (vacuum) is the state where all negative energy "quasi-particles" $(E<0)$ are occupied and no positive energy particles exist. If ϕ remains finite on the Fermi surface, the positive and negative states are separated by a gap $\sim2|\phi|$. The corresponding Green's function G now has the representation

$$G(pp_0)=i\frac{p_0+\bar{\epsilon}_p\tau_3+\phi_p\tau_1}{p_0{}^2-E_p{}^2+i\epsilon}.\quad(2.16)$$

In order to extract the diagonal and nondiagonal parts in spin space, we will use the trick

$$\begin{aligned}O_{\mathrm{D}}&=\tfrac{1}{2}\,\mathrm{Tr}\,(\Lambda O),\\ O_{\mathrm{ND}}&=-(i/2)\,\mathrm{Tr}\,(\Lambda O\tau_2),\\ \Lambda&=[E_p+H_0'(p)]/2E_p.\end{aligned}\quad(2.17)$$

Applying this to Eq. (2.13a) with Eqs. (2.12), (2.14), and (2.15), we finally obtain the following equations for χ and ϕ

$$\frac{\epsilon_p\chi_p+\phi_p^2}{E_p}=\frac{g^2\pi}{(2\pi)^4}P\int\left[\frac{E_p}{\Omega_k}+\frac{E_{p-k}+\Omega_k}{E_pE_{p-k}\Omega_k}\right.$$
$$\left.\times(\bar{\epsilon}_p\bar{\epsilon}_{p-k}-\phi_p\phi_{p-k})\right]\frac{h(k)^2}{E_p^2-(E_{p-k}-\Omega_k)^2}d^3k,$$
$$\epsilon_p\phi_p=\frac{g^2\pi}{(2\pi)^4}\int(\bar{\epsilon}_{p-k}\phi_p+\bar{\epsilon}_p\phi_{p-k})$$
$$\times\frac{h(k)^2d^3k}{E_{p-k}\Omega_k(E_{p-k}+\Omega_k)}. \quad (2.18)$$

The second equation, coming from the nondiagonal condition, has a trivial solution $\phi=0$. If a finite solution ϕ exists, it cannot follow from perturbation treatment since there is no inhomogeneous term to start with.

Equation (2.18) is equivalent to, but slightly different from, the corresponding conditions of Bogoliubov because of a slightly different definition of the nondiagonal part of the self-energy operator, which is actually due to an inherent ambiguity in approximating nonlocal operators by local ones. (This is the same kind of ambiguity as one encounters in the derivation of a potential from field theory. The difference between the local operator Σ and the nonlocal one S shows up in a situation like that in Fig. 3, and the compensation between Σ and S is not complete.) We may avoid this unpleasant situation, by extending the Hartree self-consistency conditions to all virtual matrix elements, but this would mean that ϕ (and χ) must be treated as nonlocal. We will discuss this situation in a separate section since such a generalization brings simplification in dealing with the problem of gauge invariance and collective excitations.

For the moment we consider the second equation of (2.18) and rewrite it

$$\phi_p=A_p\frac{g^2\pi}{(2\pi)^4}\int\frac{\phi_{p-k}}{E_{p-k}}\frac{h(k)d^3k}{\Omega_k(E_{p-k}+\Omega_k)},$$
$$A_p=\bar{\epsilon}_p\Big/\left(\epsilon_p-\frac{g^2\pi}{(2\pi)^4}\int\frac{\bar{\epsilon}_ph(k)^2d^3k}{E_{p-k}\Omega_k(E_{p-k}+\Omega_k)}\right). \quad (2.19)$$

This is essentially the energy gap equation of BCS if $g^2A_ph(k)^2/\Omega_k(E_{p-k}+\Omega_k)$ is identified with the effective interaction potential V, and if $\bar{\epsilon}_p\sim\epsilon_p(\chi_p\sim 0)$. It has a solution

$$\phi\sim\Omega_m\exp(-1/VN),$$

if $VN\ll 1$, N being the density of states: $N=dn/d\epsilon_p$ on the Fermi surface.

The phonon self-energy Π may be studied similarly from Eq. (2.13), which should determine the renormalization of the phonon field. It does not play an essential role in superconductivity, though it gives rise to an important correction when the Coulomb effect is taken into account. (See the following section.)

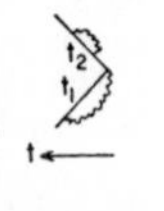

FIG. 3. An example of the situation where the cancellation of Σ_{ND} *versus* S_{ND} is not complete. The two self-energy parts overlap in time, and their centers of time t_1 and t_2 are such that $t_1>t_2$. If calculated according to the usual perturbation theory, this process will not be eliminated by the condition $\Sigma_{ND}=(S_1)_{ND}$.

From the nature of Eq. (2.12), it is clear that $\tau_1\phi$ can actually be pointed in any direction in the 1–2 plane of the τ space: $\tau_1\phi_1+\tau_2\phi_2$. It was thus sufficient to take $\phi_1\neq 0$, $\phi_2=0$. Any other solution is obtained by a transformation

$$\Psi\rightarrow\exp(i\alpha\tau_3/2)\Psi,$$
$$(\phi,0)\rightarrow(\phi\cos\alpha,\phi\sin\alpha). \quad (2.20)$$

In view of the definition of Ψ, Eq. (2.20) is a gauge transformation with a constant phase. Thus the arbitrariness in the direction of ϕ is the 1–2 plane is a reflection of the gauge invariance.

For later use, we also mention here the particle-antiparticle conjugation C of the quasi-particle field Ψ. This is defined by

$$C:\ \Psi\rightarrow\Psi^C=C\Psi^+=\tau_2\Psi^+,$$

or

$$\begin{pmatrix}\psi_1{}^C\\ \psi_2{}^{+C}\end{pmatrix}=\begin{pmatrix}-i\psi_2\\ i\psi_1{}^+\end{pmatrix}, \quad (2.21)$$

and changes quasi-particles of energy-momentum (p_0,p) into holes of energy-momentum $(-p_0,-p)$, or interchanges up-spin and down-spin electrons. Under C, the τ operators transform as

$$C:\ \tau_i\rightarrow C^{-1}\tau_iC=-\tau_i{}^T,\quad i=1,2,3 \quad (2.22)$$

where T means transposition.

As a consequence, we have also

$$C:\ L(p)\rightarrow L^C(-p)=-L(-p)^T. \quad (2.23)$$

Finally we make a remark about the Coulomb interaction. When this is taken into account, the phonon interaction factor $g^2h(k)^2\Delta(k,k_0)$ in Eq. (2.12a) has to be replaced by

$$[g^2h(k)^2\Delta(kk_0)+ie^2/k^2]/$$
$$\{1-i\Pi(kk_0)[\Delta(kk_0)+ie^2/g^2h(k)^2k^2]\}.$$

As is well known, the denominator represents the screening of the Coulomb interaction. Discussion about this point will be made later in connection with the plasma oscillations.

3. NONLOCAL (ENERGY-DEPENDENT) SELF-CONSISTENCY CONDITIONS

In the last section we remarked that the self-consistency conditions Eq. (2.13) may be extended to all virtual matrix elements, namely, not only on the energy shell (diagonal) and for the virtual pair creation out of the vacuum, but also for the self-energy effects which appear in intermediate states of any process.

This simply means that ϕ and π are now nonlocal; i.e., depend both on energy and momentum arbitrarily, and are to be completely equated with S and P, respectively,

$$\Sigma(pp_0)=S(pp_0),\quad \Pi(kk_0)=P(kk_0). \tag{3.1}$$

Actually, these self-energies can no more be incorporated in H_0' as the zeroth order Lagrangian since they contain infinite orders of time derivatives.[13] Nevertheless, Eq. (3.1) has a precise meaning in the bare particle perturbation theory. It defines the (proper) self-energy parts (in the sense of Dyson) as an infinite sum of the special class of diagrams illustrated in Fig. 2.

The earlier condition of Eq. (2.13) represented, as was noted there, only an approximation to this sum. In other words, Eqs. (2.13) and (3.1) are not exactly identical even on the energy shell.

The Hartree-Fock approximation based on Eq. (3.1) could be interpreted as a nonperturbation approximation to determine the "dressed" single particles (together with the "dressed vacuum") or the Green's function $\langle 0|T(\Psi(xt),\Psi^+(x't'))|0\rangle$ for the true interacting system. Such single particles will satisfy

$$L(p,p_0)u\cong 0,\quad M(k,k_0)\cong 0. \tag{3.2}$$

We use the approximate equality since a really stable single particle may not exist.

Let us assume that these determine the approximate renormalized dispersion law

$$p_0^2=E_r(p)^2,\quad k_0^2=\Omega_r(k)^2. \tag{3.3}$$

If we write for Σ

$$\Sigma(pp_0)=p_0\zeta(pp_0)+\chi(pp_0)\tau_3+\phi(pp_0)\tau_1, \tag{3.4}$$

where ζ, χ, ϕ are even functions of p_0, then

$$\begin{aligned}E_r^2(p)&=[\bar{\epsilon}(pp_0)^2+\phi(pp_0)^2]/[1-\zeta(pp_0)]^2|_{p_0^2=E_r(p)^2}\\&\equiv E(pp_0)^2/Z(pp_0)^2|_{p_0^2=E_r(p)^2}.\end{aligned} \tag{3.5}$$

The Green's functions G and Δ will be given by

$$\begin{aligned}G(pp_0)&=i/L(pp_0)\\&=i\int_0^\infty \frac{dx}{p_0^2-x+i\epsilon}\\&\quad\times\text{Im}\,\frac{p_0Z(px)+\bar{\epsilon}(px)\tau_3+\phi(px)\tau_1}{x^2Z(px)^2-E(px)^2},\end{aligned} \tag{3.6}$$

$$\begin{aligned}\Delta(kk_0)&=i/M(kk_0)\\&=i\int_0^\infty \frac{dx}{k_0^2-x+i\epsilon}\,\text{Im}\,\frac{1}{M(kx)}.\end{aligned}$$

This representation assumes that $G(p_0)[\Delta(k_0)]$ is analytic except for a branch cut on the real axis. The imaginary part in the integrand is expected to have a delta function or a sharp peak at $x=E_r^2(p)$ $[\Omega_r(k^2)]$. These properties are necessary in order that the vacuum is stable and the quasi-particles and phonons have a valid physical meaning as excitations.[14] In the following, we will generally consider this quasi-particle peak only, and write

$$G(pp_0)=i\frac{p_0Z(pp_0)+\bar{\epsilon}(pp_0)\tau_3+\phi(pp_0)\tau_1}{p_0Z(pp_0)^2-E(pp_0)^2+i\epsilon},\quad\text{etc.}$$

The Hartree equations now take the form

$$\Sigma(pp_0)=-i\frac{g^2}{(2\pi)^4}\int\tau_3G(p-k,\,p_0-k_0)\tau_3h(kk_0)^2d^3kdk_0,$$

$$\begin{aligned}\Pi(kk_0)=i\frac{g^2}{(2\pi)^4}\int\text{Tr}\,[\tau_3G(k-p,\,k_0-p_0)\\ \times\tau_3G(pp_0)]d^3pdp_0.\end{aligned} \tag{3.7}$$

This equation for Σ is much simpler than the previous one (2.18) since we may just equate the coefficients of 1, τ_3, τ_1 on both sides. In particular, we get the energy gap equation

$$\begin{aligned}\phi(pp_0)=-\frac{ig^2}{(2\pi)^4}\int\frac{\phi(p'p_0')}{p_0^2Z(p'p_0')^2-E(p'p_0')^2+i\epsilon}\\ \times h(\mathbf{p}-\mathbf{p}',\,p_0-p_0')^2\Delta(\mathbf{p}-\mathbf{p}',\,p_0-p_0')d^3p'dp_0',\end{aligned} \tag{3.8}$$

which is to be compared with Eq. (2.19).

Although the existence of a solution to Eq. (3.6) may be difficult to establish, the solution, if it exists, should not be much different from the older solution to Eq. (2.19). At any rate, our assumption about the analyiticity of G and Π is consistent with Eq. (3.6) or (3.7) which implies that Σ and Π are also analytic except for a cut on the real axis.

In later calculations we shall encounter various integrals which we may classify into three types regarding their sensitivity to the energy gap. First, a normal self-energy part, for example, represents the effect of the bulk of the surrounding electrons on a particular electron, and is insensitive to the change of the small fraction $\sim\phi/E_F$ of the electrons near the Fermi surface in a superconductor. Such a quantity is

[13] It would seem then that we lose the advantage of the generalization since we cannot find the Bogoliubov transformation. However, we could still start from the older solution (2.13) as the zeroth approximation to Eq. (3.1), and then calculate the correction; namely, the "radiative" correction to the Bogoliubov vacuum and the Bogoliubov quasi-particle. These corrections would take account of the single-particle transitions which remain after the Bogoliubov condition (2.13) is imposed.

[14] This is a representation of the Lehmann type [H. Lehmann, Nuovo cimento 11, 342 (1954)] which can be derived by defining the Green's functions in terms of Heisenberg operators. See also V. M. Galizkii and A. B. Migdal, J. Exptl. Theoret. Phys. U.S.S.R. 34, 139 (1958) [translation: Soviet Phys. JETP 7, 96 (1958)].

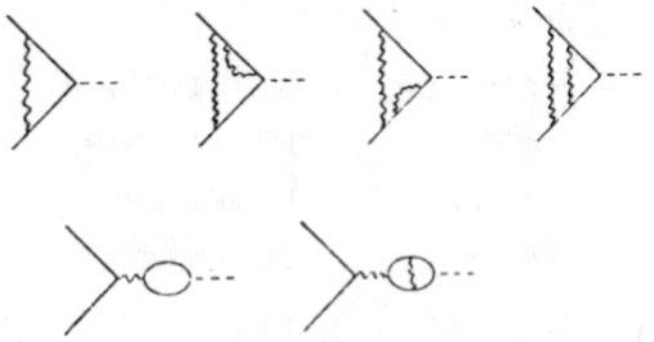

FIG. 4. Construction of the vertex part Γ in bare particle picture. The second line represents the polarization diagrams.

given by an integral like

$$g^2\int \frac{\epsilon_k}{E_k} f(\mathbf{p}-\mathbf{k})d^3k, \tag{3.9}$$

where the region $\epsilon_k \lesssim E_k=(\epsilon_k^2+\phi^2)^{\frac{1}{2}}$ makes little contribution if $f(p-k)$ is a smooth function.

Second, the energy gap itself is determined from an equation of the form

$$g^2\int \frac{d^3k}{E_k} f(\mathbf{p}-\mathbf{k}) \sim g^2 \int_{E_k \lesssim \omega_m} \frac{d^3k}{E_k} f(\mathbf{p}-\mathbf{k}) \sim 1, \tag{3.10}$$

which means that even if g^2 is small, such an expression is always of the order 1.

Finally we meet with integrals like

$$g^2\int \frac{\epsilon_k\phi}{E_k^3} f(\mathbf{p}-\mathbf{k})d^3k, \quad g^2\int \frac{\phi^2}{E_k^3} f(\mathbf{p}-\mathbf{k})d^3k, \quad \text{etc.} \tag{3.11}$$

They have an extra cutoff factor $\sim 1/E$, $1/E^2$, etc., in the integrand which restricts the contribution to an energy interval $\sim 2\phi$ near the Fermi surface. The integrals are thus of the order

$$g^2N\phi/\omega_m, \quad g^2N, \quad \text{etc.}$$

In the following, we will not be primarily concerned with the ordinary self-energy effects. We will assume that proper renormalization has been carried out, or else simply disregard it unless essential. When we carry out perturbation type calculations, we will arrange things so that quantities of the second type are taken into account rigorously, and treat quantities of the third type as small, and hence negligible ($g^2N \ll 1$).

4. INTEGRAL EQUATIONS FOR VERTEX PARTS[15]

In the presence of an electromagnetic potential, the original Lagrangian $\mathfrak{L}$ has to be modified according to the rule

$$i\frac{\partial}{\partial t} \rightarrow i\frac{\partial}{\partial t}+eA_0, \quad \mathbf{p}\rightarrow \mathbf{p}-\frac{e}{c}\mathbf{A}$$

for the electron. Going to the two-component representation, this corresponds to the prescription

$$i\frac{\partial}{\partial t} \rightarrow i\frac{\partial}{\partial t}+e\tau_3 A_0, \quad \mathbf{p}\rightarrow \mathbf{p}-\frac{e}{c}\tau_3\mathbf{A} \tag{4.1}$$

acting on Ψ. It can also be inferred from the gauge transformation $\Psi \rightarrow \exp(i\alpha\tau_3)\Psi$ as was observed previously. So the ordinary charge-current operator turns out to be in our form given by

$$\rho(x)=\frac{e}{2}([\Psi^+(x),\tau_3\Psi(x)]+\{\Psi^+(x),\Psi(x)\}),$$

$$\begin{aligned}\mathbf{j}(x)=\frac{-ie}{4m}([\Psi^+(x),\,(\nabla-ie\tau_3\mathbf{A})\Psi(x)]\\ +[(-\nabla-ie\tau_3 A)\Psi^+(x),\,\Psi(x)]\\ +\{\Psi^+(x),\,(\nabla\tau_3-ieA)\Psi(x)\}\\ -\{(\tau_3\nabla-ie\mathbf{A})\Psi^+(x),\,\Psi(x)\}).\end{aligned} \tag{4.2}$$

The second terms on the right-hand side, being infinite C numbers, arise from the rearrangement of ψ and ψ^+, and will actually be compensated for by the first terms.

This expression, however, has to be modified when we go to the quasi-particle picture.

For we have seen that the self-energy ϕ of a quasi-particle is a gauge-dependent quantity. If we want to have the quasi-particle picture and gauge invariance at the same time, then it is clear that the electromagnetic current of a quasi-particle must contain, in addition to the normal terms given by Eq. (4.2), terms which would cause a physically unobservable transformation of ϕ if the electromagnetic potential is replaced by the gradient of a scalar. In other words, the complete charge current of a quasi-particle has to satisfy the continuity equation, which Eq. (4.2) does not, since

$$\partial\rho/\partial t+\nabla\cdot\mathbf{j}=2\Psi^+\phi\tau_2\Psi.$$

In order to find such a conserving expression for charge current, it is instructive to go back to the bare electron picture, in which the self-energy is represented by a particular class of diagrams discussed in the previous sections.

It is well known[16] in quantum electrodynamics that, in any process involving electromagnetic interaction, perturbation diagrams can be grouped into gauge-invariant subsets, such that the invariance is maintained by each subset taken as a whole. Such a subset can be constructed by letting each photon line in a diagram interact with a charge of all possible places along a chain of charge-carrying particle lines. The gauge-invariant interaction of a quasi-particle with an electromagnetic potential should then be obtained by attaching a photon line at all possible places in the diagrams of Fig. 2. The result is illustrated in Fig. 4,

[15] Hereafter we will often use the four-dimensional notation $x=(\mathbf{x},t)$, $p=(\mathbf{p},p_0)$, $d^4p=d^3pdp_0$.

[16] Z. Koba, Progr. Theoret. Phys. (Kyoto) 6, 322 (1951).

which consists of the "vertex" part Γ and the self-energy part Σ.

In this way we are led to consider the modification of the vertex due to the phonon interaction in the same approximation as the self-energy effect is included in the quasi-particle. It is not difficult to see that it corresponds to a "ladder approximation" for the vertex part, and we get an integral equation[17]

$$\Gamma_i(p',p)=\gamma_i(p',p)-g^2\int \tau_3 G(p'-k)\Gamma_i(p'-k,\ p-k)$$
$$\times G(p-k)\tau_3 h(k)^2\Delta(k)d^4k, \quad (4.3)$$

where γ_i, $i=0, 1, 2, 3$ stand for the free particle charge current $[\tau_3, (1/2m)(\mathbf{p}+\mathbf{p}')]$ which follows from Eq. (4.2). Similar equations may be set up for any type of vertex interactions.

Equation (4.3) is the basis of the rest of this paper. It expresses a clear-cut approximation procedure in which the "free" charge-current operator γ_i of a quasi-particle is modified by a special class of "radiative corrections" due to H_{int}'.

As the next important step, we observe that there exist exact solutions to Eq. (4.3) for the following four types of vertex interactions

$$\begin{aligned}
&(a)\quad \gamma^{(a)}(p',p)=L_0(p')-L_0(p)\\
&\qquad\qquad =(p_0'-p_0)-\tau_3(\epsilon_{p'}-\epsilon_p),\\
&\qquad \Gamma^{(a)}(p',p)=L(p')-L(p)\\
&\qquad\qquad =\gamma_a(p')-[\Sigma(p')-\Sigma(p)],\\
&(b)\quad \gamma^{(b)}(p',p)=L_0(p')\tau_3-\tau_3L_0(p)\\
&\qquad\qquad =(p_0'-p_0)\tau_3-(\epsilon_{p'}-\epsilon_p), \qquad (4.4)\\
&\qquad \Gamma^{(b)}(p',p)=L(p')\tau_3-\tau_3L(p),\\
&(c)\quad \gamma^{(c)}(p',p)=L_0(p')\tau_1+\tau_1L_0(p'),\\
&\qquad \Gamma^{(c)}(p',p)=L(p')\tau_1+\tau_1L(p),\\
&(d)\quad \gamma^{(d)}(p',p)=L_0(p')\tau_2+\tau_2L(p),\\
&\qquad \Gamma^{(d)}(p',p)=L(p')\tau_2+\tau_2L(p).
\end{aligned}$$

The verification is straightforward by noting that $G(p)=i/L(p)$, and making use of Eq. (3.7).

The fact that there are simple solutions is not accidental. These solutions express continuity equations and other relations following from the four types of operations, which do not depend on the presence or absence of the interaction:

$$\begin{aligned}
&(a)\quad \Psi(x)\to e^{i\alpha(x)}\Psi(x), &&\Psi^+(x)\to\Psi^+(x)e^{-i\alpha(x)},\\
&(b)\quad \Psi\to e^{i\tau_3\alpha}\Psi, &&\Psi^+\to\Psi^+e^{-i\tau_3\alpha},\\
&(c)\quad \Psi\to e^{\tau_1\alpha}\Psi, &&\Psi^+\to\Psi^+e^{\tau_1\alpha},\\
&(d)\quad \Psi\to e^{\tau_2\alpha}\Psi, &&\Psi^+\to\Psi^+e^{\tau_2\alpha},
\end{aligned}\qquad (4.5)$$

where $\alpha(x)$ is an arbitrary real function.

(a) and (b) correspond, respectively, to the spin rotation around the z axis, and the gauge transformation. The entire Lagrangian is invariant under them, so that we obtain continuity equations for the z component of spin and charge, respectively:

$$\begin{aligned}
&\frac{\partial}{\partial t}\Psi^+\Psi+\nabla\cdot\Psi^+\frac{\mathbf{p}}{m}\tau_3\Psi=0,\\
&\frac{\partial}{\partial t}\Psi^+\tau_3\Psi+\nabla\cdot\Psi^+\frac{\mathbf{p}}{m}\Psi=0,
\end{aligned}\qquad (4.6)$$

where Ψ is the true Heisenberg operator.

These equations are identical with

$$\Psi^+\gamma^{(a)}\Psi=0,\quad \Psi^+\gamma^{(b)}\Psi=0. \qquad (4.7)$$

Taken between two "dressed" quasi-particle states, the left-hand side of Eq. (4.7) will become

$$\begin{aligned}
e^{-i(p'-p)\cdot x}\langle|p'|\Psi^+(x)\gamma^{(n)}\Psi(x)|p\rangle
&=u_{p'}{}^*\Gamma^{(n)}(p',p)u_p\\
&=0,\quad (n=a,\ b) \qquad (4.7')
\end{aligned}$$

where u_p, $u_{p'}$ are single-particle wave functions satisfying $L(p)u_p=u_{p'}{}^*L(p')=0$.

In this way we have shown the existence of spin and charge currents $\Gamma_i^{(a)}(p',p)$ and $\Gamma_i^{(b)}(p',p)$ for a quasi-particle, for which the continuity equations

$$(p_0'-p_0)\cdot\Gamma_0^{(n)}-\sum_{i=1}^{3}(p'-p)_i\cdot\Gamma_i^{(n)}=\Gamma^{(n)}(p',p)=0$$

will hold.

The last two transformations of Eq. (4.4) are not unitary, but mix ψ_1 and ψ_2^+ in such a way as to keep $\Psi^+\tau_3\Psi$ invariant. From infinitesimal transformations of these kinds we get

$$\begin{aligned}
&i\Psi^+\tau_1\left(\frac{\overrightarrow{\partial}}{\partial t}-\frac{\overleftarrow{\partial}}{\partial t}\right)\Psi+\nabla\cdot\Psi^+\tau_2\left(\frac{\overrightarrow{\mathbf{p}}}{m}+\frac{\overleftarrow{\mathbf{p}}}{m}\right)\Psi=0,\\
&-i\Psi^+\tau_2\left(\frac{\overrightarrow{\partial}}{\partial t}-\frac{\overleftarrow{\partial}}{\partial t}\right)\Psi+\nabla\cdot\Psi^+\tau_1\left(\frac{\overrightarrow{\mathbf{p}}}{m}+\frac{\overleftarrow{\mathbf{p}}}{m}\right)\Psi=0,
\end{aligned}\qquad (4.8)$$

which bear the same relations to $\gamma^{(c),(d)}$ and $\Gamma^{(c),(d)}$ as Eq. (4.6) did to $\gamma^{(a),(b)}$ and $\Gamma^{(a),(b)}$. Note that the above equations are unaffected by the presence of the phonon interaction.

The fact that we can find a conserved charge-current

[17] This equation may also be derived simply by considering the self-energy equation (3.7) in the presence of an external field, and expanding Σ in A. Σ should be now a function of initial and final momenta, and we define

$$\Sigma^{(A)}(p',p)\equiv\Sigma(p)\delta^4(p'-p)+\sum_{i=0}^{3}(\Gamma_i(p',p)-\gamma_i(p',p))$$
$$\times A^i(p'-p)+O(A^2).$$

In the limit $p'-p=0$, $\Gamma_i-\gamma_i=\partial\Sigma/\partial A^i$, which is the content of the Ward identity.[6] Investigation of the higher order terms in A is beyond the scope of this paper.

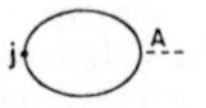

FIG. 5. The diagram for the kernel $K^{(2)}$.

for a quasi-particle is rather surprising. A quasi-particle cannot be an eigenstate of charge since it is a linear combination of an electron and a hole, tending to an electron well above the Fermi surface, and to a hole well below. We must conclude than that an accelerated wave packet of quasi-particles, whose energy is confined to a finite region of space, continuously picks up charge from, or deposits it with, the surrounding medium which extends to infinity. This situation will be studied in Sec. 7, where we will derive the charge current operators Γ_i explicitly.

5. GAUGE INVARIANCE IN THE MEISSNER EFFECT

We will next discuss how the gauge invariance is maintained in the problem of the Meissner effect when the external magnetic field is static. We calculate the Fourier component of the current $J(q)$ induced in the superconducting ground state by an external vector potential $A(q)$:

$$J_i(q)=\sum_{j=1}^{3} K_{ij}(q)A_j(q), \qquad (5.1)$$

where q is kept finite.

For free electrons, K is represented by

$$K_{ij}=K_{ij}^{(1)}+K_{ij}^{(2)}, \quad K_{ij}^{(1)}=-\delta_{ij}ne^2/m, \qquad (5.2)$$

where n is the number of electrons inside the Fermi sphere. $K^{(1)}$ comes from the expectation value of the current operator Eq. (4.2), whereas $K^{(2)}$ corresponds to the diagram in Fig. 5. [Compare also Eq. (1.1).] It is well known that in this case K_{ij} is of the form

$$K_{ij}(q)=(\delta_{ij}q^2-q_iq_j)K(q^2), \qquad (5.3)$$

so that for a longitudinal vector potential $A_i(q)\sim q_i\lambda(q)$, we have

$$J_i(q)=K_{ij}q_j\lambda(q)=0, \qquad (5.4)$$

establishing the unphysical nature of such a potential.

In the case of a superconducting state, the free electron lines in Fig. 5 will be replaced by quasi-particle lines. But then we have $K^{(2)}(q)\rightarrow 0$ as $q\rightarrow 0$ since the intermediate pair formation is suppressed due to the finite energy gap, whereas $K^{(1)}$ is essentially unaltered. Thus Eq. (5.2) takes the form of the London equation, except that even a longitudinal field creates a current.

According to our previous argument, this lack of gauge invariance should be remedied by taking account of the vertex corrections. Starting again from the free electron picture, and inserting the phonon interaction

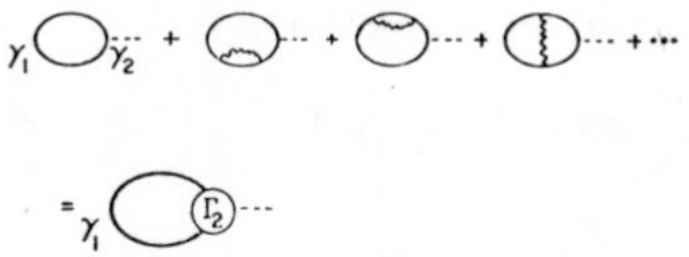

FIG. 6. Graphical derivation of Eq. (5.5). The thick lines represent quasi-particles.

effects, as indicated in Fig. 6, we arrive at the conclusion that either one of the vertices γ in Fig. 5 has to be replaced by the full Γ^{10}. In addition, there is the polarization correction represented by a string of bubbles. Let us, however, first neglect this correction. $K_{ij}^{(2)}$ is then

$$K_{ij}^{(2)}(q)=\frac{-ie^2}{(2\pi)^4}\int \mathrm{Tr}[\gamma_i(p-q/2,\,p+q/2)G(p+q/2)$$
$$\times\Gamma_j(p+q/2,\,p-q/2)G(p-q/2)]d^4p. \qquad (5.5)$$

Although we do not know $\Gamma_j(p,p)$ explicitly, we can establish Eq. (5.4) easily. For

$$-\sum_{j=1}^{3}\Gamma_j(p+q,\,q)q_j$$

is exactly the solution $\Gamma^{(b)}(p+q,\,p)$ of Eq. (4.4) where q_0 is equal to zero. Substituting this solution in Eq. (4.5) we find

$$K_{ij}^{(2)}(q)q_j$$
$$=\frac{-1}{(2\pi)^4}\int \mathrm{Tr}\{\gamma_i(p-q/2,\,p+q/2)$$
$$\times[\tau_3G(p+q/2)-G(p-q/2)\tau_3]\}d^4p$$
$$=\frac{-1}{(2\pi)^4}\int \mathrm{Tr}\{[\gamma_i(p-q,\,p)-\gamma_i(p,\,p+q)]$$
$$\times\tau_3G(p)\}d^4p$$
$$=\frac{1}{(2\pi)^4}\frac{q_i}{m}\int \mathrm{Tr}[\tau_3G(p)]d^4p, \qquad (5.6)$$

where the properties of γ_i and G under particle conjugation and a translation in p space were utilized in going from the first to the second line.

On the other hand, the part $K^{(1)}$ is, according to Eq. (4.2) given by

$$K_{ij}^{(1)}=-\delta_{ij}\frac{e^2}{2m}(\langle 0|[\Psi^+(x),\tau_3\Psi(x)]|0\rangle+\{\Psi^+(x),\Psi(x)\})$$
$$=K_{ij}^{(1a)}+K_{ij}^{(1b)}. \qquad (5.7)$$

The first term becomes further

$$-\delta_{ij}\frac{e^2}{2m}\langle 0|[\Psi^+(x),\tau_3\Psi(x)]|0\rangle$$

$$=-\delta_{ij}\frac{e^2}{m}\frac{1}{2}\lim_{\epsilon\to 0}\sum_{\pm}\langle 0|T(\Psi^+(x,t\pm\epsilon)\tau_3\Psi(x,t))|0\rangle$$

$$=-\delta_{ij}\frac{e}{m}\mathrm{Tr}[\tau_3 G(xt=0)]$$

$$=-\delta_{ij}\frac{e}{m}\frac{1}{(2\pi)^4}\int \mathrm{Tr}[\tau_3 G(p)]x^4p. \quad (5.8)$$

Thus

$$[K_{ij}^{(1a)}(q)+K_{ij}^{(2)}(q)]q_j=0. \quad (5.9)$$

The second term $K^{(1b)}$ comes from the c-number term of the current operator (4.2), and is just the anticommutator of the electron field, which does not depend on the quasi-particle picture, nor on the presence of interaction. Therefore we may write for this contribution

$$K_{ij}^{(1b)}(q)A_j(q)=\frac{-ie}{2m}\frac{1}{(2\pi)^3}\int e^{-iqx}d^3x$$
$$\times\{\Psi^+(x),(\tau_3\nabla-ieA(x))_i\Psi(x)\} \quad (5.10)$$

to show its formal gauge invariance since $\tau_3\nabla-ieA(x)$ is certainly a gauge-invariant combination for free electron field.

As for the polarization correction, we can easily show in a similar way that it vanishes for the static case ($q_0=0$) because

$$\int \mathrm{Tr}\,\Gamma_i(p-q/2,p+q/2)G(p+q/2)$$
$$\times\gamma_0(p+q/2,p-q/2)G(p-q/2)d^4p=0.$$

Thus the above proof is complete and independent of the Coulomb interaction which profoundly influences the polarization effect. Although the proof is thus rigorous, it is still somewhat disturbing since $K^{(1a)}$, $K^{(1b)}$ and $K^{(2)}$ are all infinite. Actually there is a certain ambiguity in the evaluation of $K^{(2)}$, Eq. (5.6), which is again similar to the one encountered in quantum electrodynamics.[18] An alternative way would be to expand quantities in q without making translations in p space. In this case we may write

$$-\Gamma^{(b)}(p+q/2,p-q/2)=\bar\epsilon(p+q/2)-\bar\epsilon(p-q/2)$$
$$-i\tau_2[\phi(p+q/2)+\phi(p-q/2)]$$
$$\approx p\cdot q/m-2i\tau_2\phi. \quad (5.11)$$

The first term then gives

$$\frac{e^2}{(2\pi)^3}\int\frac{\phi^2}{4E_p{}^5}\left(\frac{\mathbf{p}\cdot\mathbf{q}}{m}\right)^2 p_i\mathbf{p}\cdot\mathbf{q}d^3p\propto q^2q_i, \quad (5.12)$$

which is convergent and the same as the one obtained from Eq. (1.1) using the bare quasi-particle states. The second term also is finite and equal to

$$\frac{e^2}{(2\pi)^3}\int\frac{\phi^2}{E_p{}^3}\frac{p^2}{3m^2}q_id^3p+O(q^2)q_i\approx N\alpha^2q_i=n(e^2/m)q_i. \quad (5.13)$$

The last line follows from Eqs. (6.11) and (6.11′) below.

The calculation of $K^{(1)}$ from Eqs. (5.7) and (5.10), gives, on the other hand, the same value as Eq. (5.2), so that we get $(K_{ij}^{(1)}+K_{ij}^{(2)})q_j=0$ in the limit of small q. (The polarization correction is again zero.)

Since Eq. (5.13) is a contribution from the collective intermediate state (see Secs. 6 and 7), we may say that the collective state saves gauge invariance, as has been claimed by several people.[3,19]

It goes without saying that the effect of the vertex correction on K_{ij} will be felt also for real magnetic field. But as we shall see later, it is a small correction of order g^2N (except for the renormalization effects), and not as drastic as for the longitudinal case.

6. THE COLLECTIVE EXCITATIONS

In order to understand the mechanism by which gauge invariance was restored in the calculation of the Meissner effect, and also to solve the integral equations for general vertex interactions, it is necessary to examine the collective excitations of the quasi-particles. In fact, people[3] have shown already that the essential difference between the transversal and longitudinal vector potentials in inducing a current is due to the fact that the latter can excite collective motions of quasi-particle pairs.

We see that the existence of such collective excitations follows naturally from our vertex solutions Eq. (4.4). For taking $p=p'$, the second solution $\Gamma^{(b)}(p',p)$ becomes

$$\Gamma^{(b)}(p,p)=L(p)\tau_3-\tau_3L(p)$$
$$=2i\tau_2\phi, \quad (6.1)$$
$$\gamma^{(b)}=0.$$

In other words $\tau_2\phi(p)\equiv\Phi_0(p)$ satisfies a homogeneous integral equation:

$$\Phi_0(p)=-\frac{g^2}{(2\pi)^4}\int\tau_3G(p')\Phi(p')G(p')$$
$$\times\tau_3h(p-p')^2\Delta(p-p')d^4p. \quad (6.2)$$

We interpret this as describing a pair of a particle and an antiparticle interacting with each other to form a bound state with zero energy and momentum $q=p'-p=0$.

[18] H. Fukuda and T. Kinoshita, Progr. Theoret. Phys (Kyoto) 5, 1024 (1950).

[19] On the other hand, the way in which the collective mode accomplishes this end seems to differ from one paper to another. We will not attempt to analyze this situation here.

In fact, by defining

$$F(p,-p)\equiv -G(p)\Phi_0(p)G(p), \tag{6.3}$$

Eq. (6.2) becomes

$$L(p)F(p,-p)L(p)=-\frac{g^2}{(2\pi)^4}\int \tau_3F(p',-p')\times\tau_3h(p-p')^2\Delta(p-p')d^4p',$$

or

$$\sum_{j,l=1}^{2}L(p)_{ij}L^C(-p)_{kl}F(p,-p)_{jl}=\frac{-g^2}{(2\pi)^4}\int\sum_{j,l}(\tau_3)_{ij}(\tau_3)_{kl}F(p',-p')_{jl}\times h(p-p')^2\Delta(p-p')d^4p. \tag{6.4}$$

The particle-conjugate quantity L^C was defined in Eq. (2.23).

Equations (6.2) and (6.4) are the analog of the so-called Bethe-Salpeter equation[20] for the bound pair of quasi-particles with zero total energy-momentum. $F_{ij}(p,-p)$ is the four-dimensional wave function with the spin variables i, j and the relative energy-momentum $(p_0,\mathbf{p})$.

Since there, thus, exists a bound pair of zero momentum, there will also be pairs moving with finite momentum and kinetic energy. In other words, there will be a continuum of pair states with energies going up from zero. We have to determine their dispersion law.

For a finite total energy-momentum q, the homogeneous integral equation takes the form

$$\Phi_q(p)\equiv L(\tfrac{1}{2}q+p)F(\tfrac{1}{2}q+p,\tfrac{1}{2}q-p)L(p-\tfrac{1}{2}q)=-g^2\frac{1}{(2\pi)^4}\int\tau_3F(\tfrac{1}{2}q+p',\tfrac{1}{2}q-p')\times\tau_3h(p-p')^2\Delta(p-p')d^4p. \tag{6.5}$$

From here on we carry out perturbation calculation. Let us expand F and L in terms of the small change $L(p\pm q/2)-L(p)$, thus

$$\begin{aligned}F(\tfrac{1}{2}q+p,\tfrac{1}{2}q-p)&=F^{(0)}(p)+F^{(1)}(p,q/2)+\cdots,\\ L(p\pm q/2)&=L(p)+\Delta L(p,\pm q/2).\end{aligned} \tag{6.6}$$

Collecting terms of the first order, we get

$$\begin{aligned}L(p)F^{(1)}(p,q/2)L(p)+U^{(1)}(p,q/2)&=-g^2\frac{1}{(2\pi)^4}\int\tau_3F^{(1)}(p',q/2)\times\tau_3h(p-p')^2\Delta(p-p')d^4p',\\ U^{(1)}(p,q/2)&=\Delta L(p,q/2)F^{(0)}(p)L(p)+L(p)F^{(0)}(p)\Delta L(p,-q/2).\end{aligned} \tag{6.7}$$

[20] E. E. Salpeter and H. A. Bethe, Phys. Rev. 84, 1232 (1951).

This is an inhomogeneous integral equation for $F^{(1)}$. In order that it has a solution, the inhomogeneous term $U(p)$ must be orthogonal to the solution $\Phi_0(p)$ of the homogeneous equation. This condition can be derived as follows:

We multiply Eq. (6.7) by $F^{(0)}(p)=-G(p)\Phi_0(p)G(p)$, and integrate thus:

$$\int \mathrm{Tr}\,F^{(0)}(p)L(p)F^{(1)}(p,q/2)L(p)d^4p+\int\mathrm{Tr}\,F^{(0)}(p)U^{(1)}(p,q/2)d^4p$$
$$=-g^2\frac{1}{(2\pi)^4}\int\int\mathrm{Tr}\,F^{(0)}(p)\tau_3F^{(1)}(p',q/2)\times\tau_3h(p-p')^2\Delta(p-p')d^4pd^4p'.$$

In view of Eq. (6.5) the last line is

$$=\int\mathrm{Tr}\,L(p')F^{(0)}(p')L(p')F^{(1)}(p',q/2)d^4p',$$

so that

$$(F^{(0)},U^{(1)})\equiv\int\mathrm{Tr}\,F^{(0)}(p)U^{(1)}(p,q/2)d^4p=0. \tag{6.8}$$

This is the desired condition.

For the evaluation of Eq. (6.8), we will neglect the p dependence of the self-energy terms. Thus

$$\begin{aligned}F^{(0)}(p)&=\tau_2\phi/(p_0^2-E_p^2+i\epsilon),\quad E_p^2=\epsilon_p^2+\phi^2,\\ \Delta L(p,q/2)&=q_0/2-\tau_3(\mathbf{p}\cdot\mathbf{q}/2m+(q/2)^2/2m).\end{aligned} \tag{6.9}$$

We then obtain

$$(F^{(0)},U^{(1)})=2\pi i\int\frac{\phi^2}{E_p^3}\left[\left(\frac{q_0}{2}\right)^2-\left(\frac{\mathbf{p}\cdot\mathbf{q}}{2m}\right)^2-\frac{\epsilon_p}{m^2}\left(\frac{q}{2}\right)^2\right]d^3p=0,$$

or

$$\left(\frac{q_0}{2}\right)^2-\left(\frac{q}{2}\right)^2\left[\frac{1}{3}\frac{\bar{p}^2}{m^2}-\frac{\bar{\epsilon}_p}{m}\right]=0, \tag{6.10}$$

where the average $\bar{f}$ is defined

$$\bar{f}=\int f(p)\frac{\phi^2}{E^3}d^3p\bigg/\int\frac{\phi^2}{E^3}d^3p. \tag{6.10$'$}$$

The weight function $\phi^2/E_p^3=\phi^2/(\epsilon_p^2+\phi^2)^{\frac{3}{2}}$ peaks around the Fermi momentum, so that $p^2\sim p_F^2$, $\epsilon_p\sim 0$. Thus

$$q_0^2\approx q^2\frac{1}{3}\frac{\bar{p}^2}{m^2}\equiv\alpha^2q^2,\quad \alpha^2\approx p_F^2/3m^2, \tag{6.11}$$

which is the dispersion law for the collective excitations.[2,3] We also note, incidentally, that

$$\frac{1}{(2\pi)^3}\int \frac{\phi^2}{E^3} d^3p \approx N = mp_F/\pi^2, \tag{6.11'}$$

$$\alpha^2 N \approx p_F{}^3/3\pi^2 m = n.$$

We would like to emphasize here that these collective excitations are based on Eq. (6.2), which takes account of the phonon-Coulomb scattering of the quasi-particle pairs, but does not take into account the annihilation-creation process of the pair due to the same interaction.

It is well known that this annihilation-creation process is very important in the case of the Coulomb interaction, and plays the role of creating the plasma mode of collective oscillations. We will consider it in a later section.

As for the wave function $F^{(1)}$ itself, we have still to solve the integral equation (6.7). But this can be done by perturbation because on substituting $U^{(1)}$ in the integrand, we find that all the terms are of the type (3.11). In other words, to the zeroth order we may neglect the integral entirely and so

$$F^{(1)}(p,q/2) = -G(p)U^{(1)}(p,q/2)G(p). \tag{6.12}$$

The original function

$$\Phi_q(p) = -L(p+q/2)F(p,q/2)L(p-q/2)$$

is even simpler. We get

$$\Phi_q(p) \approx \Phi_0(p) \tag{6.13}$$

to this order.

7. CALCULATION OF THE CHARGE-CURRENT VERTEX FUNCTIONS

In this section we determine explicitly the charge-current vertex functions Γ_i, $(i=0, 1, 2, 3)$ from their integral equations. Only the particular combination $\Gamma^{(b)}$ of these was given before.

Let us first go back to the integral equation for Γ_0 generated by τ_3:

$$\Gamma_0(p+q/2, p-q/2)$$
$$= \tau_3 - g^2\int \tau\, G(p'+q/2)\Gamma_0(p'+q/2, p'-q/2) \times G(p'-q/2)\tau_3 h(p-p')^2\Delta(p-p')d^4p',$$

or

$$L(p+q/2)F_0(p+q/2, p-q/2)L(p-q/2)$$
$$= \tau_3 + g^2\int \tau_3 F_0(p'+q/2, p'-q/2) \times \tau_3 h(p-p')^2\Delta(p-p')d^4p'. \tag{7.1}$$

For small g^2, the standard approach to solve the equation would be the perturbation expansion in powers of g^2. We know, however, that there are low-lying collective excitations, discussed before, to which τ_3 can be coupled, and these excitations do not follow from perturbation.[21]

Fortunately, if we assume $q=0$, $q_0\neq 0$, then we have an exact solution to Eq. (7.1) in terms of $\Gamma^{(b)}$ of Eq. (4.4). Namely,

$$\begin{aligned}\Gamma_0(p+q/2, p-q/2) &= \Gamma^{(b)}(p+q/2, p-q/2)/q_0\\ &= \tau_3\{[Z(p+q/2)+Z(p-q/2)\}/2\\ &\quad + (p_0/q_0)[Z(p+q/2)-Z(p-q/2)]\}\\ &\quad -[\chi(p+q/2)-\chi(p-q/2)]/q_0\\ &\quad + i\tau_2[\phi(p+q/2)+\phi(p-q/2)]/q_0,\end{aligned} \tag{7.2}$$

which can readily be verified.

The second term is the result of the coupling of τ_3 to the collective mode. This can be understood in the following way. Γ_0 contains matrix elements for creation or annihilation of a pair out of the vacuum. These processes can go through the collective intermediate state with the dispersion law (6.11), so that Γ will contain terms of the form

$$R_\pm/(q_0\pm\alpha q).$$

The residues $R_\pm$ can be obtained by taking the limit

$$R_\pm = \lim_{q_0\pm\alpha q\to 0} \Gamma_0(p+q/2, p-q/2)(q_0\pm\alpha q). \tag{7.3}$$

Applying this procedure to the integral equation (7.1) for Γ_0, we find that $R_\pm$ must be a solution of the homogeneous equation; namely,

$$R_\pm = C_\pm \Phi_q(p), \tag{7.4}$$

under the condition $q_0\pm\alpha q=0$.

For the particular case $q=0$, $\Phi_q(p)$ reduces to $\tau_2\phi(p)$, which in fact agrees with Eq. (7.2) if

$$C_\pm = -2i. \tag{7.5}$$

This observation enables us to write down Γ_0 for $q\neq 0$. According to the results of Sec. 6, $\Phi_q(p)=\Phi_0(p)$ in the zeroth order in g^2N. Since corrections to the non-collective part of Γ_0 also turn out to be calculable by perturbation, we may now put

$$\Gamma_0(p+q/2, p-q/2) \approx \tau_3\bar{Z} + 2i\tau_2\bar{\phi}q_0/(q_0{}^2-\alpha^2q^2),$$
$$\bar{\phi} \equiv [\phi(p+q/2)+\phi(p-q/2)]/2,$$
$$\bar{Z} \equiv [Z(p+q/2)+Z(p-q/2)\}/2$$

to the extent that terms of order g^2N and/or the p-dependence of the renormalization constants are neglected.

In quite a similar way the current vertex $\mathbf{\Gamma}$ may be constructed. This time we start from the longitudinal

[21] If we proceeded by perturbation theory, we would find in each order terms of order 1.

component for $q_0=0$, $q\neq 0$, which has the exact solution

$$\begin{aligned}\Gamma(p+q/2,\, p-q/2)\cdot\mathbf{q}/q &= -\Gamma^{(b)}(p+q/2,\, p+q/2)/q\\ &=\frac{\mathbf{p}\cdot\mathbf{q}}{mq}\left\{1+\frac{\chi(p+q/2)-\chi(p-q/2)}{\mathbf{p}\cdot\mathbf{q}/m}\right\}\\ &\quad-\tau_3 p_0[\zeta(p+q/2)-\zeta(p-q/2)]/q\\ &\quad-2i\tau_2\frac{\phi(p+q/2)+\phi(p-q/2)}{2q}.\end{aligned} \quad (7.7)$$

For $q_0\neq 0$, then, we get

$$\begin{aligned}\mathbf{\Gamma}(p+q/2,\, p-q/2)\cdot\mathbf{q}/q &\approx (\mathbf{p}\cdot\mathbf{q}/q)\bar{Y}+2i\tau_2\phi\alpha^2 q/(q_0^2-\alpha^2q^2), \quad (7.8)\\ \bar{Y}&\equiv 1+[\chi(p+q/2)-\chi(p-q/2)]/(\mathbf{p}\cdot\mathbf{q}/m).\end{aligned}$$

Combining (7.6) and (7.8), the continuity equation takes the form

$$\begin{aligned}q_0\mathbf{\Gamma}_0-\mathbf{q}\cdot\mathbf{\Gamma}&=q_0\tau_3\bar{Z}+(\mathbf{p}\cdot\mathbf{q}/m)\bar{Y}+2i\tau_2\bar{\phi}\\ &\approx\Gamma^{(b)},\end{aligned}$$

which is indeed zero on the energy shell.

The transversal part of $\mathbf{\Gamma}$, on the other hand, is not coupled with the collective mode because the latter is a scalar wave.[22] We may, therefore, write instead of Eq. (7.8)

$$\begin{aligned}\mathbf{\Gamma}(p+q/2,\, p-q/2)&\approx(\mathbf{p}/m)\bar{Y}\\ &\quad+2i\tau_2\bar{\phi}\alpha^2 q/(q_0^2-\alpha^2q^2).\end{aligned} \quad (7.10)$$

Equations (7.6) and (7.10) for Γ_i have a very interesting structure. The noncollective part is essentially the same as the charge current for a free quasi-particle except for the renormalization $\bar{Z}$ and $\bar{Y}$, whereas the collective part is spread out both in space and time. Neglecting the momentum dependence of $\bar{Z}$, $\bar{Y}$, and ϕ, we may thus write the charge-current density (ρ,j) as

$$\begin{aligned}\rho(x,t)&\cong e\Psi^+\tau_3 Z\Psi(x,t)+\frac{1}{\alpha^2}\frac{\partial f(x,t)}{\partial t}\equiv\rho_0+\frac{1}{\alpha^2}\frac{\partial f}{\partial t},\\ \mathbf{j}(x,t)&\cong e\Psi^+(\mathbf{p}/m)Y\Psi(x,t)-\nabla f(x,t)\equiv\mathbf{j}_0-\nabla f,\end{aligned} \quad (7.11)$$

where f satisfies the wave equation

$$\left(\Delta-\frac{1}{\alpha^2}\frac{\partial^2}{\partial t^2}\right)f\approx-2e\Psi^+\tau_2\phi\Psi. \quad (7.12)$$

$(\rho_0,\mathbf{j}_0)$ is the charge-current residing in the "core" of a quasi-particle. The latter is surrounded by a cloud of the excitation field f. In a static situation, for example, f will fall off like $1/r$ from the core. When the particle is accelerated, a fraction of the charge is exchanged between the core and the cloud.

The total charge residing in a finite volume around a core is not constant because the current $-\nabla f$ reaches out to infinity.

[22] There may be transverse collective excitations (Bogoliubov, reference 2), but they do not automatically follow from the self-energy equation nor affect the energy gap structure.

8. THE PLASMA OSCILLATIONS

The inclusion of the annihilation-creation processes in the equations of the previous sections means that the vertex parts get multiplied by a string of closed loops, which represent the polarization (or shielding effect) of the surrounding medium. We will call the new quantities Λ, which now satisfy the following type of integral equations

$$\begin{aligned}\Lambda(p',p)&=\gamma-i\int\tau_3 G(p'-k)\Lambda(p'-k,\, p-k)\\ &\qquad\times G(p-k)\tau_3 D(k)d^4k\\ &\quad+iD(p'-p)\tau_3\int\mathrm{Tr}[\tau_3G(p'-k)\\ &\qquad\times\Lambda(p'-k,\, p-k)G(p-k)]d^4k,\\ D(q)&\equiv-ig^2h(q)^2\Delta(q)+e^2/q^2.\end{aligned} \quad (8.1)$$

$D(q)$ includes the effect of the Coulomb interaction [see Eq. (2.24)]. Putting

$$\begin{aligned}\bar{X}(p'-p)\equiv i\int&\mathrm{Tr}[\tau_3G(p'-k)\Lambda(p'-k,\, p'-k)\\ &\times G(p-k)]d^4k,\end{aligned} \quad (8.2)$$

Eq. (8.1) takes the same form as Eq. (4.3) for Γ with the inhomogeneous term replaced by $\gamma+\tau_3D\bar{X}$, so that Λ is a linear combination of the Γ corresponding to γ and Γ_0:

$$\Lambda=\Gamma+\Gamma_0D\bar{X}. \quad (8.3)$$

Substitution in Eq. (8.2) then yields

$$\begin{aligned}\bar{X}(p'-p)=i\int&\mathrm{Tr}[\tau_3G(p'-k)\\ &\times\Gamma(p'-k,\, p-k)G(p-k)]d^4k\\ &+iD(p'-p)\bar{X}(p'-p)\int\mathrm{Tr}[\tau_3G(p'-k)\\ &\qquad\times\Gamma_0(p'-k,\, p-k)G(p-k)]d^4k,\end{aligned}$$

or

$$\begin{aligned}\bar{X}(p'-p)=i\int&\mathrm{Tr}[\tau_3G(p'-k)\\ &\times\Gamma(p'-k,\, p-k)G(p-k)]d^4k\\ &\times\left\{1-iD(p'-p)\int\mathrm{Tr}[\tau_3G(p'-k)\right.\\ &\qquad\left.\times\Gamma_0(p'-k,\, p-k)G(p-k)]d^4k\right\}^{-1}\\ \equiv X(p'-p)/&[1-D(p'-p)X_0(p'-p)].\end{aligned} \quad (8.4)$$

Especially for $\gamma=\tau_3$, we get

$$\bar{X}_0(p'-p)=X_0(p'-p)/[1-D(p'-p)X_0(p'-p)],$$
$$\Lambda_0(p',p)=\Gamma_0(p',p)/[1-D(p'-p)X_0(p',p)]. \quad (8.5)$$

To obtain the collective excitations, let us next write down the homogeneous integral equation:

$$\Theta_q(p)=-i\int\tau_3G(p'+q/2)\Theta_q(p')\times G(p'-q/2)\tau_3D(p-p')d^4p'+i\tau_3D(q)\int\mathrm{Tr}\,[\tau_3G(p'+q/2)\Theta_q(p')\times G(p'-q/2)]d^4p', \quad (8.6)$$

which means

$$\Theta_q(p)=\Gamma_0(p+q/2,\, p-q/2)D(q)\chi(q),$$
$$\chi(q)\equiv i\int\mathrm{Tr}[\tau_3G(p'+q/2)\times\Theta_q(p')G(p'-q/2)\tau_3]d^4p'. \quad (8.7)$$

Substituting Θ_q in the second equation from the first, we get

$$1=D(q)X_0(q), \quad (8.8)$$

where $X_0(q)$ is defined in Eq. (8.4).

The solutions to Eq. (8.8) determine the new dispersion law $q_0=f(q)$ for the collective excitations.

With the solution (7.6), the quantity X_0 in Eq. (8.8) can be calculated. After some simplifications using Eq. (6.11), we obtain

$$X_0=\frac{1}{(2\pi)^3}\left[\frac{\alpha^2q^2}{q_0^2-\alpha^2q^2}\int\frac{p^2d^3p}{E_p(E_p^2-\alpha^2q^2/4)}+\frac{q^2}{4}\int\frac{p^2d^3p}{E_p(q_0^2/4-E_p^2)}\left(\frac{p^2}{3m^2E_p^2}-\frac{\alpha^2}{E_p^2-\alpha^2q^2/4}\right)\right]+O(q^4). \quad (8.9)$$

For $\alpha q\ll p$, and $q_0\gg p$ or $\ll p$, the second integral may be dropped and

$$X_0\cong\alpha^2q^2N/(q_0^2-\alpha^2q^2). \quad (8.10)$$

For small q^2, the dominant part of $D(q)$ in Eq. (8.8) is the Coulomb interaction e^2/q^2. Equation (8.8) then becomes

$$q_0^2=e^2\alpha^2N=e^2n \quad (q^2\rightarrow 0), \quad (8.11)$$

where n is the number of electrons per unit volume. This agrees with the ordinary plasma frequency for free electron gas.

We see thus that the previous collective state with $q_0^2=\alpha^2q^2$ has shifted its energy to the plasma energy as a result of the Coulomb interaction.

On the other hand, if Coulomb interaction is neglected, Eq. (8.8) leads to[23]

$$q_0^2=\alpha^2q^2[1-ig^2\Delta(q,q_0)h(q,q_0)^2N]. \quad (8.12)$$

The correction term, however, is of the order g^2N, hence should be neglected to be consistent with our approximation.

We can also study the behavior of X_0 in the limit $q_0\rightarrow 0$ for small but finite q^2:

$$X_0\approx\frac{1}{(2\pi)^3}\int\frac{p^2}{E^3}d^3p\approx N, \quad (8.13)$$

which comes entirely from the noncollective part of Γ_0, but again agrees with the free electron value.

Another observation we can make regarding $\bar{X}_0(q,q_0)$ is the following. $\bar{X}_0$ represents the charge density correlation in the ground state:

$$\bar{X}_0(q,q_0)=\int\langle 0|T(\rho(xt),\rho(0))|0\rangle e^{-i\mathbf{q}\cdot\mathbf{x}+iq_0t}d^3xdt.$$

If $|0\rangle$ is an eigenstate of charge, $\bar{X}_0$ should vanish for $q\rightarrow 0$, $q_0\neq 0$ since the right-hand side then consists of the nondiagonal matrix elements of the total charge operator Q:

$$\bar{X}_0(0,q_0)\propto\sum_n\left(\frac{1}{q_0-E_n}-\frac{1}{q_0+E_n}\right)|\langle n|Q|0\rangle|^2.$$

The converse is also true if $E_n>|q_0|$, $n\neq 0$ for some $q_0\neq 0$. Our result for $\bar{X}_0$, as is clear from Eqs. (8.5) and (8.9), has indeed the correct property in spite of the fact that the "bare" vacuum, from which we started, is not an eigenstate of charge.

9. CONCLUDING REMARKS

We have discussed here formal mathematical structure of the BCS-Bogoliubov theory. The nature of the approximation is characterized essentially as the Hartree-Fock method, and can be given a simple interpretation in terms of perturbation expansion. In the presence of external fields, the corresponding approximation insures, if treated properly, that the gauge invariance is maintained. It is interesting that the quasi-particle picture and charge conservation (or gauge invariance) can be reconciled at all. This is possible because we are taking account of the "radiative corrections" to the bare quasi-particles which are not eigenstates of charge. These corrections manifest themselves primarily through the existence of collective excitations.

There are some questions which have been left out. We would like to know, for one thing, what will happen if we seek corrections to our Hartree-Fock approximation by including processes (or diagrams) which have not been considered here. Even within our ap-

[23] Compare Anderson, reference 7.

proximation, there is an additional assumption of the weak coupling ($g^2N \ll 1$), and the importance of the neglected terms (of order g^2N and higher) is not known.

Experimentally, there has been some evidence[24] regarding the presence of spin paramagnetism in superconductors. This effect has to do with the spin density induced by a magnetic field and can be derived by means of an appropriate vertex solution. However, this does not seem to give a finite spin paramagnetism at 0°K.[25]

[24] Knight, Androes, and Hammond, Phys. Rev. **104**, 852 (1956); F. Reif, Phys. Rev. **106**, 208 (1957); G. M. Androes and W. D. Knight, Phys. Rev. Letters **2**, 386 (1959).

[25] K. Yosida, Phys. Rev. **110**, 769 (1958).

The collective excitations do not play an important role here as they are not excited by spin density. [$\Gamma^{(a)}$, Eq. (4.4), does not have the characteristic pole.]

It is desirable that both experiment and theory about spin paramagnetism be developed further since this may be a crucial test of the fundamental ideas underlying the BCS theory.

ACKNOWLEDGMENT

We wish to thank Dr. R. Schrieffer for extremely helpful discussions throughout the entire course of the work.

II. EARLY WORK

PHYSICAL REVIEW VOLUME 125, NUMBER 1 JANUARY 1, 1962

Gauge Invariance and Mass

JULIAN SCHWINGER
Harvard University, Cambridge, Massachusetts, and University of California, Los Angeles, California
(Received July 20, 1961)

It is argued that the gauge invariance of a vector field does not necessarily imply zero mass for an associated particle if the current vector coupling is sufficiently strong. This situation may permit a deeper understanding of nucleonic charge conservation as a manifestation of a gauge invariance, without the obvious conflict with experience that a massless particle entails.

DOES the requirement of gauge invariance for a vector field coupled to a dynamical current imply the existence of a corresponding particle with zero mass? Although the answer to this question is invariably given in the affirmative,[1] the author has become convinced that there is no such necessary implication, once the assumption of weak coupling is removed. Thus the path to an understanding of nucleonic (baryonic) charge conservation as an aspect of a gauge invariance, in strict analogy with electric charge,[2] may be open for the first time.

One potential source of error should be recognized at the outset. A gauge-invariant system is not the continuous limit of one that fails to admit such an arbitrary function transformation group. The discontinuous change of invariance properties produces a corresponding discontinuity of the dynamical degrees of freedom and of the operator commutation relations. No reliable conclusions about the mass spectrum of a gauge-invariant system can be drawn from the properties of an apparently neighboring system, with a smaller invariance group. Indeed, if one considers a vector field coupled to a divergenceless current, where gauge invariance is destroyed by a so-called mass term with parameter m_0, it is easily shown[3] that the mass spectrum must extend below m_0. The lowest mass value will therefore become arbitrarily small as m_0 approaches zero. Nevertheless, if m_0 is exactly zero the commutation relations, or equivalent properties, upon which this conclusion is based become entirely different and the argument fails.

If invariance under arbitrary gauge transformations is asserted, one should distinguish sharply between numerical gauge functions and operator gauge functions, for the various operator gauges are not on the same quantum footing. In each coordinate frame there is a unique operator gauge, characterized by three-dimensional transversality (radiation gauge), for which one has the standard operator construction in a vector space of positive norm, with a physical probability interpretation. When the theory is formulated with the aid of vacuum expectation values of time-ordered operator products, the Green's functions, the freedom of formal gauge transformation can be restored.[4] The Green's functions of other gauges have more complicated operator realizations, however, and will generally lack the positiveness properties of the radiation gauge.

Let us consider the simplest Green's function associated with the field $A_\mu(x)$, which can be derived from the unordered product

$$\langle A_\mu(x)A_\nu(x')\rangle = \int \frac{(dp)}{(2\pi)^3} e^{ip(x-x')} dm^2\, \eta_+(p)\delta(p^2+m^2)A_{\mu\nu}(p),$$

where the factor $\eta_+(p)\delta(p^2+m^2)$ enforces the spectral restriction to states with mass $m\geq 0$ and positive energy. The requirement of non-negativeness for the matrix $A_{\mu\nu}(p)$ is satisfied by the structure associated with the radiation gauge, in virtue of the gauge-dependent asymmetry between space and time (the time axis is specified by the unit vector n_μ):

$$A_{\mu\nu}^R(p) = B(m^2)\left[g_{\mu\nu} - \frac{(p_\mu n_\nu + p_\nu n_\mu)(np) + p_\mu p_\nu}{p^2+(np)^2}\right].$$

Here $B(m^2)$ is a real non-negative number. It obeys the sum rule

$$1 = \int_0^\infty dm^2\, B(m^2),$$

which is a full expression of all the fundamental equal-time commutation relations.

The field equations supply the analogous construction for the vacuum expectation value of current products $\langle j_\mu(x)j_\nu(x')\rangle$, in terms of the non-negative matrix

$$j_{\mu\nu}(p) = m^2 B(m^2)(p_\mu p_\nu - g_{\mu\nu}p^2).$$

The factor m^2 has the decisive consequence that $m=0$ is not contained in the current vector's spectrum of vacuum fluctuations. The latter determines $B(m^2)$ for $m>0$, but leaves unspecified a possible delta function contribution at $m=0$,

$$B(m^2) = B_0\delta(m^2) + B_1(m^2).$$

The non-negative constant B_0 is then fixed by the sum rule,

$$1 = B_0 + \int_0^\infty dm^2\, B_1(m^2).$$

[1] For example, J. Schwinger, Phys. Rev. **75**, 651 (1949).
[2] T. D. Lee and C. N. Yang, Phys. Rev. **98**, 1501 (1955).
[3] K. Johnson, Nuclear Phys. **25**, 435 (1961).
[4] J. Schwinger, Phys. Rev. **115**, 721 (1959).

We have now recognized that the vacuum fluctuations of the vector A_μ are composed of two parts. One, with $m>0$, is directly related to corresponding current fluctuations, while the other part, with $m=0$, can be associated with a pure radiation field, which is transverse in both three- and four-dimensional senses and has no accompanying current. Imagine that the current vector contains a variable numerical factor. If this is set equal to zero, we have $B_1(m^2)=0$ and $B_0=1$ or, just the radiation field. For a sufficiently small nonzero value of the parameter, B_0 will be slightly less than unity, which may be the situation for the electromagnetic field. Or it may be that the electrodynamic coupling is quite considerable and gives rise to a small value of B_0, which has the appearance of a fairly weak coupling. Can we increase further the magnitude of the variable parameter until $\int dm^2\, B_1(m^2)$ attains its limiting value of unity, at which point $B_0=0$, and $m=0$ disappears from the spectrum of A_μ? The general requirement of gauge invariance no longer seems to dispose of this essentially dynamical question.

Would the absence of a massless particle imply the existence of a stable, unit spin particle of nonzero mass? Not necessarily, since the vacuum fluctuation spectrum of A_μ becomes identical with that of j_μ, which is governed by all of the dynamical properties of the fields that contribute to this current. For the particularly interesting situation of a vector field that is coupled to the current of nucleonic charge, the relevant spectrum, in the approximate strong-interaction framework, is that of the states with $N=Y=T=0$, $R_T=-1$, $J=1$, and odd parity. This is a continuum, beginning at three pion masses.[5] It is entirely possible, of course, that $B(m^2)$ shows a more or less pronounced maximum which could be characterized approximately as an unstable particle.[6] But the essential point is embodied in the view that the observed physical world is the outcome of the dynamical play among underlying primary fields, and the relationship between these fundamental fields and the phenomenological particles can be comparatively remote, in contrast to the immediate correlation that is commonly assumed.

---

[5] The very short range of the resulting nuclear interaction together with the qualitative inference that like nucleonic charges are thereby repelled suggests that the vector field which defines nucleonic charge is also the ultimate instrument of nuclear stability.

[6] *Note added in proof*. Experimental evidence for an unstable particle of this type has recently been announced by B. C. Maglić, L. W. Alvarez, A. H. Rosenfeld, and M. L. Stevenson, in Phys. Rev. Letters 7, 178 (1961).

PHYSICAL REVIEW VOLUME 128, NUMBER 5 DECEMBER 1, 1962

Gauge Invariance and Mass. II*

JULIAN SCHWINGER
Harvard University, Cambridge, Massachusetts

(Received July 2, 1962)

The possibility that a vector gauge field can imply a nonzero mass particle is illustrated by the exact solution of a one-dimensional model.

It has been remarked[1] that the gauge invariance of a vector field does not necessarily require the existence of a massless physical particle. In this note we shall add a few related comments and give a specific model for which an exact solution affirms this logical possibility. The model is the physical, if unworldly situation of electrodynamics in one spatial dimension, where the charge-bearing Dirac field has no associated mass constant. This example is rather unique since it is a simple model for which there is an exact divergence-free solution.[2]

GENERAL DISCUSSION

The Green's function of an Abelian vector gauge field has the structure

$$\mathcal{G}_{\mu\nu}(x,x')=\pi_{\mu\nu}(-i\partial)\mathcal{G}(-i\partial)\delta(x-x'),$$

where $\pi_{\mu\nu}(p)$ is a gauge-dependent projection matrix and

$$\mathcal{G}(p)=\int_0^\infty dm^2\frac{B(m^2)}{p^2+m^2-i\epsilon},$$

which is subject to the sum rule

$$1=\int_0^\infty dm^2\, B(m^2).$$

An alternative form of $\mathcal{G}(p)$ is

$$\mathcal{G}(p)=\left[p^2+\lambda^2-i\epsilon+(p^2-i\epsilon)\int_0^\infty dm^2\frac{s(m^2)}{p^2+m^2-i\epsilon}\right]^{-1},$$

where the function $s(m^2)$ and the constant λ^2 are nonnegative. The latter has been derived[3] with the understanding that the pole at $z=0$ of the expression

$$-\frac{\lambda^2}{z}+\int_0^\infty dm^2\frac{s(m^2)}{m^2-z}=\int_0^\infty\frac{dm^2}{m^2-z}[s(m^2)+\lambda^2\delta(m^2)]$$

is completely described by the parameter λ. Accordingly,

$$\mathcal{G}(0)=\frac{1}{\lambda^2}=\int_0^\infty dm^2\frac{B(m^2)}{m^2},$$

and $\lambda^2>0$ unless $m=0$ is contained in the spectrum. Thus, it is necessary that λ vanish if $m=0$ is to appear as an isolated mass value in the physical spectum. But it is also necessary that

$$s(0)=0,$$

such that

$$\int_{\to 0}^\infty\frac{dm^2}{m^2}s(m^2)<\infty,$$

for only then do we have a pole at $p^2=0$,

$$p^2\sim 0:\quad \mathcal{G}(p)\sim B_0/(p^2-i\epsilon),\quad 0<B_0<1.$$

Under these conditions,

$$B(m^2)=B_0\delta(m^2)+B_1(m^2),$$

where

$$B_0=\left(1+\int_0^\infty\frac{dm^2}{m^2}s(m^2)\right)^{-1}$$

and

$$B_1(m^2)=[s(m^2)/m^2]\Big/\left[1+P\int_0^\infty dm'^2\frac{s(m'^2)}{m'^2-m^2}\right]^2+[\pi s(m^2)]^2.$$

The physical interpretation of $s(m^2)$ derives from the relation of the Green's function to the vacuum transformation function in the presence of sources. For sufficiently weak external currents $J_\mu(x)$,

$$\langle 0|0\rangle^J=\exp\left[\tfrac{1}{2}i\int(dx)(dx')J^\mu(x)\mathcal{G}_{\mu\nu}(x,x')J^\nu(x')\right]$$

$$=\exp\left[\tfrac{1}{2}i\int(dp)J^\mu(p)^*\mathcal{G}(p)J_\mu(p)\right],$$

which involves the reduction of the projection matrix $\pi_{\mu\nu}(p)$ to $g_{\mu\nu}$ for a conserved current, or equivalently

$$p_\mu J^\mu(p)=0.$$

We shall present this transformation function as a measure of the response to the external vector potential

$$A_\mu(p)=\mathcal{G}(p)J_\mu(p),$$

namely,

$$\langle 0|0\rangle^J=\exp\left[\tfrac{1}{2}i\int(dp)A^\mu(p)^*\mathcal{G}(p)^{*-1}A_\mu(p)\right].$$

---

* Supported in part by the Air Force Office of Scientific Research (Air Research and Development Command), under contract number AF-49(638)-589.

[1] J. Schwinger, Phys. Rev. **125**, 397 (1962).

[2] There is a divergence in the so-called Thirring model [W. E. Thirring, Ann. Phys. (New York) 3, 91 (1958)], which uses local current interactions rather than a Bose field.

[3] J. Schwinger, Ann. Phys. (New York) **9**, 169 (1960).

The probability that the vacuum state shall persist despite the disturbance is

$$|\langle 0|0\rangle^J|^2=\exp\left[-\int (dp)A_\mu(p)^*A_\mu(p)\ \mathrm{Im}\mathcal{G}(p)^{*-1}\right]$$

$$=\exp\left[-\pi\int (dp)dm^2\delta(p^2+m^2)\times s(m^2)(-\tfrac{1}{2})F^{\mu\nu}(p)^*F_{\mu\nu}(p)\right],$$

which exhibits $s(m^2)$ as a measure of the probability that an external field $F_{\mu\nu}$ will produce a vacuum excitation involving an energy-momentum transfer measured by the mass m.

The vanishing of $s(m^2)$ at $m=0$ is normal threshold behavior for an excitation function. If a zero-mass particle is not to exist, $m=0$ must be an abnormal threshold. Two possibilities can be distinguished. In the first of these, $s(m^2)$ is finite or possibly singular at $m=0$, but in such a way that

$$\lim_{z\to 0} z\int_0^\infty dm^2\frac{s(m^2)}{m^2-z}=0.$$

Then the physical mass spectrum begins at $m=0$ but there is no recognizable zero-mass particle. For the second situation, $s(m^2)$ has a delta-function singularity at $m^2=0$,

$$s(m^2)=\lambda^2\delta(m^2)+s_1(m^2),$$

and

$$s_1(m^2)=0,\quad m^2<m_0^2.$$

If the threshold mass m_0 is zero, the restriction of the previous situation applies to the function $s_1(m^2)$. Now, $m=0$ is not contained in the spectrum at all. This statement is true even if $m_0=0$ for, according to the structure of $B_1(m^2)=B(m^2)$,

$$B(m^2)=\frac{m^2s_1(m^2)}{[R(m^2)]^2+[\pi m^2s_1(m^2)]^2},$$

in which

$$R(m^2)=m^2-\lambda^2+m^2P\int_{m_0^2}^\infty dm'^2\frac{s_1(m^2)}{m'^2-m^2},$$

we have

$$\lim_{m^2\to 0}B(m^2)=\lim_{m^2\to 0}\frac{m^2s_1(m^2)}{\lambda^4}=0.$$

Let us suppose that m_0 is the threshold of a continuous spectrum. A stable particle of mass $m<m_0$ will exist if $R(m_0^2)>0$. Should both $R(m_0^2)$ and $s_1(m_0^2)$ be zero there would be a stable particle of mass m_0. No stable particle exists if $R(m_0^2)<0$. But there is always an unstable particle, in a certain sense. By this we mean that $R(m^2)$ vanishes at some mass value $m_1>m_0$, under the general restrictions required for the continuity of the function $R(m^2)$, as a consequence of this function's asymptotic approach to $+\infty$ with increasing m^2. The mass m_1 will be physically recognizable as the mass of an unstable particle if the mass width

$$\gamma=\frac{\pi m_1s_1(m_1^2)}{[dR(m_1^2)/dm_1^2]}$$

is sufficiently small. [We take the derivative of $R(m_1^2)$ to be positive, which is appropriate for the simplifying assumption that only one zero occurs.] The contribution of such a fairly sharp resonance to the sum rule for $B(m^2)$ is given by

$$\int_{m\sim m_1}dm^2B(m^2)=[dR(m_1^2)/dm_1^2]^{-1}<1.$$

SIMPLE MODELS

Some of these possibilities can be illustrated in very simple physical contexts. We consider the linear approximation to the problem of electromagnetic vacuum polarization for spaces of dimensionality $n=2$ and 1. A modification of a technique[4] previously applied to three-dimensional space yields for $m>m_0$:

$$s(m^2)=\int_0^{(1-m_0^2/m^2)^{1/2}}dv(1-v^2)(e^2/8\pi^2)\quad\text{for}\quad n=3$$

$$=\int_0^{(1-m_0^2/m^2)^{1/2}}dv(1-v^2)(e^2/4\pi^2)\times[m^2(1-v^2)-m_0^2]^{-1/2}\quad\text{for}\quad n=2$$

$$=\int_0^{(1-m_0^2/m^2)^{1/2}}dv(1-v^2)(e^2/\pi)\delta[m^2(1-v^2)-m_0^2]\quad\text{for}\quad n=1;$$

for $m<m_0$:

$$s(m^2)=0,$$

where the known result for $n=3$ has been included for comparison. The threshold mass m_0 is that for single pair creation. It should be noted that the coupling constant e^2 of electrodynamics in n-dimensional space has the dimensions of a mass raised to the power $3-n$. For $n<3$ this single pair approximation does not lead to difficulties concerning the existence of such integrals as

$$B_0^{-1}-1=\int_0^{\to\infty}\frac{dm^2}{m^2}B(m^2),$$

since, for $m\gg m_0$:

$$\begin{aligned}s(m^2)&\sim(e^2/12\pi^2)&&\text{for}\quad n=3,\\&\sim(e^2/16\pi)(1/m)&&\text{for}\quad n=2,\\&\sim(e^2/2\pi)(m_0^2/m^4)&&\text{for}\quad n=1.\end{aligned}$$

The particular situation in which we are interested appears at the limit $m_0\to 0$. Then we have

$$\begin{aligned}s(m^2)&=(e^2/16\pi)(1/m)&&\text{for}\quad n=2,\\&=(e^2/\pi)\delta(m^2)&&\text{for}\quad n=1.\end{aligned}$$

[4] *Selected Papers on Quantum Electrodynamics* (Dover Publications, New York, 1958), p. 209.

Two-dimensional electrodynamics illustrates the first of the two possibilities for an anomalous threshold at $m=0$. The spectral function $B(m^2)$ describes a purely continuous spectrum,

$$dm^2\, B(m^2)=\frac{2}{\pi}\frac{e^2}{16}\frac{dm}{m^2+(e^2/16)^2},$$

and an m integration from 0 to ∞ satisfies the sum rule. In one-dimensional electrodynamics we meet a special case of the second possibility, with

$$\lambda^2=e^2/\pi,\quad s_1(m^2)=0.$$

Accordingly,

$$B(m^2)=\delta(m^2-(e^2/\pi))$$

and the mass spectrum is localized at one point, describing a stable particle of mass $e/\pi^{1/2}$.

The basis indicated for the latter conclusion will not be very convincing, but it is an exact result. To prove this we first compute for one spatial dimension the electric current induced by an arbitrary external potential in the vacuum state of a massless charged Dirac field. The appropriate gauge-invariant expression for the current[5] is

$$j_\mu(x)=-\tfrac{1}{2}e\,\mathrm{tr}q\alpha_\mu G(x,x')\exp\left[-ieq\int_{x'}^{x}d\xi^\mu A_\mu(\xi)\right]\Bigg|_{x'\to x},$$

in which the approach of x' to x is performed from a spatial direction in order to maintain time locality. The Green's function is defined by the differential equation

$$\alpha^\mu[\partial_\mu-ieqA_\mu(x)]G(x,x')=\delta(x-x'),$$

together with the outgoing wave boundary condition, in the absence of the potential. Only two Dirac matrices appear here, $\alpha^0=-\alpha_0=1$ and $\alpha^1=\alpha_1$, which has the eigenvalues ±1. Those are also the eigenvalues of the independent charge matrix q. The Green's function equation can be satisfied by writing

$$G(x,x')=G^0(x,x')\exp\{ieq[\phi(x)-\phi(x')]\},$$

where

$$\alpha^\mu\partial_\mu\phi=\alpha^\mu A_\mu(x)$$

and

$$\alpha^\mu\partial_\mu G^0(x,x')=\delta(x-x').$$

The latter defines the free Green's function, which is given explicitly by

$$G^0(x,x')=\int_0^\infty\frac{dp}{2\pi}\exp[ip\alpha^\mu(x_\mu-x_\mu')]\quad\text{for}\quad x^0>x^{0\prime},$$

$$=-\int_{-\infty}^0\frac{dp}{2\pi}\exp[ip\alpha^\mu(x_\mu-x_\mu')]\quad\text{for}\quad x^0<x^{0\prime}.$$

At equal times, and for sufficiently small x_1-x_1', we have

$$G(x,x')\exp\left[-ieq\int_{x'}^{x}d\xi^\mu A_\mu(\xi)\right]$$

$$\cong\frac{i}{2\pi}\frac{\alpha_1}{x_1-x_1'}-\frac{eq}{2\pi}\alpha_1[\partial_1\phi(x)-A_1(x)].$$

The first term does not contribute to the vacuum current when the limit $x_1'\to x_1$ is performed symmetrically. On utilizing the relation

$$\alpha_1(\partial_1\phi-A_1)=-(\partial_0\phi-A_0),$$

we find that

$$j_\mu(x)=-\frac{e^2}{\pi}A_\mu(x)+\partial_\mu\left[\frac{e^2}{4\pi}\mathrm{tr}\phi(x)\right].$$

This expression for the induced current is Lorentz covariant, gauge invariant, and obeys the equation of conservation. It is also a linear function of the external field. To verify these statements we construct a differential equation for $\mathrm{tr}\phi(x)$ by multiplying the ϕ equation with $\partial_0-\alpha_1\partial_1$ and evaluating the trace. The result is

$$\partial^2\tfrac{1}{4}\,\mathrm{tr}\phi(x)=\partial_\mu A^\mu(x),$$

and therefore

$$\tfrac{1}{4}\,\mathrm{tr}\phi(x)=-\int(dx')D(x,x')\partial_\mu'A^\mu(x'),$$

in which D is the outgoing-wave Green's function defined by

$$-\partial^2D(x,x')=\delta(x-x').$$

By using a symbolic matrix notation for coordinates and vector indices, we can write

$$j=-(e^2/\pi)(1+\partial D\partial)A,$$

which exhibits the symmetrical projection matrix

$$\pi=1+\partial D\partial,$$
$$\partial\pi=\pi\partial=0,$$

that guarantees gauge invariance and current conservation.

We shall insert this result in the functional differential equation obeyed by the Green's functional $G[J]$, the vacuum transformation function in the presence of external currents. It is convenient to use the particular system of equations that refer to the Lorentz gauge,

$$\left\{(\partial\partial-\partial^2)\frac{1}{i}\frac{\delta}{\delta J}-(1+\partial D\partial)\left[J+j\left(\frac{1}{i}\frac{\delta}{\delta J}\right)\right]\right\}G[J]=0,$$

$$\partial\frac{\delta}{\delta J}G[J]=0,$$

[5] The necessity for the line integral factor has been noted before [J. Schwinger, Phys. Rev. Letters 3, 296 (1959)].

which also utilize a symbolic notation for vectorial coordinate functions. We have written $j(-i\delta/\delta J)$ to indicate the conversion of $j(A)$ into a functional differential operator by the substitution $A\rightarrow -i\delta/\delta J$. The functional differential equation implied by the known structure of this operator is

$$\pi\left[\left(-\partial^2+\frac{e^2}{\pi}\right)\frac{1}{i}\frac{\delta}{\delta J}-J\right]G[J]=0,$$

or, on uniting the two defining properties of the functional,

$$\left(\frac{1}{i}\frac{\delta}{\delta J}-\pi\mathcal{G}J\right)G[J]=0,$$

in which

$$[-\partial^2+(e^2/\pi)]\mathcal{G}(x,x')=\delta(x-x').$$

The Green's functional $G[J]$ is therefore given exactly by

$$G[J]=\exp\left[\tfrac{1}{2}i\int(dx)(dx')J^\mu(x)\mathcal{G}_{\mu\nu}(x,x')J^\nu(x')\right],$$

with

$$\mathcal{G}_{\mu\nu}(x,x')=\pi_{\mu\nu}(-i\partial)\mathcal{G}(-i\partial)\delta(x-x')$$

and

$$\mathcal{G}(p)=\frac{1}{p^2+(e^2/\pi)-i\epsilon}.$$

Thus, all states that can be excited by vector currents are fully described as noninteracting ensembles of Bose particles with the mass $e/\pi^{1/2}$.

Concerning the complete Green's functional including Fermi sources, $G[\eta J]$, we shall only remark that

$$G[\eta J]=\exp\left[-\frac{1}{2}\int(dx)(dx')\eta(x)\times G\left(x,x',\frac{1}{i}\frac{\delta}{\delta J}\right)\eta(x')\right]G[J],$$

in which the Green's function can be presented as

$$G(x,x',A)=G^0(x,x')\exp\left[i\int(d\xi)j^\mu(\xi,x,x')A_\mu(\xi)\right]$$

with

$$j^\mu(\xi,x,x')=eq\alpha^\mu\left(\alpha^1\frac{\partial}{\partial\xi^1}-\frac{\partial}{\partial\xi^0}\right)[D(\xi,x)-D(\xi,x')].$$

On expanding the Green's functional in even powers of the Fermi source, we encounter functional differential operators that are contained in one or more factors of the type

$$\exp\left[\int(d\xi)j^\mu(\xi,x,x')\delta/\delta J^\mu(\xi)\right],$$

the effect of which is simply to produce the translation $J\rightarrow J+j$ in $G[J]$. The first Fermi Green's function is

$$G(x,x')=G(x,x',-i\delta/\delta J)G[J]|_{J=0}$$
$$=G^0(x,x')\exp\left[\tfrac{1}{2}i\int(d\xi)(d\xi')\times j^\mu(\xi,x,x')\mathcal{G}_{\mu\nu}(\xi,\xi')j^\nu(\xi',x,x')\right].$$

The latter exponential factor is given by

$$\exp\left[-\frac{i}{4\pi}\int(dp)\left(\frac{1}{p^2-i\epsilon}-\frac{1}{p^2+(e^2/\pi)-i\epsilon}\right)\times(1-e^{ip(x-x')})\right].$$

We shall be content to note that this integral and the similar integrals encountered in more general Green's functions are completely convergent. The detailed physical interpretation of the Green's functions is rather special and apart from our main purpose.

These simple examples are quite uninformative in one important respect. They do not exhibit a critical dependence upon the coupling constant. As we have discussed previously, one can view the electromagnetic field as undercoupled and the hypothetical vector field that relates to nucleonic charge as overcoupled, in the sense of a critical value at which the massless Bose particle ceases to exist. The corresponding appearance of an anomalous zero-mass threshold must be attributed to a dynamical mechanism. We can supply an artificial mathematical model that illustrates the situation. Let the following be a contributory term in $s(m^2)$:

$$s_0(m^2)=\frac{\lambda^2}{\pi}\frac{m\gamma}{(m^2-m_0^2\kappa)^2+(m\gamma)^2},$$

in which m_0 is a characteristic physical fermion mass, and λ/m_0, γ/m_0, and κ are positive functions of the (dimensionless) coupling constant. In electrodynamics the near-resonant contributions of such a term can be identified with the creation of a unit angular momentum positronium state, while the values far below resonance refer to the creation of three-photon states (the model falsifies the latter, which should vary as m^8 for $m\ll m_0$). It is reasonable to suppose that κ decreases with increasing strength of the coupling, and we can imagine that a critical value exists for which both κ and γ reach zero, with finite λ. In that circumstance,

$$s_0(m^2)=\lambda^2\delta(m^2),$$

and the null-mass particle disappears from the spectrum. Since this argument requires that one type of excitation move down to zero mass at the critical coupling strength, it is plausible that some other types of excitation will then be located at fairly small fractions of m_0. Thus, one could anticipate that the known spin-0 bosons, for example, are secondary dynamical manifestations of strongly coupled primary fermion fields and vector gauge fields. This line of thought emphasizes that the question "Which particles are fundamental?" is incorrectly formulated. One should ask "What are the fundamental fields?"

ACKNOWLEDGMENTS

I have had the benefit of conversations on this and related topics with Kenneth Johnson and Charles Sommerfield.

Reprinted from THE PHYSICAL REVIEW, Vol. 130, No. 1, 439–442, 1 April 1963
Printed in U. S. A.

Plasmons, Gauge Invariance, and Mass

P. W. ANDERSON
Bell Telephone Laboratories, Murray Hill, New Jersey
(Received 8 November 1962)

Schwinger has pointed out that the Yang-Mills vector boson implied by associating a generalized gauge transformation with a conservation law (of baryonic charge, for instance) does not necessarily have zero mass, if a certain criterion on the vacuum fluctuations of the generalized current is satisfied. We show that the theory of plasma oscillations is a simple nonrelativistic example exhibiting all of the features of Schwinger's idea. It is also shown that Schwinger's criterion that the vector field $m\neq0$ implies that the matter spectrum before including the Yang-Mills interaction contains $m=0$, but that the example of superconductivity illustrates that the physical spectrum need not. Some comments on the relationship between these ideas and the zero-mass difficulty in theories with broken symmetries are given.

RECENTLY, Schwinger[1] has given an argument strongly suggesting that associating a gauge transformation with a local conservation law does not necessarily require the existence of a zero-mass vector boson. For instance, it had previously seemed impossible to describe the conservation of baryons in such a manner because of the absence of a zero-mass boson and of the accompanying long-range forces.[2] The problem of the mass of the bosons represents the major stumbling block in Sakurai's attempt to treat the dynamics of strongly interacting particles in terms of the Yang-Mills gauge fields which seem to be required to accompany the known conserved currents of baryon number and hypercharge.[3] (We use the term "Yang-Mills" in Sakurai's sense, to denote any generalized gauge field accompanying a local conservation law.)

The purpose of this article is to point out that the familiar plasmon theory of the free-electron gas exemplifies Schwinger's theory in a very straightforward manner. In the plasma, transverse electromagnetic waves do not propagate below the "plasma frequency," which is usually thought of as the frequency of long-wavelength longitudinal oscillation of the electron gas. At and above this frequency, three modes exist, in close analogy (except for problems of Galilean invariance implied by the inequivalent dispersion of longitudinal and transverse modes) with the massive vector boson mentioned by Schwinger. The plasma frequency is equivalent to the mass, while the finite density of electrons leading to divergent "vacuum" current fluctuations resembles the strong renormalized coupling of Schwinger's theory. In spite of the absence of low-frequency photons, gauge invariance and particle conservation are clearly satisfied in the plasma.

In fact, one can draw a direct parallel between the dielectric constant treatment of plasmon theory[4] and Schwinger's argument. Schwinger comments that the commutation relations for the gauge field A give us one sum rule for the vacuum fluctuations of A, while those for the matter field give a completely independent value for the fluctuations of matter current j. Since j is the source for A and the two are connected by field equations, the two sum rules are normally incompatible unless there is a contribution to the A rule from a free, homogeneous, weakly interacting, massless solution of the field equations. If, however, the source term is large enough, there can be no such contribution and the massless solutions cannot exist.

The usual theory of the plasmon does not treat the electromagnetic field quantum-mechanically or discuss vacuum fluctuations; yet there is a close relationship between the two arguments, and we, therefore, show that the quantum nature of the gauge field is irrelevant. Our argument is as follows:

The equation for the electromagnetic field is

$$p^2 A_\mu = (k^2-\omega^2)A_\mu(\mathbf{k},\omega) = 4\pi j_\mu(\mathbf{k},\omega).$$

[1] J. Schwinger, Phys. Rev. 125, 397 (1962).
[2] T. D. Lee and C. N. Yang, Phys. Rev. 98, 1501 (1955).
[3] J. J. Sakurai, Ann. Phys. (N. Y.) 11, 1 (1961).
[4] P. Nozières and D. Pines, Phys. Rev. 109, 741 (1958).

A given distribution of current j_μ will, therefore, lead to a response A_μ given by

$$A_\mu=\frac{4\pi}{k^2-\omega^2}j_\mu=\frac{4\pi}{p^2}j_\mu. \qquad (1)$$

(1) is merely the statement that only the electromagnetic current can be a source of the field; it is required for general gauge invariance and charge conservation according to the usual arguments.

The dynamics of the matter system—of the plasma in that case, of the vacuum in the elementary particle problem—determine a second response function, the response of the current to a given electromagnetic or Yang-Mills field. Let us call this response function

$$j_\mu(\mathbf{k},\omega)=-K_{\mu\nu}(\mathbf{k},\omega)A_\nu(\mathbf{k},\omega). \qquad (2)$$

By well-known arguments of gauge invariance, $K_{\mu\nu}$ must have a certain form: Schwinger points out that in the relativistic case it must be proportional to $p_\mu p_\nu - g_{\mu\nu}p^2$, and equivalent arguments give one the same form in superconductivity.[5] It will be convenient to consider, for simplicity, only the gauge

$$p_\mu A_\mu=0. \qquad (3)$$

Then the response is diagonal: $K_{\mu\nu}=-g_{\mu\nu}K$. For a plasma with n carriers of charge e and mass M it is simply (in the limit $p\to 0$)

$$K=ne^2/M. \qquad (4)$$

In an insulator the response is not relativistically invariant. If the insulator has magnetic polarizability α_m and electric α_e, the response equations may be written, in the gauge (3),

$$j_\mu=-\alpha_e p^2 A_\mu \quad \text{(longitudinal and time components)},$$

$$\mathbf{j}=-\alpha_m p^2\mathbf{A} \quad \text{(transverse components)}.$$

In a truly relativistic situation such as our normal picture of a vacuum, we expect

$$j_\mu=-\alpha p^2 A_\mu \qquad (5)$$

to describe normal polarizable behavior.

Since we cannot turn off the interactions, we do not actually observe the responses (1), (2), or (5). If we insert a test particle, its field $A_\mu{}^e$ induces a current j_μ which in turn acts as the source for an internal field $A_\mu{}^i$:

$$j_\mu=-K(A_\mu{}^i+A_\mu{}^e), \quad A_\mu{}^i=+4\pi j_\mu/p^2,$$

or, the total field is modified to

$$A_\mu=[p^2/(p^2+4\pi K)]A_\mu{}^e. \qquad (6)$$

The pole at which A propagates freely occurs at a mass (frequency)

$$m^2=-p^2=4\pi K, \qquad (7)$$

which in a conductor is

$$m^2=\omega^2-k^2=\omega_p{}^2. \qquad (8)$$

ω_p is the usual plasma frequency $(4\pi ne^2/M)^{1/2}$.

It is not necessary here to go in detail into the relationship between longitudinal and transverse behavior of the plasmon. In the limit $p\to 0$ both waves propagate according to (8). The longitudinal plasmon is generally thought of as entirely an attribute of the plasma, while the transverse ones are considered to result from modification of the propagation of real photons by the medium. This is reasonable in the classical case because the longitudinal plasmon disappears at a certain cutoff energy and has a different dispersion law; but in a Lorentz-covariant theory of the vacuum it would be indistinguishable from the third component of a massive vector boson of which the transverse photons are the two transverse components.

How, then, if we were confined to the plasma as we are to the vacuum and could only measure renormalized quantities, might we try to determine whether, before turning on the effects of electromagnetic interaction, A had been a massless gauge field and K had been finite? As far as we can see, this is not possible; it is, nonetheless, interesting to see what the criterion is in terms of the actual current response function to a perturbation in the Lagrangian

$$\delta L=j_\mu\delta A_\mu. \qquad (9)$$

This will turn out to be identical to Schwinger's criterion. The original "bare" response function was K:

$$j_\mu=-K_{\mu\nu}\delta A_\mu.$$

Taking into account the interaction, however, we must correct for the induced fields and currents, and we get

$$j_\mu=-K'\delta A_\mu{}^e=-K[p^2/(p^2+4\pi K)]\delta A_\mu{}^e\to -(p^2/4\pi)\delta A_\mu{}^e, \quad p^2\to 0. \qquad (10)$$

Thus, the new response to an applied perturbing field (9) is very like that of an ordinary polarizable medium. The only difference from an ordinary polarizable "vacuum" with bare response (5) is that in that case as $p\to 0$

$$K'\to -[\alpha/(1+4\pi\alpha)]p^2, \qquad (11)$$

so that the coefficient of $p^2/4\pi$ is less than unity.

This criterion is precisely the same as Schwinger's criterion

$$\int B_1(m^2)dm^2=1,$$

where $B_1 m^2$ is the weight function for the current vacuum fluctuations. This can be shown by a simple dispersion argument. Schwinger expresses the unordered

[5] M. R. Schafroth, Helv. Phys. Acta 24, 645 (1951).

product expectation value of the current as

$$\langle j_\mu(x)j_\nu(x')\rangle=\int dm^2\, m^2B_1(m^2)\int\frac{dp}{(2\pi)^3}e^{ip(x-x')}\times\eta_+(p)\delta(p^2+m^2)(p_\mu p_\nu-g_{\mu\nu}p^2).$$

The Fourier transform of the corresponding retarded Green's function is our response function:

$$K'(p)=\int\frac{dm^2\, m^2B_1(m^2)}{p^2-m^2}[p_\mu p_\nu-g_{\mu\nu}p^2],$$

and

$$\lim_{p\to 0}K'(p)=(p_\mu p_\nu-g_{\mu\nu}p^2)\int dm^2\, B_1(m^2).$$

Thus, (aside from a factor 4π which Schwinger has not used in his field equation) his criterion is also that the polarizability α', here expressed in terms of a dispersion integral, have its maximum possible value, 1.

The polarizability of the vacuum is not generally considered to be observable[6] except in its p dependence (terms of order p^4 or higher in K). In fact, we can remove (11) entirely by the conventional renormalization of the field and charge

$$A_r=AZ^{-1/2},\quad e_r=eZ^{1/2},\quad j_r=jZ^{1/2}.$$

Z, here, can be shown to be precisely

$$Z=1-4\pi\alpha'=1-\int_0^\infty dm^2\, B_1(m^2).$$

Thus, the renormalization procedure is possible for any merely polarizable "vacuum," but not for the special case of the conducting "plasma" type of vacuum. In this case, no net true charge remains localized in the region of the dressed particle; all of the charge is carried "at infinity" corresponding to the fact, well known in the theory of metals, that all the charge carried by a quasi-particle in a plasma is actually on the surface. Nonetheless, conservation of particles, if not of bare charge, is strictly maintained. Note that the situation does not resemble the case of "infinite" charge renormalization because the infinity in the vacuum polarizability need only occur at $p^2=0$.

Either in the case of the polarizable vacuum or of the "conducting" one, no low-energy experiment, and even possibly no high-energy one, seems capable of directly testing the value of the vacuum polarizability prior to renormalization. Thus, we conclude that the plasmon is a physical example demonstrating Schwinger's contention that under some circumstances the Yang-Mills type of vector boson need not have zero mass. In addition, aside from the short range of forces and the finite mass, which we might interpret without resorting to Yang-Mills, it is not obvious how to characterize such a case mathematically in terms of observable, renormalized quantities.

We can, on the other hand, try to turn the problem around and see what other conclusions we can draw about possible Yang-Mills models of strong interactions from the solid-state analogs. What properties of the vacuum are needed for it to have the analog of a conducting response to the Yang-Mills field?

Certainly the fact that the polarizability of the "matter" system, without taking into account the interaction with the gauge field, is infinite need not bother us, since that is unobservable. In physical conductors we can see it, but only because we can get outside them and apply to them true electromagnetic fields, not only internal test charges.

More serious is the implication—obviously physically from the fact that α has a pole at $p^2=0$—that the "matter" spectrum, at least for the "undressed" matter system, must extend all the way to $m^2=0$. In the normal plasma even the final spectrum extends to zero frequency, the coupling rather than the spectrum being affected by the screening. Is this necessarily always the case? The answer is no, obviously, since the superconducting electron gas has no zero-mass excitations whatever. In that case, the fermion mass is finite because of the energy gap, while the boson which appears as a result of the theorem of Goldstone[7,8] and has zero unrenormalized mass is converted into a finite-mass plasmon by interaction with the appropriate gauge field, which is the electromagnetic field. The same is true of the charged Bose gas.

It is likely, then, considering the superconducting analog, that the way is now open for a degenerate-vacuum theory of the Nambu type[9] without any difficulties involving either zero-mass Yang-Mills gauge bosons or zero-mass Goldstone bosons. These two types of bosons seem capable of "canceling each other out" and leaving finite mass bosons only. It is not at all clear that the way for a Sakurai[3] theory is equally uncluttered. The only mechanism suggested by the present work (of course, we have not discussed non-Abelian gauge groups) for giving the gauge field mass is the degenerate vacuum type of theory, in which the original symmetry is not manifest in the observable domain. Therefore, it needs to be demonstrated that the necessary conservation laws can be maintained.

I should like to close with one final remark on the Goldstone theorem. This theorem was initially conjectured, one presumes, because of the solid-state analogs, via the work of Nambu[10] and of Anderson.[11] The theorem states, essentially, that if the Lagrangian

[6] We follow here, as elsewhere, the viewpoint of W. Thirring, *Principles of Quantum Electrodynamics* (Academic Press Inc., New York, 1958), Chap. 14.

[7] J. Goldstone, Nuovo Cimento **19**, 154 (1961).

[8] J. Goldstone, A. Salam, and S. Weinberg, Phys. Rev. **127**, 965 (1962).

[9] Y. Nambu and G. Jona-Lasinio, Phys. Rev. **122**, 345 (1961).

[10] Y. Nambu, Phys. Rev. **117**, 648 (1960).

[11] P. W. Anderson, Phys. Rev. **110**, 827 (1958).

possesses a continuous symmetry group under which the ground or vacuum state is not invariant, that state is, therefore, degenerate with other ground states. This implies a zero-mass boson. Thus, the solid crystal violates translational and rotational invariance, and possesses phonons; liquid helium violates (in a certain sense only, of course) gauge invariance, and possesses a longitudinal phonon; ferro-magnetism violates spin rotation symmetry, and possesses spin waves; superconductivity violates gauge invariance, and would have a zero-mass collective mode in the absence of long-range Coulomb forces.

It is noteworthy that in most of these cases, upon closer examination, the Goldstone bosons do indeed become tangled up with Yang-Mills gauge bosons and, thus, do not in any true sense really have zero mass. Superconductivity is a familiar example, but a similar phenomenon happens with phonons; when the phonon frequency is as low as the gravitational plasma frequency, $(4\pi G\rho)^{1/2}$ (wavelength$\sim 10^4$ km in normal matter) there is a phonon-graviton interaction: in that case, because of the peculiar sign of the gravitational interaction, leading to instability rather than finite mass.[12] Utiyama[13] and Feynman have pointed out that gravity is also a Yang-Mills field. It is an amusing observation that the three phonons plus two gravitons are just enough components to make up the appropriate tensor particle which would be required for a finite-mass graviton.

Spin waves also are known to interact strongly with magnetostatic forces at very long wavelengths,[14] for rather more obscure and less satisfactory reasons. We conclude, then, that the Goldstone zero-mass difficulty is not a serious one, because we can probably cancel it off against an equal Yang-Mills zero-mass problem. What is not clear yet, on the other hand, is whether it is possible to describe a truly strong conservation law such as that of baryons with a gauge group and a Yang-Mills field having finite mass.

I should like to thank Dr. John R. Klauder for valuable conversations and, particularly, for correcting some serious misapprehensions on my part, and Dr. John G. Taylor for calling my attention to Schwinger's work.

[12] J. H. Jeans, Phil. Trans. Roy. Soc. London **101**, 157 (1903).
[13] R. Utiyama, Phys. Rev. **101**, 1597 (1956); R. P. Feynman (unpublished).
[14] L. R. Walker, Phys. Rev. **105**, 390 (1957).

SOVIET PHYSICS JETP VOLUME 24, NUMBER 1 JANUARY, 1967

SPONTANEOUS BREAKDOWN OF STRONG INTERACTION SYMMETRY AND THE ABSENCE OF MASSLESS PARTICLES

A. A. MIGDAL and A. M. POLYAKOV

Submitted to JETP editor November 30, 1965; resubmitted February 16, 1966

J. Exptl. Theoret. Physics (U.S.S.R.) **51**, 135–146 (July, 1966)

The occurrence of massless particles in the presence of spontaneous symmetry breakdown is discussed. By summing all Feynman diagrams, one obtains for the difference of the mass operators $M_a(p) - M_b(p)$ of particles a and b belonging to a supermultiplet an equation which is identical to the Bethe-Salpeter equation for the wave function of a scalar bound state of vanishing mass (a "zeron") in the annihilation channel $\bar{a}b$ of the corresponding particles. It is shown that if symmetry is spontaneously violated in a Yang–Mills type theory involving vector mesons, the zerons interact only with virtual particles and therefore unobservable. On the other hand, the vector mesons acquire a mass in spite of the generalized gauge invariance. It is shown in Appendices A and B that the asymmetrical solution corresponds to a minimal energy of the vacuum and that C-invariance of the solution implies strangeness conservation for it.

1. INTRODUCTION

SPONTANEOUS symmetry breakdown related to an instability of the system under consideration with respect to an infinitesimally weak asymmetric perturbation is often encountered in quantum statistical mechanics (ferromagnetism, superconductivity, etc.). A large number of such examples has been analyzed in Bogolyubov's review.[1]

In the theory of elementary particles the possibility of spontaneous asymmetry has been discussed by Nambu and Jona–Lasinio[2] and by Vaks and Larkin,[3] on the example of violated γ_5-invariance. In their calculations these authors ran into the existence of massless particles (zerons). Later, Goldstone has proved the theorem[4, 5] which states that bound states of zero mass must appear in systems with unstable symmetries. An example are the spin waves which occur in a ferromagnetic body and the acoustic oscillations in a boson gas. On the other hand, there are no acoustic oscillations in a superconductor. Lange[6] has shown that the cause for the inapplicability of the Goldstone theorem in this case are the long-range Coulomb forces.

In elementary particle theory, forces analogous to the Coulomb forces appear as a result of the exchange of massless Yang–Mills vector mesons[7] which guarantee SU(3) symmetry and generalized gauge invariance (cf. e.g., the paper of Glashow and Gell-Mann[8]). It is to be expected that no zerons would appear with spontaneous symmetry breakdown in such a theory, and that the vector mesons acquire a physical mass in analogy with the screening of Coulomb forces in a superconductor.

We show that the vector mesons do indeed acquire a mass and that the zerons do not interact with real particles. Such a "phantom" zeron manifests itself as an off-mass-shell pole in the scattering amplitudes. The scattering amplitudes for real particles do not have such poles and therefore the zerons are unobservable. We note that in a Yang–Mills theory with stable symmetry the vector mesons would not have a physical mass; the instability of the symmetry removes this difficulty, which has been repeatedly discussed in the literature.[9]

Our paper is organized in the following manner: In Sec. 2 the equations of Nambu[2] and Vaks and Larkin,[3] which were introduced for weak point interactions, are extended to the case of an arbitrary interaction. The Goldstone theorem[4] is confirmed for this case: a condition for the solvability of the equations so derived is the existence of a bound state of zero mass (zeron). In Sec. 3 we prove the unobservability of zerons in a theory with vector mesons. In Appendix A the necessity of choosing (from energy considerations) the asymmetric solution, whenever it exists, is justified. Appendix B contains a proof of the connection between the conservation of strangeness and C-invariance in the presence of spontaneous symmetry breaking.

2. INSTABILITY OF SYMMETRY IN QUANTUM FIELD THEORY

We explain the meaning of instability of symmetry in quantum field theory without assuming that the interaction is weak. We discuss this problem on the example of a system and n- and λ-quarks with identical initial masses and interactions invariant under the SU(2) group, involving the n and λ axes.

Instability of the symmetry means that if one adds to the Hamiltonian an infinitesimally weak asymmetric perturbation, there appears in the system a finite symmetry-breaking effect. We take as such a perturbation a splitting between the bare masses of the n and the λ, to be denoted by Δm_0. The particles n and λ will now have different mass operators $M_n(p)$ and $M_\lambda(p)$. We look for a condition which guarantees that if the initial asymmetry is removed ($\Delta m_0 \to 0$) the physical asymmetry ($M_n(p) - M_\lambda(p)$) persists., For finite Δm_0 there exists a simple equation for the difference between $M_n(p)$ and $M_\lambda(p)$:

$$\Delta M(p) \equiv M_n(p) - M_\lambda(p) = \Delta m_0$$

$$+ \int \frac{d^4p'}{(2\pi)^4} U(p, p', 0) G_n(p') \Delta M(p') G_\lambda(p') , \qquad (2.1)$$

which corresponds to the Feynmann diagram

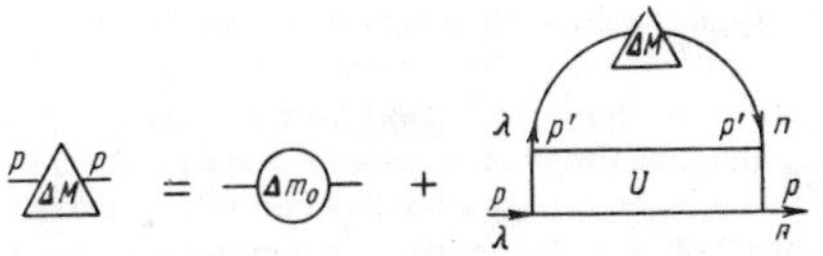

(2.1′)

Here G_n and G_λ are the exact Green's functions of n and λ:

$$G_n(p) = [p_\mu \gamma^\mu - M_n(p)]^{-1},$$

$$G_\lambda(p) = [p_\mu \gamma^\mu - M_\lambda(p)]^{-1};$$

U(p, p′, q) is the set of all $n\bar{\lambda}$-scattering graphs containing no parts connected by n or $\bar{\lambda}$ lines only.

The equation (2.1) has been derived by simple diagram subtraction. A diagrammatic derivation of this equation for an interaction with an "isoscalar" φ meson, $(\bar{n}n + \lambda\bar{\lambda})\varphi$, goes as follows:

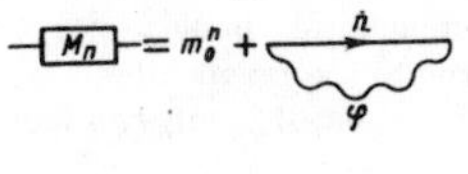

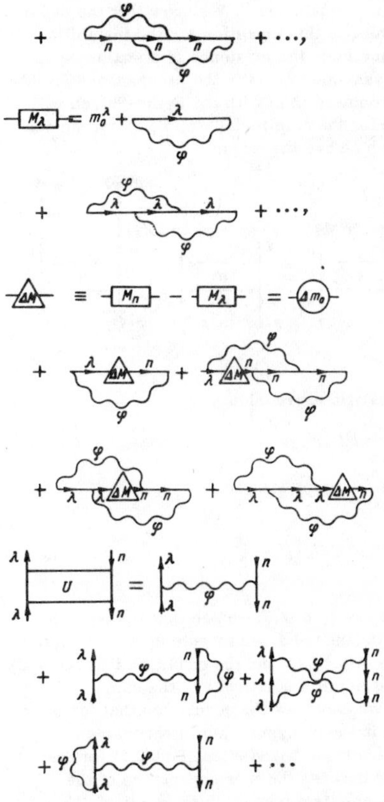

(2.1″)

Here the straight lines denote the exact Green's functions of the n and the λ, and the wavy lines are the exact Green's functions of the φ meson. Use has been made of the identity

$$A_1A_2 \ldots A_n - B_1B_2 \ldots B_n = (A_1 - B_1) B_2 \ldots B_n$$

$$+ A_1(A_2 - B_2) B_3 \ldots B_n + A_1A_2 \ldots A_{n-1}(A_n - B_n).$$

In (2.1) we go to the limit $\Delta m_0 \to 0$. Then the equation for $\Delta M(p) = M_n(p) - M_\lambda(p)$ has the form

$$\Delta M(p) = \int \frac{d^4p'}{(2\pi)^4} U(p, p', 0) G_n(p') \Delta M(p') G_\lambda(p'). \quad (2.2)$$

(This equation is a generalization to arbitrary interactions of the model equations given in [2, 3].)

The homogeneous equation (2.2) can have a nonvanishing solution corresponding to a spontaneous symmetry violation if the eigenvalue of the operator UG_nG_λ equals one.[1] We show that the physical meaning of this condition for the instability of the symmetry is the existence of a scalar bound state of vanishing mass in the $\bar{\lambda}n$ channel. To this end we compare (2.2) with the Bethe-Salpeter equation for the amplitude A(p, p′, q) for $\bar{\lambda}n$ scattering, which has the form

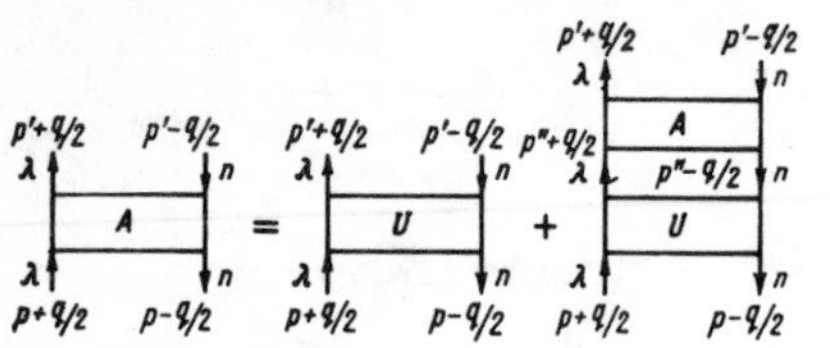

(2.3)

or the analytic expression

$$A(p,p',q) = U(p,p',q)$$
$$+\int \frac{d^4p''}{(2\pi)^4} U(p,p'',q) G_n\left(p''-\frac{q}{2}\right)$$
$$\times A(p'',p',q) G_\lambda\left(p''+\frac{q}{2}\right). \qquad (2.3')$$

Letting $q_\alpha \to 0$ in (2.3) we see that the condition for the existence of a scalar pole at $q_\alpha = 0$ in the amplitude A(p, p′, q) and the condition that Eq. (2.2) for ΔM(p) admit a solution, are the same.

Thus, we can draw the conclusion that for an interaction in the $n\bar{\lambda}$ system sufficiently strong to form a zero mass bound state, Eq. (2.2) for the difference between the mass operators of the n and the λ admits a nonvanishing solution. This solution corresponds to spontaneous breakdown of the SU(2) symmetry and can be obtained by introducing an infinitesimal symmetry violation into the initial Hamiltonian.

Since, of course, Eq. (2.2) also admits zero as a solution, corresponding to non-violated symmetry, there arises the question as to how to choose the solution. This problem is discussed in Appendix A.

If one does not pose the question as to how a mass difference can appear in a system with a symmetric Hamiltonian, the existence of the zeron in itself can be established in a simpler manner, for instance by making use of the generalized Ward identity for the vertices involving conserved currents. In our case, for the vertex $J_\alpha(p, q)$ of the transition current describing the transition $\lambda \to n$, this identity becomes

$$q_\alpha J_\alpha(p,q) = G_\lambda^{-1}(p+q/2) - G_n^{-1}(p-q/2). \qquad (2.4)$$

Let $q_\alpha \to 0$ in (2.4)

$$\lim_{q_\alpha \to 0} [q_\alpha J_\alpha(p,q)] = M_n(p) - M_\lambda(p). \qquad (2.5)$$

The left-hand side in (2.5) does not vanish if $M_\lambda(p) \neq M_n(p)$, so that $J_\alpha(p, q)$ exhibits a scalar pole at $q_\alpha = 0$. This pole corresponds to the zeron. However we have seen this pole both in J_α and in A (cf. (2.3)) at equal momenta of the n and λ, i.e., off the mass shell, since $m_n \neq m_\lambda$. In the following section we show that in theories involving vector mesons this pole does not appear on the mass shell, and thus the zerons are fictitious.

3. THE ACQUISITION OF A MASS BY THE VECTOR MESONS AND THE DISAPPEARANCE OF THE ZERONS

In this section it will be shown that in a Yang–Mills type theory with generalized gauge invariance (cf. [7,8]) the zerons which appear due to spontaneous symmetry breakdown are unobservable, and that the vector mesons acquire a physical mass, even though they have no initial mass. We shall discuss these problems using the simple example of a system consisting of a doublet of quarks n and λ, interacting with a triplet of vector mesons ρ_U^0, K^{*0}, $\bar{K}^{*0}$. The system is assumed invariant under the group SU(2) of U-spin,[10] the mesons have a vanishing bare mass and interact with the vector current of the U-spin of all particles.

As explained in Sec. 2, spontaneous symmetry breakdown in the system produces a zeron in the $\bar{\lambda}n$-channel (the Goldstone theorem). Leaving aside for the moment the problem of the observability of zerons, we show that the zeron pole in the irreducible self-energy part of the K* meson leads to a nonvanishing physical mass for the K* meson.

Denoting the amplitude for the transition of a K* meson into a zeron at momentum q by $a(q^2)q_\alpha$, the contribution of the zeron to the irreducible self-energy part of the K* meson, $\Pi_{\alpha\beta}(q)$, has the form

$$\Pi_{\alpha\beta}(q) \xrightarrow[q^2\to 0]{} -a^2(q^2) q_\alpha q_\beta / q^2. \qquad (3.1)$$

[1]One should not think that Eq. (2.2) has a solution only for a definite fixed value of the original parameters (mass, coupling constants, etc.). This is not so since (2.2) is nonlinear: UG_nG_λ depends functionally on $M_n(p)$ and $M_\lambda(p)$. Model examples and the analogy with the theory of superconductivity indicate that there is a solution if the original coupling constant corresponds to sufficiently strong attraction of the $\bar{\lambda}$ and n. As the attraction is increased, the mass difference persists and as a rule increases.

The pole (3.1) in $\Pi_{\alpha\beta}(q)$ is due to the K* meson–zeron transition (amplitude $a(q^2)q_\alpha$) followed by the propagation of the zeron (the Green's function $D(q^2) \to 1/q^2$) and the inverse zeron–K* meson transition (the factor $a(q^2)q_\alpha$).[2] In order to understand the influence exerted by the zeron pole (3.1) on the mass of the K* meson we recall that the mass m_K is defined by the equation

$$m_K^2 = \Pi(m_K^2). \tag{3.2}$$

Here the mass operator $\Pi(q^2)$ is related to $\Pi_{\alpha\beta}(q)$ as follows

$$\Pi_{\alpha\beta}(q) = \Pi(q^2)(\delta_{\alpha\beta} - q_\alpha q_\beta q^{-2}). \tag{3.3}$$

(The transversality of $\Pi_{\alpha\beta}(q)$ is a consequence of the U-spin vector current conservation.)

Comparing (3.3) and (3.1) at $q^2 \to 0$ we find

$$\Pi(0) = a^2(0) \equiv a^2 \neq 0. \tag{3.4}$$

For $m_K = 0$ we would obtain from (3.2) $\Pi(0) = m_K^2 = 0$, thus (3.4) implies $m_K \neq 0$.

We stress that $a \neq 0$ due to the spontaneous symmetry. Indeed, the Ward identity for the vertex $\Gamma_{\alpha\beta}(p, q)$ describing the K* emission in the $\bar{\lambda}$-n quark annihilation (cf. Eqs. (2.4) and (2.5)[3]) requires the existence of a pole in that vertex function, if $M_n(p) \neq M_\lambda(p)$. This pole can occur in Γ_α only if the zeron is capable of a transition into a K* meson, i.e., if $a \neq 0$. In a Yang–Mills theory with stable symmetry there is no zeron, hence

$$a^2 = \Pi(0) = m_K^2 = 0.$$

Thus one can draw the conclusion that the mass of a vector meson in a gauge invariant theory can appear if and only if the symmetry is spontaneously broken.

We go over to the proof of the assertion that the zeron does not participate in real processes and is therefore no obstacle for the theory of spontaneous symmetry breakdown in systems with generalized gauge invariance.

For the proof we consider the amplitude A for $\bar{\lambda}$n scattering and show that it does not contain zeron singularities on the mass shell (such singularities must exist off the mass shell, as seen from Eq. (2.3)). The absence of such singularities might appear strange, since they exist both in Γ_α and $\Pi_{\alpha\beta}$. However these quantities do not contain, by definition, diagrams exhibiting a K*-meson pole, whereas A contains such diagrams. If one removes these diagrams from A, the remaining amplitude $\tilde{A}$, which does not contain a K*-meson pole in the $\bar{\lambda}$n channel, will exhibit a zeron pole in the same channel. We show that the interference between the zeron pole in $\tilde{A}$ and the zeron of the other diagrams (which also exhibit K*-meson poles; the contribution of these diagrams will be denoted by $\tilde{\tilde{A}}$) cancels the resulting pole in $A = \tilde{A} = \tilde{\tilde{A}}$.

For $q^2 \to 0$ the amplitude $\tilde{A}$ has the form

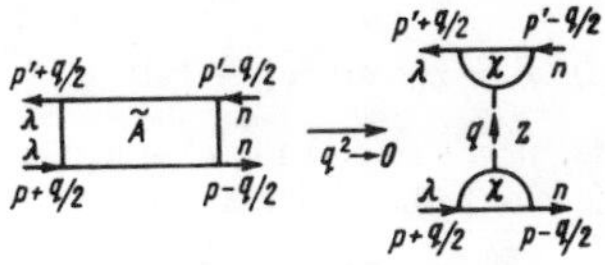

(3.5)

or

$$\tilde{A}(p, p', q) \underset{q^2 \to 0}{\longrightarrow} \frac{\chi(p, q)\chi(p', q)}{q^2}. \tag{3.5'}$$

(The amplitudes A, $\tilde{A}$, and $\tilde{\tilde{A}}$ depend on the three momenta p, p′, and q, cf. (3.5).) Here $\chi(p, q)$ denotes the sum of the diagrams which take the $\bar{\lambda}$n system into a zeron and do not exhibit a K*-pole in the $\bar{\lambda}$n-channel, since $\tilde{A}$ does not contain such a pole.

We express the residue in the zeron pole of the amplitude $\tilde{\tilde{A}}$ in terms of the same amplitude. To this end we rewrite the amplitude $\tilde{\tilde{A}}$ (before taking the limit $q^2 \to 0$) in the form

$$\tilde{\tilde{A}}(p, p', q) = \Gamma_\alpha(p, q) D_{\alpha\beta}(q) \Gamma_\beta(p', q). \tag{3.6}$$

Here $D_{\alpha\beta}(q) = \delta_{\alpha\beta}[\Pi(q^2) - q^2]^{-1}$ is the Green's function of the K* meson, $\Gamma_\alpha(p, q)$ is, as before, the sum of the $\bar{\lambda}$n–k* transition diagrams which do not exhibit a K*-meson pole.

If the $\bar{\lambda}$ and n are situated on the mass shell, the Ward identity (cf. footnote [3]) implies that $\Gamma_\alpha(p, q)$ is transverse for all q^2 ($q_\alpha \Gamma_\alpha = 0$), hence on the mass shell A does not depend on the longitudinal part of $D_{\alpha\beta}(q)$, as was to be expected

---

[2] The minus sign in (3.1) is due to the fact that all amplitudes enter with an additional factor i in the Feynman amplitude.

[3] A Ward identity of the form (2.4) can here be written for the current J_a differing from Γ_α by diagrams involving a K*-pole in the $\bar{\lambda}_n$-channel:

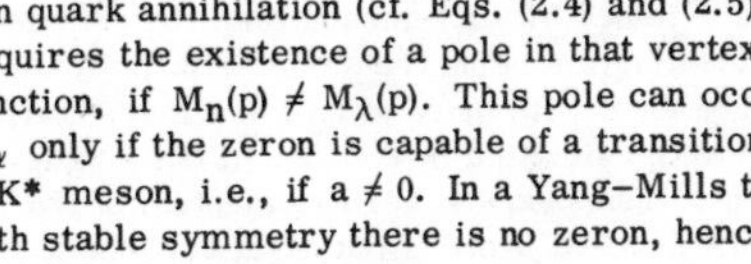

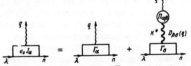

But owing to the transversality of $\Pi_{\alpha\beta}(q)$, (3.3), these diagrams do not contribute to the divergence, so that

$$e_0^{-1} q_\alpha \Gamma_\alpha(p, q) = q_\alpha J_\alpha(p, q) = G_\lambda^{-1}(p + q/2) - G_n^{-1}(p - q/2).$$

(here e_0 is the bare nλK*-coupling constant.

from gauge invariance. This has permitted us to work in the Feynman gauge for $D_{\alpha\beta}(q)$.

We now have to separate the pole in $\tilde{\tilde{A}}$ at $q^2 \to 0$. A zeron singularity in $\tilde{\tilde{A}}$ will appear, due to the zeron poles in the vertices Γ_α (we recall that $m_K \neq 0$, and therefore $D_{\alpha\beta}(q)$ does not exhibit a singularity at $q^2 \to 0$). The zeron poles in Γ_α appear as a result of the transition of the $\bar{\lambda}n$ into a zeron (described by the amplitude χ), the propagation of the zeron ($D(q^2) \to q^{-2}$), and its transition into a K* meson with the amplitude aq_α. Therefore Γ_α can be represented in the form

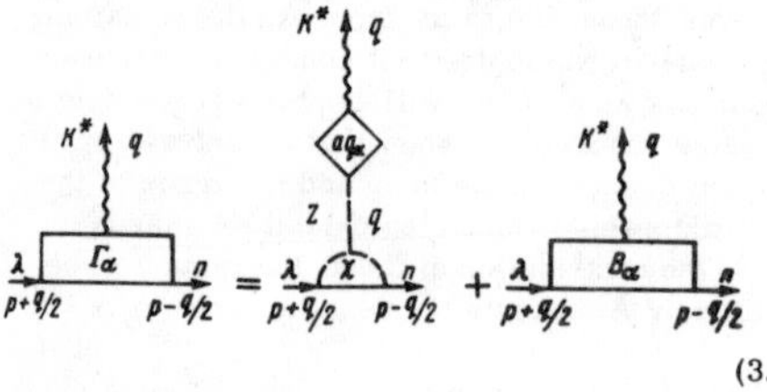

(3.7)

or

$$\Gamma_\alpha(p, q) = \chi(p, q) q^{-2} a q_\alpha + B_\alpha(p, q), \qquad (3.7')$$

where $B_\alpha(p, q)$ has no pole for $q^2 \to 0$. Substituting (3.7′) into (3.6), we obtain

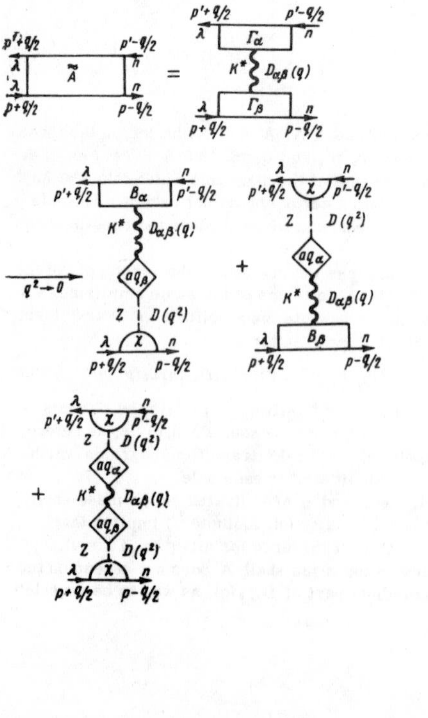

or in analytic form

$$\tilde{A}(p, p, q) = \frac{\Gamma_\alpha \Gamma_\alpha'}{\Pi(q^2) - q^2} \underset{q^2 \to 0}{\longrightarrow} \frac{\chi a q_\alpha B_\alpha' + \chi' a q_\alpha B_\alpha + a^2 \chi\chi'}{q^2 \Pi(0)},$$

$$(\chi \equiv \chi(p, q), \quad B_\alpha \equiv B_\alpha(p, q), \quad \chi' \equiv \chi(p', q),$$

$$B_\alpha' \equiv B_\alpha(p', q)). \qquad (3.8)$$

In order to express $q_\alpha B_\alpha$ in terms of χ and a we make use of the transversality condition for Γ_α:

$$q_\alpha B_\alpha(p, q) = q_\alpha \Gamma_\alpha(p, q) - a\chi(p, q) = -a\chi(p, q). \qquad (3.9)$$

Substituting (3.9) into (3.8) and remembering that $\Pi(0) = a^2$ (cf. (3.4)), we find

$$\tilde{A}(p, p', q) \underset{q^2 \to 0}{\longrightarrow} -\frac{\chi(p, q)\chi(p', q)}{q^2}. \qquad (3.10)$$

Comparing (3.10) and (3.5) we find that the pole at $q^2 \to 0$ has disappeared from $A = \tilde{A} = \tilde{\tilde{A}}$. Similarly all zeron and multi-zeron singularities disappear from any amplitude S on the mass shell.

If in some diagram $\tilde{S}$ for the amplitude S two points are joined by a zeron line, there will be another contribution $\tilde{\tilde{S}}$ to S in which the same two points are joined by a K*-meson line, the remainder of the diagram being the same as in S. Adding the diagrams $\tilde{S}$ with those of type $\tilde{\tilde{S}}$ we see that, similar to the situation in $A = \tilde{A} + \tilde{\tilde{A}}$, the singularity in S disappears at zero momentum of the zeron. (A detailed proof is not hard, but lengthy, therefore we omit it.)

As in any observable amplitude (A, J_α, etc.) the interference of zerons with K* mesons destroys all zeron poles and thresholds. This means that the zerons are fictitious particles: they can neither be emitted nor absorbed by real particles (otherwise the scattering amplitudes would exhibit zeronic singularities).

Nevertheless, a "shadow" of the zeron survives: off the mass shell all amplitudes (e.g., A(p, p′, q) in (2.3) or $J_\alpha(p, q)$ in (2.5)) have zeronic singularities, corresponding to a vanishing zeron mass. The fact that there are off-mass-shell singularities (poles or thresholds) not corresponding to physical particles is not a specific feature of the theory under consideration. If, for instance, the Green's function of the photon in ordinary quantum electrodynamics is chosen, such that its longitudinal part is singular, unphysical off-mass-shell singularities are found in all amplitudes. In distinction from quantum electrodynamics, in our case the singularities cannot be removed by changing the gauge of the K*-meson Green's function.

We note finally that we had no need to consider the initial interaction or the symmetry violation small. We have omitted only those diagrams which had no zeronic singularities and were thus inessential at $q^2 \to 0$.

4. POSSIBLE APPLICATIONS AND GENERALIZATIONS

The results derived in the preceding section can easily be separated from the specifically chosen model. In the case of spontaneous violation of an arbitrary symmetry group there will appear zerons with the quantum numbers of those group generators which are no longer conserved (cf. [5]). On the other hand, for generalized gauge invariance, the existence of a vector meson multiplet is necessary, having the quantum numbers of all the generators of the group.[8]

We have proved that through the interference of the vector meson with the zeron having the same quantum numbers, it destroys the latter and thus acquires a physical mass, in spite of the fact that its initial mass vanishes.

If the symmetry is not completely destroyed, there will be fewer zerons than there are vector mesons. Therefore, although all zerons disappear, not all vector mesons acquire a physical mass. (In the model under consideration the meson ρ^0_U remains massless, owing to strangeness conservation.) Therefore, if one considers the breakdown of SU(3) as a spontaneous effect,[5] one must admit that all the quantum numbers of this group are no longer conserved. In this case the electric charge should not be considered as directly related with these quantum numbers, i.e., there should exist supercharged particles.[10]

Another difficulty appears if one tries to identify the strangeness changing weak interaction currents with the corresponding currents associated with the generators of SU(3): the weak currents are approximately conserved and the currents of the group generators are exactly conserved, in spite of the mass differences.

In conclusion we note that under spontaneous symmetry breakdown, all the relations that follow from the current algebra[11] remain valid, since this latter method, consisting in the neglect of remote intermediate states, is related only to the smallness of the observed symmetry, but not with the mechanism which brings about this violation. Thus, in agreement with experiment, we are led to approximate universality of the vector meson interactions, the Gell-Mann–Okubo formula, etc.[4)]

We are grateful to V. G. Vaks and A. I. Larkin for useful advice and to K. A. Ter-Martirosyan for a discussion of the results.

APPENDIX A

THE ENERGETIC ADVANTAGES OF AN ASYMMETRIC VACUUM

We consider the energetic advantages of an asymmetric solution, assuming that it exists. We have seen in Sec. 2 that such a solution can be generated by an infinitesimal asymmetric perturbation of an initially symmetric Hamiltonian.

Let H_0 denote the symmetric Hamiltonian, and ϵV the symmetry breaking interaction; let E_0 be the ground state of H_0 and $E(\epsilon)$ the ground state of $H_0 + \epsilon V$. Energetic advantage means that $\lim_{\epsilon \to 0} E(\epsilon) < E_0$ for $\epsilon \to 0$. We prove that this condition is always fulfilled.

For this, we note that, owing to the variational principle,

$$E(\varepsilon) \leqslant (\Phi^* | H_0 + \varepsilon V | \Phi). \tag{A.1}$$

If one selects as a trial function Φ the eigenfunction Φ_0 of the ground state of H_0, one obtains

$$E(\varepsilon) < (\Phi_0^* | H_0 + \varepsilon V | \Phi_0) = E_0 + \varepsilon (\Phi_0^* | V | \Phi_0). \tag{A.2}$$

Since Φ_0 is symmetric and V violates the symmetry, $(\Phi_0^* | V | \Phi_0) = 0$. Consequently

$$\lim_{\varepsilon \to 0} E(\varepsilon) < E_0. \tag{A.3}$$

(This proof carries over automatically to statistical mechanics, if one makes use of the variational principle for the free energy.)

The model examples[2, 3] show what kind of contradictions appear when the symmetric solution, corresponding to an energetically disadvantageous vacuum, is used. It turns out that this "normal" solution, in distinction from the "superconducting" one, exhibits poles corresponding to "resonances" with negative lifetimes.

[4)]If the method used in Sec. 3 for the separation of the zeron pole is applied to γ_5-violation, one obtains in particular all the results usually derived from the hypothesis of a partially conserved axial vector current (PCAC). This is done in a paper by one of the authors[15].

APPENDIX B

STRANGENESS CONSERVATION IN A C-INVARIANT THEORY

In deriving the equation for $M_n(p) - M_\lambda(p)$ in Sec. 2, it was assumed that strangeness is conserved. At the same time we have seen in Appendix A that in order to assure the minimal energy for the vacuum, one must look for the solution exhibiting totally violated symmetry. We show, that if one assumes C-invariance of the solution, then it will automatically conserve strangeness. Indeed, let us not consider that strangeness is a priori conserved. Then the mass operator will involve transitions of the type $n \to \lambda$, $\lambda \to n$:

$$M(p) = \begin{pmatrix} M_{nn}(p) & M_{n\lambda}(p) \\ M_{\lambda n}(p) & M_{\lambda\lambda}(p) \end{pmatrix} \equiv M_0 + \mathbf{M}\boldsymbol{\tau}$$

$$\equiv {}^1/_2(M_{nn} + M_{\lambda\lambda}) + (M_{nn} - M_{\lambda\lambda})\tau_z + 2M_{n\lambda}\tau_x \qquad \text{(B.1)}$$

(here τ_x and τ_z are Pauli matrices; τ_y does not occur in (B.1). Since, owing to C-invariance, M(p) is symmetric in isospin indices:[5] $M_{n\lambda}(p) = M_{\lambda n}(p)$).

In place of Eq. (2.2) for $M_z = M_{nn} - M_{\lambda\lambda}$ we will have an equation for $\mathbf{M} = (M_x, 0, M_z)$ in the form

$$\mathbf{M}(p) = \int d^4p' \mathscr{K}(p, p')\mathbf{M}(p'). \qquad \text{(B.2)}$$

$\mathscr{K}(p, p')$ in turn, depends on isoscalars formed out of $\mathbf{M}(p)$:

$$\mathscr{K}(p, p') = \Phi[\mathbf{M}(p_1)\mathbf{M}(p_2)]. \qquad \text{(B.3)}$$

In general, one could have added to the right-hand side of (B.2) an isovector of the form

$$\int d^4p_1 d^4p_2 \mathscr{L}(p_1, p_2, p)\mathbf{M}(p_1) \times \mathbf{M}(p_2), \qquad \text{(B.4)}$$

but in our case, owing to C-invariance, $\mathbf{M} = (M_x, 0, M_z)$ is two-dimensional for all momenta p, and thus this term is missing.

Equation (B.2) is linear in $\mathbf{M}(p)$, therefore, in order for $\mathbf{M}(p) \neq 0$ (spontaneous symmetry violation), it is necessary that $\mathscr{K}(p, p')$ have unity as an eigenvalue. This eigenvalue can be expressed in terms of $\mathbf{M}(p)$ by means of Eq. (B.3). Then the condition for spontaneous symmetry breakdown is

$$\lambda[\mathbf{M}(p_1)\mathbf{M}(p_2)] = 1. \qquad \text{(B.5)}$$

If the eigenvalue $\lambda = 1$ of the operator $\mathscr{K}(p, p')$ is nondegenerate,[6] it follows from (B.2) that

$$\mathbf{M}(p) = \mathbf{C}\chi(p), \qquad \text{(B.6)}$$

where $\chi(p)$ is the eigenfunction of $\mathscr{K}(p, p')$, $\mathbf{C}$ is a constant vector, defined by the condition of spontaneous symmetry breakdown (B.5). But (B.5) defines only the length of $\mathbf{C}$, leaving the direction of $\mathbf{C}$ arbitrary. This is related to the fact that (B.2), (B.3) and (B.5) are invariant to rotations of $\mathbf{M}(p)$ by an angle which does not depend on the momentum p. Such a rotation implies a redefinition of the particles n and λ; by means of such a rotation one can achieve

$$\mathbf{C} = (0, 0, C_z), \quad M(p) = \begin{pmatrix} M_n(p) & 0 \\ 0 & M_\lambda(p) \end{pmatrix}. \qquad \text{(B.7)}$$

Now n and λ do not transform into each other for any momenta, and thus strangeness is conserved.[7]

Both C-conservation and the nondegeneracy of the zeron were essential for our proof. Indeed, for an accidentally degenerate zeron we would have in place of (B.6):

$$\mathbf{M}(p) = \mathbf{C}_1\chi_1(p) + \mathbf{C}_2\chi_2(p) + \ldots \qquad \text{(B.8)}$$

The solvability condition (B.5) would fix the angle between $\mathbf{C}_1$ and $\mathbf{C}_2$ and one could not direct both along the z axis. Thus, the diagonal character of $\mathbf{M}(p)$ for all p would not be possible and strangeness would not be conserved in virtual λ-n transitions. These transitions would lead to strangeness-violation in real processes of the type of $n + n \to \lambda + \lambda$ too. Without C-invariance we would also have an expression of the type (B.8) with all the consequences it implies.

Finally, we note that our assertion about the connection between conservation of charge parity and that of additive quantum numbers under spontaneous breakdown of SU(2) is confirmed by the concrete calculations of Arnowitt and Deser[12] in the electrodynamics of the muon and electron. In their approximate calculations both charge parity and the muonic charge are conserved, in analogy to strangeness in our example.

---

[5] Here and below we call the U-spin isospin, for brevity. The isospin invariance of Eqs. (B.2) and (B.3) follows from the symmetry of the Lagrangian and can be easily verified on skeleton diagrams, similar to the derivation of Eq. (2.1")

[6] It can be seen from Sec. 2 that such a degeneracy means degeneracy of the zeron with respect to some quantum number foreign to the U-spin group. Therefore the degeneracy can only be accidental, and there is no symmetry from which it results. We also note that $\chi(p)$ in (B.6) is the $\lambda \to n$ + zeron (momentum q = 0) transition amplitude.

[7] We note that Eqs. (B.2) and (B.3) are not invariant under a rotation of $\mathbf{M}(p)$ by an angle which depends on p, therefore it is nontrivial that $\mathbf{M}(p)$ can be directed along the z axis for all p. (B.8) below is an example when this is impossible.

[1] N. N. Bogolyubov, JINR Preprint D-781, Dubna, 1961.
[2] Y. Nambu and C. Jona-Lasinio, Phys. Rev. **122**, 345 (1961).
[3] V. G. Vaks and A. I. Larkin, JETP **40**, 282 (1961), Soviet Phys. JETP **13**, 192 (1961).
[4] J. Goldstone, Nuovo Cimento **19**, 154 (1961).
[5] S. Bludman and A. Klein, Phys. Rev. **131**, 2364 (1963).
[6] R. Lange, Phys. Rev. Lett. **14**, 3 (1965).
[7] C. N. Yang and R. L. Mills, Phys. Rev. **96**, 191 (1954).
[8] S. L. Glashow and M. Gell-Mann, Ann. Phys. (N. Y.) **15**, 437 (1961).
[9] J. Schwinger, Phys. Rev. **125**, 397 (1962).
[10] L. B. Okun', ITEF Preprint No. 287, Moscow, 1964.
[11] M. Gell-Mann, Phys. Rev. **125**, 1067 (1962); S. Fubini and G. Furlan, Physics **1**, 229 (1965).
[12] R. Arnowitt and S. Deser, Phys. Rev. **138**, 8712 (1965).
[13] A. M. Polyakov, JETP Letters, in press.

Translated by M. E. Mayer
18

PHYSICAL REVIEW D VOLUME 8, NUMBER 8 15 OCTOBER 1973

Dynamical Model of Spontaneously Broken Gauge Symmetries*

R. Jackiw and K. Johnson

Laboratory for Nuclear Science and Department of Physics, Massachusetts Institute of Technology, Cambridge, Massachusetts 02139

(Received 3 May 1973)

We demonstrate how a theory consisting of massless Fermi and vector-meson fields can lead to an excitation spectrum of solely massive particles. We eschew spinless Bose fields in the fundamental Lagrangian, contrary to current practice. A detailed model is presented and solved in lowest order. Fermion and axial-vector-meson masses are spontaneously generated, and the vector particle's mass is computed in terms of the fermion mass.

I. INTRODUCTION

An attractive idea for field-theoretic models involving massive vector mesons is that the mass arises from spontaneous breakdown of a gauge symmetry. Indeed the currect activity in the theory of weak interactions is centered around such a possibility.[1] In these examples, both fermions and vector mesons become massive because a canonical scalar field, already present in the Lagrangian, acquires a symmetry-violating vacuum expectation value. In a now familiar fashion, this vacuum expectation value immediately leads to a fermion mass and a massless excitation. The massless excitation, however, does not correspond to a particle, but rather combines with a massless vector gauge field to produce a massive vector-meson particle. For brevity we shall refer to this phenomenon as the Higgs mechanism.[2]

In this paper we examine the possibility that masses of fermions and vector mesons can arise spontaneously, *without* the presence of canonical scalar fields in the Lagrangian.[3] Thus we are extending the work of Nambu and Jona-Lasinio,[4] who showed that the Goldstone mechanism can take place even when the Lagrangian does not include spin-zero fields.

The fundamental reason why an apparently massless vector meson acquires a mass was given a decade ago by Schwinger.[5] The reason is that the vacuum polarization tensor, that is, the proper two-point correlation function of the conserved current J^μ to which the meson couples with strength g, acquires a pole at zero momentum transfer. To see this explicitly, consider the complete vector-meson propagator,

$$D^{\mu\nu}(q) = -i\left(g^{\mu\nu} - \frac{q^\mu q^\nu}{q^2}\right) D(q^2),$$

which is given by[6]

$$D^{\mu\nu}(q) = -i\left(g^{\mu\nu} - \frac{q^\mu q^\nu}{q^2}\right)\frac{1}{q^2 - q^2\Pi(q^2)},$$

where

$$\Pi^{\mu\nu}(q) = (g^{\mu\nu}q^2 - q^\mu q^\nu)i\Pi(q^2)$$
$$= -g^2 \int d^4x\, e^{iqx}\langle 0|T^*J^\mu(x)J^\nu(0)|0\rangle|_{pp}. \quad (1.1)$$

Here the subscript pp indicates the proper part; i.e., only one-vector-meson-irreducible graphs contribute to $\Pi^{\mu\nu}(q)$. [We insist that $\Pi^{\mu\nu}(q)$ be transverse, since the current is conserved; this entails a judicious choice of seagull terms.] Clearly, if $\Pi(q^2)$ has a pole at $q^2=0$, the vector meson is massive, even though it is massless in the absence of interactions ($g=0$, $\Pi=0$). We shall call this the Schwinger mechanism.

There is no physical principle which would prevent $\Pi(q^2)$ from acquiring a pole. Indeed, for massless spinor electrodynamics in two dimensions,[7] Schwinger found that $\Pi(q^2)$ *does* acquire a pole, as can be seen on purely dimensional grounds: In that theory g has units of mass. (The occurrence of the vector-meson mass in this model is also related to the anomaly of the axial-vector fermion current.[8] The axial-vector current will also be important in our four-dimensional analysis.) The Higgs mechanism, now seen as a special realization of the Schwinger mechanism, provides an explicit reason for a pole in $\Pi(q^2)$: The vacuum expectation value of a canonical scalar field coupled to the vector meson gives rise to tadpole contributions to $\Pi(q^2)$ which produce a pole. We show that such a pole can occur for purely dynamical reasons, even in the absence of canonical scalar fields.

In Sec. II, it is demonstrated that a pole in $\Pi(q^2)$ can arise whenever a massless *fermion* acquires a mass through spontaneous symmetry breaking. This is an elaboration on ideas of Englert and Brout.[2,3] As in Ref. 4, this symmetry breaking is not due to a vacuum expectation value of canonical scalar field; rather it is *assumed* to be a consequence of a symmetry-breaking solution to the integral equations of the theory. This gives rise to a zero-mass bound excitation in the two-fermion sector, which produces a pole in $\Pi(q^2)$. Consequently, as explained above, the vector meson also acquires a mass. Furthermore, we show that the zero-mass excitation decouples from the theory. We give in Sec. III a "phenomenological" description of the previous "fundamental" theory. The phenomenological theory, valid at low energy, is realized by a nonlinear Lagrangian, which is not renormalizable, even though the fundamental theory possesses this property. In an appendix we show how the various unrenormalized equations encountered in the text should be renormalized. Also we solve them in the lowest-order Bethe-Salpeter approximation, and exhibit explicitly the workings of the Schwinger mechanism. We find a formula for the spontaneously generated vector-meson mass.

II. SPONTANEOUS MASS GENERATION

A. Preliminaries

We consider a theory with a massless fermion field ψ and a neutral vector-meson field A^μ, interacting through a conserved current J^μ. (No internal symmetry degrees of freedom are included in this simplified discussion.) Furthermore we assume that the Schwinger-Dyson equation for the fermion mass operator $\Sigma(p)$ has a symmetry-breaking solution, $\{\gamma^5, \Sigma(p)\} \neq 0$. It is well known that this can happen if there is a massless, bound excitation in the fermion-antifermion channel. Since a fermion mass is being generated, this zero-mass excitation couples to $\bar{\psi}\psi$ or $\bar{\psi}\gamma^5\psi$. Ultimately we shall want the zero-mass state to combine with the massless vector-meson field and generate a meson mass. This can only happen if a transition between J^μ and $\bar{\psi}\psi$ or $\bar{\psi}\gamma^5\psi$ is allowed. Therefore, as long as charge-conjugation invariance is not spontaneously broken, J^μ *must contain the axial-vector current.*

We are thus led to consider a theory described by the Lagrangian

$$\mathcal{L} = i\bar{\psi}\not{\partial}\psi - \tfrac{1}{4}F^{\mu\nu}F_{\mu\nu} + gJ_5^\mu A_\mu\,,$$
$$J_5^\mu = i\bar{\psi}\gamma^\mu\gamma^5\psi\,, \qquad (2.1)$$
$$F^{\mu\nu} = \partial^\mu A^\nu - \partial^\nu A^\mu\,.$$

(A vector interaction may also be included. This complication will be discussed further in Sec. II D.) The axial-vector anomaly occurs in the above theory by virtue of the axial-vector coupling.[9] For the time being, we shall ignore this; the anomaly can be eliminated by introducing additional fermions.[10] Indeed the anomaly *must* be removed; otherwise the equation of motion

$$\partial_\nu F^{\mu\nu} = gJ_5^\mu \qquad (2.2)$$

is not consistent with the antisymmetry of $F^{\mu\nu}$. (It is interesting that the axial-vector current, together with the anomaly — which, to be sure, must be removed — is central to the present discussion, just as it is central in Schwinger's two-dimensional model.) In Sec. II E, the effect of the anomaly will be considered further.

The theory (2.1) is chirally invariant and renormalizable, in the sense that off-mass-shell Green's functions can be computed in perturbation theory. The normalization point will not be taken at $k^2=0$ because of possible infrared divergences; rather an arbitrary value $k^2 = k_0^{\,2}$ will be chosen.

B. Spontaneous Mass Generation

The proper vertex function $\Gamma_5^\mu(p, p')$ associated with J_5^μ satisfies a Ward-Takahashi identity:

$$q_\mu \Gamma_5^\mu(p, p+q) = \gamma^5 G^{-1}(p+q) + G^{-1}(p)\gamma^5\,. \qquad (2.3)$$

There must not be any anomalous exceptions to this equation, since the current, a source of the gauge field, must be always conserved; see (2.2).[11] In (2.3), $G(p)$ is the complete fermion Green's function, which is given by the following Schwinger-Dyson equation, shown diagrammatically as Eq. (2.4) in Fig. 1. We assume that a

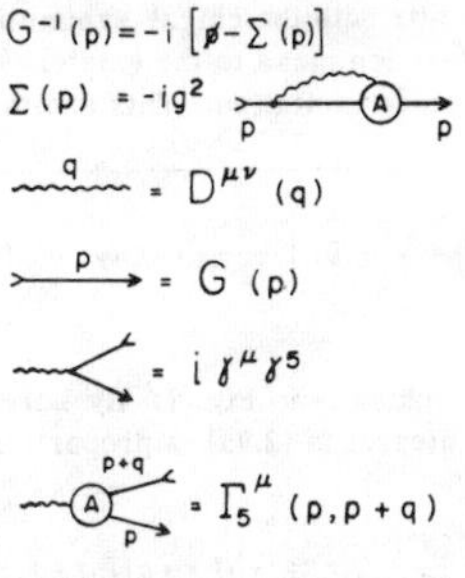

(2.4)

FIG. 1. Schwinger-Dyson equation for fermion propagator.

chiral symmetry-breaking solution for Eq. (2.4) exists such that $\{\gamma^5, \Sigma(p)\}$ is not zero. Then (2.3) implies that $\Gamma_5^\mu(p, p+q)$ has a pole at $q=0$ with residue $i\Gamma_5(p)$, where

$$\Gamma_5(p) = \{\gamma^5, \Sigma(p)\} \neq 0\,. \qquad (2.5)$$

This is our principal assumption. In the Appendix we show that to lowest order in the coupling a symmetry-breaking solution exists.

The proper vertex function is also related to a portion of the fermion-fermion scattering amplitude. The relation is exhibited as Eq. (2.6) in Fig. 2, where T' is the "one-vector-meson-irreducible" fermion-fermion scattering amplitude; i.e., the

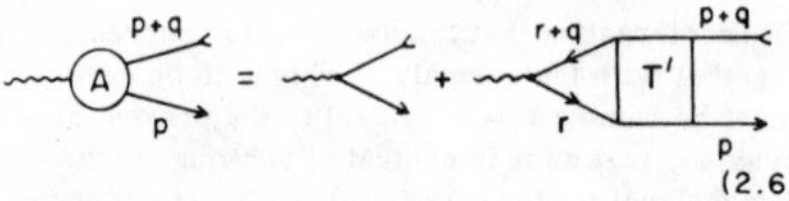

(2.6)

FIG. 2. Proper vertex function expressed in terms of the scattering amplitude.

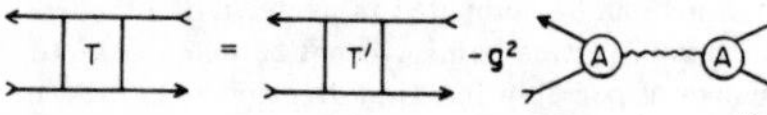

(2.7)

FIG. 3. Relation between complete and one-vector-meson-irreducible scattering amplitudes.

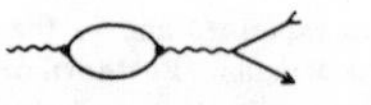

FIG. 4. Example of graph excluded from proper vertex function.

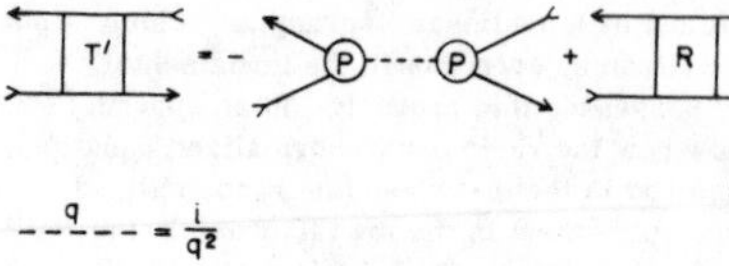

(2.8)

FIG. 5. Singular contribution to T'.

full scattering amplitude T is given by T' supplemented by one-vector-meson exchange graphs; see Eq. (2.7) in Fig. 3. The reason why T' rather than T appears in (2.6) is that Γ_5^μ is the *proper* vertex function and does not contain graphs like that of Fig. 4. [The minus sign in (2.7) arises from the fact that the graph is second-order in the coupling, i.e., $(ig)^2$ is the correct factor.]

The pole in Γ_5^μ is to be attributed to a pole in T'.[12] Thus we represent T' by a pole term plus a regular term R, as in Eq. (2.8) of Fig. 5. In other words, we attribute the chiral-symmetry breaking of the fermion mass to the existence of a massless Goldstone excitation. This excitation is a bound state in the fermion-antifermion channel; the proper vertex which describes the coupling of the bound state to $\psi\bar{\psi}$ is represented by Fig. 6 and is given by $P(p,p')$. This set of circumstances is different from that envisioned for massless spinor electrodynamics.[11] In the latter theory the electron acquires a mass spontaneously, but the Goldstone theorem is evaded due to anomalous nonconservation of the axial-vector current. In the present theory, as we have repeatedly stated, the axial-vector current must be conserved, and the Goldstone theorem holds.

Upon inserting (2.8) into (2.6), we determine the pole term in Γ_5^μ. The relation is

$$\Gamma^\mu_{5\,\text{pole}}(p,p+q) = -\left[-\operatorname{Tr}\int \frac{d^4r}{(2\pi)^4}\, G(r) i\gamma^\mu\gamma^5 G(r+q) P(r+q,r)\right] \frac{i}{q^2} P(p,p+q)\,, \tag{2.9a}$$

which is given graphically in Fig. 7. By Lorentz invariance, the integral in (2.9a) is proportional to q^μ:

$$q^\mu I(q^2) = \operatorname{Tr}\int \frac{d^4r}{(2\pi)^4} G(r) i\gamma^\mu\gamma^5 G(r+q) P(r+q,r)\,,$$

$$I(0) = \lambda\,. \tag{2.9b}$$

Thus

$$\Gamma^\mu_{5\,\text{pole}}(p,p+q) = \frac{iq^\mu}{q^2}\lambda P(p,p+q) \tag{2.9c}$$

and comparison with (2.3) yields

$$\Gamma_5(p) = \lambda P(p,p)\,. \tag{2.10}$$

Note that Eq. (2.9c) establishes the result that the singularity in Γ_5^μ is a pole in q^2, as well as a pole in q. As a consequence $P(p,p+q)$ is ambiguous up to terms of $O(q^2)$; however, terms of $O(q)$ are well defined.

It is clear that $P(p,p+q)$ must be an odd-parity vertex, and that $P(p,p)$ must be nonvanishing. Therefore the massless excitation must be a pseudoscalar, with some nonderivative coupling to the fermion antifermion channel.

Let us now examine the vacuum polarization tensor, which is given by the Schwinger-Dyson equation shown in Fig. 8, apart from seagull terms. Since the proper vertex function occurring in the equation for $\Pi^{\mu\nu}$ has a pole, $\Pi^{\mu\nu}$ also develops this singularity. The pole contribution to $\Pi^{\mu\nu}$ can be obtained by inserting the singular part of the vertex function [Fig. 7 and Eqs. (2.9)] into the Schwinger-Dyson equation of Fig. 8. The result is given diagrammatically in Fig. 9. The integrals occurring in Fig. 9 are already introduced in (2.9b). We find therefore

$$\Pi^{\mu\nu}_{\text{pole}}(q) = g^2[q^\mu I(q^2)]\frac{i}{q^2}[-q^\nu I(q^2)]$$

$$\to -ig^2\lambda^2 q^\mu q^\nu/q^2\,. \tag{2.11}$$

[The minus sign in the last factor in the first equation above arises as follows. The second pseudo-

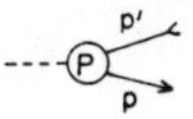

FIG. 6. Dynamical pseudoscalar vertex.

$\Gamma^{\mu}_{5\,\text{Pole}}\,(p, p+q) = -$

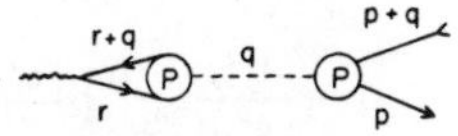

FIG. 7. Pole term in proper vertex function.

scalar vertex in Fig. 9 occurs with arguments $P(r, r+q)$, rather than $P(r+q, r)$ as in the definition $q^{\mu}I(q^2)$ in (2.9b). To make contact with (2.9b) the integration variable r is shifted to $r-q$, and q is replaced by $-q$.] Eq. (2.11) shows that $\Pi(q^2)$, defined in (1.1), has the pole $\Pi_{\text{pole}}(q^2) = g^2\lambda^2/q^2$. This indicates that the vector meson acquires a mass μ. In the approximation where only the pole term is kept, $\mu = g\lambda$. An exact formula

$$D(0) = -1/g^2\lambda^2 \tag{2.12}$$

may also be given.

C. Decoupling of the Massless Excitation

The amplitude T' contains a massless pole; nevertheless, it is true that the full on-mass shell scattering amplitude does not possess this pole. Hence the massless excitation decouples from the theory. To establish this important result, we combine (2.7) and (2.8) to obtain (2.13) as shown in Fig. 10.

Only the last two terms in (2.13) contain poles at $q^2 = 0$. The external fermions are on mass shell; thus the $q^{\mu}q^{\nu}$ term of the vector-meson propagator in the last term in (2.13) gives zero contribution since the on-mass-shell vertex is transverse. Consequently $D^{\mu\nu}(q)$ may be set equal to $-ig^{\mu\nu}D(q^2)$. Furthermore, if the leftmost vertex function in the last term in (2.13) is decomposed into a regular piece and the pole term (2.9c), one sees that the pole term does not contribute: The pole is proportional to q^{μ}, which contracts with the rightmost vertex function and annihilates it. Thus the last two terms in (2.13), which potentially contain a

$\Pi^{\mu\nu} = -g^2$

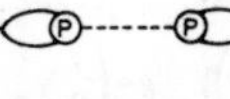

FIG. 8. Schwinger-Dyson equation for vacuum polarization tensor.

$\Pi^{\mu\nu}_{\text{Pole}} = g^2$

FIG. 9. Pole term in vacuum polarization tensor.

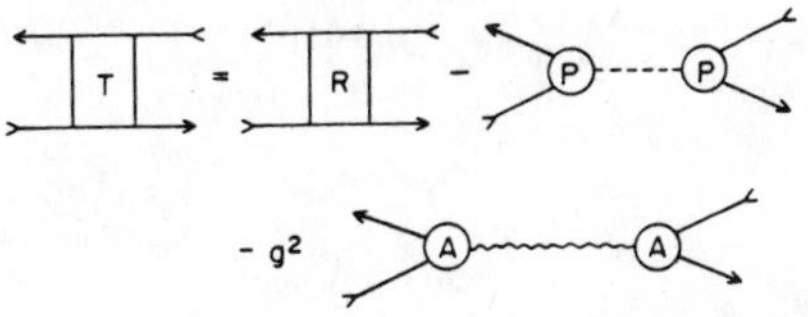

(2.13)

FIG. 10. The complete scattering amplitude.

zero-mass pole, are represented by (2.14) of Fig. 11. The right-hand part of the top diagram in Fig. 11 is the regular part of the axial-vertex function which we call $\tilde{\Gamma}^{\mu}_5$. The pole term (2.9a) in the rightmost vertex function of the last diagram in Fig. 11 has been explicitly exhibited.

We concentrate on the last term in (2.14). We may set $\tilde{\Gamma}^{\mu}_5 = \Gamma^{\mu}_5 - \Gamma^{\mu}_{5\,\text{pole}}$. Since the remaining part of the graph is proportional to q^{μ}, Γ^{μ}_5 does not contribute, and we are left with (2.15) given in Fig. 12. Evaluating this at $q^2 = 0$, and recalling that $q^2\Pi(q^2)|_{q^2=0} = g^2\lambda^2$, we find that (2.15) is

$$P(p'+q, p')\frac{i}{q^2}P(p, p+q)\,.$$

This is precisely the negative of the second term in (2.14); hence all the pole terms cancel.

Clearly a similar analysis can be performed in the crossed channel, and we conclude that T on the mass shell is free from massless poles.

D. The Problem of Other Symmetries

The theory which we have studied is formally scale-invariant. Consequently the emergence of mass terms appears to violate that symmetry and to lead to further massless excitations. It is clear that these potential Goldstone bosons are spin-zero, positive-parity objects. They could couple to the energy-momentum tensor, but not to a vector or axial-vector meson.

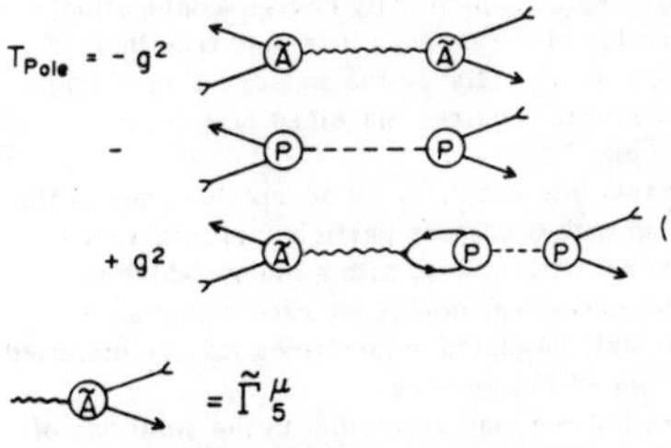

(2.14)

FIG. 11. Potentially singular portion of scattering amplitude.

$$-g^2\left[-\Gamma^{\mu}_{5\,\text{pole}}\right]\left[-ig_{\mu\nu}D(q^2)\right]\left[\Gamma^{\nu}_{5\,\text{pole}}\right]$$

$= g^2$ 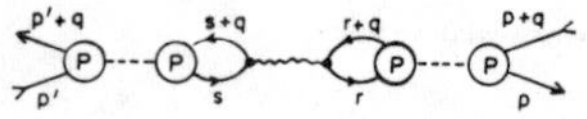

$$=-g^2\left[\frac{iq^{\mu}\lambda}{q^2}P(p'+q,p')\right]\left[\frac{-ig_{\mu\nu}}{q^2-q^2\Pi(q^2)}\right]\left[\frac{iq^{\nu}\lambda}{q^2}P(p,p+q)\right] \quad (2.15)$$

FIG. 12. Diagrammatic representation of Eq. (2.15).

However, the above circumstances very likely do not occur, since scale invariance does not appear to be realized in the solutions of a field theory, in spite of the formal symmetry of the Lagrangian. The reason is not spontaneous violation, but rather the presence of anomalies, analogous to those of the axial-vector current.[13] Thus we shall ignore any considerations of scale invariance.

Indeed it would appear that scalar massless mesons coupled to the energy-momentum tensor must be avoided. Such mesons would lead to a pole in the vacuum polarization tensor of two energy-momentum tensors. If one then considers coupling the energy-momentum tensor to a graviational gauge field, the pole would produce a mass for the graviton, in a fashion entirely analogous to our previous discussions. Since massive gravity apparently is not realized in nature, this state of affairs must be avoided.

The model in (2.1) possesses two other currents which are formally conserved. These are $\bar{\psi}\gamma^{\mu}\gamma^5\psi^c$ and its Hermitian conjugate. Here ψ^c is the charge-conjugate field $\psi_i^c = C_{ij}\bar{\psi}_j$; C is the charge-conjugation matrix $i\gamma^2\gamma^0$ which satisfies $C=-\tilde{C}=-C^{-1}=-C^{\dagger}$, and which transposes the γ matrices: $C^{-1}\gamma^{\mu}C=-\tilde{\gamma}^{\mu}$, $C^{-1}\gamma^5C=\tilde{\gamma}^5$.

If these two currents are conserved in the solutions of the theory, one obtains Ward identities which at zero momentum transfer require that $G(p)\gamma^5C+\gamma^5C\tilde{G}(-p)=0$. By charge-conjugation invariance of the theory, it is also true that $C\tilde{G}(-p)=G(p)C$. Hence the conservation of these odd currents requires massless fermions: $\{\gamma^5,\Sigma(p)\}=0$.

Clearly this result is not acceptable, nor is its evasion with Goldstone particles satisfactory. We do not wish to deal with a theory which possesses massless, doubly charged scalars. The additional, unwanted symmetries may be disposed of by one of two devices.

Firstly, one may argue that in the solutions of the theory the currents are not conserved, again because of anomalies. The envisioned situation is analogous to the discussion of massless electrodynamics, where the electron acquires a mass spontaneously.[11,14] The axial-vector current is formally conserved in that model, yet one can show that it is consistent that the equations determining the matrix elements of the axial-vector current and of its divergence admit a nonconserved solution. This comes to pass because of the divergences of perturbation theory. (Unlike the anomalies of the triangle graph and of scale invariance which occur in a low order of perturbation theory, examples of this mechanism require infinite summation of graphs.)

The second way of destroying the additional symmetry is by introducing vector couplings into the Lagrangian (2.1). This vector coupling can be a gauge coupling to the vector field A^{μ}: The interaction term in (2.1) is replaced by $\bar{\psi}\gamma^{\mu}(g'+ig\gamma^5)\psi A_{\mu}$. This leads to a parity-violating theory. If one wishes to maintain parity, one can introduce an additional (massive) vector field B_{μ} coupled to $\bar{\psi}\gamma^{\mu}\psi$. In the presence of a vector interaction, $\bar{\psi}\gamma^{\mu}\gamma^5\psi^c$ is no longer conserved. A massive vector field also destroys formal scale invariance.

E. The Problem of the Triangle Anomaly

Of course our theory (2.1) is inconsistent because of the triangle anomaly.[9] In order to remove this contradiction, the number of fermion fields must be increased.[10,15] We introduce a multiplet of n fields

$$\Psi=\begin{pmatrix}\psi_1\\ \vdots\\ \psi_n\end{pmatrix},$$

coupled to the axial-vector field by the interaction $i\bar{\Psi}\gamma^{\mu}\gamma^5\underline{g}\Psi A_{\mu}$. Provided the $n\times n$ Hermitian matrix $\underline{g}$ satisfies $\mathrm{Tr}\,\underline{g}^3=0$, the anomaly is absent. We shall limit the discussion to the simplest, two-fermion case:

$$\Psi=\begin{pmatrix}\psi_1\\ \psi_2\end{pmatrix},\quad \underline{g}=\begin{pmatrix}1 & 0\\ 0 & -1\end{pmatrix}. \quad (2.16)$$

The theory now possesses many conserved currents. Two, given by $\bar{\Psi}\gamma^\mu \underline{g}\Psi$ and $\bar{\Psi}\gamma^\mu\Psi$, or $\bar{\psi}_i\gamma^\mu\psi_i$, $i=1,2$, ensure the conservation of the individual species of fermions. In order to avoid further Goldstone bosons, this continuous symmetry must not be spontaneously broken. Fortunately even if we allow each of the fermions to acquire a *different* mass, and a *different* coupling to the massless bound state, the new symmetry remains operative. The model possesses additional, formally conserved, charged currents as discussed in the previous subsection. These must be disposed of by one of the two models mentioned above. Finally, the axial-vector current $\bar{\Psi}\gamma^\mu\gamma^5\Psi$, though beset by the triangle anomaly, would also lead to Goldstone bosons. The reason is that the anomaly does not contribute to the Ward identity at zero momentum, while it is only the zero-momentum Ward identity that is required to establish the existence of massless excitations. Thus conservation of this current must be broken, by the first method discussed above, just as in massless quantum electrodynamics.

There are also discrete symmetries present in the model; for example,

$$\psi_1 \to \psi_2, \quad \psi_2 \to \psi_1, \quad A^\mu \to -A^\mu$$

and (2.17)

$$\psi_1 \to i\psi_2, \quad \psi_2 \to -i\psi_1, \quad A^\mu \to -A^\mu.$$

(This eliminates 3-meson couplings, and the anomaly.) The discrete symmetry implies that $G_1(p)=G_2(p)$ and $P_1(p,p')=-P_2(p,p')$ [subscripts refer to fermion species]. Clearly, as a consequence of the symmetry the masses of the two fermions are the same, and the coupling of the pseudoscalar massless excitation to the fermions is equal in magnitude and opposite in sign. If we wish that the masses of the two fermions be different, this discrete symmetry must be broken. This breaking does no violence to the continuous symmetries, and since the symmetry which is being boken is discrete, no new massless excitations are required.

The discrete symmetry which we have discussed is not unlike ordinary charge-conjugation invariance. Indeed, the analogy can be made explicit by redefining the fields Ψ as follows:

$$\Psi = \tfrac{1}{2}(1+i\gamma^5)\Psi' + \tfrac{1}{2}(1-i\gamma^5)M\Psi',$$

where $M = \begin{pmatrix} 0 & 1 \\ 1 & 0 \end{pmatrix}$. The Lagrangian (2.1) becomes now

$$\mathcal{L} = i\bar{\psi}'\partial\!\!\!/\Psi' - \tfrac{1}{4}F^{\mu\nu}F_{\mu\nu} + \Psi'\gamma^\mu \underline{g}\psi' A_\mu. \qquad (2.18)$$

In the general case, when there are n fermions, care must be exercised that the spontaneously generated masses do not violate any new continuous symmetries. This would lead to massless excitations which may not decouple from the system.

F. Discussion

An interesting sum rule for the vector-meson mass may be derived in our model. Let us begin with (2.9b):

$$q^\mu I(q^2) = \mathrm{Tr}\int \frac{d^4r}{(2\pi)^4}[G(r)i\gamma^\mu\gamma^5 G(r+q)P(r+q,r)]$$

$$= -\mathrm{Tr}\int \frac{d^4r}{(2\pi)^4}[G(r+q)i\gamma^\mu\gamma^5 G(r)P(r,r+q)], \qquad (2.19a)$$

$$g^{\mu\nu}\lambda = -\mathrm{Tr}\int \frac{d^4r}{(2\pi)^4}\left[\partial^\nu G(r)i\gamma^\mu\gamma^5 G(r)P(r,r) + G(r)i\gamma^\mu\gamma^5 G(r)\frac{\partial}{\partial q_\nu}P(r,r+q)\Big|_{q=0}\right]. \qquad (2.19b)$$

We also know from (2.3) and (2.9a) that

$$q_\mu\bar{\Gamma}_5^\mu(p,p+q) = \gamma^5 G^{-1}(p+q) + G^{-1}(p)\gamma^5 - iP(p,p+q)\lambda, \qquad (2.20)$$

where $\bar{\Gamma}_5^\mu(p,p+q)$ is the vertex function, *without* its pole at $q^2=0$. Hence it is true that

$$i\lambda P(p,p) = \{\gamma^5, G^{-1}(p)\},$$

$$i\lambda\frac{\partial}{\partial q_\nu}P(p,p+q)\Big|_{q=0} = \gamma^5\partial^\nu G^{-1}(p) - \bar{\Gamma}_5^\nu(p,p). \qquad (2.21)$$

Inserting (2.21) in (2.19b) gives

$$g^{\mu\nu}\lambda^2 = -\mathrm{Tr}\int\frac{d^4r}{(2\pi)^4}[G(r)\partial^\nu G^{-1}(r)G(r)\gamma^\mu - G(r)\partial^\nu G^{-1}(r)G(r)\gamma^\mu\gamma^5 G(r)\gamma^5 G^{-1}(r)$$

$$+ G(r)\gamma^\mu\gamma^5 G(r)\gamma^5\partial^\nu G^{-1}(r) - G(r)\gamma^\mu\gamma^5 G(r)\bar{\Gamma}_5^\nu(r,r)]. \qquad (2.22a)$$

The middle two terms cancel against each other, while $\partial^\nu G^{-1}(r)$ may be related to the vertex function Γ^ν

of the vector current:

$$\partial^\nu G^{-1}(r) = -i\Gamma^\nu(r,r)\,.$$

Hence

$$g^{\mu\nu}\lambda^2 = i\,\mathrm{Tr}\int\frac{d^4r}{(2\pi)^4}[G(r)\Gamma^\nu(r,r)G(r)\gamma^\mu - G(r)\Gamma_5^\nu(r,r)G(r)i\gamma^\mu\gamma^5]\,. \tag{2.22b}$$

Recalling the definition of the vacuum polarization tensor (1.1) and its representation in terms of the Schwinger-Dyson equation of Fig. 8, we recognize that (2.22b) is equivalent to

$$g^{\mu\nu}g^2\lambda^2 = -i[\bar\Pi_A^{\mu\nu}(0) - \Pi_V^{\mu\nu}(0)]\,. \tag{2.23}$$

$\bar\Pi_A^{\mu\nu}$ is the vacuum polarization tensor associated with the nonsingular, nonconserved axial-vector current, while $\Pi_V^{\mu\nu}$ is the vacuum polarization of the conserved vector current, apart from seagull terms.

Formally, if one ignores seagull terms, $\Pi_V^{\mu\nu}(0) = 0$, since the vector current is conserved. This is also seen from (2.22a), where the integrand in the first term on the right-hand side is a total divergence $-\partial^\nu G(r)\gamma^\mu$. However, if $G(r)\to 1/r$ for large r, $\Pi_V^{\mu\nu}(0)$ is in fact a quadratically divergent constant which cancels a similar quadratic divergence in $\bar\Pi_A^{\mu\nu}(0)$. In other words, $\Pi_V^{\mu\nu}(0)$ is the seagull term, which is necessary to convert the formal Schwinger-Dyson equation for $\bar\Pi_A^{\mu\nu}$ into a correct expression. Thus we may replace (2.23) by

$$g^{\mu\nu}g^2\lambda^2 = -i\bar\Pi_A^{\mu\nu}(0)\,, \tag{2.24}$$

where $\bar\Pi_A^{\mu\nu}$ is now understood to include the requisite seagull term.

The formula (2.24) may also be understood from (1.1) and (2.11). We have

$$\Pi^{\mu\nu}(q) = (g^{\mu\nu}q^2 - q^\mu q^\nu)\left[i\bar\Pi(q^2) + i\frac{g^2\lambda^2}{q^2}\right], \tag{2.25}$$

where $\bar\Pi(q^2)$ is by definition regular. It is natural to identify $\bar\Pi_A^{\mu\nu}$ with the regular part of (2.25). Thus the following nontransverse expression is found:

$$\bar\Pi_A^{\mu\nu}(q) = ig^2\lambda^2 g^{\mu\nu} + (g^{\mu\nu}q^2 - q^\mu q^\nu)i\bar\Pi(q^2)\,, \tag{2.26}$$

and $-i\bar\Pi_A^{\mu\nu}(0) = g^{\mu\nu}g^2\lambda^2$, which agrees with (2.24). Note also that $\bar\Pi_A^{\mu\nu}(q)$, though nonconserved, has a conserved absorptive part. Of course in the approximation $\mu^2 = g^2\lambda^2$, (2.23) and (2.24) become mass formulas.

Equation (2.23) is also related to Weinberg's first sum rule.[16] If we identify $i\Pi_V^{\mu\nu}(0)$ with $g^2g^{\mu\nu}$ times the Schwinger term S_V of the vector current commutator, and similarly for the Schwinger term S_A of the *regular* part of the axial-vector current, then (2.23) reads $\lambda^2 + S_A = S_V$. This is recognized as Weinberg's first sum rule, when it is recalled that the Schwinger term of the *total* axial-vector current differs from that of the regular part by the square of the current-pseudoscalar coupling, which in our case is λ.

It is useful to estimate the various quantities that have been encountered. We set $\Sigma(p) = \not p A(p^2) + mB(p^2)$. Since $iG^{-1}(p) = \not p - \Sigma(p)$, it follows that $\Sigma(p)|_{\not p=m} = m$, or $A(m^2) + B(m^2) = 1$. From (2.5) and (2.10) we have

$$2m\gamma^5 B(m^2) = \lambda P(p,p)|_{\not p=m}\,. \tag{2.27}$$

Define $\lambda = m\bar\lambda$, where $\bar\lambda$ is a dimensionless number. Also, $P(p,p)|_{\not p=m}$ is defined to be $2\gamma^5 f$, again a dimensionless quantity. Thus

$$B(m^2) = \bar\lambda f\,. \tag{2.28}$$

The vector-meson mass satisfies $\mu^2 = g^2\lambda^2$, or

$$\mu = g\,\frac{B(m^2)}{f}\,m\,. \tag{2.29}$$

Thus if we imagine μ to be an order of magnitude larger than m, and g to be of electromagnetic strength, then $B(m^2)/f$ is of order 100. Moreover, if $B(m^2)$ is of order unity, then f is very small. Such a small coupling constant is perhaps unnatural, if it is a fundamental parameter. However, in the present context, f is of dynamical origin, and no preconceptions about its magnitude exist.

One may attempt a sort of tree approximation by setting $\Sigma(p)$ equal to m and $P(p,p+q)$ to $2\gamma^5 f$. It is then consistent, according to (2.21) that $\bar\lambda = f^{-1}$, $\mu = gmf^{-1}$, $\Gamma^\mu(r,r) = \gamma^\mu$, and $\bar\Gamma_5^\mu(r,r) = i\gamma^\mu\gamma^5$. From (2.19b) or (2.22b) one may compute $\bar\lambda$. The former gives

$$g^{\mu\nu}\bar\lambda m = -2if\,\mathrm{Tr}\int\frac{d^4r}{(2\pi)^4}\frac{(\not r+m)\gamma^\nu(\not r+m)\gamma^\mu\gamma^5(\not r+m)\gamma^5}{(r^2-m^2)^3}$$

$$= \frac{g^{\mu\nu}fm}{2\pi^2}\left(\ln\frac{\Lambda^2}{m^2} - 1\right). \tag{2.30a}$$

From (2.22b) we find

$$g^{\mu\nu}\bar\lambda^2 m^2 = -i\,\mathrm{Tr}\int\frac{d^4r}{(2\pi)^4}\frac{1}{(r^2-m^2)^2}[(\not r+m)\gamma^\nu(\not r+m)\gamma^\mu + (\not r+m)\gamma^\nu\gamma^5(\not r+m)\gamma^\mu\gamma^5]$$

$$= g^{\mu\nu}\frac{m^2}{2\pi^2}\left(\ln\frac{\Lambda^2}{m^2} - 1\right). \tag{2.30b}$$

Equations (2.30) imply that $\bar{\lambda}^2 = f^2L^2 = L$, where L is the logarithmically divergent quantity

$$\frac{1}{2\pi^2}\left(\ln\frac{\Lambda^2}{m^2} - 1\right).$$

Consistency with $\bar{\lambda}^2 = f^{-2}$ is achieved if $L = f^{-2}$.

These results are not satisfactory, since they indicate that the dynamical quantity $\bar{\lambda}$ is divergent, for which there does not appear a possibility of renormalization. In a truly consistent theory $\bar{\lambda}$ should be finite. It is not difficult to see the source of the problem. Our approximation $\Sigma(p) = m$ commits us to the result that $\lim_{p\to\infty}\Sigma(p) = m$, which would indicate that the *bare* mass is nonzero. Clearly the solution of the Schwinger-Dyson equation for $\Sigma(p)$ should have the property that, apart from terms proportional to $\not{p}$, $\Sigma(p)$ vanishes asymptotically; i.e.,

$$\lim_{p\to\infty}\lambda P(p,p) = \lim_{p\to\infty}\{\gamma^5, \Sigma(p)\} = 0 .$$

Thus we *must* not set $P(p, p+q) = 2\gamma^5 f$, and $\Sigma(p) = m$.

A more plausible approximation for $\Sigma(p)$ is $m(|p^2|/m^2)^{-\epsilon(g^2)}$,[17] where $\epsilon(g^2)$ is a positive, coupling-constant-dependent quantity. We expect that for small g^2, $\epsilon(g^2) = c^2g^2$. With this approximation, it is still consistent to leading order in g^2, to set $\Gamma^\mu(r,r) = \gamma^\mu$ and $\bar{\Gamma}_5^\mu(r,r) = i\gamma^\mu\gamma^5$. From (2.20b) we now get, instead of (2.30b), to leading order in g,

$$g^{\mu\nu}\bar{\lambda}^2 m^2 = g^{\mu\nu}\frac{m^2}{2\pi^2}\frac{1}{2\epsilon(g^2)} . \tag{2.31}$$

We thus find that $\bar{\lambda}$ is of order $1/g$, and $\mu = m/2\pi c$ is independent of g. Consequently, f becomes of order g. In the Appendix we discuss in greater detail these results, with special emphasis on the problem of renormalization. We justify the above approximation for $\Sigma(p)$, and evaluate $\epsilon(g^2)$ in lowest order.

III. PHENOMENOLOGICAL LAGRANGIANS

The mechanism which we have discussed is attractive in that it dispenses with fundamental scalar fields as the agents responsible for a pole in the vacuum polarization tensor. Furthermore, since the vector meson propagator has the form

$$-i\left(g^{\mu\nu} - \frac{q^\mu q^\nu}{q^2}\right)[q^2 - q^2\Pi(q^2)]^{-1} ,$$

its high-energy behavior is no more divergent than that of the free propagator, and the theory should be renormalizable.

An obvious shortcoming of the theory, as developed so far, is that no effective method of computation has been found. Ordinary perturbation theory will never expose a bound state: To any finite order in g we shall always have massless fermions and vector mesons.

One possible, though incomplete, approach is to describe the physical system by an effective Lagrangian. The effective Lagrangian, in tree approximation, should reproduce some of the dynamics of the complete theory. The description, necessarily limited to a low-energy domain, should take into account the following features of the complete theory: (1) the excitation spectrum, (2) couplings of the states to each other, and (3) the symmetries of the problem.

In the simple model considered in Sec. II, the excitation spectrum consists of a massive fermion, a massive axial-vector meson, and a massless pseudoscalar meson, which, however, decouples from the theory. There may be other massive bound states in the theory. Presumably these are important only at higher energies, and for a low-energy phenomenology may be ignored.

The interactions of the theory involve fermion–vector-meson couplings, fermion-pseudoscalar couplings, and vector-meson–pseudoscalar couplings. The latter two, however, are removable by an appropriate choice of fields.

Finally, the symmetries which are to be maintained are vector and axial-vector-current conservation and parity conservation. The axial-vector current should be related to a gauge symmetry.

Guided by the above considerations, we are led to the Lagrange function

$$\begin{aligned}\mathcal{L} = {} & i\bar{\psi}\not{\partial}\psi + \tfrac{1}{2}\partial^\mu\phi\partial_\mu\phi - \tfrac{1}{4}F^{\mu\nu}F_{\mu\nu} + \tfrac{1}{2}\mu^2A^2 \\ & + ig\bar{\psi}\gamma^\mu\gamma^5\psi A_\mu - \bar{\psi}\psi S(\phi) + \bar{\psi}\gamma^5\psi P(\phi) \\ & - 2gA^\mu\partial_\mu\phi a(\phi) + b(\phi) .\end{aligned} \tag{3.1}$$

The real functions $a(\phi)$, $b(\phi)$, $S(\phi)$, and $P(\phi)$ are to be determined. By parity conservation, the first three are even in ϕ; the last is odd. Also $S(0) = m$. Since we wish to realize chiral symmetry, with a *single* boson field ϕ, it is to be expected that the Lagrangian will not be a polynomial in ϕ. The equations of motion are

$$i\not{\partial}\psi = -ig\gamma^\mu\gamma^5\psi A_\mu + \psi S(\phi) - \gamma^5\psi P(\phi) , \tag{3.2}$$

$$\Box\phi = 2g\partial_\mu A^\mu a(\phi) - \bar{\psi}\psi S'(\phi) + \bar{\psi}\gamma^5\psi P'(\phi) + b'(\phi) , \tag{3.3}$$

$$\partial_\mu F^{\mu\nu} = -ig\bar{\psi}\gamma^\nu\gamma^5\psi - \mu^2A^\nu + 2g\partial^\nu\phi a(\phi) . \tag{3.4}$$

Clearly the vector current $\bar{\psi}\gamma^\mu\psi$ is conserved. The axial-vector current which must be conserved is

$$J_5^\mu = i\bar{\psi}\gamma^\mu\gamma^5\psi - 2a(\phi)\partial^\mu\phi . \tag{3.5}$$

By virtue of the equations of motion the divergence of J_5^μ is

$$\partial_\mu J_5^\mu = -2\bar\psi\gamma^5\psi[S(\phi)+a(\phi)P'(\phi)] -2\bar\psi\psi[P(\phi)-a(\phi)S'(\phi)] - 4g\partial_\mu A^\mu a^2(\phi) -2a(\phi)b'(\phi) - 2\partial^\mu\phi\partial_\mu\phi a'(\phi)\,. \tag{3.6}$$

Since ultimately we shall wish to arrive at a gauge theory, $\partial_\mu A^\mu$ may be set to zero. This is also consistent with (3.4), when it is assumed that $\partial_\mu J_5^\mu=0$. In order that (3.6) be zero, we choose $b(\phi)=0$, $a(\phi)=$constant, and

$$S(\phi) = -aP'(\phi)\,, \quad P(\phi) = aS'(\phi)\,. \tag{3.7}$$

The solution of these equations which conserves parity and leads to a properly normalized $S(0)$ is

$$S(\phi) = m\cos\phi/a\,, \quad P(\phi) = -m\sin\phi/a\,. \tag{3.8}$$

Consequently the Lagrangian (3.1) is given by

$$\mathcal{L} = i\bar\psi\partial\!\!\!/\psi + \tfrac{1}{2}\partial_\mu\phi\partial^\mu\phi - \tfrac{1}{4}F^{\mu\nu}F_{\mu\nu} + \tfrac{1}{2}\mu^2A^2 + ig\bar\psi\gamma^\mu\gamma^5\psi A_\mu - m\bar\psi\psi\cos\phi/a - m\bar\psi\gamma^5\psi\sin\phi/a - 2agA^\mu\partial_\mu\phi\,. \tag{3.9}$$

The theory (3.9) possesses a global symmetry with the transformations

$$\delta\psi=\gamma^5\psi\,,\quad \delta\bar\psi=\bar\psi\gamma^5\,,\quad \delta\phi=-2a\,,\quad \delta A^\mu=0\,. \tag{3.10a}$$

However, we require that (3.10a) should be extendible to local symmetry transformation:

$$\delta\psi=\gamma^5\psi\theta\,,\quad \delta\bar\psi=\bar\psi\gamma^5\theta\,,\quad \delta\phi=-2a\theta\,,\quad \delta A^\mu=-\frac{1}{g}\partial^\mu\theta\,. \tag{3.10b}$$

This can only happen if $2ga=\mu$. Thus the final phenomenological Lagrangian is

$$\mathcal{L} = i\bar\psi\partial\!\!\!/\psi + \tfrac{1}{2}\partial_\mu\phi\partial^\mu\phi - \tfrac{1}{4}F^{\mu\nu}F_{\mu\nu} + \tfrac{1}{2}\mu^2A^2 + ig\bar\psi\gamma^\mu\gamma^5\psi A_\mu - m\bar\psi\left(\exp\frac{2g}{\mu}\gamma^5\phi\right)\psi - \mu A^\mu\partial_\mu\phi\,. \tag{3.11}$$

The expression (3.11) may also be arrived at by a different argument. Since we seek a Lagrange function which gives a realization of a theory with chiral U(1)×U(1) symmetry, we are naturally led to a σ-type model, without isospin[18]:

$$\mathcal{L} = i\bar\psi\partial\!\!\!/\psi + \tfrac{1}{2}\partial_\mu\pi\partial^\mu\pi + \tfrac{1}{2}\partial_\mu\sigma\partial^\mu\sigma - \tfrac{1}{4}F^{\mu\nu}F_{\mu\nu} + ig\bar\psi\gamma^\mu\gamma^5\psi A_\mu - 2g(\sigma\partial^\mu\pi - \pi\partial^\mu\sigma)A_\mu + 2g^2(\sigma^2+\pi^2)A^2 - G\bar\psi(\gamma^5\pi+\sigma)\psi\,, \tag{3.12}$$

where the various couplings are chosen to be consistent with the gauge principle

$$\delta\psi=\gamma^5\psi\theta,\quad \delta\bar\psi=\bar\psi\gamma^5\theta\,,\quad \delta\pi=-2\sigma\theta,\quad \delta\sigma=2\pi\theta\,,\quad \delta A^\mu=-\frac{1}{g}\partial^\mu\theta\,. \tag{3.13}$$

However, the phenomenological Lagrangian relevant to our theory should possess only one spin-zero field, since we are ignoring the possibility of higher-mass bound states. Thus we set $\sigma^2+\pi^2=\mu^2/4g^2$, and (3.12) becomes

$$\mathcal{L} = i\bar\psi\partial\!\!\!/\psi - \tfrac{1}{4}F^{\mu\nu}F_{\mu\nu} + \tfrac{1}{2}\mu^2A^2 + \tfrac{1}{2}\frac{\mu^2}{4g^2\sigma^2}\partial_\mu\pi\partial^\mu\pi + ig\bar\psi\gamma^\mu\gamma^5\psi A_\mu - G\bar\psi(\gamma^5\pi+\sigma)\psi - \frac{\mu^2}{2g\sigma}A^\mu\partial_\mu\pi\,, \quad \sigma=\left(\frac{\mu^2}{4g^2}-\pi^2\right)^{1/2}. \tag{3.14}$$

Upon redefining the field π by

$$\pi=\frac{\mu}{2g}\sin\frac{2g}{\mu}\phi\,,\quad \sigma=\frac{\mu}{2g}\cos\frac{2g}{\mu}\phi\,, \tag{3.15}$$

we again get (3.11), with $m=G\mu/2g$.

It should be possible to derive the phenomenological Lagrangian by yet another method, which would exhibit the multiple fermion–massless-boson couplings. In this approach, which we have not studied, one would analyze the Ward identities of the underlying physical theory for n currents and two Fermi fields. The phenomenological description of the massless boson dynamics corresponds to keeping only the pole terms in these amplitudes.

It is not difficult to see that the massless field ϕ, present in (3.11), in fact decouples.[19] Change variables in (3.11):

$$\psi\to\exp\left(-\gamma^5\frac{g}{\mu}\phi\right)\psi\,,\quad \bar\psi\to\bar\psi\exp\left(-\gamma^5\frac{g}{\mu}\phi\right),\quad A^\mu\to A^\mu+\frac{1}{\mu}\partial^\mu\phi\,. \tag{3.16}$$

Then

$$\mathcal{L} = i\bar\psi\partial\!\!\!/\psi - m\bar\psi\psi - \tfrac{1}{4}F^{\mu\nu}F_{\mu\nu} + \tfrac{1}{2}\mu^2A^2 + ig\bar\psi\gamma^\mu\gamma^5\psi A_\mu\,. \tag{3.17}$$

This Lagrangian exhibits the physical spectrum of excitations: A massive fermion and vector meson interacting through the axial-vector current. Since (3.17) represents a choice of gauge, one may no longer set $\partial_\mu A^\mu$ to zero. Therefore the current $J_5^\mu = i\bar{\psi}\gamma^\mu\gamma^5\psi$ is no longer conserved. There does exist a conserved axial vector in the theory: $\mu^2 A^\mu + ig\bar{\psi}\gamma^\mu\gamma^5\psi$. However, since the time component of this object is equal to $\partial_i F^{0i}$, the "charge" vanishes, and no symmetry is generated.

The phenomenological description of our theory does not lead to a renormalizable Lagrangian. This is seen either from (3.11), where the nonrenormalizability resides in the nonpolynomial fermion-pseudoscalar interaction, or from (3.17), where it is the nonconservation of $i\bar{\psi}\gamma^\mu\gamma^5\psi$ that is responsible for the divergent, presumably unphysical, high-energy behavior of the theory. We do not consider the rapid growth at high energy as a defect of the fundamental theory. The phenomenological description is not meant to extend to high energies; in particular one does not expect to use the Lagrangians (3.11) or (3.17) for higher-order calculations.

Indeed it is quite possible that a more accurate phenomenological description of the fundamental theory can be given in terms of a renormalizable Lagrangian. For example, if there is another bound state in theory — a massive scalar — which together with the massless excitation forms a chiral multiplet, then the appropriate phenomenological Lagrangian would be (3.12) (supplemented with various chirally invariant σ and π terms) without the nonlinear constraint relating σ to π. That theory is renormalizable; it is an example of the conventional Higgs mechanism.

It is well known that the nonlinear σ model is the limiting form of a renormalizable linear σ model, where the mass of the scalar field tends to infinity. Thus we envision a hierarchy of phenomenological Lagrangians: In the low-energy domain where only the massless excitation is considered, and all other bound states are ignored, the description is in terms of (3.11). If other bound states are present, and if they should be included, then the description is as in (3.12). Renormalizability, on the phenomenological level, does not appear to be a fundamental requirement. Of course the construction of this hierarchy of phenomenological Lagrangians presupposes the possibility of solving the fundamental theory and exhibiting its complete excitation spectrum. Needless to say, at the present time, this is not possible.

In spite of our abandonment of renormalizability, the form of the phenomenological Lagrangian is nevertheless somewhat limited. For example, a nonrenormalizable Lagrangian which exhibits the Higgs mechanism could include a fermion–scalar-meson interaction of the form $G\bar{\psi}\psi(\sigma^2+\pi^2)$. Yet it does not appear that such a term can arise in a bound-state model.

It is easy to set up a correspondence between the couplings encountered in our fundamental theory and those exhibited in the phenomenological theory (3.11). In the former, the J_5^μ-ϕ transition is characterized by the strength λ, and hence the A^μ-ϕ interaction is $g\lambda$. Examining the last term in (3.11), we find an old relation $g\lambda=\mu$. Similarly, the fundamental, fermion-ϕ interaction is described by $2f$. Comparison with the next-to-last term in (3.11) gives $f=mg/\mu$, again an old result.

IV. CONCLUSION

The interesting aspect of the present investigation is the demonstration that vector particles can acquire a mass from a bound-state mechanism, rather than from a vacuum expectation value of a canonical scalar field.

The present work should be extended to include internal symmetry. More importantly, an effective computational method should be developed which bypasses ordinary perturbation theory, yet maintains renormalizability.

The physical relevance of this mechanism is not apparent at the present time. Clearly a non-Abelian version of our theory may be used for weak-interaction model building. More interesting is the possibility that the Abelian model may be relevant to pure strong interactions. A world with massless quarks and gluons which acquire their mass spontaneously without Goldstone bosons is attractive for its economy. Moreover, we see in this picture the possibility of avoiding the problem of too much symmetry in the quark model. If chiral U(3) is spontaneously broken in the fashion described here, then the troublesome ninth Goldstone boson disappears from the theory.

ACKNOWLEDGMENT

We are grateful for the interest that Steven Weinberg has shown in this work.

APPENDIX

In this appendix we return to some of the Schwinger-Dyson equations which we have encountered in the main body of the text. Our purpose is to show how they are to be written in finite, renormalized form. Also we solve them in the lowest-order Bethe-Salpeter approximation; a spontaneously generated mass is found for the fermion and vector meson.

Let us return to the equation satisfied by the proper vertex function, (2.6), and reexpress it in Bethe-Salpeter form; see Eq. (A1) of Fig. 13. In that formula K' is the Bethe-Salpeter kernel, with one-vector-meson lines deleted [compare (2.7)]. Evidently Γ_5^μ is multiplicatively renormalizable, and in the usual fashion we learn from the Ward identity (2.3) that Γ_5^μ may be renormalized by the same constant that renormalizes $G^{-1}(p)$. (The triangle anomaly is being ignored.)

Decomposing Γ_5^μ into its regular and pole parts, we have, according to (2.9c), the relation given by Eq. (A2) in Fig. 13. If (A2) is substituted into (A1), we find, by equating pole terms, that at $q^2=0$ P satisfies the homogeneous Bethe-Salpeter equation (A3) of Fig. 13. Consequently a formula for $\tilde{\Gamma}_5^\mu$ which follows from (A1)–(A3) can be found. It is Eq. (A4) of Fig. 13. The central hypothesis of this paper is that (A3) has a nontrivial solution, $P(p,p)\neq 0$. According to (2.5), (2.9), and (2.10), this allows the fermion to acquire a mass spontaneously. We shall show below that to lowest order in the coupling (A3) does indeed have a nontrivial solution.

Next we consider the equation for λ. The formula for λ^2 in terms of the vacuum polarization tensors, (2.22b), suffers from well-known overlapping divergences. Similarly the defining equation (2.9b) involves overlaps. It is our purpose to remove these overlaps and to express λ in terms of renormalized quantities.

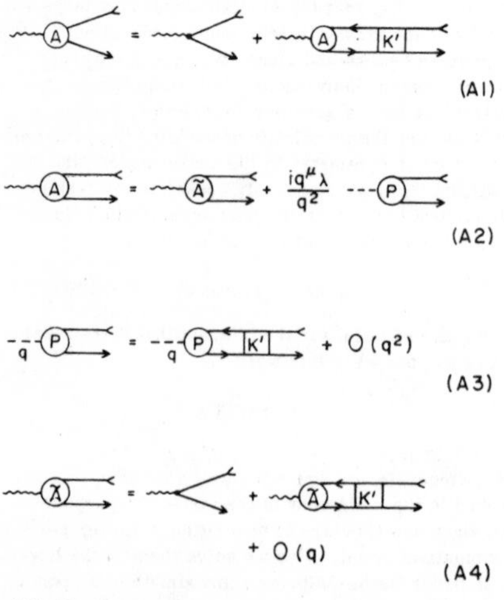

FIG. 13. Diagrammatic representation of Eqs. (A1)–(A4).

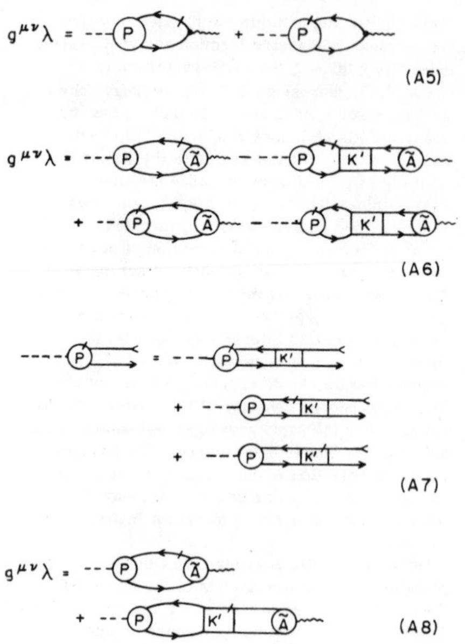

FIG. 14. Diagrammatic representation of Eqs. (A5)–(A8).

We represent (2.9b) or (2.19b) by Eq. (A5) in Fig. 14. The slash indicates differentiation with respect to q (the external momentum) and evaluation at $q=0$. The external momentum is routed through the upper fermion line. The bare vertex in (A5) may be evaluated from (A4) of Fig. 13. Hence Eq. (A6) is obtained; see Fig. 14. [The undetermined part of (A4) does not contribute, since $q=0$.] Also from (A3) of Fig. 13 we find a formula for the derivative of P. It is Eq. (A7) of Fig. 14. (Again the undetermined part disappears at $q=0$.) Upon substituting (A7) into the third term of (A6), one finds that many terms cancel and λ becomes determined by Eq. (A8) of Fig. 14.

Since P and Γ_5^μ each renormalize like G^{-1} and K' renormalizes like G^{-2}, the formula (A8) for λ is invariant under renormalization. Therefore finite, renormalized quantities P, $\tilde{\Gamma}_5^\mu$, and G can be used in evaluating λ. Furthermore, $P(r,r)$ is proportional to $\{\gamma^5,\Sigma(r)\}$; we expect it to go to zero with large r, since the bare mass is zero. Then the integrations in (A8) converge, and λ is well defined and finite. Of course if no symmetry-breaking solution exists, $P=0$ and λ vanishes.

Next we examine the lowest-order Bethe-Salpeter

equation for P. Equation (A3), written explicitly, is

$$P_{ab}(p,p+q)=\int\frac{d^4r}{(2\pi)^4}G_{cd}(r)P_{de}(r,r+q)\times G_{ef}(r+q)K_{ac,fb}(p,r;q)+O(q^2). \quad \text{(A9)}$$

(the subscripts are spinor indices). With pure axial-vector coupling, the lowest-order kernel is

$$K_{ac,fb}(p,r;q)=-g^2(i\gamma^\mu\gamma^5)_{fb}D_{\mu\nu}(p-r)(i\gamma^\nu\gamma^5)_{ac}.$$

For reasons that will emerge presently, we also imagine that there exists in the theory a massive vector meson, with vector coupling. (Recall that we found in Sec. II D that vector couplings are needed to eliminate explicitly undesirable additional symmetries.) Thus, with obvious notation, the lowest-order kernel is taken to be

$$K_{ac,fb}(p,r;q)=-g_a^2(i\gamma^\mu\gamma^5)_{fb}D^A_{\mu\nu}(p-r)(i\gamma^\nu\gamma^5)_{ac}-g_V^2(\gamma^\mu)_{fb}D^V_{\mu\nu}(p-r)(\gamma^\nu)_{ac}. \quad \text{(A10)}$$

For the propagators in (A9) and (A10) we take the lowest-order expressions. However, as we shall see, the spontaneously induced fermion mass is independent of coupling, while the vector-meson mass will not influence the result. Hence we keep all mass terms.

$$G(p)=\frac{i}{\not p-m},$$
$$D^A_{\mu\nu}(k)=-i\left(g^{\mu\nu}-\frac{k^\mu k^\nu}{k^2}\right)\frac{1}{k^2-\mu_A^2}, \quad \text{(A11)}$$
$$D^V_{\mu\nu}(k)=-i\left(g^{\mu\nu}-\frac{k^\mu k^\nu}{k^2}\right)\frac{1}{k^2-\mu_V^2}.$$

We need only determine $P(p,p)$ which is set equal to $\gamma^5\mathcal{P}(p^2)$. The Bethe-Salpeter equation (A9) at $q=0$ becomes

$$\mathcal{P}(p^2)=-3i\int\frac{d^4r}{(2\pi)^4}\frac{\mathcal{P}(r^2)}{r^2-m^2}\times\left[\frac{g_V^2}{(p-r)^2-\mu_V^2}-\frac{g_A^2}{(p-r)^2-\mu_A^2}\right]. \quad \text{(A12)}$$

The solution to (A12) for large p^2, where μ_V^2 and μ_A^2 may be ignored, is well known.[17] A solution is obtained only if $g_V^2>g_A^2$, in which case

$$\mathcal{P}(p^2)\underset{-p^2\to\infty}{\propto}\left(\frac{-p^2}{m^2}\right)^{-\epsilon},$$
$$\epsilon=\frac{3}{16\pi^2}[g_V^2-g_A^2]. \quad \text{(A13)}$$

In the case $g_A^2>g_V^2$, $\mathcal{P}(p^2)$ would increase with p^2, and the integral equation (A12) could not be satisfied. The reason for our insistence on additional vector coupling is now clear: For pure axial coupling no solution is found.

Since $\lambda P(p,p)=\lambda\gamma^5\mathcal{P}(p^2)=\{\gamma^5,\Sigma(p)\}$, it is consistent to set $\Sigma(p)$ equal to $m(|p^2|/m^2)^{-\epsilon}$, which is the formula we used in the text (Sec. II F).

According to (2.31), the axial-vector-meson mass becomes in this approximation

$$\mu_A^2=\lambda^2g_A^2=\tfrac{4}{3}m^2\frac{1}{g_V^2/g_A^2-1} \quad \text{(A14)}$$

The mass μ_A is independent of coupling strength (in the sense that it depends on the ratio of g_V^2/g_A^2), and can become arbitrarily large as $g_V^2/g_A^2\to1$.

Of course the present considerations do not hold for strong coupling, where the lowest order Bethe-Salpeter approximation is not applicable. In particular, we do not wish to imply that pure axial-vector coupling would necessarily fail in that case.

*Work supported in part through funds provided by the U. S. Atomic Energy Commission under Contract No. AT(11-1)-3069.

[1]For a summary of the pioneering work of Salam, Schwinger, 't Hooft, and Weinberg, see the review talk by B. W. Lee, in *Proceedings of the XVI International Conference on High Energy Physics, Chicago-Batavia, Ill., 1972*, edited by J. D. Jackson and A. Roberts (NAL, Batavia, Ill., 1973), Vol. 4, p. 249.

[2]P. W. Higgs, Phys. Lett. 12, 132 (1964); Phys. Rev. Lett. 13, 508 (1964); Phys. Rev. 145, 1156 (1966); F. Englert and R. Brout, Phys. Rev. Lett. 13, 321 (1964); G. S. Guralnik, C. R. Hagen, and T. W. B. Kibble, *ibid.* 13, 585 (1964).

[3]Interest in this possibility is widespread. The only investigations that we are aware of are due to F. Englert and R. Brout [Phys. Rev. Lett. 13, 321 (1964)], Y. Freundlich and D. Lurié [Nucl. Phys. B19, 557 (1970)], F. Englert, R. Brout, and M. F. Thiny [Nuovo Cimento 43, 244 (1966)], and J. M. Cornwall and R. E. Norton [UCLA report (unpublished)].

[4]Y. Nambu and G. Jona-Lasino, Phys. Rev. 122, 345 (1961).

[5]J. Schwinger, Phys. Rev. 125, 397 (1962).

[6]Our convention is $g^{00}=-g^{ii}=1$, $\gamma^5=\gamma^0\gamma^1\gamma^2\gamma^3$.

[7]J. Schwinger, Phys. Rev. 128, 2425 (1962).

[8]K. Johnson, Phys. Lett. 5, 253 (1963); D. J. Gross and R. Jackiw (unpublished); S.-S. Shei, Phys. Rev. D 6, 3469 (1972).

[9]S. L. Adler, in *Lectures on Elementary Particles and Quantum Field Theory*, edited by S. Deser, M. Grisaru, and H. Pendleton (MIT Press, Cambridge, Mass., 1970); R. Jackiw, in *Lectures on Current Algebra and Its Applications*, edited by S. Treiman, R. Jackiw, and

D. J. Gross (Princeton Univ. Press, Princeton, N. J., 1972).

[10]D. J. Gross and R. Jackiw, Phys. Rev. D 6, 477 (1972); C. Bouchiat, J. Iliopoulos, and Ph. Meyer, Phys. Lett. 38B, 519 (1972).

[11]"Anomalous exceptions" can arise not only from the triangle graph, which must be removed for consistency of the theory, but also from the absence of the symmetry in nonperturbative solutions to the theory. Such failures of Ward identities were discussed by A. Maris, V. Herscovitz, and G. Jacob [Phys. Rev. Lett. 12, 313 (1964)], K. Johnson [in *9th Latin American School of Physics, Santiago, Chile, 1967*, edited by I. Saavedra (Benjamin, New York, 1968)], M. Baker and K. Johnson [Phys. Rev. D 3, 2516 (1971)], and H. Pagels [Phys. Rev. Lett. 28, 1482 (1972); Phys. Rev. D 7, 3689 (1973)].

[12]We ignore the possibility that the pole arises from an infrared singularity in the integration implied by (2.6). This is in contrast to the pole in two-dimensional spinor electrodynamics, which is a consequence of the restricted two-dimensional phase space.

[13]K. Wilson, Phys. Rev. D 2, 1473 (1970); 2, 1478 (1970); S. Coleman and R. Jackiw, Ann. Phys. (N.Y.) 67, 552 (1971).

[14]M. Baker, K. Johnson, and B. W. Lee, Phys. Rev. 133, B209 (1964).

[15]One can construct two anomaly-free theories without increasing the number of fermions by coupling the axial vector to the real or imaginary part of $\bar{\psi}\gamma^\mu\gamma^5\psi^c$. These theories, however, do not admit a conserved vector current.

[16]S. Weinberg, Phys. Rev. Lett. 18, 507 (1967).

[17]M. Baker and K. Johnson, Phys. Rev. D 3, 2516 (1971).

[18]J. Schwinger, Ann. Phys. (N.Y.) 2, 407 (1957); M. Gell-Mann and M. Lévy, Nuovo Cimento 16, 705 (1960); J. S. Bell and R. Jackiw, *ibid.* 60, 47 (1969).

[19]The Higgs mechanism in nonpolynomial Lagrangians has been observed by S. Weinberg [Phys. Rev. 166, 1568 (1968)] and by L. Faddeev (unpublished). Weinberg has emphasized a connection between the spectral-function sum rules and the Higgs mechanism.

PHYSICAL REVIEW D VOLUME 8, NUMBER 10 15 NOVEMBER 1973

Spontaneous Symmetry Breaking Without Scalar Mesons*

John M. Cornwall† and Richard E. Norton
Department of Physics, University of California, Los Angeles, California 90024
(Received 11 June 1973)

By combining the ideas of Nambu in his study of superconductivity and of Johnson, Baker, and Willey in their approach to electrodynamics we construct a gauge theory of spontaneous symmetry breaking which is free of elementary spin-zero fields. The theory contains two fermions and two vector mesons, one of which acquires a mass via the Higgs mechanism. A formula for this vector-meson mass is derived which becomes exact, and nonzero, in the limit as the strength of interaction is appropriately scaled to zero. The vacuum energy is also discussed.

I. INTRODUCTION

The observation that a symmetry-violating solution of a Yang-Mills theory may lead via the Higgs mechanism to a renormalizable field theory of massive vector mesons coupled to nonconserved currents has aroused a flurry of efforts to apply these ideas to construct a unified theory of the weak and electromagnetic interactions.[1] To our knowledge, essentially all such efforts in this regard have utilized field theories with elementary scalar fields whose nonzero vacuum expectation values are the source of the symmetry breaking. In this sense, these theories are analogous to the many-body theory of Bose systems in the superfluid phase. However, in constrast with the situation in nonrelativistic many-body theory, where the existence of an expectation value for the Bose field follows from the repulsive nature of the potential between the particles,[2] the vacuum expectation value of the scalar field in the proposed field-theoretical models follows only after it is essentially put in by hand—by giving the mass term in the free-particle Lagrangian the wrong sign. Furthermore, since there is no suggestion from observation that scalar mesons play a significant role in weak or electromagnetic phenomena, it is generally necessary in constructing theories with scalar mesons to arrange that their observable effects are sufficiently small, i.e., that their masses are sufficiently large.

In this paper[3] we present a simple model field theory in which the spontaneous symmetry breaking and the consequent massiveness of a vector meson occur in a manner similar to the violation of electric current conservation and the consequent Meissner effect in the theory of superconductivity. Indeed, our efforts to construct a theory of this kind were inspired by the work of Nambu[4] concerning the gauge invariance of the BCS theory of superconductivity.

The specific model we study is presented in Sec. II. It will be evident that this model is not intended as a realistic theory of weak or electromagnetic interactions. Rather, it is only an example of what we feel is probably a large class of theories in which the spontaneous symmetry breaking derives from general features of an apparently symmetric interaction, without the necessity of a mass term with the wrong sign and without an elementary scalar field having a vacuum expectation value.

The model we discuss has two spin-$\frac{1}{2}$ fermions of equal bare mass and two massless vector mesons. The vector mesons are coupled to different currents, each of which generates a separate O(2) invariance of the theory. The first symmetry is an invariance with respect to rotations of the two fermion fields into each other, and it is this symmetry which is spontaneously violated, giving a mass to the associated vector meson. The other vector meson is coupled to the fermion number current, whose conservation is unbroken. This second vector meson plays the role in our model which the phonon plays in the theory of superconductivity; it provides the force which allows a nontrivial (i.e., nonzero) solution to the homogeneous Dyson equation for the symmetry-violating part of the fermion propagator.

In Sec. II we argue that the theory admits asymmetric solutions. The character of the symmetry-violating part of the fermion Green's function is similar to the chiral-symmetry-breaking part (i.e., mass term) of the electron Green's function in the Baker-Johnson[5] approach to electrodynamics. In particular, the scale of the symmetry-violating part of the propagator at asymptotically large momentum squared appears to be a free parameter, whereas the rate of decrease of this quantity as a function of momentum squared is an explicit function of the coupling constants.

In Sec. III we impose the Ward identity on the vector-meson–fermion vertex function and show that this function develops a pole at zero q^2. This pole generates a mass for the vector meson coupled to the violated current. A formula for the self-energy of this meson at zero momentum is derived.

Section IV is devoted to demonstrating that the Goldstone excitation which accompanies the symmetry breaking does not occur as a pole in the physical scattering amplitude. More specifically, we exhibit a cancellation between a pole at $q^2=0$ in the vector-meson exchange part of the scattering amplitude and a corresponding pole in the vector-meson irreducible part of this amplitude.[6,7]

In Sec. V we emphasize that the spontaneous symmetry breaking induced by the interaction leaves a finite, residual effect even when the coupling constants of this interaction are appropriately scaled to zero. The approximation of retaining only these zero-order effects of the interaction is termed the "platform approximation" to convey our impression (as yet not fully substantiated) that from this platform the remaining effects of the interaction can be calculated, in principle, as a power series in the coupling constants. The content of the theory in this approximation is essentially a free-field theory, but with the masses altered from their original symmetric values. In particular, the originally massless vector meson coupled to the violated current has a mass which, as we demonstrate, can be calculated exactly in the platform approximation.

In Sec. VI we confront the question as to which of the various solutions to the theory is the preferred one in the sense that it yields the lowest vacuum energy—which in relativistic field theory is the zero-temperature, zero-chemical-potential limit of the thermodynamic potential. Somewhat to our surprise, we are led to conclude that all the solutions, including the symmetric solution, have equal vacuum energy. We argue that this conclusion is implied whenever there is a continuous range of solutions characterized by a continuous parameter. In the present model, this parameter can be taken to be the apparently arbitrary scale (zero for the symmetric solution) of the symmetry-violating part of the fermion propagator for asymptotically large momentum. The situation is thus quite different from usual Higgs theories with scalar fields, where the difference in vacuum energy between the normal and spontaneously broken solutions is nontrivial, both at the tree level and when radiative corrections are included.[8]

II. THE MODEL

The model we study has the Lagrangian density

$$\mathcal{L}=-\bar{\psi}\left(\frac{1}{i}\gamma_\mu\partial_\mu+m_0\right)\psi-\tfrac{1}{4}F_{\mu\nu}{}^2-\tfrac{1}{4}G_{\mu\nu}{}^2+g\bar{\psi}\gamma_\mu\psi A_\mu+g'\bar{\psi}\gamma_\mu\tau_2\psi B_\mu\,, \tag{2.1a}$$

where ψ represents the two fields ψ_1 and ψ_2, where

$$F_{\mu\nu}=\partial_\mu A_\nu-\partial_\nu A_\mu\,, \tag{2.1b}$$

$$G_{\mu\nu}=\partial_\mu B_\nu-\partial_\nu B_\mu\,, \tag{2.1c}$$

and where τ_2 is the 2×2 matrix

$$\tau_2=\begin{pmatrix}0&-i\\ i&0\end{pmatrix}$$

connecting ψ_1 and ψ_2. The Lagrangian density is locally invariant under rotations through an angle θ in the plane of ψ_1 and ψ_2,

$$\psi\to e^{i\tau_2\theta}\psi\,, \tag{2.2a}$$

if simultaneously the vector field B undergoes the gauge transformation

$$B_\mu\to B_\mu+\frac{1}{g'}\partial_\mu\theta\,. \tag{2.2b}$$

We will look for solutions which violate this symmetry. The $\mathcal{L}$ in (2.1a) is also invariant under $\psi\to e^{i\alpha}\psi$, with a corresponding gauge translation of A_μ; we do not break this symmetry, except that it is always possible to add a mass term for A_μ without affecting either renormalizability or number current conservation.

The Dyson equation for the fermion propagator is

$$S^{-1}(p)=\not{p}+m_0+ig^2\int\frac{d^4k}{(2\pi)^4}\Delta^A_{\mu\nu}(k)\gamma_\mu S(p-k)\Gamma^A_\nu(p-k,p)+ig'^2\int\frac{d^4k}{(2\pi)^4}\Delta^B_{\mu\nu}(k)\gamma_\mu\tau_2 S(p-k)\Gamma^B_\nu(p-k,p)\,, \tag{2.3}$$

where the last two terms on the right refer, respectively, to the two graphs in Fig. 1, and where $\Delta^{A,B}_{\mu\nu}$ and $\Gamma^{A,B}_\nu$ are the propagators and vertex functions of the indicated mesons. We want to know if there exists a solution to this equation in which $S^{-1}(p)$ has a part proportional to the matrix

$$\tau_3=\begin{pmatrix}1&0\\0&-1\end{pmatrix}, \tag{2.4}$$

and which therefore violates the O(2) symmetry of Eq. (2.2).

Let us begin by considering Eq. (2.3) to first order in g^2 and g'^2. If we write

$$S^{-1}(p)-\not{p}-m_0=\Sigma(p)\equiv\Sigma_s(p)+\tau_3\Sigma_v(p) \tag{2.5}$$

and take $\Delta^{A,B}_{\mu\nu}$ to be in the Landau gauge, the order g^2, g'^2 version of Eq. (2.3) becomes

$$\Sigma_s(p)=i(g^2+g'^2)\int\frac{d^4k}{(2\pi)^4}\,\frac{\delta_{\mu\nu}-k_\mu k_\nu/k^2}{k^2}\,\gamma_\mu\,\frac{\not p-\not k+m_0+\Sigma_s(p-k)}{[\not p-\not k+m_0+\Sigma_s(p-k)]^2-\Sigma_v{}^2(p-k)}\gamma_\nu \tag{2.6a}$$

and

$$\Sigma_v(p)=i(g'^2-g^2)\int\frac{d^4k}{(2\pi)^4}\,\frac{\delta_{\mu\nu}-k_\mu k_\nu/k^2}{k^2}\gamma_\mu\,\frac{\Sigma_v(p-k)}{[\not p-\not k+m_0+\Sigma_s(p-k)]^2-\Sigma_v{}^2(p-k)}\gamma_\nu\,, \tag{2.6b}$$

where we have projected out separately the part of Σ proportional to the unit matrix and the part proportional to τ_3.

Equation (2.6b) for the symmetry-violating part of the fermion self-energy is similar in structure to the equation for the chirally asymmetric part of the inverse electron propagator in the Baker-Johnson-Willey[5] approach to electrodynamics. Guided by the work of these authors, we ask if there is a solution to (2.6b) which for asymptotically large p^2 behaves like

$$\Sigma_v(p^2)\underset{p^2\to\infty}{\sim}\delta m\left(\frac{p^2}{m^2}\right)^{-\epsilon}, \tag{2.7}$$

where ϵ is positive, and m is some mass. Since the large-p^2 behavior of the integral in (2.6b) comes from large $(p-k)^2$, we check the consistency of the ansatz (2.7) by substituting (2.7) into the integrand in (2.6b), ignoring all but the $(\not p-\not k)^2$ in the denominator, and performing the integration. The result is

$$\Sigma_v(p^2)\underset{p^2\to\infty}{\sim}\frac{3(g^2-g'^2)}{16\pi^2\epsilon(1-\epsilon)}\,\delta m\left(\frac{p^2}{m^2}\right)^{-\epsilon}. \tag{2.8}$$

Equations (2.7) and (2.8) can be made consistent for ϵ in the range $0<\epsilon<1$ providing

$$0<g^2-g'^2<\frac{4\pi^2}{3}\,, \tag{2.9}$$

which is not much of a restriction, except that it requires g^2 to be larger than g'^2. In particular, the choice $g=0$ in Eq. (2.1a) would not have allowed a symmetry-violating solution of the kind envisaged in Eq. (2.7).

As mentioned in Sec. I, for the asymptotic solution (2.7) the scale δm is arbitrary, and it is this feature which leads to our conclusion in Sec. VI that the vacuum energy is independent of δm. It is natural at this point to wonder if this feature is also true for the exact solution to the nonlinear equation (2.6b) and, for that matter, for the solution for the symmetry-violating part of the full Dyson equation in (2.3). We have convinced ourselves, albeit nonrigorously, that exact solutions to (2.8b) exist with the asymptotic behavior in (2.7) for a continuous range of δm including $\delta m=0$. Concerning the full Dyson equation we have little to say in this paper, except to mention that we have given preliminary consideration to the question of how the solution of the simple equation (2.6b) could be extended to include corrections to arbitrary order in g^2 and g'^2, and that it is our current impression that such an extension can be realized with the asymptotic form for Σ_v in (2.7) and the arbitrary scale for δm persisting to all orders. Further comments in this regard are included in Sec. V.

The Dyson equation in (2.6a) for the symmetric self-energy Σ_s can presumably be solved iteratively with standard techniques.[9] We do not pursue this question further here, except to observe that there appears the attractive alternative that m_0 could be chosen to be zero, and that the chirally noninvariant part of Σ_s could be generated spontaneously in a manner similar to that which we propose for Σ_v. This idea, which of course is not original here,[5] leads to a spontaneous breakdown of both chiral and scale invariance. Since the currents which generate these symmetries are generally associated with anomalies, one would not expect a zero-mass boson to accompany their violation.[10] By contrast, we pursue in the subsequent sections the point of view that the breakdown of the O(2) symmetry in Eq. (2.2) implied by a nonzero Σ_v is associated with a Goldstone excitation which via the Higgs mechanism gets decoupled from the S matrix by giving a mass to the B vector meson of Eq. (2.1).

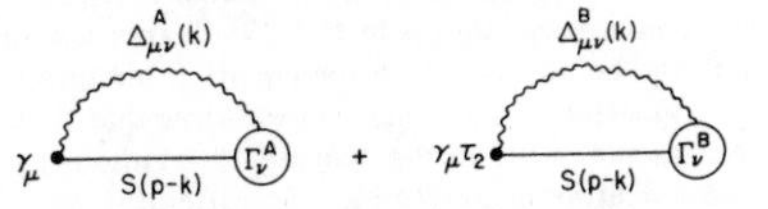

FIG. 1. The two contributions to the fermion self-energy.

III. WARD IDENTITY AND B-MESON MASS

Stimulated by the existence of a Ward identity[4] for the electric current vertex in the nonrelativistic theory of superconductivity, and also by the fact that in relativistic field theory there are no anomalies to destroy the conservation of vector[11] (as opposed to axial-vector) currents, we assume that the vertex function Γ_μ^B satisfies the conventional Ward identity,

$$k_\mu \Gamma_\mu^B(p-k,p) = \tau_2 S^{-1}(p) - S^{-1}(p-k)\tau_2, \qquad (3.1)$$

even when Σ_v is nonzero. It is evident that the right-hand side of (3.1) is not zero at $k=0$, if Σ_v in (2.5) does not vanish. This observation forces us to conclude that Γ_μ^B is singular at $k=0$ such that

$$k_\mu \Gamma_\mu^B(p-k,p)\big|_{k=0} = 2i\tau_1 \Sigma_v(p), \qquad (3.2)$$

where τ_1 is the 2×2 matrix

$$\tau_1 = \begin{pmatrix} 0 & 1 \\ 1 & 0 \end{pmatrix}. \qquad (3.3)$$

The vertex Γ_μ^B must essentially have the form

$$\Gamma_\mu^B(p-k,p) = \overline{\Gamma}_\mu^B(p-k,p) + i\tau_1 \frac{k_\mu}{k^2}\Gamma_v^B(p-k,p), \qquad (3.4)$$

where $\overline{\Gamma}_\mu^B$ is regular at $k^2=0$, and Γ_v^B satisfies

$$\Gamma_v^B(p-k,p)\big|_{k=0} = 2\Sigma_v(p). \qquad (3.5)$$

The B-meson self-energy function $\Pi_{\mu\nu}$ is shown in Fig. 2. It is given by

$$\Pi_{\mu\nu}^B(k) = -ig'^2 \int \frac{d^4p}{(2\pi)^4} \mathrm{Tr}[\Gamma_\mu^B(p-k,p)S(p)\gamma_\nu \tau_2 S(p-k)]$$
$$+ \text{contact terms}, \qquad (3.6)$$

where the trace is over the eight-dimensional direct-product space of the τ and Dirac matrices. The contact terms in (3.6) play no essential role in the following except to guarantee that as a consequence of (3.1)

$$k_\mu \Pi_{\mu\nu}^B = k_\nu \Pi_{\mu\nu}^B = 0. \qquad (3.7)$$

That is, $\Pi_{\mu\nu}^B$ must be of the form

$$\Pi_{\mu\nu}^B(k) = \left[\delta_{\mu\nu} - \frac{k_\mu k_\nu}{k^2}\right]\Pi_B(k^2), \qquad (3.8)$$

where the vector-meson propagator is given in terms of Π_B by

$$\Delta_{\mu\nu}^B(k) = \frac{\delta_{\mu\nu} - k_\mu k_\nu/k^2}{k^2 + \Pi_B(k^2)} - \lambda\frac{k_\mu k_\nu}{k^4}. \qquad (3.9)$$

Here λ is a gauge parameter which is zero in the Landau gauge.

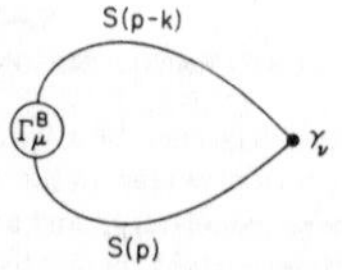

FIG. 2. The B-vector-meson self-energy $\Pi_{\mu\nu}$.

It is evident from (3.9) that the B meson acquires a mass if $\Pi_B(0)$ is unequal to zero; that is, if the coefficient of $k_\mu k_\nu$ in (3.8) has a pole at $k^2=0$. But such a pole is implied by the presence of the pole in Γ_μ^B indicated in (3.4). By inserting (3.4) into (3.6) and isolating the pole in $\Pi_{\mu\nu}$, we obtain

$$\Pi_B(0) = \lim_{k^2\to 0} -\frac{1}{k^2} g'^2 \int \frac{d^4p}{(2\pi)^4} \mathrm{Tr}[\tau_1 \Gamma_v^B(p-k,p)$$
$$\times S(p)\not{k}\tau_2 S(p-k)]. \qquad (3.10)$$

A formula for $\Gamma_v^B(p-k,p)$ valid to first order in k is obtained by differentiating (3.1) with respect to k with Γ_μ^B given by (3.4),

$$\tau_1\Gamma_v^B(p-k,p) = i[S^{-1}(p),\tau_2] - ik_\lambda \partial_\lambda S^{-1}(p)\tau_2$$
$$+ ik_\lambda \overline{\Gamma}_\lambda^B(p,p) + O(k^2). \qquad (3.11)$$

Using this relation in (3.10) we obtain after some algebra the equality

$$\Pi_B(0) = -\frac{ig'^2}{4}\int \frac{d^4p}{(2\pi)^4}\mathrm{Tr}[\gamma_\lambda \partial_\lambda S(p)$$
$$+\gamma_\lambda S(p)\overline{\Gamma}_\lambda^B(p,p)S(p)\tau_2], \qquad (3.12)$$

which is applied in Sec. V to obtain an expression for the B-meson mass. Remarkably this does not vanish in the limit where the coupling constants g^2 and g'^2 are scaled to zero [with $g^2 > g'^2$ to satisfy (2.9)], as we shall see in Sec. V.

IV. CANCELLATION OF THE GOLDSTONE EXCITATION

The assumption that the Ward identity (3.1) is valid, despite the apparent violation of the $\bar\psi\gamma_\mu\tau_2\psi$ current arising from a nonzero Σ_v, has been shown in Sec. III to imply a pole at $k^2=0$ in the vertex function Γ_μ^B. This assumed validity of the Ward identity is equivalent to the assumption that the existence of a solution to the homogeneous Bethe-Salpeter equation for the single-B-meson irreducible part of the fermion-antifermion scattering amplitude, which follows from the existence of a solution to the homogeneous Dyson equation for the symmetry-violating part of the fermion propagator,[10] reflects a pole in the solution of the corresponding inhomogeneous Bethe-Salpeter equation. Conversely, this pole in Γ_μ^B indicates a solution to the homogeneous Dyson equation for this vertex function. In this section we show that this pole does not occur in the "physical" fermion-fermion scattering amplitude. Specifically, we demonstrate that the pole in the single-B-meson irreducible part of the scattering amplitude cancels against a pole in the B-meson-exchange part of this amplitude. This cancellation is similar to the cancellation between a bound-state pole and the pole due to an elementary particle with the same quantum numbers.[7]

We denote by R and R', respectively, the coupling strengths of the $\bar\psi\gamma_\mu\tau_2\psi$ current and of the fermion-antifermion state to this pole. As indicated in Fig. 3, the single-fermion matrix element of the current $\bar\psi\gamma_\mu\tau_2\psi$, which is related to Γ_μ^B by[12]

$$\langle p'|\bar\psi\gamma_\mu\tau_2\psi|p\rangle=\bar u(p')\Gamma_\mu^B(p',p)u(p), \tag{4.1}$$

thus has a pole which in terms of R and R' is given by

$$\bar u(p')\Gamma_\mu^B(p',p)u(p)|_{\text{pole}}=iRR'\frac{k_\mu}{k^2}\bar u(p')\tau_1 u(p). \tag{4.2}$$

Here $k=p-p'$ and the $u(p)$ are the spinors for the "physical" fermions,

$$S^{-1}(p)u(p)=\bar u(p)S^{-1}(p)=0. \tag{4.3}$$

The R and R' shown in Fig. 3 can be understood in the sense of an effective Lagrangian by thinking of the current $\bar\psi\gamma_\mu\tau_2\psi$ as containing a piece proportional to $\partial_\mu\phi$,

$$\bar\psi\gamma_\mu\tau_2\psi=R\partial_\mu\phi+\cdots, \tag{4.4}$$

with ϕ the effective field for a massless scalar meson coupled to the fermions by a term in the effective Lagrangian

$$\mathcal{L}'_{\text{eff}}=R'\bar\psi\tau_1\psi\phi. \tag{4.5}$$

The effective interaction in (4.5) generates a pole at zero momentum transfer squared in the single-B-vector-meson irreducible part of the fermion-fermion scattering amplitude T_1. This pole term

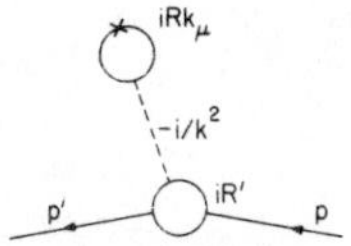

FIG. 3. Diagrammatic representation of the pole in the matrix element of Eqs. (4.1) and (4.2).

indicated in Fig. 4 is

$$\langle p_1'p_2'|T_1|p_1p_2\rangle|_{\text{pole}}=-\frac{R'^2}{k^2}[\bar u(p_1')\tau_1 u(p_1)\bar u(p_2')\tau_1 u(p_2)-(p_1'\leftrightarrow p_2')]. \tag{4.6}$$

The B-vector-meson-exchange part of the fermion-fermion scattering amplitude, which we call T_2, is shown in Fig. 5 and is equal to

$$\langle p_1'p_2'|T_2|p_1p_2\rangle=-\Delta_{\mu\nu}^B(k)[\bar u(p_2')\Gamma_\mu^B(p_2',p_2)u(p_2)\bar u(p_1')\Gamma_\nu^B(p_1',p_1)u(p_1)-(p_1'\leftrightarrow p_2')], \tag{4.7}$$

where the B-meson propagator $\Delta_{\mu\nu}^B$ is given in Eq. (3.9).

We wish to show that the pole at $k^2=0$ in (4.6) cancels against a pole in (4.7) which exists because of the pole in Γ_μ^B indicated in (4.2). As a first step in isolating the pole part of (4.7) note that the $k_\mu k_\nu$ terms in $\Delta_{\mu\nu}^B$ contribute nothing because of the Ward identity (3.1) and the equalities in (4.3). But then the pole part of $\Gamma_\mu^B(p_2',p_2)$, which is proportional to k_μ, also gives nothing for the same reasons. The pole part of (4.7) then arises only from the pole part of $\Gamma_\mu^B(p_1',p_1)$, and by appealing to (4.2) we have $[k=p_1-p_1'=-(p_2-p_2')]$

$$\langle p_1'p_2'|T_2|p_1p_2\rangle|_{\text{pole}}=-\Pi_B^{-1}(0)\left[\bar u(p_2')\left(\Gamma_\mu^B(p_2',p_2)+iRR'\frac{k_\mu}{k^2}\tau_1\right)u(p_2)\bar u(p_1')iRR'\frac{k_\mu}{k^2}\tau_1 u(p_1)-(p_1'\leftrightarrow p_2')\right], \tag{4.8}$$

with Π_B as given in (3.9). Finally, an application of the Ward identity and (4.3) once again eliminates $\Gamma_\mu^B(p_2',p_2)$ to leave

$$\langle p_1'p_2'|T_2|p_1p_2\rangle|_{\text{pole}}=\frac{R^2R'^2}{\Pi_B(0)k^2}[\bar u(p_2')\tau_1 u(p_2)\bar u(p_1')\tau_1 u(p_1)-(p_1'\leftrightarrow p_2')], \tag{4.9}$$

which shows, by comparing with (4.6), that the pole parts of T_1 and T_2 cancel, providing

$$R^2=\Pi_B(0). \tag{4.10}$$

But according to Eq. (3.8) $\Pi_B(0)$ is the coefficient of $-k_\mu k_\nu$ in the residue of the pole at $k^2=0$ in $\Pi_{\mu\nu}$. In Fig. 6 the pole part of $\Pi_{\mu\nu}(k)$ is shown in terms of R defined in (4.4), and it is easy to check that

$$\Pi_{\mu\nu}(k)|_{\text{pole}}=-\frac{k_\mu k_\nu}{k^2}R^2. \tag{4.11}$$

V. THE PLATFORM APPROXIMATION

In this section we apply Eq. (3.12) to obtain a formula for the B-vector-meson mass which becomes exact in the limit as the coupling constants g^2 and g'^2 approach zero in a manner that allows

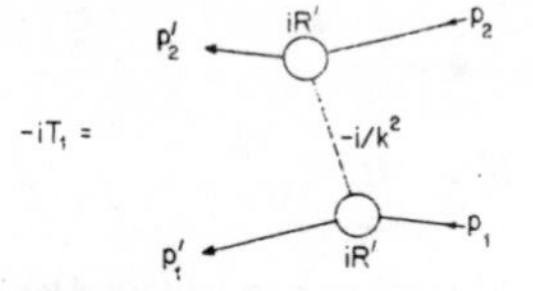

FIG. 4. The pole in the single-B-meson irreducible part of the fermion-fermion scattering amplitude.

the consistency of (2.7) and (2.8) to be maintained, that is, as $g^2 - g'^2$ approaches zero from the positive side. It is evident from Eq. (2.7) that in this limit Σ_v goes to a constant δm, and—so to speak—the only effect of the vanishing interaction is to leave the fermion masses split by an arbitrary amount $2\delta m$ and, as we shall see in Eq. (5.6), to leave the B vector meson with a finite mass whose value, for any nonzero choice of δm, can be arranged completely arbitrarily, since [see Eq. (5.6)] the ratio $g'^2(g^2 - g'^2)^{-1}$ is unrestricted in the limit as g^2 and g'^2 approach zero.

We shall refer to the "platform approximation" as the approximation of retaining only those features of the coupling between the particles which persist when g^2 and g'^2 go to zero in the sense described. This terminology reflects our impression that these zero-order effects of the interaction can serve as a platform from which the remaining effects of the coupling between the particles can be computed in a perturbation expansion in g^2 and g'^2. For example, after choosing a δm to determine the zero-order Σ_v in Eq. (2.7), one can substitute this Σ_v into the right-hand side of (2.6b) to generate the zero-order part of the left-hand side identically and, in addition, a first-order contribution to Σ_v arising from the nonasymptotic region of the integration domain. This first-order Σ_v then generates a second-order contribution when it is inserted into the right-hand side of (2.6b), etc. Although no pitfalls have yet appeared to us which would prohibit the implementation of this program to arbitrary order (i.e., in principle), we should emphasize that we have

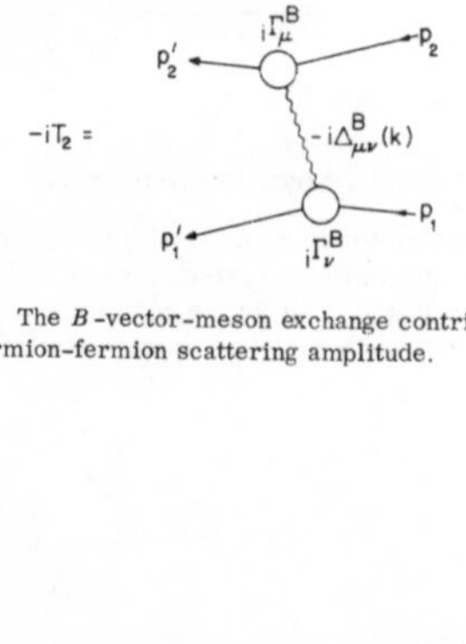

FIG. 5. The B-vector-meson exchange contribution to the fermion-fermion scattering amplitude.

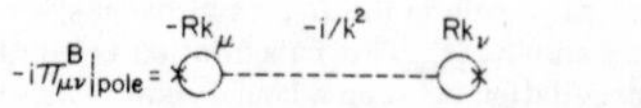

FIG. 6. The pole in the B-meson self-energy $\Pi_{\mu\nu}$.

not yet studied this problem in depth, and we cannot at present assert that the asymmetric theory discussed here is amenable to a finite perturbation calculation to all orders in g^2 and g'^2. This question is of significance, since the theory it describes—a massive vector meson coupled to an apparently nonconserved current—is certainly not renormalizable as conventionally formulated.

We now apply (3.12) to calculate $\Pi_B(0)$ in the platform approximation. Because of the explicit g'^2 in (3.12) we need only retain those parts of $S(p)$ and $\overline{\Gamma}{}^B_\lambda$ which are of zero order in the coupling constants. Thus, referring to (2.5), we take

$$S(p) = \left[m_0 + \not{p} + \tau_3 \delta m \left(\frac{p^2}{m^2}\right)^{-\epsilon}\right]^{-1}, \tag{5.1a}$$

$$\overline{\Gamma}{}^B_\lambda(p,p) = \gamma_\lambda \tau_2 \tag{5.1b}$$

in the integrand of (3.12). Since we can also ignore the derivative of the τ_3 term in $\partial_\lambda S(p)$, we have

$$\begin{aligned} \mathrm{Tr}[\gamma_\lambda \partial_\lambda S(p) + \gamma_\lambda S(p) \overline{\Gamma}{}^B_\lambda(p,p) S(p) \tau_2] &= \mathrm{Tr}[-\gamma_\lambda S(p) \gamma_\lambda S(p) + \gamma_\lambda S(p) \gamma_\lambda \tau_2 S(p) \tau_2] \qquad (5.2\mathrm{a}) \\ &= -2\,\mathrm{Tr}\left[\gamma_\lambda \gamma_\lambda \tau_3^{\,2} (\delta m)^2 \left(\frac{p^2}{m^2}\right)^{-2\epsilon} (p^2 + m_0^{\,2})^{-2}\right], \qquad (5.2\mathrm{b}) \end{aligned}$$

where in (5.2b) we have kept only the leading dependence of $[S(p)]^4$ for large p (i.e., p^{-4}) together with the term $m_0^{\,2}$ to avoid a spurious infrared divergence. Performing the trace in (5.2b) and substituting the result into (3.12) leads to

$$\Pi_B(0) = -16 i g'^2 (\delta m)^2 (m^2)^{2\epsilon} \times \int \frac{d^4p}{(2\pi)^4} \frac{1}{(p^2 + m_0^{\,2})^2 (p^2)^{2\epsilon}}, \tag{5.3}$$

which upon integration gives

$$\Pi_B(0) = \frac{g'^2(1-\epsilon)}{\pi \sin 2\pi\epsilon} \left(\frac{m^2}{m_0^{\,2}}\right)^{2\epsilon} (\delta m)^2. \tag{5.4}$$

In the platform approximation g^2, g'^2, and ϵ all go to zero such that

$$\frac{\epsilon}{g^2 - g'^2} \to \frac{3}{16\pi^2} \tag{5.5}$$

as required for the consistency of (2.4) and (2.8). Thus, since $\Pi_B(0)$ is the square of the B-meson mass M_B in this limit, and since δm is half the fermion mass splitting $m_1 - m_2$, Eq. (5.5) becomes in the platform approximation

$$M_B{}^2 = \frac{2g'^2}{3(g^2 - g'^2)}(m_1 - m_2)^2 . \tag{5.6}$$

Observe that no such result could be obtained in a conventional Higgs theory with scalar particles, because of the arbitrariness of the coupling of the scalar to the fermion fields. Indeed, one of the major advances which follows from elimination of the scalars is the large reduction in independent parameters which can occur in the Lagrangian.

VI. THE VACUUM ENERGY

Both in many-body systems and in Higgs theories with scalars, one argues that the preferred solution (either normal or spontaneously broken) is the one with the lowest ground-state energy. The potential function which is minimized by an appropriate choice of the vacuum expectation value of a Higgs scalar is essentially the vacuum energy difference for the normal and spontaneously broken vacuum; it was used by Jona-Lasinio[13] for the study of Goldstone theories, and the single-loop approximation to it was studied by Coleman and Weinberg for Higgs theories.[8]

Denote by Ω_s, Ω_v the sums of all connected vacuum graphs for the symmetry-preserving and symmetry-violating theories, respectively, divided by a four-dimensional normalization volume, and let $\Omega = \Omega_s - \Omega_v$. Then Ω is the energy difference (per unit three-dimensional volume) between the normal and symmetry-violating vacuums. Both Ω_s and Ω_v are quartically divergent; our first task is to argue that Ω is finite; otherwise, it makes no sense to compare the vacuum energy of the systems.

Let us consider Ω_s or Ω_v as a functional set of all irreducible Green's functions, that is, the proper self-energies, proper vertices, one-particle-irreducible four-point functions, etc. (as well as the vacuum expectation value of scalar fields, in the general case). It can be shown[14] that the stationarity of Ω_s or Ω_v to arbitrary variations of these Green's functions yields the complete set of Dyson equations for the normal theory or the symmetry-violating theory, respectively. There is only one symmetry-breaking parameter in our theory, namely the scale factor δm introduced in Eq. (2.7) into the symmetry-violating self-energy Σ_v. When $\delta m = 0$, we recover the normal theory. Indicating explicitly the dependence of Ω_v on δm we have

$$\Omega_s = \Omega_v(0); \ \Omega = \Omega_v(0) - \Omega_v(\delta m). \tag{6.1}$$

It is easy to see (essentially because $\mathrm{Tr}\,\tau_3 = 0$) that Ω_v is an even function of δm; we now argue that there is no infinite part to the $(\delta m)^2$ term in Ω.

The coefficient of $(\delta m)^2$ can be found by varying Ω_v with respect to S_v, Γ_v, etc., and invoking stationarity:

$$\delta\Omega_v = \frac{i}{(2\pi)^4}\mathrm{Tr}\int d^4p\, \delta S_v(p)[\text{LHS of Eq. (2.3)} - \text{RHS of Eq. (2.3)}] + \int \delta\Gamma_v \cdots \tag{6.2}$$

[recall that (2.3) is the Dyson equation for Σ_v]. For sufficiently small δm, we have $-\delta S_v = S_s \Sigma_v S_s$; moreover, at large p, the term in square brackets in (6.2) vanishes identically for any δm, because it reduces to the *linear* homogeneous equation based on (2.6b). One may now verify that the contribution from asymptotically large p to (6.2) is in fact finite.

All terms of $O((\delta m)^4)$ or higher in Ω are finite, by a naive power-counting argument, since Ω has dimension M^4. [Potential logarithmic divergences in the $(\delta m)^4$ term are saved by the asymptotic decrease of $\Sigma_v(p) \sim (p^2)^{-\epsilon}$.] Thus we conclude that Ω, the difference of two quartically divergent objects, is finite.

However, a much stronger conclusion seems warranted in the case at hand: It appears that $\Omega \equiv 0$. This conclusion is based on the following theorem: If Ω_v depends on a *continuous* symmetry-breaking parameter δm which can be chosen arbitrarily within a certain neighborhood of $\delta m = 0$, then Ω_v is independent of δm. In this case, the variation of Ω_v with respect to δm vanishes identically, because the square brackets in (6.2) vanish identically for any δm (and so for all the other Dyson equations). A simpler proof invokes the Feynman-Hellwarth theorem: Let

$$E(\delta m) = \langle \psi | H | \psi \rangle \tag{6.3}$$

be the vacuum energy (i.e., expectation) value of H in the *normalized* vacuum state $|\psi\rangle$. Then we have

$$\frac{\partial E}{\partial \delta m} = \langle \psi | \frac{\partial H}{\partial \delta m} | \psi \rangle = 0 \tag{6.4}$$

because H is independent of δm.

Thus, $\Omega = 0$ if δm can be chosen from a continuum of values. Otherwise, Ω is finite. At the moment, it is an open question whether or not the full Dyson equations [e.g., (2.3)] determine that $\delta m/m$ takes on discrete values. It is our current belief that δm can be chosen from a continuum, based on studies of equations with nonlinearities similar to those of (2.3); thus $\Omega \equiv 0$. Needless to say, this is quite different from conventional Higgs theories.

One might question whether it makes physical sense to compare the vacuum energies of two relativistic field theories, since there seems to be no physical mechanism for causing transitions between the two vacuums. However, all the remarks of this section hold for the ground-state energy of a system of particles at finite density and temperature, and in this case it is physically possible for a phase transition to occur. Of course, the required densities and temperatures must be very high; the only possible relevant circumstances seem to be the first moment of the "big-bang" model of the universe. Kirzhnits and Linde[15] have discussed the possibility of a phase transition in conventional Higgs theories, at temperatures comparable to particle masses. At the moment, it seems unlikely that consideration of finite temperature and density can change our conclusion that $\Omega \equiv 0$, but it remains an interesting and open question.

VII. CONCLUSIONS

It seems to be an important advance to rid spontaneously broken gauge theories (SBGT) of Higgs scalars, for several reasons. The first is the obvious consideration that no such scalars have been observed, and in most SBGTs the scalar masses are chosen to ensure unobservability for many years to come. Perhaps more important, there are no arbitrary parameters in $\mathcal{L}$ which characterize the coupling of scalars to themselves and to fermions, which ultimately allows for the calculation of a larger number of symmetry-breaking effects than in more conventional Higgs theories.

Clearly, the model discussed here is too restrictive in a number of respects. We have dealt explicitly only with the Abelian case, largely to avoid the tedious complications of ghost scalars which accompany closed loops of vector mesons.[1] However, it is easy to generalize the treatment of the linear, homogeneous equation for Σ_v by enlarging the B meson and the fermion to multiplets of fields. Then δm becomes a matrix which can be expanded into irreducible group representations, for each one of which there is a separate ϵ, as determined by Eq. (2.8). It is interesting to observe that the contribution to each ϵ from the B multiplet may have either sign (it is negative for an Abelian case). For example, let the fermions and B mesons both be in an $I=1$ representation of O(3); then δm can be $I=1$ or $I=2$. The B contribution is positive for $\epsilon_{I=1}$, but negative for $\epsilon_{I=2}$. Likewise, if the fermions and B mesons are each in an SU(3) octet, then ϵ_{8d} and ϵ_{8f} have the same positive contribution, while ϵ_{27} is negative. In the absence of coupling to the singlet A meson, then, in the first case $I=2$ symmetry breaking is forbidden, and in the second case only octet symmetry breaking is allowed. This is in satisfying agreement with what appears to be happening in the real world.

All our considerations have been based on the assumption of specific asymptotic forms for the propagation and vertex functions; specifically, we assumed the meson propagators to go like k^{-2} at infinity. It is well known that this is true only if the Gell-Mann–Low eigenvalue condition is satisfied,[5] namely $\beta(\alpha_0)=0$, where β is the coefficient of $k^2 \ln k^2$ in $\Pi_B(k^2)$ for large k^2, and α_0 is the bare fine-structure constant; $\beta = O(\alpha_0)$ for small α_0. If $\beta \neq 0$, the vector propagators have the asymptotic behavior $k^{-2}[\sum_0 C_N(\beta \ln k^2)^N]$, where the C_N are uninteresting numerical constants. If this asymptotic form is used in the linearized version of Eq. (2.6) for Σ_v, the resulting equation for ϵ is (schematically) $\epsilon \sim (g^2 - g'^2)F(\beta/\epsilon, \beta'/\epsilon)$. In consequence, ϵ still scales with g^2 and g'^2 if these coupling constants are small, but the actual numerical value of ϵ depends on the C_N. Thus violation of the Gell-Mann–Low eigenvalue condition does not violate the spirit of the platform approximation of Sec. V.

It remains to work out a systematic perturbation theory, starting with the linearized platform approximation and going to higher orders in the symmetry-breaking parameter δm. Such a perturbation theory will be reminiscent of the insertion of scalar tadpoles on fermion and vector lines, as in the usual Higgs theories. However, there is a difference: Our "tadpole" (e.g., Σ_v) decreases at large momenta.

Our considerations of the vacuum self-energy in Sec. VI have been very brief. It is now clear, as we shall discuss elsewhere,[14] that the techniques alluded to there allow for a full statement of relativistic statistical mechanics, including vacuum fluctuations, which is free of infinities. It is possible that this may find application in certain astrophysical processes.

---

*Work supported in part by the National Science Foundation.

†Alfred P. Sloan Foundation Fellow.

[1] For a list of references through September 1972 see B. W. Lee, in *Proceedings of the XVI International Conference on High Energy Physics, Chicago-Batavia, Ill., 1972,* edited by J. D. Jackson and A. Roberts (NAL, Batavia, Ill., 1973), Vol. 4, p. 249.

[2] See, for example, A. L. Fetter and J. D. Walecka, *Quantum Theory of Many Particle Systems* (McGraw-Hill,

New York, 1971), Chap. 14.

[3]While this paper was in preparation we received a preprint of R. Jackiw and K. Johnson [Phys. Rev. D 8, 2386 (1973)] dealing with the same issues in the context of a model with an axial vector. A variant of this model is the same as the Abelian model discussed here, upon suitable transformation of the fields. Some time ago the possibility of a dynamical origin to spontaneous symmetry breakdown was discussed by F. Englert and R. Brout [Phys. Rev. Lett. 13, 321 (1964)].

[4]Y. Nambu, Phys. Rev. 117, 648 (1960).

[5]K. Johnson, M. Baker, and R. Willey, Phys. Rev. 136, B1111 (1964); Phys. Rev. 163, 1699 (1967); M. Baker and K. Johnson, *ibid.* 183, 1292 (1969); Phys. Rev. D 3, 2516 (1971); 3, 2541 (1971); K. Johnson and M. Baker, *ibid.* 8, 1110 (1973); S. Adler, Phys. Rev. D 5, 3021 (1972); 7, 1948(E) (1973).

[6]This is also demonstrated by Jackiw and Johnson in Ref. 3.

[7]It has been known for some time that a bound state which communicates with an elementary particle channel does not appear in the S matrix. This was argued by C. J. Goebel and B. Sakita [Phys. Rev. Lett. 11, 293 (1963)] for nonrelativistic bound states and by Y. S. Jin and S. W. MacDowell [Phys. Rev. 137, B688 (1965)] for bound-state poles which move with the coupling constant—this does not take place for the Goldstone boson, which is fixed at $q^2=0$. A general proof of this cancellation, based on the Dyson equations, was given by J. M. Cornwall and D. J. Levy [Phys. Rev. 178, 2356 (1968)].

[8]S. Coleman and E. Weinberg, Phys. Rev. D 7, 1688 (1973).

[9]See, for example, S. Adler and W. A. Bardeen, Phys. Rev. D 4, 3045 (1971); 6, 734(E) (1972).

[10]M. Baker, K. Johnson, and B. W. Lee, Phys. Rev. 133, B209 (1964).

[11]W. A. Bardeen, Phys. Rev. 184, 1848 (1969).

[12]There would be an additional factor of $(Z_2Z_2')^{1/2}$ on the right-hand side of (4.1) if the spinors u were normalized to unity. Although it is of no particular significance in the present context, we note that this Z_2 dependence can be included by normalizing the spinors to $\bar{u}u=Z_2$.

[13]G. Jona-Lasinio, Nuovo Cimento 34, 1790 (1964).

[14]J. M. Luttinger and J. C. Ward [Phys. Rev. 118, 1417 (1960)] demonstrated this stationarity with regard to the Dyson equation for the fermion propagator in nonrelativistic many-body theory, and C. De Dominicis and P. C. Martin [J. Math. Phys. 5, 14 (1964); 5, 31 (1964)] extended this work to include the vertex functions. The present authors, at the time unaware of the work of Dominicis and Martin, also extended these considerations to the full set of Dyson equations in relativistic field theory. The details will be published elsewhere.

[15]D. A. Kirzhnits and A. D. Linde, Phys. Lett. 42B, 471 (1972).

PHYSICAL REVIEW D VOLUME 10, NUMBER 10 15 NOVEMBER 1974

Dynamical symmetry breaking of non-Abelian gauge symmetries*†

Estia J. Eichten
Theory Group, Stanford Linear Accelerator Center, Stanford, California 94305

Frank L. Feinberg
Department of Physics, University of California, Los Angeles, California 90024
(Received 1 May 1974)

We use the Schwinger mechanism to generate a dynamical breakdown of non-Abelian gauge symmetries. Such a breakdown is implemented by using bound-state Goldstone bosons which violate the global invariance associated with the gauge group. The usefulness of this realization of the Schwinger mechanism is that it eliminates the necessity of introducing elementary scalar particles, and, furthermore, it is a viable possibility in a pure Yang-Mills theory. Fermion and vector-meson mass relations in the pole approximation are discussed and compared with the zeroth-order mass relations in the familiar Higgs models. In addition, we obtain consistent solutions to the Bethe-Salpeter equation for the bound state in the weak-coupling limit, which yield finite vector-boson masses in the pole approximation. However, there are group-theoretical constraints which limit the possible groups and representations.

I. INTRODUCTION

Since the success of 't Hooft and Veltman[1] and Lee and Zinn-Justin[2] in showing the renormalizability of massless Yang-Mills Lagrangians, there has been considerable interest[3] in field theories possessing local non-Abelian gauge symmetries of vector mesons. Hope has been raised that weak and electromagnetic interactions may be united by using this class of field theories.[3] More recently, models have been introduced which try to incorporate strong-interaction dynamics into this scheme.[4] The key element in this approach is the introduction of spontaneous symmetry breakdown, so that S-matrix elements do not exhibit all the formal symmetry properties of the Lagrangian. Excluding anomalies, we know that such circumstances demand the appearance of massless Goldstone excitations. However, since the Lagrangians are invariant under local gauge transformations, the Goldstone excitations combine with the originally massless vector mesons associated with the gauge symmetry to produce massive vector mesons.

Some time ago, Nambu and Jona-Lasinio[5] in their pioneering work showed that a theory may possess symmetry-violating solutions by using a four-Fermi interaction to produce a chiral-symmetry-breaking bound-state Goldstone boson. The next year Schwinger[6] made the crucial observation that it is possible to have a pole in the polarization tensor of vector mesons at $q^2=0$. This pole leads to spontaneous symmetry breakdown by giving a mass[7] to otherwise massless vector mesons, a process known as the Schwinger mechanism. In Schwinger's example,[8] two-dimensional quantum electrodynamics, the model is soluble and the pole in the polarization tensor can be seen explicitly, but the Goldstone theorem is evaded due to anomalies which violate axial-vector conservation.[9] Therefore symmetry violation leading to massive particles is due to anomalies, rather than bound-state Goldstone bosons.

Most of the current resurgence of interest in spontaneous symmetry breakdown has been centered around a different realization of the Schwinger mechanism, the popular Higgs mechanism.[10] In this procedure scalar particles in the Lagrangian are assigned symmetry-violating vacuum expectation values, which supply the raison d'être for the pole in polarization tensor. However, it is desirable to produce models without fundamental scalar particles in the Lagrangian because the scalars which are needed for the Higgs case have not been observed, and furthermore, it is difficult to arrange asymptotic freedom when scalar particles are present.

Two techniques, mentioned above, can be used to avoid the use of the conventional Higgs-Kibble mechanism[11] in Yang-Mills theories. One, which may be peculiar to two dimensions,[12] is to employ anomalies, which destroy current conservation, to provide the pole in the polarization tensor of the vector particles. The other is to assume that the integral equations of the theory admit symmetry violation such that there are bound-state Goldstone bosons which generate the desired pole. Using the latter method, Jackiw and Johnson[13] and Cornwall and Norton,[14] extending the work of Nambu and Jona-Lasinio,[5] showed that it is consistent for a gauge theory of fermions interacting with massless Abelian vector mesons to have symmetry-breaking solutions such that there is a finite physical mass for the fermion and for the vector meson which generates the broken symmetry. Some time ago Englert and Brout[15] and recently Pagels[16] have suggested this approach. This model has the attractive feature that a conserved gauge current is coupled to a massive Abelian field.[17] In this paper we show that in local non-Abelian gauge theories the same mechanism is a viable possibility, with or *without* canonical fermion fields in the Lagrangian. The emerging phenomenological mass relations are similar to those obtained with the Higgs mechanism.

In Sec. II we demonstrate that it is consistent to have spontaneous mass generation in a theory composed of non-Abelian gauge mesons alone. That is, if the polarization tensor has a symmetry-violating part, then it itself acquires a pole at zero momentum transfer from a zero-mass bound state which communicates with two-vector-meson states as well as with one-vector-meson states. In the Abelian case,[13,14] such a bound state is forbidden by charge-conjugation invariance. In Sec. III fermions are included. This section is a direct extension of the work of Cornwall and Norton[14] and Jackiw and Johnson[13] to non-Abelian vector mesons. Recently Sarkar[18] has also discussed the non-Abelian problem. Here a symmetry-breaking part of the fermion self-energy or of the polarization tensor implies a zero-mass bound state in the fermion-antifermion sector as well as in the two-vector-meson sector. In both Secs. II and III it is argued that the ghost fields, which are included to maintain gauge invariance, remain massless.

Also it should be pointed out that the existence of the symmetry-violating solutions is not shown to necessarily occur, but rather that the integral equations of the theory are consistent with such solutions. Of course, these solutions, when realized, cannot be detected by making the usual perturbation expansions. We assume here that the nonperturbative masses depend analytically on the renormalized coupling constants, but that conventional perturbation theory will not reveal this structure. Another possibility, not considered here, is that the mass depends nonanalytically on the renormalized coupling constants.

In Sec. IV various mass relations are obtained in the pole approximation. Such relations may provide in a qualitative way the general nature of the symmetry-violating mass spectrum, and they are compared with the zeroth-order mass relations in the conventional Higgs models. The last section shows that consistent asymptotic solutions to the Bethe-Salpeter equation can be found in the weak-coupling limit. Such solutions exist for pure Yang-Mills Lagrangians, but when fermions are included there arise group-theoretical constraints. For SU(N) no solution is found when the fermions are in the fundamental representation, and also there is no solution for a non-Abelian chiral theory.

After completing this investigation we learned of the results of two related works. In one,[19] Cornwall has used a new approach consisting of a nonlocal nonpolynomial Lagrangian. With the assumption of asymptotic freedom he obtains results similar to those we have in Sec. V. The other work,[20] by Poggio, Tomboulis, and Tye, uses conventional Lagrangians and Ward identities to discuss symmetry violation, but attacks the problem of finding a consistent solution for the bound-state vertices with the assumption of asymptotic freedom and with the operator-product expansion.

II. DYNAMICAL SYMMETRY BREAKING IN PURE YANG-MILLS THEORIES

In previous investigations of dynamical symmetry breaking, a symmetry breakdown to generate finite masses for vector mesons was accomplished either by the Higgs mechanism[1-3] (that is, by using scalar particles) or by fermion-antifermion bound states.[13,14] However, in the massless Yang-Mills theory there is the possibility of generating finite masses for vector mesons by means of their self-interactions. Here we wish to investigate the consistency of such a scheme. We consider the following Lagrangian density for non-Abelian gauge vector bosons:

$$\mathcal{L}(x) = -\tfrac{1}{4}F_\alpha{}^{\mu\nu}(x)F_{\mu\nu\alpha}(x), \tag{2.1}$$

where

$$F_\alpha{}^{\mu\nu}(x) = \partial^\mu A^\nu_\alpha(x) - \partial^\nu A^\mu_\alpha(x) + gC_{\alpha\beta\gamma}A^\mu_\beta(x)A^\nu_\gamma(x).$$

The vector mesons transform according to the adjoint representation of a compact simple Lie Group G whose infinitesimal generators $\{T^\alpha\}$ satisfy the commutation relations

$$-i[T^\alpha, T^\beta] = C^{\alpha\beta\gamma}T^\gamma. \tag{2.2}$$

Under local infinitesimal gauge transformations the vector fields transform as

$$\delta A^\mu_\alpha(x) = C_{\alpha\beta\gamma}A^\mu_\gamma(x)\epsilon_\beta(x) + \frac{1}{g}\partial^\mu\epsilon_\alpha(x), \tag{2.3}$$

and it is easy to verify that the Lagrangian is invariant under this transformation. The conserved current associated with the gauge invariance of $\mathcal{L}(x)$ is given by

$$J_\alpha{}^\mu(x) \equiv \partial_\nu F_\alpha{}^{\mu\nu}(x). \tag{2.4}$$

The classical equations of motion

$$\partial_\nu F_\alpha^{\ \mu\nu}(x) = gC_{\alpha\beta\gamma}A^\nu_\beta F_\gamma^{\ \mu}{}_\nu(x) \tag{2.5}$$

show that this current is the current associated with Noether's theorem. We note, however, that conservation of the gauge current, Eq. (2.4), is mandatory as a consequence of the antisymmetry of $F_\alpha^{\ \mu\nu}(x)$.

In order to obtain a sensible quantum theory for gauge mesons interacting according to Eq. (2.1), we employ the method of Faddeev and Popov.[19] A well-known feature of this quantization procedure is the appearance of the Faddeev-Popov "ghost" particles in covariant gauges, that is, scalar particles obeying Fermi statistics. Also, we add to the Lagrangian density a gauge-fixing term of the form

$$-\frac{1}{2\alpha}\,[\partial_\mu A_\beta^{\ \mu}(x)]^2.$$

The Faddeev-Popov ghosts are not physical particles and will not appear in *out* states if they are not in *in* states. In perturbation theory in addition to the usual vertices there will be a ghost-ghost-vector bare vertex. Our notation is shown in Fig. 1.

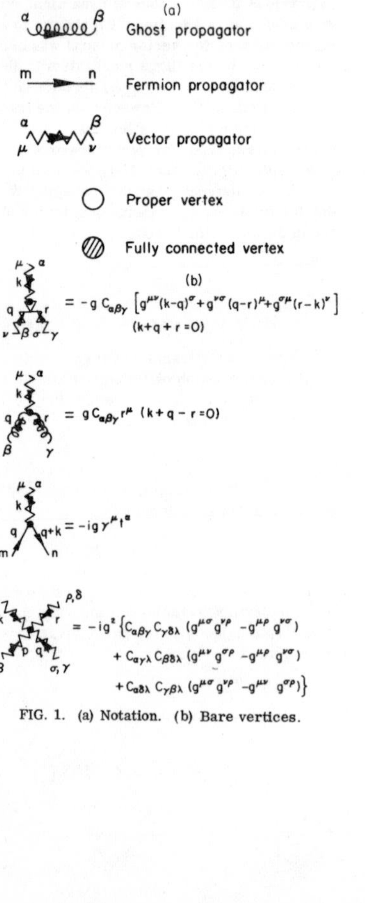

FIG. 1. (a) Notation. (b) Bare vertices.

Moreover, one may use the regulator scheme of Lee and Zinn-Justin[2] or the dimensional method of 't Hooft and Veltman[1] to render the theory free of ultraviolet divergences. We will not discuss the renormalization procedure except when necessary (see Appendix C). The infrared singularities that one would expect from perturbation theory are not present in the n-point functions by the basic hypothesis being made: The vector mesons are massive due to dynamical symmetry breaking.

To discuss dynamical symmetry breaking we will need the Dyson equations and various Ward identities, the derivations of which are left to Appendixes A and B. Also, since the current $J_\alpha^{\ \mu}(x)$ must be conserved, the dynamical symmetry breakdown must entail the appearance of bound-state Goldstone bosons, rather than anomalous current nonconservation.

Consider first the vector-meson propagator $D^{\mu\nu}_{\alpha\beta}(k)$. The bare propagator is given by

$$D^{\mu\nu}_{0\alpha\beta}(k) = -i\left[\left(g^{\mu\nu} - \frac{k^\mu k^\nu}{k^2}\right)\frac{1}{k^2+i\epsilon} + \alpha\frac{k^\mu k^\nu}{(k^2+i\epsilon)^2}\right]\delta_{\alpha\beta}. \tag{2.6}$$

The polarization tensor $\Pi^{\mu\nu}_{\alpha\beta}(k)$ is given in terms of $D^{\mu\nu}_{\alpha\beta}(k)$ and $D^{\mu\nu}_{0\alpha\beta}(k)$ by

$$\Pi^{\mu\nu}_{\alpha\beta}(k) = -i[D^{\mu\nu}_{\alpha\beta}(k)]^{-1} + i[D^{\mu\nu}_{0\alpha\beta}(k)]^{-1}. \tag{2.7}$$

Also the Ward identities [see Eq. (A14)] tell us that

$$\frac{1}{\alpha}k^\mu k^\nu D^{\mu\nu}_{\alpha\beta}(k) = \delta_{\alpha\beta}, \tag{2.8}$$

so that the full propagator may be written

$$D^{\mu\nu}_{\alpha\beta}(k) = -i\left(\frac{g^{\mu\nu} - k^\mu k^\nu/k^2}{k^2+i\epsilon}[\delta_{\alpha\beta} + \Pi_{\alpha\beta}(k^2)]^{-1} + \alpha\frac{k^\mu k^\nu}{(k^2+i\epsilon)^2}\right), \tag{2.9}$$

where

$$\Pi^{\mu\nu}_{\alpha\beta}(k) = \left(g^{\mu\nu} - \frac{k^\mu k^\nu}{k^2}\right)k^2\Pi_{\alpha\beta}(k^2),$$

since the polarization tensor is transverse.

There is a pole in the transverse part of the

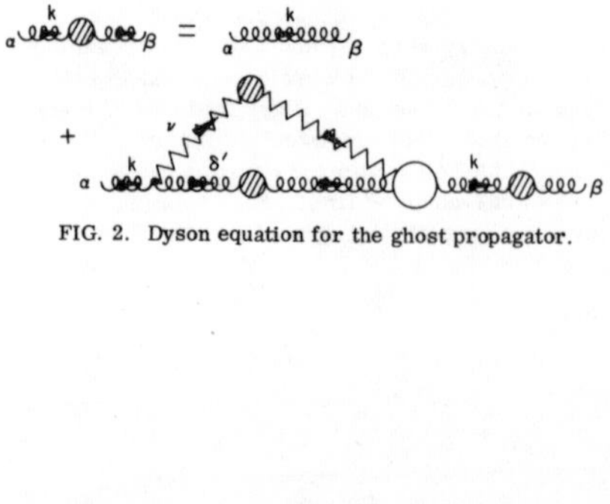

FIG. 2. Dyson equation for the ghost propagator.

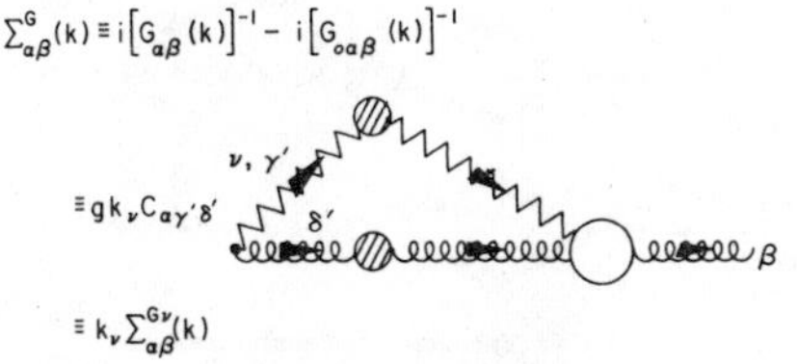

FIG. 3. Ghost self-energy.

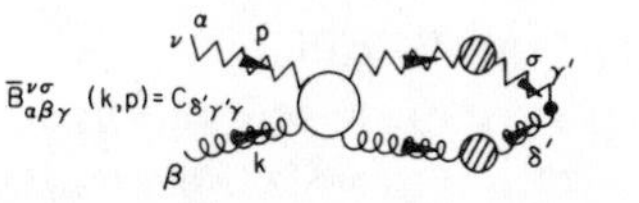

FIG. 4. $\bar B$ function for vector Ward identity.

vector-meson propagator at $k^2 = 0$ unless $\Pi_{\alpha\beta}(k^2)$ has a pole at $k^2 = 0$, in which case the vector mesons gain a mass[7] by the Schwinger mechanism.

The bare ghost propagator is given by

$$G_{0\alpha\beta}(k) = \frac{i}{k^2 + i\epsilon}\,\delta_{\alpha\beta}\,. \tag{2.10}$$

The full ghost propagator satisfies a Dyson equation (Fig. 16), shown diagrammatically as in Fig. 2. Using the equation in Fig. 2 we see that the ghost self-energy may be written as in Fig. 3. Using Lorentz invariance we may write $\Sigma^{G\nu}_{\alpha\beta}(k) = k^\nu B_{\alpha\beta}(k^2)$, where $B_{\alpha\beta}(k^2)$ will be regular in renormalized perturbation theory. Thus the full ghost propagator $G_{\alpha\beta}(k)$ may be written

$$G_{\alpha\beta}(k) = \frac{i}{k^2 + i\epsilon}\,[\delta_{\alpha\beta} + B_{\alpha\beta}(k^2)]^{-1}\,, \tag{2.11}$$

and we see that for the ghost propagator to acquire a pole other than at $k^2 = 0$, $B_{\alpha\beta}(k^2)$ must develop a nonperturbative pole. Since, as we shall see, such a pole is not necessary for dynamical symmetry breaking we will assume that $B_{\alpha\beta}(k^2)$ remains regular at $k^2 = 0$ and so the ghosts remain massless. Also, such a pole would require scalar fermion Goldstone excitations, a situation which clearly is not envisioned for the Goldstone mesons.

To see how the Goldstone bosons must appear we consider the Ward identity for the proper vector-vector-vector vertex, denoted $T^{\mu\nu\sigma}_{\alpha\beta\gamma}(k, p, -k-p)$. This Ward identity, derived in Appendix A, is

$$k_\mu T^{\mu\nu\sigma}_{\alpha\beta\gamma}(k, p, -k-p)[1-B(k^2)] = -g[g^{\nu\sigma'}C_{\alpha\beta\gamma'} - \bar B^{\nu\sigma'}_{\alpha\beta\gamma'}(k,p)][\delta_{\gamma'\gamma} + \Pi_{\gamma'\gamma}(k+p)][(p+k)^2 g^{\sigma'\sigma} - (p+k)^{\sigma'}(p+k)^{\sigma}]$$
$$+ g[g^{\sigma\nu'}C_{\alpha\beta'\gamma} - \bar B^{\sigma\nu'}_{\alpha\beta'\gamma}(k,p)][\delta_{\beta'\beta} + \Pi_{\beta'\beta}(p^2)](p^2 g^{\nu'\nu} - p^{\nu'}p^{\nu})\,, \tag{2.12}$$

where $\delta_{\alpha\beta}B(k^2) = B_{\alpha\beta}(k^2)$ and $\bar B^{\nu\sigma}_{\alpha\beta\gamma}(k,p)$ is defined as in Fig. 4.

Let the momentum $k \to 0$ in Eq. (2.12). Since $B(k^2)$ is regular and

$$G_{\alpha\beta}(k^2) = \frac{\delta_{\alpha\beta}}{k^2}\,\frac{1}{1 + B(k^2)}\,,$$

$B(0)$ may be absorbed into the renormalization of the ghost propagator. Furthermore, the vanishing of the triple divergence of the vector-vector-vector vertex [as can be seen from Eq. (2.12)] implies that $k_\mu T^{\mu\nu\sigma}_{\alpha\beta\gamma}(k,p)$ is transverse in p as $k \to 0$. Finally, we use the Lorentz decomposition of $\bar B^{\mu\nu}_{\alpha\beta\gamma}(k,p)$ as $k \to 0$:

$$\bar B^{\mu\nu}_{\alpha\beta\gamma}(k,p) \underset{k\to 0}{\longrightarrow} (p^2 g^{\mu\nu} - p^\mu p^\nu) b_{\alpha\beta\gamma}(p^2) + \frac{p^\mu p^\nu}{p^2}\, b'_{\alpha\beta\gamma}(p^2)\,. \tag{2.13}$$

Note from the definition of $\bar B^{\mu\nu}_{\alpha\beta\gamma}$, $\bar B^{\mu\nu}_{\alpha\beta\gamma}(k,p)(k+p)_\nu$ differs from the ghost-vector-ghost proper vertex, denoted by $\tilde T^{\mu}_{\alpha\beta\gamma}(k,p)$, only by the bare ghost-vector-ghost vertex. Also, $b_{\alpha\beta\gamma}(p^2)$ is proportional to $C_{\alpha\beta\gamma}$ in the absence of symmetry violation, and we assume that the antisymmetry in β and γ remains valid even when the symmetry is broken.[21] Finally, we obtain as the $k \to 0$ limit of Eq. (2.12)

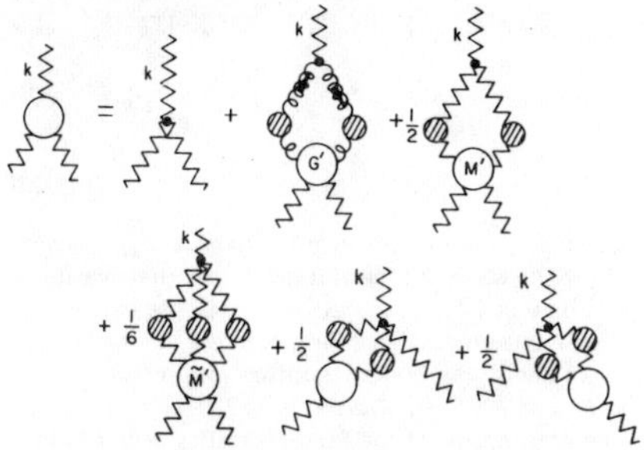

FIG. 5. Integral equation for vector vertex.

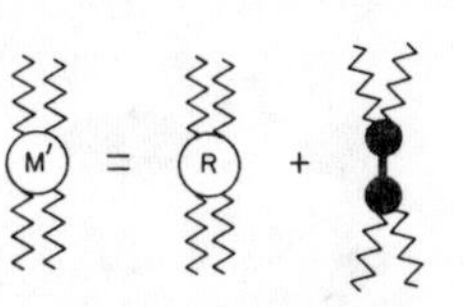

FIG. 6. Pole and regular parts of M'.

$$\lim_{k\to 0}\left[k_\mu T^{\mu\nu\sigma}_{\alpha\beta\gamma}(k,p)\frac{1}{k^2G(k^2)}\right]$$
$$=(p^2g^{\nu\sigma}-p^\nu p^\sigma)[gT^\alpha-b^\alpha(p^2),\Pi(p^2)]_{\beta\gamma}, \qquad (2.14)$$

where $[T^\alpha]_{\beta\gamma}=C_{\beta\alpha\gamma}$ and $[b^\alpha(p^2)]_{\beta\gamma}=b_{\alpha\beta\gamma}(p^2)$. Clearly if $\Pi_{\alpha\beta}(p^2)$ is symmetric under the group, then $T^{\mu\nu\sigma}_{\alpha\beta\gamma}$ is regular as $k\to 0$, and we have a trivial identity.

The basic assumption we make is that $\Pi_{\alpha\beta}(p^2)$ has a symmetry-violating part, and therefore that $T^{\mu\nu\sigma}_{\alpha\beta\gamma}$ has a pole $k\to 0$. Furthermore, we may relate the vertex function to various proper four-point and five-point functions; see Fig. 5. The prime indicates that these amplitudes are one-vector-meson irreducible in the k^2 channel. The pole in $T^{\mu\nu\sigma}_{\alpha\beta\gamma}$ as $k\to 0$ is attributed to poles in M', $\tilde{M}'$, and F'. That is, there are massless Goldstone excitations in these amplitudes and the residues of these poles factorize. Thus the four-vector-meson amplitude M' may be written as in Fig. 6. Similar expressions may be obtained for $\tilde{M}'$ and G'. The R subscript indicates that the function is regular as $k^2\to 0$, and $a(1\le a\le m)$ is summed over all Goldstone poles. Later we will find that the number of Goldstone excitations is equal to the number of massive vector mesons. The pole part of $T^{\mu\nu\sigma}_{\alpha\beta\gamma}(k,p,-p-k)$ is given by

$$T^{\mu\nu\sigma}_{\alpha\beta\gamma}(k,p,-k-p)=\sum_a k^\mu\lambda_{\alpha a}(k^2)\frac{g}{k^2}P^{\nu\sigma}_{V\beta\gamma a}(p,-k-p), \qquad (2.15)$$

where $k^\mu\lambda_{\alpha a}(k^2)$ can be written diagrammatically as in Fig. 7. We note that Eq. (2.15) implies that the pole in $T^{\mu\nu\sigma}_{\alpha\beta\gamma}$ at $k=0$ is also a pole at $k^2=0$. If we define $\lambda_{\alpha a}\equiv\lambda_{\alpha a}(0)$ we may use Eq. (2.15) and the equation in Fig. 7 to obtain

$$\lim_{k\to 0}k_\mu T^{\mu\nu\sigma}_{\alpha\beta\gamma}(k,p)=\sum_a g\lambda_{\alpha a}P^{\nu\sigma}_{V\beta\gamma a}(p). \qquad (2.16)$$

$P^{\nu\sigma}_{V\beta\gamma a}(p)$, by Eq. (2.14), is completely transverse, and we may define $P_{T\beta\gamma a}(p^2)$ by

$$P^{\nu\sigma}_{V\beta\gamma a}(p)\equiv\left(g^{\nu\sigma}-\frac{p^\nu p^\sigma}{p^2}\right)P_{T\beta\gamma a}(p^2). \qquad (2.17)$$

Also $P_{T\beta\gamma a}(p^2)$ must be symmetric in β and γ as

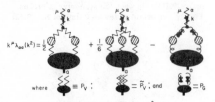

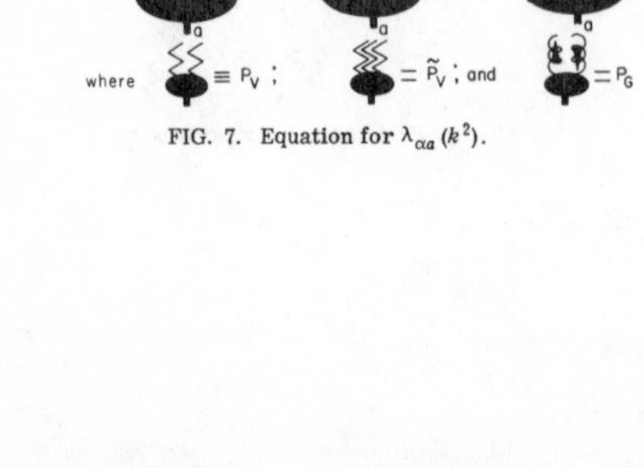

FIG. 7. Equation for $\lambda_{\alpha a}(k^2)$.

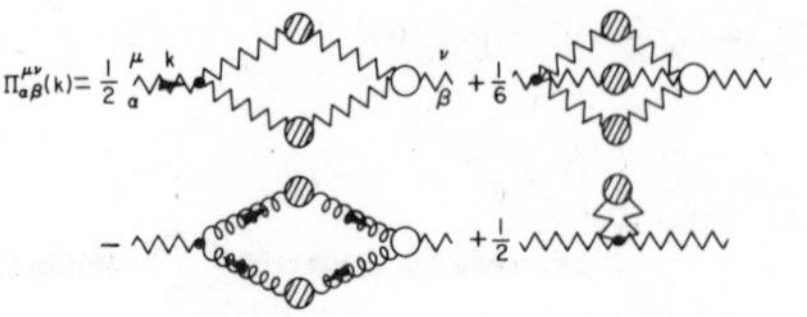

FIG. 8. Vacuum-polarization tensor.

required by the Bose symmetry of the three-vector vertex.

To determine the structure of the pole in the polarization tensor $\Pi_{\alpha\beta}(k^2)$, we consider the Feynman-Dyson equation satisfied by the vacuum-polarization tensor (Fig. 17) represented in Fig. 8. Using the equation in Fig. 7 we have

$$\Pi^{\mu\nu}_{\alpha\beta}(k)|_{\text{pole}}=\sum_a k^\mu k^\nu\lambda_{\alpha a}(k^2)\frac{1}{k^2}\lambda^*_{a\beta}(k^2), \qquad (2.18)$$

or since $\Pi^{\mu\nu}_{\alpha\beta}(k)=(k^2g^{\mu\nu}-k^\mu k^\nu)\Pi_{\alpha\beta}(k^2)$ we may write

$$\Pi^{\mu\nu}_{\alpha\beta}(k^2)|_{\text{pole}}=\sum_a\frac{1}{k^2}\lambda_{\alpha a}\lambda^*_{a\beta}. \qquad (2.19)$$

Therefore $\Pi_{\alpha\beta}(k^2)$ has a pole at $k^2=0$ with residue $\sum_a\lambda_{\alpha a}\lambda^*_{a\beta}$, and the vector mesons acquire a mass $\mu^2_{\alpha\beta}$ which in the pole approximation becomes $\mu^2_{\alpha\beta}=\sum_a\lambda_{\alpha a}\lambda^*_{a\beta}$.

Finally in this section we want to show that the full on-mass-shell Green's functions do not have poles corresponding to the Goldstone bosons. Indeed, these excitations must decouple if the theory is to be consistent.[20] Consider an arbitrary on-mass-shell scattering amplitude $T_{AB}(s=k^2,\ldots)$, and assume that the quantum numbers of the initial state (denoted by A) and the final state (denoted by B) are such that it communicates with the bound states. Then we may write $(s=k^2)$

$$T_{AB}(s=k^2,\ldots)=T^\mu_{A\alpha}(k,\ldots)D^{\alpha\beta}_{\mu\nu}(k)T^{*\nu}_{\beta B}(k,\ldots)$$
$$+\sum_a gP_{Aa}(\ldots)\frac{g}{k^2}P^*_{Ba}(\ldots)$$
$$+\text{reg}, \qquad (2.20)$$

where reg denotes the regular part of T_{AB} near $k^2=0$. Next we make the following two observations: (1) Since in $T^\mu_{A\alpha}(k,\ldots)$ all the particles except the vector meson carrying momentum k are on the mass shell, the Ward identities imply that $k_\mu T^\mu_{A\alpha}(k,\ldots)=0$. Similarly $k_\nu T^{*\nu}_{\beta B}(k,\ldots)=0$. (2) The assumption of the factorizability of the Goldstone poles and Lorentz covariance implies that

$$T^{\mu}_{A\alpha}(k,\ldots)=\frac{k^{\mu}}{k^2}[g\bar{P}_{A\alpha}(k,\ldots)]_{k=0}+T^{\mu\,\mathrm{reg}}_{A\alpha}(k,\ldots) \qquad (2.21a)$$

and

$$T^{*\nu}_{B\beta}(k,\ldots)=\frac{k^{\nu}}{k^2}[g\bar{P}^{*}_{B\beta}(k,\ldots)]_{k=0}+T^{*\nu\,\mathrm{reg}}_{B\beta}(k,\ldots). \qquad (2.21b)$$

Therefore the first term in Eq. (2.20) may be rewritten as

$$\begin{aligned}T^{\mu}_{A\alpha}(k,\ldots)D^{\alpha\beta}_{\mu\nu}(k)T^{*\nu}_{B\beta}(k,\ldots)&=T^{\mu}_{A\alpha}(k,\ldots)\frac{g^{\mu\nu}}{k^2+i\epsilon}[1+\Pi(k^2)]^{-1}{}_{\alpha\beta}T^{*\nu}_{B\beta}(k,\ldots)\\&=-\bar{P}_{A\alpha}(0,\ldots)\frac{k^{\mu}}{k^2}\frac{g^{\mu\nu}}{k^2+i\epsilon}[1+\Pi(k^2)]^{-1}{}_{\alpha\beta}\bar{P}^{*}_{B\beta}(0,\ldots)\frac{k^{\nu}}{k^2}g^2\\&\quad+T^{\mu\,\mathrm{reg}}_{A\alpha}(k,\ldots)\frac{g^{\mu\nu}}{k^2+i\epsilon}[1+\Pi(k^2)]^{-1}{}_{\alpha\beta}T^{*\nu\,\mathrm{reg}}_{B\beta}(k,\ldots),\end{aligned} \qquad (2.22)$$

and from the definitions of $\bar{P}_{Aa}$ and $\lambda_{\alpha a}$, we have

$$\bar{P}_{A\alpha}(0,\ldots)=\sum_a P_{Aa}(\ldots)\lambda^{*}_{a\alpha}. \qquad (2.23)$$

Therefore we may write

$$\begin{aligned}T_{AB}(k^2,\ldots)=&\sum_{a=1}^{m}gP_{Aa}(\ldots)\frac{g}{k^2}P^{*}_{Ba}(\ldots)-\sum_{a,b=1}^{m}gP_{Aa}(\ldots)\lambda^{*}_{a\alpha}\left(\frac{1}{k^2}\right)^2[1+\Pi(k^2)]^{-1}{}_{\alpha\beta}\lambda_{\beta b}gP^{*}_{Bb}(\ldots)\\&+T^{\mu\,\mathrm{reg}}_{A\alpha}(k,\ldots)\frac{g^{\mu\nu}}{k^2+i\epsilon}[1+\Pi(k^2)]^{-1}{}_{\alpha\beta}T^{*\nu\,\mathrm{reg}}_{B\beta}(k,\ldots)+\mathrm{reg}.\end{aligned} \qquad (2.24)$$

Now, as we have seen, the Goldstone bosons give rise to a pole contribution to $\Pi_{\alpha\beta}(k^2)$ so that

$$\Pi_{\alpha\beta}(k^2)=-\sum_a\frac{\lambda_{\alpha a}\lambda^{*}_{a\beta}}{k^2}+\Pi^{\mathrm{reg}}_{\alpha\beta}(k^2), \qquad (2.25)$$

and since $\Pi_{\alpha\beta}$ is Hermitian, there is no loss of generality in writing

$$\Pi_{\alpha\beta}(k^2)=-\delta_{\alpha\beta}\frac{\mu_{\beta}^2}{k^2}+\Pi^{\mathrm{reg}}_{\alpha\beta}(k^2). \qquad (2.26)$$

Therefore we have

$$\sum_{a=1}^{m}\lambda_{\alpha a}\lambda^{*}_{a\beta}=\mu_{\alpha}{}^2\delta_{\alpha\beta}. \qquad (2.27)$$

The number of vector mesons is equal to the dimension of the group $d(G)$. If we suppose that N vector mesons get masses and the rest remain massless we have that

$$\left[\frac{1}{k^2[1+\Pi(k^2)]}\right]_{\alpha\beta}=\begin{cases}\dfrac{1}{k^2}\delta_{\alpha\beta}, & \alpha>N \text{ (massless vector meson)}\\[2mm] \dfrac{1}{\mu_{\alpha}{}^2}\delta_{\alpha\beta}, & \alpha\leq N \text{ (massive vector meson)}.\end{cases} \quad \text{near } k^2=0 \qquad (2.28)$$

Note that condition (2.27) requires $m\geq N$.

The cancellation of the Goldstone poles in Eq. (2.24) requires as necessary and sufficient conditions

$$\frac{1}{k^2}\delta_{ab}=\frac{1}{k^2}\sum_{\alpha=1}^{N}\lambda^{*}_{a\alpha}\frac{1}{\mu_{\alpha}{}^2}\lambda_{\alpha b}, \qquad (2.29a)$$

$$0=\sum_{\alpha=N+1}^{d(G)}\lambda^{*}_{a\alpha}\lambda_{\alpha b}. \qquad (2.29b)$$

In particular, we have $N\geq m$, and therefore $N=m$; that is, the number of Goldstone bosons equals the number of massive vector mesons. Also, $\lambda_{a\alpha}=0$, $N+1\leq\alpha\leq d$. Under these conditions the amplitude T_{AB} becomes

$$T_{AB}(k^2,\ldots)=\sum_{\alpha=N+1}^{d(G)}T^{\mu\,\mathrm{reg}}_{A\alpha}(k,\ldots)\frac{g^{\mu\nu}}{k^2+i\epsilon}\times\delta_{\alpha\beta}T^{*\nu\,\mathrm{reg}}_{B\beta}(k,\ldots), \qquad (2.30)$$

so that the only poles remaining at $k^2=0$ are those associated with physical vector-meson poles. We will find more constraints on the symmetry-breaking masses in Sec. IV.

III. SYMMETRY BREAKING WITH FERMIONS

Now we want to discuss dynamical symmetry breaking in a theory with both fermions and vector mesons. The Lagrangian density one begins with is

$$\mathcal{L}(x) = -\tfrac{1}{4}F^{\alpha}{}_{\mu\nu}F^{\alpha\mu\nu} - i\bar{\psi}D^{\mu}_{F}\gamma_{\mu}\psi - \bar{\psi}m_0\psi\,, \tag{3.1}$$

where m_0 is the bare fermion mass and where

$$(D^{\mu}_{F}\psi)_n = \partial^{\mu}\psi_n + i(t^{\alpha})_{mn}\psi_m A^{\mu}_{\alpha}g\,, \tag{3.2}$$

$$-i[t^{\alpha}, t^{\beta}] = C^{\alpha\beta\gamma}t_{\gamma}\,. \tag{3.3}$$

Under a gauge transformation the fermion field transforms as follows [the vectors transform as in Eq. (2.3)]:

$$\delta\psi_n = -i(t^{\alpha}_{mn})\psi_m g\,. \tag{3.4}$$

$F^{\mu\nu}_{\alpha}$ and $C^{\alpha\beta\gamma}$ have been defined in Sec. II. t^{α} may depend on γ_5 as well as 1 in spinor space, and invariance under G implies that $[t^{\alpha}, \gamma^0 m_0] = 0$. Writing t^{α} as $t^{L}_{\alpha} + t^{R}_{\alpha}$, where $t^{R,L}_{\alpha} = \tfrac{1}{2}(1\pm\gamma_5)t_{\alpha}$, we have that $m_0 = 0$ unless $D^{*}_{L}\otimes D_R$ and $D^{*}_{R}\otimes D_L$ have the identity as an irreducible representation. This

FIG. 9. Dyson equation for fermion self-energy.

Lagrangian density must be modified in order to have a sensible quantum theory, and we use the Faddeev-Popov technique as in Sec. II.

First, since we will use a method of analysis here similar to that of Sec. II, with minor changes, we will not repeat all of the technical complications here. The new ingredient is the fact that symmetry violation will occur both in the fermion self-energy $\Sigma_{mn}(p)$ and in the polarization tensor $\Pi^{\mu\nu}_{\alpha\beta}(p)$. The Dyson equations for $\Sigma_{mn}(k)$ and $\Pi^{\mu\nu}_{\alpha\beta}(k)$ are shown in Figs. 9 and 10.

Also, we need the Ward identities for the vector-vector-vector vertex and for the vector-fermion-fermion vertex. The former is given by Eq. (2.12) and the latter is (see Appendix A)

$$k_{\mu}\Gamma^{\mu}_{m,n;\alpha}(k,p)[1+B(k^2)] = [gt^{\alpha}_{mp} - \bar{B}^{\alpha}_{mp}(k,p)]S_{pn}{}^{-1}(p+k) - S_{mp}{}^{-1}(p)\gamma^0[gt^{\alpha}_{pn} - \bar{B}^{\alpha}_{pn}(k,p)]\gamma^0\,, \tag{3.5}$$

where $S_{mn}(p)$, $\Gamma^{\mu}_{m,n;\alpha}(k,p)$, and $\bar{B}^{rm}_{\alpha}(k,p)$ are as shown in Fig. 11. Now we take the limit $k\to 0$ in these equations to obtain

$$\lim_{k\to 0}[k_{\mu}\Gamma^{\mu}_{m,n;\alpha}(k,p)][1+B(0)] = \Big[(gt^{\alpha} + \bar{b}^{\alpha}(p))\Sigma(p)\Big]_{mn} - \Big[\Sigma(p)\gamma^0(gt^{\alpha} + \bar{b}^{\alpha}(p))\gamma^0\Big]_{mn}\,, \tag{3.6a}$$

$$\lim_{k\to 0}[k_{\mu}T^{\mu\nu\lambda}_{\alpha\beta\gamma}(k,p)][1+B(0)] = [gT^{\alpha} + \bar{b}^{\alpha}(p), \Pi(p)]_{\beta\gamma}(g^{\nu\lambda} - p^{\nu}p^{\lambda}/p^2)\,. \tag{3.6b}$$

Suppose $\Sigma_{mn}(p)$ has a symmetry-violating part. Then the left-hand side of Eq. (3.6a) is nonzero and therefore $\Gamma^{\mu}_{m,n;\alpha}(k,p)$ has a pole at $k=0$. Attributing this pole to poles in the relevant one-vector-meson irreducible four- and five-point functions, as in Sec. II, leads to the conclusion that $\Gamma^{\mu}_{m,n;\alpha}(k,p)$ has a pole at $k^2=0$. Here the matrix $\lambda_{\alpha a}(k^2)$ which describes the coupling of the vector meson to the Goldstone boson is given in Fig. 12. $P_{a\,mn}(k,p)$ is the vertex for a fermion, antifermion, and Goldstone boson. The pole part of $\Gamma^{\mu}_{m,n;\alpha}(k,p)$ may be written as

$$\Gamma^{\mu}_{m,n;\alpha}(k,p)\big|_{\text{pole}} = \frac{k^{\mu}}{k^2}\lambda_{\alpha a}(k)\,gP_{a\,mn}(k,p)\,. \tag{3.7}$$

Although the Goldstone poles will shift the masses of the fermions, no pole will develop in $\Sigma_{mn}(p)$, since the Goldstone excitations are scalars. These properties of $\Sigma_{mn}(p)$ are immediately obvious from the Dyson equation for $\Sigma_{mn}(p)$ in Fig. 9. Furthermore, unless there is a superselection rule, the Goldstone poles must also appear on the left-hand side of Eq. (3.6b), implying that $\Pi_{\alpha\beta}(p^2)$ has a symmetry-violating part, and then the Dyson equation for the polarization tensor (see Fig. 10) implies that $T^{\mu\nu\lambda}_{\alpha\beta\gamma}(k,p)$ has a pole at $k^2=0$, giving masses to the vector mesons. As in Sec. II, the pole approximation to the masses is $\mu^2_{\alpha\beta} = \sum_a \lambda_{\alpha a}\lambda^{*}_{a\beta}$, where $\lambda_{\alpha a}$ is given by Fig. 12. Note also that the above argument is reversible in that symmetry violation in $\Pi_{\alpha\beta}(p^2)$ implies symmetry violation in $\Sigma_{mn}(p)$. Finally we observe that the cancellation of the Goldstone poles from on-mass-shell S-matrix elements proceeds as in Sec. II, and we obtain the obvious result that there is one

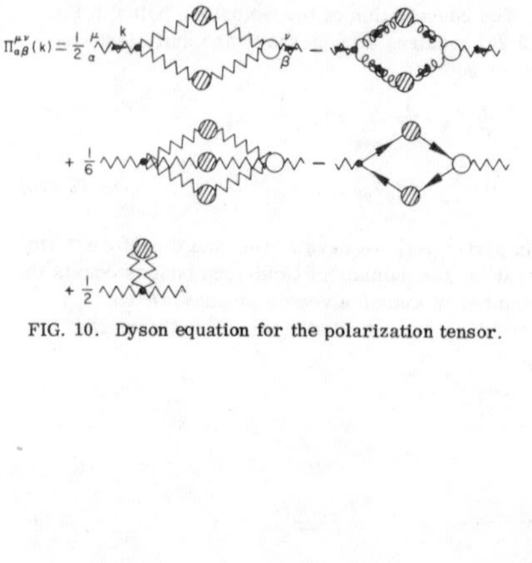

FIG. 10. Dyson equation for the polarization tensor.

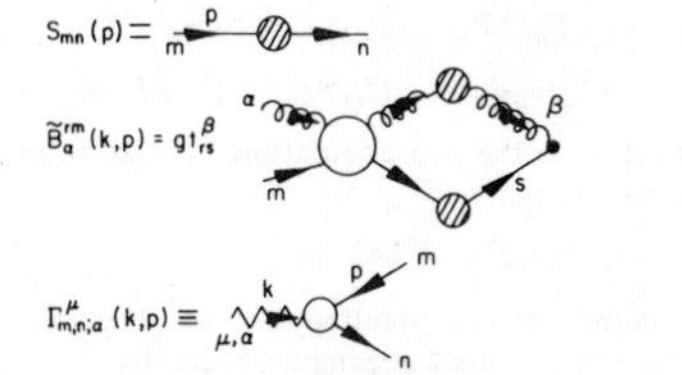

FIG. 11. Notation for fermion vector Ward identity.

and only one Goldstone excitation for each massive vector meson.

IV. MASS RELATIONS

In this section we investigate the constraints on the form of symmetry breaking masses implied by the Ward identities. It is hoped that the approximations made here can be used to develop a platform approximation of the exact theory in the sense of Cornwall and Norton.[14] Within these approximations it is desirable that the masses approach a nonzero limit as the coupling constants go to zero. In this case, the existence of the interaction, no matter how small, changes the limiting theory. Such a process actually occurs in the theory of superconductivity, where the electron-phonon interaction produces an effective electron-electron interaction which is responsible for the energy gap at the Fermi surface. The existence of this gap does not depend on the strength of the electron-phonon interaction. From the Bethe-Salpeter approximation, to be considered in the next section, we will see that it is consistent that $\mu_\alpha^2(g^2) \to \mu_\alpha^2(0) \neq 0$ as $g^2 \to 0$.

In the spirit of this approximation we will drop B, $\overline{B}$, and $\tilde{B}$ terms because they are higher order in g (see Sec. V). The Ward identity, Eq. (2.12), then becomes

$$\lambda_{\alpha a}(0) P^T_{\beta\gamma a}(p^2) = p^2 C^{\beta\alpha\lambda}\Pi_{\lambda\gamma}(p^2) - p^2\Pi_{\beta\lambda}(p^2) C^{\lambda\alpha\gamma}. \qquad (4.1)$$

The Dyson equations for the polarization tensor can be written

$$\Pi_{\alpha\beta}(k^2) = \sum_{a=1}^{m} \frac{\lambda_{\alpha a}(0)\lambda^*_{a\beta}(0)}{k^2} + \text{reg}, \qquad (4.2)$$

so in the pole approximation the mass matrix is

$$\mu^2_{\alpha\beta} = \sum_{a=1}^{m} \lambda_{\alpha a}(0)\lambda^*_{a\beta}(0). \qquad (4.3)$$

Also we know that the $\lambda_{\alpha a}$ matrices satisfy (see Sec. II) the following:

(1) $m \leq d(G)$;

(2) $\lambda_{a\alpha}(0)\lambda^*_{\alpha b}(0)$ is diagonalizable;

(3) $\lambda_{a\alpha}(0) = 0$ if $\mu_\alpha^2 = 0$.

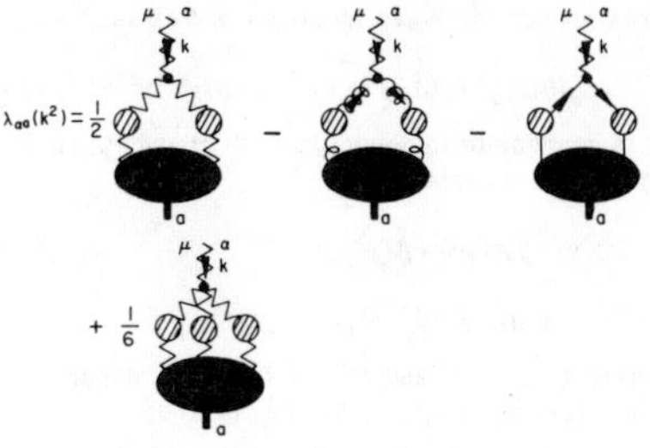

FIG. 12. $\lambda_{\alpha a}(k^2)$ with fermions.

Therefore the Ward identity, Eq. (4.1), becomes

$$\lambda_{\alpha a}(0) P^T_{\sigma\rho a}(p^2) = C^{\alpha\sigma\rho}(\mu_\sigma^2 - \mu_\rho^2) + \text{reg}. \qquad (4.4)$$

In the pole approximation we find that

$$C^{\alpha\sigma\rho}(\mu_\sigma^2 - \mu_\rho^2) = 0, \ \mu_\alpha^2 = 0 \text{ (no sum on } \sigma \text{ or } \rho) \qquad (4.5a)$$

$$(\lambda_{\alpha a})^{-1} C^{\alpha\sigma\rho}(\mu_\sigma^2 - \mu_\rho^2) = P^T_{\sigma\rho a}(p^2), \ \mu_\alpha^2 \neq 0. \qquad (4.5b)$$

A simple example which satisfies these conditions can be given for SU(2). Assume that $\mu_3 = 0$. Then $\mu_1 = \mu_2$ according to Eqs. (4.5).

In general in any simple compact Lie group the generators can be written in the Cartan-Weyl form:

$$[H_i, H_j] = 0, \quad i, j = 1, \ldots, N, \quad N = \text{rank of group}$$

$$[E_\alpha, H_i] = \alpha_i E_\alpha,$$

$$[E_\alpha, E_{-\alpha}] = \sum_i N^i_\alpha H_i,$$

$$[E_\alpha, E_\beta] = C^{\alpha+\beta}_{\alpha\,\beta} E_{\alpha+\beta}.$$

Clearly one solution for Eqs. (4.5) is that some (or all) of the vector-meson masses corresponding to the H_i generators are zero. Also, if $\mu_{E_\alpha}{}^2 = 0$, then $\mu_{H_i}{}^2 = 0$ for all H_i such that $[E_\alpha, H_i] \neq 0$, and therefore that $\mu_{E_{-\alpha}}{}^2 = 0$; i.e., there is at least an unbroken SU(2) subgroup. In general if some set of the E_α's has zero mass, then condition (4.5) implies that the smallest subgroup containing this set of E_α's has zero mass. Therefore Eqs. (4.5) imply that the massless vector mesons form a semisimple compact Lie group. These equations will be discussed further in the next section when the masses are calculated in the lowest order Bethe-Salpeter approximation.

When we include fermions there is the additional Ward identity for the fermion-fermion-vector vertex which relates the symmetry breaking masses to the residue of the pole in this vertex [that is, to the form factors $P_{Fa}(p, k)$]. Within our approximation the Ward identity for $k=0$ is

$$\lambda_{\alpha a}(0)P_{Fa}(p,0)=t^{\alpha}\Sigma(p)-\Sigma(p)\gamma^0 t^{\alpha}\gamma^0 . \tag{4.6}$$

It is convenient to decompose $\Sigma(p)$ and $P_{Fa}(p,0)$ into Dirac invariants:

$$\Sigma(p)=\not{p}A(p^2)+B(p^2), \tag{4.7a}$$

$$P_{Fa}(p,0)=\not{p}P^A_{Fa}(p^2)+P^B_{Fa}(p^2), \tag{4.7b}$$

where A, B, P^A, and P^B can have a γ_5 dependence. Therefore we may rewrite Eq. (4.6) as

$$\lambda_{\alpha a}(0)P^A_{Fa}(p^2)=\gamma^0 t^{\alpha}\gamma^0 A(p^2)-A(p^2)\gamma^0 t^{\alpha}\gamma^0, \tag{4.8a}$$

$$\lambda_{\alpha a}(0)P^B_{Fa}(p^2)=t^{\alpha}B(p^2)-B(p^2)\gamma^0 t^{\alpha}\gamma^0 . \tag{4.8b}$$

Now we can make the following transformation to diagonalize $\lambda_{\alpha a}(0)$:

$$U_{\alpha\alpha'}\lambda_{\alpha' a'}(0)V^{\dagger}_{a' a}=\mu_{\alpha}\delta_{\alpha a}, \tag{4.9}$$

where U is the unitary transformation which diagonalizes the mass matrix, i.e., $U_{\alpha\alpha'}\lambda_{\alpha' a}\lambda^*_{a\beta'}U^{\dagger}_{\beta' B}=\delta_{\alpha\beta}\mu_{\alpha}{}^2$. V is the unitary transformation which diagonalizes $\lambda_{a\alpha}\lambda^*_{\alpha b}$. Since the columns of $\lambda_{\alpha a}(0)$ are orthogonal vectors by Eqs. (2.29) such a unitary transformation exists. So we may reexpress Eqs. (4.8) in the following way for the massive gauge bosons and fermions:

$$\mu_{\alpha}\delta_{\alpha a}[V_{aa'}P^A_{Fa'}(p^2)]=(U_{\alpha\beta}\gamma^0 t^{\beta}\gamma^0)A(p^2)-A(p^2)(U_{\alpha\beta}\gamma^0 t^{\beta}\gamma^0), \tag{4.10a}$$

$$\mu_{\alpha}\delta_{\alpha a}[V_{aa'}P^B_{Fa'}(p^2)]=(U_{\alpha\beta}t^{\beta})B(p^2)-B(p^2)(U_{\alpha\beta}\gamma^0 t^{\beta}\gamma^0). \tag{4.10b}$$

For the remaining massless gauge bosons, which form a subgroup of the original group, we have

$$0=\gamma^0(U_{\alpha\beta}t^{\beta})\gamma^0 A(p^2)-A(p^2)\gamma^0(U_{\alpha\beta}t^{\beta})\gamma^0, \tag{4.10c}$$

$$0=U_{\alpha\beta}t^{\beta}B(p^2)-B(p^2)\gamma^0(U_{\alpha\beta}t^{\beta})\gamma^0 . \tag{4.10d}$$

Therefore the fermions transform as a representation of the unbroken subgroup associated with the massless gauge bosons. The representation matrices are

$$(U_{\alpha\beta}t^{\beta}) \tag{4.11}$$

for α such that $\mu_{\alpha}=0$.

In order to understand the mass relations (4.10a), (4.10b), and (4.5b) it is useful to compare them to the usual Higgs mechanism for generating symmetry breaking masses. There the Lagrangian is given by

$$\mathcal{L}(x)=-\tfrac{1}{4}F_{\alpha}{}^{\mu\nu}F_{\alpha\mu\nu}-\bar{\psi}(-i\gamma^{\mu}\partial_{\mu}+g\theta^{\alpha}_{F}\gamma^{\mu}A_{\mu\alpha}+m_0)\psi+\tfrac{1}{2}[(\partial_{\mu}\delta_{ij}+ig\theta^{\alpha}_{sij}A_{\mu\alpha})\phi_j]^2+\bar{\psi}\Gamma^i\psi\phi_i-V(\phi),$$

where θ_s is the representation matrix of the scalar fields, so that

$$-i[\theta^{\alpha}_s,\theta^{\beta}_s]=C^{\alpha\beta\gamma}\theta^{\gamma}_s . \tag{4.12a}$$

Γ^i is the Yukawa coupling matrix. The gauge symmetry of the Lagrangian requires

$$-i[\theta^{\alpha}_F,\gamma^0\Gamma^i]=\gamma^0\theta^{\alpha ij}_s\Gamma^j . \tag{4.12b}$$

Symmetry breaking is introduced by giving the scalar fields a nonzero vacuum expectation value: $\langle\phi_i\rangle\equiv v_i$ and the vector-meson mass in lowest order is given by

$$\mu^2_{\alpha\beta}=g^2 v_i\theta^{\alpha}_{sik}\theta^{\beta}_{skj}v_j . \tag{4.13}$$

Comparison with Eq. (4.3) shows we should identify $\lambda_{\alpha i}(0)$ with $g(\theta^{\alpha}_{sij}v_j)$ in the Higgs case. Now from Eq. (4.12a) we have

$$-i(v_h\theta^{\alpha}_{shi})\tfrac{1}{2}(v_i\theta^{\beta}_{slj}\theta^{\gamma}_{slj}+\theta^{\beta}_{sij}\theta^{\gamma}_{slj}v_l)=C^{\alpha\beta\gamma'}v_h\theta^{\gamma'}_{shj}\theta^{\gamma}_{sij}v_i-C^{\alpha\beta'\gamma}v_h\theta^{\beta}_{shj}\theta^{\beta'}_{sij}v_i, \tag{4.14}$$

or identifying $\frac{1}{2}(v_i\theta^{\beta}_{ij}\theta^{\gamma}_{lj}+\theta^{\beta}_{ij}\theta^{\gamma}_{lj}v_l)$ with $P^{\alpha\beta}_{Ti}(0)$ we see Eq. (4.13) is simply Eq. (4.5b). Finally from Eq. (4.12b) we have

$$-i[\theta^{\alpha}_F,\gamma^0\Gamma^i v_i]=\gamma^0\theta^{\alpha ij}_s v_i\Gamma^j, \tag{4.15}$$

and identifying Γ^j with $P^j_F(m_0{}^2)$ and $M-m_0\equiv\Gamma^i v_i$ with

$$\Sigma(p)\Big|_{\not{p}=m_0}$$

we see that Eq. (4.15) is simply Eqs. (4.10a) and (4.10b).

Thus there is a straightforward analogy between the zeroth-order mass relations in the Higgs case and the mass relations in the pole approximation for dynamical symmetry breaking. This behavior is not unexpected since both conditions follow from the Ward identities and current conservation. The general structure of such mass relations following from current-algebra constraints has been investigated by Weinstein.[22]

We have seen in this section that the Ward identities [Eqs. (4.5) and (4.10)] imply that the remaining massless bosons must be associated with an unbroken gauge symmetry of the theory. We also see by Eq. (4.10) that the Ward identities relate gauge boson and symmetry-breaking fermion masses to the Goldstone-boson form factors in a way similar to the Higgs case. In the next section we will attempt to calculate these form factors to obtain different expressions for the masses. There is no analogy for these types of relations in the Higgs case. For example, Γ^i is arbitrary in the Higgs case, but in principle is a calculable quantity here; also, v (in the Higgs case) is determined by independent scalar coupling constants, while in this case it is not arbitrary.

V. WEAK-COUPLING LIMIT

A. General formulation

In the previous sections we have been concerned with formal properties of the symmetry-breaking solutions. Here we want to show that a sensible limit exists in which there are nontrivial solutions for the symmetry-breaking quantities $\lambda_{\alpha a}(0)$, $\Sigma^{\nu}(p)$, and for the form factors for the Goldstone-boson couplings to two vectors $[P^{\mu\nu}_{\nu\alpha\beta}(k,p+k)]$, to a fermion-antifermion pair $[P_F(k,p+k)]$, and to a ghost-antighost pair $[P_{G\alpha\beta}(k,p+k)]$. The approximation scheme consists of calculating to lowest order in the coupling constants (g_i) with the modification that the basic nonperturbative quantities (gauge boson masses, fermion symmetry-breaking masses) as determined by their "pole approximations," are treated as order $(g_i)^0$ in these calculations. Of course, it is a requirement of the approach that the calculations we make be consistent with these basic assumptions. This will require that the masses we calculate do not go to zero as $g_i \to 0$.

In order to calculate the gauge boson masses in the pole approximation we must first determine $\lambda_{\alpha a}(0)$. From Eqs. (3.6) we have the result shown in Fig. 13.

The equation in Fig. 13, as written, suffers from two defects: First, it involves the derivatives of the Goldstone vertices with respect to k^{ν} at $k=0$; second, the equation involves overlapping divergences and is therefore not obviously invariant under renormalization. In Appendix C we show how a renormalized form for λ involving no derivatives of Goldstone vertices may be written. Within our approximation scheme the term which involves the triple vector-Goldstone vertex will not contribute. The other Goldstone vertices in the equation in Fig. 13 appear in the pole parts of the relevant three-point proper vertices; e.g. see Fig. 14. Again the term involving the triple vector-Goldstone vertex does not contribute here. Note that the separation in the equation in Fig. 14 into pole and regular parts is unique only up to terms of order k^2, so that the Goldstone vertices can be determined from the three-point vertices only up to terms of order k^2. However, this will be sufficient for our purposes. From the integral equations satisfied by the various three-point vertices (see Appendix B) it is easy to derive the following set of homogeneous equations for the bound-state vertices:

$$P^{\mu\nu}_{V\alpha\beta a}(p,-k+p)=\int\frac{d^4q}{(2\pi)^4}\Big[\tfrac{1}{2}K^{\mu\nu\sigma\rho}_{VV\alpha\beta\gamma\delta}(p,-k+p,q,-k+q)D^{\sigma\sigma'}_{F\gamma\gamma'}(q)D^{\rho\rho'}_{F\delta\delta'}(q-k)\,P^{\sigma'\rho'}_{V\gamma'\delta'}(q,q-k)$$
$$-K^{\mu\nu}_{VG\alpha\beta\gamma\delta}(p,p-k,q,q-k)G_{F\gamma\gamma'}(q)\,G_{F\delta\delta'}(q-k)P_{G\gamma'\delta'}(q,q-k)$$
$$-K^{\mu\nu}_{VF\alpha\beta}(p,p-k,q,q-k)S_F(q)S_F(q-k)P_F(q,q-k)\Big]\,, \tag{5.1a}$$

$$P_{G\alpha\beta}(p,p-k)=\int\frac{d^4q}{(2\pi)^4}\Big[\tfrac{1}{2}K^{\sigma\rho}_{GV\alpha\beta\gamma\delta}(p,p-k,q,q-k)D^{\sigma\sigma'}_{F\gamma\gamma'}(q)D^{\rho\rho'}_{F\delta\delta'}(q-k)\,P^{\sigma'\rho'}_{V\gamma'\delta'}(q,q-k)$$
$$-K_{GG\alpha\beta\gamma\delta}(p,p-k,q,q-k)G_{F\gamma\gamma'}(q)G_{F\delta'\delta}(q-k)\,P_{G\gamma'\delta'}(q,q-k)$$
$$-K_{GF\alpha\beta}(p,p-k,q,q-k)S_F(q)P_F(q,q-k)S_F(q-k)\Big]\,, \tag{5.1b}$$

$$P_F(p,p-k)=\int\frac{d^4q}{(2\pi)^4}\Big[\tfrac{1}{2}K^{\sigma\rho}_{GV\alpha\beta\gamma\delta}(p,p-k,q,q-k)D^{\sigma\sigma'}_{F\gamma\gamma'}(q)D^{\rho\rho'}_{F\delta\delta'}(k+q)P^{\sigma'\rho'}_{V\gamma'\delta'}(q,q-k)$$
$$-K_{FG\gamma\delta}(p,p-k,q,q-k)G_{F\gamma\gamma'}(q)G_{F\delta'\delta}(k+q)\,P_{G\gamma'\delta'}(q,q-k)$$
$$-K_{FF}(p,p-k,q,q-k)S_F(q)P_F(q,q-k)S_F(k+q)\Big]\,. \tag{5.1c}$$

This is a set of coupled equations for the Goldstone vertices (or form factors). The triple vector coupling has been omitted since we will use these equations only in the weak-coupling limit.

Our objective now is to find solutions to Eqs. (5.1) at $k=0$ in the weak-coupling limit for large p^2 ($p^2 \gg \mu^2, M^2$). Since the kernels with which we are concerned are not Fredholm kernels it is to be expected that solutions will exist for some range of coupling constants, and that they will depend on the coupling constant explicitly. The integrations will be performed by continuing to Euclidean momenta and then writing the form factors in terms of Lorentz invariants. The angular integration may be performed to reduce integrals over d^4q to integrals over a single variable $(-q^2/-p^2)\equiv x$, where integrals are from $x=0$ to $x=\infty$. So, for example,

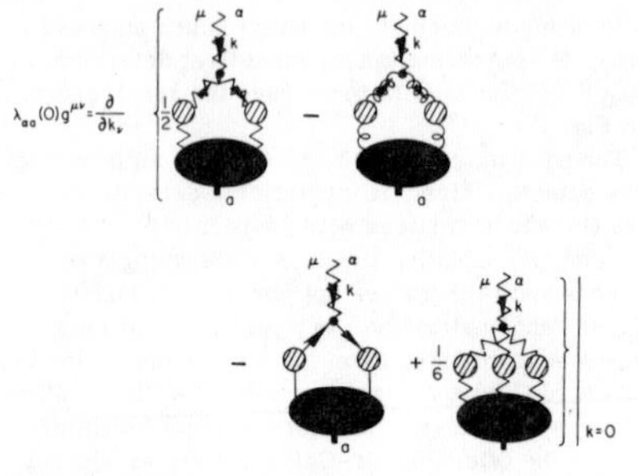

FIG. 13. Vector-meson–Goldstone-boson coupling matrix at $k=0$.

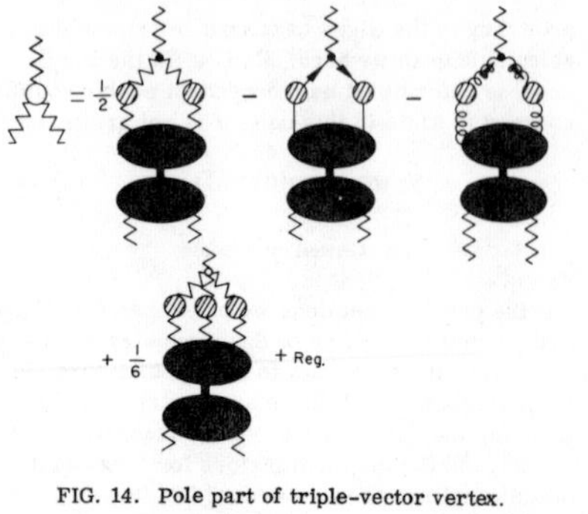

FIG. 14. Pole part of triple-vector vertex.

$$H^{\mu\nu}(p)\tfrac{1}{2}\int\frac{d^4q}{(2\pi)^4}K^{\mu\nu\sigma\rho}_{VV\alpha\beta\gamma\delta}(p,p,q,q)\times D^{\sigma\sigma'}_{\gamma\gamma'}(q)D^{\rho\rho'}_{\delta\delta'}(q)P^{\sigma'\rho'}(q)=3\int_0^\infty\frac{dx}{x}f^{TT}_{\alpha\beta\gamma'\delta'}\left(x,\frac{\mu^2}{p^2},g\right),\tag{5.2}$$

where

$$H^{\mu\nu}(p)\equiv\left(g^{\mu\nu}-\frac{p^\mu p^\nu}{p^2}\right).$$

To any finite order of perturbation theory $f(x,0,g)\sim\ln^n(x)$, where n depends on the order of perturbation theory. We will calculate the functions f to leading order in μ^2/p^2 within our approximation scheme. Since we are interested in the equations only in the weak-coupling limit the following substitutions may be made:

$$D^{\mu\nu}_{\beta\gamma}(q)=\left(g^{\mu\nu}-\frac{q^\mu q^\nu}{q^2}\right)\left(\frac{1}{q^2-\mu^2+i\epsilon}\right)_{\beta\gamma}+\alpha_R\frac{q^\mu q^\nu}{(q^2+i\epsilon)^2}\delta_{\alpha\beta},\tag{5.3}$$

$$S_F(q)=\frac{1}{\not q-M+i\epsilon},\tag{5.4}$$

$$G_F(q)=\frac{\delta_{\alpha\beta}}{q^2+i\epsilon},\tag{5.5}$$

i.e., propagators are given their lowest order form except for the nonperturbative masses $\mu^2{}_{\alpha\beta}$ and M. For the two-particle irreducible kernels we use the order-g_i^2 terms. In order to proceed with the calculation we separate the form factors into invariants:

$$P^{\mu\nu}_{V\alpha\beta a}(p,p)\equiv\tfrac{1}{2}H^{\mu\nu}(p)P^S_{T\{\alpha,\beta\}}(p^2)+\tfrac{1}{2}H^{\mu\nu}(p)P^A_{T[\alpha,\beta]}(p^2),\tag{5.6a}$$

$$P_{G\alpha\beta a}(p,p)\equiv\tfrac{1}{2}P^S_{G\{\alpha,\beta\}a}(p^2)+\tfrac{1}{2}P^A_{G[\alpha,\beta]a}(p^2),\tag{5.6b}$$

$$P_{Fa}(p,p)\equiv\frac{\not p}{p^2}P^A_{Fa}(p^2)+P^B_{Fa}(p^2).\tag{5.6c}$$

In Eq. (5.6c), P^A_{Fa} and P^B_{Fa} may contain γ^5's as well as unit Dirac matrices. Also we know that $P^A_{T[\alpha,\beta]}(p^2)=0$ as a consequence of the Bose symmetry of the gauge boson fields. So we are left with a coupled system of five equations in five unknowns. We will work in the Landau gauge $\alpha_R=0$ for convenience. Continuing to Euclidean space and performing all angular integrations as discussed above, we arrive at the following system of coupled equations:

$$P_i(p^2)=\sum_{j=1}^{5}\int_0^\infty\frac{dx}{x}f_{ij}\left(x,\frac{\mu^2}{p^2},g\right)P_j(xp^2),\tag{5.7}$$

where

$$P_1(p^2)\equiv P^S_{Ta\{\alpha,\beta\}}(p^2),\quad P_2(p^2)\equiv\tfrac{1}{2}P^S_{Ga\{\alpha,\beta\}}(p^2),$$
$$P_3(p^2)\equiv\tfrac{1}{2}P^A_{G[\alpha,\beta]}(p^2),\quad P_4(p^2)\equiv P^A_{Fa}(p^2),\tag{5.8}$$
$$P_5(p^2)\equiv P^B_{Fa}(p^2).$$

Appendix D lists all the lowest order kernels and the various f_{ij}'s.

B. Pure Yang-Mills theories

We consider first the pure Yang-Mills Lagrangians. The asymptotic solutions obtained from the weak-coupling limit of the Bethe-Salpeter equation yield a finite mass matrix for the vector mesons and give an acceptable behavior for $\Pi_{\alpha\beta}(p^2)$ at large p^2 via the Ward identities. $P_1(p^2)$ and $P_2(p^2)$ satisfy a coupled system of equations which has no nontrivial solutions since the region of integration $k^2\ll p^2$ in the integral equations can be as important as the asymptotic region $k^2\gg p^2$ (see the relevant kernels in Appendix D). However, these terms vanish by symmetry in the calculation of $\lambda_{\alpha a}(0)$ and therefore do not contribute to the mass. The Ward identity, Fig. 10, in the weak-coupling limit with $P_1(p^2)=0$ becomes

$$0=[T^\alpha,\Pi(p^2)]_{\beta\gamma}\quad (p^2 \text{ large}). \tag{5.9}$$

Since $p^2\Pi_{\alpha\beta}(p^2)$ is finite at $p^2=0$ and does not commute with T^α if there is global symmetry breaking Eq. (5.9) cannot be valid for small p^2. $P_1(p^2)=0$ is consistent in integral expressions for, e.g., $\lambda_{\alpha a}(0)$, but in equations such as the Ward identities, a consistent approximation is found by using the asymptotic solutions only for $p^2>m^2$ (m^2 is some relevant mass scale). Although $P_1(p^2)=0$ for large p^2 there is a nonzero asymptotic coupling of the vector mesons to the Goldstone boson, a coupling which contributes to the vector-meson masses. These terms arise from the fact that the pole parts of bound-state vertices may be determined to parts linear in the Goldstone boson momentum. Thus we write (with $P_1=0$)

$$P^{\mu\nu}_{V\alpha\beta}\left(p+\frac{k}{2},p-\frac{k}{2}\right)=\left(g^{\mu\nu}-\frac{p^\mu p^\nu}{p^2}\right)p\cdot kP'_{T[\alpha,\beta]}(p^2)+\frac{p^\mu p^\nu}{p^2}p\cdot kP'_{L[\alpha,\beta]}(p^2)+(p^\mu k^\nu+p^\nu k^\mu)P'_{S[\alpha,\beta]}(p^2)+O(k^2). \tag{5.10}$$

Also, the ghost-ghost-Goldstone vertex may be expanded similarly about $k=0$. The Bethe-Salpeter equation for $P'_{T[\alpha,\beta]}(p^2)$ is [see Eqs. (D2h), (D2i), and (D2j)]

$$\begin{aligned}3P'_{T[\alpha,\beta]}\left(\frac{p^2}{m^2}\right)=\frac{g^2}{16\pi^2}(C_{\alpha\beta\lambda}C_{\gamma\delta\lambda})\tfrac{1}{2}\int_0^\infty\frac{dx}{x}\Bigg\{&\left[\left(3+\frac{25}{4x}-\frac{3}{4x^2}\right)\theta(x-1)+(3+\tfrac{25}{4}x-\tfrac{3}{4}x^2)x\theta(1-x)\right]P'_{T[\gamma,\delta]}\left(\frac{xp^2}{m^2}\right)\\&+\left[\left(\frac{1}{2}-\frac{1}{4x}\right)\theta(x-1)+\left(\frac{1}{2}-\frac{1}{4}x\right)x\theta(1-x)\right]P'_{G[\gamma,\delta]}\left(\frac{xp^2}{m^2}\right)\\&+\frac{1}{xp^2}\left[\left(\frac{1}{4}-\frac{1}{4x}\right)\theta(x-1)-(\tfrac{1}{4}-\tfrac{1}{4}x)x\theta(1-x)\right]P_{G[\gamma,\delta]}\left(\frac{xp^2}{m^2}\right)\Bigg\}.\end{aligned} \tag{5.11}$$

$P'_{L[\alpha,\beta]}(p^2)$ and $P'_{S[\alpha,\beta]}(p^2)$, determined in terms of $P_{G[\alpha,\beta]}(p^2)$, $P_{G'[\alpha,\beta]}(p^2)$, and $P'_{T[\alpha,\beta]}(p^2)$, can be shown to satisfy the following relations which follow from the Ward identity, Fig. 10:

$$\begin{aligned}\lambda_{\alpha a}(0)P'_{Sa[\beta,\gamma]}(p^2)&=\tfrac{1}{2}P'_{La[\beta,\gamma]}(p^2)\lambda_{\alpha a}(0)\\&=\tfrac{1}{4}P'_{Ta[\beta,\gamma]}\lambda_{\alpha a}(0)\quad(\text{for all } p^2)\\&=C_{\alpha\beta\gamma}\Pi(p^2)\quad(p^2 \text{ large}).\end{aligned} \tag{5.12}$$

Equation (5.9) implies that for large p^2 the polarization tensor is symmetric under the global gauge group. Therefore it may be written as $\delta_{\alpha\beta}\Pi(p^2)$, and asymptotically the global symmetry is not violated. Such behavior suggests the possibility of dynamical symmetry breaking of only the local gauge group leading to a common mass for all gauge bosons. Smit[24] has considered such a possibility.

To solve Eq. (5.11) we first must consider $P_3^{\alpha\beta}(p^2)\equiv P_{G[\alpha,\beta]}(p^2)$. $P_3(p^2)$ satisfies the equation [see Eq. (D2d)]

$$P_3^{\alpha\beta}\left(\frac{p^2}{m^2}\right)=\frac{g^2}{16\pi^2}\tfrac{3}{8}C_{\alpha\beta\lambda}C_{\lambda\gamma\delta}\times\int_0^\infty dx\left[\frac{1}{x^2}\theta(x-1)+\theta(1-x)\right]\times P_3^{\gamma\delta}\left(x\frac{p^2}{m^2}\right). \tag{5.13}$$

This equation can be solved by making a Mellin transformation, or, equivalently, by converting to a differential equation given by

$$\frac{d}{dy}\left(y^3\frac{d}{dy}\frac{P_3^{\alpha\beta}(y)}{y}\right)=-\frac{2g^2}{16\pi^2}C_{\alpha\beta\lambda}C_{\lambda\gamma\delta}\tfrac{3}{8}P_3^{\gamma\delta}(y). \tag{5.14}$$

To solve this equation we must know the symmetry structure of $P_3(p^2)$. Writing $P_3^{\alpha\beta}(p^2)=X_0^{\alpha\beta}(p^2/\mu^2)^\lambda$ we have the solutions given by eigenvalues of the following equation:

$$\det\left[\frac{2g_{(\lambda)}^2}{16\pi^2}C_{\alpha\beta\lambda}C_{\lambda\gamma\delta}\tfrac{3}{8}-\lambda(\lambda+2)(\delta^\gamma_\alpha\delta^\delta_\beta-\delta^\delta_\alpha\delta^\gamma_\beta)\right]=0, \tag{5.15}$$

and the eigenvectors $X^{\alpha\beta}$ associated with each eigenvalue are obtained in the standard manner. Since we are concerned with the leading behavior as $p^2\to\infty$ we choose the solution associated with the largest acceptable eigenvalue. For purposes of calculating vector-meson masses we need $C_{\alpha\beta\gamma}P_G^{\beta\gamma}(p^2/m^2)$. This expression may be evaluated easily from Eq. (5.14). We obtain

$$C_{\alpha\beta\gamma}P_{Ga}^{\beta\gamma}(p^2)=A_{\alpha a}\left(\frac{p^2}{m^2}\right)^\lambda,\quad \lambda=1-\tfrac{3}{8}g^2\frac{1}{16\pi^2}C_2(G), \tag{5.16}$$

$$\delta_{\alpha\beta}C_2(G)\equiv C_{\alpha\gamma\delta}C_{\beta\gamma\delta}.$$

To determine $P'_{G\alpha\beta}(p^2/\mu^2)$ and $P'_{T\alpha\beta}(p^2/\mu^2)$ we need one more equation, that satisfied by $P'_{G\alpha\beta}(p^2/\mu^2)$, which is [see Eqs. (D2k) and (D2l)]

$$P'_{G[\alpha,\beta]}\left(\frac{p^2}{m^2}\right)=\frac{g^2}{16\pi^2}\frac{1}{2}C_{\alpha\beta\lambda}C_{\gamma\delta\lambda}\int_0^\infty\frac{dx}{x}\left\{\left[-\frac{1}{2x}\theta(x-1)-\frac{x^2}{2}\theta(1-x)\right]P'_{G[\gamma,\delta]}\left(x\frac{p^2}{m^2}\right)-\left[\left(\frac{3}{4}-\frac{1}{x}\right)\theta(x-1)-\frac{x^2}{4}\theta(1-x)\right]\frac{P_{G[\gamma,\delta]}(xp^2/m^2)}{xp^2}\right\}. \tag{5.17}$$

Using Eq. (5.16) in Eqs. (5.17) and (5.11) we find that

$$C_{\alpha\beta\gamma}P'_{G[\beta,\gamma]}=-\frac{A_{\alpha a}}{m^2}\left(\frac{p^2}{m^2}\right)^{-(3/8)g^2(1/16\pi^2)C_2(G)},$$

$$C_{\alpha\beta\gamma}P'_{T[\beta,\gamma]}=\frac{1}{3}\frac{A_{\alpha a}}{m^2}\left(\frac{p^2}{m^2}\right)^{-(3/8)g^2(1/16\pi^2)C_2(G)} \tag{5.18}$$

Finally we want to determine $\lambda_{\alpha a}(0)$ from Eqs. (5.13) and (5.17). The integrals involved are asymptotically dominated so that the problem of small-p^2 behavior, discussed above, is irrelevant. Using the equation in Fig. 13 with the external momentum routed equally (and oppositely) through the two internal vector lines and entirely through one of the two internal ghost lines (see Appendix C), we obtain

$$\lambda_{\alpha a}(0)g^{\mu\nu}=g^2\int\frac{d^4p}{(2\pi)^4}3\frac{p^\mu p^\nu C_{\alpha\beta\gamma}}{(p^2-\mu_{\beta\beta'}{}^2+i\epsilon)(p^2-\mu_{\gamma\gamma'}{}^2+i\epsilon)}P'_{T\beta'\gamma'}\left(\frac{p^2}{m^2}\right)-g^2\int\frac{d^4p}{(2\pi)^4}\frac{4p^\mu p^\nu C_{\alpha\beta\gamma}}{(p^2)^3}P_{G\beta\gamma a}\left(\frac{p^2}{m^2}\right)-2g^2\int\frac{d^4p}{(2\pi)^4}\frac{p^\mu p^\nu C_{\alpha\beta\gamma}}{(p^2)^2}P'_{G\beta\gamma a}\left(\frac{p^2}{m^2}\right). \tag{5.19a}$$

We may isolate the divergent parts of this equation to obtain

$$(\lambda_{\alpha a}g^{\mu\nu})_{\text{divergent}}=g^2\int\frac{d^4p}{(2\pi)^4}\frac{p^\mu p^\nu}{(p^2)^2}\left[3C_{\alpha\beta\gamma}P'_{T\beta\gamma a}\left(\frac{p^2}{m^2}\right)-2C_{\alpha\beta\gamma}P_{G\beta\gamma}\left(\frac{p^2}{m^2}\right)\frac{1}{p^2}-C_{\alpha\beta\gamma}P'_{G\beta\gamma}\left(\frac{p^2}{m^2}\right)\right]$$
$$=3g^2\int\frac{d^4p}{(2\pi)^4}\frac{p^\mu p^\nu}{(p^2)^2}\left[-\frac{1}{3}\frac{A_{\alpha a}}{m^2}\left(\frac{p^2}{m^2}\right)^{-\epsilon G}+\frac{1}{3}\frac{A_{\alpha a}}{m^2}\left(\frac{p^2}{m^2}\right)^{-\epsilon G}\right]=0. \tag{5.19b}$$

Therefore the divergences cancel and the masses of the vector mesons are finite in the pole approximation. To calculate the finite part explicitly we would have to keep terms to next order in (m^2/q^2). We do not attempt to calculate these nonasymptotic terms here, but simply note that keeping such terms we expect to obtain an expression of the form

$$\lambda_{\alpha a}=a_1\frac{g^2}{16\pi^2}C_2(G)\times\int_{m^2}^\infty dq^2\frac{1}{q^2}\bar{A}_{\alpha\mu}\left(\frac{p^2}{m^2}\right)^{-a_2g^2C_2(G)(1/16\pi^2)}\approx\frac{a_1}{a_2}\bar{A}_{\alpha a}, \tag{5.20}$$

where $\bar{A}_{\alpha a}$ is a constant independent of g. In the Abelian case considered by Cornwall and Norton[14] and by Jackiw and Johnson,[13] $\lambda(0)$ can be found explicitly, but here, on the other hand, part of the leading finite contribution to the mass comes from nonleading parts of the Goldstone couplings.

C. Including fermions

When we include fermions into the Lagrangian, we obtain a larger system of coupled equation for determining consistent asymptotic solutions for the form factors P_T, P_G, and P_F. $P_{Fa}(p,p)$ is given by

$$P_{Fa}(p,p)=(\not{p}/p^2)P_{4a}(p^2)+P_{5a}(p^2). \tag{5.21}$$

The coupled system of equations for P_4 and P_5 is given by Eq. (5.7), with f's given in Appendix D. Since f_{44} and f_{45} are zero to leading order in m^2/p^2 we have $P_{4a}(p^2)=0$ in our approximation scheme. $P_{5a}(p^2)$ satisfies the equation [see Eq. (D2g)]

$$P_{5a}(p^2/m^2)=\frac{3}{16\pi^2}g^2\int_0^\infty\frac{dx}{x}[\theta(x-1)+x\theta(1-x)]\times[\gamma^0\theta^\alpha\gamma^0P_{5a}(xp^2/m^2)\theta_\alpha]. \tag{5.22}$$

The inclusion of fermions also allows the possibility of a nonzero solution for $P_{T\alpha\beta}(p^2)$, which satisfies the equation

$$P_{T\alpha\beta}(p^2)=\frac{g^2}{16\pi^2}\int_1^\infty\frac{dx}{x}\left(2\,\mathrm{tr}[\theta_\alpha P_{5a}(xp^2/m^2)\theta_\beta M]-\mathrm{tr}[\tfrac{1}{2}\{\theta_\alpha,\theta_\beta\}\times\{M,P_{5a}(xp^2/m^2)\}]\right). \tag{5.23}$$

We now consider the solutions of Eqs. (5.22) and (5.23) for $P_{5a}(p^2/m^2)$ in some simple cases.

(i) Vector gauge bosons belonging to a simple Lie group. Now

$$P_{5a}\left(\frac{p^2}{m^2}\right)=\frac{3}{16\pi^2}g_V{}^2\int_0^\infty\frac{dx}{x}[\theta(x-1)+x\theta(1-x)]\times\theta^\alpha P_5\left(x\frac{p^2}{m^2}\right)\theta_\alpha. \tag{5.24}$$

To obtain solutions we convert this equation to the differential equation

$$\frac{d}{dp^2}\left\{(p^2)^2\frac{d}{dp^2}\left[P_{5a}\left(\frac{p^2}{m^2}\right)\right]\right\}=\frac{-3}{16\pi^2}g_V{}^2\theta^\alpha P_5\left(\frac{p^2}{m^2}\right)\theta_\alpha. \tag{5.25}$$

The solutions to this equation are of the form

$$P_{5a}\left(\frac{p^2}{m^2}\right)=N_a\left(\frac{p^2}{m^2}\right)^{-\lambda}, \tag{5.26}$$

where $0<\lambda<1$ in order for the original integral to converge. The eigenvalue condition is

$$-\lambda(-\lambda+1)N_a=\frac{-3}{16\pi^2}g_V{}^2\theta^\alpha N_a\theta_\alpha. \tag{5.27}$$

Thus the existence of a consistent solution requires that

$$\theta^\alpha N_a\theta_\alpha=AN_a,\quad A>0. \tag{5.28}$$

If Eq. (5.28) is satisfied, then

$$\lambda=\frac{3g_V{}^2}{16\pi^2}A. \tag{5.29}$$

To investigate this constraint consider the simple cases of fermions belonging to the fundamental representation of SU(N) or O(N). For SU(N) the representation matrices θ_F^α are given by N^2-1 Hermitian traceless matrices denoted $\lambda^\alpha/2$. From the completeness of the λ^α matrices, it is easy to show

$$\sum_\alpha\frac{\lambda_{ij}^\alpha}{2}(N_a)_{jk}\left(\frac{\lambda^\alpha}{2}\right)_{kl}=-\frac{1}{2N}(N_a)_{il}+\tfrac{1}{2}\delta_{il}\,\mathrm{tr}[N_a]. \tag{5.30}$$

We may take the trace of the mass relation (4.6) to show that $\mathrm{tr}[N_a]$ vanishes, and therefore Eq. (5.29) gives $A=-\frac{1}{2}N$. Similarly in O(N) the constant $A=-\frac{1}{2}$ for fermions in the fundamental representation. Thus for fermions in the fundamental representation of SU(N) or O(N) there is no consistent asymptotic solution for symmetry breaking in the weak-coupling limit.

(ii) Non-Abelian gauge vector bosons and an Abelian axial-vector gauge symmetry. This case is very similar to the first case:

$$P_{5a}\left(\frac{p^2}{m^2}\right)=\frac{3}{16\pi^2}\int_0^\infty\frac{dx}{x}[\theta(x-1)+x\theta(1-x)]\times\left[g_V{}^2\theta^\alpha P_5\left(x\frac{p^2}{m^2}\right)\theta_\alpha-g_A{}^2P_5\left(\frac{p^2}{m^2}\right)\right]. \tag{5.31}$$

Now we have the following conditions for a consistent solution:

$$\theta^\alpha N_a\theta_\alpha=AN_a,\quad A>0\text{ and }Ag_V{}^2-g_A{}^2>0 \tag{5.32}$$

and with these conditions the solution is given by

$$P_{5a}\left(\frac{p^2}{m^2}\right)=N_a\left(\frac{p^2}{m^2}\right)^{-(3/16\pi^2)(Ag_V{}^2-g_A{}^2)}. \tag{5.33}$$

(iii) Chiral SU(N)×SU(M). In the case of non-Abelian chiral groups we do not have a consistent scheme for calculating $P_{5a}(p^2/\mu^2)$ because θ^α is given by

$$\theta_{V+A}^{\alpha_1}\left(\frac{1+\gamma_5}{2}\right)\oplus\theta_{V-A}^{\alpha_2}\left(\frac{1-\gamma_5}{2}\right),$$

and therefore

$$\gamma^0\theta_\alpha\gamma^0P_{5a}\theta^\alpha=\theta_{V+A}^{\alpha_1}\left(\frac{1-\gamma^5}{2}\right)P_{5a}\left(\frac{1+\gamma^5}{2}\right)\theta_{V+A,\alpha_1}+\theta_{V-A}^{\alpha_2}\left(\frac{1+\gamma^5}{2}\right)P_{5a}\left(\frac{1-\gamma^5}{2}\right)\theta_{V-A,\alpha_2}=0. \tag{5.34}$$

The solutions for P_{5a} here are not dominated by the asymptotic behavior. Basically the lack of an asymptotic solution in the weak-coupling limit can be attributed to the Fredholm nature of the fermion-fermion kernels in a chiral theory.[23]

We now calculate the contribution of fermions to $\lambda_a^\alpha(0)$. The calculation is straightforward as in the Abelian case[13,14] if the necessary consistency condition discussed above is satisfied. Consider case (a) above. Then

$$P_{5a}\left(\frac{p^2}{m^2}\right)=N_a\left(\frac{p^2}{m^2}\right)^{-(3g_V{}^2/16\pi^2)A} \tag{5.35}$$

for $A=(\theta^\alpha N_a\theta_\alpha/N_a)>0$. The contribution from the fermions to $\lambda_a^\alpha(0)$ then becomes (Fig. 22)

$$\lambda_a^\alpha(0)|_{\text{fermion}}=4g_{V\,i}{}^2\int_{m^2}^\infty\frac{d^4q}{(2\pi)^4}\mathrm{tr}[\theta^\alpha N_aM]\times\frac{1}{(q^2)^2}\left(\frac{q^2}{m^2}\right)^{-(3g_V{}^2/16\pi^2)A}=-\frac{4}{3A}\mathrm{tr}[\theta^\alpha MN_a]. \tag{5.36}$$

Thus the fermions give a contribution to the vector masses which is finite as $g\to0$ as required for consistency of the weak-coupling scheme.

Finally we note that the assumption made in Sec. V, that the terms B, $\overline{B}$, and $\bar{\bar{B}}$ [defined by

Figs. 3, 4, and Eq. (3.5), respectively] are higher order in g, can be verified within our approximation scheme. Since in perturbation theory B, $\overline{B}$, and $\tilde{B}$ are of order g^2 to lowest order, the only way for these terms to have contributions of order $(g)^0$ is from the contributions of the Goldstone form factors to B, $\overline{B}$, $\tilde{B}$. Such contributions give factors of $1/g^2$ from the loop integrations.

To see that such terms do not arise consider B ($\overline{B}$ and $\tilde{B}$ can be analyzed similarly), which is given by

$$k_\nu B^{\alpha\beta}(k^2) = igC_{\alpha\gamma\delta}\int \frac{d^4q}{(2\pi)^4} D^{\nu\nu'}_{\gamma\gamma'}(q)G_{\delta\delta'}(k-q) \times \tilde{T}^{\nu'}_{\gamma'\delta'\beta}(q,k). \tag{5.37}$$

In our approximation we have

$$k_\nu B^{\alpha\beta}(k^2) = igC_{\alpha\gamma\delta}\int \frac{d^4q}{(2\pi)^4}\frac{g^{\nu\nu'}-q^\nu q^{\nu'}/q^2}{q^2-\mu_{\gamma\gamma'}{}^2+i\epsilon}\frac{1}{(k-q)^2}\left[g\lambda_{\gamma'a}(q^2)\frac{q^{\nu'}}{q^2}P_{G\delta\beta a}(q,k)+\tilde{T}^{\nu'}_{\gamma'\delta\beta}(q,k)_{\text{reg}}\right]. \tag{5.38}$$

The pole contribution vanishes due to the transversality of the vector propagator, and $\tilde{T}^{\nu'}_{\gamma'\delta\beta}(q,k)_{\text{reg}}$ gives contributions at least of order g. Therefore $B^{\alpha\beta}(k^2)$ has no term of zeroth order in g.

VI. COMMENTS AND CONCLUSIONS

The main result of this investigation is that dynamical symmetry breaking by means of bound-state Goldstone bosons is a viable possibility in Yang-Mills theories, with or without the presence of fermions. The global symmetry of the vector mesons is broken, so that symmetry-violating masses are obtained for the fermions and vector mesons. Although explicit computations are not feasible yet, the consistency of this scheme is shown in the lowest order Bethe-Salpeter approximation. There are several techniques which may be used to develop a perturbative expansion in which explicit calculations are possible. One fruitful approach could be to consider a canonical transformation which would exhibit the presence of dynamically broken solutions in a transparent manner, similar to the Bogoliubov transformation (in superconductivity). Clearly such a transformation must be very complicated.

The pattern of symmetry breaking obtained is similar to that found in the Higgs models. For example, the remaining massless gauge bosons are associated with a gauge symmetry that is a subgroup of the original group. However, the weak-coupling limit (Sec. V) fails for chiral groups, and constraints are put on the representation content of the fermions in nonchiral groups. If one does not use the weak-coupling approximation, the above restrictions need not apply. Also, the gauge dependence of our results needs further investigation.

Many of the Lagrangians we have been investigating have the property of asymptotic freedom within the context of the Callan-Symanzik equation and the usual perturbative calculations. However, the essentially nonperturbative nature of our assumption for symmetry-breaking solutions does not allow us easily to establish properties such as asymptotic freedom. In fact we expect that the $g\to 0$ limit of the full theory is not equivalent to the $g=0$ theory, at least in that the gauge bosons gain symmetry-breaking masses which survive in the $g\to 0$ limit. Moreover, within the Higgs mechanism no theory which has suitable infrared behavior has been found which also has asymptotic freedom.[25] Of course the mechanism suggested in this paper differs dynamically from the Higgs mechanism. Whether asymptotic freedom is a true property of the theories considered in this paper deserves further investigation.

ACKNOWLEDGMENTS

We would like to thank E. Poggio, E. Tomboulis, and S.-H. Tye for communicating their results to us prior to publication and for constructive comments on the preliminary version of our manuscript.

APPENDIX A: DERIVATION OF THE WARD IDENTITIES

We derive the Ward identities by using the functional integration techniques of Faddeev and Popov as applied by Lee and Zinn-Justin.[2] Standard notation is followed (see Fig. 1).

It is well known that the generating functional

$$W \equiv W(J^\alpha_\mu, \eta, \overline{\eta})$$

for the Lagrangian density of Eq. (3.1) has infinities associated with the non-Abelian gauge invariance of the Lagrangian. Using the Faddeev-Popov technique one may remove these infinities to obtain a finite generating functional given by

$$W = e^Z = \int d\omega\, d\omega^* \exp\left\{ i\int d^4x\, \omega^*_\alpha(x)\left[-\delta^{\alpha\beta} + ig\partial_\mu C_{\alpha\beta\gamma}\frac{\delta}{\delta J^\gamma_\mu(x)}\right]\omega_\beta(x)\right\} e^{Z_0}, \tag{A1}$$

where

$$e^{Z_0} \equiv \int d\bar\psi\, d\psi\, dA^\mu_\alpha \exp\left\{ i \int d^4x \left[\mathcal{L}(x) - \frac{1}{2\alpha}(\partial_\mu A^\mu_\alpha)^2 + \bar\eta\psi + \bar\psi\eta + J^\mu_\alpha A^\alpha_\mu \right]\right\} . \tag{A2}$$

We may define a ghost propagator in an "external field," $G^{\alpha\beta}(x, y; (\delta/\delta J))$, by

$$\left[-\Box_x \delta^{\alpha\beta} + ig\partial^\mu C_{\alpha\gamma'\beta} \frac{\delta}{\delta J^\mu_{\gamma'}(x)} \right] G^{\beta\gamma}\left(x, y; \frac{\delta}{\delta J}\right) = \delta^{\alpha\gamma}\delta^4(x-y) . \tag{A3}$$

Here $G^{\beta\gamma}(x-y; 0)$ is the free ghost propagator and $e^{-Z_0} G(x, y; (\delta/\delta J)) e^{Z_0}$ is the full ghost propagator of the gauge theory.

e^{Z_0} is invariant under the gauge transformations in Eqs. (2.3) and (3.3). If we make an infinitesimal transformation characterized by $\epsilon_\alpha(x)$, then the terms in δe^{Z_0} linear in $\epsilon_\alpha(x)$ must vanish. Therefore

$$\left[g\left(\bar\eta t^\alpha \frac{\delta}{\delta\bar\eta} - \eta(\gamma^0 t^\alpha \gamma^0)^T \frac{\delta}{\delta\eta} \right) + J^\beta_\mu \left(\delta^{\alpha\beta}\partial^\mu - gC^{\alpha\beta\gamma} \frac{\delta}{\delta J^\gamma_\mu} \right) + \frac{i}{\alpha}[G^{-1}]^{\alpha\beta}\left(\partial_\mu \frac{\delta}{\delta J^\mu_\beta} \right) \right] e^{Z_0} = 0 . \tag{A4}$$

In order to reexpress Eq. (A4) in terms of e^Z we introduce Δ, defined by $e^Z \equiv \Delta e^{Z_0}$ so that

$$\Delta = \int d\omega^* d\omega \exp\left(i \int d^4x\, \omega^*_\alpha G^{-1}{}_{\alpha\beta}\,\omega_\beta \right) = e^{\mathrm{tr}\ln G^{-1}{}_{\alpha\beta}} . \tag{A5}$$

Operating on Eq. (A4) with $\int d^4y\, \Delta G$ and doing some algebra give

$$\left\{ \frac{i}{\alpha}\partial_\mu \frac{\delta}{\delta J_{\alpha\mu}(x)} + g\int d^4y \left[\bar\eta(y) t_\beta \frac{\delta}{\delta\bar\eta(y)} - \eta(y)(\gamma^0 t^\beta \gamma^0)^T \frac{\delta}{\delta\eta(y)} \right] G^{\beta\alpha}\left(y, x; \frac{\delta}{\delta J}\right) + \int d^4y\, J^\gamma_\nu(y) \left[\partial^\nu \delta_{\gamma\beta} - igC_{\gamma\delta\beta} \frac{\delta}{\delta J^\nu_\delta(y)} \right] G^{\beta\alpha}\left(y, x; \frac{\delta}{\delta J}\right) \right\} e^Z = 0 . \tag{A6}$$

Equation (A6) is the generation equation for the Ward identities of the fully connected Green's functions.

It is more useful, however, to obtain the generating functional equation for Ward identities of proper vertices.[26] The generator of proper vertices, Γ, is given by a Legendre transformation of Z:

$$\Gamma(\mathcal{A}; \underline{\psi}, \underline{\bar\psi}) = Z(J, \eta, \bar\eta) - i\int d^4x [J^\mu_\alpha \mathcal{A}^\alpha_\mu + \bar\eta(x)\underline{\psi}(x) + \underline{\bar\psi}(x)\eta(x)] , \tag{A7}$$

where

$$\mathcal{A}^\mu_\alpha = -i\frac{\delta Z}{\delta J^\mu_\alpha} , \quad \underline{\psi} = -i\frac{\delta Z}{\delta\bar\eta} , \quad \underline{\bar\psi} = -i\frac{\delta Z}{\delta\eta} .$$

We also define Γ_0 by

$$\Gamma_0 \equiv \Gamma - \frac{i}{2\alpha}\int d^4y [\partial_\mu \mathcal{A}^\mu_\alpha(y)]^2 . \tag{A8}$$

In terms of these variables, Eq. (A6) becomes

$$\int d^4y \left\{ ig\frac{\delta\Gamma_0}{\delta\underline{\psi}(y)} t_\beta \left(i\underline{\psi}(y) + \frac{\delta}{\delta\bar\eta(y)} \right) - ig\frac{\delta\Gamma_0}{\delta\underline{\bar\psi}(y)} (\gamma^0 t_\beta \gamma^0)^T \left(i\underline{\bar\psi} + \frac{\delta}{\delta\eta} \right) + \frac{\delta\Gamma_0}{\delta\mathcal{A}^\nu_\gamma(y)} \left[i\partial^\nu_y \delta_{\gamma\beta} + gC_{\gamma\delta\beta}\left(i\mathcal{A}^\nu_\delta + \frac{\delta}{\delta J^\delta_\nu} \right) \right] \right\} \times G^{\beta\alpha}\left(y, x; i\mathcal{A}^\alpha_\mu + \frac{\delta}{\delta J^\mu_\alpha} \right) = 0 . \tag{A9}$$

Note that the ghost propagator may be written as

$$G_{\alpha\beta}(x, y; \mathcal{A}^\mu_\alpha) = G_{\alpha\beta}\left(x, y; \frac{\delta}{\delta J^\mu_\alpha} + i\mathcal{A}^\alpha_\mu \right) 1 . \tag{A10}$$

Next we define certain proper vertices in terms of the ghost functional. It is easy to verify that these definitions are correct.

ghost-ghost-vector proper vertex:

$$\mathcal{T}^{\nu\mu}_{\alpha\beta,\gamma}(x, y, z; \mathcal{A}, \underline{\bar\psi}) = -\frac{\delta}{\delta\mathcal{A}^\gamma_\mu(z)} G^{-1}{}_{\alpha\beta} ; \tag{A11a}$$

proper four-point ghost-ghost-fermion-fermion amplitude $H_{\alpha\beta,mn}$:

$$\int d^4\omega H_{\alpha\beta,mn}(s, y; z, \omega)\underline{\psi}_n(\omega) = -\frac{\delta}{\delta\underline{\overline{\psi}}_m(z)} G^{-1}{}_{\alpha\beta}\,, \tag{A11b}$$

$$\int d^4z\,\underline{\overline{\psi}}_m(z) H_{\alpha\beta,mn}(x, y, z, \omega) = -\frac{\delta}{\delta\underline{\psi}_n(\omega)} G^{-1}{}_{\alpha\beta}\,. \tag{A11c}$$

Finally we use the chain rule to reexpress functional derivatives with respect to $(\eta, \bar{\eta}, J)$ to functional derivatives with respect to $(\underline{\overline{\psi}}, \underline{\psi}, \mathcal{A})$. Then multiplying on the right by G^{-1} yields (explicit functional dependence is suppressed)

$$\begin{aligned}
& g\underline{\overline{\psi}}(x)t_\alpha \frac{\delta\Gamma_0}{\delta\underline{\overline{\psi}}(x)} - g\frac{\delta\Gamma_0}{\delta\underline{\psi}(x)}\gamma^0 t_\alpha\gamma^0\underline{\psi} - \left[i\partial^\mu\delta^{\alpha\beta} + igC^{\alpha\delta\beta}\mathcal{A}^\mu_\delta(x)\right]\frac{\delta\Gamma_0}{\delta\mathcal{A}^\mu_\beta(x)} \\
&\quad + ig\int d^4y\,d^4y'\,d^4z\,\tilde{T}^{\nu\mu}_{\alpha\beta}(x, y, z)G_{\beta\beta'}(y, y')\left[-i\frac{\delta\mathcal{A}^\delta_\mu(z)}{\delta J^{\mu'}_{\delta'}(y')}C_{\beta'\delta''\delta'}\frac{\delta\Gamma_0}{\delta\mathcal{A}^{\delta''}_{\mu'}(y')}\right] \\
&\quad + ig\int d^4y\,d^4y'\,d^4z\,d^4\omega\,H_{\alpha\beta,mn}(x, y; z, \omega)G_{\beta\beta'}(y, y')\left[-i\underline{\psi}_m(\omega)\frac{\delta\underline{\overline{\psi}}}{\delta J^\nu_\delta(y')}C_{\gamma\delta\beta'}\frac{\delta\Gamma_0}{\delta\mathcal{A}^\gamma_\mu(y')}\right. \\
&\qquad\qquad\left. - i\underline{\overline{\psi}}_n(n)\frac{\delta\underline{\psi}_m(z)}{\delta J^\nu_\delta(y)}C_{\gamma\delta\beta}\frac{\delta\Gamma_0}{\delta\mathcal{A}^\gamma_\delta(y')}\right] \\
&\quad + ig\int d^4y\,d^4y'\,d^4z\left[G^{\beta\beta'}(y, y')\tilde{T}^{\nu\mu}_{\alpha\beta,\gamma}(x, y, z)\left(\frac{\delta\Gamma_0}{\delta\underline{\psi}_r(y')}t^{rs}_{\beta'}\frac{\delta\mathcal{A}^\mu_\gamma(z)}{\delta\bar{\eta}_s(y')} - \frac{\delta\Gamma_0}{\delta\underline{\overline{\psi}}_r(y')}\gamma^0 t^{sr}_{\beta'}\gamma^0\frac{\delta\mathcal{A}^\gamma_\mu(z)}{\delta\eta_s(y')}\right)\right] \\
&\quad + ig\int d^4y\,d^4y'\,d^4z\,d^4\omega\,G^{\beta\beta'}(y, y')H_{\alpha\beta,mn}(x, y', z, \omega) \\
&\qquad\times\left[\frac{\delta\Gamma_0}{\delta\underline{\overline{\psi}}_r(y')}t^{rs}_{\beta'}\frac{\delta\underline{\overline{\psi}}_m(z)}{\delta\bar{\eta}_s(y')}\underline{\psi}_n(\omega) - \frac{\delta\Gamma_0}{\delta\underline{\overline{\psi}}_r(y')}\gamma^0 t^{sr}_{\beta'}\gamma^0\frac{\delta\underline{\overline{\psi}}_m(z)}{\delta\eta_s(y')}\underline{\psi}_n(\omega) + \frac{\delta\Gamma_0}{\delta\underline{\psi}_r(y)}t^{rs}_{\beta'}\frac{\delta\underline{\psi}_n(z)}{\delta\bar{\eta}_s(y')}\underline{\overline{\psi}}_m(\omega)\right. \\
&\qquad\left. - \frac{\delta\Gamma_0}{\delta\underline{\overline{\psi}}_r(y')}\gamma^0 t^{sr}_{\beta'}\gamma^0\frac{\delta\underline{\psi}_n(z)}{\delta\bar{\eta}_s(y')}\underline{\overline{\psi}}_n(\omega)\right] = 0\,.
\end{aligned} \tag{A12}$$

Equation (A12) is the generating equation for Ward identities involving proper vertices. Functional derivatives with respect to the appropriate variables of Eq. (A12) evaluated in the limit of zero external fields yield the Ward identities associated with the gauge invariance of the original Lagrangian.

Finally, we want to write explicitly those Ward identities that are relevant for the discussion of symmetry breaking. First Eq. (B7) evaluated at zero external fields gives

$$k^\mu\langle 0|\mathcal{A}^\alpha_\mu(0)|0\rangle = 0\,. \tag{A13}$$

Therefore there are no tadpole contributions. Taking the functional derivative of Eq. (A12) with respect to $\mathcal{A}^\mu_\alpha(x)$ and using the Dyson equation for the ghost (see Fig. 16) we have

$$\frac{1}{\alpha}k^\mu k^\nu D^{\alpha\beta}_{\mu\nu}(k) = \delta_{\alpha\beta}\,. \tag{A14}$$

The longitudinal part of the vector propagator is unrenormalized and the full vector propagator can be written as

$$D^{\mu\nu}_{\alpha\beta}(k) = \left(g^{\mu\nu} - \frac{k^\mu k^\nu}{k^2}\right)D_{\alpha\beta}(k^2) + \frac{k^\mu k^\nu}{k^2}\frac{\alpha}{k^2}\delta_{\alpha\beta}\,. \tag{A15}$$

This equation is also valid if scalar particles are included. Next we take two functional derivatives on Eq. (A12) with respect to $\mathcal{A}$ to obtain Eq. (2.12). Note that the triple divergence of this vertex vanishes. Therefore, suppressing internal-symmetry indices, we have by Lorentz invariance that

$$\begin{aligned}
k_\mu T^{\mu\nu\gamma} = {} & \left[g^{\nu\gamma} - \frac{(p+k)^\nu(p+k)^\gamma}{(p+k)^2}\right]V_1 + \left[(p+k)^\nu(p+k)^\gamma - \frac{p\cdot(p+k)}{p^2}p^\nu(p+k)^\gamma\right]V_2 \\
& + \left[p^\nu p^\gamma - \frac{p\cdot(p+k)}{(p+k)^2}p^\nu(p+k)^\gamma\right]V_3 + \left[\frac{(p+k)^\nu p^\gamma}{p\cdot(p+k)} - p^\nu(p+k)^\gamma\frac{p\cdot(p+k)}{p^2(p+k)^2}\right]V_4\,.
\end{aligned} \tag{A16}$$

Therefore, assuming regularity as k or $p \to 0$,

$$k_\mu T^{\mu\nu\gamma} \underset{k\to 0}{\longrightarrow} \left(g^{\nu\gamma} - \frac{p^\nu p^\gamma}{p^2}\right) V_1 , \tag{A17a}$$

$$p_\nu k_\mu T^{\mu\nu\gamma} \to 0 \text{ as } p \to 0 \text{ or } k \to 0 . \tag{A17b}$$

Finally, taking the functional derivative of Eq. (A14) with respect to ψ, and with respect to $\overline{\psi}$ yields Eq. (3.5).

APPENDIX B: DERIVATION OF THE DYSON EQUATIONS

In order to derive the Dyson equations of interest we start from the Schwinger functional equations for the fields ψ, A^μ_α, and ω^*_α coupled to external sources $\overline{\eta}$, J^μ_α, and S_α, respectively.[27] These equations are given by

$$\left[i\gamma^\mu\left(\partial_\mu + g\theta^\alpha_F \frac{\delta}{\delta J^\alpha_\mu}\right) - m\right]\left(-i\frac{\delta}{\delta\overline{\eta}(x)}\right)e^Z = \eta(x)e^Z , \tag{B1a}$$

$$\begin{aligned}\left[g^{\mu\nu}\Box - \left(\frac{\alpha-1}{\alpha}\right)\partial^\mu\partial^\nu\right]&\left(-i\frac{\delta}{\delta J^\alpha_\nu(x)}\right)e^Z - i\partial_\nu\left(C_{\alpha\beta\gamma}\frac{\delta}{\delta J^\beta_\mu(x)}\frac{\delta}{\delta J^\gamma_\nu(x)}\right)e^Z \\ &= -ig\frac{\delta}{\delta\eta(x)}\gamma^\mu\theta^\alpha_F\frac{\delta}{\delta\overline{\eta}(x)} + ig\frac{\delta}{\delta J^\nu_\beta(x)}\left[\partial^\mu\frac{\delta}{\delta J^\gamma_\nu(x)} - \partial^\nu\frac{\delta}{\delta J^\gamma_\mu(x)} + gC_{\lambda\gamma\delta}\frac{\delta}{\delta J^\lambda_\mu(x)}\frac{\delta}{\delta J^\delta_\nu(x)}\right]C_{\alpha\beta\gamma}e^Z \\ &\quad - g\left(\partial^\mu\frac{\delta}{\delta s_\beta(x)}\right)\frac{\delta}{\delta s^*_\gamma(x)}C_{\alpha\beta\gamma}e^Z ,\end{aligned} \tag{B1b}$$

and

$$\Box\left(-i\frac{\delta}{\delta s^*_\alpha(x)}\right)e^Z - igC_{\alpha\gamma\beta}\partial^\mu\left(\frac{\delta}{\delta J^\mu_\gamma(x)}\frac{\delta}{\delta s^*_\beta(x)}\right)e^Z = s_\alpha(x)e^Z . \tag{B1c}$$

The various two-point functions in an external potential $J^\mu_\alpha(x)$ are defined by

$$S_F(x,y;J) \equiv -\left.\frac{\delta^2 Z}{\delta\overline{\eta}(x)\,\delta\eta(y)}\right|_{\eta=\overline{\eta}=s=s^*=0} , \tag{B2a}$$

$$D^{\mu\nu}_{F\alpha\beta}(x,y;J) \equiv -\left.\frac{\delta^2 Z}{\delta J^\alpha_\mu(x)\,\delta J^\beta_\nu(y)}\right|_{\eta=\overline{\eta}=s=s^*=0} , \tag{B2b}$$

$$G_{F\alpha\beta}(x,y;J) \equiv -\left.\frac{\delta^2 Z}{\delta s^*_\alpha(x)\,\delta s_\beta(y)}\right|_{\eta=\overline{\eta}=s=s^*=0} \tag{B2c}$$

The Dyson equations for these two-point functions follow directly by applying

$$\frac{\delta}{\delta\eta(y)} , \quad \frac{\delta}{\delta J^\lambda_\rho(y)} ,$$

and

$$\frac{\delta}{\delta s^\lambda(y)}$$

to Eqs. (B1a), (B1b), and (B1c), respectively, and then setting $\overline{\eta} = \eta = s = s^* = 0$.

For example, from the fermion equation we have

$$\begin{aligned}(i\gamma^\mu\partial_\mu - m)S_F(x,y;J) &+ g\gamma^\mu\theta^\alpha_F\langle A_{\alpha\mu}\rangle S_F(x,y;J) \\ &+ g\gamma^\mu\theta^\alpha_F\int d^4x'\,d^4y'\,d^4z'\,S_F(x,x';J)\,\Gamma^\nu_\beta(x',y',z';J) \\ &\times S_F(y',y;J)\,D^{\alpha\beta}_{F\mu\nu}(x,z';J) = \delta^4(x-y) .\end{aligned} \tag{B3}$$

$\Gamma^\mu_\alpha(x',y',z';J)$ is the proper fermion-fermion vector three-point function. Using the definition of the free fermion propagator $S^0_F(x-y)$, i.e., $(i\gamma^\mu\partial_\mu - m)S^0_F(x-y) = \delta^4(x-y)$, we obtain the Dyson equation for the fermion propagator in the external potential $J^\mu_\alpha(x)$ as the integral form of the differential equation for $S_F(x,y;J)$ (B3). Defining the fermion self-energy $\Sigma_F(x,y;J)$ by

$$\begin{aligned}S_F(x,y;J) \equiv S^0_F(x-y) &+ \int d^4z\,d^4z'\,S^0_F(x-z) \\ &\times \Sigma_F(z,z';J)\,S_F(z',y;J) ,\end{aligned} \tag{B4a}$$

we obtain an equation for the self-energy by substituting (B4a) into (B3), which yields the diagrammatic equation shown in Fig. 15, which may be written

$$\begin{aligned}\Sigma_F(x,y;J) = &-ig\gamma^\mu\theta_\alpha\langle A^\alpha_\mu(x)\rangle\delta^4(x-y) \\ &- i\int d^4x'\,d^4z\,g\gamma^\mu\theta_\alpha S_F(x,x';J) \\ &\times \Gamma^\nu_\beta(x',y,z;J)\,D^{\alpha\beta}_{F\mu\nu}(x,z;J) .\end{aligned} \tag{B4b}$$

The Dyson equations for the ghost propagator and the vector-boson propagator follow by the same method. The result for the ghost propagator is

$\Sigma_F(x,y;J) =$ 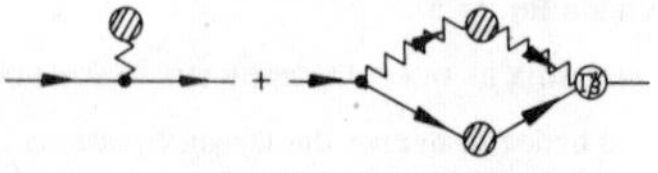

FIG. 15. Fermion self-energy in external field.

$\Sigma_G^{\alpha\beta}(x,y;J) =$ 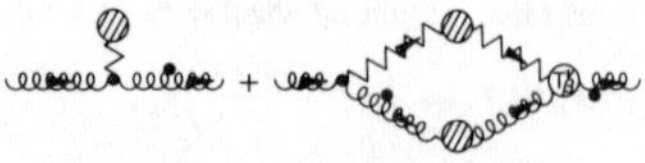

FIG. 16. Ghost self-energy in external field.

$$G_F^{\alpha\beta}(x, y; J) = G_F^{0\alpha\beta}(x-y) + \int d^4z\, d^4z'\, G_F^{0\alpha\alpha'}(x-z) \times \Sigma_G^{\alpha'\beta'}(z, z'; J)\, G_F^{\beta'\beta}(z', y; J), \quad \text{(B5)}$$

where the ghost self-energy Σ_G is given in Fig. 16, where $\tilde{T}^{\mu}_{\alpha\beta\gamma}(x, y, z; J)$ is the proper ghost-ghost-vector three-point function.

For the vector-boson propagator

$$D^{\mu\nu}_{F\alpha\beta}(x, y; J) = D^{\mu\nu}_{0\alpha\beta}(x-y) + \int d^4z\, d^4z'\, D^{\mu\mu'}_{0\alpha\alpha'}(x-z)\, \Pi^{\mu'\nu'}_{\alpha'\beta'}(z, z'; J) \times D^{\nu'\nu}_{F\beta'\beta}(z', y; J), \quad \text{(B6)}$$

where the gauge boson polarization tensor $\Pi^{\mu\nu}_{\alpha\beta}(x, y; J)$ is given in Fig. 17, where $T^{\mu\nu\sigma}_{\alpha\beta\gamma}(x,y,z;J)$ is the proper three-vector three-point function and and $M^{\mu\nu\sigma\rho}_{\alpha\nu\gamma\delta}(x, y, z; \omega; J)$ is the proper four-vector four-point function. T_0 and M_0 denote the associated bare vertices.

Since the propagators and self-energies depend on the external potential $J^{\mu}_{\alpha}(x)$ only through the vacuum expectation values of the gauge boson fields $\langle A^{\mu}_{\alpha}(x)\rangle$ we can consider Σ_F, Σ_G, and $\Pi^{\mu\nu}_{\alpha\beta}$ as functionals of $\langle A^{\mu}_{\alpha}(x)\rangle$ instead of $J^{\mu}_{\alpha}(x)$. Then the proper three-point function can be expressed as follows:

$$\frac{\delta\Sigma_F(x, y; \langle A\rangle)}{\delta\langle A^{\mu}_{\alpha}(z)\rangle} = \frac{\delta S_F^{-1}(x, y; \langle A\rangle)}{\delta\langle A^{\mu}_{\alpha}(z)\rangle}$$

$$= \int d^4x'\, d^4y'\, d^4z'\, S_F^{-1}(x, x'; \langle A\rangle)(-i)^3 \frac{\delta^3 Z}{\delta\eta(x')\,\delta\bar{\eta}(y')\,\delta J^{\nu}_{\beta}(z')} S_F^{-1}(y', y; \langle A\rangle) D^{-1\beta\alpha}_{\nu\mu}(z', z; \langle A\rangle)$$

$$= \Gamma^{\mu}_{\alpha}(x, y, z; \langle A\rangle). \quad \text{(B7a)}$$

Similarly

$$\frac{\delta\Sigma_G^{\beta\gamma}(x, y; \langle A\rangle)}{\delta\langle A^{\alpha}_{\mu}(z)\rangle} = \tilde{T}^{\mu}_{\beta\gamma\alpha}(x, y, z; \langle A\rangle) \quad \text{(B7b)}$$

and

$$\frac{\delta\Pi^{\mu\nu}_{\alpha\beta}(x, y; \langle A\rangle)}{\delta\langle A^{\gamma}_{\sigma}(z)\rangle} = T^{\mu\nu\sigma}_{\alpha\beta\gamma}(x, y, z; \langle A\rangle). \quad \text{(B7c)}$$

We may also derive the Dyson equations for the three-point proper vertices. We use the following identities:

$$g\gamma^{\mu}\theta^{\alpha}_{F} \int d^4x'\, d^4y'\, D^{\alpha\beta}_{F\mu\nu}(x, x'; \langle A\rangle) \frac{\delta}{\delta\langle A^{\beta}_{\nu}(y)\rangle} [S_F(x, y; \langle A\rangle)\, S_F^{-1}(y, z; \langle A\rangle)] = 0, \quad \text{(B8a)}$$

$$gC_{\alpha\beta\gamma} \int d^4x'\, d^4y'\, D^{\mu\nu}_{F\gamma\delta}(x, x'; \langle A\rangle) \frac{\delta}{\delta\langle A^{\delta}_{\nu}(x')\rangle} \partial^x_{\mu}[G_F(x, y'; \langle A\rangle)\, G_F^{-1}(y', y; \langle A\rangle)]_{\beta\lambda} = 0, \quad \text{(B8b)}$$

and

$$\frac{1}{2} \int d^4y'\, d^4z\, d^4z'\, d^4\omega\, T^{\mu\sigma\rho}_{\alpha\gamma\delta}(x, z, \omega) \frac{\delta}{\delta\langle A^{\beta}_{\nu}(y)\rangle} [D^{\sigma\sigma'}_{F\delta\delta'}(\omega, y'; \langle A\rangle)\, D^{-1\sigma'\rho'}_{F\delta'\gamma'}(y', z'; \langle A\rangle)]\, D^{\rho'\rho}_{F\gamma'\gamma}(z', z; \langle A\rangle)$$

$$+ \frac{1}{6} \int d^4x'\, d^4x''\, d^4z\, d^4z'\, d^4z''\, d^4y'\, d^4y''\, M^{\mu\sigma\rho\xi}_{\alpha\gamma\delta\lambda}(x, x', x'', z) \frac{\delta}{\delta\langle A^{\beta}_{\nu}(y)\rangle}$$

$$\times \left[D^{\xi\xi'}_{F\lambda\lambda'}(z, z'; \langle A\rangle) \frac{\delta}{\delta\langle A^{\xi'}_{\lambda'}(z')\rangle} (D^{\sigma'\rho'}_{F\gamma\delta'}(x', y''; \langle A\rangle)\, D^{-1\rho'\rho''}_{F\delta'\delta''}(y'', y'; \langle A\rangle)) \right] D^{\rho''\rho}_{F\delta''\delta}(y', x''; \langle A\rangle)$$

$$+\frac{1}{2}\int d^4x'\, d^4x''\, d^4z\, d^4y'\, d^4y''\, M^{\mu\sigma\rho\xi}_{\alpha\gamma\delta\lambda}(x,x',x'',z)\langle A^{\xi}_{\lambda}(z)\rangle \frac{\delta}{\delta\langle A^{\beta}_{\nu}(y)\rangle}$$

$$\times[D^{\sigma\sigma'}_{F\gamma\gamma'}(x',y';\langle A\rangle)\, D^{-1\sigma'\rho'}_{F\gamma'\delta'}(y',y'';\langle A\rangle)]\, D^{\rho'\rho}_{F\delta'\delta}(y'',x'';\langle A\rangle)$$

$$-g\gamma^{\mu}\theta^{F}_{\alpha}\int d^4x'\, d^4y'\, S_F(x,y;\langle A\rangle)\frac{\delta}{\delta\langle A^{\beta}_{\nu}(y)\rangle}[S_F{}^{-1}(y,x';\langle A\rangle)\, S_F(x',x;\langle A\rangle)]$$

$$+igC_{\alpha\gamma\delta}\int d^4z'\, d^4z''\lim_{x\to z}[\partial^x_{\mu}G_{\delta\delta'}(x,y;\langle A\rangle)]\frac{\delta}{\delta\langle A^{\beta}_{\nu}(y)\rangle}[G_{F\gamma\gamma'}{}^{-1}(z'',z';\langle A\rangle)\, G_{F\gamma'\delta'}(z',z;\langle A\rangle)]=0\,. \quad \text{(B8c)}$$

Now using Eqs. (B7a) and (B4b) in Eq. (B8a), and Eq. (B7b) and Fig. 16 in Eq. (B8b), and finally Eq. (B7c) and Fig. 17 in Eq. (B8c) we obtain a system of functional equations for the self-energies Σ_F, Σ_G, and Π. We do not write the full system down here as it is very complicated. However, to show what these equations look like we give the equation for the fermion-self energy:

$$\Sigma'_F(x,y;D,S_F,G_F)=ig^2\gamma^{\mu}\theta^F_{\alpha}D^{\alpha\beta}_{\mu\nu}(x,y;\langle A\rangle)\, S_F(x,y;\langle A\rangle)\,\gamma^{\nu}\theta^F_{\beta}$$

$$-ig^2\gamma_{\mu}\theta^{\alpha}_F\int d^4x'\, d^4x''\, D^{\alpha\beta}_{\mu\nu}(x,x';\langle A\rangle)\, S_F(x,x'';\langle A\rangle)\frac{\delta\Sigma'_F(x'',y;D,S_F,G_F)}{\delta\langle A^{\beta}_{\nu}(x')\rangle}\,, \quad \text{(B9a)}$$

where

$$\Sigma_F(x,y;D,S_F,G_F;\langle A\rangle_{\text{ext}})\equiv g\gamma_{\mu}\theta^{\alpha}_F\langle A^{\mu}_{\alpha}(x)\rangle\,\delta^4(x-y)$$
$$+\Sigma'_F(x,y;D,S_F,G_F)\,. \quad \text{(B9b)}$$

Graphically the iterative solution of the coupled system of equations looks like those in Fig. 18.

We may now define four-point proper kernels which are two-particle irreducible in the s channel by functional derivatives with respect to S, D, and G of the self-energy terms at $\langle A\rangle=0$:

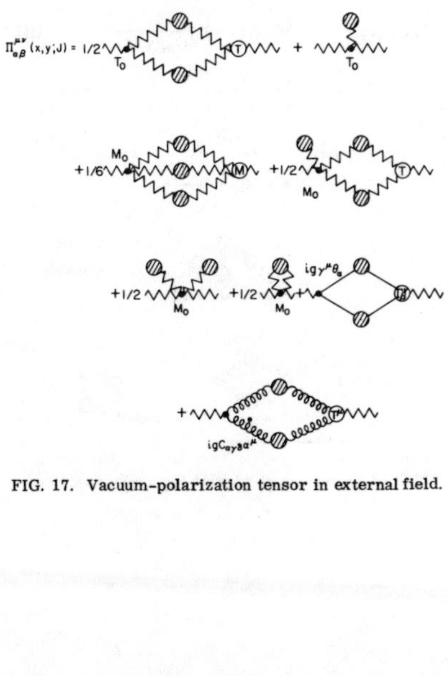

FIG. 17. Vacuum-polarization tensor in external field.

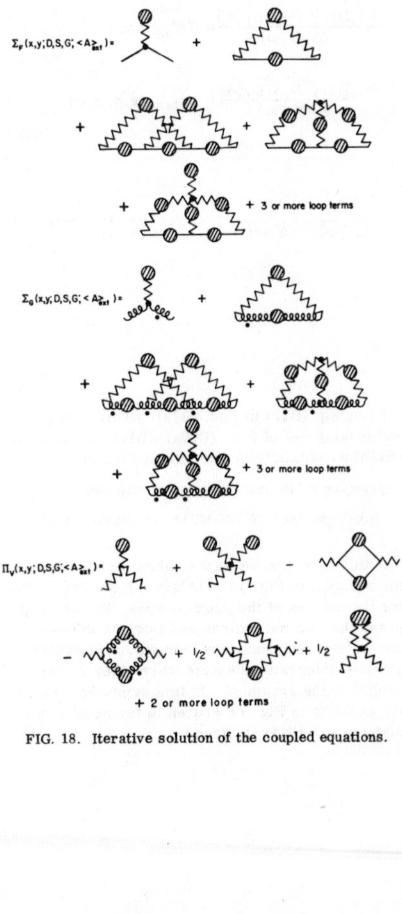

FIG. 18. Iterative solution of the coupled equations.

$$\frac{\delta\Sigma_F(x,y;D,S,G,\langle A\rangle)}{\delta S_F(z,\omega)} \equiv K_{FF}(x,y,z,\omega), \tag{B10a}$$

$$\frac{\delta\Sigma_F(x,y;D,S,G,\langle A\rangle)}{\delta G_F^{\alpha\beta}(z,\omega)} \equiv K_{FG}^{\alpha\beta}(x,y,z,\omega), \tag{B10b}$$

$$\frac{\delta\Sigma_F(x,y;D,S,G,\langle A\rangle)}{\delta D_{\alpha\beta}^{\mu\nu}(z,\omega)} \equiv K_{FV\alpha\beta}^{\mu\nu}(x,y,z,\omega), \tag{B10c}$$

$$\frac{\delta\Sigma_G^{\alpha\beta}(x,y;D,S,G,\langle A\rangle)}{\delta S_F(z,\omega)} \equiv K_{GF}^{\alpha\beta}(x,y,z,\omega), \tag{B10d}$$

$$\frac{\delta\Sigma_G^{\alpha\beta}(x,y;D,S,G,\langle A\rangle)}{\delta G^{\gamma\delta}(z,\omega)} \equiv K_{GG}^{\alpha\beta\gamma\delta}(x,y,z,\omega), \tag{B10e}$$

$$\frac{\delta\Sigma_G^{\alpha\beta}(x,y;D,S,G,\langle A\rangle)}{\delta D_{\gamma\delta}^{\mu\nu}(z,\omega)} \equiv K_{GV\mu\nu}^{\alpha\beta\gamma\delta}(x,y,z,\omega), \tag{B10f}$$

$$\frac{\delta\Pi_{\alpha\beta}^{\mu\nu}(x,y;D,S,G;\langle A\rangle)}{\delta S_F(z,\omega)} \equiv K_{VF\alpha\beta}^{\mu\nu}(x,y,z,\omega), \tag{B10g}$$

$$\frac{\delta\Pi_{\alpha\beta}^{\mu\nu}(x,y;D,S,G,\langle A\rangle)}{\delta G^{\gamma\delta}(z,\omega)} \equiv K_{VG\alpha\beta\gamma\delta}^{\mu\nu}(x,y,z,\omega), \tag{B10h}$$

$$\frac{\delta\Pi_{\alpha\beta}^{\mu\nu}(x,y;D,S,G,\langle A\rangle)}{\delta D_{\alpha\rho}^{\gamma\delta}(z,\omega)} \equiv K_{VV\alpha\beta\gamma\delta}^{\mu\nu\alpha\rho}(x,y,z,\omega). \tag{B10i}$$

We also define two-particle irreducible (in the z channel) proper three-point functions by the explicit variation of the self-energies with respect to $\delta/\delta\langle A\rangle$ at $\langle A\rangle = 0$:

$$\left(\frac{\delta}{\delta\langle A_\alpha^\mu(z)\rangle}\right)_{\text{explicit}} \Sigma_F(x,y;D,S,G;\langle A\rangle)\bigg|_{\langle A\rangle=0} \equiv -ig\gamma_\mu\theta_F^\alpha\delta''(x-y)+\overline{\Gamma}_\mu^\alpha(x,y,z), \tag{B11a}$$

$$\left(\frac{\delta}{\delta\langle A_\gamma^\mu(z)\rangle}\right)_{\text{explicit}} \Sigma_G(x,y;D,S,G,\langle A\rangle)\bigg|_{\langle A\rangle=0} \equiv gC_{\alpha\beta\gamma}\partial_y^\mu\delta^4(x-y)+\hat{T}_{\alpha\beta\gamma}^\mu(x,y,z), \tag{B11b}$$

$$\left(\frac{\delta}{\delta\langle A_\gamma^\sigma(z)\rangle}\right)_{\text{explicit}} \Pi_{\alpha\beta}^{\mu\nu}(x,y;D,S,G,\langle A\rangle)\bigg|_{\langle A\rangle=0} \equiv T_{\alpha\alpha\beta\gamma}^{\mu\nu\sigma}(x,y,z)+\overline{T}_{\alpha\beta\gamma}^{\mu\nu\sigma}(x,y,z). \tag{B11c}$$

Graphically the series for $\overline{\Gamma}$, $\hat{T}$, and $\overline{T}$ can be obtained from the series for Σ_F, Σ_G, and Π. For example, see the equations in Fig. 19.

We finally write the Dyson equations for the various three-point functions. We note

$$\begin{aligned}\frac{\delta}{\delta\langle A_\mu^\alpha(x)\rangle} = \left(\frac{\delta}{\delta\langle A_\alpha^\mu(x)\rangle}\right)_{\text{explicit}} + \int d^4y\,d^4z\,d^4y'\,d^4z'\bigg[&S_F(y,y')\Gamma_\alpha^\mu(y',z',x)S_F(z',z)\frac{\delta}{\delta S_F(y,z)}\\ &+D_{F\beta\beta'}^{\nu\nu'}(y,y')T_{\beta'\alpha\gamma'}^{\nu'\mu\sigma'}(y',x,z')D_{F\gamma'\gamma}^{\sigma'\sigma}(z',z)\frac{\delta}{\delta D_{F\nu\sigma}^{\beta\gamma}(y,z)}\\ &+G_{\beta\beta'}(y,y')\tilde{T}_{\beta'\gamma'\alpha}^{\mu}(y',z',x)G_{\gamma'\gamma}(z',z)\frac{\delta}{\delta G_{\beta\gamma}(y,z)}\bigg].\end{aligned} \tag{B12}$$

Using Eq. (B12) in Eqs. (B7a), (B7b), and (B7c) and making use of Eqs. (B10a)–(B10i) and (B11a)–(B11c) we obtain the equations in Fig. 20.

APPENDIX C: FINITE EXPRESSION FOR THE GAUGE BOSON MASSES IN THE WEAK-COUPLING LIMIT

In this appendix we want to show how to rewrite the equation in Fig. 13 to obtain a finite expression for the masses of the gauge bosons. We will suppress the internal indices and Lorentz indices, and we will also suppress momentum variables and loop integrations except where they are essential to the argument. In this symbolic notation the equation in Fig. 13 written in terms of renormalized quantities is given by

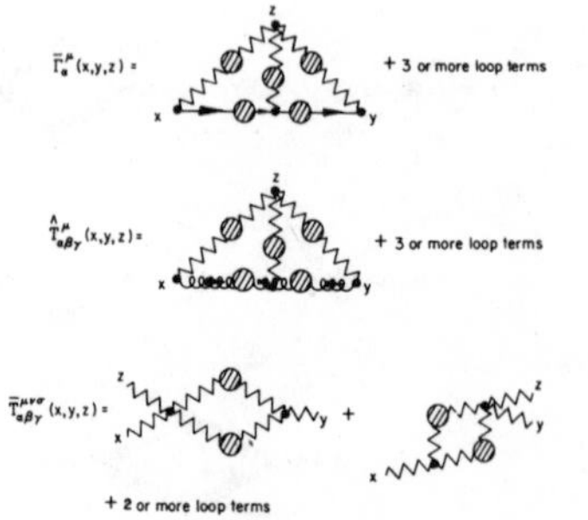

FIG. 19. Lowest-order terms for various vertices.

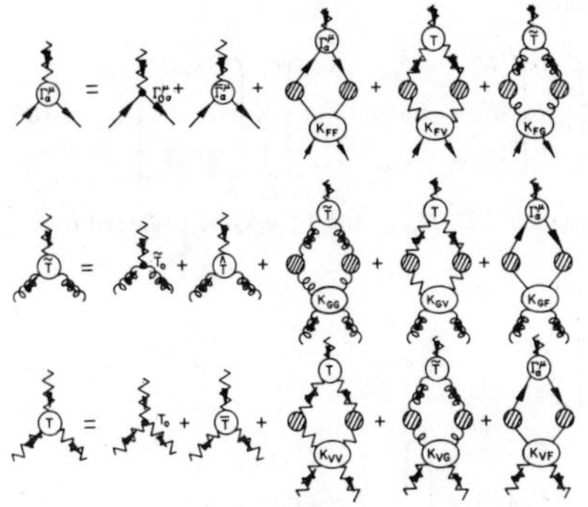

FIG. 20. Integral equations for various vertices.

$$\lim_{k\to 0} g^{\mu\nu}\lambda_{\alpha a}(0) = \lim_{k\to 0} -\frac{\partial}{\partial k^\nu}[-\tfrac{1}{2}Z_1 T^\mu_{0\alpha} \;\; \tilde{Z}_1\tilde{T}^{0\mu}_{\alpha} \;\; Z_{1F}\Gamma^\mu_{0\alpha}]\begin{bmatrix} DgP_{Va}D \\ GgP_{Ga}G \\ SgP_{Fa}S \end{bmatrix}, \tag{C1}$$

where Z_1, $\tilde{Z}_1$, and Z_{1F} are the vertex function renormalization constants for the three-vector, ghost-vector-ghost, and fermion-vector-fermion vertices, respectively, and $\lambda_{\alpha a}P_{Va}$, $\lambda_{\alpha a}P_{Ga}$, $\lambda_{\alpha a}P_{Fa}$ renormalize like their associated vertex functions. For simplicity we have omitted terms involving the four-vector vertex. The argument given here can easily be extended to include this term, but we are mainly interested in the weak-coupling limit where the term is irrelevant. Here it is not indicated how the external momentum, k, is routed through the loops. This routing is important since the bare vertices $T^{\mu\nu\sigma}_{0\alpha\beta\gamma}$ and $\tilde{T}^{\mu}_{0\alpha\beta\gamma}$ depend on momentum, and the separation of the Dyson equations into pole regular parts is unique only up to terms of order k^2. The proper choice of routing for our purposes will be shown below. The full Dyson equations for the vertex functions is, in symbolic notation,

$$\begin{bmatrix} T \\ \tilde{T} \\ \Gamma \end{bmatrix} = \begin{bmatrix} \frac{1}{2}K_{VV} & -K_{VG} & -K_{VF} \\ \frac{1}{2}K_{GV} & K_{GG} & -K_{GF} \\ \frac{1}{2}K_{FV} & -K_{FG} & K_{FF} \end{bmatrix}\begin{bmatrix} D & T & D \\ G & \tilde{T} & G \\ S & \Gamma & S \end{bmatrix} + \begin{bmatrix} Z_1 T_0 \\ \tilde{Z}_1\tilde{T}_0 \\ Z_{1F}\Gamma_0 \end{bmatrix}, \tag{C2}$$

and to order k^2

$$\begin{bmatrix} P_V \\ P_G \\ P_F \end{bmatrix} = \begin{bmatrix} \frac{1}{2}K_{VV} & -K_{VG} & -K_{VF} \\ \frac{1}{2}K_{GV} & K_{GG} & -K_{GF} \\ \frac{1}{2}K_{FV} & -K_{FG} & K_{FF} \end{bmatrix}\begin{bmatrix} D & P_V & D \\ G & P_G & G \\ S & P_F & S \end{bmatrix}. \tag{C3}$$

Substituting Eq. (C2) into Eq. (C1) we obtain

$$g^{\mu\nu}\lambda_{\alpha a}(0) = \lim_{k^\mu\to 0}\frac{\partial}{\partial k^\mu}\left\{[T^\mu_\alpha \;\; \tilde{T}^\mu_\alpha \;\; \Gamma^\mu_\alpha] - [DT^\mu_\alpha D \;\; G\tilde{T}^\mu_\alpha G \;\; S\Gamma^\mu_\alpha S]\begin{bmatrix} \frac{1}{2}K_{VV} & \frac{1}{2}K_{VG} & \frac{1}{2}K_{VF} \\ -K_{GV} & K_{GG} & -K_{GF} \\ -K_{FV} & -K_{FG} & K_{FF} \end{bmatrix}\right\}\begin{bmatrix} \frac{1}{2}DgP_{Va}D \\ -GgP_{Ga}G \\ -SgP_{Fa}S \end{bmatrix}. \tag{C4}$$

Now all explicit dependence on the renormalization constants vanishes. In Eq. (C4) we must have an expression which is automatically finite, as all subtractions have been performed and we are not free to make any additional overall subtraction for the pole part of Π. Now we choose the routing of the external momentum so that the term where $\partial/\partial k_\mu$ acts on the expression in parentheses vanishes at $k=0$. This can be done by routing momentum as shown in Fig. 21. $\Gamma^\mu_{0\alpha}$ has no dependence on the momentum k. By routing the external momentum through the incoming ghost leg, $\tilde{T}^\mu_{0\alpha\beta\gamma}$ does not depend explicitly on the external momentum. Also noting that in the Landau gauge the vector-boson propagators are transverse, the explicit dependence of $T^{\mu\nu\sigma}_{0\alpha\beta\gamma}$ on the external momentum k will drop out. Therefore, we write

$$g^{\mu\nu}\lambda_{\alpha a}(0) = \left\{[T^\mu_\alpha \;\; \tilde{T}^\mu_\alpha \;\; \Gamma^\mu_\alpha] - [DT^\mu_\alpha D \;\; G\tilde{T}^\mu_\alpha G \;\; S\Gamma^\mu_\alpha S]\begin{bmatrix} \frac{1}{2}K_{VV} & \frac{1}{2}K_{VG} & \frac{1}{2}K_{VF} \\ -K_{GV} & K_{GG} & -K_{GF} \\ -K_{FV} & -K_{FG} & K_{FF} \end{bmatrix}\right\}_{k=0}\frac{\partial}{\partial k^\nu}\left\{\begin{bmatrix} \frac{1}{2}DgP_VD \\ -GgP_GG \\ -SgP_FS \end{bmatrix}\right\}_{k=0}. \tag{C5}$$

Now we differentiate Eq. (C3) with respect to k^ν to obtain

$$\frac{\partial}{\partial k^\nu}\begin{bmatrix}P_V\\ Pg\\ P_F\end{bmatrix} - \begin{bmatrix}\frac{1}{2}K_{VV} & -K_{VG} & -K_{VF}\\ \frac{1}{2}K_{GV} & K_{GG} & -K_{GF}\\ \frac{1}{2}K_{FV} & -K_{FG} & K_{FF}\end{bmatrix}\left\{\frac{\partial}{\partial k^\nu}\begin{bmatrix}DP_VD\\ GP_GG\\ SP_FS\end{bmatrix}\right\} = \left\{\frac{\partial}{\partial k^\nu}\begin{bmatrix}\frac{1}{2}K_{VV} & -K_{VG} & -K_{VF}\\ \frac{1}{2}K_{GV} & K_{GG} & -K_{GF}\\ \frac{1}{2}K_{FV} & -K_{FG} & K_{FF}\end{bmatrix}\right\}\begin{bmatrix}DP_VD\\ GP_GG\\ SP_FS\end{bmatrix} \qquad \text{(C6)}$$

Substituting (C6) into (C5) and noting by (C3) that the vertices $T^{\mu\nu\sigma}_{\alpha\beta\gamma}$, $\tilde{T}^{\mu}_{\alpha\beta\gamma}$, and Γ^{μ}_{α} may be replaced by their regular parts we have

$$g^{\mu\nu}\lambda_{\alpha\alpha}(0) = [\tfrac{1}{2}T^{\mu}_{R\alpha} \;\; -\tilde{T}^{\mu}_{R\alpha} \;\; -\Gamma^{\mu}_{R\alpha}]_{k=0}\left\{\frac{\partial}{\partial k^\nu}\begin{bmatrix}D_gP_V(k=0)D\\ GgP_G(k=0)G\\ SgP_F(k=0)S\end{bmatrix}\right\}_{k=0}$$

$$+[\tfrac{1}{2}DT^{\mu}_{R\alpha}D \;\; -G\tilde{T}^{\mu}_{R\alpha}G \;\; -S\Gamma^{\mu}_{R\alpha}S]_{k=0}\left\{\frac{\partial}{\partial k^\nu}\begin{bmatrix}K_{VV} & K_{VG} & K_{VF}\\ K_{GV} & K_{GG} & K_{GF}\\ K_{FV} & K_{FG} & K_{FF}\end{bmatrix}\right\}_{k=0}\begin{bmatrix}\frac{1}{2}DgP_VD\\ -GgP_GG\\ -SgP_FS\end{bmatrix}_{k=0}. \qquad \text{(C7)}$$

Now in the weak-coupling limit we replace the propagators and the kernels by their weak-coupling limit expressions [see Eqs. (5.3)–(5.5) and Appendix D]. The regular parts of the three-point vertices cannot be replaced by their bare vertices because terms involving iterations of the kernel (although they are higher order in g in the expansion of the regular part of the vertices) may give contributions of zeroth order in g to $\lambda_{\alpha\alpha}(0)$ because they involve additional loop integrations which may provide appropriate factors of $1/g$. As we have seen in Sec. V such factors can occur since the power dependence of the Goldstone form factors contains g^2 explicitly.

To investigate exactly which terms can contribute to the weak-coupling limit for $\lambda_{\alpha\alpha}(0)$, we use the results of Sec. V that

$$P_G \sim \left(\frac{p^2}{m^2}\right)^{1-\epsilon_G}, \quad P_T \sim \left(\frac{p^2}{m^2}\right)^{-\epsilon_V}$$

and

$$P_F \sim \left(\frac{p^2}{m^2}\right)^{-\epsilon_F},$$

where $\epsilon_{G,F,V} \sim O(g^2)$. With these forms it is straightforward to check for the three-point vertices that iterations of kernels which produce intermediate states between kernels, other than two vector states, do not contribute to the expression (C7) for $\lambda_{\alpha\alpha}(0)$, in the weak-coupling limit. We may reexpress those terms in Eq. (C7) which do contribute to $\lambda_{\alpha\alpha}(0)$ as shown in Fig. 22. The prime denotes differentiation with respect to k^ν, and $A^{\mu\nu\sigma}_{\alpha\beta\gamma}(0,q)$, part of the three-vector vertex, is a solution to the integral equation

$$A^{\mu\nu\sigma}_{\alpha\beta\gamma}(0,q) = T^{\mu\nu\sigma}_{0\alpha\beta\gamma}(0,q) + \int \frac{d^4k}{(2\pi)^4} A^{\mu\nu'\sigma'}_{\alpha\beta'\gamma'}(0,q) D^{\nu'\lambda}_{\beta'\epsilon}(k) D^{\sigma'\lambda'}_{\gamma'\epsilon'}(k) \times K^{\lambda\lambda'\nu\sigma}_{VV\epsilon\epsilon'\beta\gamma}(k,q). \qquad \text{(C8)}$$

Finally we note that terms which are iterations of the kernel in Eq. (C8) require counterterms to make the integrals finite. However, these counterterms are at least $O(g^2)$, and there are no loop

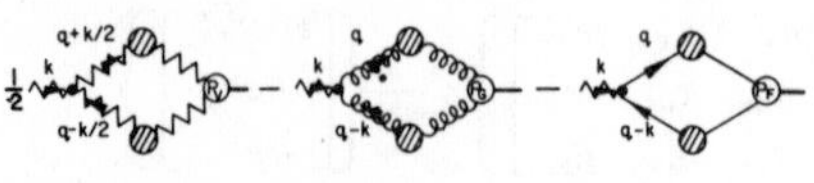

FIG. 21. Convenient routing of loop momenta.

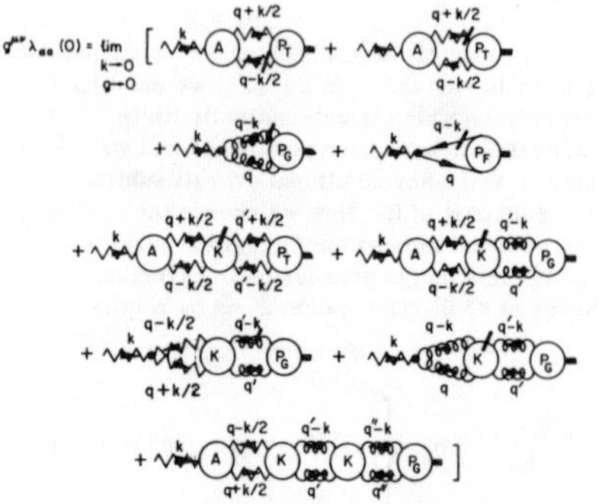

FIG. 22. Masses in weak-coupling limit.

integrals associated with them which might make them finite. Therefore they may be ignored in the calculation of $\lambda_{\alpha\alpha}(0)$ to lowest order in g. For example, consider the term in $A^{\mu\nu\sigma}_{\alpha\beta\gamma}(0,q)$ which has one iteration of the kernel. This term, $A^{\mu\nu\sigma}_{1\alpha\beta\gamma}(0,q)$ has the form

$$A^{\mu\nu\sigma}_{1\alpha\beta\gamma}(0,q)=\int\frac{d^4k}{(2\pi)^4}T^{\mu\nu'\sigma'}_{0\alpha\beta'\gamma'}(0,k)D^{\nu'\lambda}_{\beta'\epsilon}(k)D^{\sigma'\lambda'}_{\gamma'\epsilon'}(k)K^{\lambda\lambda'\nu\sigma}_{VV\epsilon\epsilon'\beta\gamma}(k,q)-g^2a\ln\frac{\Lambda^2}{m^2}T^{\mu\nu\sigma}_{0\alpha\beta\gamma}(0,q), \tag{C9}$$

where Λ is a cutoff and the constant a is independent of g. When Eq. (C9) is inserted into the expression for $\lambda_{\alpha\alpha}(0)$ we obtain an integral of the form I, where

$$I=g^2\int\frac{d^4k}{(2\pi)^4}A^{\mu\nu'\sigma'}_{1\alpha\beta'\gamma'}(0,k)D^{\nu'\nu}_{\beta'\beta}(k)D^{\sigma'\sigma}_{\gamma'\gamma}(k)\left(\frac{k^2}{m^2}\right)^{-bg^2} \tag{C10a}$$

$$=g^2\int\frac{d^4k}{(2\pi)^4}\left\{\left[\int\frac{d^4q}{(2\pi)^4}T^{\mu\nu'\sigma'}_{0\alpha\beta'\gamma'}(0,k)D^{\nu'\lambda}_{\beta'\epsilon}(k)D^{\sigma'\lambda'}_{\gamma'\epsilon'}K^{\lambda\lambda'\eta\eta'}_{VV\epsilon\epsilon'\delta\delta'}(k,q)D^{\eta\nu}_{\delta\beta}(q)D^{\eta'\sigma}_{\delta'\gamma}(q)\left(\frac{q^2}{m^2}\right)^{-bg^2}\right]\right.$$

$$\left.-g^2a\ln\frac{\Lambda^2}{m^2}T^{\mu\nu'\sigma'}_{0\alpha\beta'\gamma'}(0,k)D^{\nu'\nu}_{\beta'\beta}(k)D^{\sigma'\sigma}_{\gamma'\gamma}(k)\left(\frac{k^2}{m^2}\right)^{-bg^2}\right\}. \tag{C10b}$$

b does not depend on g. The integration over q can be performed using the methods of Sec. V to give

$$I=g^2\int\frac{d^4k}{(2\pi)^4}\left[T^{\mu\nu'\sigma'}_{0\alpha\beta'\gamma'}(0,k)D^{\nu'\nu}_{\beta'\beta}(k)D^{\sigma'\sigma}_{\gamma'\gamma}(k)\left(\frac{c}{b}-g^2a\ln\frac{\Lambda^2}{m^2}\right)\left(\frac{k^2}{m^2}\right)^{-bg^2}\right], \tag{C11}$$

where c is a constant independent of g. Therefore, the cutoff dependent part of I does not contribute to $\lambda_{\alpha\alpha}(0)$ in the weak-coupling limit. Since the counterterms do not contribute to $\lambda_{\alpha\alpha}(0)$, the equation in Fig. 22 is equal, in our approximation, to Eq. (C1) with $Z_1=\tilde{Z}_1=Z_{1F}=1$. The resulting expression is the most convenient one for the explicit calculations performed in Sec. V.

APPENDIX D: LIST OF LOWEST-ORDER BETHE-SALPETER KERNELS AND THE REDUCED KERNELS f_{ij}

Here we list the various kernels needed in Sec. V:

$$K^{\mu\nu\sigma\rho}_{VV\alpha\beta\gamma\delta}(p,p,q,q)=-ig^2\left\{\tfrac{1}{2}C_{\alpha\gamma\lambda}C_{\delta\beta\lambda'}[g^{\mu\xi}(2p-q)^\sigma+g^{\xi\sigma}(2q-p)^\mu+g^{\sigma\mu}(-q-p)^\xi]\right.$$

$$\times\left[\left(g^{\xi\xi'}-\frac{(q-p)^\xi(q-p)^{\xi'}}{(q-p)^2}\right)\left(\frac{1}{(p-q)^2-\mu^2+i\epsilon}\right)_{\lambda\lambda'}+\alpha_R\frac{(q-p)^\xi(q-p)^{\xi'}}{[(q-p)^2+i\epsilon]^2}\delta_{\lambda\lambda'}\right]$$

$$\times[g^{\nu\rho}(-p-q)^{\xi'}+g^{\rho\xi'}(2q-p)^\nu+g^{\xi'\nu}(2p-q)^\rho]$$

$$\left.+(p\to-p,\ \alpha\leftrightarrow\beta,\ \mu\leftrightarrow\nu)\right\}$$

$$+\tfrac{1}{2}g[C_{\alpha\beta\lambda}C_{\delta\gamma\lambda}(g^{\mu\rho}g^{\nu\sigma}-g^{\mu\sigma}g^{\nu\rho})+C_{\alpha\delta\lambda}C_{\beta\gamma\lambda}(g^{\mu\nu}g^{\sigma\rho}-g^{\mu\sigma}g^{\nu\rho})$$

$$+C_{\alpha\gamma\lambda}C_{\delta\beta\lambda}(g^{\mu\rho}g^{\nu\sigma}-g^{\mu\nu}g^{\sigma\rho})], \tag{D1a}$$

$$K^{\mu\nu}_{VG\alpha\beta\gamma\delta}(p,p,q,q)=-ig^2\left[\tfrac{1}{2}\left(C_{\alpha\gamma\lambda}C_{\delta\beta\lambda}(p-q)^\nu q^\mu\frac{1}{(p-q)^2+i\epsilon}\right)+\tfrac{1}{2}(p\to-p,\ \alpha\leftrightarrow\beta,\ \mu\leftrightarrow\nu)\right], \tag{D1b}$$

$$K^{\sigma\rho}_{GV\alpha\beta\gamma\delta}(p,p,q,q)=K^{\rho\sigma}_{VG\delta\gamma\alpha\beta}(q,q,p,p), \tag{D1c}$$

$$K_{GG\alpha\beta\gamma\delta}(p,p,q,q)=-ig^2C_{\alpha\gamma\lambda}C_{\delta\beta\lambda'}p^{\xi'}q^\xi\left\{[g^{\xi\xi'}-(p-q)^\xi(p-q)^{\xi'}/(p-q)^2][(q-p)^2-\mu^2+i\epsilon]^{-1}{}_{\lambda\lambda'}\right.$$

$$\left.+\alpha_R\frac{(p-q)^\xi(p-q)^{\xi'}}{(p-q)^2}\right\}, \tag{D1d}$$

$$K^{\mu\nu}_{VF\alpha\beta\gamma\delta}(p,p,q,q)=-ig^2[\gamma^\nu\theta_\beta S_F(q-p)\gamma^\mu\theta_\alpha+\gamma^\mu\theta_\alpha S_F(q+p)\gamma^\nu\theta_\beta], \tag{D1e}$$

$$K^{\sigma\rho}_{FV\gamma\delta}(p,p,q,q)=K^{\rho\sigma}_{VF\delta\gamma}(q,q,p,p), \tag{D1f}$$

$$K_{FG\gamma\delta}(p,p,q,q)=K_{GF\delta\gamma}(q,q,p,p)=0\,, \tag{D1g}$$

$$K_{FF}^{mm'nn'}(p,p,q,q)=ig^2(\gamma^\mu\theta_\alpha)^{mm'}(\gamma^\nu\theta_\beta)^{n'n}\left\{\left[g_{\mu\nu}-\frac{(p-q)^\mu(p-q)^\nu}{(p-q)^2}\right]\left(\frac{1}{(p-q)^2-\mu^2+i\epsilon}\right)_{\alpha\beta}+\alpha_R\frac{(p-q)^\mu(p-q)^\nu}{((p-q)^2+i\epsilon)^2}\delta_{\alpha\beta}\right\}. \tag{D1h}$$

Also we list the various f_{ij}'s used in Sec. V. f_{ij}^{10} represents the coupling of a first-order vertex in the Goldstone boson momentum to a zeroth-order one and f_{ij}^{11} represents the coupling of a first-order vertex to a first-order vertex:

$$f_{11}^{\alpha\beta\gamma\delta}\left(x,\frac{\mu^2}{p^2},g\right)=\frac{-1}{16\pi^2}g^2\tfrac{1}{6}[C_{\alpha\gamma\lambda}C_{\lambda\delta\beta}+C_{\beta\gamma\lambda}C_{\lambda\delta\alpha}]\left[\theta(x-1)(9x^2+50x-7)\frac{1}{4x^2}+\theta(1-x)(9+50x-7x^2)\tfrac{1}{4}\right]+\theta\left(\frac{\mu^2}{p^2}\right), \tag{D2a}$$

$$\begin{aligned}f_{12}^{\alpha\beta\gamma\delta}(x,0,g)&=\frac{1}{16\pi^2}g^2\tfrac{1}{2}[C_{\alpha\gamma\lambda}C_{\lambda\delta\beta}+C_{\beta\gamma\lambda}C_{\lambda\delta\alpha}]\left[\theta(x-1)\frac{1-3x}{12x}+\theta(1-x)\frac{x^2-3x}{12}\right]\\&=\frac{x}{3}f_{21}^{\alpha\beta\gamma\delta}(x,0,g),\end{aligned} \tag{D2b}$$

$$f_{22}^{\alpha\beta\gamma\delta}(x,0,g)=\frac{-1}{16\pi^2}g^2\tfrac{1}{2}[C_{\alpha\gamma\lambda}C_{\lambda\delta\beta}+C_{\beta\gamma\lambda}C_{\lambda\delta\alpha}]\left[\frac{3}{4x}\theta(x-1)+\tfrac{3}{4}x\theta(1-x)\right], \tag{D2c}$$

$$f_{33}^{\alpha\beta\gamma\delta}(x,0,g)=\frac{1}{16\pi^2}\frac{g^2}{2}C_{\alpha\beta\lambda}C_{\lambda\gamma\delta}\frac{3}{4}\left[\frac{1}{x}\theta(x-1)+x\theta(1-x)\right], \tag{D2d}$$

$$\begin{aligned}f_{14}^{\alpha\beta}\left(x^2,\frac{\mu^2}{p^2},\frac{M^2}{p^2},g\right)=\frac{1}{12\pi^2}g^2\Bigg\{&\left[\theta(x-1)\left(-\frac{3}{2}+\frac{1}{x}\right)-\theta(1-x)\frac{x^2}{2}\right]\tfrac{1}{2}\gamma^0\{\theta_\beta,\theta_\alpha\}\gamma^0\\&+\left[\theta(x-1)\frac{-3}{2x}+\theta(1-x)\frac{x}{2}\right]\frac{1}{2p^2}(\gamma^0\theta_\beta M\gamma^0\theta_\alpha\gamma^0M\gamma^0+M\theta_\beta M\gamma^0\theta_\alpha\gamma^0+\alpha\leftrightarrow\beta)\\&+\left[\theta(x-1)\left(\frac{-3}{2x}+\frac{1}{x^2}\right)+\theta(1-x)\frac{-x}{2}\right]\\&\times\frac{1}{2p^2}(\gamma^0\{\theta_\beta,\theta_\alpha\}\gamma^0M\gamma^0M\gamma^0+M\{\theta_\beta,\theta_\alpha\}\gamma^0M\gamma^0+M\gamma^0M\{\theta_\beta,\theta_\alpha\}\gamma^0)\Bigg\}\\&+O\left(\left(\frac{\mu^2}{p^2}\right)^2,\frac{\mu^2M^2}{(p^2)^2},\left(\frac{M^2}{p^2}\right)^2\right),\end{aligned} \tag{D2e}$$

$$\begin{aligned}f_{15}^{\alpha\beta}\left(x^2,\frac{\mu^2}{p^2},\frac{M^2}{p^2},g\right)=\frac{1}{12\pi^2}g^2\Bigg\{&[3\theta(x-1)+3x\theta(1-x)\tfrac{1}{2}(\theta_\beta M\gamma^0\theta_\alpha\gamma^0+\alpha\leftrightarrow\beta)\\&+\left[\left(-\frac{3}{2}+\frac{1}{x}\right)\theta(x-1)+\frac{x^2}{2}\theta(1-x)\right]\tfrac{1}{2}(\{\theta_\beta,\theta_\alpha\}\gamma^0M\gamma^0+\gamma^0M\{\theta_\beta,\theta_\alpha\}\gamma^0)\Bigg\}\\&+O\left(\frac{\mu^2}{p^2},\frac{M^2}{p^2}\right),\end{aligned} \tag{D2f}$$

$$f_{55}^{nn'mm'}\left(x,\frac{\mu^2}{p^2},\frac{M^2}{p^2}g\right)=\frac{3}{16\pi^2}g^2\{[\theta(x-1)+x\theta(1-x)][(\gamma^0\theta^\alpha\gamma^0)^{nn'}(\theta^\beta)^{m'm}\delta_{\alpha\beta}]\}+O\left(\frac{\mu^2}{p^2},\frac{M^2}{p^2}\right), \tag{D2g}$$

$$f_{11}^{11}(x)=\frac{g^2}{16\pi^2}\tfrac{1}{6}C_{\alpha\beta\lambda}C_{\gamma\delta\lambda}\left[\left(3+\frac{25}{4x}-\frac{3}{4x^2}\right)\theta(x-1)+\left(3+\frac{25}{4x}-\frac{3}{4x^2}\right)x\theta(1-x)\right], \tag{D2h}$$

$$f_{13}^{11}=\frac{g^2}{16\pi^2}\tfrac{1}{6}C_{\alpha\beta\lambda}C_{\gamma\delta\lambda}\left[\left(\frac{1}{2}-\frac{1}{4x}\right)\theta(x-1)+(\tfrac{1}{2}-\tfrac{1}{4}x)x^2\theta(1-x)\right], \tag{D2i}$$

$$f_{13}^{10}=\frac{g^2}{16\pi^2}\tfrac{1}{6}C_{\alpha\beta\lambda}C_{\gamma\delta\lambda}\left[\left(\frac{1}{4}-\frac{1}{4x}\right)\theta(x-1)-\left(\frac{x}{4}-\frac{x^2}{4}\right)\theta(1-x)\right]\frac{1}{xp^2}, \tag{D2j}$$

$$f_{33}^{11}=\frac{g^2}{16\pi^2}\tfrac{1}{2}C_{\alpha\beta\gamma}C_{\gamma\delta\lambda}\left[\frac{1}{2x}\theta(x-1)+\frac{x^2}{2}\theta(1-x)\right], \tag{D2k}$$

$$f_{33}^{10}=-\frac{g^2}{16\pi^2}\tfrac{1}{2}C_{\alpha\beta\gamma}C_{\gamma\delta\lambda}\left[\left(\frac{3}{4}-\frac{1}{x}\right)\theta(x-1)-\frac{x^2}{4}\theta(1-x)\right]\frac{1}{xp^2}. \tag{D2l}$$

*Work supported in part by the National Science Foundation.

†Work supported in part by the U. S. Atomic Energy Commission.

[1]G. 't Hooft, Nucl. Phys. B33, 173 (1971).

[2]B. W. Lee and J. Zinn-Justin, Phys. Rev. D 5, 3121 (1972).

[3]A comprehensive review of this work may be found in B. W. Lee, in *Proceedings of the XVI International Conference on High Energy Physics, Chicago-Batavia, Ill., 1972,* edited by J. D. Jackson and A. Roberts (NAL, Batavia, Ill., 1973), Vol. 4, p. 249.

[4]See S. Weinberg, Phys. Rev. Lett. 31, 494 (1973), and the references therein. Also I. Bars, M. B. Halpern, and M. Yoshimura, *ibid.* 29, 969 (1972); H. Georgi and S. Glashow, *ibid.* 32, 438 (1974).

[5]Y. Nambu and G. Jona-Lasinio, Phys. Rev. 122, 345 (1961). Also Y. Freundlich and D. Lurié, Nucl. Phys. B19, 557 (1970).

[6]J. Schwinger, Phys. Rev. 125, 397 (1962).

[7]In reality the Schwinger mechanism only eliminates the pole at $q^2=0$. It is possible that the pole does not occur at any finite q^2 and that the particle does not appear in the spectrum.

[8]J. Schwinger, Phys. Rev. 128, 2325 (1962).

[9]K. Johnson, Phys. Lett. 5, 253 (1963); D. Gross and R. Jackiw (unpublished); S.-S. Shei, Phys. Rev. D 6, 3469 (1972).

[10]P. W. Higgs, Phys. Lett. 12, 132 (1964); Phys. Rev. Lett. 13, 508 (1964); Phys. Rev. 145, 1156 (1966); F. Englert and R. Brout, Phys. Rev. Lett. 13, 321 (1964); G. S. Guralnik, C. R. Hagen, and T. W. B. Kibble, *ibid.* 13, 585 (1964).

[11]One may also consider the Higgs mechanism with composite scalars; see T. Goldman and P. Vinciarelli, this issue, Phys. Rev. D 10, 3431 (1974).

[12]K. Milton (unpublished).

[13]R. Jackiw and K. Johnson, Phys. Rev. D 8, 2386 (1973).

[14]J. M. Cornwall and R. Norton, Phys. Rev. D 8, 3338 (1973).

[15]F. Englert and R. Brout, Phys. Rev. Lett. 13, 321 (1964).

[16]H. Pagels, Phys. Rev. D 7, 3689 (1973).

[17]The relevance of these ideas to strong interactions has been pointed out by J. J. Sakurai, Ann. Phys. (N.Y.) 11, 1 (1960).

[18]S. Sarkar (unpublished).

[19]J. M. Cornwall, Phys. Rev. D 10, 500 (1974).

[20]E. Poggio, E. Tomboulis, and S. Tye (unpublished).

[21]This assumption is made to simplify our equations, but is not essential to the discussion.

[22]M. Weinstein, Phys. Rev. D 8, 2511 (1973).

[23]N. Nakanishi, Prog. Theor. Phys. Suppl. 43, 1 (1969).

[24]J. Smit, Phys. Rev. D 10, 2473 (1974).

[25]D. Gross and F. Wilczek, Phys. Rev. D 8, 3633 (1973). Also see T. P. Cheng, E. Eichten, and L.-F. Li, *ibid.* 9, 2259 (1974).

[26]B. W. Lee, Phys. Lett. 46B, 214 (1963).

[27]K. Johnson, *Ninth Latin American School of Physics,* edited by I. Saavedra (Benjamin, Reading, Mass., 1968).

PHYSICAL REVIEW D VOLUME 10, NUMBER 8 15 OCTOBER 1974

Effective action for composite operators*

John M. Cornwall†
Department of Physics, University of California, Los Angeles, California 90024

R. Jackiw and E. Tomboulis
Laboratory for Nuclear Science and Department of Physics, Massachusetts Institute of Technology, Cambridge, Massachusetts 02139
(Received 24 June 1974)

An effective action and potential for composite operators is obtained. The formalism is used to analyze, by variational techniques, dynamical symmetry breaking and coherent solutions to field theory. A Rayleigh-Ritz procedure is introduced which replaces arbitrary variations with parametric variations. Previously unsolved nonlinear equations become, in the Rayleigh-Ritz approximation, solvable algebraic equations.

I. INTRODUCTION

Field-theoretic descriptions of natural processes suffer from a serious shortcoming: The only available method for solving dynamical equations is the perturbative expansion. Yet it is clear that there exist phenomena, apparently physically important, which cannot be easily seen in the perturbation series. Examples are spontaneous symmetry violation, bound states, entrapment of various experimentally unobserved excitations, etc. What is needed is an approximation scheme that preserves some of the nonlinear features of field theory, which presumably lead to these cooperative and coherent effects.

Recently in the course of various investigations of spontaneous symmetry violation at zero and finite temperature, it became possible to sum large classes of ordinary perturbation-series diagrams which contribute to the effective action $\Gamma(\phi)$ (the generating functional of single-particle irreducible n-point functions), and which preserve a much richer nonlinear structure than the familiar classical (tree) approximation.[1-3] In the present paper we continue that development. We study a generalization of the effective action, $\Gamma(\phi, G)$, which depends not only on $\phi(x)$—a possible expectation value of the quantum field $\Phi(x)$—but also on $G(x,y)$—a possible expectation value of $T\Phi(x)\Phi(y)$. Physical solutions require

$$\frac{\delta\Gamma(\phi,G)}{\delta\phi(x)}=0, \tag{1.1a}$$

$$\frac{\delta\Gamma(\phi,G)}{\delta G(x,y)}=0. \tag{1.1b}$$

Hence the formalism is especially appropriate for the study of dynamical symmetry violation, which is characterized by the fact that even though (1.1a) has only the symmetric solution $\phi(x)=0$, symmetry-breaking solutions exist for (1.1b).

In Sec. II we define $\Gamma(\phi, G)$ and derive a series expansion for it which is analogous to the WKB loop expansion previously obtained for $\Gamma(\phi)$.[4] In Sec. III we show that the bubble sum for $\Gamma(\phi)$, which has been recently performed[1-3] and which is dominant in an O(N)-invariant spinless theory for large N, is trivially obtained in the present formalism. It corresponds to a single graph—that of the Hartree-Fock approximation. Section IV is devoted to spontaneous symmetry violation by bound states. We show that the Hartree-Fock approximation to $\Gamma(\phi, G)$ leads to a gap equation for G, which upon linearization gives the ladder Bethe-Salpeter model that recently has been offered as an example of dynamical symmetry violation.[5] We also demonstrate how the nonlinear aspects can be analyzed. Rather than considering the arbitrary variation (1.1), which leads to an intractable nonlinear integral equation, we perform a Rayleigh-Ritz variation which gives a nonlinear algebraic equation. In Sec. V we adopt our formalism to the study of time-independent but position-dependent solutions to (1.1). We show that $\Gamma(\phi, G)$ in that case corresponds to the stationary expectation of the Hamiltonian in a normalized state $|\psi\rangle$ for which

$$\begin{aligned}&\langle\psi|\Phi(x)|\psi\rangle=\phi(\vec{x}),\\&\langle\psi|\Phi(x)\Phi(y)|\psi\rangle|_{x_0=y_0}=\phi(\vec{x})\phi(\vec{y})+\hbar G(\vec{x},\vec{y}).\end{aligned} \tag{1.2}$$

Now, in the Hartree-Fock approximation, (1.1) becomes equivalent to the variational equations derived by Kuti,[6] in his interesting development of a functional Schrödinger picture for field theory.

Studies of the sort here presented were initiated years ago by Lee and Yang,[7] and others.[8] These authors concerned themselves with nonrelativistic statistical mechanics. With the exception of a few isolated works,[9] little has been done to extend and apply these techniques to relativistic field theories. We hope that our use of functional methods to replace combinatorial analysis makes the general proofs more transparent, and that our sample applications demonstrate the utility of these ideas for practical calculations.

II. THE LOOP EXPANSION FOR $\Gamma(\phi,G)$

A. Definitions

We define $Z(J,K)$, the generating functional for Green's functions of nonlocal, composite fields (the derivation will be carried out for Bose fields; the generalization to fermions is trivial, and will be indicated in Sec. IV):

$$Z(J,K)=e^{(i/\hbar)W(J,K)}$$
$$=\int d\Phi \exp\left\{\frac{i}{\hbar}\left[I(\Phi)+\int d^4x\,\Phi(x)J(x)+\tfrac{1}{2}\int d^4x\,d^4y\,\Phi(x)K(x,y)\Phi(y)\right]\right\}. \tag{2.1}$$

The Φ integration is functional. $I(\Phi)$ is the classical effective action

$$I(\Phi)=\int d^4x\,\mathcal{L}(x). \tag{2.2a}$$

$\mathcal{L}$ is the effective Lagrangian, containing gauge and ghost terms if a gauge theory is discussed. The field $\Phi(x)$ can possess components; the specifying index is suppressed. The classical action (2.2a) may also be written as

$$I(\Phi)=\int d^4x\,d^4y\,\Phi(x)iD^{-1}(x-y)\Phi(y)+I_{\rm int}(\Phi),$$
$$I_{\rm int}(\Phi)=\int d^4x\,\mathcal{L}_{\rm int}(x), \tag{2.2b}$$

where $D(x-y)$ is the free propagator

$$iD^{-1}(x-y)=-(\Box+m^2)\delta^4(x-y) \tag{2.3}$$

and the interaction Lagrangian $\mathcal{L}_{\rm int}$ is at least cubic in the fields.

$\Gamma(\phi,G)$ is a double Legendre transform of $W(J,K)$. We define

$$\frac{\delta W(J,K)}{\delta J(x)}=\phi(x), \tag{2.4a}$$

$$\frac{\delta W(J,K)}{\delta K(x,y)}=\tfrac{1}{2}[\phi(x)\phi(y)+\hbar G(x,y)]. \tag{2.4b}$$

Eliminate J and K in favor of ϕ and G and set

$$\Gamma(\phi,G)=W(J,K)-\int d^4x\,\phi(x)J(x)-\tfrac{1}{2}\int d^4x\,d^4y\,\phi(x)K(x,y)\phi(y)-\tfrac{1}{2}\hbar\int d^4x\,d^4y\,G(x,y)K(y,x). \tag{2.5}$$

Evidently it is also true that

$$\frac{\delta\Gamma(\phi,G)}{\delta\phi(x)}=-J(x)-\int d^4y\,K(x,y)\phi(y), \tag{2.6a}$$

$$\frac{\delta\Gamma(\phi,G)}{\delta G(x,y)}=-\tfrac{1}{2}\hbar K(x,y). \tag{2.6b}$$

Since physical processes correspond to vanishing sources J and K, Eqs. (2.6) provide a derivation of the stationarity requirement (1.1).

Let us observe that the conventional effective action is merely $\Gamma(\phi,G)$ at $K=0$, or equivalently from (2.6b), it is $\Gamma(\phi,G)$ at that value of $G(x,y)$ for which (2.6b) vanishes:

$$\Gamma(\phi)=\Gamma(\phi,G_0),\qquad \frac{\delta\Gamma(\phi,G_0)}{\delta G_0(x,y)}=0. \tag{2.7}$$

Furthermore it is known that $\Gamma(\phi,G)$ is the generating functional in ϕ for *two*-particle irreducible Green's functions expressed in terms of the propagator G.[10] For example the diagrammatic expansion of

$$\left.\frac{\delta\Gamma(\phi,G)}{\delta\phi(x)\delta\phi(y)}\right|_{\phi=0}$$

is the Feynman-Dyson series for the inverse 2-point function of the theory, with *two*-particle reducible graphs deleted, and with lines representing $\hbar G(x,y)$. (The reader may convince himself of this by working out all the differentiations.[11])

We now describe the series expansion for $\Gamma(\phi,G)$. We introduce the functional operator $\mathfrak{D}^{-1}(\phi)$ by the definition

$$i\mathfrak{D}^{-1}\{\phi;x,y\}=\frac{\delta^2 I(\phi)}{\delta\phi(x)\delta\phi(y)}=iD^{-1}(x-y)+\frac{\delta^2 I_{\rm int}(\phi)}{\delta\phi(x)\delta\phi(y)}. \tag{2.8}$$

The required series is

$$\Gamma(\phi,G)=I(\phi)+\tfrac{1}{2}i\hbar\,{\rm Tr\,Ln}G^{-1}+\tfrac{1}{2}i\hbar\,{\rm Tr}\,\mathfrak{D}^{-1}(\phi)G+\Gamma_2(\phi,G)+{\rm const}. \tag{2.9a}$$

The trace, the logarithm, and the product $\mathfrak{D}^{-1}G$ in the second and third terms are taken in the functional sense. The constant, independent of ϕ and G, is evaluated so that (2.7) is satisfied:

$$\Gamma(\phi,G)=I(\phi)+\tfrac{1}{2}i\hbar\,{\rm Tr\,Ln}DG^{-1}+\tfrac{1}{2}i\hbar\,{\rm Tr}\mathfrak{D}^{-1}(\phi)G+\Gamma_2(\phi,G)-\tfrac{1}{2}i\hbar\,{\rm Tr}1. \tag{2.9b}$$

$\Gamma_2(\phi,G)$ is computed as follows. In the classical action $I(\Phi)$ shift the field Φ by $\phi(x)$. The new action $I(\Phi+\phi)$ possesses terms cubic and higher in Φ; these define an "interaction" part $I_{\rm int}(\phi;\Phi)$ where the vertices depend on $\phi(x)$. $\Gamma_2(\phi,G)$ is given by all the two-particle irreducible vacuum graphs in a "theory" with vertices determined by $I_{\text{int}}(\phi;\Phi)$ and propagators set equal to $G(x,y)$. Another way to say it is that only those vacuum graphs are kept which, upon opening one line, yield proper self-energy graphs. Since the vertices depend on $\phi(x)$, and $G(x,y)$ is not merely a function of $x-y$, this is not a translationally invariant theory. Nevertheless, $\Gamma_2(\phi,G)$ is easily determined by the usual Feynman-Dyson-Wick expansion. $\Gamma_2(\phi,G)$ is of order $\hbar^2$, and the number of loops corresponds to powers of $\hbar$.

In subsequent sections, formula (2.9) will be evaluated in illustrative examples and applied in various contexts. The remainder of Sec. II is devoted to the derivation.

B. Derivation of (2.9)

The derivation of (2.9) is facilitated by two observations. First, note that inasmuch as $\Gamma(\phi,G)$ is the generating functional (in ϕ) for two-particle irreducible n-point functions for the theory governed by the action $I(\Phi)$ (and lines set equal to $\hbar G$), it follows that $\Gamma(0,G)$ is the sum of all two-particle irreducible vacuum graphs of the same theory. According to (2.5) and (2.6) this is also given by

$$\Gamma(0,G)=\mathrm{Tr}\,G\frac{\partial\Gamma(0,G)}{\partial G}-i\hbar\ln\int d\Phi\exp\left\{\frac{i}{\hbar}\left[\tfrac{1}{2}\Phi iD^{-1}\Phi+I_{\text{int}}(\Phi)+\Phi J^0-\frac{1}{\hbar}\Phi\frac{\partial\Gamma(0,G)}{\partial G}\Phi\right]\right\}$$

$$+i\hbar\ln\int d\Phi\exp\left\{\frac{i}{\hbar}\left[\tfrac{1}{2}\Phi iD^{-1}\Phi\right]\right\}$$

$$=\text{two-particle irreducible vacuum graphs of a theory governed by } I(\Phi),\text{ with lines representing } \hbar G. \tag{2.10}$$

(In the remainder of Sec. II we use a compact notation where all integrations are suppressed and derivatives are functional.) In (2.10) J^0 is that value of J which makes $\partial W(J,K)/\partial J=\phi$ vanish, i.e., all tadpoles are removed. The normalization factor, frequently omitted since it is a constant, is here explicitly exhibited.

Second, observe that the double Legendre transform (2.4) and (2.5) can also be performed sequentially. Thus we may set, at fixed K,

$$\Gamma^K(\phi)=W(J,K)-\phi J. \tag{2.11}$$

Then we define

$$\frac{\partial\Gamma^K(\phi)}{\partial K}=\tfrac{1}{2}(\phi\phi+\hbar G), \tag{2.12}$$

$$\Gamma(\phi,G)=\Gamma^K(\phi)-\tfrac{1}{2}\phi K\phi-\tfrac{1}{2}\hbar\,\mathrm{Tr}\,GK. \tag{2.13}$$

That the definition of $\Gamma(\phi,G)$ in (2.12) and (2.13) is equivalent to (2.4) and (2.5) follows from the equality

$$\frac{\partial\Gamma^K(\phi)}{\partial K}=\frac{\partial W(J,K)}{\partial J}\frac{\partial J}{\partial K}\bigg|_\phi+\frac{\partial W(J,K)}{\partial K}-\frac{\partial J}{\partial K}\bigg|_\phi\phi$$

$$=\frac{\partial W(J,K)}{\partial K}. \tag{2.14}$$

To establish (2.9) we use (2.13). According to (2.1) and (2.13), $\Gamma^K(\phi)$ is the effective action for a theory governed by the classical action

$$I^K(\Phi)=I(\Phi)+\tfrac{1}{2}\Phi K\Phi. \tag{2.15}$$

Hence according to previous analysis[4]

$$\Gamma^K(\phi)=I^K(\phi)+\Gamma_1^K(\phi), \tag{2.16a}$$

where $\Gamma_1^K(\phi)$ has the representation

$$\Gamma_1^K(\phi)=-i\hbar\ln\int d\Phi\exp\left\{\frac{i}{\hbar}\left[I^K(\Phi+\phi)-I^K(\phi)-\Phi\frac{\partial I^K(\phi)}{\partial\phi}-\Phi\frac{\partial\Gamma_1^K(\phi)}{\partial\phi}\right]\right\}$$

$$=-i\hbar\ln\int d\Phi\exp\left\{\frac{i}{\hbar}\left[\tfrac{1}{2}\Phi[i\mathcal{D}^{-1}(\phi)+K]\Phi+I_{\text{int}}(\phi,\Phi)-\Phi\frac{\partial\Gamma_1^K(\phi)}{\partial\phi}\right]\right\}. \tag{2.16b}$$

Upon equating (2.9) [which is here viewed as defining $\Gamma_2(\phi,G)$] with (2.13) and using (2.15) and (2.16), we find

$$\Gamma_2(\phi,G)+\text{const}=-\tfrac{1}{2}\hbar\,\mathrm{Tr}[i\mathcal{D}^{-1}(\phi)+K]G-\tfrac{1}{2}i\hbar\,\mathrm{Tr}\,\mathrm{Ln}G^{-1}+\Gamma_1^K(\phi). \tag{2.17}$$

The proof of (2.9) will now follow, if it can be shown that $\Gamma_2(\phi,G)$, as given by (2.17), is the sum of two-particle irreducible vacuum graphs governed by vertices of $I_{\text{int}}(\phi;\Phi)$ and propagators G.

We proceed by eliminating K in (2.17). According to (2.6) and (2.9), K satisfies

$$-\tfrac{1}{2}\hbar K=-\tfrac{1}{2}i\hbar G^{-1}+\tfrac{1}{2}i\hbar\mathfrak{D}^{-1}(\phi)+\frac{\partial\Gamma_2(G)}{\partial G}\,. \tag{2.18}$$

Hence (2.17) is

$$\Gamma_2(\phi,G)+\text{const}=\text{Tr}G\frac{\partial\Gamma_2(\phi,G)}{\partial G}-i\hbar\ln\int d\Phi\exp\left\{\frac{i}{\hbar}\left[\tfrac{1}{2}\Phi iG^{-1}\Phi+I_{\text{int}}(\phi;\Phi)-\Phi\frac{\partial\Gamma_1^K(\phi)}{\partial\phi}-\frac{1}{\hbar}\Phi\frac{\partial\Gamma_2(\phi,G)}{\partial G}\Phi\right]\right\}$$
$$+i\hbar\ln\int d\Phi\exp\left\{\frac{i}{\hbar}\left[\tfrac{1}{2}\Phi G^{-1}\Phi\right]\right\}. \tag{2.19}$$

Comparing (2.19) with (2.10), we see that $\Gamma_2(\phi,G)$ is precisely the sum of two-particle irreducible vacuum graphs in a theory governed by the action $\frac{1}{2}\Phi iG^{-1}\Phi+I_{\text{int}}(\phi;\Phi)$, since it has already been previously shown[4] that $\partial\Gamma_1^K(\phi)/\partial\phi$ is precisely that value of an external current which makes tadpoles vanish. This completes the proof.

C. Discussion of (2.9)

The series for $\Gamma(\phi,G)$ may also be understood in the following way. According to its definition, $\Gamma(\phi,G)$ is given by

$$e^{i\Gamma(\phi,G)/\hbar}=\int d\Phi\exp\left\{\frac{i}{\hbar}\left[I(\Phi)+(\Phi-\phi)J+\tfrac{1}{2}\Phi K\Phi-\tfrac{1}{2}\phi K\phi-\tfrac{1}{2}\hbar\,\text{Tr}GK\right]\right\}$$
$$=\int d\Phi\exp\left\{\frac{i}{\hbar}\left[I(\Phi)-(\Phi-\phi)\frac{\partial\Gamma(\phi,G)}{\partial\phi}-\frac{1}{\hbar}(\Phi-\phi)\frac{\partial\Gamma(\phi,G)}{\partial G}(\Phi-\phi)+\text{Tr}G\frac{\partial\Gamma(\phi,G)}{\partial G}\right]\right\}. \tag{2.20a}$$

Upon shifting by ϕ, which removes the one-particle reducible graphs, we find

$$\Gamma(\phi,G)-\text{Tr}G\frac{\partial\Gamma(\phi,G)}{\partial G}=-i\hbar\ln\int d\Phi\exp\left\{\frac{i}{\hbar}I(\phi,G;\Phi)\right\},$$
$$I(\phi,G;\Phi)=I(\Phi+\phi)-\Phi\frac{\partial\Gamma(\phi,G)}{\partial\phi}-\frac{1}{\hbar}\Phi\frac{\partial\Gamma(\phi,G)}{\partial G}\Phi. \tag{2.20b}$$

Varying this with respect to G gives

$$-\text{Tr}G\frac{\partial^2\Gamma(\phi,G)}{\partial G\partial G}\int d\Phi\exp\left\{\frac{i}{\hbar}I(\phi,G;\Phi)\right\}=-\frac{\partial\Gamma(\phi,G)}{\partial\phi\partial G}\int d\Phi\,\Phi\exp\left\{\frac{i}{\hbar}I(\phi,G;\Phi)\right\}$$
$$-\frac{1}{\hbar}\int d\Phi\,\Phi\frac{\partial^2\Gamma(\phi,G)}{\partial G\partial G}\Phi\exp\left\{\frac{i}{\hbar}I(\phi,G;\Phi)\right\}. \tag{2.21a}$$

The first term on the right-hand side vanishes, since the expectation of the field Φ in the theory with the action $I(\phi,G;\Phi)$ is zero. [This is a consequence of the shift which was performed in passing from (2.20a) to (2.20b) and was established explicitly previously.[4]] The remaining terms in (2.21a) therefore imply

$$\hbar G=\frac{\int d\Phi\,\Phi\Phi\exp\{(i/\hbar)I(\phi,G;\Phi)\}}{\int d\Phi\exp\{(i/\hbar)I(\phi,G;\Phi)\}}\,. \tag{2.21b}$$

This means that G is the exact connected propagator of the theory.

Turning now to (2.18), we see that

$$G^{-1}=\mathfrak{D}^{-1}(\phi)-iK-\Sigma(\phi,G), \tag{2.22a}$$

where

$$\Sigma(\phi,G)=\frac{2i}{\hbar}\frac{\partial\Gamma_2(\phi,G)}{\partial G}\,. \tag{2.22b}$$

But (2.22a) is just the Schwinger-Dyson equation for the propagator and $\Sigma(\phi,G)$ is to be interpreted as the proper self-energy part, with *no* propagator insertions. However, since Σ is also given by a derivative with respect to G of Γ_2, Γ_2 must be two-particle irreducible. For if Γ_2 has a two-particle reducible contribution of the form $\bar{\Gamma}GG\bar{\Gamma}'$, then Σ would have a contribution of the form $\bar{\Gamma}G\bar{\Gamma}'$; but such structures do not belong in Σ. The absence of two-particle reducible contributions to Γ_2 is a consequence of the fact that the propagator G has no radiative corrections, and is exact.

D. Concluding remarks

Frequently one is interested only in translation-invariant solutions. In that case, we set $\phi(x)$ to a constant ϕ, and take $G(x,y)$ to be a function only of $x-y$. A generalization of the effective potential may be defined by

$$V(\phi,G)\int d^4x=-\Gamma(\phi,G)\big|_{\text{translation invariant}}\,. \tag{2.23}$$

The series for $V(\phi, G)$ can be easily obtained from (2.9). We define the Fourier-transformed propagators

$$G(p)=\int d^4x\, e^{ip(x-y)}G(x-y), \tag{2.24a}$$

$$\mathfrak{D}\{\phi;p\}=\int d^4x\, e^{ip(x-y)}\mathfrak{D}\{\phi;x-y\}, \tag{2.24b}$$

$$D(p)=\int d^4x\, e^{ip(x-y)}D(x-y). \tag{2.24c}$$

Equation (2.9) reduces to

$$V(\phi,G)=U(\phi)-\tfrac{1}{2}i\hbar\int\frac{d^4p}{(2\pi)^4}\ln\det D(p)G^{-1}(p)$$
$$-\tfrac{1}{2}i\hbar\int\frac{d^4p}{(2\pi)^4}\operatorname{tr}[\mathfrak{D}^{-1}\{\phi;p\}G(p)-1]$$
$$+V_2(\phi,G). \tag{2.25}$$

The determinant and the trace are no longer functional; they apply to the component degrees of freedom. $U(\phi)$ is the classical potential; $-V_2(\phi, G)$ is the sum of all the two-particle irreducible vacuum graphs of the theory with vertices given by $I_{\text{int}}(\phi; \Phi)$ and propagator $G(p)$. The vertices still depend on ϕ, but this is now a constant parameter. Since translation invariance is maintained, an overall factor of space-time volume must be removed.

In terms of $V(\phi, G)$, the stationarity requirements become

$$\frac{\partial V(\phi,G)}{\partial\phi}=0, \tag{2.26a}$$

$$\frac{\partial V(\phi,G)}{\partial G}=0. \tag{2.26b}$$

$V(\phi, G)$ is a function of ϕ and a functional of $G(p)$. Hence only the second derivative in (2.26) is functional.

The sequences of Legendre transforms may be continued. In this way an effective action can be defined which depends functionally not only on ϕ and G, but also on irreducible 3-point, 4-point, etc. functions. The obvious generalization of (1.1) is the requirement that the effective action is stationary with respect to independent variations of any irreducible Green's function. Details of this will be given elsewhere.[12]

III. AN O(N)-INVARIANT SPINLESS MODEL

As an illustration of the general formalism, we now present an analysis of an O(N)-invariant spinless model, governed by the Lagrangian

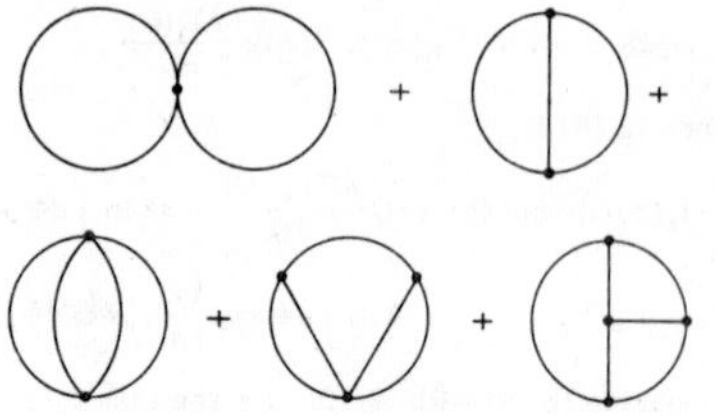

FIG. 1. Two-particle irreducible graphs contributing to $\Gamma_2(\phi, G)$ up to the three-loop level in a $\lambda\Phi^4$ theory. The solid line represents the propagator $\hbar G(x,y)$. There are two kinds of vertices: a three-point vertex proportional to $\lambda\phi$ and a four-point vertex.

$$\mathcal{L}=\tfrac{1}{2}\partial_\mu\Phi_a\partial^\mu\Phi_a-\tfrac{1}{2}m^2\Phi^2-\frac{\lambda}{4!N}\Phi^4,$$
$$\Phi^2=\Phi_a\Phi_a,\quad \Phi^4=(\Phi^2)^2,\quad a=1,\ldots,N. \tag{3.1}$$

The propagator $i\mathfrak{D}^{-1}(\phi)$ is

$$i\mathfrak{D}_{ab}^{-1}\{\phi;x,y\}=-\left[\Box+m^2+\frac{\lambda}{6N}\phi^2(x)\right]\delta_{ab}\delta^4(x-y)$$
$$-\frac{\lambda}{3N}\phi_a(x)\phi_b(x)\delta^4(x-y). \tag{3.2}$$

Vertices of the shifted theory are given by the interaction Lagrangian

$$\mathcal{L}_{\text{int}}(\phi;\Phi)=-\frac{\lambda}{6N}\phi_a(x)\Phi_a(x)\Phi^2(x)$$
$$-\frac{\lambda}{4!N}\Phi^4(x). \tag{3.3}$$

Consequently the diagrams contributing to $\Gamma_2(\phi, G)$ are as depicted in Fig. 1, up to three-loop contributions. Each line represents the propagator $\hbar G_{ab}(x,y)$, and there are two kinds of vertices:

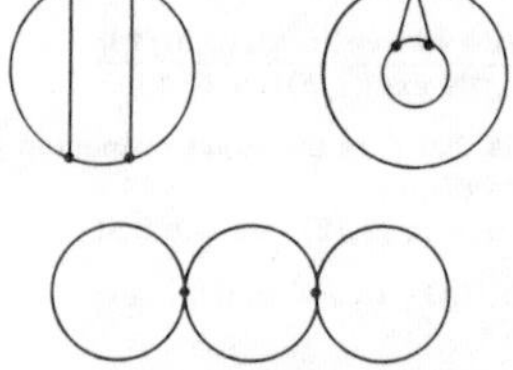

FIG. 2. Two-particle reducible graphs which do not contribute to $\Gamma_2(\phi, G)$ in a $\lambda\Phi^4$ theory. [These graphs do contribute to the ordinary effection action $\Gamma(\phi)$. In that instance the solid line represents the propagator $\hbar\mathfrak{D}\{\phi;x,y\}$.]

a four-point vertex proportional to λ and a three-point vertex, arising from the shift, proportional to $\lambda\phi_a(x)$. The numerical factors are not indicated; they are determined in the usual fashion by Wick's theorem.

[If we were computing the ordinary effective action $\Gamma(\phi)$, then the lines would represent the propagator $\hbar\mathfrak{D}_{ab}\{\phi;x,y\}$ and there would be additional contributions which are two-particle reducible. On the three-loop level these are depicted in Fig. 2.]

We shall evaluate $\Gamma(\phi, G)$ in the Hartree-Fock approximation, which corresponds to retaining only that contribution to $\Gamma_2(\phi, G)$ which is lowest-order in coupling constant. The relevant graph is the first entry, the double bubble, of Fig. 1:

$$\Gamma(\phi, G)+\text{const}=I(\phi)+\tfrac{1}{2}i\hbar\,\mathrm{Tr}\,\mathrm{Ln}G^{-1}+\tfrac{1}{2}i\hbar\,\mathrm{Tr}\mathfrak{D}^{-1}(\phi)G-\frac{\lambda}{4!N}\hbar^2\int d^4x[G_{aa}(x,x)G_{bb}(x,x)+2G_{ab}(x,x)G_{ba}(x,x)]\,. \tag{3.4}$$

From this $\Gamma(\phi)$ can be obtained by solving for $G_{ab}(x,y)$:

$$\begin{aligned}\frac{\delta\Gamma(\phi,G)}{\delta G_{ab}(x,y)}&=0\\&=-\tfrac{1}{2}i\hbar G_{ab}^{-1}(x,y)+\tfrac{1}{2}i\hbar\,\mathfrak{D}_{ab}^{-1}\{\phi;x,y\}-\frac{\lambda\hbar^2}{12N}[\delta_{ab}G_{cc}(x,x)+2G_{ab}(x,x)]\,\delta^4(x-y),\end{aligned} \tag{3.5}$$

$$G_{ab}^{-1}(x,y)=\mathfrak{D}_{ab}^{-1}\{\phi;x,y\}+\frac{i\lambda\hbar}{6N}[\delta_{ab}G_{cc}(x,x)+2G_{ab}(x,x)]\,\delta^4(x-y).$$

Eliminating $\mathfrak{D}_{ab}^{-1}$ between (3.4) and (3.5) finally gives, apart from constants,

$$\begin{aligned}\Gamma(\phi)=I(\phi)&+\tfrac{1}{2}i\hbar\,\mathrm{Tr}\,\mathrm{Ln}G^{-1}\\&+\frac{\lambda\hbar^2}{4!N}\int d^4x[G_{aa}(x,x)G_{bb}(x,x)\\&+2G_{ab}(x,x)G_{ba}(x,x)],\end{aligned} \tag{3.6}$$

where $G_{ab}(x,y)$ satisfies the equation (3.5).

It is of course impossible to solve (3.5) for $G_{ab}(x,y)$ exactly. However, a simplification occurs if we consider the large-N limit, and keep terms dominant in N. In that limit ϕ_a is to be taken to be $O(\sqrt{N})$, hence $\mathfrak{D}_{ab}$ and G_{ab} are $O(1)$. We decompose

$$G_{ab}(x,y)=\delta_{ab}g(x,y)+\tilde{G}_{ab}(x,y), \tag{3.7}$$

where $\tilde{G}_{ab}(x,y)$ is traceless. To retain the $O(1)$ parts of (3.5) it is sufficient to keep only the first term in the bracket. Thus

$$g^{-1}(x,y)=i\left[\Box+m^2+\frac{\lambda}{6N}\phi^2(x)+\tfrac{1}{6}\lambda\hbar g(x,x)\right]\times\delta^4(x-y). \tag{3.8}$$

Upon defining

$$\chi(x)=m^2+\frac{\lambda}{6N}\phi^2(x)+\tfrac{1}{6}\lambda\hbar g(x,x) \tag{3.9a}$$

we find that

$$g^{-1}(x,y)=i[\Box+\chi(x)]\,\delta^4(x-y) \tag{3.9b}$$

and

$$\Box+\chi=\Box+m^2+\frac{\lambda}{6N}\phi^2-\tfrac{1}{6}i\lambda\hbar(\Box+\chi)^{-1}. \tag{3.10}$$

From (3.6)

$$\begin{aligned}\Gamma(\phi)&=\int d^4x\left(\tfrac{1}{2}\partial_\mu\phi_a\partial^\mu\phi_a-\tfrac{1}{2}m^2\phi^2-\frac{\lambda}{4!N}\phi^4\right)+\tfrac{1}{2}i\hbar N\,\mathrm{Tr}\,\mathrm{Ln}(\Box+\chi)+\frac{3N}{2\lambda}\int d^4x\left(\chi-m^2-\frac{\lambda}{6N}\phi^2\right)^2\\&=\int d^4x\left(\tfrac{1}{2}\partial_\mu\phi_a\partial^\mu\phi_a+\frac{3N}{2\lambda}\chi^2-\frac{3N}{\lambda}m^2\chi-\tfrac{1}{2}\phi^2\chi\right)+\tfrac{1}{2}i\hbar N\,\mathrm{Tr}\,\mathrm{Ln}(\Box+\chi).\end{aligned} \tag{3.11}$$

The result (3.11) has been previously obtained.[1-3] The power of the present formalism is now apparent: Only one graph has to be evaluated. Our method in this example is related to the combinatorial trick which has been previously utilized in the analysis of this problem.[3]

IV. DYNAMICAL SYMMETRY BREAKING

We construct the Hartree-Fock approximation to the generalized effective potential for an Abelian gauge theory of fermions and vector mesons which has recently been studied as an example of dynamical symmetry violation.[5] A gap equation is derived; in linearized form it coincides with the Bethe-Salpeter ladder equation which was previously solved.[5] A Rayleigh-Ritz procedure is developed to study the nonlinear aspects of the problem.

A. Effective potential

The Lagrangian is

$$\mathcal{L}=\bar{\psi}(i\not{\partial}-m-g_A\gamma_\mu A^\mu-g_B\tau_2\gamma_\mu B^\mu)\psi$$
$$-\tfrac{1}{4}A_{\mu\nu}A^{\mu\nu}-\tfrac{1}{4}B_{\mu\nu}B^{\mu\nu},$$
$$A^{\mu\nu}=\partial^\mu A^\nu-\partial^\nu A^\mu,$$
$$B^{\mu\nu}=\partial^\mu B^\nu-\partial^\nu B^\mu, \qquad (4.1)$$

where $\psi=(\psi_1,\psi_2)$ is a two-component field in "isospin" space and τ_2 is the usual Pauli matrix. When the gauge symmetry

$$\psi\rightarrow e^{i\theta\tau_2}\psi,$$
$$B_\mu\rightarrow B_\mu-\frac{1}{g_B}\partial_\mu\theta \qquad (4.2)$$

is spontaneously broken, the B meson picks up a mass M_B, and the masses of the fermions split from the symmetric value m by an amount $\pm\delta m$. There is a gauge symmetry for the A meson, also, which remains unbroken.

The generalized effective action for this problem will depend only on the complete propagators of the theory: $G(x,y)$ for the fermions and $\Delta_i^{\mu\nu}(x,y)$, $i=A,B$, for the vector mesons. A field dependence is not included, since we do not expect that any of the fields acquire a vacuum expectation value in the absence of sources. The formula (2.9) is now applicable, with the field variable eliminated and with the following modification which reflects Fermi statistics: All factors of $-\frac{1}{2}$ appearing in (2.9) are replaced by 1.[13] Moreover, since we seek a translation-invariant solution, only the effective potential is of interest.

Thus for our problem we have

$$V(G,\Delta_i)=-i\int\frac{d^4p}{(2\pi)^4}\,\mathrm{tr}[\ln S^{-1}(p)G(p)-S^{-1}(p)G(p)+1]$$
$$+\tfrac{1}{2}i\sum_{i=A,B}\int\frac{d^4p}{(2\pi)^4}\,\mathrm{tr}[\ln D^{-1}(p)\Delta_i(p)-D^{-1}(p)\Delta_i(p)+1]+V_2(G,\Delta_i). \qquad (4.3)$$

Space-time indices on the boson propagators have been suppressed, and in Sec. IV $\hbar=1$. $S(p)$ and $D^{\mu\nu}(p)$ are the free propagators:

$$S(p)=\frac{i}{\not{p}-m},$$
$$D^{\mu\nu}(p)=\frac{-i}{p^2}P^{\mu\nu}, \qquad (4.4)$$
$$P^{\mu\nu}=g^{\mu\nu}-p^\mu p^\nu/p^2.$$

As is discussed below, the Landau gauge must be used for consistency with the subsequent analysis. In the Hartree-Fock approximation $V_2(G,\Delta_i)$ is given by the graphs of Fig. 3, where the solid lines represent $G(p)$, the wavy line represents $\Delta_A^{\mu\nu}(p)$, and the zigzag line represents $\Delta_B^{\mu\nu}(p)$. The analytic expression is

$$V_2(G,\Delta_i)=\tfrac{1}{2}i\sum_{i=A,B}\int\frac{d^4p\,d^4k}{(2\pi)^8}\,\mathrm{tr}\Gamma_i^\mu G(p)\Gamma_i^\nu G(p+k)$$
$$\times\Delta_{i,\mu\nu}(k),$$
$$\Gamma_A^\mu=g_A\gamma^\mu,\quad \Gamma_B^\mu=g_B\gamma^\mu\tau_2. \qquad (4.5)$$

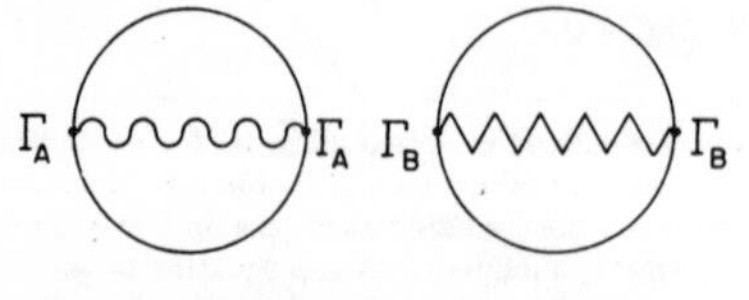

FIG. 3. Hartree-Fock approximation to $V_2(G,\Delta_i)$. The solid line is the fermion propagator G; the wavy line is the boson propagator Δ_A; the zigzag line is the boson propagator Δ_B. Γ_A and Γ_B represent the fermion-boson interactions and are defined in (4.5).

The reason for using the Landau gauge can now be explained. Our Hartree-Fock approximation replaces the complete $B^\mu\psi\bar{\psi}$ vertex by the free vertex. But in a spontaneously broken theory, this vertex has a Goldstone pole, which certainly is not in the bare vertex. The approximation therefore makes sense only in the Landau gauge, since then the Goldstone pole is annihilated in all vacuum graphs.

B. The gap equation

Demanding that $V(G,\Delta_i)$ be stationary against variations of G gives from (4.3) and (4.5)

$$G^{-1}=S^{-1}+\sum_{i=A,B}\Gamma_i G\Gamma_i\Delta_i. \qquad (4.6)$$

For notational simplicity, all integrations are suppressed. Equation (4.6) is also represented pictorially in Fig. 4. The symmetry-breaking part of G^{-1} is proportional to τ_3. Since S^{-1} has no such contribution, the symmetry-breaking part

FIG. 4. Equation satisfied by the fermion propagator. S is the free fermion propagator, given in (4.4).

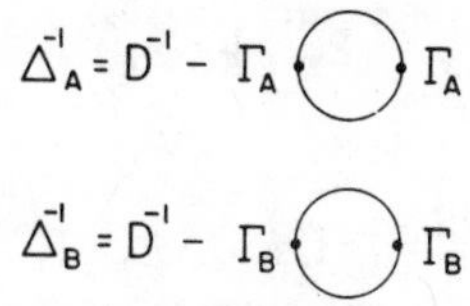

FIG. 5. Equations satisfied by the boson propagators. D is the free boson propagator, given in (4.4).

satisfies a homogeneous, nonlinear equation. Even when the meson propagators are replaced with their free form, Eq. (4.6) remains nonlinear and intractable. [In that approximation (4.6) is the gap equation proposed years ago by Johnson and others.[14]] When the equation is further simplified by linearizing the Fermi propagator, i.e., setting $G^{-1}=S^{-1}-\Sigma$, $G=S+S\Sigma S$, the symmetry-violating part of (4.6) coincides with the ladder Bethe-Salpeter equation, which, as has been shown previously,[5] possesses symmetry-breaking as well as symmetry-preserving solutions.

The equations for the Bose propagators are obtained by varying $V(G,\Delta_i)$ with respect to Δ_i. One finds

$$\Delta_i^{-1}=D^{-1}-\Gamma_i G\Gamma_i G. \tag{4.7}$$

Graphically this is represented in Fig. 5. In its linearized form, the equation, together with (4.6), implies that the B meson acquires a mass (see below).

We shall need the results of the linearized theory for the subsequent analysis. Hence we summarize them now.

C. Summary of linearized theory

The linearized theory is analyzed for $g_A^2, g_B^2 \ll 1$. For the fermion propagator, we define

$$G^{-1}=S^{-1}-\Sigma, \tag{4.8a}$$

$$\Sigma=\Sigma_N+\Sigma_S+\Sigma_V. \tag{4.8b}$$

Σ_V is the symmetry-violating part, proportional to τ_3. $\Sigma_N+\Sigma_S$ is symmetric, where Σ_N is that part of the self-energy which is also present in the normal solution while Σ_S, though symmetric, arises from the symmetry-violating properties of the theory. The linearized equation for Σ_V, which follows from (4.6), is given below and in Fig. 6, where the insertion V represents Σ_V (dashed lines are free propagators):

$$\Sigma_V=-\sum_{i=A,B}\Gamma_i S\Sigma_V S\Gamma_i D. \tag{4.9}$$

The solution to (4.9) is[5]

$$\Sigma_V(p^2) \underset{|-p^2|\gg m^2}{=} -i\delta m\left(\frac{-p^2}{m^2}\right)^{-\epsilon}\tau_3, \tag{4.10a}$$

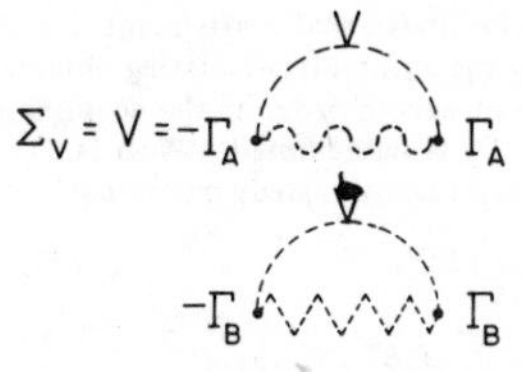

FIG. 6. Linearized gap equation for Σ_V, the symmetry-breaking part of the proper fermion self-energy. Σ_V is also denoted by V, and dashed lines represent free propagators.

$$\epsilon=\frac{3}{16\pi^2}(g_A^2-g_B^2)+O(g_A^2,g_B^2). \tag{4.10b}$$

Because ϵ is small, it is a good approximation to take

$$\Sigma_V(p^2) \underset{|-p^2|\le m^2}{=} -i\delta m\tau_3. \tag{4.10c}$$

(This ignores various threshold effects, which in any event are washed out in subsequent integrations.) The normal part Σ_N is quadratic in the coupling constant and is set to zero. Before discussing Σ_S, we turn to the meson propagators.

We define

$$\Delta_i^{\mu\nu}(p)=\frac{-iP^{\mu\nu}}{p^2-\Pi^i(p^2)}. \tag{4.11}$$

For the A propagator, which has no symmetry breaking, $\Pi^A(p^2)$ is dropped since it is $O(g_A^2)$. For the B propagator, G in (4.7) is expanded in powers of Σ, from (4.8a). We seek the symmetry-violating part, hence only Σ_V is kept. It is sufficient to go to second order in Σ_V. All other terms which are not kept are quadratic in the coupling constant. It was found[5] that the symmetry-violating part of $\Pi^B(p^2)$, $\Pi_V^B(p^2)$, behaves like a constant near $p^2=0$; and more recently it was shown that $\Pi_V^B(p^2)$ decreases at infinity[15]:

$$\begin{aligned}\Pi_V^B(p^2) &\underset{|-p^2|\gg m^2}{=} M_B^2\left(\frac{-p^2}{m^2}\right)^{-2\epsilon},\\ \Pi_V^B(p^2) &\underset{|-p^2|\le m^2}{=} M_B^2.\end{aligned} \tag{4.12}$$

The B-meson mass is also calculable[5]:

$$M_B^2=\frac{g_B^2}{2\pi^2\epsilon}(\delta m)^2+O(g_A^2,g_B^2). \tag{4.13}$$

We shall not use this formula for M_B^2; rather it shall be derived below. [The approximation scheme which yields (4.12) is not gauge-invariant in that the vacuum-polarization tensor for the B meson is not transverse. However, since we are in the Landau gauge, only the $g^{\mu\nu}$ part of that tensor survives when the complete propagator (4.11) is formed. The final results are gauge-invariant.[15]]

Note that the symmetry-violating objects Σ_V and $\Pi_V^B(p^2)$ are of zeroth order in the coupling constant.

Finally, Σ_S is determined. When (4.8) is inserted in (4.6) and iterated, one finds[16]

$$\Sigma_S = \Sigma_S^{(1)} + \Sigma_S^{(2)},$$

$$\Sigma_S^{(1)} = -\sum_{i=A,B} \Gamma_i S \Sigma_V S \Sigma_V S, \tag{4.14}$$

$$\Sigma_S^{(2)} = -\Gamma_B S \Gamma_B D \Pi_V^B D.$$

The graphical representation is in Fig. 7, where the large dot represents the symmetry-breaking B-meson self-energy given by (4.12). Note that Σ_S is proportional to the symmetry-breaking parameters $(\delta m)^2$ and $M_B{}^2$ and is second order in the coupling. For large p, $\Sigma_S(p)$ decreases like $(\not{p}/p^2)(-p^2/m^2)^{-2\epsilon}$.

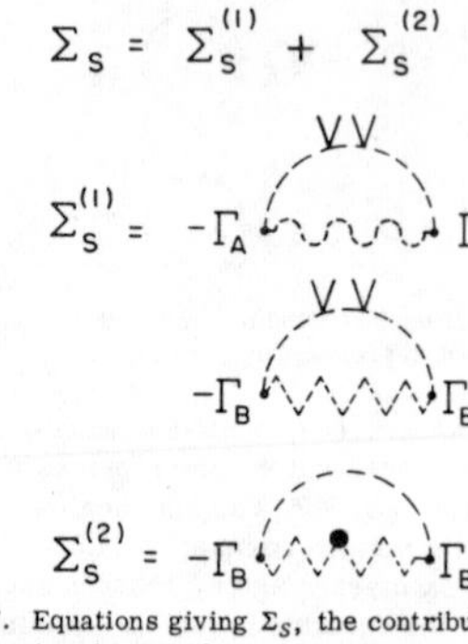

FIG. 7. Equations giving Σ_S, the contribution of symmetry breaking to the symmetric part of the proper fermion self-energy. The large black circle represents Π_V^B.

D. Rayleigh-Ritz approximation

By performing an arbitrary variation on $V(G,\Delta_i)$ nonlinear equations emerge which can only be solved in a linearized approximation. We now develop a strategy for analyzing the nonlinear aspects of the problem. Rather than performing arbitrary variations, we evaluate $V(G,\Delta_i)$ with specific, parameter-dependent expressions for G and Δ_i, and then vary these parameters. The forms for the propagators that we shall use are solutions of the linear theory,[5] summarized in Sec. IV C. They depend on the symmetry-breaking quantities δm, $M_B{}^2$, and ϵ. Our goal therefore is to write down an effective potential which depends on the numbers δm, $M_B{}^2$, and ϵ, and whose minimum approximately determines them. [Clearly, δm plays the role of some suitably regularized expectation value $\langle\bar\psi(x)\tau_3\psi(x)\rangle$.] However, we must insure that our formula is free of divergences. *A priori* one can expect to encounter quartic, quadratic, and logarithmic divergences. In fact all these divergences are absent for the following reasons.

The quartic divergences disappear if we subtract from $V(G,\Delta_i)$ the same expression evaluated at *symmetric* forms for the propagators, denoted by the subscript N. [In scalar field theories this is accomplished by setting $V(\phi)|_{\phi=0}=0$.] We therefore consider

$$\Omega = V(G,\Delta_i) - V(G_N,\Delta_{iN}). \tag{4.15}$$

It was shown previously that the quadratic divergences are also absent, provided the propagators solve the linearized theory; specifically provided that ϵ is given by (4.10b).[17] [It is unnecessary to require that $M_B{}^2$ be given by (4.13).] This is because any quadratic divergence in Ω is proportional to $(\delta m)^2$ or $M_B{}^2$. (Ω has no dependence on odd powers of δm or Σ_V since $\operatorname{tr}\tau_3=0$.) It is easy to show that these terms vanish when the propagators satisfy the equations of the linearized theory:

$$\Omega = \tfrac{1}{2}\delta m \left.\frac{\partial\Omega}{\partial\delta m}\right|_{M_B{}^2=0} + M_B{}^2 \left.\frac{\partial\Omega}{\partial M_B{}^2}\right|_{M_B{}^2=0;\ \delta m=0}$$

$$+\text{higher powers of the masses.} \tag{4.16a}$$

In $\partial\Omega/\partial\delta m|_{M_B{}^2=0}$ only terms linear in δm are kept. But it is also true that

$$\left.\frac{\partial\Omega}{\partial\delta m}\right|_{M_B{}^2=0} = \int\frac{d^4p}{(2\pi)^4}\operatorname{tr}\frac{\partial G(p)}{\partial\delta m}\left.\frac{\delta V(G,\Delta_i)}{\delta G(p)}\right|_{M_B{}^2=0}, \tag{4.16b}$$

$$\left.\frac{\partial\Omega}{\partial M_B{}^2}\right|_{M_B{}^2=0;\ \delta m=0} = \int\frac{d^4p}{(2\pi)^4}\left[\frac{\partial\Delta_B^{\mu\nu}(p)}{\partial M_B^2}\left.\frac{\delta V(G,\Delta_i)}{\delta\Delta_B^{\mu\nu}(p)}\right|_{M_B^2=0;\ \delta m=0} + \operatorname{tr}\frac{\partial G(p)}{\partial M_B^2}\left.\frac{\delta V(G,\Delta_i)}{\delta G(p)}\right|_{M_B^2=0;\ \delta m=0}\right]. \tag{4.16c}$$

Since only terms linear in δm matter in (4.16b), the symmetry-breaking part of $G(p)$ must satisfy only the linearized equation for $\delta V(G,\Delta_i)/\delta G(p)$ to vanish. In (4.16c) all symmetry-breaking parameters are set to zero. Hence that term will vanish provided $\Delta_B^{\mu\nu}(p)$ satisfies its equation up to symmetry-breaking terms. Therefore when ϵ is fixed by (4.10b), and is *not* viewed as a variational parameter, there are no $(\delta m)^2$ or M_B^2 terms and no quadratic divergences.

Finally, the logarithmic divergences are removed by the power-law falloff in the mass operators (4.10) and (4.12). What could be a possible logarithmic infinity becomes a finite term proportional to an inverse power of ϵ. Such terms necessarily involve $(\delta m)^4$, M_B^4, and $(\delta m)^2 M_B^2$. It is the inverse powers of ϵ that spoil ordinary perturbation theory, since ϵ is $O(g_A^2, g_B^2)$. However, once they have been isolated, the weak-coupling limit may be taken with impunity.

According to (4.3) and (4.5), Ω is given by the following symbolic expression:

$$\Omega(\delta m, M_B^2) = -i\,\mathrm{tr}[\ln G_N^{-1}G - G_N^{-1}G + 1] + \tfrac{1}{2}i \sum_{i=A,B} \mathrm{tr}[\ln \Delta_{iN}^{-1}\Delta_i - \Delta_{iN}^{-1}\Delta_i + 1]$$
$$+ i\,\mathrm{tr}\Sigma_N[G - G_N] - \tfrac{1}{2}i \sum_{i=A,B} \mathrm{tr}\Pi_N^i[\Delta_i - \Delta_{iN}] + \tfrac{1}{2}i \sum_{i=A,B} \mathrm{tr}[\Gamma_i G \Gamma_i G \Delta_i - \Gamma_i G_N \Gamma_i G_N \Delta_{iN}]. \qquad (4.17)$$

The exact variational equations for the *normal* propagators have been used to eliminate S^{-1} and D^{-1} in (4.3) in favor of the complete, normal propagators G_N and Δ_{iN}. These equations are, of course,

$$G_N^{-1} = S^{-1} - \Sigma_N, \qquad (4.18a)$$

$$\Sigma_N = -\sum_{i=A,B} \Gamma_i G_N \Gamma_i \Delta_{iN}, \qquad (4.18b)$$

$$\Delta_{iN}^{-1} = D^{-1} - \Pi_N^i, \qquad (4.18c)$$

$$\Pi_N^i = \Gamma_i G_N \Gamma_i G_N. \qquad (4.18d)$$

The last three terms in (4.17) may be combined. Observe that the last term may be rewritten, with the help of (4.18), as

$$\tfrac{1}{2}i \sum_{i=A,B} \mathrm{tr}[\Gamma_i G \Gamma_i G \Delta_i - \Gamma_i G_N \Gamma_i G_N \Delta_{iN}]$$
$$= -i\,\mathrm{tr}\Sigma_N[G - G_N] + \tfrac{1}{2}i \sum_{i=A,B} \mathrm{tr}\Pi_N^i[\Delta_i - \Delta_{iN}] + \bar{\Omega}, \qquad (4.19)$$
$$\bar{\Omega} = \tfrac{1}{2}i \sum_{i=A,B} \mathrm{tr}[\Gamma_i G \Gamma_i G - \Gamma_i G_N \Gamma_i G_N][\Delta_i - \Delta_{iN}]$$
$$+ \tfrac{1}{2}i \sum_{i=A,B} \mathrm{tr}\Gamma_i[G - G_N]\Gamma_i[G - G_N]\Delta_{iN}.$$

Hence the last three terms in (4.17) can be replaced by $\bar{\Omega}$. [It is important to appreciate that it is entirely legitimate to use Eqs. (4.18) to simplify the dependence of Ω on *normal* propagators. It would be illegitimate to make corresponding reductions on the symmetry-breaking propagators, since they contain variational parameters.]

Equations (4.17) to (4.19) for Ω are exact (in the Hartree-Fock approximation). They will now be approximated. First we set $G_N = S$, $\Sigma_N = 0$, $\Delta_{iN} = \Delta_A = D$, $\Pi^A = 0$. Consequently,

$$\Omega(\delta m, M_B^2) = -i\,\mathrm{tr}[\ln S^{-1}G - S^{-1}G + 1] + \tfrac{1}{2}i \sum_{i=A,B} \mathrm{tr}\Gamma_i[G - S]\Gamma_i[G - S]D$$
$$+ \tfrac{1}{2}i\,\mathrm{tr}[\ln D^{-1}\Delta_B - D^{-1}\Delta_B + 1] + \tfrac{1}{2}i\,\mathrm{tr}[\Gamma_B G \Gamma_B G - \Gamma_B S \Gamma_B S][\Delta_B - \Delta]. \qquad (4.20)$$

The evaluation proceeds by inserting Eq. (4.8) for G, with $\Sigma_N = 0$, and Σ_V and Σ_S given by (4.10) and (4.14), respectively, while Δ_B is set equal to (4.11), with $\Pi^B(p^2) = \Pi_V^B(p^2)$ given by (4.12). We keep only terms that are proportional to inverse powers of the coupling (these come from inverse powers of ϵ), as well as terms of zeroth order in ϵ and coupling. That is, we set ϵ to zero everywhere as long as no divergence arises; if $\epsilon = 0$ is not allowed, (4.10) and (4.11) are used. Moreover, even if a divergence is present at $\epsilon = 0$, terms that are multiplied by higher powers of the coupling constant are dropped.

Rather than integrating immediately, it is convenient to simplify. We rewrite the logarithm of the first term in (4.20):

$$i\,\mathrm{tr}\ln G^{-1}S = i\,\mathrm{tr}\ln(1 - \Sigma_V S - \Sigma_S S). \qquad (4.21a)$$

This is expanded in powers of Σ_S. Only the first power is significant; higher powers give contributions $O(g_A^2, g_B^2)$. Thus we may replace (4.21a) by

$$i\,\mathrm{tr}\ln(1 - \Sigma_V S) - i\,\mathrm{tr}\frac{\Sigma_S S}{1 - \Sigma_V S}$$
$$= i\,\mathrm{tr}\ln(1 - \Sigma_V S) - i\,\mathrm{tr}\Sigma_S G. \qquad (4.21b)$$

Next we analyze the second term in (4.20), which is a two-loop integral. Note that the Γ_i's provide two powers of the coupling strength. Thus for present purposes it is sufficient to keep only terms which diverge when $\epsilon=0$ (it is easy to verify that no ϵ^{-2} terms are present):

$$\tfrac{1}{2}i\sum_{i=A,B}\mathrm{tr}\Gamma_i[G-S]\Gamma_i[G-S]D=i\sum_{i=A,B}\mathrm{tr}[G-S]\Gamma_i[G-S]\Gamma_i D-\tfrac{1}{2}i\sum_{i=A,B}\mathrm{tr}[G-S]\Gamma_i[G-S]\Gamma_i D$$
$$=i\sum_{i=A,B}\mathrm{tr}[G-S]\Gamma_i[S\Sigma_V S+S\Sigma_V S\Sigma_V S]\Gamma_i D-\tfrac{1}{2}i\sum_{i=A,B}\mathrm{tr}[S\Sigma_V S]\Gamma_i[S\Sigma_V S]\Gamma_i D$$
$$-\tfrac{1}{2}i\sum_{i=A,B}\mathrm{tr}[S\Sigma_V S\Sigma_V S]\Gamma_i[S\Sigma_V S\Sigma_V S]\Gamma_i D. \tag{4.22a}$$

In expanding $G-S$, Σ_S has been ignored compared to Σ_V, and only the first two powers of Σ_V are kept. All terms that are dropped lead to convergent integrals at $\epsilon=0$, hence they are at least quadratic in the coupling. Also, only even powers of Σ_V survive the trace. Equations (4.9) and (4.14) allow us to recognize that (4.22a) is equivalently

$$-i\,\mathrm{tr}[G-S][\Sigma_V+\Sigma_S^{(1)}]+\tfrac{1}{2}i\,\mathrm{tr}S\Sigma_V S\Sigma_V+\tfrac{1}{2}i\,\mathrm{tr}S\Sigma_V S\Sigma_V S\Sigma_S^{(1)}$$
$$=-i\,\mathrm{tr}G[S^{-1}-G^{-1}-\Sigma_S^{(2)}]+i\,\mathrm{tr}S\Sigma_S^{(1)}+\tfrac{1}{2}i\,\mathrm{tr}S\Sigma_V S\Sigma_V+\tfrac{1}{2}i\,\mathrm{tr}S\Sigma_V S\Sigma_V S\Sigma_S^{(1)}. \tag{4.22b}$$

Consequently the first two terms in (4.20) reduce to the following, with the help of (4.21) and (4.22):

$$i\,\mathrm{tr}\ln(1-\Sigma_V S)+\tfrac{1}{2}i\,\mathrm{tr}\Sigma_V S\Sigma_V S+i\,\mathrm{tr}\Sigma_S^{(2)}G$$
$$-i\,\mathrm{tr}\Sigma_S G+i\,\mathrm{tr}\Sigma_S^{(1)}S+\tfrac{1}{2}i\,\mathrm{tr}S\Sigma_V S\Sigma_V S\Sigma_S^{(1)}. \tag{4.23a}$$

The last four terms in (4.23a) combine to

$$i\,\mathrm{tr}\Sigma_S^{(1)}[S-G]+\tfrac{1}{2}i\,\mathrm{tr}S\Sigma_V S\Sigma_V S\Sigma_S^{(1)}$$
$$=-\tfrac{1}{2}i\,\mathrm{tr}S\Sigma_V S\Sigma_V S\Sigma_S^{(1)}. \tag{4.23b}$$

Therefore the final, simplified expression for the first two terms in Ω is

$$i\,\mathrm{tr}\ln(1-\Sigma_V S)+\tfrac{1}{2}i\,\mathrm{tr}\Sigma_V S\Sigma_V S$$
$$+\tfrac{1}{2}i\sum_{i=A,B}\mathrm{tr}[S\Sigma_V S\Sigma_V S]\Gamma_i[S\Sigma_V S\Sigma_V S]\Gamma_i D. \tag{4.24}$$

The last term in (4.20) is also expanded in powers of Σ_V. The significant contribution is

$$\tfrac{1}{2}i\,\mathrm{tr}[S\Sigma_V S]\Gamma_B[S\Sigma_V S]\Gamma_B[\Delta_B-D]$$
$$+i\,\mathrm{tr}[S]\Gamma_B[S\Sigma_V S\Sigma_V S]\Gamma_B[\Delta_B-D]. \tag{4.25}$$

We thus arrive at a completely reduced formula for Ω:

$$\Omega(\delta m,M_B^2)=i\,\mathrm{tr}\ln(1-\Sigma_V S)+\tfrac{1}{2}i\,\mathrm{tr}\Sigma_V S\Sigma_V S+\tfrac{1}{2}i\sum_{i=A,B}\mathrm{tr}[S\Sigma_V S\Sigma_V S]\Gamma_i[S\Sigma_V S\Sigma_V S]\Gamma_i[D]$$
$$+\tfrac{1}{2}i\,\mathrm{tr}[\ln D^{-1}\Delta_B-D^{-1}\Delta_B+1]+\tfrac{1}{2}i\,\mathrm{tr}[S\Sigma_V S]\Gamma_B[S\Sigma_V S]\Gamma_B[\Delta_B\Pi_V^B D]$$
$$+i\,\mathrm{tr}[S]\Gamma_B[S\Sigma_V S\Sigma_V S]\Gamma_B[\Delta_B\Pi_V^B D]. \tag{4.26}$$

We have replaced Δ_B-D by $\Delta_B\Pi_V^B D$. A graphical representation for (4.26) is given in Fig. 8.

The integrations are straightforward, subject to a minor ambiguity mentioned below. The final result is

$$\Omega(\delta m,M_B^2)=\frac{(\delta m)^4}{32\pi^2\epsilon}-\frac{m^4}{8\pi^2}\left\{\left(1+\frac{\delta m}{m}\right)^4\ln\left(1+\frac{\delta m}{m}\right)+\left(1-\frac{\delta m}{m}\right)^4\ln\left(1-\frac{\delta m}{m}\right)-7\left(\frac{\delta m}{m}\right)^2\right\}$$
$$+\frac{3M_B^4}{128\pi^2\epsilon}\left[1-4\epsilon\ln\frac{M_B^2}{m^2}-2\epsilon\right]-\frac{3g_B^2M_B^2(\delta m)^2}{128\pi^4\epsilon^2}\left[1-4\epsilon\ln\frac{M_B^2}{m^2}+\epsilon\right]+O(g_A^2,g_B^2). \tag{4.27}$$

All divergences are absent, as anticipated by our general argument. The first two terms in (4.27) come from the first two terms in (4.26). The third term in (4.26) gives no significant value. The $O(M_B^4)$ term in (4.27) arises from the fourth term in (4.26). The $O(M_B^2(\delta m)^2)$ contribution to Ω results from the last term in (4.26). The ambiguity occurs in the terms that are zeroth order in the coupling. They receive a contribution both from asymptotically large and from finite momenta. This ambiguity could be resolved by specifying more exactly the transition between the power-law behavior (4.10a), (4.12a) for Σ_V, Π_V^B and the low-energy behavior (4.10c), (4.12b). However, various reasonable transition behaviors make only a small difference in the numerical coefficients.

Variation of M_B^2 yields

$$M_B^2=\tfrac{8}{3}\frac{(\delta m)^2}{\gamma^2-1}(1-\epsilon),$$
$$\gamma^2=g_A^2/g_B^2>1. \tag{4.28}$$

In deriving (4.28), we have assumed that $|4\epsilon\ln(M_B^2/m^2)|\ll 1$. This means that M_B^2 cannot become arbitrarily small or large compared to m^2. Equation (4.28) coincides [up to $O(\epsilon)$ terms] with (4.13), giving us the promised derivation of that result. We substitute the value (4.28) for M_B^2 into (4.27) and obtain an effective potential which depends only on δm:

$$\Omega(\delta m)=\frac{(\delta m)^4}{32\pi^2\epsilon}-\frac{m^4}{8\pi^2}\left\{\left(1+\frac{\delta m}{m}\right)^4\ln\left(1+\frac{\delta m}{m}\right)+\left(1-\frac{\delta m}{m}\right)^4\ln\left(1-\frac{\delta m}{m}\right)-7\left(\frac{\delta m}{m}\right)^2\right\}$$
$$-\frac{(\delta m)^4}{6\pi^2\epsilon(r^2-1)^2}\left\{1-4\epsilon\ln\frac{\frac{8}{3}}{r^2-1}\left(\frac{\delta m}{m}\right)^2\right\}. \tag{4.29}$$

Because the underlying physical model is unrealistic, we shall not pursue a detailed study of (4.29), beyond noting the form for small $\delta m/m$ ($|\delta m/m|$ must be less than 1; otherwise Ω becomes complex):

$$\Omega(\delta m)\approx\frac{(\delta m)^4}{32\pi^2\epsilon}-\frac{25(\delta m)^4}{48\pi^2}+\frac{(\delta m)^6}{120\pi^2 m^2}$$
$$-\frac{(\delta m)^4}{6\pi^2\epsilon(r^2-1)^2}\left[1-4\epsilon\ln\frac{\frac{8}{3}}{r^2-1}\left(\frac{\delta m}{m}\right)^2\right]. \tag{4.30}$$

This has a minimum at

$$\left(\frac{\delta m}{m}\right)^2=\frac{125}{3}\left[1-\frac{0.06}{\epsilon}+\frac{0.32}{\epsilon(r^2-1)^2}-\frac{0.64}{(r^2-1)^2}\right] \tag{4.31}$$

provided that quantity is positive; also it must be less than 1. Furthermore, we have dropped $4\epsilon\ln[\frac{8}{3}/(r^2-1)](\delta m/m)^2$ compared to 1 in (4.31); this approximation was already appealed to in the calculation of M_B^2. There are two kinds of solution to (4.31). First, when r is large compared to 1, the last two terms in the brackets of (4.31) may be ignored and ϵ comes out to be ≈ 0.06. In this case $M_B^2\ll m^2$. As r approaches 1, the second term in the brackets becomes negligible compared to the third. Then $(r^2-1)^2\approx 0.64(1-1/2\epsilon)$ and it must be that $\epsilon\geq\frac{1}{2}$. It is not clear whether such a large ϵ is really consistent with our approximation scheme. In any case for these solutions $M_B^2\approx m^2$. [The requirement that $|4\epsilon\ln(M_B^2/m^2)|\ll 1$ is not very restrictive and can be easily met.]

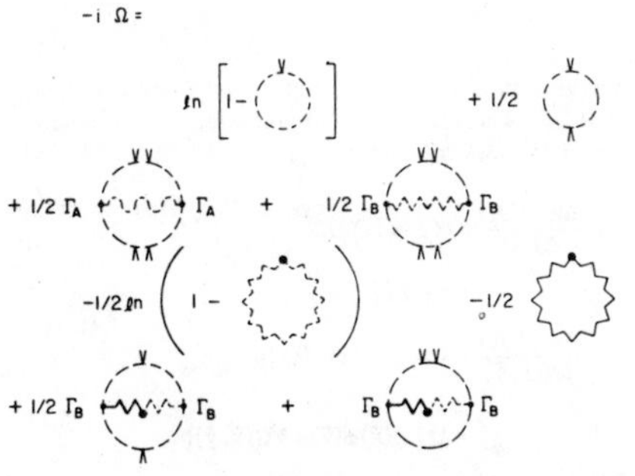

FIG. 8. Formula for $\Omega(\delta m, M_B^2)$.

E. Discussion

Observe that the form (4.29) is reminiscent of the effective potential in a theory with scalar mesons, where δm plays the role of the scalar fields.[18] There is a quartic term in δm, followed by a logarithmic one. The main differences are the appearance of inverse powers of the coupling constant, and the fact that Ω becomes complex for $|\delta m/m|>1$.

It is interesting to contrast the present results with the linear theory. Now δm is calculable and there are constraints on the coupling constants (ϵ and r). In the linear theory, neither of the two conditions is present.

In a further investigation, we hope to survey graphs that contribute beyond the Hartree-Fock approximation. It is most important to ascertain whether it continues to be possible to select the dominant contribution, for small coupling, and whether our present results are stable against higher-order corrections.

V. STATIC, POSITION-DEPENDENT SOLUTIONS

There is considerable interest in finding solutions to field theory which correspond to an energy eigenstate $|\psi\rangle$ in which the expectation of the field Φ is nonvanishing and non-translation-invariant,

$$\langle\psi|\Phi(x)|\psi\rangle=\phi(x). \tag{5.1}$$

Since $|\psi\rangle$ is an energy eigenstate, $\phi(x)$ is time-independent, $\phi(x)=\phi(\vec{x})$. Although at the present time the physical interpretation of these states has not been firmly fixed, physical intuition suggests that they correspond to coherent excitations, not unlike the familiar Thomas-Fermi nucleus of conventional theory. One might suppose that the observed, physical particles correspond to such states, while the excitations associated with the underlying fields are trapped for some marvelous, as yet ununderstood reason.

We demonstrate that the formalism, which we have here developed, provides a natural framework for a study of these questions. We first show that $\Gamma(\phi, G)$, for static ϕ and G, determines the stationary expectation value of the Hamiltonian H in a normalized state $|\psi\rangle$ for which

$$\langle\psi|\Phi(x)|\psi\rangle = \phi(\bar{x}), \tag{5.2a}$$

$$\langle\psi|\Phi(x)\Phi(y)|\psi\rangle|_{x_0=y_0} = \phi(\bar{x})\phi(\bar{y}) + \hbar G(\bar{x},\bar{y}), \tag{5.2b}$$

$$\langle\psi|H|\psi\rangle = E(\phi, G). \tag{5.3}$$

(The operators are all evaluated at a fixed time x^0.)

For static solutions, $\Gamma(\phi, G)$ is time-translation-invariant, and has an overall factor of time "volume." The relation between $\Gamma(\phi, G)$ and $E(\phi, G)$ will be shown to be

$$-E(\phi, G)\int dt = \Gamma(\phi, G)|_{\text{static}}. \tag{5.4}$$

[The precise meaning of $\Gamma(\phi, G)|_{\text{static}}$ is spelled out below.] Thus we see that the stationary requirements on $\Gamma(\phi, G)$, (1.1), are merely instances of the quantum-mechanical variation principle.

$E(\phi, G)$ is computed for a scalar self-interacting field theory in the Hartree-Fock approximation and the equations for ϕ and G are derived. Next, following Kuti,[6] a Schrödinger picture is introduced, and the abstract variational principle is realized in a functional Schrödinger equation. Kuti's Rayleigh-Ritz method of solving this equation[6] is shown to be equivalent to our Hartree-Fock approximation.

A. Physical interpretation of $\Gamma(\phi,G)|_{\text{static}}$

The problem of finding all the bound states of a general quantum theory, and specifically of a field theory, may of course be formulated as a variational principle. One seeks the states $|\psi\rangle$ which make $\langle\psi|H|\psi\rangle$ stationary against arbitrary variation of $|\psi\rangle$, subject to the normalization constraint $\langle\psi|\psi\rangle = 1$. Rather than performing the arbitrary variation in one fell swoop, it is convenient first to perform a restricted variation, where certain quantities are held fixed, and then to vary these quantities.

We choose to require additional constraints: The expectations of $\Phi(x)$ and of $\Phi(x)\Phi(y)|_{x_0=y_0}$ are fixed. Imposing all conditions with the help of Lagrange multipliers, one is led to consider the variation of

$$\langle\psi|H|\psi\rangle - \epsilon\langle\psi|\psi\rangle - \int d\bar{x}\, J(\bar{x})\langle\psi|\Phi(x)|\psi\rangle - \tfrac{1}{2}\int d\bar{x}\, d\bar{y}\, K(\bar{x},\bar{y})\langle\psi|\Phi(x)\Phi(y)|\psi\rangle|_{x_0=y_0}. \tag{5.5}$$

Clearly $|\psi\rangle$ satisfies the equation

$$\left[H - \int d\bar{x}\, J(\bar{x})\Phi(x) - \tfrac{1}{2}\int d\bar{x}\, d\bar{y}\, K(\bar{x},\bar{y})\Phi(x)\Phi(y)|_{x_0=y_0}\right]|\psi\rangle = \epsilon|\psi\rangle. \tag{5.6}$$

Thus $|\psi\rangle$ is an energy eigenstate of a theory governed by a modified Hamiltonian or, equivalently, by an action with source terms added to it,

$$\int d^4x\, \mathcal{L}(x) + \int d^4x\, J(x)\Phi(x) + \tfrac{1}{2}\int d^4x\, d^4y\, K(x,y)\Phi(x)\Phi(y),$$
$$J(x) = J(\bar{x}), \quad K(x,y) = \delta(x_0 - y_0)K(\bar{x},\bar{y}). \tag{5.7}$$

The energy eigenvalue of this problem is ϵ. We restrict the discussion to the lowest energy eigenstate.

It is known that the energy of the lowest state is also given by

$$W(J,K) = -\epsilon\int dt, \tag{5.8}$$

where $W(J,K)$ is defined in (2.1). (The time infinity may be removed if the sources are considered to be acting over a large but finite time interval.) Hence we conclude that

$$-\frac{1}{T}W(J,K) = \langle\psi|H|\psi\rangle - \int d\bar{x}\, J(\bar{x})\langle\psi|\Phi(x)|\psi\rangle - \tfrac{1}{2}\int d\bar{x}\, d\bar{y}\, K(\bar{x},\bar{y})\langle\psi|\Phi(x)\Phi(y)|\psi\rangle|_{x_0=y_0},$$
$$T = \int dt. \tag{5.9}$$

Varying this with respect to J and K and recalling that $|\psi\rangle$ is a normalized eigenstate of the modified Hamiltonian, we find

$$\frac{\delta W(J,K)}{\delta J(\bar{x})} = T\langle\psi|\Phi(x)|\psi\rangle = T\phi(\bar{x}),$$
$$\frac{\delta W(J,K)}{\delta K(\bar{x},\bar{y})} = \tfrac{1}{2}T\langle\psi|\Phi(x)\Phi(y)|\psi\rangle|_{x_0=y_0} = \tfrac{1}{2}T[\phi(\bar{x})\phi(\bar{y}) + \hbar G(\bar{x},\bar{y})] \tag{5.10a}$$

Thus

$$\langle\psi|H|\psi\rangle = E(\phi, G)$$
$$= -\frac{1}{\tau}\left\{W(J,K) - \int d^4x J(x)\phi(\vec{x}) - \tfrac{1}{2}\int d^4x\, d^4y K(x,y)[\phi(\vec{x})\phi(\vec{y}) + \hbar G(\vec{x},\vec{y})]\right\} . \quad (5.10b)$$

The right-hand side of (5.10b) is $-(1/\tau)\Gamma(\phi, G)|_{\text{static}}$ [compare with (2.4) and (2.5)]. The precise meaning of $\Gamma(\phi, G)|_{\text{static}}$ is as follows. First $\Gamma(\phi, G)$ is evaluated with time-translation-invariant forms $\phi(x) = \phi(\vec{x})$; $G(x, y) = G(x_0 - y_0; \vec{x}, \vec{y})$. This does not yet give $\Gamma(\phi, G)|_{\text{static}}$, since we still must express $G(x_0 - y_0; \vec{x}, \vec{y})$ in terms of $G(\vec{x}, \vec{y})$. From (2.21b) and (5.10a) we see that $G(\vec{x}, \vec{y}) = G(0; \vec{x}, \vec{y})$. The desired relation is obtained from the equation

$$\frac{\delta\Gamma(\phi, G)}{\delta G(x_0 - y_0; \vec{x}, \vec{y})} = -\tfrac{1}{2}\hbar K(x, y)$$
$$= -\tfrac{1}{2}\hbar\delta(x_0 - y_0)K(\vec{x}, \vec{y}) .$$

Once $G(x_0 - y_0; \vec{x}, \vec{y})$ is known in terms of $G(\vec{x}, \vec{y})$, $\Gamma(\phi, G)$ can be expressed as a functional of $\phi(\vec{x})$ and $G(\vec{x}, \vec{y})$. This then is $\Gamma(\phi, G)|_{\text{static}}$.[19]

The arbitrary variation is now completed by varying $E(\phi, G)$ with respect to ϕ and G, and determining values for ϕ and G which render $E(\phi, G)$ stationary. For a physically sensible system, there will always be a solution with constant ϕ corresponding to the vacuum expectation value of $\Phi(x)$.

There may also be a solution with a position-dependent $\phi(\vec{x})$. This then corresponds to a non-translation-invariant energy eigenstate $|\psi\rangle$, with eigenvalue $E(\phi, G)$. $|\psi\rangle$ is *not* the vacuum, since we do not expect translation invariance to be spontaneously broken. Since the underlying theory is translationally invariant, the state $|\psi\rangle$ is infinitely degenerate with respect to energy. One can construct other states by an application of the momentum operator $\vec{P}$,

$$|\psi; \vec{a}\rangle = \exp\left(-\frac{i}{\hbar}\vec{a}\cdot\vec{P}\right)|\psi\rangle . \quad (5.11)$$

Each of the states $|\psi, \vec{a}\rangle$ has the same energy, and

$$\langle\psi; \vec{a}|\Phi(x)|\psi; \vec{a}\rangle = \phi(\vec{x} + \vec{a}) . \quad (5.12)$$

Momentum eigenstates may also be formed:

$$|\vec{q}\rangle = \int d\vec{a} \exp\left(\frac{i}{\hbar}\vec{q}\cdot\vec{a}\right)|\psi; \vec{a}\rangle . \quad (5.13)$$

B. Sample calculation

We compute $E(\phi, G)$ for the self-interacting Bose field considered in Sec. III. For simplicity, and to make contact with other work, we consider only one field, $N = 1$. From (3.4)

$$\Gamma(\phi, G) + \text{const} = I(\phi) + \tfrac{1}{2}i\hbar \operatorname{Tr} \operatorname{Ln} G^{-1} + \tfrac{1}{2}i\hbar \operatorname{Tr} \mathfrak{D}^{-1}(\phi)G - \tfrac{1}{8}\lambda\hbar^2 \int d^4x\, G(x,x)G(x,x) , \quad (5.14)$$

where

$$i\mathfrak{D}^{-1}\{\phi; x, y\} = -[\Box + m^2 + \tfrac{1}{2}\lambda\phi^2(\vec{x})]\,\delta^4(x - y). \quad (5.15)$$

The equation

$$\frac{\delta\Gamma(\phi, G)}{\delta G(x, y)} = -\tfrac{1}{2}\hbar\delta(x_0 - y_0)K(\vec{x}, \vec{y})$$

implies

$$G^{-1}(x, y) = -i\delta(x^0 - y^0)K(\vec{x}, \vec{y}) + \mathfrak{D}^{-1}\{\phi; x, y\} + \tfrac{1}{2}i\lambda\hbar G(x, x)\delta^4(x - y). \quad (5.16)$$

Time-translation-invariant solutions to (5.16) are clearly of the form

$$G^{-1}(x_0 - y_0; \vec{x}, \vec{y}) = i\delta''(x_0 - y_0)\delta(\vec{x} - \vec{y}) + i\delta(x_0 - y_0)f(\vec{x}, \vec{y}) . \quad (5.17)$$

Hence if we define the Fourier transform

$$\tilde{G}^{-1}(\omega; \vec{x}, \vec{y}) = \int_{-\infty}^{\infty} dx_0\, e^{i\omega x_0} G^{-1}(x_0; \vec{x}, \vec{y}),$$

we find

$$\tilde{G}^{-1}(\omega; \vec{x}, \vec{y}) = -i\omega^2\delta(\vec{x} - \vec{y}) + if(\vec{x}, \vec{y}), \quad (5.18a)$$

$$\tilde{G}(\omega; \vec{x}, \vec{y}) = \frac{i}{\omega^2 - f}(\vec{x}, \vec{y}) , \quad (5.18b)$$

where the inverse in (5.18b) is taken in the functional sense in the $\vec{x}, \vec{y}$ variables. It follows that

$$G(\vec{x}, \vec{y}) = G(0; \vec{x}, \vec{y})$$
$$= \int_{-\infty}^{\infty} \frac{d\omega}{2\pi} \frac{i}{\omega^2 - f}(\vec{x}, \vec{y})$$
$$= \tfrac{1}{2}f^{-1/2}(\vec{x}, \vec{y})$$

or

$$f(\vec{x}, \vec{y}) = \tfrac{1}{4}G^{-2}(\vec{x}, \vec{y}) . \quad (5.19)$$

Hence $G(x_0 - y_0; \vec{x}, \vec{y})$ is expressed in terms of $G(\vec{x}, \vec{y})$ by

$$G^{-1}(x_0 - y_0; \vec{x}, \vec{y}) = i\delta''(x_0 - y_0)\delta(\vec{x} - \vec{y}) + i\delta(x_0 - y_0)\tfrac{1}{4}G^{-2}(\vec{x}, \vec{y}) , \quad (5.20a)$$

$$\tilde{G}^{-1}(\omega;\vec{x},\vec{y}) = -i\omega^2\delta(\vec{x}-\vec{y}) + \tfrac{1}{4}iG^{-2}(\vec{x},\vec{y}). \quad (5.20b)$$

We can now evaluate $\Gamma(\phi,G)|_{\text{static}}$. The first term on the right-hand side of (5.14) gives simply

$$-\int dx_0 \int d\vec{x}\left(\tfrac{1}{2}[\vec{\nabla}\phi(\vec{x})]^2 + \tfrac{1}{2}m^2\phi^2(\vec{x}) + \frac{\lambda}{4!}\phi^4(\vec{x})\right). \quad (5.21a)$$

The second term is evaluated in the Fourier representation:

$$\tfrac{1}{2}i\hbar\int dx_0 \int_{-\infty}^{\infty}\frac{d\omega}{2\pi}\,\mathrm{Tr\,Ln}(-\omega^2+\tfrac{1}{4}G^{-2}) = -\tfrac{1}{4}\hbar\int dx_0\,\mathrm{Tr}G^{-1}(\vec{x},\vec{y}) = -\tfrac{1}{4}\hbar\int dx_0\int d\vec{x}\,G^{-1}(\vec{x},\vec{x}). \quad (5.21b)$$

Constants have been dropped and both Tr and Ln refer to functional operations in $\vec{x},\vec{y}$ variables. The third term is

$$\begin{aligned}
\tfrac{1}{2}i\hbar\int d^4x\, d^4y\, \mathfrak{D}^{-1}\{\phi;\vec{x},\vec{y}\}G(x_0-y_0;\vec{x},\vec{y})
&= \tfrac{1}{2}\hbar\int d^4x\left\{-\frac{\partial^2}{\partial x_0^2}G(x_0;\vec{x},\vec{y})\Big|_{x_0=0} + \vec{\nabla}_x^{\,2}G(0;\vec{x},\vec{y})\big|_{\vec{x}=\vec{y}} - [m^2+\tfrac{1}{2}\lambda\phi^2(\vec{x})]G(0;\vec{x},\vec{x})\right\}\\
&= \tfrac{1}{2}\hbar\int dx_0\int d\vec{x}\left\{\int_{-\infty}^{\infty}\frac{d\omega}{2\pi}\omega^2\tilde{G}(\omega;\vec{x},\vec{x}) + \vec{\nabla}_x^{\,2}G(\vec{x},\vec{y})\big|_{\vec{x}=\vec{y}} - [m^2+\tfrac{1}{2}\lambda\phi^2(\vec{x})]G(\vec{x},\vec{x})\right\}\\
&= \tfrac{1}{2}\hbar\int dx_0\int d\vec{x}\left\{\tfrac{1}{4}G^{-1}(\vec{x},\vec{x}) + \vec{\nabla}_x^{\,2}G(\vec{x},\vec{y})\big|_{\vec{x}=\vec{y}} - [m^2+\tfrac{1}{2}\lambda\phi^2(\vec{x})]G(\vec{x},\vec{x})\right\}. \quad (5.21c)
\end{aligned}$$

Again constants have been dropped when (5.20b) was used to evaluate the ω integration. Finally, the last term in (5.14) is

$$-\tfrac{1}{8}\lambda\hbar^2\int dx_0\int d\vec{x}\,G(0;\vec{x},\vec{x})G(0;\vec{x},\vec{x}) = -\tfrac{1}{8}\lambda\hbar^2\int dx_0\int d\vec{x}\,G(\vec{x},\vec{x})G(\vec{x},\vec{x}). \quad (5.21d)$$

Collecting all the terms in (5.21) we get $\Gamma(\phi,G)|_{\text{static}}$, hence $E(\phi,G)$:

$$\begin{aligned}
E(\phi,G) = \int d\vec{x}\Big\{&\tfrac{1}{2}[\vec{\nabla}\phi(\vec{x})]^2 + \tfrac{1}{2}m^2\phi^2(\vec{x}) + \frac{\lambda}{4!}\phi^4(\vec{x})\\
&+ \tfrac{1}{2}\hbar[-\vec{\nabla}_x^{\,2}G(\vec{x},\vec{y})|_{\vec{x}=\vec{y}} + m^2G(\vec{x},\vec{x}) + \tfrac{1}{2}\lambda\phi^2(\vec{x})G(\vec{x},\vec{x})] + \tfrac{1}{8}\hbar G^{-1}(\vec{x},\vec{x}) + \tfrac{1}{8}\lambda\hbar^2G(\vec{x},\vec{x})G(\vec{x},\vec{x})\Big\}. \quad (5.22)
\end{aligned}$$

The equations which are obtained from varying $\phi(\vec{x})$ and $G(\vec{x},\vec{y})$ in $E(\phi,G)$ are

$$0 = -\vec{\nabla}^2\phi(\vec{x}) + m^2\phi(\vec{x}) + \tfrac{1}{6}\lambda\phi^3(\vec{x}) + \hbar\tfrac{1}{2}\lambda\phi(\vec{x})G(\vec{x},\vec{x}), \quad (5.23a)$$

$$\tfrac{1}{4}G^{-2}(\vec{x},\vec{y}) = [-\vec{\nabla}^2 + m^2 + \tfrac{1}{2}\lambda\phi^2(\vec{x}) + \hbar\tfrac{1}{2}\lambda G(\vec{x},\vec{x})]\,\delta(\vec{x}-\vec{y}). \quad (5.23b)$$

Consequently the energy of the physical state $|\psi\rangle$ is

$$E = -\tfrac{1}{4}\int d\vec{x}[\tfrac{1}{6}\lambda\phi^4(\vec{x}) + \hbar\lambda\phi^2(\vec{x})G(\vec{x},\vec{x}) - \hbar G^{-1}(\vec{x},\vec{x}) + \hbar^2\tfrac{1}{2}\lambda G(\vec{x},\vec{x})G(\vec{x},\vec{x})], \quad (5.24)$$

where $\phi(\vec{x})$ and $G(\vec{x},\vec{y})$ satisfy (5.23).

In the classical limit $\hbar\to 0$, $\phi(\vec{x})$ is a solution to the classical equation of motion, and $G(\vec{x},\vec{y})$ is the classical propagator. The quantum corrections are the self-consistent corrections of the Hartree-Fock approximation which modify the mass term m^2 by the position-dependent quantity $\hbar\tfrac{1}{2}\lambda G(\vec{x},\vec{x})$.

C. Schrödinger representation

An entirely different approach to the study of unconventional solutions of a field theory has been developed by Kuti.[6] Here we review his development and show that it leads to the same results—Eqs. (5.22), (5.23), and (5.24)—as our method.

An abstract quantum-mechanical state may be realized by a "wave function." For a field theory involving the field operator $\Phi(x)$, the wave function is a functional of a c number $\Phi(\vec{x})$ (the time is fixed, hence suppressed):

$$|\psi\rangle \to \Psi\{\Phi\}. \quad (5.25a)$$

The action of the operator $\Phi(x)$ on $|\psi\rangle$ is realized by multiplying $\Psi\{\Phi\}$ by $\Phi(\vec{x})$:

$$\Phi(x)|\psi\rangle \to \Phi(\vec{x})\Psi\{\Phi\}. \quad (5.25b)$$

The only other independent operator in the theory

is the canonical momentum $\Pi(x)$. The action of that operator on $|\psi\rangle$ is realized by functional differentiation

$$\Pi(x)|\psi\rangle \to \frac{\hbar}{i}\frac{\delta}{\delta\Phi(\vec{x})}\Psi\{\Phi\}. \quad (5.25c)$$

Finally, the inner product is defined by functional integration:

$$\langle\psi_1|\psi_2\rangle \to \int d\Phi\, \Psi_1^*\{\Phi\}\Psi_2\{\Phi\}. \quad (5.25d)$$

More precisely what is being done is introducing eigenstates at fixed time of the field $\Phi(x)$. When these states are denoted by $|\Phi\rangle$, the wave function $\Psi\{\Phi\}$ is merely

$$\langle\Phi|\psi\rangle = \Psi\{\Phi\}.$$

The analogy with ordinary quantum mechanics is clear.

Energy eigenstates satisfy the Schrödinger equation

$$\int d\vec{x}\, \mathcal{H}\left\{\frac{\hbar}{i}\frac{\delta}{\delta\Phi(\vec{x})}, \Phi(\vec{x})\right\}\Psi\{\Phi\} = E\Psi\{\Phi\}, \quad (5.26)$$

where $\mathcal{H}\{\Pi(x), \Phi(x)\}$ is the Hamiltonian density. The time development is

$$\Psi\{\Phi; t\} = e^{-iEt/\hbar}\Psi\{\Phi\}.$$

For example, in our Bose model (5.26) is

$$\int d\vec{x}\left[-\frac{\hbar^2}{2}\frac{\delta^2}{\delta\Phi(\vec{x})\delta\Phi(\vec{x})} + \tfrac{1}{2}[\vec{\nabla}\Phi(\vec{x})]^2 + \tfrac{1}{2}m^2\Phi^2(\vec{x}) + \frac{\lambda}{4!}\Phi^4(\vec{x})\right]\Psi\{\Phi\} = E\Psi\{\Phi\}. \quad (5.27)$$

A direct solution of the functional integro-differential equation is of course impossible. Let us return, however, to the variational principle which can be used to derive (5.26). We demand that

$$\frac{\langle\psi|H|\psi\rangle}{\langle\psi|\psi\rangle} = \frac{\int d\Phi \int d\vec{x}\, \Psi^*\{\Phi\}\mathcal{H}\Psi\{\Phi\}}{\int d\Phi\, |\Psi\{\Phi\}|^2} \quad (5.28)$$

be stationary against arbitrary variations of $\Psi\{\Phi\}$. In the usual fashion, this yields (5.26). The form of $\langle\psi|H|\psi\rangle$ can be computed as follows. We define

$$\phi(\vec{x}) = \frac{\int d\Phi\, \Phi(\vec{x})|\Psi\{\Phi\}|^2}{\int d\Phi\, |\Psi\{\Phi\}|^2}, \quad (5.29a)$$

$$\phi(\vec{x})\phi(\vec{y}) + \hbar G(\vec{x},\vec{y}) = \frac{\int d\Phi\, \Phi(\vec{x})\Phi(\vec{y})|\Psi\{\Phi\}|^2}{\int d\Phi\, |\Psi\{\Phi\}|^2}. \quad (5.29b)$$

It follows that

$$\int d\Phi\, \Phi(\vec{x})|\Psi\{\Phi+\phi\}|^2 = 0, \quad (5.30a)$$

$$\int d\Phi\, \Phi(\vec{x})\Phi(\vec{y})|\Psi\{\Phi+\phi\}|^2 = \hbar G(\vec{x},\vec{y})\int d\Phi\, |\Psi\{\Phi\}|^2. \quad (5.30b)$$

Hence $\langle\psi|H|\psi\rangle/\langle\psi|\psi\rangle$ for our Bose model is given by

$$\int d\vec{x}\left\{\tfrac{1}{2}[\vec{\nabla}\phi(\vec{x})]^2 + \tfrac{1}{2}m^2\phi^2(\vec{x}) + \frac{\lambda}{4!}\phi^4(\vec{x}) + \tfrac{1}{2}\hbar[-\vec{\nabla}_x^2 G(\vec{x},\vec{y})|_{\vec{x}=\vec{y}} + m^2 G(\vec{x},\vec{x}) + \tfrac{1}{2}\lambda\phi^2(\vec{x})G(\vec{x},\vec{x})]\right\}$$
$$+\left[\int d\Phi\, |\Psi\{\Phi\}|^2\right]^{-1}\int d\Phi \int d\vec{x}\left\{\tfrac{1}{2}\hbar^2\left|\frac{\delta\Psi\{\Phi\}}{\delta\Phi(\vec{x})}\right|^2 + \frac{\lambda}{4!}[\Phi^4(\vec{x}) + 4\phi(\vec{x})\Phi^3(\vec{x})]\,|\Psi(\Phi+\phi)|^2\right\}. \quad (5.31)$$

[We recognize that the first integral in the exact formula (5.31) coincides with the corresponding terms in our approximate expression for $E(\phi, G)$ in (5.22). The second integral in (5.31) is approximated in (5.22) by the last two terms in that equation.]

Rather than applying the variational principle in an arbitrary way, which would merely reproduce the intractable exact equation (5.27), a Rayleigh-Ritz type ansatz is made. Following Kuti[6] we take as a trial function

$$\Psi\{\Phi\} = \exp\left\{-\frac{1}{4\hbar}\int d\vec{x}\, d\vec{y}[\Phi(\vec{x}) - \phi(\vec{x})] \times G^{-1}(\vec{x},\vec{y})[\Phi(\vec{y}) - \phi(\vec{y})]\right\} \quad (5.32)$$

and view ϕ and G as variational parameters. With this choice (5.30) is obviously satisfied, while the second integral in (5.31) gives

$$\int d\vec{x}[\tfrac{1}{8}\hbar G^{-1}(\vec{x},\vec{x}) + \hbar^2\tfrac{1}{8}\lambda G(\vec{x},\vec{x})G(\vec{x},\vec{x})]. \quad (5.33)$$

This reproduces the last two terms in (5.22). Hence Kuti's Rayleigh-Ritz approximation is entirely equivalent to our Hartree-Fock calculation.

This set of equations has been studied in two-dimensional space-time by Kuti,[6] Dashen, Hasslacher, and Neveu.[20]

D. Comments

Although the two approaches to the problem of static solutions in a field theory yield the same equations for $\phi(\vec{x})$ and $G(\vec{x},\vec{y})$, the following differences should be noted. It is difficult to study the systematics of the corrections to the Rayleigh-Ritz method. In our method, the effective action $\Gamma(\phi, G)$ and the energy $E(\phi, G)$ can be expanded in the series (2.9). On the other hand, Kuti's method gives an explicit expression for the wave function. Hence one can evaluate off-diagonal matrix elements of arbitrary operators, not just diagonal expectations of products of fields, which is all that the effective action provides.

VI. CONCLUSION

We have illustrated how the generalized effective action compactly probes the nonlinear structure of field theory. Moreover, by a suitable parametrization, the exact nonlinear integral equations can be replaced by approximate ordinary equations involving numerical parameters. The former give a precise point-by-point description of the theory, but are intractable. The latter summarize only average properties of the theory, but can be studied by conventional techniques, just as the exact Schrödinger equation can be well analyzed by the Rayleigh-Ritz variational principle. That this can be done for field theory in principle has been known since Schwinger's work in the early fifties, but as far as we know the present work contains the only application of this method.

ACKNOWLEDGMENT

J. Kuti shared with us his unpublished researches, and we are most grateful to him. One author (R. J.) benefited from conversations with R. Dashen, while another (J. M. C.) acknowledges the hospitality at the Center for Theoretical Physics at M.I.T. Finally, we all are grateful to A. Kerman who explained to us how methods, similar to those studied here, are used in many-body physics.

---

*This work supported in part through funds provided by the Atomic Energy Commission under Contract No. AT(11-1)-3069.

†Alfred P. Sloan Research Fellow.

[1]L. Dolan and R. Jackiw, Phys. Rev. D 9, 3320 (1974).

[2]H. Schnitzer, Phys. Rev. D 10, 1800 (1974).

[3]S. Coleman, R. Jackiw, and H. D. Politzer, this issue, Phys. Rev. D 10, 2491 (1974).

[4]R. Jackiw, Phys. Rev. D 9, 1686 (1974).

[5]R. Jackiw and K. Johnson, Phys. Rev. D 8, 2386 (1973); J. M. Cornwall and R. E. Norton, *ibid.* 8, 3338 (1973).

[6]J. Kuti, unpublished.

[7]T. D. Lee and C. N. Yang, Phys. Rev. 117, 22 (1960).

[8]J. M. Luttinger and J. C. Ward, Phys. Rev. 118, 1417 (1960). P. Martin and C. De Dominicis, J. Math. Phys. 5, 14 (1964); 5, 31 (1964).

[9]H. D. Dahmen and G. Jona-Lasinio, Nuovo Cimento 52A, 807 (1967); A. N. Vasil'ev and A. K. Kazanskii, Teor. Mat. Fiz. 12, 352 (1972) [Theoret. Math. Phys. 12, 875 (1972)]. Also T. D. Lee presented, in the course of a seminar at M.I.T., a relativistic generalization of the results of Ref. 7. One of us (J.M.C.) has been informed by P. Mannheim that he is working on an effective potential for composite operators in quantum electrodynamics.

[10]A graph is said to be "two-particle irreducible" if it does not become disconnected upon opening two lines. Otherwise it is "two-particle reducible."

[11]In this connection it is important to note the relationship that exists between derivatives of $\Gamma(\phi, G)$ and $W(J,K)$. That formula is a generalization of the result relevant to the ordinary effection action $\Gamma(\phi)$:

$$\int d^4z \frac{\delta^2\Gamma(\phi)}{\delta\phi(x)\delta\phi(z)} \frac{\delta^2 W(J)}{\delta J(z)\delta J(y)} = -\delta^4(x-y)\,.$$

Using a self-explanatory compact notation the formulas are

$$\left(\Gamma_{\phi\phi} - \frac{2}{\hbar}\Gamma_G - \frac{4}{\hbar}\phi\Gamma_{\phi G} + \frac{4}{\hbar^2}\varphi^2\,\Gamma_{GG}\right)W_{JJ} + \left(\frac{2}{\hbar}\Gamma_{\phi G} - \frac{4}{\hbar^2}\phi\,\Gamma_{GG}\right)W_{JK} = -1\,,$$

$$\left(\frac{2}{\hbar}\Gamma_{\phi G} - \frac{4}{\hbar^2}\phi\,\Gamma_{GG}\right)W_{KJ} + \frac{4}{\hbar^2}\Gamma_{GG}W_{KK} = -1\,,$$

$$\left(\frac{2}{\hbar}\Gamma_{\phi G} - \frac{4}{\hbar^2}\phi\,\Gamma_{GG}\right)W_{JJ} + \frac{4}{\hbar^2}\Gamma_{GG}W_{JK} = 0\,,$$

$$\left(\Gamma_{\phi\phi} - \frac{2}{\hbar}\Gamma_G - \frac{4}{\hbar}\phi\Gamma_{\phi G} + \frac{4}{\hbar^2}\phi\phi\,\Gamma_{GG}\right)W_{KJ} + \left(\frac{2}{\hbar}\Gamma_{\phi G} - \frac{4}{\hbar^2}\phi\,\Gamma_{GG}\right)W_{KK} = 0\,.$$

[12]R. E. Norton and J. M. Cornwall, unpublished. See also Dahmen and Jona-Lasinio, Ref. 9.

[13]This is most easily understood by recalling that for fermions the functional integral $\int d\psi\, d\bar{\psi}\, \exp(\frac{1}{2}i\bar{\psi}A\psi)$ gives $\mathrm{Det}A$, while for a boson $\int d\phi\, \exp(\frac{1}{2}i\phi A\phi) = \mathrm{Det}^{-1/2}A$.

[14]K. Johnson, in proceedings of the Seminar on Unified Theories of Elementary Particles, 1963, Rochester, N.Y. (unpublished); Th. A. J. Maris, V. E. Herscovitz, and G. Jacob, Phys. Rev. Lett. 12, 313 (1964).

[15]J. M. Cornwall, Phys. Rev. D 10, 500 (1974).

[16]We are using a similar notation, Π, for two different objects. In order to avoid confusion we explain in detail. When an argument is explicitly indicated, viz. $\Pi(k^2)$, as in (4.11) and (4.12), we mean the Lorentz-invariant part of the vacuum-polarization tensor. In symbolic equations like (4.14), Π occurs without an argument; in that case it denotes the full Lorentz-covariant vacuum-polarization tensor. The connection is $\Pi \to \Pi^{\mu\nu}(k) = iP^{\mu\nu}\Pi(k^2)$.

[17]Cornwall and Norton, Ref. 5.

[18]S. Coleman and E. Weinberg, Phys. Rev. D 7, 1888

(1973); S. Weinberg, *ibid.* 7, 2887 (1973); R. Jackiw, Ref. 4.

[19]The analysis of the physical interpretation of $\Gamma(\phi, G)|_{\text{static}}$ is an adaptation to the present context of the corresponding argument for $\Gamma(\phi)|_{\text{static}}$. That discussion is due to K. Symanzik, Commun. Math. Phys. 16, 48 (1970). We learned it from S. Coleman, in proceedings of the Lectures given at the International Summer School of Physics "Ettore Majorana," 1973 (unpublished).

[20]R. Dashen, B. Hasslacher, and A. Neveu, Phys. Rev. D (to be published).

III. EARLY HYPERCOLOR

PHYSICAL REVIEW D VOLUME 13, NUMBER 4 15 FEBRUARY 1976

Implications of dynamical symmetry breaking

Steven Weinberg*
Lyman Laboratory of Physics, Harvard University, Cambridge, Massachusetts
(Received 8 September 1975)

An analysis is presented of the physical implications of theories in which the masses of the intermediate vector bosons arise from a dynamical symmetry breaking. In the absence of elementary spin-zero fields or bare fermion masses, such theories are necessarily invariant to zeroth order in the weak and electromagnetic gauge interactions under a global $U(N)\otimes U(N)$ symmetry, where N is the number of fermion types, not counting color. This symmetry is broken both intrinsically by the weak and electromagnetic interactions and spontaneously by dynamical effects of the strong interactions. An effective Lagrangian is constructed which allows the calculation of leading terms in matrix elements at low energy; this effective Lagrangian is used to analyze the relative direction of the intrinsic and spontaneous symmetry breakdown and to construct a unitarity gauge. Spontaneously broken symmetries which belong to the gauge group of the weak and electromagnetic interactions correspond to fictitious Goldstone bosons which are removed by the Higgs mechanism. Spontaneously broken symmetries of the weak and electromagnetic interactions which are not members of the gauge group correspond to true Goldstone bosons of zero mass; their presence makes it difficult to construct realistic models of this sort. Spontaneously broken elements of $U(N)\otimes U(N)$ which are not symmetries of the weak and electromagnetic interactions correspond to pseudo-Goldstone bosons, with mass comparable to that of the intermediate vector bosons and weak couplings at ordinary energies. Quark masses in these theories are typically less than 300 GeV by factors of order α. These theories require the existence of "extra-strong" gauge interactions which are not felt at energies below 300 GeV.

I. INTRODUCTION

When unified gauge theories of the weak and electromagnetic interactions were first proposed, it was assumed[1] that the spontaneous symmetry breakdown responsible for the intermediate-vector-boson masses is due to the vacuum expectation values of a set of spin-zero fields. For a variety of reasons, the attention of theorists has since been increasingly drawn to the possibility that this symmetry breaking is of a purely dynamical nature.[2] That is, it is supposed that there may be no elementary spin-zero fields in the Lagrangian, and that the Goldstone bosons associated with the spontaneous symmetry breakdown are bound states.

Almost all the effort that has been put into analyses of dynamical symmetry breaking has been directed to the difficult mathematical problem, of whether and how this phenomenon can occur in a variety of field-theoretic models. In this article I would like to address quite a different question: Assuming that dynamical symmetry breaking is a mathematical possibility in gauge field theories, what are the consequences for the real world?

Why should we believe that the masses of the intermediate vector bosons arise from dynamical symmetry breaking? The absence of *strongly* interacting elementary spin-zero fields is indicated by a number of requirements: asymptotic freedom,[3] electroproduction sum rules,[4] and the naturalness of order-α parity and strangeness conservation.[5] On the other hand, the absence of *weakly* interacting elementary spin-zero fields is much less certain. Apart from simplicity, the best reason for this assumption comes from the requirement for a natural hierarchy of gauge symmetry breaking.[6] In order to put together the observed weak and electromagnetic interactions into a simple gauge group, it is necessary to suppose[7] that in the spontaneous breakdown of this simple group to the nonsimple gauge group of the observed interactions, vector-boson masses are generated that are much larger than the masses expected for the W and Z; this conclusion is even stronger if we try to include the strong interactions as well.[8] This superstrong symmetry breakdown may well be due to the vacuum expectation values of elementary spin-zero fields. However, at ordinary energies, far below the superheavy vector-boson masses, physics is described by an effective field theory[9] involving those fermions and vector bosons that did not get masses from the superstrong spontaneous symmetry breakdown, but no spin-zero fields. Likewise, the gauge group of this effective field theory consists of a direct product of those simple and U(1) subgroups of the simple gauge group that were not broken at the superstrong level. The only way that the non-superheavy fermions and vector bosons can then acquire masses is from a dynamical breakdown of this remaining gauge group. Futhermore, the mass scale determined by the dynamical symmetry breakdown is expected to be of the order of magnitude of the renormalization point at which the largest of the gauge couplings of the effective field theory reaches a value of order unity; this mass scale is in general enormously different from the mass scale of the superstrong symmetry breakdown.

For the purposes of this article, we will not need to commit ourselves to the general picture described above. However, our assumptions are those inspired by this picture: We assume that weak, strong, and electromagnetic interactions are described by a gauge field theory (perhaps an "effective" field theory[9]) involving fermions and gauge fields but no elementary spin-zero fields or bare fermion masses, and we suppose that the vector-boson and fermion masses arise from a dynamical breakdown of this gauge group. These assumptions are spelled out more precisely in Sec. II.

In further support of these assumptions, it should be mentioned that it is the absence of bare fermion masses that makes it natural for a spontaneous dynamical symmetry breakdown to occur. Any spontaneous symmetry breaking requires the appearance of massless Goldstone bosons,[10] whether or not they are eventually eliminated by the Higgs mechanism.[11] For dynamical symmetry breaking, these Goldstone bosons would have to be bound states. However, we would normally expect that any bound state at zero mass would move away from zero mass if we changed the strength of the binding interactions, in which case dynamical symmetry breaking could only occur for a discrete set of coupling strengths,[12] and could not be considered "natural." In the theories considered here, with zero bare fermion mass, there is only one mass scale, defined by the renormalization point at which the gauge couplings are specified; thus a small change in the gauge coupling constant corresponds to a general change of mass scale,[13] and cannot shift a massless bound state away from zero mass.

The consequences of our assumptions turn out to be quite striking. Before turning on the weak and electromagnetic interactions, the strong interactions are necessarily invariant, not only under the strong gauge group, but also, as shown in Sec. III, under a global $U(N)\otimes U(N)$ group, where N is the number of fermion types, not counting color or possible other strong gauge indices. This global group is broken in two different ways, described in Sec. IV. It is *spontaneously* broken down to some subgroup H by dynamical effects of the strong interactions. It is also *intrinsically* broken to that subgroup S_W of $U(N)\otimes U(N)$ which leaves the weak and electromagnetic interactions invariant. And of course the gauge group of the weak and electromagnetic interactions is a subgroup G_W of S_W.

It is this double breakdown of $U(N)\otimes U(N)$, shown symbolically in Fig. 1, that will occupy most of our attention in this paper. Indeed, aside from the final section, the bulk of this paper can be regarded as a mathematical analysis of general theories in which there is both a strong spontaneous symmetry breaking and a weak intrinsic symmetry breaking induced by gauge interactions. The property that is specific to theories without spinless fields is that the over-all global group is $U(N)\otimes U(N)$, but most of our discussion would apply to any other global group.

Our analysis is complicated by three factors:

(1) We cannot use perturbation theory to describe the strong interactions responsible for the spontaneous symmetry breaking. This problem is evaded here by restricting ourselves to processes at relatively "low" energies, not greater than the expected masses of the intermediate vector bosons, or roughly $e\times 300$ GeV. It is shown in Sec.

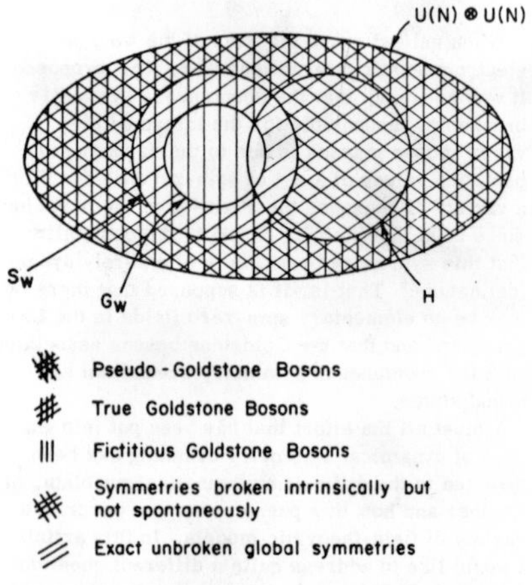

FIG. 1. Schematic representation of the various subgroups of $U(N)\otimes U(N)$. Cross hatchings indicate the various ways that global or local symmetries are broken; the unhatched lens represents the unbroken exact local symmetries, such as electromagnetic gauge invariance. As discussed in the text, H is the subgroup of $U(N)\otimes U(N)$ which is not spontaneously broken; S_W is the global symmetry group of the weak and electromagnetic interactions; and G_W is the weak and electromagnetic gauge group.

V that at such energies the terms of leading order in e in the matrix element for any process are given by calculating tree graphs, using an effective Lagrangian[14] of reasonably simple structure. The effective Lagrangian involves those fermions and strong gauge vector bosons that did not acquire masses from the spontaneous dynamical symmetry breaking (which we identify as the usual quarks and gluons) plus the Goldstone bosons that accompany the spontaneous symmetry breakdown, and the gauge bosons of the weak and electromagnetic interactions.

(2) In defining a particular theory, it is not enough to specify the group structure of the non-spontaneously broken subgroup H and the gauge subgroup G_W; we must also say how these subgroups line up with each other[15] within the overall group $U(N)\otimes U(N)$. In Sec. VI it is shown that the alignment of these subgroups is determined by the condition that the Goldstone bosons must not have tadpoles; otherwise perturbation theory breaks down.

(3) There is a Goldstone boson for every independent broken symmetry[10] in $U(N)\otimes U(N)$, but those Goldstone bosons that correspond to generators of the gauge group G_W are "fictitious" Goldstone bosons, which are eliminated by the Higgs mechanism.[11] In Sec. VII we show how in general to define a unitarity gauge[16] in which these fictitious Goldstone bosons are absent. The masses of the intermediate vector bosons can be determined (to leading order in e) by inspecting the effective Lagrangian in the unitarity gauge. The other Goldstone bosons which are not eliminated by the Higgs phenomenon are studied in Sec. VIII. This class consists of "true" Goldstone bosons of zero mass, corresponding to broken symmetries in S_W but not G_W, and "pseudo"-Goldstone bosons[17] with mass of order $e\times 300$ GeV, corresponding to broken symmetries of $U(N)\otimes U(N)$ which are neither in S_W nor G_W (see Fig. 1).

Different aspects of this analysis have been discussed before, but not to the best of my knowledge all together. Thus, effective Lagrangians for both broken global[14] and broken gauge symmetries[18] are an old story, but not for the case where the broken-symmetry group consists of a group of approximate global symmetries with an exact gauge subgroup. Also, the problem of subgroup alignment mentioned in item (2) above has been studied in the presence of strong interactions,[15] but with a nongauge perturbation, and also with a gauge perturbation,[19] but in the absence of strong interactions. Finally, previous attempts at a general definition of the unitarity gauge[16] and the pseudo-Goldstone bosons[17] dealt only with a spontaneous symmetry breakdown produced by vacuum expectation values of elementary spin-zero fields.

In Sec. IX we take up an unrealistic example which is designed to show how this analysis can be applied to specific theories. As usual in models with spontaneously broken symmetries, we can obtain quite detailed information about the interaction of soft Goldstone bosons with quarks and vector bosons, despite the presence of strong interactions. Surprisingly, one can also solve the subgroup alignment problem explicitly. There are just two possible ways that the subgroups can line up, corresponding to the possible signs of a single unknown parameter. In one case there are two massive vector bosons, one "photon," one true Goldstone boson, two pseudo-Goldstone bosons, and a finite quark mass of second order in e; in the other case there are three massive vector bosons, no photons, no true Goldstone bosons, two pseudo-Goldstone bosons, and an exact symmetry which keeps the quark mass zero to all orders in e. Evidently the subgroup alignment is crucial in determining the physical content of theories with a given group structure. It is striking that for both alignments the theory contains unwelcome massless particles: in one case a true Goldstone boson, in the other a massless quark. This is a common problem in theories with dynamical symmetry breaking.

The last section offers a series of remarks about the application of the formalism developed in this article to models of the real world.

This article is not intended as an argument that the masses of the intermediate vector bosons actually do arise from dynamical symmetry breaking. Indeed, some of the difficulties of constructing realistic models based on dynamical symmetry breaking are emphasized in Sec. X. However, it would be wise at least to keep in mind that the experiments designed to find intermediate vector bosons may discover pseudo-Goldstone bosons as well.

II. GENERAL ASSUMPTIONS

The theories to be discussed in this paper are governed by the following general assumptions:

(a) The Lagrangian is locally invariant under a gauge group

$$G_S\otimes G_W. \tag{2.1}$$

Here G_S describes the strong interactions, and has gauge couplings roughly of order unity; G_W describes the weak and electromagnetic interactions, and has gauge couplings roughly of order e. [Both G_S and G_W may themselves be direct products of simple and/or U(1) gauge groups.] As discussed in Sec. X, it is likely that G_S is larger than the usual color SU(3) gauge group.

(b) In addition to the vector gauge fields required under (a) there is a set of fermion fields $\psi_{nm}(x)$. Here n is an N-valued row or "flavor" index labeling fermion type $\mathcal{P}$, $\mathcal{N}$, λ, $\mathcal{P}'$, etc., on which G_W acts, and m is a column or "color" index, on which G_S acts. (The status of the leptons is considered briefly in Sec. X.)

(c) The Lagrangian contains no fermion mass terms and no elementary spin-zero fields.

(d) The theory is renormalizable.

These assumptions are made here because they seem to be required[6] in simple gauge theories with hierarchies of symmetry breaking, of the type described in the Introduction. However, for the purpose of this paper it will not be necessary to suppose that the strong, weak, and electromagnetic interactions arise from a superstrongly broken simple gauge group; it will only be assumed that physics at "ordinary" energies (say, up to a few thousand GeV) is governed by assumptions (a)–(d).

Under these assumptions, the Lagrangian must take the form[20]

$$\mathcal{L}=-\bar{\psi}\gamma^\mu\mathfrak{D}_\mu\psi-\tfrac{1}{4}\mathfrak{F}_{\alpha\mu\nu}\mathfrak{F}_\alpha^{\mu\nu}-\tfrac{1}{4}G_{\sigma\mu\nu}G_\sigma^{\mu\nu}, \qquad (2.2)$$

where $\mathfrak{D}_\mu\psi$ is the gauge-covariant derivative of the fermion field

$$(\mathfrak{D}_\mu\psi)_{nm}=\partial_\mu\psi_{nm}-i\sum_{n'\alpha}(w_\alpha)_{nn'}\psi_{n'm}W_{\alpha\mu} -i\sum_{m'\sigma}(s_\sigma)_{mm'}\psi_{nm'}S_{\sigma\mu}, \qquad (2.3)$$

with $W_{\alpha\nu}$ and $S_{\sigma\mu}$ the G_W and G_S gauge fields, and w_α and s_σ the matrices representing the corresponding group generators. (The gauge coupling constants are included as factors in w_α and s_σ.) Also, $F_{\alpha\mu\nu}$ and $G_{\sigma\mu\nu}$ are the usual covariant curls of $W_{\alpha\mu}$ and $S_{\sigma\mu}$, respectively.

III. U(N)⊗U(N) SYMMETRY

The most striking consequence of the general assumptions outlined in the last section is the existence of an "accidental" approximate global symmetry of the Lagrangian. In the limit $e\to 0$ the Lagrangian is automatically invariant not only under the *local* G_S transformations on the fermion column (i.e., color) indices, but also under a group U(N)⊗U(N) of *global* transformations on the N-valued fermion row (i.e., $\mathcal{P}$, $\mathcal{N}$, λ, $\mathcal{P}'$, . . .) index. That is, for each of the $2N^2$ independent Hermitian matrices λ_A (with Dirac matrix factor of 1 or γ_5) there is a vector or axial-vector current

$$J_A^\mu=-i\sum_{mn'n}\bar{\psi}_{n'm}\gamma^\mu(\lambda_A)_{n'n}\psi_{nm}. \qquad (3.1)$$

Apart from triangle anomalies (about which more will be said later) these are all conserved:

$$\partial_\mu J_A^\mu=0 \quad (\text{for } e=0). \qquad (3.2)$$

In what follows it will be convenient to normalize the λ_A and hence the J_A so that

$$\mathrm{Tr}(\lambda_A\lambda_B)=8\delta_{AB}. \qquad (3.3)$$

(An unusual extra factor of 4 appears here because the trace includes a trace on Dirac indices; this is necessary because half the λ_A are proportional to the Dirac matrix γ_5.)

It should be emphasized that the U(N)⊗U(N) symmetry arises only because of our assumptions that the Lagrangian contains no fermion mass terms $m\bar{\psi}\psi$, no scalar field couplings $\phi\bar{\psi}\psi$, and no nonrenormalizable interactions, such as a Fermi interaction $\bar{\psi}\psi\bar{\psi}\psi$. Any one of these terms might in general destroy the U(N)⊗U(N) symmetry. On the other hand, once we make these assumptions, the U(N)⊗U(N) symmetry is inescapable—the fermion fields must enter the Lagrangian only in the form $\bar{\psi}\gamma^\mu\mathfrak{D}_\mu\psi$, and in the limit $e\to 0$ the covariant derivative $\mathfrak{D}_\mu$ contains only matrices which commute with all λ_A.

IV. SYMMETRY BREAKING

The U(N)⊗U(N) symmetry is in general broken by the weak and electromagnetic interactions. This intrinsic symmetry breaking can be quantitatively described by writing the generators w_α of the weak and electromagnetic gauge group as linear combinations of the U(N)⊗U(N) operators λ_A:

$$w_\alpha=\sum_A e_{\alpha A}\lambda_A. \qquad (4.1)$$

In accordance with our previous assumptions, the coefficients $e_{\alpha A}$ are all of order e. Emission and absorption of virtual W bosons will produce order-e^2 perturbations which are not expected to be U(N)⊗U(N) invariant.

We shall assume that in addition to this intrinsic symmetry breaking, even in the limit $e\to 0$, there is a *spontaneous* breakdown of the symmetry group G_S⊗U(N)⊗U(N) of the strong interactions, caused by strong forces among fermions, antifermions, and G_S gauge bosons. As usual in any spontaneous symmetry breaking, it is perfectly natural for some subgroup U to be left unbroken. For the sake of simplicity and definiteness, we shall assume that U does not mix the strong gauge group G_S with the accidental global symmetry group U(N)⊗U(N); that is, the unbroken symmetry group for $e\to 0$ is a direct product

$$U=H_S\otimes H, \qquad (4.2)$$

where H_S is local and a subgroup of G_S, while H is global and a subgroup of U(N)⊗U(N). Most of the considerations below would also apply to the more general case where the unbroken subgroup is not of the form (4.2), but our discussion would have to be made considerably more elaborate to deal with this case.

For any independent generator of G_S which is not a generator of the unbroken gauge subgroup H_S, the corresponding strongly interacting vector boson gets a mass from the Higgs phenomenon. Also, depending on the nature of the unbroken global subgroup H, some of the fermions will acquire a mass from the spontaneous breakdown of U(N) ⊗ U(N) to H. In the limit $e \to 0$ this theory contains no very large or very small dimensionless parameters, so we would expect all these masses to be of the same order of magnitude, say M. The only dimensional parameter in the theory for $e \to 0$ is the scale characterizing the renormalization point of the G_S gauge couplings, so we would also expect M to be determined by the condition that the largest G_S gauge coupling reaches a critical value of order unity at a renormalization point characterized by momenta of order M.

We will see below that M is likely to be quite large, of order 300 GeV. The physics of strong interactions at lower energies $E \ll M$ can therefore be described in terms of those fermions and G_S gauge bosons which do *not* pick up masses of order M from the spontaneous symmetry breakdown, and hence remain massless in the limit $e \to 0$. These may be identified as the ordinary quarks and gluons, respectively. (Of course, we do not at this point rule out the possibility that H_S is just G_S, so that *all* of the G_S gauge bosons remain massless.) As we shall see, turning on the weak and electromagnetic interactions will give the quarks masses of order $e^2 M$, while the gluons will remain massless.

In addition to quarks and gluons, this theory necessarily contains one other class of hadrons with masses which vanish for $e \to 0$, the Goldstone bosons. For every linearly independent generator of U(N) ⊗ U(N) which is not a generator of the unbroken subgroup H, there must appear a Goldstone boson Π_a. Since U(N) ⊗ U(N) is not a gauge group, there is no Higgs phenomena which can eliminate these Goldstone bosons in the limit $e \to 0$.

The coupling of the ath Goldstone boson to the Ath U(N) ⊗ U(N) current is described by a parameter F_{aA}, defined by

$$\langle 0|J_A^\mu(0)|\Pi_a\rangle = F_{aA} P_\Pi^\mu (2\pi)^{-3/2}(2E_\Pi)^{-1/2}. \qquad (4.3)$$

These F_{aA} have the dimensions of a mass, and will play a role here like that played by the parameter F_π of current algebra. We expect that all F_{aA} are of the order of the mass M introduced earlier,

$$F_{aA} \approx M,$$

because in the limit $e \to 0$ this is the only mass in the theory.

It will be very convenient to adapt the basis for U(N) ⊗ U(N) to the pattern of symmetry breaking. We may define the generators of the unbroken subgroup H as linear combinations of the $2N^2$ generators λ_A of U(N) ⊗ U(N),

$$t_i = \sum_A C_{iA} \lambda_A, \qquad (4.4)$$

with the C_{iA} chosen as orthonormal vectors so that

$$\sum_A C_{iA} C_{jA} = \delta_{ij}, \qquad (4.5)$$

$$\mathrm{Tr}(t_i t_j) = 8\delta_{ij}. \qquad (4.6)$$

The unbroken-symmetry currents have no couplings to the Goldstone bosons, so that

$$\sum_A C_{iA} F_{aA} = 0. \qquad (4.7)$$

Further, by a suitable unitary transformation we can always choose the Π_a states so as to diagonalize the positive Hermitian matrix

$$\sum_A F_{aA} F_{bA}.$$

If the element with $b=a$ is denoted F_a^2, we have then

$$F_{aA} = F_a B_{aA}, \qquad (4.8)$$

with

$$\sum_A B_{aA} B_{bA} = \delta_{ab}, \qquad (4.9)$$

and also

$$\sum_A B_{aA} C_{iA} = 0. \qquad (4.10)$$

There is one Goldstone boson for each independent broken symmetry, so the B's and C's form a complete orthonormal set of vectors

$$\sum_a B_{aA} B_{aB} + \sum_i C_{iA} C_{iB} = \delta_{AB}. \qquad (4.11)$$

Correspondingly, we can define a set of broken symmetry generators

$$x_a \equiv \sum_A B_{aA} \lambda_A, \qquad (4.12)$$

with

$$\mathrm{Tr}(t_i x_a) = 0, \qquad (4.13)$$

$$\mathrm{Tr}(x_a x_b) = 8\delta_{ab}. \qquad (4.14)$$

The generators t_i and x_a span the algebra of U(N) ⊗ U(N).

V. EFFECTIVE LAGRANGIAN

We now want to consider the special phenomena which arise because U(*N*) ⊗ U(*N*) is simultaneously broken both intrinsically by the weak and electromagnetic interactions and also spontaneously by dynamical effects of the strong interactions. This is not a mere matter of expanding in powers of e, because the theory has infrared singularities which, at momenta p much less than the characteristic mass M, introduce factors M/p which can compensate for factors e.

In order to explore this problem, let us consider the leading pole singularities produced by soft virtual quarks, Goldstone bosons, *and* G_W vector bosons in a general Green's function with external quark, gluon, Goldstone boson, and/or G_W vector bosons carrying momenta of order $p \ll M$.[21] These singularities can be calculated from the sum of all tree graphs for this Green's function, constructed from an effective Lagrangian[14] involving quark, G_W vector boson, gluon, and Goldstone boson fields. [For the moment, we are ignoring the effects of loops containing *hard* virtual G_W vector bosons. These produce U(*N*) ⊗ U(*N*)-breaking corrections of order e^2 in the effective Lagrangian, and will be considered at the end of this section.]

The effective Lagrangian here takes the general form

$$\mathfrak{L} = -\tfrac{1}{4} F_{\alpha\mu\nu} F^{\mu\nu}_{\alpha} + \mathfrak{L}_1. \tag{5.1}$$

The term $\mathfrak{L}_1$ is subject to three conditions:

(1) In the limit $e \to 0$ the W dependence drops out, and $\mathfrak{L}_1$ becomes invariant under U(*N*) ⊗ U(*N*), with fields transforming according to one of the usual nonlinear realizations of U(*N*) ⊗ U(*N*), in which the unbroken subgroup H is realized algebraically. It will be convenient to define the fields so that Π transforms like the so-called exponential parametrization[22] of the cosets in U(*N*) ⊗ U(*N*)/H. That is, a general U(*N*) ⊗ U(*N*) transformation, which is represented on the fermion fields in the original Lagrangian by a matrix g, induces on the fields in the effective Lagrangian the nonlinear transformations[22]

$$\Pi_a \to \Pi'_a(\Pi, g), \tag{5.2}$$

$$q \to \exp\Big(i\sum_i \mu_i(\Pi, g)t_i\Big)q, \tag{5.3}$$

where q is the quark field multiplet (with components corresponding to those fermion fields that do not acquire masses of order M from the spontaneous symmetry breaking), and Π' and μ are functions defined by the relation

$$g\exp\Big(i\sum_a \Pi_a x_a/F_a\Big) = \exp\Big(i\sum_a \Pi'_a(\Pi, g)x_a/F_a\Big) \times \exp\Big(i\sum_i \mu_i(\Pi, g)t_i\Big). \tag{5.4}$$

The gluon fields are of course invariant under these transformations.

(2) The currents to which the products $W_{\alpha\mu}W_{\beta\nu}\cdots$ of G_W gauge fields (or their derivatives) couple in $\mathfrak{L}_1$ take the form

$$\sum_{A,B,\ldots} e_{\alpha A} e_{\beta B} \cdots T_{AB\ldots}, \tag{5.5}$$

where $T_{AB\ldots}$ is a quantity, formed out of quark, gluon, and Goldstone boson fields and their derivatives, which transforms like a U(*N*) ⊗ U(*N*) tensor (i.e., like $\lambda_A\lambda_B\cdots$) when Π and q undergo the transformations (5.2) and (5.3).

(3) $\mathfrak{L}_1$ is locally as well as globally invariant under the unbroken subgroup H_S of the strong gauge group and under the weak and electromagnetic gauge group G_W.

It is shown in Appendix A that these conditions require $\mathfrak{L}_1$ to be constructed from just the following ingredients:

(i) The quark fields q.

(ii) Their covariant derivatives (for notation, see below):

$$D_\mu q \equiv \partial_\mu q - i\sum_{ai} t_i q E_{ai}(\Pi)\partial_\mu \Pi_a F_a^{-1} - i\sum_{BA\alpha i} t_i\, q\Lambda_{AB}(\Pi) e_{\alpha A} W_{\alpha\mu} C_{iB} + \text{gluon terms}. \tag{5.6}$$

(iii) Covariant derivatives of the Goldstone boson fields:

$$D_\mu \Pi_a \equiv F_a\Big[\sum_b D_{ba}(\Pi)\partial_\mu \Pi_b F_b^{-1} + \sum_{A\alpha B} \Lambda_{AB}(\Pi) e_{\alpha A} W_{\alpha\mu} B_{aB}\Big]. \tag{5.7}$$

(iv) A covariant curl of the W field:

$$\mathfrak{F}_{A\mu\nu} = \sum_{B\alpha} \Lambda_{BA}(\Pi) e_{\alpha A} F_{\alpha\mu\nu}. \tag{5.8}$$

(v) The usual covariant curl of the gluon field.

Here $F_{\alpha\mu\nu}$ is the usual Yang-Mills G_W-covariant curl,[20] and the functions D, E, and Λ are defined by the formulas

$$S^{-1}(\Pi)\frac{\partial}{\partial \Pi_a} S(\Pi) = iF_a^{-1}\Big[\sum_b D_{ab}(\Pi)x_b + \sum_i E_{ai}(\Pi)t_i\Big], \tag{5.9}$$

$$S^{-1}(\Pi)\lambda_A S(\Pi) = \sum_B \Lambda_{AB}(\Pi)\lambda_B, \tag{5.10}$$

with

$$S(\Pi) \equiv \exp\Big(i\sum_b \Pi_b x_b/F_b\Big). \tag{5.11}$$

With our normalization convention (3.3), the Λ matrices are orthogonal,

$$\Lambda^T(\Pi) = \Lambda^{-1}(\Pi). \tag{5.12}$$

[At this point, the reader may wish to be reminded that indices A, B, etc. label all generators of $U(N)\otimes U(N)$; i, j, etc. label the unbroken generators; and a, b, etc. label the broken generators.]

In addition to these limitations on the ingredients in $\mathcal{L}_1$, the conditions (1)–(3) also require that $\mathcal{L}_1$ must be invariant under formal global H transformations, with $D_\mu\Pi$, $D_\mu q$, and $\mathfrak{F}_{A\mu\nu}$ transforming according to whatever (linear) representations of H they happen to contain.

At this point the effective Lagrangian we have derived still has an extremely complicated structure, involving unlimited numbers of q, $D_\mu q$, $D_\mu\Pi$, $G_{\sigma\mu\nu}$, and $\mathfrak{F}_{A\mu\nu}$ functions. Indeed, if we were to take this Lagrangian seriously as a basis for higher-order calculations, we would have to keep all these interactions in order to provide counterterms for the infinite number of primitive divergents that would arise. However, the structure of the effective Lagrangian can be very much simplified if we use it only to determine the matrix elements to lowest order in e and p/M.

Ordinary dimensional analysis leads us to expect that an interaction v appearing in the effective Lagrangian with n_{vq} quark fields, n_{vs} gluon fields, $n_{v\Pi}$ Goldstone boson fields, n_{vw} G_W-gauge boson fields, and n_{vd} derivatives, will have a coupling constant of order

$$g_v \approx e^{n_{vw}} M^{n_{vm}}, \tag{5.13}$$

where M is the characteristic mass introduced in the last section (of order F_a) and

$$n_{vm} = 4 - \tfrac{3}{2}n_{vq} - n_{vs} - n_{v\Pi} - n_{vd}. \tag{5.14}$$

If a Green's function has external lines with momenta of order

$$p \approx eM, \tag{5.15}$$

then each power of M counts like a factor $1/e$. The total number of factors of e or p/M is

$$N_e = \sum_v N_v(n_{vw} - n_{vm}), \tag{5.16}$$

where N_v is the number of vertices of type v. However, a tree graph with E_q external quark lines, E_Π external Goldstone boson lines, and E_W external G_W-gauge bosons will obey the topological relation

$$\sum_v (n_{vq} + n_{v\Pi} + n_{vw} - 2)N_v = E_q + E_\Pi + E_W - 2. \tag{5.17}$$

(Gluon terms on both sides cancel, because we exclude *internal* gluon lines.[21]) Therefore, the total number of factors of e or p/M is

$$N_e = E_q + E_\Pi + E_W - 2 + \sum_v N_v\Delta_v, \tag{5.18}$$

where

$$\Delta_v \equiv n_{vw} + \tfrac{1}{2}n_{vq} + n_{vd} + n_{vs} - 2. \tag{5.19}$$

Thus for any given set of external lines, the terms of lowest order in e or p/M will be given by graphs composed of vertices with the smallest possible values of Δ_v.

In fact, the smallest values of Δ_v for any allowed interaction is $\Delta_v = 0$. Keeping only terms with $\Delta_v = 0$, and normalizing fields appropriately, gives an effective Lagrangian of the general form

$$\mathcal{L}_1 = -\bar{q}\gamma^\mu D_\mu q - \tfrac{1}{2}\sum_a D_\mu\Pi_a D^\mu\Pi_a - \sum_a \bar{q}\gamma^\mu\Gamma_a q D_\mu\Pi_a + \text{Fermi interactions}. \tag{5.20}$$

Here Γ_a is a constant matrix, proportional to γ_5 and/or 1, and of order $1/M$, which has the same H-transformation properties as Π_a; also, the "Fermi interactions" are $\bar{q}q\bar{q}q$ terms of order $1/M^2$ which are H invariant, but may involve any of the 16 Dirac covariants. Note that $D_\mu q$ and $D_\mu\Pi$ have just the right dependence on quark, gluon, G_W vector boson, and Goldstone boson fields, so that it is possible to construct an effective Lagrangian obeying all necessary symmetry conditions with only $\Delta_v = 0$ terms. (To the extent that a graph with loops is dominated by states with energy $E \ll M$, it can also be calculated with this effective Lagrangian.)

There is, in fact, one other term which in effect has $\Delta_v = 0$ and therefore should be added to the effective Lagrangian (5.20). The emission and reabsorption of a hard G_W vector boson produces an effective interaction of second order in e. For any such term, the Δ_v in Eq. (5.19) should be increased by two units, so that it becomes

$$\Delta_v = n_{vw} + \tfrac{1}{2}n_{vq} + n_{vd} + n_{vs}. \tag{5.21}$$

Thus we get a term with $\Delta_v = 0$ of second order in e if it is constructed *solely* from Goldstone boson fields, with no G_W gauge fields, quark fields, gluon fields, or derivatives. (See Fig. 2.)

This term will take the form

$$\mathcal{L}_2(\Pi) = -\sum_{\alpha AB} e_{\alpha A} e_{\alpha B} J_{AB}(\Pi), \tag{5.22}$$

where J_{AB} is the time-ordered product of two $U(N)\otimes U(N)$ currents J_A^μ and J_B^ν, integrated over momenta (including a $1/k^2$ weight factor from the W_α^μ propagator) and contracted over space-time indices. This two-point function is a $U(N)\otimes U(N)$ tensor, i.e., it transforms like $\lambda_A\lambda_B$. As shown in Appendix B, the most general form of such a tensor is

$$J_{AB}(\Pi) = -\sum_{CD}\Lambda_{AC}(\Pi)\Lambda_{BD}(\Pi)I_{CD}, \qquad (5.23)$$

with I a Π-independent quantity of order M^4 which behaves as a tensor under the unbroken subgroup H. Thus, the $O(e^2)$ term with $\Delta_\nu=0$ is

$$\mathcal{L}_2(\Pi) = -\sum_{\alpha ABCD} e_{\alpha A}e_{\alpha B}\Lambda_{AC}(\Pi)\Lambda_{BD}(\Pi)I_{CD}. \qquad (5.24)$$

This contains both a Π mass term (with some Goldstone boson masses of order eM) and a nonderivative Π self-interaction, like that in pion-pion scattering.

Incidentally, the integral in J_{AB} is expected to receive its major contribution from momenta of order M, because in these theories there is no reason why the integral should start to converge at any lower momentum. This is why we do not include the effects of G_W vector-boson masses here; these masses will turn out to be of order eM, so their effect is a higher-order correction.[23]

So far, we have seen that the leading ($\Delta_\nu=0$) part of the effective Lagrangian takes the form

$$\mathcal{L}_{\rm eff} = -\tfrac{1}{4}F_{\alpha\mu\nu}F_\alpha^{\mu\nu} + \mathcal{L}_1 + \mathcal{L}_2. \qquad (5.25)$$

We could go on and describe the structure of higher terms with $\Delta_\nu=1$, $\Delta_\nu=2$, etc. However, we will content ourselves with describing one term of particular interest, associated with the quark masses.

The quarks are *defined* as the fermions that do not pick up a mass of order M from the spontaneous breakdown of $U(N)\otimes U(N)$ to H, so there is no quark mass term $\bar{q}q$ with $\Delta_\nu=-1$ in the effective Lagrangian. However, emission and absorption of a G_W vector boson can produce a term of order e^2 with $\Delta_\nu=+1$. Following the same reasoning as for $\mathcal{L}_2$, this term must take the form (see Fig. 3)

$$\mathcal{L}_m = -\sum_{\alpha AB}\bar{q}N_{AB}(\Pi)q e_{\alpha A}e_{\alpha B}, \qquad (5.26)$$

where $\bar{q}Nq$ transforms as a tensor under $U(N)\otimes U(N)$. As shown in Appendix C, the most general form of N is

$$N_{AB}(\Pi) = \sum_{CD}\Lambda_{AC}(\Pi)\Lambda_{BD}(\Pi)Q_{CD}, \qquad (5.27)$$

where Q_{CD} is a Π-independent matrix of order M which behaves as a tensor under the unbroken subgroup H. Thus the quark mass term is

$$\mathcal{L}_m = -\sum_{\alpha ABCD}\bar{q}Q_{CD}q\Lambda_{AC}(\Pi)\Lambda_{BD}(\Pi)e_{\alpha A}e_{\alpha B}. \qquad (5.28)$$

The quark mass matrix is then

$$\mathfrak{M} = \sum_\alpha Q_{AB}e_{\alpha A}e_{\alpha B}, \qquad (5.29)$$

so quark masses are expected to be of order e^2M. In addition, there are multilinear interactions of Goldstone bosons with quarks, including a $\Pi\bar{q}q$ coupling of order $e^2M/F\sim e^2$.

VI. ALIGNMENT OF SUBGROUPS

There is one further step that must be taken before the effective Lagrangian can be used for actual calculations. The symmetry $U(N)\otimes U(N)$ is supposed to be spontaneously broken to some subgroup H by dynamical effects of the strong interactions. However, although we can presume that the structure of H is determined by dynamical considerations, the strong interactions alone do not determine *which* subgroup of $U(N)\otimes U(N)$ with this structure is left unbroken. Given any solution of the strong-interaction dynamics with a subgroup H left invariant, we can find another solution in which the subgroup left invariant is

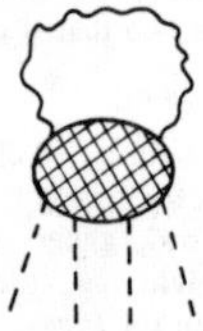

FIG. 2. Diagrams which contribute to the term $\mathcal{L}_2$ in the effective Lagrangian. Wavy lines are intermediate vector bosons and dashed lines are Goldstone bosons.

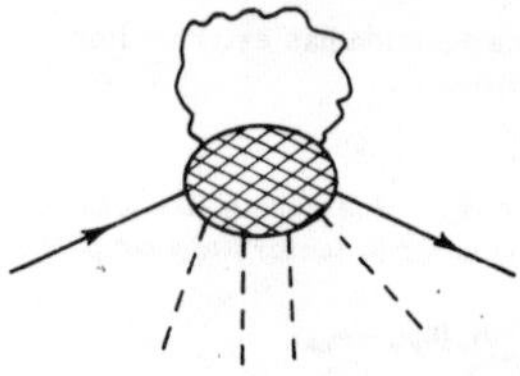

FIG. 3. Diagrams which contribute to the term $\mathcal{L}_m$ in the effective Lagrangian. Wavy lines are intermediate vector bosons, dashed lines are Goldstone bosons, and straight lines are quarks.

$$H(g) = gHg^{-1}, \tag{6.1}$$

where g is any element of $U(N) \otimes U(N)$.

Normally we do not concern ourselves with this sort of ambiguity, for these different solutions are usually physically equivalent. However, in our case, the theory contains a perturbation, the weak and electromagnetic interactions, which also break $U(N) \otimes U(N)$ down to some fixed subgroup S_W. (Of course, S_W contains the gauge group G_W of the weak and electromagnetic interactions, but as shown in Sec. X, in general it is larger.) Thus the different solutions corresponding to the different unbroken subgroups (6.1) are physically inequivalent, and we must decide which is the correct one.

This problem was encountered in a different context some time ago by Dashen,[15] who gave a general solution. It is necessary to construct a potential, given to lowest order as

$$V(g) = \langle 0, g | \mathcal{H}' | 0, g \rangle, \tag{6.2}$$

where $\mathcal{H}'$ is the symmetry-breaking perturbation (in our case an operator of second order in the weak and electromagnetic interactions) and $|0,g\rangle$ is the vacuum corresponding to the solution that has unbroken subgroup $H(g)$:

$$H(g)|0,g\rangle = |0,g\rangle. \tag{6.3}$$

The g that defines the "correct" solution in the presence of the perturbation $\mathcal{H}'$ is defined by the condition that $V(g)$ be a *minimum*.

For our purposes, it will be much more convenient to keep the solution of the strong- interaction symmetry breaking fixed, and instead vary the way that the weak and electromagnetic gauge group is inserted into the larger $U(N) \otimes U(N)$ global group. That is, we now fix the vacuum and the unbroken subgroup H, and instead let the gauge group be

$$G_W(g) = g^{-1} G_W g, \tag{6.4}$$

where g runs over all elements of $U(N) \otimes U(N)$. This has the advantage that we can fix the choice of the generators t_i and x_a from the beginning; the whole effect of varying the gauge group is that the "charges" $e_{\alpha A}$ are replaced with

$$e_{\alpha A}(g) = \sum_B R_{BA}(g) e_{\alpha B}, \tag{6.5}$$

where $R(g)$ is the regular representation of $U(N) \otimes U(N)$.

Using $e_{\alpha A}(g)$ in place of $e_{\alpha A}$ in Eq. (5.24), we see that the $O(e^2)$ term in the leading part of the effective Lagrangian is

$$\mathcal{L}_2(\Pi, g) = -\sum_{\alpha ABCD} e_{\alpha A}(g) e_{\alpha B}(g) \Lambda_{AC}(\Pi) \Lambda_{BD}(\Pi) I_{CD}. \tag{6.6}$$

The potential (6.2) is just given by the vacuum-fluctuation part of $\mathcal{L}_2$:

$$V(g) = -\mathcal{L}_2(0, g) \tag{6.7}$$

or more explicitly

$$V(g) = -\sum_{\alpha AB} e_{\alpha A}(g) e_{\alpha B}(g) I_{AB}. \tag{6.8}$$

Another interpretation can now be put on the condition that $V(g)$ be a minimum. Using Eqs. (6.5), (B2), (B6), and (5.2) in (6.6), we have

$$\mathcal{L}_2(\Pi, g_1 g) = \mathcal{L}_2(\Pi'(\Pi, g_1), g), \tag{6.9}$$

where g_1 and g are arbitrary elements of $U(N) \otimes U(N)$, and Π' is the image of Π under g_1. Thus the variation of $\mathcal{L}_2(\Pi, g)$ with respect to g may be determined from its variation with respect to Π. In particular, for $\Pi = 0$ and g_1 infinitesimal, Eq. (6.7) and (6.9) give

$$V\left(\left(1 + i\sum_a \frac{\epsilon_a x_a}{F_a}\right) g\right) = -\mathcal{L}_2(\epsilon, g), \tag{6.10}$$

so the condition that $V(g)$ be stationary with respect to g is equivalent to the condition that

$$\left.\frac{\partial \mathcal{L}_2(\Pi, g)}{\partial \Pi_a}\right|_{\Pi=0} = 0. \tag{6.11}$$

Graphically, this says that tadpole graphs, in which a single Goldstone boson disappears into the vacuum, necessarily vanish. The rationale for this condition is that otherwise perturbation theory in e would break down; the dominator of the propagator of a Goldstone boson at zero four-momentum is at most of order e^2 (see below) so a tadpole produced by second-order effects of the weak and electromagnetic interactions would be of zeroth order in e.[24]

Not only must we choose the gauge group $G_W(g)$ so that $V(g)$ is stationary; we must choose it so that $V(g)$ is at least a local *minimum*. The reason is again to be found in Eq. (6.9); the condition that $V(g)$ be a minimum is equivalent to the condition that $\Pi = 0$ be a minimum of $-\mathcal{L}_2(\Pi, g)$, and this in turn ensures the positivity of the mass matrix

$$m^2_{ab}(g) = -\left.\frac{\partial^2}{\partial \Pi_a \partial \Pi_b} \mathcal{L}_2(\Pi, g)\right|_{\Pi=0}. \tag{6.12}$$

From now on we will assume that g has been chosen from the beginning so that (6.11) is satisfied and (6.12) is positive. We can therefore drop the explicit argument g everywhere.

In order to put these conditions in a more useful form, we note that

$$\Lambda(\Pi) = \exp\left(i \sum_a X_a \Pi_a / F_a\right), \tag{6.13}$$

where $(X_a)_{AB}$ is the matrix representing the generator x_a in the adjoint representation of $U(N)\otimes U(N)$. Hence (5.24) gives

$$\mathcal{L}_2(\Pi) = -\operatorname{Tr}\left[E\exp\left(i\sum_a \frac{X_a\Pi_a}{F_a}\right)I\exp\left(-i\sum_a \frac{X_a\Pi_a}{F_a}\right)\right], \tag{6.14}$$

where E is the matrix

$$E_{AB} \equiv \sum_\alpha e_{\alpha A}e_{\alpha B}. \tag{6.15}$$

Condition (6.11) therefore reads

$$\operatorname{Tr}\{E[X_a, I]\} = 0, \tag{6.16}$$

and (6.12) gives the mass matrix as

$$m^2{}_{ab} = -\frac{1}{2F_aF_b}\operatorname{Tr}\{E[X_a,[X_b,I]] + E[X_b,[X_a,I]]\}. \tag{6.17}$$

Since I_{AB} is H invariant, we can generalize (6.16) to

$$\operatorname{Tr}\{E[\Lambda_A, I]\} = 0 \tag{6.18}$$

where Λ_A is the adjoint representation of an arbitrary generator λ_A of $U(N)\otimes U(N)$. Also, Eq. (6.17) can be simplified because the two terms on the right are equal; the Jacobi identity gives their difference as

$$\operatorname{Tr}\{E[X_a,[X_b,I]]\} - \operatorname{Tr}\{E[X_b,[X_a,I]]\} = \operatorname{Tr}\{E[[X_a,X_b,],I]\},$$

and this vanishes according to (6.18). Thus, (6.17) may be written

$$m^2{}_{ab} = m^2{}_{ba} = -\frac{1}{F_aF_b}\operatorname{Tr}\{E[X_a,[X_b,I]]\}. \tag{6.19}$$

VII. UNITARITY GAUGE AND VECTOR BOSON MASSES

The effective Lagrangian derived in Sec. V is still locally invariant under the gauge group G_W of the weak and electromagnetic interactions. We are therefore free to adopt a "unitarity gauge," in which the particle content of the theory is explicitly displayed.[16]

Suppose for a moment that the gauge group G_W were a sufficiently large subgroup of $U(N)\otimes U(N)$, so that its generators w_α, together with the generators t_i of the unbroken subgroup H, would completely span the algebra of $U(N)\otimes U(N)$. (This includes the case usually discussed, where G_W is the whole of the original symmetry group.) Then any element of $U(N)\otimes U(N)$ could be written as a product of an element of G_W times an element of H. This would in particular be true of the element $\exp(i\Sigma_a x_a\Pi_a/F_a)$, and therefore we could write

$$\exp\left(i\sum_a x_a\Pi_a/F_a\right) = \exp\left(-i\sum_\alpha \Theta_\alpha(\Pi)w_\alpha\right) \times \exp\left(i\sum_i \mu_i(\Pi)t_i\right)$$

for some real parameters $\theta_\alpha(\Pi)$ and $\mu_i(\Pi)$. But by comparing this with Eq. (5.4), we see that the gauge transformation $\exp(i\Sigma_\alpha\theta_\alpha w_\alpha)$ would carry the Goldstone boson field Π_a into $\Pi_a' = 0$. Thus, in this case there would be a choice of gauge which eliminates all Goldstone bosons.

In the general case, we do *not* expect the generators of G_W and H to span the algebra of $U(N)\otimes U(N)$. Note that if G_W is too large, then the weak and electromagnetic interactions will not break the symmetries of H sufficiently; we would then find that any quarks which do not get masses of order M from the spontaneous breakdown of $U(N)\otimes U(N)$ to H will remain massless to all orders in e (see Sec. X).

However, as shown in Appendix D, we can always write a general element λ of the algebra of $U(N)\otimes U(N)$ in the form

$$\lambda = -\sum_\alpha \theta_\alpha w_\alpha + \sum_a \phi_a x_a + \sum_i \mu_i t_i, \tag{7.1}$$

with ϕ_a constrained by the condition that

$$0 = \operatorname{Tr}\left(\theta_\alpha \sum_a \phi_a F_a^2 x_a\right) = \sum_{aA} e_{\alpha A}B_{aA}F_a^2\phi_a. \tag{7.2}$$

(The reason for adopting this particular constraint, and in particular for inserting the factor F_a^2, will be made clear below.) Hence, since every element of $U(N)\otimes U(N)$ that is infinitesimally close to the identity may be expressed in the form

$$\exp\left(-i\sum_\alpha \theta_\alpha w_\alpha\right)\exp\left(i\sum_a \phi_a x_a\right)\exp\left(i\sum_i \mu_i t_i\right)$$

[with ϕ_a satisfying (7.2)], it follows that every element of $U(N)\otimes U(N)$ in at least some finite region around the identity may be written in this form. In particular we may write

$$\exp\left(i\sum_a x_a\Pi_a/F_a\right) = \exp\left(-i\sum_\alpha \theta_\alpha(\Pi)w_\alpha\right) \times \exp\left(i\sum_a \phi_a(\Pi)x_a\right) \times \exp\left(i\sum_i \mu_i(\Pi)t_i\right). \tag{7.3}$$

But Eq. (5.4) then tells us that the gauge transformation $\exp[+i\Sigma_\alpha\theta_\alpha(\Pi)w_\alpha]$ carries the Goldstone boson field Π_a into

$$\Pi_a'/F_a = \phi_a(\Pi). \tag{7.4}$$

That is, we can adopt a gauge in which, according to Eq. (7.2),

$$\sum_{aA} e_{\alpha A} B_{aA} F_a \Pi'_a = 0 \tag{7.5}$$

or equivalently

$$\mathrm{Tr}\Big(w_\alpha \sum_a x_a F_a \Pi'_a\Big) = 0. \tag{7.6}$$

This is the unitarity gauge.

The unitarity gauge as we have defined it has the crucial property of eliminating the zeroth-order mixing between the Goldstone bosons and the G_W vector bosons. From Eqs. (5.7) and (5.15), we see that the part of the effective Lagrangian that is quadratic in gauge and/or Goldstone fields is

$$-\tfrac{1}{2}\sum_a (D_\mu \Pi'_a)_{\rm lin}(D^\mu \Pi'_a)_{\rm lin}, \tag{7.7}$$

with $(D_\mu \Pi')_{\rm lin}$ the linear part of the covariant derivative

$$(D_\mu \Pi'_a)_{\rm lin} = \partial_\mu \Pi'_a + F_a \sum_{A\alpha} e_{\alpha A} W'_{\alpha\mu} B_{aA}. \tag{7.8}$$

Hence Eq. (7.5) has the effect of insuring that the Π-W cross terms drop out in the quadratic part of the effective Lagrangian. It was to bring this about that we inserted the factor F_a^2 in Eq. (7.2), and it is this feature of the unitarity gauge that justifies the statement that it correctly displays the particle content of the theory.

The same result can be obtained by a simple generalization of the method of Jackiw and Johnson and Cornwall and Norton.[2] In a general gauge there are "black boxes," connecting a single G_W-gauge boson with a single Goldstone boson line. If we sum up the pole singularities produced by a linear chain of Π and W lines, we find that the only poles in the sum which correspond to particles of zero spin and zero mass are those in channels described by precisely the condition (7.5). It is also easy to see that the unitarity gauge as usually defined in theories with elementary scalar fields does satisfy Eq. (7.5).

Now that we have eliminated the Π-W cross terms, the mass of the G_W vector bosons may be read off from the effective Lagrangian. Equation (7.7) contains a term quadratic in W

$$-\tfrac{1}{2}\sum_{\alpha\beta} \mu^2{}_{\alpha\beta}\, W'_{\alpha\mu} W'^{\mu}_{\beta}. \tag{7.9}$$

with a vector-boson mass

$$\mu^2{}_{\alpha\beta} = \sum_{aAB} F_a{}^2 B_{aA} B_{aB} e_{\alpha A} e_{\alpha B} \tag{7.10}$$

or equivalently

$$\mu^2{}_{\alpha\beta} = \tfrac{1}{64}\sum_a F_a{}^2\, \mathrm{Tr}(x_a w_\alpha)\, \mathrm{Tr}(x_a w_\beta). \tag{7.11}$$

It is immediately apparent from (7.11) that $\mu^2{}_{\alpha\beta}$ vanishes if either w_α or w_β is the generator of a symmetry that is not spontaneously broken. Also, the μ's that are not zero are of order eF, and since the Fermi coupling constant G_F must be of order e^2/μ^2, we can conclude that

$$M \approx F \approx G_F{}^{-1/2} \approx 300\ \mathrm{GeV}$$

as previously indicated.

VIII. CLASSIFICATION OF THE GOLDSTONE BOSONS

The other side of the Higgs phenomenon, complementary to the appearance of vector-boson masses, is the disappearance of Goldstone bosons.[11] We will now consider the nature and the mass spectrum of the Goldstone bosons that are *not* eliminated by the Higgs phenomenon.

It is useful to begin by studying the eigenvectors and eigenvalues of the formal Goldstone boson mass matrix $m^2{}_{ab}$, before transformation to the unitarity gauge. First, note that there is an eigenvector of $m^2{}_{ab}$ with eigenvalue zero for every independent broken gauge generator. Every generator w_α of G_W may be written in the form

$$w_\alpha = \sum_a \frac{x_a u_{a\alpha}}{F_a} + h_\alpha, \tag{8.1}$$

where h_α is a linear combination of the generators t_i of the unbroken symmetry subgroup H. But then Eq. (6.19) gives

$$\sum_b m^2{}_{ab} u_{b\alpha} = -\frac{1}{F_a}\, \mathrm{Tr}\{E[(W_\alpha - H_\alpha), [X_a, I]]\},$$

where W_α and H_α are the matrices representing w_α and h_α in the regular representation of U(N) $\otimes$ U(N). The W term can be rewritten in terms of $[W, E]$, which vanishes because the sum $\sum_\alpha e_{\alpha A} e_{\alpha B}$ is G_W invariant. The H term can be rewritten in terms of the double commutator $[X_a, [H_\alpha, I]]$, which vanishes because I is invariant under H, plus the double commutator $[[H_\alpha, X_\alpha], I]$, which gives no contribution because of the "subgroup-alignment" condition (6.18). Thus, we see that $u_{a\alpha}$ is our eigenvector

$$\sum_b m^2{}_{ab} u_{b\alpha} = 0. \tag{8.2}$$

These will be called the *fictitious Goldstone bosons*, because as we shall see, it is just these that are eliminated by the unitarity gauge condition.

The number of independent fictitious Goldstone bosons is evidently equal to the dimensionality of G_W minus the dimensionality of that subgroup H_W of G_W which is unbroken by the spontaneous symmetry breakdown of $U(N)\otimes U(N)$ to H. Of course, for every generator of H_W there will be a vector boson whose mass remains zero to all orders in e, so the number of fictitious Goldstone bosons equals the number of *massive* vector bosons. (In the real world, H_W would presumably consist solely of electromagnetic gauge transformations.)

The argument that led from (8.1) to (8.2) did not actually depend on the fact that W_α is a generator of G_W, but only on the fact that it is a symmetry of the weak and electromagnetic interactions, so that W_α commutes with E. But in general there will be broken symmetries of the weak interactions that are not themselves generators of G_W. Exactly the same reasoning tells us that for each of these there will be another eigenvector of $m^2{}_{ab}$ with eigenvalue zero. These will be called the *true Goldstone bosons*, because they remain massless to all orders in e but, as we shall see, they are not eliminated by the Higgs phenomenon. The number of true and fictitious Goldstone bosons is equal to the dimensionality of the complete global symmetry group S_W of the weak and electromagnetic interactions, minus the dimensionality of that subgroup of S_W which is left unbroken by the spontaneous breakdown of $U(N)\otimes U(N)$ to H. Of course, the occurrence of true Goldstone bosons would present grave difficulties for any theory that has pretensions of providing a realistic model of the actual world. These problems are further discussed in Sec. X.

Finally, there will in general be eigenvectors u_a of $m^2{}_{ab}$ for which the quantity $\sum_a u_a x_a/F_a$ *cannot* be expressed as a linear combination of generators of S_W and generators of H. There is no reason in this case why the eigenvalue should vanish, so we expect a mass m of order

$$m^2 \approx e^2 I/F^2 \approx e^2 M^2 \tag{8.3}$$

about the same as for the vector bosons. These are called the *pseudo-Goldstone bosons*,[17] because they are not the Goldstone bosons of any true symmetry of the whole theory, but only of an accidental approximate symmetry which appears exact in the limit $e\to 0$. As we shall see, the pseudo-Goldstone bosons, like the true Goldstone bosons, are not eliminated by the Higgs phenomenon. The total number of all Goldstone bosons, fictitious, true and pseudo, is simply equal to the dimensionality $2N^2$ of $U(N)\otimes U(N)$ minus the dimensionality of the unbroken subgroup H.

By use of the familiar Schmidt orthogonalization technique, we can choose an orthonormal set of eigenvectors of $m^2{}_{ab}$

$$\sum_b m^2{}_{ab} u_b^n = m_n{}^2 u_a^n , \tag{8.4}$$

$$\sum_a u_a^n u_a^m = \delta_{nm} , \tag{8.5}$$

with each u^n representing either a fictitious, true, or pseudo-Goldstone bosons. That is, we first choose an orthonormal set of u_a^n vectors corresponding to fictitious Goldstone bosons, for which $\sum u_a^n x_a/F_a$ is a linear combination of H and G_W generators; then add an orthonormal set corresponding to true Goldstone bosons, for which $\sum u_a^n x_a/F_a$ is a linear combination of H and S_W but not H and G_W generators; and finally add an orthonormal set corresponding to pseudo-Goldstone bosons, for which $\sum u_a^n x_a/F_a$ is not a linear combination of H and S_W generators. With a set of orthonormal vectors u^n constructed in this way, we can define a corresponding set of Goldstone boson fields

$$\Pi^n \equiv \sum_a u_a^n \Pi_a , \tag{8.6}$$

each of definite mass and type (fictitious, true, or pseudo). Further, since the u^n form a complete set, we can also write

$$\Pi_a \equiv \sum_n u_a^n \Pi^n . \tag{8.7}$$

In particular, we have

$$\sum_a \partial_\mu \Pi^a \partial^\mu \Pi^a = \sum_n \partial_\mu \Pi^n \partial^\mu \Pi^n ,$$

so the Π^n are canonically normalized.

Now let us impose the condition (7.6) that defines the unitarity gauge. A vector u^n which corresponds to a true or a pseudo-Goldstone boson will be orthogonal to all u_a^n corresponding to the fictitious Goldstone bosons, and hence also to the vectors $u_{a\alpha}$ defined by Eq. (8.1). But it follows then that

$$\begin{aligned}\mathrm{Tr}\Big(w_\alpha \sum_a x_a u_a^n F_a\Big) &= \mathrm{Tr}\Big(\sum_b \frac{x_b u_{b\alpha}}{F_b} \sum_a x_a u_a^n F_a\Big)\\ &= 8\sum_a u_{a\alpha} u_a^n = 0 .\end{aligned}$$

Hence the unitarity gauge condition (7.6) imposes no constraint on the fields Π^n representing true or pseudo-Goldstone bosons. On the other hand, a u_a^n which corresponds to a fictitious Goldstone boson allows the decomposition

$$\sum_n \frac{u_a^n x_a}{F_a} = w^n + h^n ,$$

where w^n and h^n are generators of G_W and H, respectively. It follows that

$$\begin{aligned}\mathrm{Tr}\Big(w^n \sum_a x_a u_a^m F_a\Big) &= \mathrm{Tr}\Big(\sum_b \frac{x_b u_b^n}{F_b} \sum_a x_a u_a^m F_a\Big)\\ &= 8\sum_a u_a^n u_a^m = 8\delta_{nm} .\end{aligned}$$

Thus (7.5) requires that

$$0 = \mathrm{Tr}\left(w^n \sum_a x_a \Pi_a' F_a\right) = 8\Pi^{n\prime}.$$

We conclude that *the whole effect of the condition of unitarity gauge is to eliminate the* Π^n *corresponding to fictitious Goldstone bosons,* leaving the masses and fields of the true and the pseudo-Goldstone bosons unchanged.

IX. AN EXAMPLE

We shall now descend from the generality of the previous discussion to the consideration of one specific example. It probably is unnecessary to remark that this model is totally unrealistic as a theory of real particles or interactions; it is presented solely for the purposes of illustration.

Our model contains two color multiplets of fermions, called q and h. (Color indices are dropped everywhere.) In the limit $e \to 0$, the strong interactions are necessarily invariant under a group $U(2) \otimes U(2)$ of global transformations

$$\begin{pmatrix} q \\ h \end{pmatrix} \to U \begin{pmatrix} q \\ h \end{pmatrix}, \tag{9.1}$$

with U a unitary matrix (commuting with color), involving both the Dirac matrices 1 and γ_5. The generators of this algebra are defined as

$$\vec{\lambda}_L \equiv \frac{1}{\sqrt{2}}(1+\gamma_5)\vec{\tau}, \quad \lambda_{0L} \equiv \frac{1}{\sqrt{2}}(1+\gamma_5),$$
$$\vec{\lambda}_R \equiv \frac{1}{\sqrt{2}}(1-\gamma_5)\vec{\tau}, \quad \lambda_{0R} \equiv \frac{1}{\sqrt{2}}(1-\gamma_5), \tag{9.2}$$

with $\vec{\tau}$ the usual 2×2 Pauli matrices.

We assume that the $U(2)\otimes U(2)$ symmetry is dynamically broken down to the largest subgroup which will allow *one* of the two fermion multiplets to acquire a mass:

$$H = U(1)\otimes U(1) \otimes U(1). \tag{9.3}$$

(The color gauge group G_S is assumed to remain unbroken.) We can always define the fermion fields so that it is q that remains massless; the labels "q" and "h" thus stand for "quark" and "heavy fermion." With this definition, the generators of H are

$$t_L \equiv \frac{1}{\sqrt{2}}(\lambda_{0L}+\lambda_{3L}) = \tfrac{1}{2}(1+\gamma_5)(1+\tau_3),$$
$$t_R \equiv \frac{1}{\sqrt{2}}(\lambda_{0R}+\lambda_{3R}) = \tfrac{1}{2}(1-\gamma_5)(1+\tau_3), \tag{9.4}$$
$$t_0 \equiv \tfrac{1}{2}(\lambda_{0L}+\lambda_{0R}-\lambda_{3L}-\lambda_{3R}) = \frac{1}{\sqrt{2}}(1-\tau_3).$$

We can complete an orthonormal basis for $U(N)\otimes U(N)$ with the five additional generators

$$x_{1L} \equiv \lambda_{1L} = \frac{1}{\sqrt{2}}(1+\gamma_5)\tau_1,$$
$$x_{2L} \equiv \lambda_{2L} = \frac{1}{\sqrt{2}}(1+\gamma_5)\tau_2,$$
$$x_{1R} \equiv \lambda_{1R} = \frac{1}{\sqrt{2}}(1-\gamma_5)\tau_1, \tag{9.5}$$
$$x_{2R} \equiv \lambda_{2R} = \frac{1}{\sqrt{2}}(1-\gamma_5)\tau_2,$$
$$x_0 = \tfrac{1}{2}(\lambda_{0L}-\lambda_{0R}-\lambda_{3L}+\lambda_{3R}) = \frac{1}{\sqrt{2}}\gamma_5(1-\tau_3).$$

With respect to the $O(2)\otimes O(2)$ group generated by t_L and t_R, these generators transform according to the representations

$$\{x_{1L}, x_{2L}\} \quad (2,1),$$
$$\{x_{1R}, x_{2R}\} \quad (1,2),$$
$$x_0 \quad (1,1).$$

The Goldstone bosons Π_a may be chosen to belong to corresponding $O(2)\otimes O(2)$ multiplets. With this definition, we automatically have

$$\langle 0|J_A^\mu|\Pi_a\rangle \propto B_{aA}F_a, \tag{9.6}$$

where B_{aA} are the coefficients in (9.5),

$$x_a = \sum_A B_{aA}F_a\lambda_A,$$

and the F's are equal within irreducible H multiplets,

$$F_{1L} = F_{2L} \equiv F_L, \quad F_{1R} = F_{2R} \equiv F_R. \tag{9.7}$$

We expect that the F's are of the same order of magnitude as the heavy quark mass,

$$F_L \approx F_R \approx F_0 \approx M_h. \tag{9.8}$$

It is straightforward to calculate the covariant derivatives (5.6) and (5.7) for $e=0$ as power series in the Π_a. For instance, the effective Lagrangian contains a bilinear interaction of Goldstone bosons with quarks

$$-\frac{1}{F_L^2}\bar{q}\gamma^\mu(1+\gamma_5)q(\Pi_{1L}\partial_\mu\Pi_{2L}-\Pi_{2L}\partial_\mu\Pi_{1L}) - \frac{1}{F_R^2}\bar{q}\gamma^\mu(1-\gamma_5)q(\Pi_{1R}\partial_\mu\Pi_{2R}-\Pi_{2R}\partial_\mu\Pi_{1R}) \tag{9.9}$$

and a trilinear self-interaction of Goldstone bosons

$$-\frac{1}{\sqrt{2}F_0}\left(1-\frac{F_0^2}{F_L^2}\right)\partial_\mu\Pi_0(\Pi_{1L}\partial^\mu\Pi_{2L}-\Pi_{2L}\partial^\mu\Pi_{1L}) + \frac{1}{\sqrt{2}F_0}\left(1-\frac{F_0^2}{F_R^2}\right)\partial_\mu\Pi_0(\Pi_{1R}\partial^\mu\Pi_{2R}-\Pi_{2R}\partial^\mu\Pi_{1R}). \tag{9.10}$$

The vector and axial-vector bilinear covariants that can be formed from the q field do not share the same H transformation properties as any of the Π_a, so there is no $\bar{q}qD\Pi$ term in the leading part of the effective Lagrangian. There is, however, a Fermi interaction, which after Fierz transformations may be put in the form

$$-G_{LL}[\bar{q}\gamma^\mu(1+\gamma_5)q][\bar{q}\gamma_\mu(1+\gamma_5)q]-G_{LR}[\bar{q}\gamma^\mu(1+\gamma_5)q][\bar{q}\gamma_\mu(1-\gamma_5)q]-G_{RR}[\bar{q}\gamma_\mu(1-\gamma_5)q][\bar{q}\gamma_\mu(1-\gamma_5)q]\ . \quad (9.11)$$

We expect the constants G_{LL}, G_{LR}, and G_{RR} to be of order $1/M_h^2$.

Now let us turn on the "weak" interactions. We will assume that the weak gauge group is

$$G_W = \mathrm{SU}(2) \quad (9.12)$$

with both left- and right-handed fermion fields $[(1\pm\gamma_5)q, (1\pm\gamma_5)h]$ transforming as G_W doublets. However, we do not immediately know *which* SU(2) subgroup of U(2)⊗ U(2) generates the weak interactions. In general, the generators of SU(2) might be any matrices of the form

$$w_\alpha = e\sum_\beta[(g_L)_{\alpha\beta}\lambda_{L\beta}+(g_R)_{\alpha\beta}\lambda_{R\beta}], \quad (9.13)$$

where e is the SU(2) gauge coupling constant; α and β run over the values 1, 2, 3; $\vec{\lambda}_L$ and $\vec{\lambda}_R$ are the matrices (9.2); and g_L and g_R are *unknown* 3×3 orthogonal matrices. Nor is it arbitrary which g matrices we choose; the definition of the fermions has been fixed (up to an H transformation) by our convention that the spontaneous breakdown of U(2)⊗ U(2) to H gives a mass to h, not q.

In order to settle this question, we must examine the "potential" term in the effective Lagrangian. In general, this has the form (6.8):

$$V(g) = -\sum_{\alpha AB} e_{\alpha A}(g)e_{\alpha B}(g)I_{AB}, \quad (9.14)$$

where I_{AB} is some unknown H invariant of order M_h^4, and $e_{\alpha A}(g)$ are the coefficients which give the G_W generators as linear combinations of the U(2)⊗ U(2) generators

$$e_{\alpha,\beta L}(g) = e(g_L)_{\alpha\beta}, \qquad e_{\alpha,\beta R}(g) = e(g_R)_{\alpha\beta}. \quad (9.15)$$

Thus V here takes the form

$$V(g) = -e^2\sum_{\alpha\beta\gamma}[(g_L)_{\alpha\beta}(g_L)_{\alpha\gamma}I_{L\beta,L\gamma} + 2(g_L)_{\alpha\beta}(g_R)_{\alpha\gamma}I_{L\beta,R\gamma} + (g_R)_{\alpha\beta}(g_R)_{\alpha\gamma}I_{R\beta,R\gamma}].$$

But the g matrices are orthogonal, so this immediately simplifies to

$$V(g) = -2e^2\sum_{\beta\gamma}(g_L^{-1}g_R)_{\beta\gamma}I_{L\beta,R\gamma} + \text{constant}.$$

In addition, the H invariance of I_{AB} requires that it be invariant under *independent* rotations around the 3 axis on either the $L\alpha$ and/or $R\alpha$ indices; thus, in particular,

$$I_{L\beta,R\gamma} = In_\beta n_\gamma, \quad (9.16)$$

where $\vec{n}$ is a unit vector pointing in the 3-direction

$$\vec{n} = (0,0,1),$$

and I is some unknown constant of order M_h^4. The potential has now become simply

$$V(g) = -2e^2 I(n, g_L^{-1}g_R n). \quad (9.17)$$

This is to be minimized over the whole range of orthogonal matrices g_L, g_R. The location of such a minimum is quite obvious:

$$\text{(A) For } I>0,\ g_L^{-1}g_R n = +n. \quad (9.18a)$$

$$\text{(B) For } I<0,\ g_L^{-1}g_R n = -n. \quad (9.18b)$$

This does not, of course, entirely determine g_L and g_R; given any solution, we can find another of the form

$$g_L' = g_1 g_L g_2, \quad g_R' = g_1 g_R g_3,$$

where g_1 is an arbitrary orthogonal matrix, and g_2 and g_3 are orthogonal matrices representing arbitrary independent rotations around the 3 axis. But g_1 represents a redefinition of the weak gauge couplings by an SU(2) transformation belonging to the gauge group G_W, while g_2 and g_3 represent a redefinition of the fermion fields by a transformation belonging to that subgroup H of U(2)⊗ U(2) which is not spontaneously broken. Clearly, there is no way that this remaining ambiguity in the g's could ever be resolved, nor is there any reason why we would wish to do so. Thus, we can freely choose *any* orthogonal g_L and g_R matrices which satisfy the condition for a minimum, Eq. (9.18).

We will now need to consider the two cases separately.

A. $I>0$

Here it is convenient to choose g_L and g_R as unit matrices

$$g_L = g_R = 1. \quad (9.19)$$

The generators of the weak gauge group are then given by (9.13) as

$$\vec{w} = e(\vec{\lambda}_L + \vec{\lambda}_R) = e\sqrt{2}\,\vec{\tau}, \qquad (9.20)$$

with $\vec{\tau}$ the usual 2×2 Pauli matrices. There is a single 2×2 matrix which is a generator of *both* a G_W gauge transformation and an unbroken H transformation

$$e\tau_3 = \frac{1}{\sqrt{2}} w_3 = \frac{e}{2}(t_L + t_R) - \frac{e}{\sqrt{2}} t_0 . \qquad (9.21)$$

This corresponds to a "photon," which keeps zero mass despite the spontaneous symmetry breaking. The gauge bosons corresponding to the other two generators of G_W, w_1, and w_2, acquire a mass by the Higgs mechanism, given by (7.10) is

$$\mu_1{}^2 = \mu_2{}^2 = e^2(F_L{}^2 + F_R{}^2). \qquad (9.22)$$

These massive gauge bosons have w_3 charges +1 and −1. Also, q has w_3 charge +1, while the Goldstone bosons $\Pi_{1L} \pm i\Pi_{2L}$, $\Pi_{1R} \pm i\Pi_{2R}$, and Π_0 have w_3 charges ±1, ±1, and 0, respectively.

In order to classify the Goldstone bosons, we note first that there are two independent linear combinations of the x_a that can be expressed as linear combinations of generators of H and G_W:

$$e(x_{1L} + x_{1R}) = w_1 , \qquad (9.23)$$

$$e(x_{2L} + x_{2R}) = w_2 , \qquad (9.24)$$

and there is also one other linear combination of the x_a that can be expressed as a linear combination of a generator of H and the generator of the exact global symmetry $\psi \to \exp(i\gamma_5\epsilon)\psi$ of the whole Lagrangian

$$x_0 = \sqrt{2}\,\gamma_5 1 - \frac{1}{\sqrt{2}}(t_L + t_R). \qquad (9.25)$$

In accordance with the conclusions of Sec. VIII, we must therefore expect that this theory has two fictitious Goldstone bosons, one true Goldstone boson, and 5 − 2 − 1 = 2 pseudo-Goldstone bosons. Their fields are of the form

$$\Pi^n = \sum_a u_a^n \Pi_a ,$$

with u^n a set of *orthonormal* vectors subject to certain conditions: For the fictitious Goldstone bosons $\sum_a u_a^n x_a/F_a$ must be a linear combination of (9.23) and (9.24), so the fields are

$$\Pi^{\mathrm{I}} \equiv (F_L{}^2 + F_R{}^2)^{-1/2}(F_L\Pi_{1L} + F_R\Pi_{1R}), \qquad (9.26)$$

$$\Pi^{\mathrm{II}} \equiv (F_L{}^2 + F_R{}^2)^{-1/2}(F_L\Pi_{2L} + F_R\Pi_{2R}). \qquad (9.27)$$

For the true Goldstone boson, $\sum_a u_a^n x_a/F_a$ must be proportional to (9.25), so the field is simply

$$\Pi^0 \equiv \Pi_0 . \qquad (9.28)$$

For the pseudo-Goldstone bosons the u_a^n need only be orthogonal to all the others, so the fields are

$$\Pi^1 \equiv (F_L{}^2 + F_R{}^2)^{-1/2}(-F_R\Pi_{1L} + F_L\Pi_{1R}), \qquad (9.29)$$

$$\Pi^2 \equiv (F_L{}^2 + F_R{}^2)^{-1/2}(-F_R\Pi_{2L} + F_L\Pi_{2R}). \qquad (9.30)$$

The masses of Π^{I}, Π^{II}, and Π^0 are zero, while second-order weak (and "electromagnetic") effects will give Π^1 and Π^2 masses that are equal (because w_3 invariance is unbroken) and of order eM_h. These masses are proportional to the constant I in Eq. (9.16), but I is unknown, so the calculation is not worthwhile. The effect of the transformation to unitarity gauge is just to eliminate the fields of the fictitious Goldstone bosons:

$$\Pi^{\mathrm{I}\prime} = \Pi^{\mathrm{II}\prime} = 0 .$$

The five "old" fields Π_a' may then be expressed in terms of the three "new" fields $\Pi^{n\prime}$ as

$$\Pi_{1L}' = -\frac{F_R}{(F_L{}^2 + F_R{}^2)^{1/2}}\Pi^{1\prime},$$

$$\Pi_{1R}' = \frac{F_L}{(F_L{}^2 + F_R{}^2)^{1/2}}\Pi^{1\prime},$$

$$\Pi_{2L}' = -\frac{F_R}{(F_L{}^2 + F_R{}^2)^{1/2}}\Pi^{2\prime},$$

$$\Pi_{2R}' = \frac{F_L}{(F_L{}^2 + F_R{}^2)^{1/2}}\Pi^{2\prime},$$

and, of course

$$\Pi_0' = \Pi^{0\prime} .$$

In particular, the bilinear interaction (9.9) of the nonfictitious Goldstone bosons with the quarks is (now dropping primes)

$$(F_L{}^2 + F_R{}^2)^{-1/2}\left[-\frac{F_R}{F_L{}^2}\bar{q}\gamma^\mu(1+\gamma_5)q - \frac{F_L}{F_R{}^2}\bar{q}\gamma^\mu(1-\gamma_5)q\right](\Pi^1\partial_\mu\Pi^2 - \Pi^2\partial_\mu\Pi^1), \qquad (9.31)$$

and their trilinear self-interaction (9.10) is

$$2^{-1/2}(F_L{}^2 + F_R{}^2)^{-1/2}F_0{}^{-1}\left[F_R\left(1 - \frac{F_0{}^2}{F_L{}^2}\right) + F_L\left(1 - \frac{F_0{}^2}{F_R{}^2}\right)\right]\partial_\mu\Pi^0(\Pi^1\partial^\mu\Pi^2 - \Pi^2\partial^\mu\Pi^1). \qquad (9.32)$$

The quark mass matrix is given to order e^2M_h by Eq. (5.29), which here becomes

$$M_q = e^2 \sum_\alpha (Q_{\alpha L,\alpha L} + 2Q_{\alpha L,\alpha R} + Q_{\alpha R,\alpha R}),$$

where Q_{AB} is some constant matrix of order M_h which transforms as a tensor under H, in the sense of Eq. (C4). (Again, we let α and β run over the values 1, 2, 3.) It is straightforward to show that the most general such tensor has

$$Q_{\alpha L,\beta L} = Q_{\alpha R,\beta R} = 0,$$

and also (with a suitable choice of relative phase for the left- and right-handed quark fields)

$$Q_{\alpha L,\beta R} = Q[\tfrac{1}{2}(1+\gamma_5)n^+_\alpha n^-_\beta + \tfrac{1}{2}(1-\gamma_5)n^-_\alpha n^+_\beta],$$

where $Q \approx M_h$ is some unknown constant, and

$$\vec{n}^\pm \equiv \frac{1}{\sqrt{2}}(1, \pm i, 0).$$

Thus, the quark here does acquire a mass of order e^2M_h

$$M_q = 2e^2Q. \tag{9.33}$$

Also, using Eq. (5.28), the Yukawa coupling here is

$$\frac{4iM_q}{\sqrt{2}F_0}\bar{q}\gamma_5 q\Pi^0 \tag{9.34}$$

as required by a Goldberger-Treiman relation.

It happens that in this model either the massive vector bosons or the pseudo-Goldstone bosons are absolutely stable. This is just because they have w_3 charges ± 1; the only lighter states into which they could decay consist of "photons," true Goldstone bosons, and quark-antiquark pairs, all of which are w_3 neutral.

B. $I<0$

Here it is convenient to choose g_L and g_R as equal and opposite rotations of 90° about the 1 axis. The generators of the weak gauge group are then

$$w_1 = e(\lambda_{1L} + \lambda_{1R}) = e\sqrt{2}\,\tau_1,$$

$$w_2 = e(\lambda_{3L} - \lambda_{3R}) = e\sqrt{2}\,\gamma_5\tau_3,$$

$$w_3 = e(-\lambda_{2L} + \lambda_{2R}) = -e\sqrt{2}\,\gamma_5\tau_2,$$

with $\vec{\tau}$ again the 2×2 Pauli matrices. No linear combination of these generators is a generator of the unbroken subgroup H, so there is no "photon" here—every vector boson gets a mass. From (7.11), we find that the masses are

$$\mu_1^2 = \mu_3^2 = e^2(F_L^2 + F_R^2),$$

$$\mu_2^2 = F_0^2.$$

In order to classify the Goldstone bosons, we note first that there are three independent linear combinations of the x_a that can be expressed as linear combinations of the generators of H and G_W:

$$x_{1L} + x_{1R} = w_1/e, \tag{9.35}$$

$$x_0 = \frac{1}{\sqrt{2}}(t_L - t_R) - w_2/e, \tag{9.36}$$

$$x_{2L} - x_{2R} = -w_3/e, \tag{9.37}$$

while there is no other linear combination of the x_a that can be expressed as a linear combination of a generator of H and the generator of any exact global symmetry of the whole Lagrangian. According to Sec. VIII, we must now expect that this theory has three fictitious Goldstone bosons, no true Goldstone bosons, and $5-3=2$ pseudo-Goldstone bosons. Their fields are of the form

$$\Pi^n = \sum_\alpha u^n_a \Pi_a, \tag{9.38}$$

with u^n_a a set of orthonormal vectors subject to certain conditions: For the fictitious Goldstone bosons $\sum_a u^n_a x_a/F_a$ must be a linear combination of (9.35)–(9.37), so the fields are

$$\Pi^{\text{I}} = (F_L^2 + F_R^2)^{-1/2}(F_L\Pi_{1L} + F_R\Pi_{1R}), \tag{9.39}$$

$$\Pi^0 = \Pi_0, \tag{9.40}$$

$$\Pi^{\text{II}} = (F_L^2 + F_R^2)^{-1/2}(F_L\Pi_{2L} - F_R\Pi_{2R}). \tag{9.41}$$

For the pseudo-Goldstone bosons the u^n must simply be orthogonal to the others, so the fields are

$$\Pi^1 = (F_L^2 + F_R^2)^{-1/2}(-F_R\Pi_{1L} + F_L\Pi_{1R}), \tag{9.42}$$

$$\Pi^2 = (F_L^2 + F_R^2)^{-1/2}(F_R\Pi_{2L} + F_L\Pi_{2R}). \tag{9.43}$$

The masses of Π^{I}, Π^{II}, and Π^0 are zero, while second-order weak effects give Π^1 and Π^2 masses proportional to I and of order eM_h. These latter masses are equal, because the whole Lagrangian has an exact global symmetry which is not spontaneously broken, generated by

$$\lambda_{0L} - \lambda_{0R} + \lambda_{3L} - \lambda_{3R} = t_L - t_R = \lambda_{0L} - \lambda_{0R} + w_2/e, \tag{9.44}$$

and this rotates Π^1 and Π^2 into each other.

The effect of the transformation to unitarity gauge is to eliminate the fields of the fictitious Goldstone bosons

$$\Pi^{\text{I}\prime} = \Pi^{0\prime} = \Pi^{\text{II}\prime} = 0.$$

The five fields Π_a' may be expressed in terms of the two remaining Π^n, fields, as

$$\Pi'_{1L} = -(F_L{}^2 + F_R{}^2)^{-1/2} F_R \Pi^{1\prime},$$
$$\Pi'_{1R} = (F_L{}^2 + F_R{}^2)^{-1/2} F_L \Pi^{1\prime},$$
$$\Pi'_{2L} = +(F_L{}^2 + F_R{}^2)^{-1/2} F_R \Pi^{2\prime},$$
$$\Pi'_{2R} = (F_L{}^2 + F_R{}^2)^{-1/2} F_L \Pi^{2\prime},$$
$$\Pi'_0 = 0.$$

In particular, the bilinear interaction (9.9) of the nonfictitious Goldstone bosons with the quarks is (now dropping primes)

$$\left[\frac{F_R}{F_L{}^2(F_L{}^2+F_R{}^2)^{1/2}}\,\bar{q}\gamma^\mu(1+\gamma_5)q - \frac{F_L}{F_R{}^2(F_L{}^2+F_R{}^2)^{1/2}}\,\bar{q}\gamma^\mu(1-\gamma_5)q\right](\Pi^1\partial_\mu\Pi^2 - \Pi^2\partial_\mu\Pi^1). \tag{9.45}$$

The trilinear coupling (9.10) now vanishes. Also, as a consequence of the exact unbroken symmetry generated by (9.44), the quark mass remains zero to all orders in e. The Yukawa $\Pi\bar{q}q$ coupling vanishes, as required by a Goldberger-Treiman relation, even though it is not forbidden by the symmetry generated by (9.44).

The moral of this analysis is twofold. First, a theory with a given group-theoretic character and a given field content can have enormously different physical consequences depending on how the subgroups G_W and H line up with each other. The differences between the two cases found in this section are summarized in Table I. In addition, although these theories do not have the predictive power of a theory in which the spontaneous symmetry breaking is due to vacuum expectation values of weakly coupled scalar fields, the predictive power of theories with dynamical symmetry breaking is by no means negligible.

X. IMPLICATIONS

The foregoing analysis has been chiefly concerned with mathematical formalism rather than physical applications. We close with some remarks about the implications of this analysis for real particles and interactions.

The weak interactions in this class of theories arise both from the exchange of intermediate vector bosons and also from a direct Fermi interaction in the effective Lagrangian. Both are of the same order of magnitude; the direct Fermi interaction has a coupling constant of order M^{-2} (where M is the scale associated with the dynamical symmetry breaking), while the exchange of vector bosons of mass $\mu \approx eM$ produces an effective Fermi coupling of order $e^2/\mu^2 \approx M^{-2}$. Either way, we are led to the estimate that $M \approx 300$ GeV.

The two kinds of weak interactions can of course be distinguished by their energy dependence at energies of order eM or greater. They may also be distinguished even at lower energies by their symmetry properties; the direct Fermi interaction is invariant under the unbroken subgroup H of $U(N) \otimes U(N)$, while the vector-boson exchange interaction is not.

Some of the fermions of these theories may get masses of order M from the dynamical symmetry breaking. However, the ordinary quarks $\mathcal{P}, \mathfrak{N}, \lambda$, $\mathcal{P}'$ (etc.?) can hardly be this heavy,[25] so we must suppose that the unbroken subgroup H must be large enough to prevent the appearance of masses of order M. The masses of the ordinary quarks would then have to arise from higher-order corrections, which would presumably give them values of order e^2M. This is a gratifying result, for it offers at least a qualitative explanation of the mysterious fact that the ratio of the mass scale of the hadrons (say, 1 GeV) to that of the Fermi interaction (300 GeV) is roughly of order α.

Of course, in order to produce quark masses of order e^2M, the unbroken symmetry group H must not be too large. Specifically, there must be no chiral symmetries in H which are also symmetries of the weak and electromagnetic interactions. The breakdown of $U(N) \otimes U(N)$ to H may be signaled by the appearance of fermion masses of order M, but

TABLE I. Summary of the properties of the model discussed in Section IX, in two cases corresponding to the two different possible minima of the potential $V(g)$.

	A	B
Massive vector bosons	2 degenerate	3 (2 degenerate)
Massless vector bosons	1	0
Unbroken global symmetries (including fermion conservation)	1	2
True Goldstone bosons	1	0
Pseudo-Goldstone bosons	2 degenerate	2 degenerate
Quark mass	$\approx e^2 M_h$	0

it is also possible that H forbids all fermion masses of order M while allowing other nonchiral interactions, such as scalar, tensor, or pseudoscalar Fermi interactions of order M^{-2}.

In all cases that I have examined, H must be sufficiently small so that there are some broken symmetries, not in H, that are also not in the weak and electromagnetic gauge group G_W. In particular, in a theory with N' heavy fermions and $N-N'$ ordinary quarks, the broken symmetry generators include all Hermitian matrices, chiral or nonchiral, which connect ordinary quark and heavy fermion fields. If these were all generators of the weak and electromagnetic gauge group, their multiple commutators would be also gauge generators. But these commutators span the algebra of $SU(N) \otimes SU(N)$. This is not possible because then all the unbroken chiral symmetries in $SU(N) \otimes SU(N)$, which keep the ordinary quarks from getting masses of order M, would also be symmetries of the weak and electromagnetic interactions, so that the ordinary quarks could not get masses of any order in e. Also, the weak and electromagnetic gauge group cannot include $SU(N) \otimes SU(N)$; triangle anomalies would make such a theory nonrenormalizable.[26]

For every broken symmetry which is not a symmetry of the weak and electromagnetic interactions, there is a pseudo-Goldstone boson that is not eliminated by the Higgs phenomenon. These particles have masses of order $e \times 300$ GeV, and do not interact strongly at ordinary energies, but the charged pseudo-Goldstone bosons could of course be pair-produced by the electromagnetic interactions. Thus, it will be important to distinguish carefully between pseudo-Goldstone bosons and intermediate vector bosons when colliding beams reach energies adequate to produce such particles in pairs.

It makes a great difference in the description of the decay modes and interactions of pseudo-Goldstone bosons whether they can interact with ordinary hadrons and leptons singly, or only in pairs. The broken symmetry generator corresponding to a given pseudo-Goldstone boson might have no matrix elements between quark or lepton states, but only between states of which one is a heavy (≈ 300 GeV) fermion. The pseudo-Goldstone bosons would still interact in pairs with ordinary hadrons, as in Eqs. (9.31) and (9.45), but they could only decay into each other. On the other hand, the quarks might not be entirely neutral under the various broken symmetry generators in $U(N) \otimes U(N)$, in which case some of the pseudo-Goldstone bosons would be able to decay into ordinary hadrons. In the absence of a candidate for a realistic model, it is not worth pursuing these various possibilities in great detail.

In addition to pseudo-Goldstone bosons, such theories will usually have true Goldstone bosons of zero mass which also are not eliminated by the Higgs mechanism. This is because there are always some broken global symmetries of the weak and electromagnetic interactions which are not themselves elements of the weak and electromagnetic gauge group. For instance, one such global symmetry is the U(1) chiral transformation which multiplies all fermion fields with a common factor $\exp(i\gamma_5\theta)$; this must not be a member of the weak and electromagnetic gauge group, because if it were then triangle anomalies would make the theory nonrenormalizable, and it must be spontaneously broken, because otherwise none of the fermions in the theory could pick up any mass. If the generator x of any such broken exact global symmetry could be written as a sum of a gauge symmetry generator w and a nonspontaneously broken symmetry generator h, then the corresponding Goldstone boson would be eliminated by the Higgs mechanism; however, in this case the theory would have an extra exact nonspontaneously broken symmetry (apart from fermion conservation) generated by $x-w=h$. This is what happens in case B of the model discussed in Sec. IX; the extra symmetry there keeps the quark massless to all orders in e. If we do not want to allow such extra exact symmetries then the generators of broken nongauge symmetries of the weak and electromagnetic interactions must not be linear combinations of gauge and unbroken symmetry generators, and the corresponding Goldstone bosons cannot be eliminated by the Higgs mechanism.

In the particular case of the chiral U(1) symmetry mentioned above, it is possible that the massless true Goldstone boson, although not eliminated by the Higgs mechanism, would nevertheless not be observable as a free particle. There is a triangle anomaly connecting one $\gamma_\mu\gamma_5$ vertex to two colored gluons; this anomaly forces us to include gluon terms in the conserved chiral current, which make it not gluon-gauge invariant. Since the corresponding true Goldstone boson cannot be proved to appear as a pole in any gluon-gauge-invariant operator, there is at least a chance that it is a trapped particle, like the unobserved ninth pseudoscalar meson[27] with mass $<\sqrt{3}m_\pi$.

In a variety of models there are also true Goldstone bosons which *could* be observed as free particles. For instance, in the familiar four-quark version[28] of the $SU(2) \otimes U(1)$ model there is (in the absence of elementary spin-zero fields) an exact global symmetry of the weak and electromagnetic as well as the strong interactions, of the form

$$\mathfrak{N}_R \to \cos\phi\, \mathfrak{N}_R + \sin\phi\, \lambda_R,$$
$$\lambda_R \to -\sin\phi\, \mathfrak{N}_R + \cos\phi\, \lambda_R,$$
$$\mathcal{P}_R, \mathcal{P}'_R, \mathcal{P}_L, \mathfrak{N}_L, \mathcal{P}'_L, \lambda_L \quad \text{invariant.}$$

[As usual a subscript L or R denotes multiplication with $(1+\gamma_5)$ or $(1-\gamma_5)$, respectively.] This must be spontaneously broken, for otherwise λ and $\mathfrak{N}$ quarks could not have any mass, and for the same reason it cannot be expressed as a sum of a gauge and an unbroken generator. Also, in this case the symmetry current is both conserved and gluon-gauge-invariant. Thus, a four-quark SU(2) $\otimes$ U(1) model with purely dynamical symmetry breaking will have an untrapped massless true Goldstone boson.

I do not know whether present experiments rule out the possibility of electrically neutral and weakly interacting spin-zero bosons of zero mass. However, if we assume (as seems reasonable) that such particles do not in fact exist, then their absence puts a strong constraint on theories of dynamical symmetry breaking. In particular, it is probably necessary to have weak interactions that connect the ordinary quarks with heavy (~ 300-GeV) fermions, not only as a means of giving masses of order $e^2 \times 300$ GeV to the ordinary quarks, but also to avoid the unwanted anomaly-free global symmetries of the weak interactions.

We now come to one of the most puzzling and unsatisfactory features of dynamical symmetry breaking. In the currently popular gauge theories of strong interactions, the strong gauge coupling constant is fairly small at a renormalization point of order 2–3 GeV, and decreases further with increasing energy.[3] How then can the strong interaction produce a spontaneous symmetry breaking characterized by parameters F_a of order 300 GeV? Indeed, we believe that the strong interactions do induce a spontaneous symmetry breakdown, with the pion octet playing the role of Goldstone bosons, but the parameter F_π is 190 MeV, not 300 GeV.

Another difficulty arises when we try to include the leptons. If it is the ordinary strong interactions that produce the dynamical symmetry breaking discussed in this article, then can the color-neutral leptons get a mass in any order of e?

One way to approach these problems is to suppose that in addition to the color SU(3) associated with the observed strong interactions, there is another gauge group whose generators commute with color SU(3), associated with a new class of "extra-strong" interactions, which act on leptons as well as other fermions. If the gauge coupling constant of the extra-strong interactions reaches a value of order unity at a renormalization point of scale 300 GeV, then the extra-strong interactions could produce the dynamical symmetry breaking discussed in this article. Also, we would not observe direct effects of the extra-strong interactions at ordinary energies, provided that this dynamical symmetry breaking left no subgroup of the extra-strong gauge group unbroken, so that all vector bosons of the extra strong interactions got masses of order 300 GeV.

At first sight this possibility seems quite natural in the framework of the unified simple gauge theories of weak electromagnetic, and strong interactions discussed in the Introduction.[c] The spontaneous superstrong breakdown of the original simple gauge group can leave any number of subgroups unbroken, and some of these may have gauge couplings which grow faster with decreasing renormalization-point energy than the coupling constant of the ordinary strong interactions. However, this naturalness disappears on closer examination. Within the realm of validity of perturbation theory, the gauge couplings g_i of the various simple subgroups of the original unified simple group are given by

$$g_i^2(\mu) = \frac{g^2(\overline{M})}{1 + 2b_i g^2(\overline{M}) \ln(\overline{M}/\mu)},$$

where μ is the scale of a variable renormalization point; $\overline{M}$ is the superlarge mass at which all the $g_i(\mu)$ become equal; and b_i is the coefficient of g_i^3 in the Gell-Mann–Low function $\beta_i(g)$. Suppose we identify the onset of strong coupling for any simple subgroup as the point μ_i at which $g_i(\mu)$ reaches some definite value, of order unity, but taken sufficiently small so that perturbation theory is still valid. Then the ratio of the μ's for two subgroups will be given by

$$\mu_j/\mu_i = (\mu_i/\overline{M})^{b_i/b_j - 1}.$$

But $\overline{M}$ is likely to be enormous,[6] perhaps as large as 10^{19} GeV. Thus, unless b_i and b_j are unreasonably close, the onset of strong interaction will differ by many orders of magnitude for different simple subgroups. From this point of view, it is hard to understand how the onset of strong coupling for the ordinary strong interactions (a few hundred MeV) and the extra-strong interactions (300 GeV) could be so close.

Another possibility is that the color SU(3) gauge group is a subgroup of a larger gauge group which acts on leptons as well as on other fermions, and whose coupling constant reaches a value of order unity at a renormalization point of order 300 GeV. This could produce a dynamical symmetry breakdown of the larger group to color SU(3). In the effective field theory[9] which describes physics below 300 GeV, there could be a color SU(3) gauge symmetry, but since perturbation theory breaks down at 300 GeV, the strong gauge coupling in this effective theory would have no simple relation to the strong gauge coupling above 300 GeV, and might well be somewhat smaller. It would rise very slowly with decreasing renormalization-point energy, and even if it started just under 300 GeV at a value only a little less than its value above 300 GeV, it would not regain this value until much smaller renormalization scales were reached.

In either case, it is not the ordinary color SU(3) gluons that could produce the dynamical symmetry breaking which gives masses to the intermediate vector bosons. These gluons presumably *do* produce the dynamical breakdown of the previously unbroken subgroup H to SU(3) or SU(4) at energies of order $F_\pi = 190$ MeV, with the pion octet as Goldstone bosons, and with the quark masses of order $e^2 \times 300$ GeV furnishing the intrinsic H breaking which gives masses to the pion octet.

In closing, it is interesting to compare the conclusions of this article with the results obtained in theories with elementary spin-zero fields. In order to give the weak interactions the right strength, the vacuum expectation values of some of these fields must be of order 300 GeV. However, we can still distinguish between three kinds of theory:

I. It may be that the spin-zero fields have weak [say, $O(e^2)$] couplings to themselves and to the fermions. This is the case originally considered,[1] and it is the sort of theory with by far the greatest predictive power. The quark and lepton masses in such theories could arise directly from vacuum expectation values of the spin-zero fields, so there would be no need for heavy fermions. A characteristic feature of these theories is the appearance of Higgs scalars with masses that are less than 300 GeV by a factor of order $\sqrt{f}$, where f is the coupling constant of the quartic interaction.

II. It may be that the spin-zero fields have weak couplings to fermions, but strong interactions to themselves. In this case much of the gauge-theory phenomenology would survive, but it would be impossible to relate the vector-boson mass ratio to mixing angles, or to say anything at all about the existence of Higgs scalars. Again, there would be no need for heavy fermions.

III. It may be that the spin-zero fields have strong couplings both to fermions and to themselves. In general, such a theory would have very little predictive power; we would not even be able to say that weak processes like β decay arise from exchange of single vector bosons rather than from complicated higher-order effects.

Viewed in this way, gauge theories with dynamical symmetry breaking seem hardly distinguishable from theories of type III. The one significant difference, which gives theories of the type discussed in this article much greater predictive power than theories of type III, is the occurrence of a natural accidental symmetry, $U(N)\otimes U(N)$.

APPENDIX A: STRUCTURE OF COVARIANT DERIVATIVES

First, we note that under the requirements (1) and (2) of Sec. V, the effective Lagrangian $\mathcal{L}_1$ can be made formally invariant under $U(N)\otimes U(N)$ by introducing the G_W gauge fields

$$W_{A\mu} = \sum_\alpha e_{\alpha A} W_{\alpha\mu}, \tag{A1}$$

and imagining that these fields transform under $U(N)\otimes U(N)$ like λ_A. That is, we give $W_{A\mu}$ the formal transformation rule

$$W_{A\mu} \to W'_{A\mu} = \sum_B R_{AB}(g) W_{B\mu}, \tag{A2}$$

where g is an arbitrary element of $U(N)\otimes U(N)$, and $R_{AB}(g)$ is the corresponding orthogonal matrix in the regular representation of $U(N)\otimes U(N)$:

$$g^{-1}\lambda_A g = \sum_B R_{AB}(g)\lambda_B. \tag{A3}$$

[This is just like the well-known trick of introducing a fictitious octet of "photons" in order to study the SU(3) properties of electromagnetic corrections.]

Equation (A2) is a linear transformation rule, while $\mathcal{L}_1$ also contains quark and Goldstone boson fields which transform according to the nonlinear rules (5.2) and (5.3). It is therefore convenient, in order to make the whole of $\mathcal{L}_1$ invariant under $U(n)\otimes U(N)$, to replace $W_{A\mu}$ with a field which belongs to the same sort of nonlinear realization of $U(N)\otimes U(N)$ as does q:

$$\tilde{W}_{A\mu} \equiv \sum_B \Lambda_{BA}(\Pi) W_{B\mu} = \sum_{B,\alpha} \Lambda_{BA}(\Pi) e_{\alpha B} W_{\alpha\mu}, \tag{A4}$$

where Λ_{AB} is an orthogonal matrix defined by

$$\Lambda_{AB}(\Pi) = R_{AB}\left(\exp\left(i \sum_a \frac{\Pi_a x_a}{F_a}\right)\right). \tag{A5}$$

It is straightforward, using the transformation rules (A2) and (5.2)–(5.4), to check that $\tilde{W}_{A\mu}$ undergoes the $U(N)\otimes U(N)$ transformation

$$\tilde{W}_{A\mu} \to \tilde{W}'_{A\mu} = \sum_B R_{AB}\left(\exp\left(i \sum_i \mu_i(\Pi, g) t_i\right)\right) \tilde{W}_{B\mu}. \tag{A6}$$

The field $\tilde{W}_{A\mu}$ thus behaves under $U(N)\otimes U(N)$ transformations just like q, except of course that it belongs to a different linear representation of the unbroken subgroup H.

The Lagrangian $\mathcal{L}_1$ will thus be globally $U(N)$

$\otimes$U(N) invariant if it is algebraically invariant under the unbroken subgroup H and if it is composed of just the following ingredients: quark fields q, their U(N)$\otimes$U(N) covariant derivatives

$$\partial_\mu q - i\sum_{i,a} t_i q E_{ai}(\Pi)\partial_\mu \Pi, \qquad \text{(A7)}$$

together with Goldstone boson fields Π_a and their covariant derivatives

$$D_\mu \Pi_a \equiv F_a \sum_b D_{ba}(\Pi)\partial_\mu \Pi_b F_b^{-1}, \qquad \text{(A8)}$$

and $\tilde{W}_{A\mu}$ fields and their covariant derivatives

$$\partial_\mu \tilde{W}_{A\nu} - i\sum_{i,B}(t_i)_{AB}\tilde{W}_{B\nu}E_{ai}(\Pi)\partial_\mu \Pi_a,$$

plus gluon fields and their derivatives. [The functions D_{ba} and E_{ai} are defined by Eq. (5.9).] Aside from terms which are separately U(N)$\otimes$U(N) invariant and hence may be dropped, the covariant derivative of the gauge field may also be written

$$D_\mu \tilde{W}_{A\nu} = \sum_B \Lambda_{BA}(\Pi)\partial_\mu W_{B\nu}. \qquad \text{(A9)}$$

We now must impose requirement (3) of Sec. V, that the Lagrangian be locally as well as globally invariant under G_W and H_S. For a space-time dependent U(N)$\otimes$U(N) transformation $g(x)$, the gauge field $W_{\alpha\mu}$ transforms according to the rule

$$w_\alpha W'_{\alpha\mu} = g(w_\alpha W_{\alpha\mu})g^{-1} - (\partial_\mu g)g^{-1}. \qquad \text{(A10)}$$

Also, derivatives of g appear in the G_W transformation rules for the quantities (A7)–(A9). By using the derivative of Eq. (5.4) to evaluate these g derivatives, we can easily see that in order to cancel them, we must add gauge field terms to (A7) and (A8), so that those derivatives become G_W-covariant quantities[18]:

$$\partial_\mu q - i\sum_{i,a} t_i q E_{ai}(\Pi)\partial_\mu \Pi_a - i\sum_{A,i} t_i q C_{iA}\tilde{W}_{A\mu}, \qquad \text{(A11)}$$

$$\sum_b D_{ba}(\Pi)\partial_\mu \Pi_b + F_a \sum_a C_{aA}\tilde{W}_{A\mu}. \qquad \text{(A12)}$$

Note that these quantities are still formally locally U(N)$\otimes$U(N) covariant, as well as globally G_W covariant.

Also, G_W invariance requires that derivatives of the G_W gauge field only appear in the Yang-Mills curl[20]

$$F_{\alpha\mu\nu} \equiv \partial_\mu W_{\alpha\nu} - \partial_\nu W_{\alpha\mu} - \sum_{\beta\gamma} C_{\alpha\beta\gamma} W_{\beta\mu} W_{\gamma\nu}, \qquad \text{(A13)}$$

where $C_{\alpha\beta\gamma}$ are the structure constants of G_W. It is elementary to show that

$$\sum_{B\alpha} \Lambda_{BA}(\Pi) e_{\alpha B} F_{\alpha\mu\nu} = D_\mu \tilde{W}_{A\nu} - D_\nu \tilde{W}_{A\mu} - \sum_{BC} f_{ABC}\tilde{W}_{B\mu}\tilde{W}_{C\nu}, \qquad \text{(A14)}$$

where f_{ABC} are the U(N)$\otimes$U(N) structure constants. Thus, this quantity is both locally G_W covariant and (formally) globally U(N)$\otimes$U(N) covariant.

Finally, we must impose invariance under the unbroken strong gauge group H_S. According to our assumptions, it is only q that transforms nontrivially under H_S, so we must simply add a gluon term in (A11). This, together with (A12) and (A14), comprise the three sorts of fully covariant derivatives allowed in the effective Lagrangian $\mathcal{L}_1$.

APPENDIX B: STRUCTURE OF $\mathcal{L}_2$

The quantity $J_{AB}(\Pi)$ in Eq. (5.10) must be an U(N)$\otimes$U(N) tensor in the sense that for any element g of U(N)$\otimes$U(N), we have

$$J_{AB}(\Pi') = \sum_{CD} R_{AC}(g)R_{BD}(g)J_{CD}(\Pi), \qquad \text{(B1)}$$

with Π' and R defined by Eqs. (5.4) and (A3). But it follows from (A4) and (A5) that

$$R(g)\Lambda(\Pi) = \Lambda(\Pi')R\Big(\exp\Big(i\sum_i \mu_i t_i\Big)\Big), \qquad \text{(B2)}$$

so contracting Eq. (B1) on the left-hand side with $\Lambda^{-1}(\Pi')$ gives

$$I_{AB}(\Pi') = \sum_{CD} R_{AC}\Big(\exp\Big(i\sum_i \mu_i t_i\Big)\Big) \times R_{BD}\Big(\exp\Big(i\sum_i \mu_i t_i\Big)\Big) I_{CD}(\Pi), \qquad \text{(B3)}$$

where

$$I_{AB}(\Pi) \equiv \sum_{CD} \Lambda^{-1}_{AC}(\Pi)\Lambda^{-1}_{BD}(\Pi)J_{CD}(\Pi). \qquad \text{(B4)}$$

If we now choose $\Pi_a = 0$ and

$$g = \exp\Big(i\sum_a x_a \pi_a / F_a\Big),$$

we find from (5.4) that

$$\Pi'_a = \pi_a, \quad \mu'_i = 0,$$

so Eq. (B3) reads here

$$I_{AB}(\pi) = I_{AB}(0) \equiv I_{AB}. \qquad \text{(B5)}$$

Hence I_{AB} is a constant. Equation (B3) then says that it is a constant tensor under H. That is,

$$I_{AB} = \sum_{CD} R_{AC}(h)R_{BD}(h)I_{CD} \qquad \text{(B6)}$$

for arbitrary elements $h \in H$.

APPENDIX C: STRUCTURE OF $\mathcal{L}_m$

The quantity $N_{AB}(\Pi)$ in Eq. (5.26) is an $U(N) \otimes U(N)$ tensor, in the sense that for an arbitrary element g of $U(N)\otimes U(N)$, we have

$$\exp\left(-i\sum_i t_i\mu_i(\Pi, g)\right)N_{AB}(\Pi')\exp\left(i\sum_i t_i\mu_i(\Pi, g)\right) = \sum_{CD} R_{AC}(g)R_{BD}(g)N_{CD}(\Pi), \tag{C1}$$

with μ_i, Π', and R defined by Eqs. (5.4) and (A3). Using Eq. (B2) and contracting Eq. (C1) on the left with $\Lambda^{-1}(\Pi')$ gives

$$\exp\left(-i\sum_i t_i\mu_i\right)Q_{AB}(\Pi')\exp\left(i\sum_i t_i\mu_i\right) = \sum_{CD} R_{AC}\left(\exp\left(i\sum_i t_i\mu_i\right)\right)R_{BD}\left(\exp\left(i\sum_i t_i\mu_i\right)\right)Q_{CD}(\Pi'). \tag{C2}$$

If we choose $\Pi_a = 0$ and

$$g = \exp\left(i\sum_a x_a\pi_a/F_a\right),$$

Then Eq. (5.4) gives

$$\Pi'_a = \pi_a, \quad \mu_i = 0,$$

so Eq. (C2) in this case just tells us that $Q(\pi)$ is π independent,

$$Q_{AB}(\pi) = Q_{AB}(0) \equiv Q_{AB}. \tag{C3}$$

Equation (C2) may thus be written

$$h^{-1}Q_{AB}h = \sum_{CD} R_{AC}(h)R_{BD}(h)Q_{CD} \tag{C4}$$

for arbitrary elements $h \in H$.

APPENDIX D: CONSTRUCTION OF UNITARITY GAUGE

Recall first that the x_a and t_i are defined to span the algebra of $U(N)\otimes U(N)$, so that any element λ of this algebra may be written

$$\lambda = \sum_i \mu_i^0 t_i + \sum_a \phi_a^0 x_a,$$

with μ_i^0 and ϕ_a^0 so far unconstrained. We then define

$$\phi_a(\theta) \equiv \phi_a^0 + \sum_{\alpha A} \theta_\alpha e_{\alpha A} B_{aA},$$

and choose θ to minimize the positive continuous function

$$\sum_a F_a^2 \phi_a^2(\theta).$$

Denoting the value of $\phi_a(\theta)$ at this minimum as ϕ_a, we then have

$$0 = 2\sum_a F_a^2\phi_a \sum_A e_{\alpha A}B_{aA},$$

so ϕ_a satisfies the constraint (7.2). Also, we can rewrite λ as

$$\lambda = \sum_i \mu_i^0 t_i + \sum_a \phi_a x_a - \sum_{a\alpha A} \theta_\alpha e_{\alpha A} B_{aA} x_a$$

$$= \sum_i \mu_i^0 t_i + \sum_a \phi_a x_a - \sum_\alpha \theta_\alpha w_\alpha + \sum_{i\alpha A} \theta_\alpha e_{\alpha A} C_{iA} t_i,$$

so that λ may be decomposed as in Eq. (7.1), with

$$\mu_i = \mu_i^0 + \sum_{\alpha A} \theta_\alpha e_{\alpha A} C_{iA}.$$

ACKNOWLEDGMENTS

I am grateful for frequent valuable discussions on this subject with colleagues at Harvard and M.I.T. I also thank D. J. Gross and H. Pagels for enlightening conversations.

*Work supported in part by the National Science Foundation under grant No. NSF-GP-13547X.

[1]S. Weinberg, Phys. Rev. Lett. 19, 1264 (1967); A. Salam, in *Elementary Particle Physics*, edited by N. Svartholm (Almqvist and Wiksells, Stockholm, 1968), p. 367.

[2]Dynamical mechanisms for spontaneous symmetry breaking were first discussed by Y. Nambu and G. Jona-Lasinio, Phys. Rev. 122, 345 (1961); J. Schwinger, *ibid.* 125, 397 (1962); 128, 2425 (1962). The subject was revived in the context of modern gauge theories by R. Jackiw and K. Johnson, Phys. Rev. D 8, 2386 (1973); J. M. Cornwall and R. E. Norton, *ibid.* 8, 3338 (1973). More recent references include D. J. Gross and A. Neveu, *ibid.* 10, 3235 (1974); E. J. Eichten and F. L. Feinberg, *ibid.* 10, 3254 (1974); J. M. Cornwall, R. Jackiw, and E. Tomboulis, *ibid.* 10, 2438 (1975); E. C. Poggio, E. Tomboulis, and S-H. H. Tye, *ibid.* 11, 2839 (1975); K. Lane, *ibid.* 10, 1353 (1974); 10, 2605 (1974); S. Weinberg, in *Proceedings of the XVII International Conference on High Energy Physics, London, 1974*, edited by J. R. Smith (Rutherford Laboratory, Chilton, Didcot, Berkshire, England, 1974), p. III–59.

[3]D. J. Gross and F. Wilczek, Phys. Rev. Lett. 30, 1343 (1973); H. D. Politzer, *ibid.* 30, 1346 (1973). The problem of maintaining asymptotic freedom in theories with strongly interacting spin-zero fields has been discussed by D. J. Gross and F. J. Wilczek, Phys. Rev. D 8, 3633 (1973); T. P. Cheng, E. Eichten, and L. F. Li, *ibid.* 9, 2259 (1974).

[4]C. Callan and D. Gross, Phys. Rev. Lett. 22, 156 (1969).

[5]S. Weinberg, Phys. Rev. Lett. 31, 494 (1973); Phys. Rev. D 8, 4482 (1973).

[6]H. Georgi, H. Quinn, and S. Weinberg, Phys. Rev. Lett. 33, 451 (1974).

[7]S. Weinberg, Phys. Rev. D 5, 1962 (1972).

[8]See Ref. 6. Specific models which unify the weak, electromagnetic, and strong interactions have been proposed by J. C. Pati and A. Salam, Phys. Rev. D 8, 1240 (1973); H. Georgi and S. L. Glashow, Phys. Rev. Lett. 32, 438 (1974); H. Fritzsch and P. Minkowski (unpublished).

[9]At this point we use the term "effective field theory" in the sense of T. Appelquist and J. Carrazone, Phys. Rev. D 11, 2856 (1975).

[10]J. Goldstone, Nuovo Cimento 19, 154 (1961); J. Goldstone, A. Salam, and S. Weinberg, Phys. Rev. 127, 965 (1965). Also see Ref. 2.

[11]P. W. Higgs, Phys. Lett. 12, 132 (1965); Phys. Rev. Lett. 13, 508 (1964); Phys. Rev. 145, 1156 (1966); F. Englert and R. Brout, Phys. Rev. Lett. 13, 321 (1964); G. S. Guralnik, C. R. Hagen, and T. W. B. Kibble, *ibid.* 13, 585 (1964); T. W. B. Kibble, Phys. Rev. 155, 1554 (1967).

[12]For instance, in the model of Nambu and Jona-Lasinio, Ref. 2, the dynamical symmetry breakdown can occur for only one value of the Fermi coupling constant.

[13]This is the "dimensional transmutation" of S. Coleman and E. Weinberg, Phys. Rev. D 7, 1888 (1973).

[14]The use of effective Lagrangians, to reproduce the low-energy theorems associated with a spontaneously broken symmetry, was initiated by S. Weinberg, Phys. Rev. Lett. 18, 188 (1967). The formalism was further developed by J. Schwinger, Phys. Lett. 24B, 473 (1967); S. Weinberg, Phys. Rev. 166, 1568 (1968); S. Coleman, J. Wess, and B. Zumino, *ibid.* 177, 2239 (1968); C. G. Callan, S. Coleman, J. Wess, and B. Zumino, *ibid.* 177, 2247 (1968). An explicit construction of an effective Lagrangian involving composite Goldstone boson fields has been given by S. Coleman (unpublished).

[15]R. F. Dashen, Phys. Rev. D 3, 1879 (1971). For a recent application, see H. Pagels, *ibid.* 11, 1213 (1975).

[16]See Ref. 11. For a more general definition of the unitarity gauge, see S. Weinberg, Phys. Rev. D 7, 1068 (1973).

[17]Pseudo-Goldstone bosons are the Goldstone bosons associated with the spontaneous breakdown of any "accidental" symmetry of the $e=0$ terms in the Lagrangian. They were originally discussed in theories with elementary spin-zero fields; see S. Weinberg, Phys. Rev. Lett. 29, 1698 (1972); Phys. Rev. D 7, 2887 (1973). For a generalization, see H. Georgi and A. Pais, Phys. Rev. D 12, 508 (1975).

[18]See Callan, Coleman, Wess, and Zumino, Ref. 14, Sec. IV. Also see Jackiw and Johnson, Ref. 2, Sec. III.

[19]S. Coleman and E. Weinberg, Ref. 13; S. Weinberg, Ref. 17.

[20]C. N. Yang and R. L. Mills, Phys. Rev. 96, 191 (1954); R. Utiyama, *ibid.* 101, 1597 (1956); M. Gell-Mann and S. Glashow, Ann. Phys. (N.Y.) 15, 437 (1961).

[21]Note that we exclude pole terms produced by internal gluons here. These can be included by stitching together the black boxes calculated from this effective Lagrangian with soft gluon lines.

[22]See Coleman, Wess, and Zumino, Ref. 14.

[23]This may be contrasted with the calculation described in Ref. 17. There, the integrals began to converge for virtual momenta of the order of the intermediate-vector-meson mass, and the pseudo-Goldstone boson mass was of order e^2M, not eM.

[24]This is essentially the same argument that forces us to diagonalize the matrix elements of a perturbation between degenerate states before beginning to construct a perturbation series. Also, compare S. Weinberg, Phys. Rev. Lett. 31, 494 (1973), Sec. III.

[25]It is not so clear what meaning should be attached to quark masses if quarks are not observable as free particles. One interpretation is that the quark masses are the inputs to current-algebra calculations, as used by M. Gell-Mann, R. J. Oakes, and B. Renner, Phys. Rev. 175, 2195 (1968); S. L. Glashow and S. Weinberg, Phys. Rev. Lett. 20, 224 (1968). The success of these calculations indicates that the quark masses are small compared with some scale characteristic of the strong interactions.

[26]I thank H. Georgi for this remark.

[27]J. Kogut and L. Susskind, Phys. Rev. D 10, 3468 (1974); K. Lane, *ibid.* 10, 1353 (1974); 10, 2605 (1974); S. Weinberg, Ref. 2; J. Kogut and L. Susskind, Phys. Rev. D 11, 3594 (1975); S. Weinberg, *ibid.*, 11, 3583 (1975).

[28]S. L. Glashow, J. Iliopoulos, and L. Maiani, Phys. Rev. D 2, 1285 (1970); also see S. Weinberg, *ibid.* 5, 1412 (1972).

PHYSICAL REVIEW D VOLUME 19, NUMBER 4 15 FEBRUARY 1979

Implications of dynamical symmetry breaking: An addendum

S. Weinberg
Lyman Laboratory of Physics, Harvard University, Cambridge, Massachusetts 02138
(Received 2 June 1978)

It is shown that the dynamical symmetry breakdown of a gauge symmetry can in some cases lead to simple relations among the masses of intermediate vector bosons.

This note is an addendum to a general survey[1] of the physical implications of dynamical symmetry breaking.[2] By a "dynamical" symmetry breaking is meant any spontaneous breakdown of a global or gauge symmetry for which the associated Goldstone bosons are composite rather than elementary particles. Formation of such bound states requires strong forces among the constituent particles, and for this reason it has generally been supposed that a dynamical breakdown of gauge symmetries would not naturally provide any simple relations among the intermediate-vector-boson masses generated by the symmetry breakdown. The possibility of such mass relations was not considered in Ref. 1.

In this note I wish to show that a dynamical breakdown of a gauge symmetry can indeed lead to simple relations among intermediate-vector-boson masses. In particular, a dynamical breakdown of SU(2)× U(1) in the gauge theory[3] of weak and electromagnetic interactions can lead to the same successful relation $M_Z/M_W = \sec\theta$ that is found when the symmetry breakdown is due to vacuum expectation values of scalar-field doublets. For a dynamical symmetry breakdown this is not as automatic as for symmetry breaking by scalar-doublet vacuum expectation values, but depends on the assumption of a specific pattern of spontaneous symmetry breaking. The source of this mass relation can be traced, both for symmetry breaking by scalar doublets and in the dynamical case to be considered here, to the same simple property of the Goldstone bosons which mix with the SU(2)× U(1) gauge fields. However, there remain severe difficulties in developing realistic detailed models of elementary particles in which the symmetry breaking is purely dynamical.

First, let us consider a gauge model that is illustrative, though its quark content is quite unrealistic. The gauge group of the weak, electromagnetic, and strong interactions is as usual taken as $SU(2)\times U(1)\times G_S$, but G_S is arbitrary. The coupling constants g, g', and g_s associated with the subgroups SU(2), U(1), and G_S are assumed to have the orders of magnitude $g\approx g'\approx e$ and $g_s\approx 1$. [Strictly speaking, g_s becomes of order unity at a renormalization scale Λ, which, as we shall see in this example, would have to be of the order of 200 GeV. Hence, G_S could not consist solely of the usual color group SU(3).] The model contains just two G_S multiplets of quarks, U and D, which form a left-handed SU(2)× U(1) doublet $(1+\gamma_5)(U,D)$ and right-handed singlets $(1-\gamma_5)U$ and $(1-\gamma_5)D$, but no scalar fields. (The "color" indices associated with G_S are dropped everywhere.) In the limit $e\to 0$, the strong interactions will automatically be invariant not only under the gauge group G_S, but also under an "accidental" global SU(2)× SU(2) symmetry,[4] consisting of independent unitary unimodular transformations on the doublets $(1+\gamma_5)(U,D)$ and $(1-\gamma_5)(U,D)$. We assume that the strong forces associated with G_S produce a dynamical breakdown of SU(2)× SU(2), which for $e=0$ leaves the global "isospin" subgroup SU(2) unbroken. [At the same time, G_S itself may also break down to some gauge subgroup, perhaps SU(3).] It is the residual global invariance of the strong interactions for $e=0$ that leads in this model to a simple relation between M_W and M_Z.

To see this, we use the general lowest-order formula[5] for the intermediate-vector-boson matrix

$$\mu^2_{\alpha\beta}=\tfrac{1}{64}\sum_a F_a^{\,2}\,\mathrm{Tr}(t_\alpha x_a)\,\mathrm{Tr}(t_\beta x_a)\,. \qquad (1)$$

Here x_a are the generators of all spontaneously broken global symmetries, in a suitably orthonormalized basis[6]; F_a are the couplings of the corresponding Goldstone bosons to the associated currents; and t_α are the generators of the weak and electromagnetic gauge groups, including all coupling-constant factors. In the present case, we have

$$\begin{aligned} &x_a=\gamma_5\tau_a,\quad a=1,2,3\\ &t_i=\tfrac{1}{4}g(1+\gamma_5)\tau_i,\quad i=1,2,3 \qquad (2)\\ &t_0=-g'[\tfrac{1}{4}(1-\gamma_5)\tau_3+\tfrac{1}{6}]\,,\end{aligned}$$

with τ_a the Pauli isospin matrices. The residual global SU(2) symmetry prevents the appearance of any positive-parity Goldstone bosons, and also imposes a relation among the F_a,

$$F_1=F_2=F_3\equiv F\,. \qquad (3)$$

This is the same relation as is satisfied by the F's in the nondynamical case, where the symmetry breakdown is due to vacuum expectation values of scalar doublets, and it leads to the same relation among intermediate-vector-boson masses. From Eq. (1), we then find the nonvanishing elements of the intermediate-vector-boson mass matrix,

$$\begin{aligned} \mu^2{}_{11} &= \mu^2{}_{22} = \mu^2{}_{33} = \tfrac{1}{16}F^2g^2 , \\ \mu^2{}_{30} &= \mu^2{}_{03} = \tfrac{1}{16}F^2gg' , \\ \mu^2{}_{00} &= \tfrac{1}{16}F^2g'^2 . \end{aligned} \tag{4}$$

(All Goldstone bosons are eliminated by the Higgs mechanism here.) It is easy to see that the nonvanishing mass eigenvalues are

$$M_W{}^2 = \tfrac{1}{16}F^2g^2, \quad M_Z{}^2 = \tfrac{1}{16}F^2(g^2+g'^2) . \tag{5}$$

With $g'/g \equiv \tan\theta$, these have the usual ratio $M_Z/M_W = \sec\theta$.

The U and D of this model cannot, of course, be identified with any known particles. Their mass must be of order F, because for $e=0$ there are no free parameters in the theory except for the G_S renormalization scale Λ, so that M_U, M_D, and F must all be of order Λ. But if we identify $\sqrt{2}\,g^2/8M_W{}^2$ with the usual Fermi coupling constant G_F, then F takes the value $2^{-3/4}G_F{}^{-1/2}$, or 175 GeV. Also, U and D are nearly degenerate, since their masses are split by the weak and electromagnetic interactions only in order α.

In order to construct a slightly more realistic model, we consider the same gauge group $SU(2) \times U(1) \times G_S$, but now we suppose that there are *four* quark flavors U_1, D_1, U_2, D_2, which form two left-handed $SU(2)\times U(1)$ doublets $(1+\gamma_5)(U_1,D_1)$ and $(1+\gamma_5)(U_2,D_2)$, plus four right-handed singlets. In this model, the strong interactions for $e\to 0$ will automatically have an accidental global $SU(4)\times SU(4)$ symmetry.[4] We assume that this symmetry suffers a spontaneous dynamical breakdown in such a way that two quarks, U and D, receive equal masses, while the other two, u and d, remain massless. The subgroup of $SU(4)\times SU(4)$ which remains unbroken is assumed to be the largest subgroup consistent with such masses, with the generators (in a U,D,u,d basis) given by

$$\begin{pmatrix} \vec{\tau} & 0 \\ 0 & 0 \end{pmatrix}, \quad \begin{pmatrix} 0 & 0 \\ 0 & \vec{\tau} \end{pmatrix}, \quad \gamma_5 \begin{pmatrix} 0 & 0 \\ 0 & \vec{\tau} \end{pmatrix}, \quad \begin{pmatrix} 1 & 0 \\ 0 & -1 \end{pmatrix} . \tag{6}$$

(Also, parity is assumed to be not spontaneously broken.) The orthonormalized generators of the broken part of $SU(4)\times SU(4)$ can then be taken as

$$\begin{aligned} \vec{x}_A &= \gamma_5 \begin{pmatrix} \vec{\tau} & 0 \\ 0 & 0 \end{pmatrix}, \quad x_B = \frac{\gamma_5}{\sqrt{2}} \begin{pmatrix} 1 & 0 \\ 0 & -1 \end{pmatrix}, \\ \vec{x}_C &= \frac{1}{\sqrt{2}} \begin{pmatrix} 0 & -i\vec{\tau} \\ i\vec{\tau} & 0 \end{pmatrix}, \quad \vec{x}_C' = \gamma_5 \vec{x}_C , \\ \vec{x}_D &= \frac{1}{\sqrt{2}} \begin{pmatrix} 0 & \vec{\tau} \\ \vec{\tau} & 0 \end{pmatrix}, \quad \vec{x}_D' = \gamma_5 \vec{x}_D , \\ x_E &= \frac{1}{\sqrt{2}} \begin{pmatrix} 0 & 1 \\ 1 & 0 \end{pmatrix}, \quad x_E' = \gamma_5 x_E , \\ x_F &= \frac{1}{\sqrt{2}} \begin{pmatrix} 0 & -i \\ i & 0 \end{pmatrix}, \quad x_F' = \gamma_5 x_F . \end{aligned} \tag{7}$$

Using the unbroken part of $SU(4)\times SU(4)$, we easily see that there are only three independent F_a parameters; they are $F_{Ai} \equiv F_A, F_B$ and $F_{Ci} = F'_{Ci} = F_{Di} = F'_{Di} = F_E = F'_E = F_F = F'_F \equiv F_C$. (Here $i=1,2,3$.) The quark fields U,D,u,d of definite mass are not necessarily the same as those in the original weak doublets $(1+\gamma_5)(U_1,D_1)$ and $(1+\gamma_5)(U_2,D_2)$. However, we can always put the weak doublet into the form

$$\begin{aligned} &(1+\gamma_5)\begin{pmatrix} u \\ d\cos\phi + D\sin\phi \end{pmatrix}, \\ &(1+\gamma_5)\begin{pmatrix} U \\ -d\sin\phi + D\cos\phi \end{pmatrix}. \end{aligned} \tag{8}$$

The angle ϕ must be determined by minimizing a "potential" $V(\phi)$, given to lowest order in e by the sum of graphs in which W or Z is emitted and absorbed by a strong-interaction vacuum fluctuation.[7] By using the unbroken $SU(2)\times SU(2)\times SU(2)\times U(1)$ subgroup of $SU(4)\times SU(4)$, we easily see that the Z contribution is ϕ independent, while the W contribution is a sum of terms proportional to $\cos^2\phi$ or $\sin^2\phi$. The whole potential therefore has the ϕ dependence $V(\phi) = A + B\cos^2\phi$. Thus, depending on the sign of B, the angle ϕ at which $V(\phi)$ is a minimum must take the values $\phi=\pi/2$ or $\phi=0$. Let us consider these two cases in turn:

(a) $\phi = \pi/2$: The gauge generators (in a U,D,u,d basis) here take the form

$$\begin{aligned} t_1 &= \tfrac{1}{4}g(1+\gamma_5)\begin{pmatrix} 0 & -i\tau_2 \\ i\tau_2 & 0 \end{pmatrix}, \\ t_2 &= \tfrac{1}{4}g(1+\gamma_5)\begin{pmatrix} 0 & i\tau_1 \\ -i\tau_1 & 0 \end{pmatrix}, \\ t_3 &= \tfrac{1}{4}g(1+\gamma_5)\begin{pmatrix} \tau_3 & 0 \\ 0 & \tau_3 \end{pmatrix}, \\ t_0 &= -g'\left\{\tfrac{1}{4}(1-\gamma_5)\begin{pmatrix} \tau_3 & 0 \\ 0 & \tau_3 \end{pmatrix} + \tfrac{1}{6}\right\}. \end{aligned} \tag{9}$$

From Eq. (1), we find that the nonvanishing elements of the intermediate-vector-boson mass matrix are

$$\begin{aligned}\mu^2{}_{11}&=\mu^2{}_{22}=\tfrac{1}{4}F_C{}^2g^2\,,\\ \mu^2{}_{33}&=\tfrac{1}{16}F_A{}^2g^2\,,\\ \mu^2{}_{30}&=\mu^2{}_{03}=\tfrac{1}{16}F_A{}^2gg'\,,\\ \mu^2{}_{00}&=\tfrac{1}{16}F_A{}^2g'^2\,.\end{aligned}\tag{10}$$

The nonvanishing intermediate-vector-boson masses are then

$$M_W{}^2=\tfrac{1}{4}F_C{}^2g^2,\quad M_Z{}^2=\tfrac{1}{16}F_A{}^2(g^2+g'^2)\,.\tag{11}$$

Hence no simple mass relation arises in this case.

(b) $\phi=0$: The gauge generators (in a U,D,u,d basis) here take the form

$$\begin{aligned}t_i&=\tfrac{1}{4}g(1+\gamma_5)\begin{pmatrix}\tau_i&0\\0&\tau_i\end{pmatrix},\\ t_0&=-g'\left\{\tfrac{1}{4}(1-\gamma_5)\begin{pmatrix}\tau_3&0\\0&\tau_3\end{pmatrix}+\tfrac{1}{6}\right\}.\end{aligned}\tag{12}$$

From Eq. (1), we find the nonvanishing elements of the intermediate-vector-boson mass matrix are

$$\begin{aligned}&\mu^2{}_{11}=\mu^2{}_{22}=\mu^2{}_{33}=\tfrac{1}{16}F_A{}^2g^2\,,\\ &\mu^2{}_{30}=\mu^2{}_{03}=\tfrac{1}{16}F_A{}^2gg',\quad \mu^2{}_{00}=\tfrac{1}{16}F_A{}^2g'^2\,,\end{aligned}\tag{13}$$

so the masses are

$$M_W{}^2=\tfrac{1}{16}F_A{}^2g^2,\quad M_Z{}^2=\tfrac{1}{16}F_A{}^2(g^2+g'^2)\tag{14}$$

and have the same ratio $M_Z/M_W=\sec\theta$ as in the simpler two-flavor model. The reason for this is just that the only Goldstone bosons which mix with W and Z are those associated with $\vec{x}_A$, and the unbroken subgroup of SU(4) × SU(4) requires these to have equal F_a values.

This model is still far from realistic, whether $\phi=\pi/2$ or $\phi=0$. In both cases, the u and d quarks remain massless to all orders in e. In addition for $\phi=\pi/2$, the two light quarks u,d are in SU(2) × U(1) doublets with D and U, not with each other. Furthermore, in both cases the model contains 17 physical Goldstone bosons, some of them "true" Goldstone bosons of zero mass. It is not clear how the light quarks could get reasonable masses or how the "true" Goldstone bosons could be eliminated with a dynamical symmetry breaking.

Note added in proof. After this paper was submitted for publication, I received a report by L. Susskind, which deals with similar questions. [The motivation in his paper for dynamical symmetry breaking, in terms of grand unified gauge theories, is the same as that described by H. Georgi, H. Quinn, and S. Weinberg, Phys. Rev. Lett. 33, 451 (1974).] The undiscovered new strong interaction of Susskind is a special case of what was called an "extra strong" interaction in Ref. 1, and a "superstrong" interaction by S. Weinberg, Phys. Today 30 (No. 4), 42 (1977). Susskind independently observes that the relation $M_Z/M_W=\sec\theta$ follows if dynamical symmetry breaking leaves an "isospin" subgroup unbroken. In addition, he points out that the origin for this relation is essentially the same as that for symmetry breaking by vacuum expectation values of scalar doublets. In the latter case, the part of the Lagrangian which is relevant to the calculation of gauge boson masses is the "kinematic" Lagrangian $\mathcal{L}_\phi=-\frac{1}{2}\sum_n(D_\mu\phi_n)^\dagger(D_\mu\phi_n)$, the sum running over N scalar doublets (ϕ_n^0,ϕ_n^-). By setting $\phi_n^0=\phi_{n1}+i\phi_{n2}$, $\phi_n^-=\phi_{n3}+i\phi_{n4}$, one finds that in the limit $e=0$, $\mathcal{L}_\phi$ has an $O(4)^N=[SU(2)\times SU(2)]^N$ symmetry, with an O(3) = SU(2) subgroup which is automatically left unbroken by the vacuum expectation values of the ϕ_{n1} fields, and which transforms the weak SU(2) generators as a three-vector. As shown both here and in Susskind's paper, this is the same feature that allows one to derive the Z-W mass ratio in the case of dynamical symmetry breaking.

I also wish to comment here on the problems of developing a grand unified theory of strong as well as weak and electromagnetic interactions in which the spontaneous symmetry breaking at all levels is due to vacuum expectation values of elementary scalar fields. As is well known, such theories require constraints on the parameters in the Lagrangian. However, it is *not* true that these constraints necessarily incorporate extremely small parameters, such as 10^{-19}, or that they need to involve quadratic divergences at all. It is only necessary to suppose that at a stationary point of the potential where the SU(3) × SU(2) × U(1) subgroup is unbroken, some of the non-Goldstone eigenvalues of the scalar mass matrix vanish. Nonperturbative effects will then produce a minimum of the potential very near this stationary point, at which SU(2) × U(1) is spontaneously broken, with W and Z masses which are automatically less than the superheavy gauge boson masses by a factor $\exp(-C/e^2)$, where C is a numerical constant of order unity. (Also, any quadratic divergences are always an artifact of the cutoff procedure; they do not appear if we use dimensional regularization.) These matters are discussed by E. Gildener and S. Weinberg, Phys.

Rev. D 13, 3333 (1976), Sec. VI, and in a paper now in preparation.

I am very grateful to Kenneth Lane for remarks which led me to take up this problem, and for valuable conversations on various aspects of this subject. This research was supported in part by the National Science Foundation under Grant No. PHY77-22864.

---

[1]S. Weinberg, Phys. Rev. D 13, 974 (1976).

[2]Dynamical mechanisms for spontaneous symmetry breaking were first discussed by Y. Nambu and G. Jona-Lasinio, Phys. Rev. 122, 345 (1961); J. Schwinger, *ibid.* 125, 397 (1962); 128, 2425 (1962). The application to modern gauge theories is due to R. Jackiw and K. Johnson, Phys. Rev. D 8, 2386 (1973); J. M. Cornwall and R. E. Norton, *ibid.* 8, 3338 (1973).

[4]It is of course understood that the whole theory is also invariant under the global U(1) symmetry associated with fermion number conservation. However, the full global symmetry group for N quark flavors is $SU(N) \times SU(N) \times U(1)$, and not $U(N) \times U(N)$ (as supposed in Ref. 1), because invariance under the *chiral* U(1) symmetry is broken by triangle anomalies in the presence of instantons; see G. 't Hooft, Phys. Rev. Lett. 37, 8 (1976), and references cited therein.

[5]See Ref. 1, Eq. (7.11). The factor $\frac{1}{64}$ appears here because all traces include sums over Dirac indices.

[6]See Ref. 1, Sec. IV.

[7]See Ref. 1, Sec. VI.

PHYSICAL REVIEW D VOLUME 20, NUMBER 10 15 NOVEMBER 1979

Dynamics of spontaneous symmetry breaking in the Weinberg-Salam theory

Leonard Susskind*
Stanford Linear Accelerator Center, Stanford University, Stanford, California 94305
(Received 5 July 1978)

We argue that the existence of fundamental scalar fields constitutes a serious flaw of the Weinberg-Salam theory. A possible scheme without such fields is described. The symmetry breaking is induced by a new strongly interacting sector whose natural scale is of the order of a few TeV.

I. WHY NOT FUNDAMENTAL SCALARS?

The need for fundamental scalar fields in the theory of weak and electromagnetic forces[1] is a serious flaw. Aside from the subjective esthetic argument, there exists a real difficulty connected with the quadratic mass divergences which always accompany scalar fields.[2] These divergences violate a concept of naturalness which requires the observable properties of a theory to be stable against minute variations of the fundamental parameters.

The basic underlying framework of discussion of naturalness assumes the existence of a fundamental length scale κ^{-1} which serves as a real cutoff. Many authors[3] have speculated that κ should be of order 10^{19} GeV corresponding to the Planck gravitational length. The basic parameters of such a theory are some set of dimensionless bare couplings g_0 and masses. The dimensionless bare masses are defined as the ratio of bare mass to cutoff:

$$\mu_0 = m_0/\kappa\,. \tag{1}$$

The principle of naturalness requires the physical properties of the output at low energy to be stable against very small variations of g_0 and μ_0. One such striking property is the existence of a "light" mass spectrum of order 1 GeV. From a dimensionless viewpoint the light spectrum has mass 10^{19} times smaller than the fundamental scale. It is in order to ask what kind of special adjustments of parameters must be made in order to ensure such a gigantic ratio of mass scales.

To illustrate a case of an unnatural adjustment, consider a particle which receives a self-energy which is quadratic in κ. To make the discussion simple, suppose the form of the mass correction is

$$\begin{aligned} m^2 &= m_0^{\,2} + \Delta m^2 \\ &= m_0^{\,2} + \kappa^2 g_0^{\,2}\,. \end{aligned} \tag{2}$$

Solving for $\mu_0^{\,2}$ gives

$$\mu_0^{\,2} = \frac{m_0^{\,2}}{\kappa^2} = \frac{m^2}{\kappa^2} - g_0^{\,2}\,. \tag{3}$$

Now if m is a physical mass of order 1 GeV and $\kappa \sim 10^{19}$ GeV, then

$$\mu_0^{\,2} = -g_0^{\,2}(1 - 10^{-38})\,. \tag{4}$$

Equation (4) means that $\mu_0^{\,2}$ must be adjusted to the 38th decimal place. What happens if it is not? Then the mass will come out to be of order 10^{19} GeV.

Such adjustments are unnatural and will be assumed absent in the correct theory. Unfortunately, all present theories contain such unnatural adjustments because of the quadratic divergences in the scalar-particle masses.

Not all theories in which the physical and cutoff scales are vastly different are unnatural. Fortunately there exists a class of non-Abelian gauge theories where enormous scale ratios may occur naturally. These are the asymptotically free theories.[4]

In asymptotically free theories the scale-dependent "running" coupling constant satisfies

$$q\frac{\partial g}{\partial q} = -Cg^3 + O(g^5)\,, \tag{5}$$

where q is the momentum scale at which the coupling is measured. The constant C is positive so that the coupling increases toward the infrared. It is believed that such theories spontaneously generate masses corresponding to values of q for which g becomes large. Integrating Eq. (5) gives

$$\frac{1}{g^2(q)} = 2C\ln(q/\kappa) + \frac{1}{g_0^{\,2}}\,, \tag{6}$$

where $g^2(K)$ is identified with the bare coupling g_0. Of course, Eq. (6) is inaccurate when g becomes large, but we may use it as a guide to where the coupling becomes large. It is evident that the value of m which makes $g(m)$ become large satisfies

$$\frac{m}{\kappa}=\exp\left(-\frac{1}{2Cg_0^2}\right). \tag{7}$$

Evidently, to make $m/\kappa \sim 10^{-19}$ requires the bare coupling g_0 to be

$$g_0^2=\frac{1}{38C\ln 10}=\frac{0.012}{C}. \tag{8}$$

As an example, consider pure non-Abelian SU_3 Yang-Mills theory. The constant C is given by[4]

$$C=\frac{11}{16\pi^2} \tag{9}$$

so that

$$g_0^2 \sim 0.2.$$

This is hardly an unnatural value for g_0^2. Furthermore, the value of m/κ is not violently sensitive to small variations of g_0.

II. A NATURAL SCENARIO

Let us assume that at the smallest distances (Planck length) nature is described by a very symmetric "grand unified" theory. The grand unifying group is called G. Let us suppose that G is spontaneously broken. This might occur for a variety of reasons, including the existence of scalar fields or gravitational attraction. Since we shall forbid unnatural adjustments of constants we must assume that any masses which are generated in the first round of symmetry breaking are of order 10^{19} GeV.

At a somewhat larger distance scale, say 10^{17} GeV, a phenomenological description should exist. It will contain those survivors of the first symmetry breakdown which gained no mass. Furthermore, it will have a symmetry group

$$G_1\otimes G_2\otimes G_3\otimes\cdots$$

consisting of factors which are not broken by the first breakdown.

The survivor fields will include

(1) the gauge bosons for the group $G_1\otimes G_2\otimes\cdots$,

(2) some subset of fermions which were protected by unbroken γ_5 symmetries,

(3) some Goldstone bosons. In what follows we assume no such Goldstone bosons are present.

If not for the fermions, the different G_i would define uncoupled gauge sectors. These sectors are, in general, coupled by fermions having nontrivial transformation properties under more than one G_i. For example, quarks form the bridge which couples quantum-chromodynamic (QCD) gluons with the photon, intermediate-vector-boson sector in the standard theory.

Thus, to specify a theory we must give a set of G_i and a set of fermion fields along with their transformation properties under all G_i. Furthermore, we will also need a set of coupling constants g_i. These may be taken to be the running couplings at a low enough energy so that the effects of the very heavy masses have disappeared. Henceforth we assume this to be 10^{+17} GeV. Henceforth we define $K=10^{17}$.

Consider next the evolution of the running couplings. Some of them will increase and some will decrease as the energy scale is lowered. From studying examples it is clear that different g_i may blow up and produce masses at rather different scales. To see why this is so consider the case of two uncoupled gauge theories G_1 and G_2. Each will have its bare coupling g_1, g_2 and will evolve to give a mass scale

$$\frac{m_1}{K}\approx\exp\left(-\frac{1}{2C_1g_1^2}\right),\quad \frac{m_2}{K}\approx\exp\left(-\frac{1}{2C_2g_2^2}\right), \tag{10}$$

and

$$\frac{m_1}{m_2}=\exp\left[-\frac{1}{2}\left(\frac{1}{C_2g_2^2}-\frac{1}{C_1g_1^2}\right)\right]. \tag{11}$$

If we now assume $1/C_ig_i^2$ is large enough to make m_i/K very small then a few present differences between C_1 and C_2 or g_1 and g_2 can easily make $m_1/m_2 \sim 10^{-3}$.

Thus our expectation is for a $G_1\otimes G_2\otimes G_3\cdots$ gauge theory with a set of Fermi fields connecting the G_i and a set of dynamically produced mass scales fairly well separated. The question to which this paper is addressed is: Can this type of theory produce the required kinds of spontaneous symmetry breakdown needed to understand weak, electromagnetic, and strong interactions?

III. A WARMUP EXAMPLE

The set of subgroups G_i must include SU_3 (color) and the electromagnetic-weak group $SU_2\otimes U_1$ which we will call flavor. The fermion content must include quarks and leptons. As our simplest example we consider a theory with the massless flavor doublet (u, d) of color-triplet quarks. The quarks interact with an octet of color gluons and the four flavor gauge fields W^α and B. The coupling constants are chosen as they would be in realistic models so that the QCD coupling becomes ~1 at 1 GeV and the electromagnetic charge is $\sim\frac{1}{3}$. We call the SU_3, SU_2, and U_1 coupling constants g_3, g_2, and g_1. This theory is the standard theory of a single quark doublet with the exception that no fundamental scalar Higgs field are included.

The Lagrangian for our warmup model is

$$\mathcal{L} = -\tfrac{1}{4}F_{\mu\nu}F_{\mu\nu} - \tfrac{1}{4}W_{\mu\gamma}W_{\mu\gamma} - \tfrac{1}{4}B_{\mu\nu}B_{\mu\nu} + i\bar{\psi}\gamma_\mu(\partial_\mu + ig_3\hat{f} + ig_2\hat{W}_\mu + ig_1\hat{B}_\mu)\psi . \quad (12)$$

The objects $\hat{F}_\mu$, $\hat{W}_\mu$, and $\hat{B}_\mu$ are constructed from the SU_3, SU_2, U_1 vector potentials, Dirac matrices, 3×3 color matrices, and 2×2 flavor matrices. For example,

$$\hat{W}_\mu = W_\mu^\alpha \frac{\tau^\alpha}{2}\left(\frac{1-\gamma_5}{2}\right), \quad (13)$$

where τ^α are flavor Pauli matrices. Similarly,

$$\hat{B}_\mu = \left[\frac{1-\gamma_5}{4} - \frac{1}{3} + \left(\frac{1+\tau_3}{2}\right)\left(\frac{1+\gamma_5}{2}\right)\right]B_\mu . \quad (14)$$

In analyzing the above model we will make use of a number of standard assumptions. We now list them:

(1) The weak-electromagnetic sector can be treated as a small perturbation. The remaining assumptions apply to the pure SU_3 (color) sector when g_1 and g_2 are switched off.

(2) The strong interactions are invariant under chiral $SU_2 \otimes SU_2$ in the limit of vanishing bare quark mass.[6]

(3) Chiral $SU_2\times SU_2$ is spontaneously broken and realized in the Nambu-Goldstone mode. The pion is the Goldstone boson. The "order parameter" signaling the spontaneous breakdown is $\langle 0|\bar{\psi}\psi|0\rangle = \langle 0|\bar{u}u + \bar{d}d|0\rangle$, which is nonzero.

(4) The chiral limit ($m_\pi^2 \to 0$) is a smooth one in which all strong-interaction quantities (other than m_π) change by only a few percent. In particular, this includes f_π—the pion decay constant.

Our problem is to determine the behavior of the W, Z, and photon masses in this theory. In particular, we would like to know if the strong interactions can somehow replace the Higgs scalars and provide masses for the intermediate vector bosons. To this end we must examine the effects of quarks and SU_3 gluons on the W and B propagators. The relevant processes are shown in Figs. 1(a)–1(c).

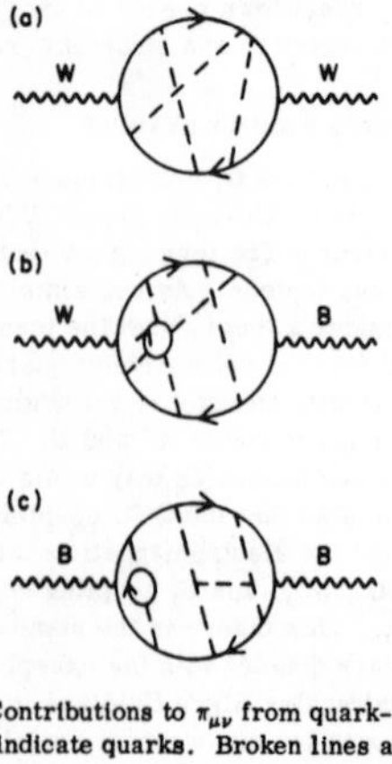

FIG. 1. Contributions to $\pi_{\mu\nu}$ from quark-gluon states. Solid lines indicate quarks. Broken lines are gluons.

Let us first ignore the B field and concentrate on the class of processes illustrated in Fig. 1(a). Invoking the familiar arguments of gauge invariance, we write the one-particle-irreducible vacuum polarization as

$$\pi_{\mu\nu}^{\alpha\beta} = \delta^{\alpha\beta}\left(\frac{g_2}{4}\right)^2 (k^2 g_{\mu\nu} - k_\mu k_\nu)\pi(k^2) , \quad (15)$$

where α,β indicate SU_2 indices. Evidently the W propagator is modified from

$$\delta^{\alpha\beta}\left(\frac{k^2 g_{\mu\nu} - k_\mu k_\nu}{k^4}\right) \quad (16)$$

to

$$\frac{\delta^{\alpha\beta}(g_{\mu\nu} - k_\mu k_\nu/k^2)}{k^2[1 + g_2^{\,2}\pi(k^2)/4]} . \quad (17)$$

Unless $\pi(k^2)$ is singular at $k^2=0$ the W propagator will have a pole at $k^2=0$ indicating a massless vector boson. From this point of view, the role of the fundamental scalar Goldstone bosons is to provide a pole in π at $k^2=0$.

Now consider the contribution of the pion to $\pi(k^2)$. Since no explicit scalars are included, the quarks must be massless. (Recall that the only source of quark mass in the Weinberg-Salam theory is the Yukawa couplings.) It then follows that the pion is massless, at least insofar as it is regarded as an unperturbed state of the pure strong interaction. Thus we can immediately write the pion contribution to π as a massless pole in the vicinity of $k^2=0$:

$$\pi(k^2) \approx f_\pi^{\,2}/k^2 . \quad (18)$$

Accordingly, the pion replaces the usual scalar fields and shifts the mass of the W to

$$M_W^{\,2} = \left(\frac{g_2}{2}\right)^2 f_\pi^{\,2} \approx (30\ \text{MeV})^2 . \quad (19)$$

Next consider the contribution of the pion to the processes in Figs. 1(b) and 1(c). For this we need to know the coupling of the pion to the Abelian U_1 current. From Eq. (14) we see that this current is

$$\bar{\psi}\gamma_\mu\left[\frac{1}{2}\left(\frac{1-\gamma_5}{2}\right) - \frac{1}{3} + \left(\frac{1+\tau_3}{2}\right)\left(\frac{1+\tau_5}{2}\right)\right]\psi . \quad (20)$$

The term which couples to the neutral pion is

$$-\tfrac{1}{4}\bar{\psi}\tau_3\gamma_5\gamma_\mu\psi . \quad (21)$$

Thus Figs. 1(b) and 1(c) receive pion-pole contributions

$$\pi_{WB}=\frac{g_1g_2}{4k^2}f_\pi^2, \quad (22)$$

$$\pi_{BB}=\left(\frac{g_1}{4k^2}\right)^2 f_\pi^2. \quad (23)$$

All of this is summarized by a mass matrix

$$M^2=\begin{bmatrix} g_2^2 & 0 & 0 & 0 \\ 0 & g_2^2 & 0 & 0 \\ 0 & 0 & g_2^2 & g_1g_2 \\ 0 & 0 & g_1g_2 & g_1^2 \end{bmatrix}\tfrac{1}{4}f_\pi^2, \quad (24)$$

where the labeling of the rows and columns is (W^+, W^-, W^0, B).

The mass matrix in Eq. (24) is identical to that in the Weinberg-Salam (WS) theory with the exception that f_π would be replaced by the vacuum expectation value of the scalar field ϕ. Thus the masses of Z and $W^\pm$ are in the same ratio as in the WS theory but are scaled down by the factor

$$\frac{f_\pi}{\langle\phi\rangle}\approx\frac{1}{3000}. \quad (25)$$

Naturally, in the model we are considering the pion is absent from the real spectrum, being replaced by the longitudinal $W^\pm$ and Z.

The correspondence between the pion and the usual scalar doublet ϕ may be made manifest. Define

$$\pi^\alpha=\bar\psi\gamma_5\tau^\alpha\psi, \qquad \sigma=\bar\psi\psi. \quad (26)$$

The two component field ϕ of WS may be replaced by

$$\begin{pmatrix}\phi_1\\ \phi_2\end{pmatrix}\longrightarrow\begin{pmatrix}\pi_1+i\pi_2\\ \pi_3+i\sigma\end{pmatrix}. \quad (27)$$

It is easily seen that such a two-component object transforms as a spinor under left-handed SU_2 and has the same Abelian charge as ϕ. Lastly, the spontaneous breaking of the symmetry is accomplished by the usual strong interactions which (we believe) give rise to $\langle\sigma\rangle\neq 0$.

An interesting point we wish to emphasize before attempting a realistic example involves the $SU_2\times SU_2$ symmetry of the hadron sector (before g_2 and g_1 are switched on). In general, to consistently couple a sector to the weak-electromagnetic interaction that sector need only have $SU_2\times U_1$ symmetry. The extra symmetry under $SU_2\times SU_2$ is also present in the WS model. To see it we write

$$\begin{pmatrix}\phi_1\\ \phi_2\end{pmatrix}=\begin{pmatrix}\alpha_1+i\alpha_2\\ \alpha_3+i\alpha_4\end{pmatrix} \quad (28)$$

and then note that the scalar field Lagrangian in WS has symmetry under the four-dimensional rotations $(=SU_2\times SU_2)$ in the $(\alpha_1,\alpha_2,\alpha_3,\alpha_4)$ space.

In the WS model the additional symmetry is accidental and may be eliminated if nonrenormalizable interaction or additional scalar multiplets are introduced. In our case it is entirely natural, following from the symmetries of a multiplet of Dirac fermions.

It is interesting to ask what evidence exists for the $SU_2\times SU_2$ symmetry. In our example the extra symmetry implies ordinary isospin [SU_2 (left) + SU_2 (right)] symmetry and guarantees that f_π is the same for neutral and charged pions. If the symmetry were reduced to $SU_2\times U_1$ then in general $f_{\pi^\pm}\neq f_{\pi^0}$. The result would be a modification of the structure of the mass matrix in Eq. (24). The success of the WS model in neutral-current phenomenology is rather sensitive to this structure. Therefore, a large deviation from $SU_2\times SU_2$ symmetry in the scalar field sector will be inconsistent with observed neutral currents.

A final point involves the existence of more than one quark multiplet. If the number of quark doublets is increased from one to N, the hadronic chiral symmetry becomes $SU_{2N}\times SU_{2N}$. Since mass terms are forbidden when the fundamental scalars are absent this will necessarily be a symmetry of the hadronic sector. The number of Goldstone bosons will be N^2-1. The longitudinal Z and $W^\pm$ will again absorb three of these, leaving N^2-4 spin-zero objects. These objects will gain mass because the weak interactions explicitly violate the symmetries which correspond to them. In other words, they are what Weinberg calls pseudo-Goldstone bosons. Their mass will in general be of the same order of magnitude as that of Z and W.

IV. A MORE REALISTIC EXAMPLE

Let us now consider the possible existence of a new undiscovered strongly interacting sector, similar to ordinary strong interactions except with a mass scale of order 10^3 GeV. To be specific we introduce a new family of fermions called "heavy-color" quarks and an associated field χ. The heavy-color quarks form a flavor $SU_2\times U_1$ doublet and an n-tuple in a new SU_n symmetry space called heavycolor. Heavycolor is a gauge symmetry requiring a multiplet of gauge bosons G_μ. The symmetry of the theory is then

$$SU_n(HC)\otimes SU_3(C)\otimes SU_2\otimes U_1$$

with couplings g_n, g_3, g_2, g_1. The fermion content includes

(1) *Leptons*. These are flavor doublets and color–heavy-color singlets.

(2) *Quarks*. These are flavor doublets, color triplets, heavy-color singlets.

(3) *Heavy-color quarks*. These are flavor doublets, color singlets, and heavy-color n-tuples.

The coupling g_n is chosen so that a mass scale of order 1 TeV—the heavy-color (HC) interaction—becomes strong. To make this precise we first consider the pure HC theory ignoring quarks, leptons, color, and flavor. The bare g_n is then adjusted so that the lightest nonzero mass of a heavy-color hadron is ~1 TeV.

The Lagrangian of our model is

$$\mathcal{L} = -\tfrac{1}{4}G_{\mu\nu}G_{\mu\nu} - \tfrac{1}{4}F_{\mu\nu}F_{\mu\nu} - \tfrac{1}{4}W_{\mu\nu}W_{\mu\nu} - \tfrac{1}{4}B_{\mu\nu}B_{\mu\nu} + \bar{\chi}\gamma_\mu(\partial_\mu + ig_n\hat{G}_\mu + ig_2\hat{W}_\mu + ig_1\hat{B}_\mu)\chi + \bar{\psi}\gamma_\mu(\partial_\mu + ig_3\hat{F}_\mu + ig_2\hat{W}_\mu + ig_1\hat{B}_\mu)\psi\,. \qquad (29)$$

In speculating about the solution of this model we will make use of the following observations and assumptions.

(1) The evolutions of the heavy-color and color couplings with scale are only slightly different from what they would be if each sector were completely isolated. The justification for this is that heavy color and color are only coupled by their *weak* interactions with B and W. If g_1 and g_2 were zero, g_n and g_3 would evolve completely separately. In fact, the quark-gluon and heavy-color-quark–heavy-color-gluon worlds would be completely noninteracting.

(2) The isolated heavy-color sector is essentially similar to the color sector except scaled up in energy by ~3000. This means that $\langle 0|\bar{\chi}\chi|0\rangle \neq 0$. This implies the existence of a family of massless heavy-color pions with decay constants $F_\pi \sim f_\pi \times 3000$. It also means that there exists a rich spectrum of heavy-color hadrons.

As in our warmup example the Z and $W^\pm$ gain a mass. This time the mass is mainly due to the mixing of the HC-pion with W and B. The ordinary pion becomes very slightly mixed with the HC-pion but remains exactly massless. This is so because the ordinary and HC axial-vector currents are separately conserved. The longitudinal components of Z and W can only absorb one linear combination of the Goldstone bosons associated with these currents.

The model described here is certainly incomplete. As it stands it cannot account for the masses of leptons and quarks. We shall discuss this further in Sec. V.

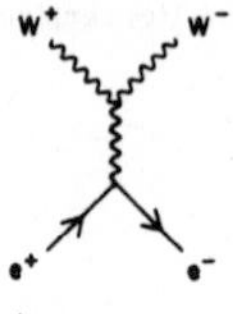

FIG. 2. The process $e^+e^- \to W^+W^-$.

V. IMPLICATIONS OF HEAVY COLOR

The behavior of processes at and above the TeV range is very different in the usual and present theories. By the usual theory I will always mean two things. First, symmetry breaking is caused by fundamental scalar fields. Second, all coupling constants including the scalar self-coupling are small so that perturbation theory is applicable.

Our first problem is to determine the mass scale of the heavy-color hadrons. To this end we observe that Eq. (19) will be replaced by

$$M_W^2 = \frac{g_2^2}{4}F_\pi^2\,, \qquad (30)$$

where F_π is the HC-pion decay constant. Since we know that $M_W \sim 90$ GeV, we find

$$F_\pi \sim 250 \text{ GeV}\,. \qquad (31)$$

Since we have assumed that the heavy-color sector is simply a scaled-up version of the usual strong interactions, it follows that heavy-color-hadron masses are F_π/f_π times their hadronic counterparts. Since $F_\pi/f_\pi \sim 3\times10^3$, the mass of a low-lying HC-hadron will be ~3 TeV.

The main differences in behavior of this and the usual model involve processes in which longitudinally polarized Z's and W's are produced at energies above a TeV. For example, consider e^+e^- annihilation. As in the usual theory the e^+e^- pair can form a virtual photon or Z boson which can then materialize as a pair of transverse $W^\pm$ bosons. This process is illustrated in Fig. 2. The transverse $W^\pm$ are weakly coupled and therefore contribute a smooth nonresonant contribution to R that can be computed in perturbation theory. (See Fig. 3.)

In the usual weakly coupled scalar version of the Weinberg-Salam theory the virtual γ or Z can also materialize as a pair of charged scalars disguised as longitudinal $W^\pm$. Since the scalars are also weakly coupled the contribution to R is smooth, nonresonant, and similar to Fig. 3.

In the present theory the scalar sector is replaced by the heavy-color sector and the virtual γ-Z may decay into a pair of heavy-color quarks. The resulting behavior of R will exhibit all the

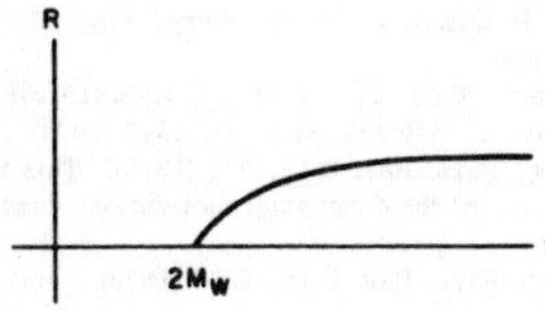

FIG. 3. R as a function of energy for transverse gluon production.

characteristics of resonances and final-state interaction which characterized R at ordinary energies ~0–3 GeV. It should exhibit the bumps of the HC-ρ, HC-Ω, and so on. (See Fig. 4.) The only difference is that the entire scale of masses, widths, and level separations will be of order 1 TeV instead of 1 GeV.

The final states of such processes will involve increasing multiplicities of HC-hadrons. If our experience in hadron physics is a good guide then most of the final HC-hadrons will be HC-pions. Of course real HC-pions do not exist, having been replaced by the longitudinal $W^{\pm}$, Z states. Indeed, the following theorem is easy to prove: To lowest order in α the amplitude for producing a given state including some set of Z_{long} and W_{long} is equal to the amplitude for a state in which the Z_L, W_L are replaced by HC-pions.

The longitudinal bosons decay in a conventional way to leptons and hadrons. However, the distribution of the longitudinal bosons will not resemble the usual theory. In the usual theory each boson, longitudinal or transverse, costs a factor of α since all couplings are small. In the present theory longitudinal bosons proliferate like pions once the energy exceeds a few TeV.

Perhaps the most interesting consequence of the new theory is the existence of a new conservation law—heavy-color baryon number $= \int \chi^* \chi \, d^3x$. The lightest HC-hadron carrying HC-baryon number will be stable and have a mass ~1–2 TeV. If the heavy-color group SU_n has odd (even) n this particle will be a fermion (boson). Its only interaction with ordinary matter will be weak-electromagnetic. It may be charged like the proton or neutral like the neutron. If, like the proton, it is charged and found in any abundance in the universe, it may be detectable as a component of cosmic rays.

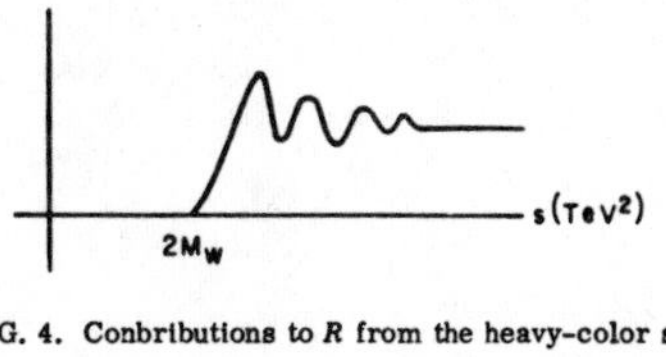

FIG. 4. Conbributions to R from the heavy-color sector.

VI. CONCLUSIONS

One aspect of the scalar-boson problem has not been mentioned in this paper. The usual scalar-boson mechanism provides masses not only for the vectors but also the leptons and quarks. In the present example the only way to mimic the fermion mass mechanism would involve four-Fermi nonrenormalizable interactions. Indeed, if the scalar field ϕ is replaced by HC-quark bilinears in the Yukawa couplings, a quartic coupling of the form $\chi^\dagger\chi\,\psi^\dagger\psi$ is generated. This coupling produces the conventional fermion mass matrix when $\langle\bar{\chi}\chi\rangle$ gets a nonvanishing value.

The inability to generate mass without four-Fermi couplings is due to the chiral γ_5 symmetry of vector couplings. In the present theory this symmetry is a continuous symmetry if we ignore weak instantons. Therefore, any dynamical fermion mass generation would require massless Goldstone bosons.

In general, by adding more sectors, including a gauge group which mixes e and μ as well as strange and nonstrange quarks, we can reduce the γ_5 symmetry to a discrete symmetry. In order to do this we must make use of the instantons of this new sector which means the coupling must be significantly greater than α. If this theory can be made to work then no Goldstone bosons would be required by dynamical mass generation.

Notes added.

(1) After submitting this paper for publication I became aware of the work of S. Weinberg in which many of the motivations for heavy-color are described.[5]

(2) Very recently Weinberg[6] and Georgi, Lane, and Eichten have been led to consider a very similar proposal for replacing Higgs bosons by dynamically bound pionlike objects originating in a strong interaction whose scale is ~1 TeV. I would like to thank S. Weinberg for a copy of his report and H. Georgi for communicating his interest in the problem to me.

ACKNOWLEDGMENT

I would like to thank K. Wilson for explaining the reasons why scalar fields require unnatural adjustments of bare constants.

*Permanent address: Dept. of Physics, Stanford University.

[1]For a review of the conventional theory see S. Weinberg, in *Proceedings of the International Symposium on Lepton and Photon Interactions at High Energies, Hamburg, 1977*, edited by F. Gutbrod (DESY, Hamburg, 1977).

[2]The particular concept of naturalness and the objections to scalar fields described in this paper are due to K. Wilson, private communication.

[3]H. Georgi, H. Quinn, and S. Weinberg, Phys. Rev. Lett. 33, 451 (1974).

[4]H. D. Politzer, Phys. Rev. Lett. 30, 1346 (1973); D. Gross and F. Wilczek, *ibid*. 30, 1343 (1973).

[5]S. Weinberg, Phys. Rev. D 13, 974 (1975). This paper contains many of the conceptual motivations described here.

[6]S. Weinberg, Phys. Rev. D 19, 1277 (1979).

IV. EXTENDED HYPERCOLOR

Nuclear Physics B155 (1979) 237–252
North-Holland Publishing Company

MASS WITHOUT SCALARS

Savas DIMOPOULOS
Physics Dept., Columbia University, New York, NY 10027, USA

Leonard SUSSKIND
Theoretical Institute, Physics Dept., Stanford University, Stanford, CA 94305, USA

Received 20 February 1979

We attempt to show that fundamental scalar fields can be eliminated from the theory of weak and electromagnetic interactions. We do this by constructing an explicit example in which the scalar field sectors are replaced by strongly interacting gauge systems. Unlike previous examples, our present work gives a natural explanation for fermion masses. The cost is a significant expansion of the size of the gauge group.

1. Introduction

It has been argued that the presence of scalar fields in unified weak-electromagnetic theories requires certain fundamental parameters to be adjusted to absurd precesion [1,2]. Indeed it seems that to establish a hierarchy of mass scales, beginning at the Planck mass (10^{18} GeV) and ending ar ordinary particle masses requires fundamental unrenormalized masses to be adjusted [1,2] to 30 decimal places! Perhaps in some future theory such adjustments will appear natural, but at present divine intervention is the only available explanation. For this reason we have sought an alternative mechanism for generating fermion and gauge boson masses in which no fundamental scalars are ever invoked.

In a previous paper [2], one of us showed how the generation of intermediate vector boson ($W^{\pm}$, Z) masses could arise without fundamental scalars or unnatural adjustments. However, this paper offered no explanation for the origin of quark and lepton masses. It is the purpose of this paper to fill that gap.

We will present some examples which we argue are capable of generating a set of mass scales with realistic orders of magnitude for both fermions and bosons. The examples in their present form are too simple to take completely seriously. We offer them as an "existence proof" for a family of theories which can produce reasonable scales without unnatural adjustments. We also feel that the examples offer valuable clues to the various mechanisms which may be needed.

Before entering into a technical discussion of models, we must warn the reader that we will deal with complicated systems of interacting gauge sectors, some of which are strongly interacting. Thus we can only guess the patterns of spontaneous symmetry breaking which occur. Indeed our guesses are modelled on the usually assumed chiral behavior of QCD which is incompletely understood [3].

The reader should keep in mind that by *quark mass generation* we mean "current-masses" which are present in the absence of ordinary QCD strong interactions.

2. Technicolor

In this section we will review the ideas of ref. [2]. There the standard SU(2) × U(1) theory of weak and electromagnetic forces in the absence of fundamental scalar fields was considered. The ordinary quarks and leptons are minimally coupled to the gauge fields $W^{\pm}$, W^0, B. In this paper we ignore the usual strong interactions and color degrees of freedom which can easily be restored.

To replace the scalar field sector, a new family of *massless* fermions called "Techniquarks" are introduced. The T-quark fields from electroweak doublets and N-tuplets under a new "Technicolor" group $SU(N)_{TC}$. They are also assumed to be color singlets.

The T-quarks interact *via* an unbroken $SU(N)_{TC}$ gauge interaction which we assume behaves in a manner which parallel QCD. The main difference is that the QTD running coupling becomes strong at a mass of order 1 TeV instead of 1 GeV. Thus the scale on which T-hadrons exist is 10^3 times heavier than ordinary hadrons.

The T-color binding forces generate a spontaneous breakdown of chiral flavor SU(2) × SU(2) leading to the existence of massless T-pion Nambu-Goldstone bosons [3].

Now in ref. [2] it was shown how these massless T-pions replace the usual fundamental scalar fields in producing a mass matrix for the intermediate vector bosons. The graphs responsible for the shifts of mass and mixing of the boson fields are shown in fig. 1. The resulting mass matrix is identical to that in the Weinberg-Salam theory [4] except that the vacuum expectation value of the scalar field is replaced by F_π, the technicolor analog of the pion decay constant. Its value must be 250 GeV in order to give the Z, $W^{\pm}$ their expirical masses. This represents a rescaling of QCD by a factor of ~2000.

The most interesting point about this mechanism is that it provides a simple explanation of the observed neutral current behavior. In general, the consistency of the Weinberg-Salam theory requires the scalar field sector to have SU(2) × U(1) symmetry only. The larger SU(2) × SU(2) symmetry of the T-quark sector insures a remaining unbroken T-isospin invariance after spontaneous breakdown. The resulting equality of the F_π's for the neutral and charged T-pions gives the boson mass-matrix the same special form as in the usual theory [4]. Thus

$$\cos\theta_W M_Z = M_W\,, \tag{2.1}$$

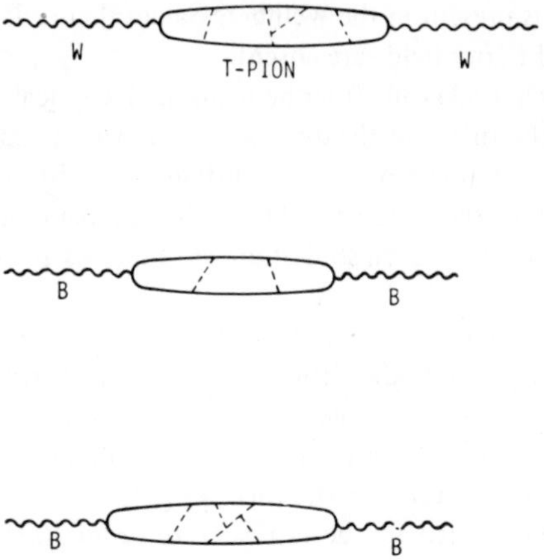

Fig. 1. Origin of W-S mass matrix.

where θ_W is the usual weak angle.

The reader may wonder about the origin of the enormous mass scale ~1 TeV. In particular, why is the ratio of scales of QCD and QTD of order 10^3? To understand this, let us recall some facts from the renormalization group. For an asymptotically free theory, the mass scale at which the interaction becomes strong is given by [2,5]

$$m = k \exp - \frac{c}{g_0^2} , \tag{2.2}$$

where k is the cutoff and g_0 the bare coupling. It is well-known [6] that in QCD if k is of order the Planck mass and $g_0 \sim$ the electric charge then $m \sim 1$ GeV.

Now suppose that a second sector exists with a somewhat different value of c/g_0^2. The nature of eq. (1.2) is to magnify a small difference in c/g_0^2 into an enormous difference in mass scales. Thus our expectation is that the strongly interacting sectors of a theory will have very different scales.

In addition to generating gauge boson masses, the scalar fields in the Weinberg-Salam theory are also responsible for the fermion mass matrix. Unfortunately, the T-sector as described in ref. [2] is not capable of replacing the scalar fields for this purpose. In order to see why this is so let us consider the simple case of a single doublet of ordinary quarks interacting with the above system through its SU(2) × U(1) charges. The system has an Abelian γ_5 symmetry * in the absence of strong interactions.

$$q \rightarrow \exp i\theta\gamma_5\, q . \tag{2.3}$$

* We emphasize that by quark masses we mean current algebra masses which are present in the absence of ordinary strong interactions. Note also that electroweak instantons give minute masses to the NGB of the U(1) axial symmetry (2.3).

This same symmetry is present in the Weinberg-Salam theory if the Yukawa couplings of the scalar and Fermi fields are absent.

The symmetry in eq. (2.3) can either be realized algebraically or in the Nambu-Goldstone mode. In the first case the quarks remain exactly massless while in the second a massless η-like boson would result. Either possibility is unacceptable. Actually the weakness of the $SU(2) \times U(1)$ couplings make it very unlikely that the symmetry is spontaneously broken so that massless quarks are almost certainly the result.

The purpose of this paper is to argue that the above situation is not inevitable. We would like to construct a model with the following features.

(i) Naturalness: no parameter needs to be adjusted to unreasonable accuracy. This presumably means the elimination of scalars *. All mass scales should emerge from infrared instabilities of the massless theories [2].

(ii) Masses for the intermediate vector bosons of order 100 GeV with the usual relationship between M_Z, M_W $\sin \theta_W$. This means that the dynamical symmetry breaking mechanisms must have $SU(2)_L \times SU(2)_R$ flavor symmetry [2].

(iii) Dynamically induced quark and lepton masses. Furthermore it should occur naturally that these particles should be 10^{-2}–10^{-3} times lighter than Z and $W^{\pm}$.

(iv) No massless or almost massless Nambu-Goldstone bosons (NGB) should exist in the observable spectrum of particles [5]

3. General hypotheses

In this section we will formulate some general hypotheses about dynamical symmetry breaking. These hypotheses are suggested by analogy with the current theory of strong interactions: QCD. There, although also unproven, the assumptions about chiral symmetry breaking are supported by empirical evidence [7].

For the sake of definiteness we consider an $SU(N > 2)$ gauge theory with n left-handed fermion fields l^i_α $(i = 1, 2, ..., n)$ where α labels the fundamental representation of $SU(N)$ and n right-handed fermions $r^j_\alpha (j = k, ..., n)$ **. In addition to the gauge symmetry the theory has an $SU(n) \times SU(n)$ "spectator" global chiral symmetry. (By spectator symmetry we mean that it commutes with the

* In globally supersymmetric theories scalars are protected from receiving large masses. If supersymmetry is broken at a mass scale ~1 TeV then scalar mass scales ~100 GeV are not unnatural. If however, the breaking of supersymmetry occurs at a much higher scale, say $\sim 10^{12}$ TeV, then absurdly unnatural adjustments are again needed to ensure scalar scales ~100 GeV.

** A non-trivial constraint in all of the models that we discuss is the absence of anomalies in the strongly interacting gauge sectors. For future reference note that the 5 and 10 representations of SU(5) have equal and opposite anomalies. This follows because, together with the singlet of SU(5), they constitute the 16-dimensional representation of the anomaly-free group SO(10). See ref. [10].

gauge interaction under consideration. At some later stage we may want to gauge some part of a spectator symmetry.)

At the scale M when the coupling constant becomes large, non-perturbative phenomena occur. We shall assume the following hypotheses for the non-perturbative behavior.

(i) Resistance to self-breaking: the non-Abelian gauge symmetry remains unbroken. The phenomenon of confinement occurs. The phsyical states are gauge singlets.

(ii) Bilinear condensation: following Nambu and Jona Lasinio [3] we assume that the vacuum state rearranges by the formation of a condensate of pairs. The condensate is of the form

$$\langle l_{\alpha}^{i\dagger} f_{ij} r_{\alpha}^{j} \rangle \neq 0 \,. \tag{3.1}$$

(iii) The condensates forms in such a way as to give equal masses to n Dirac fermions. This means that f_{ij} is a unitary matrix.

(iv) If the spectator symmetry or any part of it is gauged with a sufficiently weak coupling then the above pattern does not change. However those NGB's which correspond to gauged broken generators become the longitudinal components of massive spectator gauge bosons. The above hypotheses which are well-known for QCD become more subtle when the strong group is SU(2). In this case the $2n$ chiral fermions can all be considered left-handed. This is because the antiparticle representation of SU(2) is unitarily equivalent to the particle representation. Accordingly the theory contains $2n$ left-handed fermions in the fundamental representation. Therefore the theory has an SU($2n$) spectator symmetry instead of the $SU(n)_L \times SU(n)_R$ symmetry. Note that the U(1) symmetry is destroyed by the Adler-Bell-Jackiw anomaly.

In analysing the SU(2) model we shall assume that the hypotheses (i)–(iv) generalize. Using a notation in which we have n left-handed and n right-handed fields, the condensate can be taken as

$$\langle l_{\alpha}^{i\dagger} r_{\alpha}^{i} \rangle \neq 0 \,. \tag{3.2}$$

The fields r_{α}^{i} may be replaced by their n left-handed antiparticle fields. Thus let i run from 1 to $2n$ and define (where c means charge conjugation)

$$l_{\alpha}^{i+n} = \epsilon_{\alpha\beta} (r_{\beta}^{c})^{i} \,. \tag{3.3}$$

Thus the condensate in eq. (3.2) becomes

$$\langle \epsilon_{\alpha\beta}\, l_{\alpha}^{i}(1)\, l_{\beta}^{n+i}(2) \rangle \neq 0 \,, \tag{3.4}$$

where the notations (1), (2) indicate upper and lower spin components of the left-handed Weyl spinors. The expression in eq. (3.4) may be written in the compact form

$$\langle \epsilon_{\alpha\beta}\, l_{\alpha}(1)\, \eta l_{\beta}(2) \rangle \,, \tag{3.5}$$

where the matrix η is defined by

$$\eta = \begin{bmatrix} 0 & I \\ I & 0 \end{bmatrix} \quad (\leftarrow n \rightarrow \ \leftarrow n \rightarrow) \tag{3.6}$$

The choice of condensate is ambiguous up to an SU($2n$) transformation of the fields l^i_α under which

$$\eta \rightarrow U^{\mathrm{T}} \eta U . \tag{3.7}$$

The subgroup of SU($2n$) that leaves the condensate invariant is the unitary symplectic group Sp($2n$). This is to be contrasted with the situation when the strong group is SU($N > 2$). In that case the spectator SU(n) $\times$ SU(n) breaks down to SU(n).

In the SU(2) gauge theory an additional case is possible. Namely, the total number of chiral Weyl fields may be odd, equal to $2n + 1$. (Anomalies forbid this for $N > 2$.) In this case the $2n + 1$ left-handed Weyl fields cannot completely pair to form massive Dirac fields. The simplest assumption is that one field remains massless and the remaining $2n$ fields behave as above. That is they pair to form n massive Dirac fermions, thus breaking the spectator symmetry from SU($2n + 1$) down to Sp($2n$). One massless confined chiral field remains.

Consider now the situation where a subgroup of the spectator group is gauged with a weaker coupling. According to hypothesis (iv) the symmetry breaking condensate persists. The orientation of the condensate in spectator space is not in general arbitrary. This is because the weaker interactions may define a direction in spectator space. The remaining freedom in the condensate orientation is the subgroup of the spectator group not explicitly broken by the weaker interactions. Determining the condensate is in general a difficult strong interaction problem [5,8].

Happily this problem does not arise from certain interesting special cases. For example if the weaker gauge group is the entire spectator group the orientation is arbitrary. Another example arises in the SU($N > 2$) gauge theory if the weaker gauge group includes SU(n)$_{\mathrm{L}}$ or SU(n)$_{\mathrm{R}}$. In both cases the (assuming hypothesis (iii) the weaker gauge group is sufficient to rotate the condensate to a standard form.

In general the condensate may break the spectator symmetry down to a subgroup of the weaker gauge group. In this case the broken generators acquire a mass by eating the corresponding Nambu-Goldstone bosons.

The subgroup of the weaker gauge group which is left invariant by the condensate will behave at lower energies like an unbroken gauge theory. The non-Abelian components may become strong and confining at a lower mass scale. Thus the process may repeat itself. In the following sections we shall apply these techniques to some primitive models of fermion and W-boson mass generation.

4. A model

In this section we will describe a model which we believe satisfies the four requirements listed in sect. 2.

The model contains the following degrees of freedom.

(a) The usual SU(2) × U(1) electroweak gauge bosons W^A_μ, B_μ. A is a weak isovector index.

(b) A multiplet of Dirac fermions Q^i_α. The index α is an electroweak doublet index. The index i labels another internal space called *extended technicolor* (ETC). In the present example ETC is an SU(3) space with i indexing the 3 representation

$$Q^i_\alpha \equiv \begin{pmatrix} Q^1_\alpha \\ Q^2_\alpha \\ q_\alpha \end{pmatrix} \equiv \begin{pmatrix} U^1 \\ U^2 \\ u \end{pmatrix}, \qquad \begin{pmatrix} D^1 \\ D^2 \\ d \end{pmatrix}. \tag{4.1}$$

After spontaneous breakdown of ETC Q^1_α and Q^2_α will become techniquarks and q_α an ordinary quark doublet.

(c) ETC gauge bosons E^I_μ where I is an ETC adjoint index.

(d) Gauge bosons L_X and R_Y of two new strong SU(2) gauge groups $SU(2)_{(L)}$ and $SU(2)_{(R)}$. The generators are called left-spin and right-spin. The subscripts X, Y label L and R spin adjoint representations.

(e) Chiral Weyl fermions ℓ^i_x, r^i_y where ℓ and r are left- and right-handed. They are both ETC triplets and L or R spin doublets respectively.

The usual minimal couplings between gauge bosons and fermions are assumed. This history has no triangle anomalies in the L, R and ETC sectors. Anomalies in the electroweak sector can be eliminated in one of two ways. The charges of the quark doublets can be chosen $\pm\frac{1}{2}$ or additional multiplets can be invoked.

The coupling constants of the theory are g_1, g_2, g_E, g_L, g_R corresponding to the electroweak SU(2) × U(1), ETC, L and R gauge groups. They can be defined as their running values at some mass scale. It is convenient to choose this scale to be larger than the largest symmetry breaking scale appearing in the models ($\gtrsim$ 100 TeV). The couplings g_L, g_R will be chosen equal and large enough so that they induce strong interactions at a scale ~100 TeV *. The ETC coupling is somewhat smaller so that it becomes large at a few TeV. The entire structure is summarized in the diagram of fig. 2. The lines joining fermions to gauge groups indicate non-trivial transformation properties. The particular representation content of a fermion under a particular group is indicated by an integer labelling the bond.

The next few sections explain how we believe this complicated system works.

* 100 TeV is a buzzword for a new mass scale much bigger than 1 TeV.

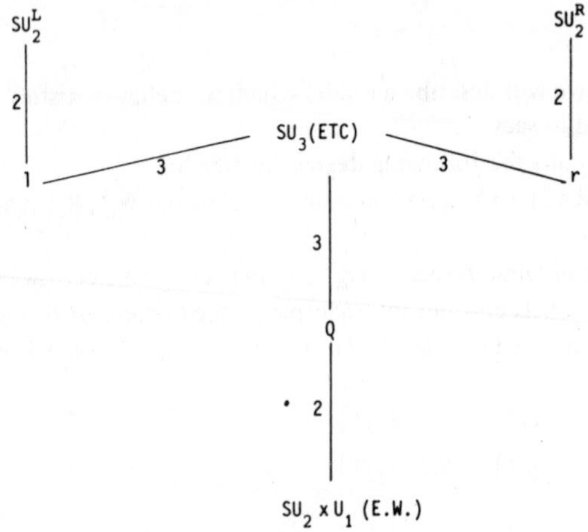

Fig. 2. A simple model of mass generation.

5. The strongest sector

In this section we will consider the isolated system of ℓ fermions and L gauge bosons. (The considerations for the right sector are completely parallel.) The isolated L sector consists of L gauge bosons and 3 L spin doublets ℓ^i_x. We assume the coupling g_L becomes large at about 100 TeV. At this scale a condensate forms. In particular

$$\langle\phi^{ij}_L\rangle = \epsilon_{xx'}\langle l^i_x(1)\, l^j_{x'}(2) - l^i_x(2)\, l^j_{x'}(1)\rangle \neq 0 \,, \tag{5.1}$$

where 1 and 2 indicate the two spin components of the Weyl spinors. The condensate ϕ^{ij}_L transform under the spectator SU(3) group as a $\overline{3}$. Without loss of generality we may assume

$$\langle\phi^{1,2}_L\rangle \neq 0 \,, \qquad \langle\phi^{2,3}_L\rangle = \langle\phi^{1,3}_L\rangle = 0 \,. \tag{5.2}$$

That is, the 1 and 2 components pair to form a massive Dirac fermion and ℓ^3 remains massless.

Here we must digress to discuss the fate of the massless field l^3. We shall assume that the strong L gauge group confines all non-singlet states. Thus the object ℓ^3 is not found in the physical space of states.

Evidently, the surviving spectator symmetry is the SU(2) subgroup which leaves l^3 invariant. Five generators of the original SU(3) are broken. This results in the appearance of 5 NGB's. Note that the apparent global U(1) symmetry

$$l^i \to e^{i\theta}\, l^i \tag{5.3}$$

is broken by triangle anomalies.

The right-hand sector R behaves in an analogous fashion. In particular if $g_R = g_L$

there exists an exact discrete symmetry L $\leftrightarrow$ R. The internal SU(3) directions of the condensates ϕ_L and ϕ_R are entirely arbitrary and will only become correlated when SU(3) is gauged. At that point we shall assume that the SU(3) dynamics aligns these directions.

6. ETC

Now consider the SU(3) gauge interactions of the ETC group. The coupling g_E is smaller than $g_{L,R}$. Therefore it makes sense to treat ETC as a perturbation in the 100 TeV region.

The invariant SU(2) subgroup of ETC plays the role of ordinary T-color. Renormalization effects will increase the strength of the $SU(2)_{TC}$ gauge coupling as the mass scale dimenishes. Thus at about 1 TeV the $SU(2)_{TC}$ becomes a strong interaction.

7. Quarks

The quark multiplets

$$Q'_\alpha = \begin{pmatrix} Q'_\alpha \\ Q^a_\alpha \\ q_\alpha \end{pmatrix} = \begin{pmatrix} U'_{L,R} \\ U^2_{L,R} \\ u_{L,R} \end{pmatrix}, \quad \begin{pmatrix} D'_{L,R} \\ D^2_{L,R} \\ d_{L,R} \end{pmatrix}, \tag{7.1}$$

interact *via* their vector $SU(3)_{ETC}$ charges. When $SU(3)_{ETC}$ breaks down to $SU(2)_{TC}$ the T-quarks $Q^{1,2}_\alpha$ become strongly interacting. The quark q, being an $SU(2)_{TC}$ singlet communicates with the TC sector very weakly. This is because the 5 broken generators of SU(3) are extremely massive ($\sim$100 TeV). Thus it is appropriate to first study the isolated TC sector consisting of the three gauge boson $E^{1,2,3}$ and $Q^{1,2}_\alpha$.

According to the discussion in sect. 3, the new spectator symmetry is SU(4). The four chiral components are

$$U = \begin{pmatrix} U_1 \\ U_2 \end{pmatrix}_L, \quad D = \begin{pmatrix} D_1 \\ D_2 \end{pmatrix}_L, \quad \sigma \uparrow \;\; \rightarrow \tau . \tag{7.2}$$

$$U^c = \begin{pmatrix} U^c_2 \\ -U^c_1 \end{pmatrix}_L \quad D = \begin{pmatrix} D^c_2 \\ -D^c_1 \end{pmatrix}$$

The generators of this SU(4) are 1, σ_i, τ_α, $\sigma_i \cdot \tau_\alpha$ where $\boldsymbol{\tau}$ changes U's to D's and $\boldsymbol{\sigma}$ changes U to U^c's and D to D^c's. When the $SU(2)_{TC}$ forces become strong at a

TeV, a condensate forms. Up to an SU(4) rotation the condensate has the form

$$\langle U^c U + D^c D \rangle \neq 0\ . \tag{7.3}$$

The condensate can be rotated in two different ways: baryon conserving and baryon violating. The subgroup of SU(4) which conserves baryon number is the SU(2) × SU(2) × U(1) group generated by $\boldsymbol{\tau}$ and, $\boldsymbol{\tau}\sigma_3$, and σ_3. The remaining baryon number violating generators are $\boldsymbol{\tau}\sigma_1$, $\boldsymbol{\tau}\sigma_2$, σ_1, σ_2.

We wish to show that the condensate (7.3) is an extremum of the energy when the electroweak interactions are turned on. Thus consider the eight-dimensional space spanned by the baryon number violating generators of SU(4). When there generators act they rotate the condensate changing the electroweak energy. This change in energy is easily shown to be independent of the direction in the eight-dimensional space of baryon violating condensates. All the remaining rotations of the condensate can be accomplished by using the electroweak SU(2) × U(1). Thus we conclude:

(a) the condensate (7.3) lies on a trajectory of extrema;

(b) the only ambiguity in the choice of the condensate is the usual SU(2) × U(1).

8. Pseudos

In this section we will discuss the various Nambu-Goldstone bosons of the spectrum before the electroweak interactions are turned on. These correspond to the five generators broken by the condensate (7.3). Three of them have the quantum numbers of the pion and two of them have diquark quantum number UD, U^cD^c. When the electroweak interactions are turned on the diquark massless bosons acquire a finite mass of order $\sqrt{\alpha}$ TeV ~ 100 GeV. Note the couplings of the broken ETC generators also give masses to the diquark bosons. This contribution to the mass is small. The three pion-like bosons are eaten by the weak vector bosons as in the standard technicolor picture, giving the usual pattern of masses to the $W^{\pm}$ and Z. The photon remains massless.

Let us consider the 10 NGB's associated with the condensates ϕ_L and ϕ_R. Five of these bosons correspond to $SU(3)_{(L+R)}$ (generated by the sums of left and right spin). These give mass to the corresponding 5 ETC gauge bosons.

The symmetries generated by the differences of L and R spin are explicitly broken. This has two effects of interest. First it gives mass to the remaining 5 NGB's. We expect this mass to be of order $g_E \times 100$ TeV. Secondly, the relative orientations of the condensates ϕ_L and ϕ_R become fixed. It is easy to see that parallel orientations in the ETC space are extrema of the energy.

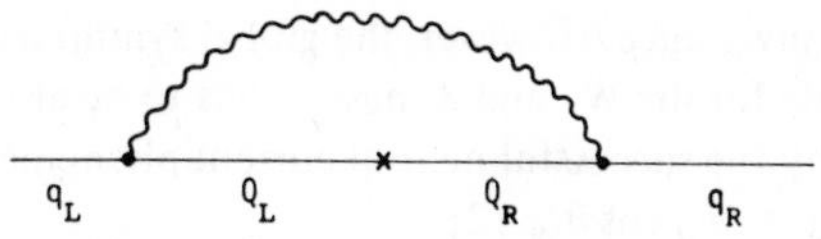

Fig. 3. The mass feed-down mechanism.

9. Quark masses

In this section we describe a mechanism by which heavy techniquarks feed down masses to the light quarks. This mass feed-down mechanism is effected by the heavy (~100 TeV) generators that connect light quarks to techniquarks. To see how this happens consider the graph of fig. 3. The cross X indicates the chirality mixing caused by the condensate. The magnitude of this mixing is characterized by the 1 TeV mass scale. When loop momenta q^2 much bigger than (1 TeV)^2 flow through the internal fermion propagator of Q this soft chirality mixing goes away and the graph of fig. 3 vanishes. Thus the 1 TeV mass scale serves as an ultraviolet cutoff which renders the graph of fig. 3 finite. Since $m_{\mathrm{E}} \sim 100$ TeV $\gg m_{\mathrm{Q}} \sim 1$ TeV, we can compute the corrections by approximating the ETC gauge boson exchange by a 4-Fermi type coupling with a loop cutoff of order 1 TeV $\sim m_{\mathrm{Q}}$. Thus we find

$$m_{\mathrm{q}} = \frac{g_{\mathrm{E}}^2}{8\pi^2}\frac{1}{M_{\mathrm{E}}^2}\int_0^{M_{\mathrm{Q}}} \frac{\mathrm{d}^4 l}{l^2 + m_{\mathrm{Q}}^2}\, m_{\mathrm{Q}}$$

$$\sim \frac{g_{\mathrm{E}}^2}{8\pi^2}\frac{m_{\mathrm{Q}}^3}{m_{\mathrm{E}}^2}\,, \tag{9.1}$$

where $g_{\mathrm{E}}^2/8\pi^2$ is the coupling of the heavy ETC bosons and is expected to be somewhat smaller than 1.

As stated in sect. 2, in general, the ratios of different strong interaction scales will be very large. For example, if $M_{\mathrm{Q}} \sim 1$ TeV and $M_{\mathrm{E}} \sim 100$ TeV we obtain $m_{\mathrm{q}} \sim 100$ MeV.

10. The problem of vertical splittings

The previous model demonstrated the possibility of generating quark masses. This model however can not explain the large vertical splittings such as u-d, c-s, b-t and e-ν, μ-ν. To be sure, the model would generate electroweak mass differences but these typically of order α. In fact this is a general problem. It is caused by the vertical SU(2) $\times$ SU(2) symmetry of the world without electroweak interactions. In order to consistently introduce electroweak interaction we only need a theory with global SU(2) × U(1) invariance. However, the global symmetry of the techniquarks (which are responsible for the $W^{\pm}$ and Z masses) has to be at least SU(2) × SU(2). This is necessitated by the successful neutral current phenomenology and in particular the relation $M_W = M_Z \cos\theta_W$ [2].

11. A model of leptons

In this section we construct a model with the following desirable features.

(i) The light fermions exhibit vertical splittings. In particular the neutrino remains massless while the electon gains a mass.

(ii) The heavy T-lepton world remains SU(2) × SU(2) symmetric thus guaranteeing a standard Z, $W^{\pm}$ mass matrix.

The ETC group for this model will be SU(5). The breaking pattern needed is SU(5) → SU(4). In subsect. 11.1 we shall show how this can be done by using the principles of sect. 3.

11.1. Breaking of ETC

Consider a strong group $SO(10)_{(L)}$ or any other group ($SO(4N+2), N > 1$) which is anomaly free and has complex representations [9]. Introduce chiral fermions ℓ_α and r_α^i which belong to a complex representation of $SO(10)_{(L)}$ (index α). ℓ_α's are $SU(5)_{ETC}$ singlets while r_α^i's are $SU(5)_{ETC}$ quintets. Since α labels a complex representation of $SO(10)_{(L)}$ the global spectator symmetry of the theory is SU(5) × U(1) and *not* SU(6). At the scale ~100 TeV when the $SO(10)_{(L)}$ forces become strong the following condensate forms:

$$\langle \bar{l}_\alpha r_\alpha^i \rangle \neq 0 . \tag{11.1}$$

This breaks the gauge $SU(5)_{ETC}$ symmetry down to $SU(4)_{TC}$. In order to avoid anomalies in the SU(5) sector we introduce a completely parallel strong $SO(10)_{(R)}$ sector with fermions r_α and ℓ_α^i and a condensate

$$\langle \bar{r}_\alpha l_\alpha^i \rangle \neq 0 . \tag{11.2}$$

When the condensates (11.1) and (11.2) align (L ↔ R symmetry) the desirable breaking pattern occurs.

Another way to obtain the same breaking pattern is to replace the strong $SO(10)_{(L)}$ by an $SU(5)_{(L)}$ with chiral fermion content: ℓ^α, r_i^α, $r^{\alpha\beta}$, $\ell_i^{\alpha\beta}$. α and $\alpha\beta$ label the 5- and 10-dimensional representations of $SU(5)_{(L)}$ and i labels the fundamental representation of $SU(5)_{ETC}$. The condensates are: $\langle \bar{l}^\alpha r_i^\alpha \rangle \neq 0$ and $\langle \bar{r}^{\alpha\beta} l_i^{\alpha\beta} \rangle \neq 0$. To avoid $SU(5)_{ETC}$ anomalies we introduce an $SU(5)_{(R)}$ sector with fermions: r^α, ℓ_i^α, $\ell^{\alpha\beta}$, $r_i^{\alpha\beta}$.

The condensates are $\langle \bar{r}^{\alpha} l_i^{\alpha} \rangle \neq 0$ and $\langle \bar{l}^{\alpha\beta} r_i^{\alpha\beta} \rangle \neq 0$. The desirable symmetry breaking occurs when all condensates align *.

The spectrum of the strong L and R sector consists of some massive ~100 TeV fermions together with massless and confined fermions and vector bosons. For the case when the strong groups L and R are SO(10) the model is illustrated in fig. 4.

11.2. *Light spectrum of the model*

The fermion content of the model includes the following $SU(5)_{ETC}$ multiplets:

$$\begin{pmatrix} N_1 \\ N_2 \\ N_3 \\ N_4 \\ V \end{pmatrix}_{L,5}, \quad \begin{pmatrix} 0 & N_{12} & N_{13} & N_{14} & N_{15} \\ & 0 & N_{23} & N_{24} & N_{25} \\ & & 0 & N_{34} & N_{35} \\ & & & 0 & N_{45} \\ & & & & 0 \end{pmatrix}_{R,10},$$

$$\begin{pmatrix} E_1 \\ E_2 \\ E_3 \\ E_4 \\ e \end{pmatrix}_{L,5}, \quad \begin{pmatrix} E_1 \\ E_2 \\ E_3 \\ E_4 \\ e \end{pmatrix}_{R,5}$$

The multiplets N_{iL}, E_{iL} and E_{iR} are 5's under SU(5) while N_{ijR} is a 10. The notations are intended to suggest neutrino and electron. The left-handed multiplets form weak isospin doublets and carry weak hypercharge −1. The right-handed multiplets are isospin singlets. The multiplet E_R has hypercharge −2 while N_{R10} is neutral.

The SU(4) content of the fermions is
4 quartets: N_i, N_{i5}, $E_{i,L}$, $E_{i,R}$, $(i = 1–4)$,
1 sextet: $N_{i,j}$, $(i, j = 1–4)$,
3 singlets: ν, e_L, e_R.

* There are very many interesting patterns of symmetry breaking that can be realized dynamically by using the principles of sect. 3. Consider, for example, a strong SU(2) sector (Greek indices) and the chiral fermion doublets ℓ_{α}^{a} transforming like the 10-dimensional representation of a weaker SO(10) gauge group. This model has no anomalies. Furthermore the condensate $\langle \epsilon_{\alpha\beta} l_{\alpha}^{a}(1)\, l_{\beta}^{b}(2) - (1 \leftrightarrow 2) \rangle \neq 0$ breaks SO(10) down to SU(5). This is what happens in the first stage of breaking in the recently proposed Georgi-Nanapoulos unifoed model [10a]. Another academic exercise for the reader is to find strong sectors for the breaking of SU(5) down to SU(1) × SU(2) × U(1) which occurs in the Georgi-Glashow model [10b]. Notice that, in contrast to our examples in the text, the breaking SO(10) → SU(5) was aone through only one strongly interacting sector.

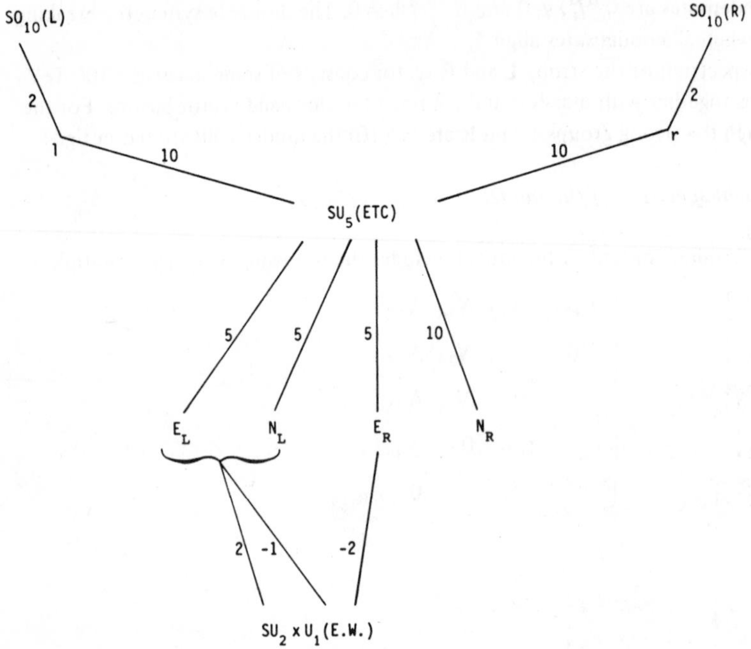

Fig. 4. A model of leptons.

We assume that the non-Abelian SU(4) sector becomes a strong interaction at a scale ~1 TeV. The corresponding spectator symmetry of the four quartets and the sextet is SU(2) × SU(2) × U(1). The U(1) symmetry multiplies the quartet fields by a a phase and the sectet fields by another phase. The SU(2) × SU(2) chiral symmetry is the one needed to ensure the correct Z, $W^{\pm}$ mass matrix. Note that this full spectator symmetry is not exact. It is violated by the heavy broken ETC couplings. The only exact symmetry is SU(2) × U(1) electroweak. The violations are very small. Their magnitude is of order $(M_Q/M_E)^2$ and they can be treated as small perturbations. By an SU(2) × U(1) electroweak rotation the condensate can always be brought in the form:

$$\langle \overline{N}_{i5,\mathrm{R}} N_{i\mathrm{L}} + \overline{E}_{i\mathrm{R}} E_{i\mathrm{L}} \rangle \neq 0 \,.$$

This condensate yields the correct vector boson mass matrix. There are no pseudo-Goldstone bosons left over.

The lepton spectrum is as follows: the quartet N's and E's acquire constitituent masses of order 1 TeV. The sextet of N_{ij}'s remains massless and confined. The neutrino ν_L remains massless because there is no SU(4) singlet right-handed N particle that it can mix with. The electron acquires a mass by the mechanism of sect. 9 *. The model in general contains a dangerous anomaly in the hypercharge current arising from the strong ETC sector. The anomalies divergence is proportional to

$$\partial_\mu Y_\mu \sim g_E^2 E_{\mu\nu} \tilde{E}_{\mu\nu} \{\sum_L Y(i)\, C_2(i) - \sum_R Y(j)\, C_2(j)\} ,$$

where L, R means sum over the left (right) SU(5) multiplets of fermions, $Y(i)$ is the hypercharge of a multiplet and C_2 is the quadratic Casimir operator for the representation. With the standard Y assignments corresponding to an electrically uncharged neutrino, the anomaly vanishes.

12. Conclusions

In this paper we have exhibited a class of models with dynamical symmetry breaking which induces both fermion and vector boson masses of reasonable magnitudes. This was accomplished without the appearance of true Nambu-Goldstone bosons or other unwanted light objects. Unlike the standard theory no unnatural adjustments need to be made.

The models, as presently constructed are unrealistic on several accounts: the "XEROX" replication of e, μ, τ ... is not accounted for, color has not been incorporated. *CP* violation is absent. However, we feel that there is sufficient richness in this approach to include all the desirable features. The xerox duplication, for example, can be added onto the ETC sector **.

We wish to thank Ken Lane for stimulating conversations. We are also indebted to J. Bjorken for helpful comments.

References

[1] K. Wilson, private communication.
[2] L. Susskind, Dynamics of spontaneous symmetry breaking in the Weinberg-Salam theory, SLAC-PUB-2142.
[3] Y. Nambu and G. Jona Lasino, Phys. Rev. 122 (1961) 345.
[4] S. Weinberg, Phys. Rev. Lett. 19 (1967) 1264;

* Had we not changed the strong L and R sectors of this model from SU_2 to SO(10) (or SU(5) the breaking pattern would have been $SU(5)_{ETC} \to Sp(4)_{TC}$. Since Sp(4) has only real representations [9] it is easy to see that neutrino would become a Majorana fermion with a mass of order 1 eV. A Majorana neutrino would mediate double beta decay with an affective Fermi coupling of order $G_F^2 m_\nu \Delta M$. Here G_F is the usual Fermi coupling, m_ν is the neutrino mass ($\sim$1 eV) and ΔM the Q-value for the process.

** We thank Ken Lane for a conversation on this point.

A. Salam, *in* Elementary particle physics, ed. N. Svartholm (Almqvist and Wiksell, Stockholm, 1968) p. 367.
[5] S. Weinberg, Phys. Rev. D13 (1976) 974.
[6] H. Georgi, H. Quinn and S. Weinberg, Phys. Rev. Lett. 33 (1974) 451.
[7] S. Adler and R. Dashen, Current algebras (Benjamin, New York, 1968);
M. Gell-Mann, R.J. Oakes and B. Renner, Phys. Rev. 175 (1968) 2195.
[8] R.F. Dashen, Phys. Rev. D3 (1971) 1879;
H. Pagels, Phys. Rev. D11 (1975) 1213
S. Weinberg, ref. [5];
F. Wilczek and A. Zee, Phys. Rev. D15 (1977) 3701.
[9] H. Georgi and S.L. Glashow, Phys. Rev. D6 (1972) 429.
[10] (a) H. Georgi and D. Nanopoulos, Harvard preprint (1978);
(b) H. Georgi and S.L. Glashow, Phys. Rev. Lett. 32 (1974) 438.

Volume 90B, number 1,2 PHYSICS LETTERS 11 February 1980

DYNAMICAL BREAKING OF WEAK INTERACTION SYMMETRIES

Estia EICHTEN[1,2] and Kenneth LANE[1]
Lyman Laboratory of Physics, Harvard University, Cambridge, MA 02138, USA

Received 28 September 1979

A realistic theory of dynamically broken weak interactions not only requires fermions with a new strong interaction at 1 TeV, but also additional gauge couplings which explicitly break quark and lepton chiral symmetries. This "sideways" gauge interaction connects light fermions to the 1 TeV fermions and, very likely, to themselves as well. It is dynamically broken near 100 TeV to a subgroup containing the 1 TeV interactions. This dynamical scheme is subject to definite low-energy tests: (1) Charged and neutral spinless pseudo-Goldstone bosons with mass expected between 10 and 40 GeV; the charged ones are readily detectible in e^+e^- annihilation. (2) Flavor-changing neutral-current interactions which mediate rare decays at rates not far below present upper limits.

This letter presents the results of our model-independent investigation of dynamical symmetry breaking (DSB) for the weak interactions. In addition to the previously discussed new gauge interaction [1,2] whose running coupling becomes O(1) near 1 TeV (we call this "hypercolor" and the corresponding gauge group $\mathcal{G}_H$), we conclude that there must be yet another gauge interaction which is dynamically broken down to hypercolor near 100 TeV. This near interaction is required for a realistic theory of weak $(SU(2) \times U(1))_W$ interactions without elementary Higgs fields. We call it the "sideways" interaction, based on a gauge group $\mathcal{G}_S$, because it gauges generations, or families, of quarks and leptons. It produces their "current-algebraic" masses by explicitly breaking their chiral symmetries. We deduce several constraints on the $\mathcal{G}_S$ representations of fermions. The most important of these is that quarks and leptons having the same helicity and weak isospin, $(T_3)_W$, must belong to the same irreducible representation of $\mathcal{G}_S$, thereby implying a quark–lepton unification at moderate energies ($\ll 10^{16}$ GeV).

Equally important, these theories have testable low-energy ($\ll 1$ TeV) consequences. They are:

(1) The existence of electrically charged "pseudo-Goldstone bosons" [3], readily detectable in e^+e^- annihilation. Their mass will be between 10 and 40 GeV. They couple to quark–antiquark and lepton–antilepton pairs much like Higgs mesons, i.e., to the masses of their pair, and have a width of order 10^{-5} to 10^{-3} times a typical hadronic width. They will appear pointlike at present energies, and each charged pair asymptotically contributes $\frac{1}{4}$ unit to

$$R = \sigma(e^+e^- \to \text{hadrons})/\sigma(e^+e^- \to \mu^+\mu^-).$$

(2) Flavor-changing neutral current interactions which may mediate such rare processes as $K_L \to \mu e$, $K^+ \to \pi^+\mu e$, $\mu \to e\gamma$, etc. Naively, these processes are expected to have an effective Fermi constant $\sim (100\ \text{TeV})^{-2} \sim 10^{-5}\, G_F$, and branching fractions not far below present upper limits.

The formalism of dynamically broken gauge theories was worked out long ago [4] and later applied to weak interactions by Weinberg [1]. Motivated in part by his analysis, in part by experience with color SU(3), we assume the following basic scenario:

(i) The fundamental lagrangian of the theory is a gauge interaction, containing fermions and gauge vector bosons as the only constituents. There are *no* elementary scalar fields, at least none whose characteristic energy scale is less than the so-called grand-unification mass $\gtrsim 10^{16}$ GeV.

(ii) The lagrangian contains *no* fermion bare mass terms and, in the neglect of weakly-coupled or strong-

[1] Research supported in part by the National Science Foundation under Grant No. PHY77-22864.
[2] Research supported in part by the Sloan Foundation.

ly-broken interactions, it is invariant under a group of chiral transformations.

(iii) Fermions whose gauge interactions become strong at scale M have their chiral symmetries dynamically "broken", provided their representation content allows it. They acquire dynamical mass of $O(M)$. More precisely, their propagator has a momentum-dependent, symmetry-breaking mass term which behaves like M^3/p^2 (when $p^2 \gg M^2$) for asymptotically-free strong interactions [5]. The corresponding Goldstone bosons have decay constant $F \sim M$. This is what happens in QCD, where the quark chiral symmetry $SU(n) \times SU(n)$ is dynamically broken to $SU(n)$; there $M \approx 300$ MeV, $F = f_\pi \approx 100$ MeV.

(iv) Spontaneously-broken chiral symmetries which are also gauge symmetries result in the dynamical Higgs mechanism [4]. The corresponding gauge boson "eats" the Goldstone boson and acquires mass $\approx gF$, where g is the gauge coupling. We emphasize that all gauge symmetry breaking (at least below $\sim 10^{16}$ GeV) is to proceed dynamically.

The implications of applying this scenario to the weak interactions are:

(a) Weak gauge boson masses will be $\mu_{W,Z} \approx eF$, where e is the electric charge. Thus, $F \approx G_F^{-1/2} \approx 300$ GeV and the fermion dynamical mass M presumably a few times F, say 1 TeV [1].

(b) The new fermions must have asymptotically-free gauge interactions becoming strong near 1 TeV [1,2]. This cannot be $\mathcal{G}_3 \equiv$ ordinary color SU(3) interactions, but a new interaction we call hypercolor, with unbroken non-abelian gauge group $\mathcal{G}_H$. Fermions transforming non-trivially under $\mathcal{G}_H$ will be called hyperfermions, while the ordinary quarks and leptons are $\mathcal{G}_H$-singlets.

(c) By analogy with QCD, we expect hyperfermions to be confined, bound into hyperhadrons whose typical mass scale is 1 TeV. Note, especially, that the physical Higgs scalar of the standard weak-interaction model will be a 1 TeV hyperhadron.

Because of its many successes, we shall assume the weak gauge group to be $\mathcal{G}_W = (SU(2) \times U(1))_W$, with quarks and leptons transforming as weak doublets and singlets in the usual way [6]. $\mathcal{G}_W$ is to be dynamically broken to $U(1)_{EM}$. It is known from experiment that [7]

$$\rho = \mu_W^2/\mu_Z^2 \cos^2\theta_W = 0.985 \pm 0.023, \qquad (1)$$

where $\sin^2\theta_W \equiv g_1^2/(g_1^2 + g_2^2) = 0.218 \pm 0.020$, and g_2 and g_1 are the $SU(2)_W$ and $U(1)_W$ coupling constants, respectively.

Susskind and Weinberg [2] have shown, independently, how to obtain $\rho = 1$ in a dynamical scheme [8] [‡1]. They introduce (at least) one pair of hyperfermions Q = (U,D) with a chiral $SU(2) \times SU(2)$ symmetry. Note that this requires vectorial $\mathcal{G}_H$ interactions for hyperfermions. $Q_L = ((1-\gamma_5)/2)\,Q$ is a weak isodoublet, with hypercharge Y_{Q_L}. The right-handed U_R and D_R are weak isosinglets with $Y_{U_R} = Y_{Q_L} + 1/2$ and $Y_{D_R} = Y_{Q_L} - 1/2$. It is assumed that, when $\mathcal{G}_H$ interactions become strong near 1 TeV, $SU(2) \times SU(2)$ is dynamically broken down to the diagonal (vectorial) SU(2) subgroup; i.e., $|\langle\Omega|\bar{U}U|\Omega\rangle| = |\langle\Omega|\bar{D}D|\Omega\rangle| \sim M^3$. This implies three pseudoscalar Goldstone bosons, Π_a, with *equal* decay constants, F_π:

$$\langle\Omega|\bar{Q}\gamma_\mu\gamma_5 \tfrac{1}{2}\tau_a Q|\Pi_b(k)\rangle = ik_\mu F_\pi \delta_{ab}, \quad a,b = 1,2,3. \qquad (2)$$

The Π_a are eaten by $W^\pm$ and Z, and a straightforward calculation yields

$$\mu_W^2/\mu_Z^2 = \tfrac{1}{4} g_2^2 F_\pi^2 / \tfrac{1}{4}(g_1^2 + g_2^2) F_\pi^2 \equiv \cos^2\theta_W. \qquad (3)$$

Note that preservation of the diagonal SU(2) symmetry is crucial to this result. Finally, if the N-dimensional $\mathcal{G}_H$ representation $\mathcal{D}^H(N)$ of hyperfermions contains r_Q irreducible representations $\mathcal{D}^H(N_r)$, $\sum_{r=1}^{r_Q} N_r = N$, then there are r_Q hyperdoublets, and the decay constants F_r satisfy

$$F_\pi \equiv \left(\sum_{r=1}^{r_Q} F_r^2\right)^{1/2} = 2^{-1/4} G_F^{-1/2} \approx 250 \text{ GeV}. \qquad (4)$$

All quark and lepton chiral symmetries must be explicitly broken in a realistic theory. This may be accomplished in a dynamical theory only by introducing a new gauge interaction (group $\mathcal{G}_S$). The mass-generating effective lagrangian thus produced must be $\mathcal{G}_W \times \mathcal{G}_3 \times \mathcal{G}_H$ invariant. With quarks, leptons, and hyperfermions transforming under $\mathcal{G}_W$ as specified above,

‡1 To our knowledge, this was the first paper to point out that the value of ρ depends only on the $\mathcal{G}_W$-representation content of the unphysical Goldstone bosons, not on whether they are elementary or dynamical.

Volume 90B, number 1,2 PHYSICS LETTERS 11 Feburary 1980

the most general form is [‡2]

$$\mathcal{L}^{\mathrm{m}}_{\mathrm{eff}} = -\frac{1}{F_{\mathrm{S}}^2}\sum_{i,j=1}^{n}\sum_{r=1}^{r_{\mathrm{Q}}}\sum_{k=1}^{N_r}\{[\bar{\mathrm{u}}_{i\mathrm{L}}\gamma_\mu \mathrm{U}^{(r)}_{k\mathrm{L}} + \bar{\mathrm{d}}_{i\mathrm{L}}\gamma_\mu \mathrm{D}^{(r)}_{k\mathrm{L}}]$$

$$\times [\Gamma^{\mathrm{u}}_{ij;r}\bar{\mathrm{U}}^{(r)}_{k\mathrm{R}}\gamma^\mu \mathrm{u}_{j\mathrm{R}} + \Gamma^{\mathrm{d}}_{ij;r}\bar{\mathrm{D}}^{(r)}_{k\mathrm{R}}\gamma^\mu \mathrm{d}_{j\mathrm{R}}]$$

$$+ [\bar{\nu}_{i\mathrm{L}}\gamma_\mu \mathrm{U}^{(r)}_{k\mathrm{L}} + \bar{\mathrm{e}}_{i\mathrm{L}}\gamma_\mu \mathrm{D}^{(r)}_{k\mathrm{L}}][\Gamma^{\mathrm{e}}_{ij;r}\bar{\mathrm{D}}^{(r)}_{k\mathrm{R}}\gamma^\mu \mathrm{e}_{j\mathrm{R}}] + \mathrm{h.c.}\} . \quad (5)$$

Here, we have assumed n generations of quarks and leptons, with weak isodoublets $\mathrm{q}_{i\mathrm{L}} = (\mathrm{u}_{i\mathrm{L}}, \mathrm{d}_{i\mathrm{L}})$ and $\ell_{i\mathrm{L}} = (\nu_{i\mathrm{L}}, \mathrm{e}_{i\mathrm{L}})$, and isosinglets $\mathrm{u}_{j\mathrm{R}}, \mathrm{d}_{j\mathrm{R}}$, and $\mathrm{e}_{j\mathrm{R}}$. $\mathcal{E}_3$ indices have been suppressed in eq. (5) to allow for the possibility that hyperfermions need not be $\mathcal{E}_3$ singlets so long as $\mathcal{L}^{\mathrm{m}}_{\mathrm{eff}}$ is.

Eq. (5) is the low-energy ($M \sim 1$ TeV $\ll F_{\mathrm{S}}$) approximation to a $\mathcal{G}_{\mathrm{S}}$ interaction which links quarks and leptons to hyperfermions. Intermediate $\mathcal{E}_{\mathrm{S}}$ bosons have mass $\approx g_{\mathrm{S}}F_{\mathrm{S}}$. The matrices $\Gamma^{\mathrm{u,d,e}}$ are determined by $\mathcal{G}_{\mathrm{S}}$ couplings of light fermions to hyperfermions and by the mass matrix of the corresponding $\mathcal{G}_{\mathrm{S}}$ bosons; we expect $|\Gamma^{\mathrm{u,d,e}}_{ij;r}| = \mathrm{O}(1)$. Mass generation by $\mathcal{L}^{\mathrm{m}}_{\mathrm{eff}}$ is equivalent to the self-energy graph in fig. 1. Thus, current-algebraic masses have the same momentum dependence (up to logs) as the dynamical hyperfermions mass; this is a "hard" mass until $p > M \sim 1$ TeV. To estimate F_{S}, we approximate the self-energy graph by

$$(m^{\mathrm{u}}_{\mathrm{LR}})_{ij} \approx \sum_{r,k}\Gamma^{\mathrm{u}}_{ij;r}\langle\Omega|\bar{\mathrm{U}}^{(r)}_{k\mathrm{R}}\mathrm{U}^{(r)}_{k\mathrm{L}}|\Omega\rangle/F_{\mathrm{S}}^2$$

$$\approx (1\ \mathrm{TeV})^3/F_{\mathrm{S}}^2, \quad (6)$$

with similar expressions for m^{d} and m^{e}. Assuming typical elements of $m^{\mathrm{u,d,e}}$ to be O(1 GeV), we obtain [‡3]

$$F_{\mathrm{S}} \approx 30\ \mathrm{TeV}. \quad (7)$$

‡2 If we assumed instead that $(\mathrm{U_R}, \mathrm{D_R})$ is a weak isodoublet while $\mathrm{U_L}$ and $\mathrm{D_L}$ are isosinglets, $\mathcal{L}^{\mathrm{m}}_{\mathrm{eff}}$ would have an apparently different form than in eq. (5). However, the exhibited form can always be recovered by redefinition of hyperfermion fields and Fierz rearrangements. Thus, we cannot unambiguously deduce the structure of $\mathcal{G}_{\mathrm{S}}$ currents from the form of $\mathcal{L}^{\mathrm{m}}_{\mathrm{eff}}$ alone.

‡3 The large range of quark and lepton masses, from O(1 MeV) to O(10 GeV), may arise from diagonalizing large mass matrices. It is also possible that the smallest u, d, and e masses arise sequentially, i.e., from higher-order $\mathcal{G}_{\mathrm{S}}$ interactions. In either case, the "typical" 1 GeV mass is a reasonable choice for estimating F_{S}.

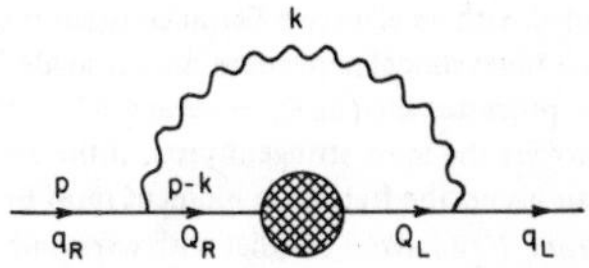

Fig. 1. Typical self-energy graph for a light quark q. Straight lines are fermions, the wavy line is a $\mathcal{G}_{\mathrm{S}}$ boson, and the blob represents a hyperfermion dynamical mass insertion. $\mathcal{G}_3$ and $\mathcal{G}_{\mathrm{H}}$ gluon corrections are not shown.

The form of $\mathcal{L}^{\mathrm{m}}_{\mathrm{eff}}$ in eq. (5) suggests, but does not imply, that $[\mathcal{G}_{\mathrm{S}}, \mathrm{SU(2)_W}] = 0$ (see footnote 2). Since we have not had to specify the charge or color of hyperfermions, $\mathcal{G}_{\mathrm{S}}$ need not commute with $\mathrm{U(1)_W}$ or $\mathcal{G}_3$. In fact, we shall show that it cannot commute with either. Obviously, $[\mathcal{G}_{\mathrm{S}}, \mathcal{G}_{\mathrm{H}}] \neq 0$. At an energy scale Λ_{S}, $\mathcal{G}_{\mathrm{S}}$ is dynamically broken down to a subgroup containing $\mathcal{G}_{\mathrm{H}}$. This DSB is analogous to that responsible for $\mathcal{G}_{\mathrm{W}} \to \mathrm{U(1)_{EM}}$, and requires yet another strong interaction whose characteristic scale is $\Lambda_{\mathrm{S}} \sim$ few $F_{\mathrm{S}} \sim 100$ TeV[‡4]! Fermions which get dynamical mass of $\mathrm{O}(\Lambda_{\mathrm{S}})$ cannot contribute to $\mathcal{G}_{\mathrm{W}}$ breakdown. If they have weak interactions at all, they must be parity-conserving. Fortunately, to discuss physics below 1 TeV, we do not need to know details of the 100 TeV world and the DSB of $\mathcal{G}_{\mathrm{S}}$ to $\mathcal{G}_{\mathrm{H}}$ [9] [‡5].

Whether $\mathcal{G}_{\mathrm{S}}$ currents contributing to $\mathcal{L}^{\mathrm{m}}_{\mathrm{eff}}$ are just as shown in eq. (5) or can be read off only after Fierz rearrangement, their algebra almost certainly contains currents linking different generations of light fermions. Hence the name "sideways" interactions: $\mathcal{G}_{\mathrm{S}}$ gauges the generations, i.e., quantum numbers commuting with the "vertical" $\mathrm{SU(2)_W}$. Quantum numbers of the unbroken, non-abelian subgroup $\mathcal{G}_{\mathrm{H}}$ are confined, while broken $\mathcal{G}_{\mathrm{S}}$ quantum numbers correspond to the distinguishable fermion generations.

Thus, flavor-changing neutral current interactions

‡4 We do not mean to exclude the possibility that $\mathcal{G}_{\mathrm{S}}$ contains this strong interaction, i.e., that $\mathcal{G}_{\mathrm{S}}$ "self-destructs" at 100 TeV. Whether or not this can happen depends on the representation content of fermions transforming under $\mathcal{G}_{\mathrm{S}}$ as well as on the size of the $\mathcal{G}_{\mathrm{S}}$ gauge coupling(s) at 100 TeV.

‡5 These authors, as well as S. Weinberg (unpublished), have independently realized that a strongly broken "sideways" interaction is needed to break chiral symmetries of light fermions.

are expected with an effective Fermi constant $\sim\frac{1}{4}F_S^{-2}$ $\sim 10^{-5}G_F$ times model-dependent mixing angle factors. Rare processes such as $K_L \to \mu e$ and $K^+ \to \pi^+\mu e$ should provide the most stringent tests. If the decay $\mu \to e\gamma$ can occur, the transition moment must be $\lesssim m_\mu^2 M^2/2m_\mu F_S^4$ to avoid conflict with experiment. The moment will be at least this small if all hyperfermions to which μ and e jointly couple have the same electric charge [10]. If $\mathcal{G}_S$ interactions exist which mediate $|\Delta S| = 2$ processes, the K_L-K_S mass difference puts a strong limit on the effective Fermi constant, $G_{|\Delta S|=2}$. A crude estimate gives

$$(G_{|\Delta S|=2})^{-1} \gtrsim f_K^2 m_{K^0}/2\Delta m_{K^0} \approx (10^3\,\text{TeV})^2, \qquad (8)$$

where $f_K \approx f_\pi \approx 100$ MeV.

There also are strong constraints on the fermion representation in $\mathcal{G}_S$ and its subgroup $\mathcal{G}_H$. For example, if $[\mathcal{G}_S, SU(2)_W] = 0$, the $\mathcal{G}_S$ representation $\mathcal{D}_L^S$ containing all weak isodoublet fermions with $(T_3)_W = +1/2$ must be identical to, but distinct from, the one containing all those with $(T_3)_W = -1/2$. If $[\mathcal{D}_S, SU(2)_W] \neq 0$, fermions with $(T_3)_W = \pm 1/2$ will belong to the same (possibly reducible) $\mathcal{G}_S$ representation, which we again denote by $\mathcal{D}_L^S$. In either case, U_L, D_L, U_R, and D_R are in identical representations, $\mathcal{D}^H$, of the subgroup $\mathcal{G}_H$, and $\mathcal{D}^H \subset \mathcal{D}_L^S$ (at least once). Denote the possibly reducible $\mathcal{G}_S$ representation containing all (right-handed) weak isosinglet fermions by $\mathcal{D}_R^S$. If the u_{iR} and hyperfermions to which they couple form a distinct representation from the d_{iR} and their associated hyperfermions, we shall label these $\mathcal{D}_{u_R}^S$ and $\mathcal{D}_{d_R}^S$, respectively. Whether $[\mathcal{G}_S, SU(2)_W] = 0$ or not, $\mathcal{D}_{u_R}^S$ and $\mathcal{D}_{d_R}^S$ cannot be identical; otherwise, $\Gamma^u = \Gamma^d$, u and d quarks are degenerate in pairs, and all mixing angles vanish. Obviously, $\mathcal{D}_L^S$ and $\mathcal{D}_R^S$ must be such that the weak hypercharge current has no $\mathcal{G}_S$-anomalous divergence.

The most important constraint on representations regards their reducibility. With at least three generations of light fermions, the $\mathcal{G}_S$ representations are large and it seems extremely likely that $\mathcal{D}^H$ will be a reducible representation of $\mathcal{G}_H$. If $\mathcal{D}^H$ decomposes into r_Q irreducible representations (IR's), the chiral symmetry group of hyperfermions which is respected by $\mathcal{G}_H \times \mathcal{G}_3$ interactions contains $\Pi_{r=1}^{r_Q}$ $[SU(2) \times SU(2)]_r$, where the left-handed part of each $(SU(2) \times SU(2)]_r$ is generated by currents which couple to $\mathcal{G}_W$ bosons. After DSB, there will be r_Q triplets of Goldstone "hyperpions", Π_a^r, with decay constant $F_r \approx F_\pi r_Q^{-1/2}$ ($F_\pi = 250$ GeV). The combination $\Sigma_{r=1}^{r_Q} F_r \Pi_a^r$ is eaten by $W^\pm$ and Z ‡6. Orthogonal triplet combinations, corresponding to "extra" chiral symmetries, are physical pseudo-Goldstone bosons, acquiring mass from $\mathcal{G}_W$ and broken $\mathcal{G}_S$ interactions. We shall collectively denote the pseudo-Goldstone bosons by P's, the charged ones by $P^\pm$, and the neutral ones by P^0. We shall show that, in *any* realistic theory, their mass will lie between 10 and 40 GeV. This estimate, as well as the very existence of P's, does not depend on whether $[\mathcal{G}_S, SU(2)_W] = 0$ or not.

Suppose, first, that some combinations of the extra chiral generators are conserved by $\mathcal{G}_S$ interactions; this will require extending the symmetry to quarks and/or leptons that belong to the same irreducible representation of $\mathcal{G}_S$ as the hyperfermions involved. This situation occurs if $[\mathcal{G}_S, U(1)_W] = [\mathcal{G}_S, \mathcal{G}_3] = 0$, so that quarks and leptons belong to distinct IR's of $\mathcal{G}_S$. More generally, it happens whenever $\mathcal{D}_L^S$ *and* $\mathcal{D}_R^S$ are reducible. Because $\mathcal{G}_S$ treats u_{iR} and d_{iR} differently, the extra chiral symmetry respected by $\mathcal{G}_S$ is likely to be products of only $SU(2)_L \times U(1)_R$ or $SU(2)_L$. This is unimportant for calculating the mass of P's, because we shall interpolate them with appropriate combinations of the left-handed weak isospin current $(j_{L\mu}^a)_t$ for the tth IR in $\mathcal{D}_L^S$; here $a = 1,2,3$. The mass arises solely from $\mathcal{G}_W$ interactions, and may be calculated by standard PCAC techniques. For $P^\pm$, we obtain $m_\pm \approx \sqrt{\alpha}\mu_Z \approx 5$ GeV. $m_\pm$ vanishes with μ_Z because momenta up to ~ 1 TeV are important in the PCAC integral and, for such large momenta, there is a partial cancellation between photon and Z contributions which becomes complete when $\mathcal{G}_W$ symmetry is restored.

Neutral P^0's for which $\mathcal{G}_S$-singlet currents can be constructed do *not* acquire mass from $\mathcal{G}_W$ interactions alone. This follows from a generalization of Dashen's theorem [11]. The $\mathcal{G}_W$ contribution to the mass of such a P^0 is given by

‡6 Strictly speaking, the unphysical Goldstone bosons, which are the *only* ones that couple to the weak currents

$$j_{L\mu}^a = \Sigma_1^n(\bar{q}_{iL}\gamma_\mu \tfrac{1}{2}\tau_a q_{iL} + \bar{\ell}_{iL}\gamma_\mu \tfrac{1}{2}\tau_a \ell_{iL}) + \Sigma_1^{r_Q}\Sigma_1^{N_r}\bar{Q}_{kL}\gamma_\mu \tfrac{1}{2}\tau_a Q_{kL},$$

are

$$(\sqrt{n}f_\pi\pi_a + \Sigma_{r=1}^{r_0} F_r\Pi_a^r)(n f_\pi^2 + F_\pi^2)^{-1/2},$$

where π_a is the obvious combination of Goldstone bosons in the $\bar{q}q$ sector.

Volume 90B, number 1,2 PHYSICS LETTERS 11 February 1980

$$F_0^2 m_0^2 = \langle\Omega|[\mathcal{F}_0, [\mathcal{H}_0, \mathcal{H}_W]]|\Omega\rangle. \quad (9)$$

Here, $F_0 \sim F_r$ is the decay constant, $\mathcal{H}_W$ is the weak hamiltonian, and $\mathcal{F}_0$ is the chiral charge – a linear combination of $\int d^3x\,(j^3_{L0})_t$. Since charged weak currents are all left-handed, $\mathcal{F}_0$ can be replaced by the corresponding combination of electric charges, $\int d^3x\,(j_0^{EM})_t$. In general, these annihilate the vacuum giving $m_0 = 0$ ‡7. Finally, since the P^0 current involves light as well as heavy fermions, the chiral Ward identity tells us that P^0 couples with strength $(m_1+m_2)/F_0$ to a light fermion–antifermion pair whose current-algebraic masses are m_1 and m_2. In all respects, these mesons are like the axion and must be avoided [12] ‡8.

There are two immediate corollaries: (1) Quarks and leptons must not be in entirely separate representations of $\mathcal{G}_S$. Thus, there is some sort of quark–lepton unification at 100 TeV, although this certainly cannot lead to proton decay. (2) Grand unified gauge theories incorporating DSB cannot contain more than one IR of the full group. Otherwise in either case, $\mathcal{D}_L^S$ and $\mathcal{D}_R^S$ are reducible and axion-like mesons result ‡9.

The remaining possibility, now a necessity, is that the extra chiral symmetries are explicitly broken. This requires (strongly broken) $\mathcal{G}_S$ interactions linking the IR's in $\mathcal{D}^H$ to one another. A necessary condition for this is that $\mathcal{D}_L^S$ be irreducible and, if $\mathcal{D}_R^S$ is reducible, that it contains at most two IR's, $\mathcal{D}^S_{u_R}$ and $\mathcal{D}^S_{d_R}$, with weak isosinglet leptons belonging to one or both of these. The symmetry-breaking effective lagrangian will have the same form as eq. (5), involving hyperfermions only, and will be invariant under $\mathcal{G}_W \times \mathcal{G}_3 \times \mathcal{G}_H$. If the effective Fermi constant is still F_S^{-2}, the $\mathcal{G}_S$ contribution to charged *and* neutral P masses is

$$F_r^2(m_\pm^2)_S \approx F_r^2(m_0^2)_S \approx \Gamma^Q F_S^{-2}\langle\bar{Q}Q\bar{Q}Q\rangle. \quad (10)$$

Using $\Gamma^Q = O(1)$, $F_r \approx F_\pi r_Q^{-1/2}$, and $\langle\bar{Q}Q\rangle/F_r^3 \approx \langle\bar{q}q\rangle/f_\pi^3 \approx 15$ [15], we obtain

$$(m_\pm)_S \approx (m_0)_S \approx 15 F_\pi^2/r_Q F_S \approx 30\, r_Q^{-1}\,\mathrm{GeV}. \quad (11)$$

The largest uncertainties in this formula are the value of r_Q ($\geqslant 2$) and the magnitude of Γ^Q matrix elements. The larger r_Q is, the greater the range of P masses we expect. Including the $\mathcal{G}_W$ contribution to $m_\pm$, about 5 GeV, the lightest $P^\pm$ and P^0 should have a mass in the range 10 to 40 GeV.

The coupling of $P^\pm$ and P^0 to a light fermion–antifermion pair will be $\approx (m_1+m_2)/F_r$ times Cabibbo-like mixing angle factors. Thus, P^0's will not be easy to produce or detect. One promising way to search for them is $\Upsilon(b\bar{b}) \to \gamma + P^0$ and $\xi(t\bar{t}) \to \gamma + P^0$. The $P^\pm$ can be pair-produced in e^+e^- annihilation. Because they appear pointlike up to ~1 TeV, each pair contributes $R_\pm = \frac{1}{4}(1-4m_\pm^2/S)^{3/2}$ to R. For $m_\pm \approx 10$ GeV, the principal decay modes of P^+ are expected to be to $c\bar{b}$ or $c\bar{s}$ and to $\tau^+\nu_\tau$ or $\mu^+\nu_\mu$, depending on mixing angles. The obvious signal is a hadron jet with roughly the beam energy, a muon (or electron, if $\tau^+\nu_\tau$) carrying 1/6 to 1/2 the beam energy, and a large amount of missing neutral energy.

Charged and neutral P-mesons with mass between 10 and 40 GeV are a necessary feature of any realistic dynamical theory of $\mathcal{G}_W$-symmetry breaking. To our knowledge, no model with elementary Higgs bosons requires charged scalars in this mass range. Thus, the case for dynamically broken weak symmetries will be particularly compelling if they are found as expected.

We are especially indebted to Steven Weinberg for discussions and encouragement and to Sidney Coleman for continually prodding us to worry about Goldstone bosons. We also wish to express our appreciation for useful conversations and assistance to V. Baluni, A. Bohr, K. Cox, S. Dimopoulous, E. Farhi, H. Georgi, S. Glashow, H. Harari, M. Peskin, L. Susskind and M. Weinstein. Finally, we gratefully acknowledge the support of Lawrence Berkeley Laboratory (for E.E.) and of NORDITA and Stanford University (for K.L.) during the final stages of this work.

‡7 There is one loophole through which P^0 may escape masslessness: individual electric currents $(j_\mu^{EM})_t$ may have $\mathcal{G}_S$-anomalous divergences; only the total electric currents must be anomaly free. Then, P^0 is expected to get a mass $F_0^2 m_0^2 \propto \exp(-8\pi^2/g_S^2)\, m^4 M^4/F_S^4$, or $m_0 \lesssim 10^{-2}$ MeV, for $m \approx 1$ GeV, $M \sim 1$ TeV. See 't Hooft [16].

‡8 A comprehensive review of experimental limits on axions is given by Donnelly et al. [13].

‡9 For examples with reducible representations see ref. [14].

References

[1] S. Weinberg, Phys. Rev. D13 (1976) 974.
[2] L. Susskind, Dynamics of spontaneous symmetry breaking in the Weinberg–Salam Theory, Phys. Rev. D, to be published;

S. Weinberg, Phys. Rev. D19 (1978) 1277.
[3] S. Weinberg, Phys. Rev. Lett. 29 (1972) 1698; Phys. Rev. D7 (1973) 2887.
[4] R. Jackiw and K. Johnson, Phys. Rev. D8 (1973) 2386; J.M. Cornwall and R.E. Norton, Phys. Rev. D8 (1973) 3338;
E. Eichten and F. Feinberg, Phys. Rev. D10 (1974) 3254.
[5] K.D. Lane, Phys. Rev. D10 (1974) 2605.
[6] S. Weinberg, Phys. Rev. Lett. 19 (1967) 1264;
A. Salam, in: Elementary particle physics, ed. N. Svartholm (Almquist and Wiksells, Stockholm, 1968) p. 367.
[7] See P. Langacker et al., Univ. of Pennsylvania preprint COO-3071-243 (1979), and references contained therein.
[8] M. Weinstein, Phys. Rev. D8 (1973) 2511.
[9] S. Dimopoulous and L. Susskind, Nucl. Phys. B155 (1979) 237.
[10] J. Bjorken, K.D. Lane and S. Weinberg, Phys. Rev. D16 (1977) 1474.
[11] R. Dashen, Phys. Rev. 183 (1969) 1245.
[12] S. Weinberg, Phys. Rev. Lett. 40 (1978) 223;
F. Wilczek, Phys. Rev. Lett. 40 (1978) 279.
[13] T.W. Donnelly et al., Phys. Rev. D18 (1978) 1607.
[14] E. Farhi and L. Susskind, A Technicolored G.U.T., SLAC preprint (1979);
V. Baluni, Quark mass spectrum and *CP* nonconservation, IAS preprint in preparation.
[15] S. Weinberg, The problem of mass, in: Trans. N.Y. Acad. Sci., Series II, 38 (1977) 185.
[16] G. 't Hooft, Phys. Rev. D14 (1976) 3432.

V. HYPERCOLOR AND GUTS

PHYSICAL REVIEW D VOLUME 20, NUMBER 12 15 DECEMBER 1979

Grand unified theory with heavy color

E. Farhi
Stanford Linear Accelerator Center, Stanford University, Stanford, California 94305

L. Susskind
Department of Physics, Stanford University, Stanford, California 94305
(Received 6 August 1979)

An example of the unification of electroweak, color, and heavy-color forces in the unifying group SU(7) is presented. This simple toy model predicts a nontrivial mass spectrum for two families of quarks and leptons. The usual Higgs scalar sector is replaced by the strong interaction heavy-color sector at ~1 TeV.

I. INTRODUCTION

In this paper we demonstrate the possibility of combining the ideas of the Georgi-Glashow (GG) grand unified $SU(5)_{GG}$ model[1] with the heavy-color mechanism[2-5] which has been suggested as a replacement for the fundamental Higgs fields of the Weinberg-Salam model. There are a number of features of the $SU(5)_{GG}$ group which make it a particularly plausible candidate for a unifying group. The organization of particle multiplets is very elegant and naturally explains the observed quantum numbers of the quarks and leptons. The quantization of electric charge is a straightforward consequence. Furthermore the unrenormalized Weinberg angle satisfies $\sin^2\theta = \frac{3}{8}$. This is modified by renormalization[6] so that the real angle satisfies $\sin^2\theta \approx 0.2$. Another nontrivial consequence is that the lifetime of the proton is greater than the current lower bound.[6]

On the other hand, the theory also has bad features. Among them is the existence of two vastly different scales of Higgs expectation values.[7] The parameters of the Higgs sector must be tuned to ridiculous precision to maintain this "gauge hierarchy." Furthermore, the number of parameters involving the lower-mass Higgs sector is excessive. In addition to the Higgs self-coupling there are a large number of Yukawa couplings. No natural explanation for their extremely small magnitude has been given. Finally the simplest Higgs assignments lead to the unacceptable bare-mass relation

$$m_e = m_d .$$

Apparently the successes of the theory involve those features which are independent of the light Higgs sector. The failures suggest that the 100-GeV symmetry breaking is being incorrectly treated. It therefore seems reasonable to explore alternatives to the usual Higgs sector. In this paper a toy model is used to illustrate how the lowest-mass Higgs sector of $SU(5)_{GG}$ can be replaced by heavy color in a simple and elegant unification.

II. HEAVY COLOR

In this section we will briefly review the salient features of Refs. 2 and 3, in which heavy color (HC) was introduced. The reader is urged to consult the original papers for more complete explanations.

(1) HC is an unbroken gauge group with a running coupling which becomes strong at a scale ~1 TeV. It is essentially a scaled-up version of quantum chromodynamics (QCD) involving heavy-color fermions U and D which parallel the ordinary u, d quarks. These heavy-color quarks may or may not have color (C) but definitely have conventional electroweak (EW) interactions.

(2) The strong interactions at 1 TeV cause a spontaneous breaking of the flavor-chiral $SU(2)^{\text{left}} \times SU(2)^{\text{right}}$ of heavy-color quarks.[2] The result is massless Goldstone heavy-color pions. These objects replace the Higgs fields and induce a mass for the Z and $W^{\pm}$.

(3) Global isospin conservation of the heavy-color-quark sector is sufficient to guarantee the empirical relation

$$\frac{M_W}{M_Z} = \cos\theta_W . \qquad (2.1)$$

(4) Heavy color should be part of a bigger group[3] with ordinary quarks and heavy-color quarks in the same multiplet. This is to allow ordinary quarks to gain mass through radiative corrections as shown in Fig. 1. The gauge bosons b which mediate transitions from ordinary quarks q to heavy-color quarks Q have mass[3] ~10–100 TeV.

One serious deficiency of Refs. 2 and 3 was that no consistent example was offered in which both leptons and quarks gain mass. Two obvious

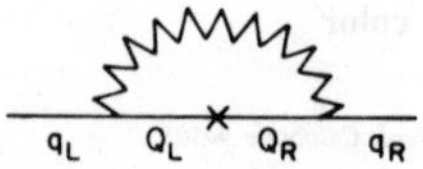

FIG. 1. Light quarks q gain mass by coupling to heavy-color quarks Q through emission of b bosons.

possibilities come to mind. In the first, additional heavy-color particles called heavy-color leptons are introduced. These feed mass to the ordinary leptons as shown in the graph of Fig. 2.

Unfortunately this type of model has too much symmetry. In particular the weak hypercharges of leptons and of quarks would be separately conserved[8] leading to two Goldstone bosons. Only one of these could be absorbed by the Higgs effect, leaving a massless "axion." This is empirically unacceptable.

A second possibility would be to allow both leptons and quarks to couple to heavy-color quarks so that leptons would gain mass from the graph shown in Fig. 3. This would allow lepton-quark transitions as in Fig. 4, thus risking baryon violation by 100-TeV bosons.

One other potential danger implicit in Refs. 2 and 3 is the existence of stable heavy-color-quark bound states analogous to protons in QCD. While these are very heavy (1 TeV) they could lead to unpleasant astrophysical or cosmological consequences.

In this paper we shall see that all of the above difficulties are surmounted in our toy model, while the good features of $SU(5)_{GG}$ are preserved.

III. THE MODEL

For the purpose of simplicity we will choose to build our toy model out of the simplest possible parts. In particular we shall choose the HC group to be the smallest possible non-Abelian group—$SU(2)_{HC}$. We shall ultimately pay a price for the smallness of the HC group.

The minimal extension of Georgi-Glashow SU(5) to include HC is SU(7). This is our choice of unifying group. The components of a fundamental seven-dimensional representation are labeled

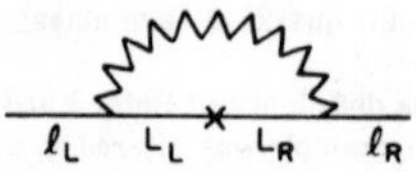

FIG. 2. Leptons l gain mass by coupling to heavy-color leptons L.

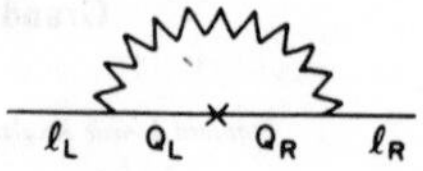

FIG. 3. Leptons l gain mass by coupling to heavy-color quarks.

$$SU(5)_S \left\{ \begin{pmatrix} \left.\begin{matrix} \psi_{T_1} \\ \psi_{T_2} \end{matrix}\right\} \text{HC} \\ \left.\begin{matrix} \psi_{d_R} \\ \psi_{d_Y} \\ \psi_{d_B} \end{matrix}\right\} \text{C} \\ \left.\begin{matrix} \psi_{\bar{e}} \\ \psi_{\bar{\nu}} \end{matrix}\right\} \text{EW} \end{pmatrix} \right. \quad \left.\begin{matrix} \\ \\ \\ \\ \\ \end{matrix}\right\} SU(5)_{GG} \tag{3.1}$$

(HC = heavy color, C = color, EW = electroweak, GG = Georgi-Glashow, S = strong, R, Y, B = red, yellow, blue).

Following arguments of Georgi,[9] we choose an anomaly-free set of representations formed from antisymmetric products of seven-dimensional representations. *All fermions are two-component (Weyl) left-handed fermions.* Our particular choice is $[2]+[4]+[6]$, where m means the antisymmetric products of m seven-dimensional representations. For example, the $[2]$ has representation vectors $\psi_{ij}=-\psi_{ji}$. The choice $[2]+[4]+[6]$ is not *ad hoc* but follows from the requirement of no anomalies.

Under the breakdown $SU(2)_{HC}\times SU(5)_{GG}$ these representations transform as follows:

$$\begin{aligned} &[2]=(1,10)+(2,5)+(1,1), \\ &[4]=(1,10)+(2,\overline{10})+(1,\overline{5}), \\ &[6]=(1,\overline{5})+(2,1). \end{aligned} \tag{3.2}$$

For example, the $(2,\overline{10})$ in the $[4]$ consists of tensors with one HC index (called p) and three $SU(5)_{GG}$ indices

$$(2,\overline{10})=\psi_{p\alpha\beta\gamma}. \tag{3.3}$$

According to Georgi,[9] the number of observable families [a family means a $(\overline{5}+10)$ of $SU(5)_{GG}$] is given by the total number of $\overline{5}$'s minus the total number of 5's. For the example under consideration, this would mean zero. The point according

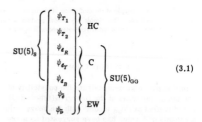

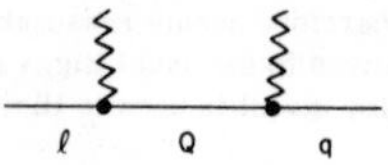

FIG. 4. Lepton goes to quark plus b mesons.

to Georgi is that a left-handed $\underline{\bar{5}}$ and $\underline{5}$ can be paired into a single Dirac fermion with a large mass term which does not violate $SU(5)_{GG}$. Such particles could well have superheavy masses. The same is true for the 10's and $\overline{10}$'s.

However, in our case all 5's and $\overline{10}$'s belong to HC doublets. This prevents them from combining with the heavy-color-free $\bar{5}$'s and 10's. Thus we expect two families of ordinary particles and a doublet of heavy-color families formed from $(2,\overline{10})$ and $(2,5)$.

We now display the particle identification in terms of $SU(5)_{GG}$ representations. A subscript p means that each entry is a HC doublet. A capital letter indicates a heavy-colored particle.

The [2]:

$$(1,10)=\begin{pmatrix} 0 & \bar{c}_B & -\bar{c}_Y & u_R & d_R \\ -\bar{c}_B & 0 & \bar{c}_R & u_Y & d_Y \\ c_Y & -\bar{c}_R & 0 & u_B & d_B \\ -u_R & -u_Y & -u_B & 0 & \bar{e} \\ -d_R & -d_Y & -d_B & -\bar{e} & 0 \end{pmatrix}, \quad (3.4a)$$

$$(2,5)=\begin{pmatrix} D_R \\ D_Y \\ D_B \\ \bar{E} \\ \bar{N} \end{pmatrix}_p, \quad (3.4b)$$

$$(1,1)=(\bar{\nu}_\mu). \quad (3.4c)$$

The [4]:

$$(1,10)=\begin{pmatrix} 0 & \bar{u}_B & -\bar{u}_Y & c_R & s_R \\ -\bar{u}_B & 0 & \bar{u}_R & c_Y & s_Y \\ \bar{u}_Y & -\bar{u}_R & 0 & c_B & s_B \\ -c_R & -c_Y & -c_B & 0 & \bar{\mu} \\ -s_R & -s_Y & -s_B & -\bar{\mu} & 0 \end{pmatrix}, \quad (3.5a)$$

$$(2,\overline{10})=\begin{pmatrix} 0 & U_B & -U_Y & \bar{U}_R & \bar{D}_R \\ -U_B & 0 & U_R & \bar{U}_Y & \bar{D}_Y \\ U_Y & -U_R & 0 & \bar{U}_B & \bar{D}_B \\ -\bar{U}_R & -\bar{U}_Y & -\bar{U}_B & 0 & E \\ -\bar{D}_R & -\bar{D}_Y & -\bar{D}_B & -E & 0 \end{pmatrix}_p, \quad (3.5b)$$

$$(1,\bar{5})=\begin{pmatrix} \bar{s}_R \\ \bar{s}_Y \\ \bar{s}_B \\ \mu \\ \nu_\mu \end{pmatrix} \quad (3.5c)$$

The [6]:

$$(1,\bar{5})=\begin{pmatrix} \bar{d}_R \\ \bar{d}_Y \\ \bar{d}_B \\ e \\ \nu_e \end{pmatrix}, \quad (3.6a)$$

$$(2,1)=(N)_p. \quad (3.6b)$$

The objects $D, U, \bar{D}, \bar{U}$ are colored, heavy-colored HC-quarks. $\bar{E}, E, \bar{N}$, and N are heavy-colored HC-leptons with the ordinary quantum numbers of $\bar{e}$, $e, \bar{\nu}$, and ν. One especially interesting feature of (3.4a) and (3.5a) is the interchange of the roles of $\bar{u}$ and $\bar{c}$ between the two multiplets. We will see that this identification keeps the mass matrix diagonal.

The 48 gauge bosons of SU(7) can be classified into several groups. First of all there are the usual 8 color gluons and 4 electroweak bosons. Twenty generators connect the 6 and 7 components $\psi_{\bar{e}}$, $\psi_{\bar{\nu}}$ to the remaining 5 components. These generators change an $SU(2)_{EW}$ index to a color or HC index. They will become superheavy ($\sim 10^{16}$ GeV) and will be ignored for the most part. There are 3 HC generators which remain unbroken. The corresponding heavy-color gluons mediate a confining force with a scale ~1 TeV. The 12 bosons which can connect the components ψ_T and ψ_d are called b bosons.

Finally a diagonal generator, orthogonal to hypercharge,

$$\frac{1}{\sqrt{140}}\begin{pmatrix} -5 & & & & & & \\ & -5 & & & & & \\ & & 2 & & & & \\ & & & 2 & & & \\ & & & & 2 & & \\ & & & & & 2 & \\ & & & & & & 2 \end{pmatrix}$$

is coupled to the gauge boson b'. The bosons b, the color and heavy-color gluons, and one linear combination of b' and hypercharge together correspond to a subgroup of $SU(5)_S$. The breaking of $SU(5)_S$ down to $SU(3)_G \times SU(2)_{HC}$ will cause b and b' to become massive.

IV. SYMMETRY BREAKING

At some very high energy ($\sim 10^{16}$ GeV) called the grand unification mass (GUM) SU(7) is a good symmetry. We shall require three separate stages of symmetry breakdown. The first occurs near the GUM and breaks SU(7) to $[SU(2)\times U(1)]_{EW} \times SU(5)_S$. [Note that we do *not* go through a stage where we break down to $SU(5)_{GG}$.] The group $SU(5)_S$ contains both color and heavy color. We shall not speculate further about this stage.

As energy comes down from the GUM the electroweak and $SU(5)_S$ coupling constants evolve independently according to the renormalization group.[6] This is depicted in Fig. 5. The second stage is also at a presently inaccessible energy of order[3] 100 TeV and breaks $SU(5)_S$ to $SU(3)_C \times SU(2)_{HC}$. (The unification of color and HC above 100 TeV has been suggested as a solution to the strong *CP* problem.[4,5]) We shall not speculate about the origin of this breakdown. However, it can be parametrized phenomenologically by a Higgs multiplet ϕ in the conjugate of the [2]. Indeed if

$$\langle\phi^{12}\rangle = -\langle\phi^{21}\rangle \neq 0\,, \tag{4.1}$$

then the required breakdown occurs. The 12 gauge bosons b receive equal masses and the boson b' also becomes massive.

During the first stage of symmetry breaking no fermion can gain mass. The invariance $[SU(2)\times U(1)]_{EW} \times SU(5)_S$ protects all fermions from mass generation. The second symmetry-breaking stage allows only one mass term to occur. For example, if a Higgs field ϕ^{ij} is used for the second stage, then the coupling (σ is a 2×2 spin matrix needed to make a scalar)

$$\psi_{ij}\,\sigma\,\psi_{kl}\,\phi^{ij}\phi^{kl} \tag{4.2}$$

between the Higgs field and the [2] is allowed by

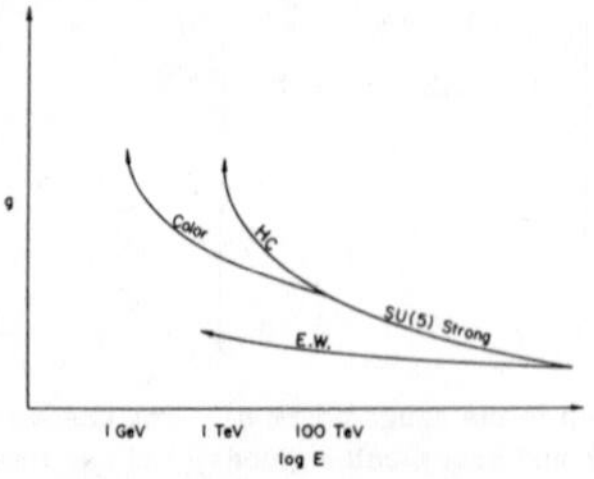

FIG. 5. Evolution of coupling constants with energy.

group theory. If $\langle\phi^{12}\rangle \neq 0$ then (4.2) generates the Majorana mass for $\bar{\nu}_\mu$. Thus we can assume that the $\bar{\nu}_\mu$ becomes a massive Majorana particle with $m \sim 100$ TeV.

After this second symmetry breaking the coupling constants of $SU(2)_{HC}$ and $SU(3)_C$ evolve independently. We assume the HC coupling becomes large at ~1 TeV while the color coupling remains weak until ~1 GeV. [In fact, this is contrary to the expected renormalization behavior. This is of course due to the simplifying assumption that HC is SU(2).]

The strong HC interaction at 1 TeV precipitates the last stage of symmetry breakdown. At this energy only the HC interactions are large so it makes sense to study the HC world in isolation from the other interactions. Consider a closed world of HC gluons interacting with $U, \bar{U}, D, \bar{D}$ and $E, \bar{E}, N, \bar{N}$. As long as the other interactions are ignored the system has a global SU(16) symmetry which mixes the 16 left-handed fields among themselves. (Remember that heavy-color quarks come in three colors.)

A subgroup of this symmetry is ordinary chiral $SU(2)_{left} \times SU(2)_{right}$. To see this it is convenient to replace all the barred fields by their right-handed charge conjugates U_R, D_R, E_R, N_R. Because of the reality of the representations of $SU(2)_{HC}$ the right-handed particles transform equivalently to the left-handed ones under HC. The $SU(2)_{left} \times SU(2)_{right}$ subgroup is defined to act on the doublets $(U,D)_{left}$, $(U,D)_{right}$ and $(N,E)_{left}$, $(N,E)_{right}$.

As in QCD, we expect the HC interactions to cause a spontaneous breakdown of chiral symmetry by precipitating vacuum condensates which up to an SU(16) rotation have the form[2]

$$\langle\bar{U}U\rangle = \langle\bar{D}D\rangle = S$$

$$\text{(no color sum)} \tag{4.3}$$

$$\langle\bar{E}E\rangle = \langle\bar{N}N\rangle = S\,.$$

When we turn on the color and electroweak interactions the condensates becomes determined up to an $SU(2)_{left} \times U(1)$ rotation. Requiring color invariance and electric charge conservation fixes the condensates to have the form (4.3).

In fact when ordinary color interactions as well as b' exchange are accounted for the HC-lepton and HC-quark condensates need not be equal. Thus we write

$$\begin{aligned}\langle\bar{U}U\rangle &= \langle\bar{D}D\rangle = S_Q\,,\\ \langle\bar{E}E\rangle &= \langle\bar{N}N\rangle = S_L\,.\end{aligned} \tag{4.4}$$

These condensates violate $[SU(2)\times U(1)]_{EW}$ according to the standard pattern and give mass to

Z and W bosons leaving the photon massless. The isospin $[SU(2)_{\rm left}+SU(2)_{\rm right}]$ invariance of the HC world insures[2] the empirically successful relation

$$\frac{M_W}{M_Z}=\cos\theta_W\,. \tag{4.5}$$

The nonvanishing expectation values of $\bar{U}U$, $\bar{D}D$, $\bar{N}N$, $\bar{E}E$ spontaneously violate 119 of the 255 generators of SU(16). Among these, 3 have the quantum numbers of the π^+, π^-, π^0 and these are absorbed by the $W^\pm$, Z. The remaining 116 are pseudo-Goldstone bosons which receive mass when the color and/or $[SU(2)\times U(1)]_{EW}$ interactions are turned on.

V. INTERACTIONS MEDIATED BY b BOSONS

New interactions connecting heavy-colored states to heavy-color singlets are mediated by the heavy b bosons. The new interaction vertices are always between particles in the same SU(7) representation. The transitions can occur between two states if they are related by changing a color index ($i=3,4,5=R,Y,B$) to a heavy-color index ($i=1,2$). Thus, for example, we identify [see Eqs. (3.4a) and (3.4b)]

$$D_{Rp}=\psi_{3p}\quad(p=1,2)\,.$$

Changing the p index to a color index, say 4, gives the transition

$$\psi_{3p}\to\psi_{34}+\bar{b} \tag{5.1}$$

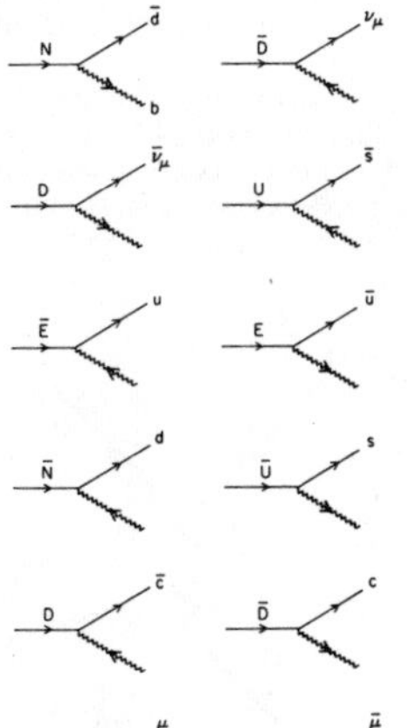

FIG. 6. Vertices involving the b meson.

or

$$D_{Rp}\to\bar{c}_B+\bar{b}\,. \tag{5.2}$$

The new vertices are listed in Fig. 6 (by convention we label the negatively charged heavy boson b and its antiparticle $\bar{b}$).

Labeling HC-quarks, HC-leptons, quarks, and leptons by Q, L, q, l we see that three types of processes occur:

$$\begin{aligned}&Q\to\bar{q}+\bar{b}\,,\\&\bar{L}\to q+\bar{b}\,,\\&\bar{Q}\to l+\bar{b}\,.\end{aligned} \tag{5.3}$$

Evidently some exotic kinds of interactions can be mediated by exchange of two b's or $\bar{b}$'s. For example,

$$\bar{l}\to\bar{q}+2\bar{b}\,. \tag{5.4}$$

One might worry that baryon number might be violated by such processes. However, it is easy to see that there are two conserved quantities

$$\begin{aligned}&N_1=N_q-\tfrac{3}{2}N_L-\tfrac{1}{2}N_Q-\tfrac{1}{2}N_b\,,\\&N_2=N_l+\tfrac{1}{2}N_L-\tfrac{1}{2}N_Q+\tfrac{1}{2}N_b\,,\end{aligned} \tag{5.5}$$

where N_i is the number of particles minus antiparticles of type i. Since no Q's, $\bar{Q}$'s, L's, $\bar{L}$'s, or b's occur in final states of low-energy reactions, the conservation of N_1 and N_2 guarantee baryon and lepton conservation. Of course the 20 superheavy bosons mediate baryon violation as in $SU(5)_{GG}$.

VI. IMPLICATIONS

The main important effect mediated by b exchange is the mass generation of the ordinary fermions. This mass generation is a kind of radiative feeddown of the HC-quark and HC-lepton masses. The relevant graphs are shown in Fig. 7. The crosses on the HC-fermion lines in Fig. 7 indicate the absorption by vacuum condensates of a pair.[3] The induced masses will approximately be of the form

$$m_e=m_{\nu_e}=0\,, \tag{6.1a}$$

$$m_\mu\simeq\frac{3S_Q}{m_b^2}g_5^2\,, \tag{6.1b}$$

$$m_c=m_s\simeq\frac{2S_Q}{m_b^2}g_5^2\,, \tag{6.1c}$$

$$m_u=m_d\simeq 1\frac{S_L}{m_b^2}g_5^2\,. \tag{6.1d}$$

g_5^2 is the $SU(5)_S$ coupling constant at 100 TeV. It is common to all graphs and has no reason to be far from unity. The factors 3, 2, 1 are group

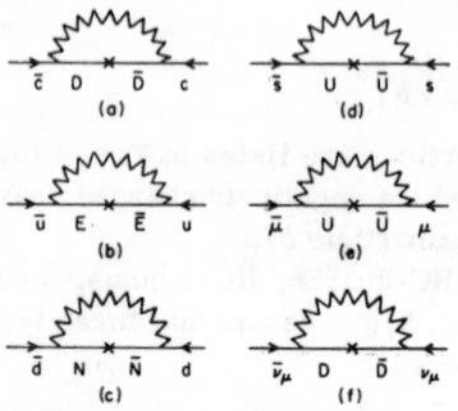

FIG. 7. Ordinary fermion mass generations. Note that an incoming left-handed particle is equivalent to an outgoing right-handed charge conjugate.

theoretic in origin. Taking $m_b \sim 100$ TeV would give ordinary masses in the hundreds of MeV.

At this point the reader can see why we made the unusual identification of particles in the $\underline{10}$'s. If we interchanged the $\bar{u}$ and $\bar{c}$ quarks the up-quark mass matrix would have been off diagonal.

The mass matrix is deficient in two ways. First of all the Cabibbo angle is zero. Secondly there are disappointing u-d and c-s degeneracies.

However, the mass matrix has two interesting features. The standard $SU(5)_{GG}$ mass relations $m_e = m_d$ and $m_\mu = m_s$ following from the simplest choice of Higgs fields do not follow. Also the radiative character of the mass mechanism naturally accounts for the smallness of fermion masses relative to electroweak bosons.

The muon neutrino ν_μ is coupled through the graph of Fig. 7(f) to a particle we have called $\bar{\nu}_\mu$. As discussed previously the $\bar{\nu}_\mu$ gets a Majorana mass ~100 TeV. The mixing of Fig. 7(f) results in a small Majorana mass order $m_\mu{}^2/m_{\bar{\nu}}$ for the ν_μ.

As mentioned previously, there are various pseudo-Goldstone bosons in the spectrum connected with the SU(16) chiral group of the HC sector. In particular, one of these bosons with $\bar{L}L$ quantum numbers only receives mass from b's and $U(1)_{EW}$. This spinless boson has the quantum numbers of $\bar{E}N$. The vertices in Fig. 6 permit it to decay to a muon antineutrino and a charmed baryon composed of u, d, c. The leading process is shown in Fig. 8.

The U(1) contribution to the mass of this $\bar{L}L$ pseudo-Goldstone boson is analogous to the electromagnetic shift of the pion mass. Roughly speaking we should expect the mass of the $\bar{L}L$ state to be

$$m_{\bar{L}L}{}^2 \sim (m_{\pi^+}{}^2 - m_{\pi^0}{}^2)\frac{F_\pi{}^2}{f_\pi{}^2}\,, \qquad (6.4)$$

where F_π and f_π are the axial-vector couplings (pion decay constants) for HC-pions and pions.

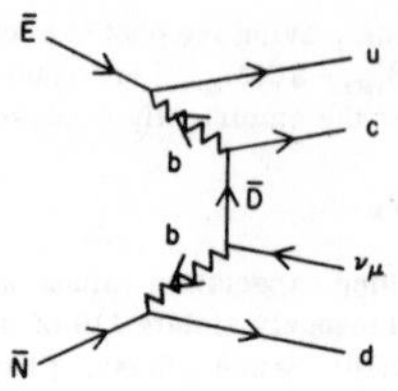

FIG. 8. The decay of the $\overline{EN}$ state.

From the relation

$$M_W = g_{EW}\frac{F_\pi}{2}$$

we know[2] $F_\pi/f_\pi \sim 2000$. This gives

$$m_{\bar{L}L} \sim 70 \text{ GeV}.$$

The contribution arising from b exchange may be comparable. Accordingly $m_{\bar{L}L}$ ought to be comparable to the Z and W masses.

An additional amusing feature of the toy model is the absence of nucleon decay. The usual $SU(5)_{GG}$ proton decay process is shown in Fig. 9. Because of the peculiar interchange of $\bar{c}$ and $\bar{u}$ in the $\underline{10}$'s the top vertex of Fig. 9 has $\bar{u}$ replaced by $\bar{c}$, thus eliminating proton decay. This corresponds to a 90° right-handed "Cabibbo-type rotation" as discussed by Jarlskog.[10]

We do not suggest that proton decay is readily forbidden. In a more realistic model with nonvanishing Cabibbo angle, the natural interpretation would be that proton decay is Cabibbo suppressed.

Similarly a nonvanishing θ_c might also allow the c quark in Fig. 8 to be replaced by a u quark. This would lead to the interesting prediction of an $\overline{EN}$ narrow resonance coupled to the proton-muon-antineutrino channel.

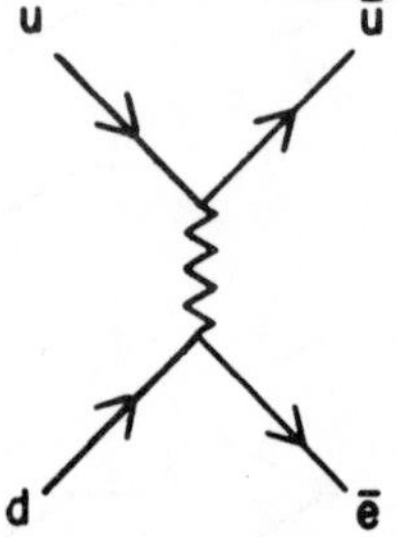

FIG. 9. Proton decay in ordinary $SU(5)_{GG}$.

VII. COMMENTS

We have exhibited a toy model of a grand unified theory including the known forces and heavy color. The representation content of the model yields two ordinary fermion families and a heavy-color family. The HC sector generates masses both for electroweak bosons and the ordinary quarks and leptons.

One of the most significant features of our model is the way in which the relation

$$\cos\theta_W = M_W/M_Z$$

arises. This relation is not a general consequence of the $SU(2)\times U(1)_{EW}$ structure. It follows from the higher $SU(2)_{left}\times SU(2)_{right}$ symmetry of the heavy color world. This symmetry requires the somewhat surprising occurrence of right-handed heavy-color neutrinos N_R. In general, a different group structure or choice of fermion representation would not reproduce this result.

The shortcomings of the model are serious. No mechanism was offered to explain the 100-TeV symmetry breaking of the strong SU(5) down to color and HC. Furthermore, the subsequent evolution of the $SU(3)_C$ and $SU(2)_{HC}$ couplings is required to be perverse in that the HC coupling must become strong before color. Both of these features and the lack of another family (t, b, τ, ν_τ) point toward a larger group structure.

A candidate model can easily be constructed to include three families of fermions. Consider the group SU(9) and the anomaly-free representation $[2]+[4]+[6]+[8]$. At the grand unification mass it breaks down to $SU(2)\times U(1)_{EW}\times SU(7)_S$. At 100 TeV $SU(7)_S$ breaks down to $SU(3)_C\times SP(4)_{HC}$ where SP(4) is the subgroup of SU(4) matrices which have the property

$$U\eta U^T = \eta \quad (T = \text{transpose}).$$

Here η is the symplectic metric

$$\eta = \begin{pmatrix} 0 & 1 & 0 & 0 \\ -1 & 0 & 0 & 0 \\ 0 & 0 & 0 & 1 \\ 0 & 0 & -1 & 0 \end{pmatrix}.$$

The breakdown can be induced by a Higgs field in the [2] with expectation value

$$\phi_{ij} = \begin{pmatrix} \eta & 0 & 0 \\ 0 & 0 & 0 \\ 0 & 0 & 0 \end{pmatrix}. \qquad (\text{columns: HC, C, EW})$$

The representations of SP(4) are all real like those of SU(2). The $\underline{4}$ and $\bar{\underline{4}}$ of SU(4) both transform as $\underline{4}$'s of SP(4). The antisymmetric tensor which is a $\underline{6}$ in SU(4) becomes a $\underline{5}+\underline{1}$ in SP(4). The breakdown of the fermion content into $SP(4)_{HC}\times SU(5)_{GG}$ is as follows:

$$[2] = (1,10)+(4,5)+(1,1)+(5,1),$$

$$[4] = (1,\bar{5})+(4,\overline{10})+(5,10)+(1,10)+(4,5)+(1,1),$$

$$[6] = (4,1)+(5,\bar{5})+(1,\bar{5})+(4,\overline{10})+(1,10),$$

$$[8] = (4,1)+(1,\bar{5}).$$

This material is just sufficient to assemble two families of heavy-color families in the $\underline{4}$ and one in the $\underline{5}$, $\underline{3}$ ordinary families and 2 heavy Majorana neutrinos.

This model and our toy model belong to a sequence in which all of the good features of our toy model are preserved. This sequence, discovered by Georgi,[9] is defined by the group $SU(2n+1)$ with fermions in the representations $[2], [4], \ldots, [2n]$. The strong group $SU(2n-1)$ is broken to SU(3) $\times SP(2n-4)$. These models contain $n-1$ ordinary families. A paper discussing these points in detail is in preparation.

Notes added:

(1) Closely related models based on O(18) have been considered by H. Georgi and E. Witten and discussed by H. Georgi at the Lepton-Quark meeting in Hamburg in 1978.

(2) Ken Lane has discovered a pair of true Goldstone bosons in our model. To see them let $\chi[m]$ be the field for fermions in the $[m]$ of SU(7). Then the current densities

$$j_0[m] = \chi^\dagger[m]\chi[m]$$

are conserved up to anomalies. Since $j[m]$ is SU(7) invariant, its anomaly is determined up to a numerical factor

$$\partial_\mu j_\mu[m] = c(m)\tilde{F}F$$

Evidently two anomaly-free linear combinations fail to commute with the condensates and therefore are realized in the Goldstone-Nambu mode. The problem is obviously due to the reducibility of the fermion representation. Fortunately an elegant escape is available. Georgi, and independently, Bjorken, have pointed out to us that our fermion representation is actually a single spinor representation of O(14) in which SU(7) is embedded. There are 43 new generators and a subset of these can mediate transitions between the different SU(7) multiplets, thus eliminating the unwanted conserved currents.

(3) P. Ramond has emphasized the possible occurrence of Majorana neutrino masses in models based on orthogonal groups. See his lecture at the

Sanibel Symposium in Florida in 1979, Caltech Report No. CALT-68-709.

ACKNOWLEDGMENTS

We wish to thank S. Raby, P. Sikivie, and J. Bjorken for helpful discussions. We would like to thank Ken Lane for his continuing effort to root out all the difficulties with our model. This work was supported in part by the Department of Energy under Contract No. DE-AC03-76SF00515. The work of L.S. was supported in part by the National Science Foundation.

[1]H. Georgi and S. L. Glashow, Phys. Rev. Lett. 32, 438 (1974).

[2]L. Susskind, Phys. Rev. D 20, 2619 (1979). Similar speculations have been made by S. Weinberg, Phys. Rev. D 19, 1277 (1979).

[3]S. Dimopoulos and L. Susskind, Nucl. Phys. B155, 237 (1979); E. Eichten and K. Lane, Harvard report, 1979 (unpublished).

[4]S. Dimopoulos and L. Susskind, Columbia University Report No. Print-79-0196 (unpublished).

[5]W. Fischler, Phys. Rev. D 20, 3399 (1979).

[6]H. Georgi, H. K. Quinn, and S. Weinberg, Phys. Rev. Lett. 33, 451 (1974); A. J. Buras, J. Ellis, M. K. Gaillard, and D. V. Nanopoulos, Nucl. Phys. B135, 66 (1978).

[7]E. Gildener and A. Patrascioiu, Phys. Rev. D 16, 423 (1977).

[8]We thank Ken Lane and Savas Dimopoulos for helpful conversations on this point.

[9]H. Georgi, Nucl. Phys. B156, 126 (1979).

[10]C. Jarlskog, Phys. Lett. 82B, 401 (1979).

VI. PSEUDO GOLDSTONE BOSONS

VOLUME 43, NUMBER 23 PHYSICAL REVIEW LETTERS 3 DECEMBER 1979

Experimental Characteristics of Dynamical Pseudo Goldstone Bosons

M. A. B. Bég
The Rockefeller University, New York, New York 10021
and
H. D. Politzer and P. Ramond
California Institute of Technology, Pasadena, California 91125
(Received 15 October 1979)

The hypothetical existence of new color interactions, which participate in the spontaneous breaking of the weak-interaction group, will in general lead to relatively light composite pseudo Goldstone bosons. Their production and decay characteristics are analyzed to be close to, yet actually distinguishable from, those of the elementary Higgs bosons of the Weinberg-Salam model.

The usual implementation of the Goldstone-Higgs mechanism of spontaneous symmetry breaking, via elementary spin-0 fields, is one of the less attractive features of conventional quantum flavordynamics (QFD). And the situation is acute in attempts to unify QFD and quantum chromodynamics (QCD) into a single gauge theory. In such grand unified theories, one is obliged to introduce a multitude of Higgs fields with judiciously contrived couplings; the essential simplicity of the gauge-theoretic approach is, thereby, irretrievably lost. It has been suggested,[1-4] therefore, that one discard elementary Higgs fields altogether, and seek a dynamical mechanism for symmetry breakdown.

In the simplest dynamical mechanism, the requisite Goldstone bosons, which furnish the longitudinal degrees of freedom for massive gauge fields, are bound states of a new species of quark, whose superstrong gauge interactions (generated by gauging a color′ degree of freedom and described by a theory hereinafter called QC′D) spontaneously break chiral symmetry.

The color′ quarks are likely to come in several flavors, in which case there will be several light pseudo Goldstone bosons as well. In this Letter we observe that these particles may be as light as 10 GeV, that they will be relatively pointlike (of size 1 TeV^{-1}), and will have production and decay modes that are determined by partial conservation of axial-vector-current arguments and hence are fairly model independent. Their signatures are, crudely speaking, similar to those of elementary Higgs particles; consequently, we stress the differences. If spin-0 weakly interacting particles are discovered with masses of tens of GeV, the question of composite versus elementary need not wait until energies of 1 TeV probe the possible bound-state structure.

The color′ degree of freedom, first introduced by Weinberg,[3] may be dubbed "hypercolor."[4] The terminology is convenient, with words such as hyperquark, hyperpion, and hyper-σ having an obvious meaning. We take the weak and electromagnetic interactions of the hyperquarks to be isomorphic to those of ordinary quarks so that a flavor doublet such as (u', d') transforms as (u, d) under the electroweak group.

The existence of hyperquarks is, of course, logically independent of the existence of elementary Higgs particles; we do not rule out the possibility that there may exist both hyperpions and elementary weak scalars. In such a case, $f_{\pi'}$ (the hyperpion decay constant) need not be as large as suggested below, and hypercolor would be even more accessible at low energies. However, careful attention must be paid to mixing to identify the physical states and analyze their decays.[5] Our discussion of hyperpions would still be applicable and would be an aid in deciphering which was which.

Three of the hyperpions must be exactly massless to be absorbed by the $W^{\pm}$ and Z, and in the process they contribute to the W and Z masses:

$$m_Z \cos\xi = m_W \cong \tfrac{1}{2} e f_{\pi'} \sin\xi, \qquad (1)$$

where e is the unit of electric charge and ξ is the Glashow-Weinberg-Salam angle.[1] Hence $f_{\pi'} \cong 300$ GeV. (Note that the relationship between W and Z masses, embodying the so-called $\Delta I_{\rm weak} = \frac{1}{2}$ rule, emerges naturally.) In the following, we shall assume that QC′D may be regarded as scaled-up QCD, albeit with a different number of colors; the characteristic mass scales, Λ' and Λ, in the two theories may, therefore, be related via

$$f_{\pi'}/\Lambda'(N_{C'})^{1/2} = f_\pi/\Lambda(N_C)^{1/2}, \qquad (2)$$

where N_C is the number of quark colors. Since $\Lambda \sim 300$ MeV, we have $\Lambda' \sim (3/N_{C'})^{1/2}$ TeV.

No completely satisfactory hypercolor model of dynamical symmetry breaking exists as yet. There are several outstanding problems:
(i) While $f_{\pi'}$ gives mass to the W's, quarks and leptons must get their masses elsewhere.[6]
(ii) Lifting the mass degeneracy within weak isospin quark doublets appears—at least in the simplest models for generating fermion mass—to destroy the $\Delta I_{\rm weak} = \frac{1}{2}$ relation between W and Z masses.[7] (iii) Unified models typically have many hyperquarks,[8] and hence many massless, unabsorbed hyperpions.

We proceed assuming that these problems are solvable and, in particular, that the unabsorbed hyperpions acquire masses; the order of magnitude of these masses may be estimated as follows. Current-algebra considerations imply that the near-Goldstone modes in QC′D and QCD satisfy

$$m_{\pi'}^2 = m_\pi^2 \frac{m_{q'}}{m_q} \frac{\langle 0|\bar{q}'q'|0\rangle}{f_{\pi'}^2} \frac{f_\pi^2}{\langle 0|\bar{q}q|0\rangle}. \qquad (3)$$

Also, since QC′D is presumed to be scaled-up QCD, we may equate the dimensionless ratios

$$\frac{\langle 0|\bar{q}'q'|0\rangle}{N_{C'}\Lambda'^3} = \frac{\langle 0|\bar{q}q|0\rangle}{N_C\Lambda^3}. \qquad (4)$$

Equations (2)–(4) yield

$$m_{\pi'}^2 = m_\pi^2 \frac{m_{q'}}{m_q} \frac{f_{\pi'}}{f_\pi} (\tfrac{3}{4})^{1/2}, \qquad (5)$$

using three colors for QCD and assuming four for QC′D, respectively. For $m_{q'} \sim m_q$, e.g., if the current quark masses are of a similar origin, Eq. (5) implies $m_{\pi'} \sim 7$ GeV. This estimate is to be contrasted with the naive expectation for pseudo-Goldstone-boson[9] masses arising at the one-loop level: $m_{\pi'} \sim \alpha^{1/2} f_{\pi'} \cong 25$ GeV. (Later, where a specific numerical value is required, we take $m_{\pi'} \sim 15$ GeV.)

As pseudo Goldstone bosons, the hyperpions have fairly model-independent decay modes since they couple primarily to the divergences of currents. To the extent that these currents contain contributions from the known quarks and leptons, these divergences will be of the form

$$\partial_\mu j_5^{\,\mu} = \ldots + (m_i + m_j)\bar{f}_i \gamma_5 f_j, \qquad (6)$$

$$\partial_\mu j^\mu = \ldots + (m_i - m_j)\bar{f}_i f_j, \qquad (7)$$

where m_i is the mass of the fermion f_i. The hyperpion couplings are to be contrasted with those of the left-over elementary Higgs scalar φ of the Weinberg-Salam model:

$$\mathcal{L} = \ldots + (1/\langle\varphi\rangle)\varphi m_i \bar{f}_i f_i. \qquad (8)$$

Thus both types of bosons have the strongest couplings to the heaviest fermion available, but there are measurable differences. For instance, a flavor-diagonal decay of the π' is S wave, while that of φ is P wave leading to the difference in branching ratios:

$$\frac{\Gamma(\varphi \to b\bar{b})}{\Gamma(\varphi \to \tau\bar{\tau})} \cong 3\left(\frac{m_b}{m_\tau}\right)^2 \left(\frac{m_\varphi^2 - 4m_b^2}{m_\varphi^2 - 4m_\tau^2}\right)^{3/2}, \qquad (9)$$

$$\frac{\Gamma(\pi' \to b\bar{b})}{\Gamma(\pi' \to \tau\bar{\tau})} \cong 3\left(\frac{m_b}{m_\tau}\right)^2 \left(\frac{m_{\pi'}^2 - 4m_b^2}{m_{\pi'}^2 - 4m_\tau^2}\right)^{1/2}. \qquad (10)$$

Furthermore, if we look at exclusive channels, processes such as $\varphi \to D\bar{D}$ and $\pi' \to D\bar{D}\pi$ are allowed, whereas $\varphi \to D\bar{D}\pi$ and $\pi' \to D\bar{D}$ are forbidden.

The 2γ decay of neutral hyperpions can be computed, as for π^0, from the triangle anomaly:

$$\frac{\Gamma(\pi'^0 \to 2\gamma)}{\Gamma(\pi^0 \to 2\gamma)} = \left(\frac{f_\pi}{f_{\pi'}}\right)^2 \left(\frac{m_{\pi'}}{m_\pi}\right)^3 \frac{4}{3}, \qquad (11)$$

modulo possible differences in q' and q charges. However, the branching ratio to 2γ is likely to be very small compared to the decay to massive fermions $f\bar{f}$:

$$\frac{\Gamma(\pi'^0\to 2\gamma)}{\Gamma(\pi'^0\to f\bar{f})} \propto e^4\left(\frac{m_{\pi'}}{m_f}\right)^2 \times(\text{phase-space ratio}). \quad (12)$$

The hyperpions are of size Λ'^{-1} or $f_{\pi'}^{-1}$. The charged π''s, therefore, will be produced liberally in e^+e^- annihilation; the deviation from

$$R_{e^+e^-}(s)=\tfrac{1}{4}\sum_{\pi'} Q_{\pi'}^2(1-4m_{\pi'}^2/s)^{3/2} \quad (13)$$

is of order $N_{\pi'}s/f_{\pi'}^2$, where $N_{\pi'}$ is the number of charged hyperpions, $Q_{\pi'}$ are their electric charges, and $s\equiv(\text{c.m. energy})^2 \ll f_{\pi'}^2$. That the departure from the pointlike limit, Eq. (13), is of the order stated may be verified by writing an *effective* chiral Lagrangian[3] for the π''s.

In fact, all of the couplings of π''s can be deduced using an effective-Lagrangian approach, as long as the π''s are soft, on the scale set by $f_{\pi'}$. For simplicity we imagine using linear representations with the pseudoscalar π''s accompanied by scalar σ''s. From the implicit q' content of the π' and from the W and Z masses that now come from σ' vacuum-expectation values, it is evident that the chiral quadruplets transform as complex doublets, Φ's, under the electroweak group:

$$\Phi=\begin{pmatrix} i\pi'^+ \\ \tfrac{1}{2}\sqrt{2}(\sigma'-i\pi'^0)\end{pmatrix}. \quad (14)$$

The relevant couplings can be read off from

$$\mathcal{L}_{\rm eff}=(D^\mu\varphi)^\dagger(D_\mu\varphi)+\ldots, \quad (15)$$

where D_μ is the appropriate gauge-covariant derivative. Note that now

$$\langle\sigma'\rangle=f_{\pi'}\approx 300\ \text{GeV}. \quad (16)$$

From this standpoint, the "soft" couplings of π' and σ' resemble a model with several elementary Higgs fields, where the physical spin-0 particles include not only the scalar Higgs particle of the minimal Weinberg-Salam model but also additional scalars and pseudo-Goldstone pseudoscalars. The distinction lies in the expected masses. The masses of fundamental spinless particles are typically comparable to each other. In contrast, a dynamical σ' would have a mass of order Λ', ~1–3 TeV by analogy with QCD, and its decays into light π''s would be so fast as to make it virtually unrecognizable as a distinct particle. Hence, the prominent light particles are all pseudoscalars.

We exhibit the contrasts in some simple models. The coupling of π''s to a lepton or quark doublet (a,b), mediated by $W^\pm$ and Z, are described by the effective interaction

$$\mathcal{L}_{\rm eff}=(G_F/\sqrt{2})f_{\pi'}\sqrt{2}\,\bar{a}[(m_a-m_b)-\gamma_5(m_a+m_b)]b(-i)\pi'^+ +\text{H.c.} +(G_F/\sqrt{2})f_{\pi'}2(m_a\bar{a}i\gamma_5a-m_b\bar{b}i\gamma_5b)\pi'^0. \quad (17)$$

Note that the σ' does not couple directly to a or b and, therefore, it is not responsible for their masses. Note also the *universality* of π' couplings to any doublet: Apart from the trivial mass rescaling, the coupling to (c,d) is the same as that to (a,b).

The contrast with the minimal standard model, in which

$$\mathcal{L}_1=+\frac{e}{2m_W\sin\xi}(m_a\bar{a}a+m_b\bar{b}b)\varphi^0, \quad (18)$$

is discussed above [e.g., Eqs. (9) and (10)]. Consider then a variant with an additional Higgs doublet (a total of five physical spin-0 fields):

$$\mathcal{L}_2=\mathcal{L}_1+2^{-1/2}(f_1\bar{a}a+f_2\bar{b}b)\eta^0+2^{-1/2}(-f_1\bar{a}i\gamma_5a+f_2\bar{b}i\gamma_5b)\chi^0+\tfrac{1}{2}\bar{a}[(f_2-f_1)+\gamma_5(f_2+f_1)]\chi^++\text{H.c.} \quad (19)$$

There is no universality here relating the f's of one fermion doublet to those of another. A further contrast to the dynamical scheme is the presence of two relatively light neutrals which undergo P-wave decays. Needless to say, the distinctions would be blurred if, for some reason, the f's are universal and equal to the would-be π' couplings and if φ and η not only have masses of order 1 TeV but also manage to decay *strongly* into pairs of pseudoscalar mesons.

We have discussed some experimental consequences of hypercolor interactions that would be manifest at energies well below typical hyperhadron masses; these low-energy consequences, follow from partial conservation of axial-vector-

TABLE I. Some characteristics of spin-0 particles which occur in theories with a dynamical Higgs mechanism [σ', π'^0, $\pi'^\pm$] and in theories with elementary Higgs fields [φ^0, χ^0, $\chi^\pm$]. The φ decay rates quoted are for m_φ = 15 GeV. Note that σ', the dynamical analog of φ^0, decays rapidly to $\pi'^+\pi'^-$, etc., via *strong* interactions.

PARTICLE	EXPECTED MASS (in GeV)	FEASIBLE PRODUCTION MECHANISM	2γ DECAY		f f̄ DECAYS				PSEUDOSCALAR MESON CHANNELS	
			Rate (in sec^{-1})	Photon Pols.	Universal Coupling	$\nu_\tau\tau$ Rate (in sec^{-1})	$\tau\bar{\tau}$ Rate (in sec^{-1})	$\Gamma(b\bar{b})/\Gamma(\tau\bar{\tau})$	$D\bar{D}$	$D\bar{D}\pi$
σ'	~2000	?	-	-	-	-	-	-	-	-
π'^0	~15	Via Heavy Flavors in $e^+e^-\to$. . . or $pp\to$ . . . . / $W^\pm\to\pi'^\pm\pi'^0$	$\sim 2\times10^{15}$	⊥	Yes	-	$\sim10^{20}$	~15	No	Yes
$\pi'^\pm$	~15	$e^+e^-\to\pi'^+\pi'^-$ / $W^\pm\to\pi'^\pm\pi'^0$	-	-	Yes	$\sim3\times10^{19}$	-	-	Yes	Yes
ϕ^0	$\gtrsim10$	Via Heavy Flavors and Virtual W-Pairs in $e^+e^-\to$.. or $pp\to$ ...	$\lesssim10^{17}$	∥	Yes	-	$\sim4\times10^{19}$	~10	Yes	No
χ^0	Could be ~15	Same as for π'^0	?	⊥	No	-	?	?	No	Yes
$\chi^\pm$	Could be ~15	Same as for $\pi'^\pm$	-	-	No	?	-	-	Yes	Yes

current arguments and are therefore fairly insensitive to how hypercolor is implemented. We have also contrasted the experimental profile of relatively low-lying dynamical bosons, in the hypercolor scenario, with that of elementary Higgs particles. (Some of our results are summarized in Table I.) It is our hope that our experimental colleagues will be able to shed some light on the nature of the Higgs mechanism in the not-too-distant future.

This work was partially supported by the U. S. Department of Energy under Contract Grant No. EY-76-C-02-2232B.*000 and also by the U. S. Department of Energy under Contract No. De-Ac-79-ER0068.

Two of us (M.A.B.B. and P.R.) are indebted to the Aspen Center for Physics for the hospitality which led to our collaboration; one of us (M.A.B. B.) is grateful to Brookhaven National Laboratory for hospitality during the course of this work. It is a pleasure to acknowledge stimulating conversations with M. Gell-Mann.

[1]M. A. B. Bég and A. Sirlin, Annu. Rev. Nucl. Sci. 24, 379 (1974).

[2]M. A. B. Bég, in *New Frontiers in High Energy Physics, Proceedings of Orbis Scientiae, 1978, Coral Gables*, edited by A. Perlmutter and L. Scott (Plenum, New York, 1978).

[3]S. Weinberg, Phys. Rev. D 13, 974 (1976).

[4]Compare L. Susskind, SLAC Report No. 2142, 1978 (Phys. Rev. D, to be published).

[5]See, for example, M. Weinstein, Phys. Rev. D 8, 2511 (1973). We thank E. Eichten and K. Lane for discussions on this point.

[6]S. Dimopoulos and L. Susskind, Nucl. Phys. B155, 237 (1979).

[7]E. Eichten and K. Lane, to be published; M. Voloshin and V. Zakharov, private communication.

[8]M. Gell-Mann, P. Ramond, and R. Slansky, to be published; E. Farhi and L. Susskind, to be published; M. A. B. Bég, to be published.

[9]S. Weinberg, Phys. Rev. Lett. 29, 1698 (1972).

M.A.B. Bég, H.D. Politzer and P. Ramond, Phys. Rev. Lett. 43, 1701 (1979).

Added Notes:

(i) Our discussion assumes that there exists a gauge invariant mechanism for generation of what is effectively "current mass" for the fermions. Whether this can be done without running into any conflict with low energy phenomenology is not yet fully clear. [For a recent review, see: M.A.B. Bég, Proc. of EPS International Conf. on High Energy Physics, Lisbon, Portugal (1981)].

(ii) Note that the tests, for distinguishing hyperpions from physical Higgs fields of the canonical theory, described as parity tests in this paper, actually follow from CP invariance if the hyperpions are flavor-neutral -- an expectation in all existing models for the lightest neutral pseudo-Goldstone bosons. [See, for example, E. Farhi and L. Susskind, CERN Report TH-2975 (1980)]. The point is relevant because some models of mass generation violate parity. Other tests are proposed in the paper of Ali and Bég [Phys. Lett. 103B, 376 (1981)].

Nuclear Physics B168 (1980) 69–92

TECHNICOLOURED SIGNATURES*

Savas DIMOPOULOS
Department of Physics, Stanford University, Stanford, CA 94305, USA

Received 3 December 1979

Scalarless theories with extended technicolour have several new consequences at energies of a few hundred GeV, which will be accessible in future accelerators. A large class of such theories have several (60, 116 or 132) pseudo-Nambu–Goldstone bosons with masses from a few GeV to ~300 GeV. These bosons are classified and their model-independent properties are discussed. Their most important properties are: (i) they can be singly produced; (ii) their decays give rise to some exotic final states involving jets; (iii) more than 90% of them are coloured. Therefore, they can be produced in the future p̄p and pp machines, predominantly *via* 2-gluon collisions.

1. Introduction

1.1. Brief motivation

The extended technicolour [1, 2] (ETC) models are the only known candidates for theories with the following desirable features:

(i) they do not have elementary scalar fields;

(ii) they can generate masses for quarks, leptons, and vector bosons;

(iii) they offer the possibility of introducing a small *CP* violation without the appearance of the notorious strong *CP* problem [3].

We should remind the reader that scalars are unwanted because: (a) They carry several couplings (Yukawa couplings, quartic couplings, and vacuum expectation values) about which gauge theories have nothing intelligent to say. Thus they spoil the predictive power of gauge theories. Even grand unified 3 family models need ~20 undetermined parameters [4, 5]. Often these couplings have unnaturally small values (e.g., $g_{\text{Yukawa}} \sim 10^{-6}$). These facts suggest that elementary scalars are nothing but a good way to parametrize low energy ($E \ll 300$ GeV) phenomenology. (b) They cannot naturally account for the huge difference between the electroweak scale (~300 GeV) and the grand unified scale [6, 7] $\gtrsim 10^{15}$ GeV [5,8]. There exist extremely strong acausal correlations between the small distance ($r \lesssim 10^{-28}$ cm) and large-distance physics ($r \gtrsim 10^{-16}$ cm). Thus, for example, a 1% change of a

* Based on an invited talk given at the European Physical Society (EPS) meeting in Geneva, July 4, 1979.

small-distance quantity can change the values of quantities of the large-distance physics by 25 orders of magnitude [7]!

1.2. *Quick review*

Historically the ETC models came into existence [1] in an effort to fix up a big drawback of the old technicoloured [9, 7] (TC) models: the fact that TC models could not generate masses for quarks and leptons. They did generate masses only for the electroweak vector bosons and naturally accounted for the heavy isospin symmetry relation $M_W = M_Z \cos \Theta_W$ [7]. The old TC models consisted of (a) a new strong interaction, called technicolour, which becomes strong at a mass scale of ~1 TeV; and (b) *one* doublet of massless "techniquarks," denoted by U and D. The techniquarks bind *via* the technicolour forces and form technihadrons with masses of order of $\gtrsim 1$ TeV. Some bosons, for symmetry reasons, escape from receiving any mass. These are the 3 technipions. They are the Nambu–Goldstone bosons of the spontaneously broken chiral $SU(2)_L \times SU(2)_R$ associated with the two massless techniquarks U and D. The massless technipions have exactly the right quantum numbers to play the role of the usual Nambu–Goldstone bosons of the Higgs mechanism of the standard $SU(2)_L \times U(1)$ electroweak model [10]. Thus they can be eaten and give masses to the $W^{\pm}$ and Z^0.

The most important new feature of the ETC models is that they allow transitions between ordinary fermions and technifermions. These transitions are mediated by a new set of heavy (mass ~1–100 TeV) ETC vector bosons. It is these transitions that allow the quarks and leptons to acquire their masses.

1.3. *Some consequences*

An extremely important feature of the ETC models [1, 2] that is not shared by the old TC models is the apparent necessity for having more than one electroweak doublet of technifermions, denoted by $U_1, D_2; U_2, D_2; U_3, D_3$, etc. This appears to be a necessary feature of all known quasirealistic models [11, 12] and it is almost certain that it should be true for realistic models. This feature has the amusing consequence that there exist several technipion-like almost massless bosons with masses ranging from about ~1 GeV to ~300 GeV. They are called pseudo-Nambu–Goldstone bosons or, for short, pseudos. Many quasirealistic ETC models have either 60, 116 or 132 such bosons. Thus it appears difficult to miss such an overpopulated pseudo-world living less than an order of magnitude of energy away from us. In fact, the upcoming generation of accelerators, especially the FNAL TEVATRON ($p\bar{p}$ 1000 + 1000 GeV; 1984), should find these particles.

Since more than 90% of these pseudo-Nambu–Goldstone bosons carry ordinary $SU(3)_{colour}$ (see sect. 3) we expect that the $p\bar{p}$ and pp machines will be very useful for finding them and studying their properties.

2. Scaleontology

In this section we discuss the masses of the new types of hadrons that exist in ETC models. There are at least two new strong gauge forces in ETC models [1, 2]. (a) The first one becomes strong at a mass scale of order of ~10 TeV. Thus we expect to find several hadrons at this mass range. This sector of the theory is the least understood so we shall not discuss it any further. We should, however, emphasize that this sector may produce some light pseudos or true Nambu–Goldstone particles with exotic properties. (b) The second strong gauge force is technicolour [9, 7]. It becomes strong at a scale of order of 1 TeV and gives mass to the $W^{\pm}$ and Z^0. The exact scale at which the technicolour forces become strong is very important, for it is this scale that determines the masses of the technihadrons and the numerous pseudos mentioned before. The value of this scale μ_{TC} depends crucially on the number of technifermion electroweak doublets. μ_{TC} decreases as the number of technifermion doublets goes up. An important feature of the ETC models is that the number of technifermion doublets almost inevitably has to go up. This comes about, in all quasirealistic models [11, 12] because the technifermions have to give masses to several quarks and leptons.

For the sake of definiteness, from now on we shall assume that we have one family of technifermions consisting of 16 chiral members*. This comes about naturally when we attempt to construct grand unified models which contain technicolour [11, 12]. Let us denote these $SU(3)_{colour}$ technifermions by $U_c^L, D_c^L, E^L, N^L; U_c^R, D_c^R, E^R, N^R$, where $c = 1, 2, 3$ is a colour index. Let us for the moment turn off all the ordinary $SU(3)_{colour} \times SU(2)_L \times U(1)$ and forces mediated by the heavy ETC vector bosons [1, 2] and keep only the technicoloured forces. Let us furthermore assume that all the technifermions belong to the same complex representation of the technicolour group. (The case of real representation of the TC group will be considered later.) Then the theory has a global $SU(8)_L \times SU(8)_R$ chiral symmetry. At the scale of $\mu_{TC} \sim$ TeV the technicoloured forces become strong and form the following 8 condensates [1, 2] ($c = 1, 2, 3$; no sum on c):

$$\langle \bar{U}_c^R U_c^L \rangle = \langle \bar{D}_c^R D_c^L \rangle = \langle \bar{N}^R N^L \rangle = \langle \bar{E}^R E^L \rangle . \tag{2.1}$$

The point that we wish to emphasize here is that there exist four electroweak isodoublets of technifermions which form condensates and therefore effectively four technipions. Let $F_{T\pi}$ be the T-pion decay constant associated with *each* electroweak doublet. Then this corresponds to a Higgs-type model with four Higgs fields ϕ_1, ϕ_2, ϕ_3, ϕ_4 and with vacuum expectation values $\langle\phi_1\rangle = \langle\phi_2\rangle = \langle\phi_3\rangle = \langle\phi_4\rangle = F_{T\pi}$ [7]. Thus the W mass is given by $M_W^2 = (1/4g_2^2)4F_{T\pi}^2$ and therefore $2F_{T\pi} \simeq 250$ GeV or $F_{T\pi} \simeq 125$ GeV. In general, if we have N electroweak doublets of technifermions

* In some quasirealistic models (see refs. [11, 12] with three or four families of ordinary fermions one gets more than one family of technifermions. This has the effect of decreasing the masses and increasing the number of pseudos.

then the corresponding $F_{T\pi}$ will be $F_{T\pi} \simeq N^{-1/2}$ 250 GeV. This is a useful fact because it means that as the technihadron world becomes more populated it becomes lighter and therefore more accessible to the future accelerators. The same holds true for pseudos. The pseudos, as we shall discuss, owe their existence as well as their relatively light masses to the existence of several electroweak doublets of fermions which is necessary in ETC models*.

To estimate the masses of the technihadrons, let us assume that they scale naively, i.e., in proportion to $F_{T\pi}/f_\pi$. This yields a mass for the technirho of order

$$M_{T\rho} \sim \frac{F_{T\pi}}{f_\pi} m_\rho \sim \frac{125\ \mathrm{GeV}}{93\ \mathrm{MeV}} m_\rho \sim 1039\ \mathrm{GeV}\,. \tag{2.2}$$

These estimates are too naive since they assume that the ratio m_ρ/f_π is group independent. Note that the mass of the technirho would be twice as large if we had only one electroweak doublet of T-fermions. Having four electroweak doublets also increased the number of technirhos. Now instead of four technivector mesons (technirho and techniomega) we shall have 64 of them all with (naive) masses of order of ~1040 GeV. They will have the same colour and electroweak quantum numbers as the pseudos shown in table 1a.

3. Pseudotypes

3.1. The 16 colour pseudomultiplets that occur when the technifermion representation is complex

In this section we wish to discuss the pseudo-Nambu–Goldstones in these theories. As in sect. 2, we assume for definiteness that we have one family of technifermions. As mentioned, this occurs naturally in GUT models which include technicolour [11, 12]. All the groundwork for discussing the pseudos has been done in the part of sect. 2 preceding eq. (2.1).

Let us begin our discussion of pseudos from eq. (2.1) where the form of the condensates was explicitly displayed. These condensates break the chiral $SU(8)_L \times SU(8)_R$ symmetry down to the $SU(8)_{(L+R)}$ symmetry. The broken 63 axial generators give rise to 63 massless Nambu-Goldstone pseudoscalars. They correspond to all but one (the $U(1)_A$ piece) of the ground-state pseudoscalars that can be formed by combining a T-fermion and an anti-T-fermion. They are all shown in table 1a, except for charge conjugates, together with their colour, electric charge, and masses (to be estimated in sect. 4). Three of these are eaten by the $W^{\pm}$ and the Z^0. The remaining 60 receive masses when the $SU(3) \times SU(2) \times U(1)$ forces and the forces mediated by

* See previous footnote.

TABLE 1

(a) The 16 multiplets of pseudos that occur when the technifermion representation is complex

Pseudotype	Colour	Charge	~Mass (GeV)	Name
$\bar{U}_c U_c, \bar{D}_c D_c \ldots$	1	$0, \pm 1$	0	technipions
$\bar{U}_c D_{c'}, \ldots$	8	$0, \pm 1, 0$	300	coloured T-pions and T-eta
$\bar{E}U$	3	$\frac{5}{3}$	240	T-leptoquark
$\bar{E}D$	3	$\frac{2}{3}$	205	T-leptoquark
$\bar{N}U$	3	$\frac{2}{3}$	205	T-leptoquark
$\bar{N}D$	3	$-\frac{1}{3}$	200	T-leptoquark
$\bar{N}E - \bar{U}D$	1	1	few–100	charged axions
$\bar{E}E - \bar{N}N + \bar{U}U - \bar{D}D$	1	0	few–100	axion
$\bar{N}N + \bar{E}E - \bar{U}U - \bar{D}D$	1	0	few–100	paraxion

(b) The 18 additional colour multiplets that occur when the technifermions belong to a real representation with an antisymmetric invariant symbol

Pseudotype	Colour	Charge	~Mass (GeV)	Name
UU	$\bar{3}$	$\frac{4}{3}$	225	ditechniquarks
[UD]	$\bar{3}$	$\frac{1}{3}$	200	ditechniquarks
DD	$\bar{3}$	$-\frac{2}{3}$	205	ditechniquarks
{UD}	6	$\frac{1}{3}$	320	ditechniquarks
UN	3	$\frac{2}{3}$	205	
UE	3	$-\frac{1}{3}$	200	
DN	3	$-\frac{1}{3}$	200	
DE	3	$-\frac{4}{3}$	225	
EN	1	-1	few–100	ditechnilepton

(c) The 22 additional colour multiplets that occur when the technifermions belong to a real representation with a symmetric metric

Pseudotype	Colour	Charge	~Mass (GeV)	Name
UU	6	$\frac{4}{3}$	340	
DD	6	$-\frac{2}{3}$	330	
{UD}	6	$\frac{1}{3}$	320	
[UD]	$\bar{3}$	$\frac{1}{3}$	200	
UN	3	$\frac{2}{3}$	205	
UE	3	$-\frac{1}{3}$	200	
DN	3	$-\frac{1}{3}$	200	
DE	3	$-\frac{4}{3}$	225	
EN	1	-1	few–100	ditechnilepton triplet
EE	1	-2	few–100	ditechnilepton triplet
NN	1	0	few–100	ditechnilepton triplet

We indicate the pseudos that appear in the various types of models with one complete family of technifermions. The masses shown are only rough approximate estimates. The reason for the great ambiguity in the masses of the colourless pseudos is the fact that the contribution of the ETC generators is quite model dependent (see sect. 4). In models with more families of technifermions we shall have more replicas of these pseudos. Their masses will be smaller in proportion to the square root of the number of families.

the heavy ETC generators are turned on. They can be divided into the following types:

(1) Coloured pseudos. All but four of the pseudos are coloured. There exist four octets, four triplets, and four antitriplets.

(1a) The four colour octets consist of an isotroplet and an isosinglet. They can be written

$$\Pi_f^a = \bar{Q}\gamma_5\lambda^a\tau_f Q, \qquad (a=1,2,\ldots 8; f=0,1,2,3). \tag{3.1}$$

Here $a=1,2,\ldots 8$ labels a colour SU(3) Gell–Mann matrix, $f=1,2,3$ labels a Pauli matrix, and $f=0$ labels the identity matrix, $Q=\begin{pmatrix}U\\D\end{pmatrix}$. Π_f^a, $f=1,2,3$, are just like coloured and heavy (mass $\lesssim 300$ GeV) versions of the pion. We call them coloured T-pions. Π_0^a is called coloured T-eta. They are quite amusing for several reasons. For one thing, they, together with the gluons (and colour pseudosextets), would be the strongest carriers of ordinary colour. Thus they should be abundantly produced at very high energy $p\bar{p}$ collisions ($E_{cm} \gtrsim 10$ TeV). In fact, each colour component proliferates as much as the $W_L^\pm$ and Z_L^0 [7] (which proliferate because they too are like pions). Just like all other pseudos they can be singly produced. More on these later.

(1b) The four colour triplets also consist of an isotriplet and an isosinglet. They can be written:

$$K_f^c = \bar{L}\gamma_5\tau_f Q^c. \tag{3.2}$$

Here $L=\begin{pmatrix}N\\E\end{pmatrix}$, $c=1,2,3=$ colour index. We shall call them technileptoquarks. They are expected to have masses of order of $\lesssim 200$ GeV. They are expected to be quite narrow ($\Gamma \sim 1$ MeV) in general. Details later.

(2) Colourless pseudos. There are four colourless pseudos again consisting of an isotriplet and an isosinglet. They can be written:

$$\Pi_f^{TL} = \bar{L}\gamma_5\tau_f L, \qquad (f=0,1,2,3). \tag{3.3}$$

Let us introduce also

$$\Pi_f^{TQ} = \bar{Q}^c\gamma_5\tau_f Q^c, \qquad (f=0,1,2,3), \qquad (c=1,2,3). \tag{3.4}$$

Then it is clear that:

$$\Pi_i^{TQ} + \Pi_i^{TL}\ (i=1,2,3) \text{ are eaten by } W^\pm \text{ and } Z^0;$$

$$\Pi_0^{TQ} + \Pi_0^{TL} \text{ is eaten by the technicoloured anomaly}.$$

$\Pi_f^{TF} \equiv \Pi_f^{TQ} - \Pi_f^{TL}$ $(f=0,1,2,3)$ are pseudos. They consist of an isosinglet and an isotriplet. We shall refer to them as the axion family.

$\Pi_1^{TF} \pm i\Pi_2^{TF} \equiv \Pi_\pm^{TF}$ is an electrically charged pseudo and therefore it receives mass from the electroweak interactions.

The neutral pseudos Π_0^{TF} and Π_3^{TF} work as follows. The linear combination of Π_0^{TF} and Π_3^{TF} which is the Nambu–Goldstone boson of the difference $Y_\mu^{TQ} - Y_\mu^{TL}$ of the techniquark and technilepton hypercharges does *not* receive any mass from the electroweak interactions since $Y_\mu^{TQ} - Y_\mu^{TL}$ commutes with $SU(2)_L \times U(1)_Y$. This Nambu–Goldstone boson is the axion [13] of these dynamical models [3, 14, 2]. This axion is strongly analogous to the neutral Π^0 of QCD which is also massless in the chiral limit. The only place where the axion can receive mass is the ~10 TeV heavy extended technicolour generators which are also responsible for quark and lepton mass generation [1, 2]. This is an important constraint for ETC model building. It means that, for example, the quarks and the leptons should *not* get their masses from two *disjoint* sets of technifermions (called T-quarks and T-leptons respectively). For in this case we could define an exactly conserved ETC singlet current, namely $Y_\mu^{TQ} + Y_\mu^{quarks} - Y_\mu^{TL} - Y_\mu^{leptons}$, which would be spontaneously broken and therefore would yield a massless Nambu–Goldstone boson. In general, this constraint means that we should *not* have several reducible ETC multiplets [14, 2]. We will assume that these constraints are satisfied and that the axion receives a mass from the ~10 TeV ETC generators [1, 2]. Similar discussion applies to the Nambu–Goldstone boson orthogonal to the axion, which we call paraxion. The paraxion is essentially the Nambu–Goldstone boson coupling to the difference of the T-quarks' and T-leptons' U(1) (axial) currents [3, 14] in the same way that the axion essentially corresponds to the difference of the $I(3)_R$'s of T-quarks and T-leptons.

3.2. *There exist 18 additional colour pseudomultiplets when the technifermions belong to a real representation with an antisymmetric metric*

When the technicolour group is real and we turn off all other interactions (i.e., $SU(3)_c \times SU(2)_L \times U(1) \times ETC/TC$) then handedness has no dynamical meaning. All particles can be considered left-handed. Thus, instead of an $SU(8)_L \times SU(8)_R$ chiral symmetry we have an $SU(16)_L$ symmetry. If the metric of the technifermions' representation is antisymmetric then the condensates of eq. (2.1) break this $SU(16)_L$ down to Sp(16) leaving 119 massless Nambu–Goldstone bosons (on the absence of all other forces). Of these, three are the technipions that will be eaten by $W^\pm$ and Z^0, 60 are those that also arise in models with complex technicolour groups and have been discussed in subsect. 3.1. The remaining 56 exist only because of the reality of the technicolour group and are shown in table 1b. We call them pseudodifermions. From this table we see that there are three colour antitriplets, one sextet, four triplets, one singlet, and their antiparticles, i.e., 18 colour multiplets of pseudos.

3.3. *There exist 22 additional colour multiplets when the technifermions belong to a real representation with a symmetric metric*

As we mentioned in subsect. 3.2, when the technicolour group is real the global chiral symmetry associated with one technifamily is not $SU(8)_L \times SU(8)_R$ but $SU_L(16)$. If the technirepresentation is characterized by a symmetric invariant tensor, δ_{ab}, then the condensates of eq. (2.1) break the symmetry down to SO(16), leaving us with 135 Nambu–Goldstone bosons. Three of these are the technipions that are eaten by $W^{\pm}$ and Z^0. Sixty of them are identical to the ones that appear in models with complex technirepresentation and are shown in table 1a. The remaining 72 pseudos signal the fact that the technirepresentation is real and has a symmetric metric. The 72 pseudos form 22 colour multiplets and are shown in table 1c (except for charge conjugates). The differences with the 18 additional pseudotypes that occur in the case of an antisymmetric metric (see table 1b) are the following. In table 1c we have six (instead of two) colour sextet pseudos. In particular the isotriplet of pseudos, consisting of UU, UD, DD are now colour sextets instead of colour $\bar{3}$'s.

Very important new additions in table 1c are the four colourless ditechnileptonic pseudos: EE, NN, $\bar{\text{E}}\bar{\text{E}}$, $\bar{\text{N}}\bar{\text{N}}$. EE, EN, and NN form an electroweak triplet of colour singlet pseudos. These pseudos can in fact be quite light in a large class of models with possible masses between a few and 100 GeV. The most amusing one is the doubly charged EE. It contributes one unit to $R(e^+e^- \to \gamma \to \text{anything})$. Another interesting feature of these possibly light, colourless, dileptopseudos is the fact that in a large class of models they are very narrow and can decay only *via* emission of four quarks and/or leptons [11]. This comes about because they can easily be protected by lepton and baryon number type conservation laws against decays into two ordinary fermions [11] (see subsect. 7.6).

4. Pseudomasses

In this section we discuss some approximate estimates for the masses of the various pseudos. The pseudos receive their masses because the symmetries to which they correspond are explicitly broken by the ordinary $SU(3)_{\text{colour}}$ interactions, electroweak interactions, and the heavy ~10 TeV ETC generators. We begin with a discussion of the masses of the coloured pseudos.

The coloured pseudos receive their mass *via* the graphs of fig. 1. The computation of these graphs is isomorphic to that of the electromagnetic mass difference of the π ions. All we need to do is scale up all masses by $F_{T\pi}/f_\pi$ and replace α by $\alpha_c(F_{T\pi})C_2(\text{R})$, where $C_2(\text{R})$ is the quadratic Casimir operator of the pseudo's representation and $\alpha_c(F_{T\pi}) \sim 0.1$ the $SU(3)_{\text{colour}}$ coupling constant at $F_{T\pi}$. To do this note that $m^2_{\pi^+} - m^2_{\pi^0}$ does not change much in the chiral limit where $m_{\text{quark}} = 0$ and $m_{\pi^0} = 0$. Thus the mass of π in the chiral limit is purely electromagnetic and equal to $\simeq (m^2_{\pi^+} - m^2_{\pi^0})^{1/2} \simeq 35.5$ MeV. Thus the $SU(3)_{\text{colour}}$ contribution to the mass of a coloured pseudo is of order

$$M_c \sim \left(\frac{C_2(\text{R})\alpha_c(F_{T\pi})}{\alpha}\right)^{1/2} \frac{F_{T\pi}}{f_\pi} 35.5\ \text{MeV} \simeq 170\sqrt{C_2(\text{R})}\ \text{GeV}\,. \tag{4.1}$$

Here $f_\pi \simeq 95$ MeV.

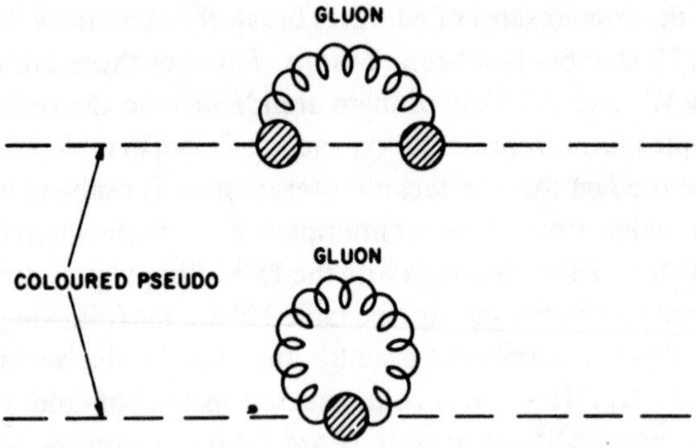

Fig. 1. Graphs that contribute mass to the coloured pseudos.

Next we discuss the electroweak contribution to the mass of a pseudo. The electromagnetic contribution to the mass of a pseudo is

$$M_{\rm EM} \sim e_{\rm ps} \frac{F_{\rm T\pi}}{f_\pi} 35.5\ {\rm MeV} \simeq e_{\rm ps}\, 47\ {\rm GeV}\,. \tag{4.2}$$

Here $e_{\rm ps}$ is the electric charge of the pseudo in units of the pion's electric charge. The contribution of the complete electroweak gauge group to the mass of a pseudo (which is more relevant at these scales $\sim F_{\rm T\pi}$), barring possible cancellations [2], can be of order of 100 GeV if the pseudo carries $SU(2)_{\rm L}$ quantum numbers. If a pseudo carries both $SU(3)_{\rm colour}$ and electroweak charges then its mass is given by $m \sim \sqrt{m_{\rm c}^2 + m_{\rm EW}^2}$.

The third source from which pseudos receive a contribution to their mass is the heavy ~10 TeV ETC generators. This mass contribution (called $m_{\rm ETC}$) is quite model dependent and pseudo dependent. It can be anywhere from 0 GeV to ~100 GeV, depending on the model and on the pseudo. For example, in the Fahri–Susskind SU(7) model there is a massless neutral axion [11]. This is a general feature of models with several decoupled representations [2,14].

If the group is enlarged to O(14), then the representations become coupled at ~100 TeV and the axion receives a mass which is roughly of order of 1 GeV. In general, the ETC models have to have ETC vector bosons with masses as small as 1 TeV if there exists a quark or lepton with mass of 30 GeV (see subsect. 7.1). If these relatively light ETC vector bosons explicitly break the symmetry associated with a given pseudo then it is easily seen that this pseudo receives a contribution of $M_{\rm ETC}^2 \sim (100\ {\rm GeV})^2$ to its mass squared. The reason for this relatively large contribution is that the scale of technicoloured spontaneous symmetry breaking of order of $\langle \bar{E}E\rangle^{1/3} \sim 300$ GeV (see subsect. 7.1) and the masses of the lightest ~1 TeV ETC vector bosons which explicitly break the chiral symmetry corresponding to the pseudo are *not* very far apart. Therefore the chiral symmetry is broken relatively strongly and the corresponding mass^2 contribution to the pseudo is not small

$\sim(100\ \text{GeV})^2$. This contribution is *not* in general very important for the pseudos that already have mass^2 contributions from $SU(3)_{\text{colour}}$ or the electroweak $SU(2)_L$ interactions. However, this contribution is very important for the axionic objects and for the charged partners $\Pi_{\pm}^{TF}$ of the axion that were shown, by Eichten and Lane [2], to receive only a small mass from the electroweak interactions. The only way for axions or their charged partners to avoid getting large masses ~100 GeV from the 1 TeV ETC generators is if they correspond to currents which are singlets of a subgroup of the ETC group which is connected to the remainder of the ETC group by generators that have masses of order of 3 TeV or heavier.

5. Decays of pseudos

In this section we wish to make some comments on the dominant decays of pseudos. We have two types of dominant possibilities. The pseudo can decay into $SU(3)\times SU(2)_L\times U(1)$ gauge vector bosons or into ordinary fermions. The dominant decays of the coloured technipions and technieta are into pairs of gauge bosons. The dominant decays of the majority of pseudos are into pairs of ordinary fermions. Occasionally global conservation laws forbid such a decay [11]. In this case a pseudo will have to decay into four ordinary fermions and it will be quite narrow.

Let us begin our discussion with the coloured T-pions and T-eta. The dominant decay modes of these pseudos are (see fig. 2):

$$\Pi^{c\pm}\to W^{\pm}+\text{gluon}\,,$$
$$\Pi^{c3}\to Z\ \text{or}\ \gamma+\text{gluon}\,,$$
$$\Pi^{c0}\to 2\ \text{gluons}\,. \tag{5.1}$$

To give a very rough estimate of the orders of magnitude of the widths of these particles based on scaling arguments we simply note that the relevant mass scales in the problem are $\langle\bar{E}E\rangle^{1/3}\sim 300$ GeV, pseudo-masses ~300 GeV, and $F_{T\pi}\sim 125$. They are all of the same order. Thus we expect an overestimate of all these widths to be of order

$$\Gamma(\Pi\text{'s})\sim\alpha_s^2\ 300\ \text{GeV}\sim\frac{\alpha_s\alpha}{\sin^2\theta_W}\ 300\ \text{GeV}\sim 3\ \text{GeV}\,. \tag{5.2}$$

The majority of the remaining pseudos decay by converting the technifermions into a pair of ordinary fermions. This generally happens by the exchange of a heavy ETC gauge boson whose mass we denote by μ_{ETC} (see fig. 3). μ_{ETC} can range from 1 TeV to ~250 TeV, depending on the transition that it mediates and/or the ETC model under discussion (see subsect. 7.1). A very rough order of magnitude estimate for the width of a pseudo decaying in this way is $\Gamma\sim\mu_{\text{ETC}}^{-4}M^5$ where M is the pseudo-mass. This yields a variety of widths bounded from above by approximately $\Gamma\sim 3$ GeV (for $M\simeq 310$ GeV, $\mu_{\text{ETC}}\sim 1$ TeV). More characteristic are width of

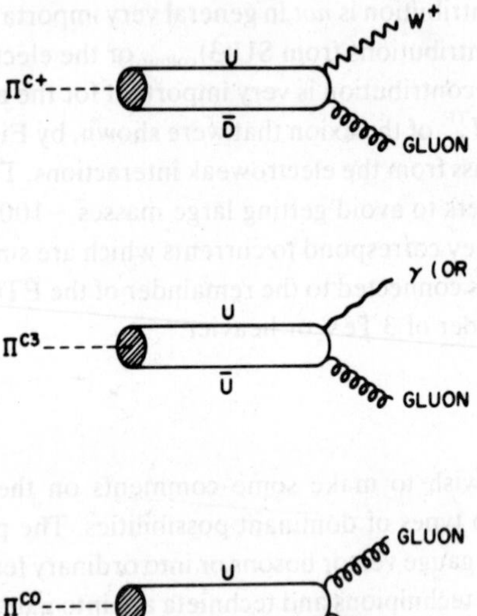

Fig. 2. Important decay modes of the coloured technipions (Π^{c+}, Π^{c3}) and the coloured technieta.

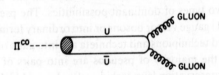

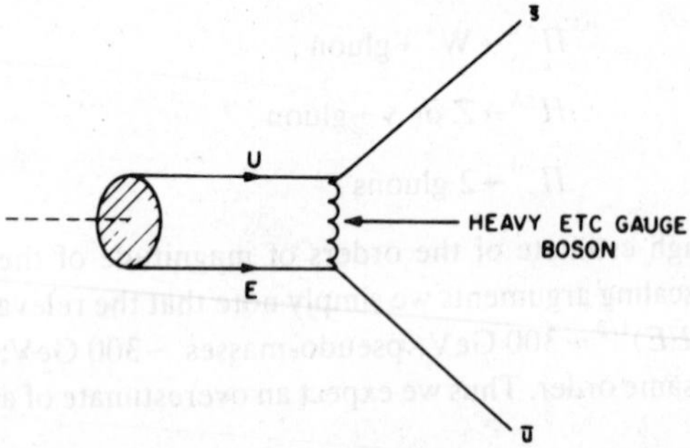

Fig. 3. A usual decay mode of a common pseudo into two ordinary fermions.

order $\Gamma \sim 20$ MeV corresponding to $M \sim 200$ GeV and $\mu_{\rm ETC} \sim 2$ TeV). The width can be as small as a few eV for a pseudo of $M \sim 200$ GeV decaying *via* $\mu_{\rm ETC} \sim 100$ TeV generators only.

Thus we see that the widths are quite pseudo dependent and model dependent*. In general we expect the pseudos that can decay into a pair of *heavy* ordinary fermions to be wider because these transitions are expected to be mediated by the lighter ETC

* Much of the model dependence arises because ordinary fermions may either all get their masses from technifermions *via* ETC gauge bosons of varying masses (see fig. 11) or the light fermions get their masses from heavy ordinary fermions *via* not so heavy technicolourless ETC gauge bosons.

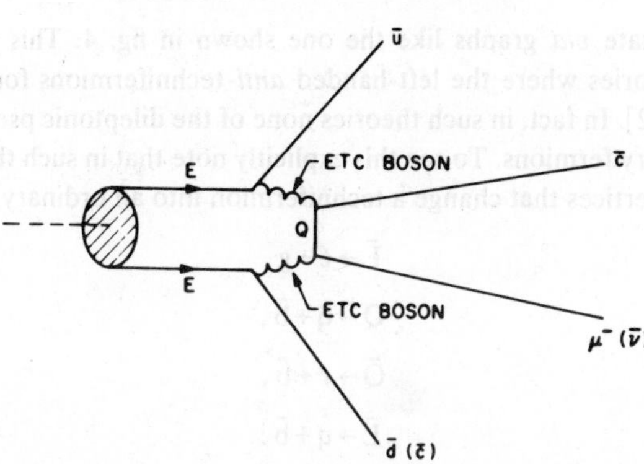

Fig. 4. A (dileptonic) pseudo decays into four ordinary fermions. This is the dominant decay mode of pseudos whch are forbidden to decay into two fermions by some conservation laws.

generators ($\mu_{\rm ETC} \sim 1$ TeV) which are responsible for the heavy fermions' masses (see subsect. 7.1 and ref. [1]). Thus in this respect these pseudos are higgsomimes, i.e., they tend to couple in proportion to the mass of the fermions.

Another model-dependent and very interesting feature occurs when a pseudo cannot decay into a pair of ordinary fermions and instead it decays into, say, a quartet of ordinary fermions [11] by exchange of two heavy ETC vector bosons (see fig. 4). In this case the width is of order of $\Gamma \sim M^9/\mu_{\rm ETC}^8$. In the Fahri–Susskind model this occurs for the ditechnilepton EN($\rightarrow$ cud $\bar{\nu}_\mu$). This pseudo is protected against 2-body decays by an exact fermion number conservation. In general, such conservation laws are quite natural and necessary in ETC models. They are nothing but the statements of baryon and lepton number conservations generalized to include T-coloured and ETC-coloured particles.

In the case of the doubly charged ditechnielectron EE (see table 1c) it is easy to find sufficient conditions under which it is forbidden to decay into a pair of ordinary fermions. The pair of ordinary fermions into which EE decays would have to consist of two negatively charged leptons because these are the only two fermion states with charge −2. Thus if the theory does not have ETC vertices (see fig. 5) in which a technielectron E becomes a negatively charged lepton by the emission of an ETC gauge boson then EE cannot decay to any two-fermion state. It can decay into a

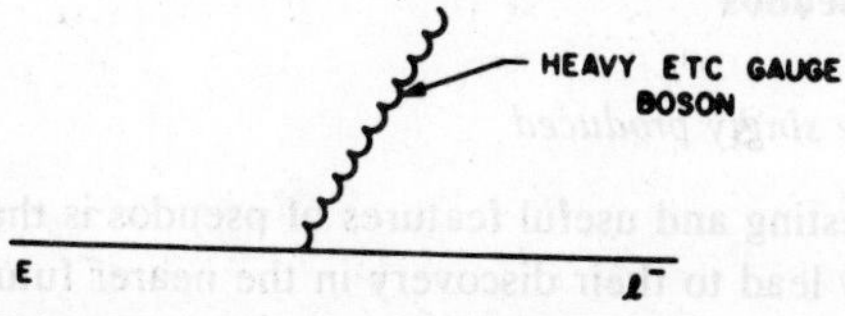

Fig. 5. A technielectron E turns into a negative lepton ℓ^- by the emission of an ETC gauge boson. The absence of such graphs is sufficient to ensure that the ditechnielectron EE decays in higher order *via* graphs like fig. 4. These graphs are absent in theories where the *anti*-techniquarks form left-handed electroweak doublets.

four-fermion state *via* graphs like the one shown in fig. 4. This is exactly what happens in theories where the left-handed *anti*-technifermions form electroweak doublets [11, 12]. In fact, in such theories none of the dileptonic pseudos can decay into two ordinary fermions. To see this explicitly note that in such theories the only allowed ETC vertices that change a technifermion into an ordinary fermion are

$$\begin{aligned} \bar{L} &\to \ell + g\,, \\ Q &\to \bar{q} + \bar{b}\,, \\ \bar{Q} &\to \ell + \bar{b}\,, \\ \bar{L} &\to q + \bar{b}\,. \end{aligned} \tag{5.3}$$

Here $\bar{b}$ and g denote ETC gauge bosons. The reason that no other such processes are allowed follows from the fact that the chirality and the $SU(2)_{electroweak}$ (EW) quantum numbers of the fermions are conserved in the processes (5.3). The latter follows because the ETC gauge bosons do not carry $SU(2)_{EW}$ quantum numbers. (The $SU(2)_{EW}$ group breaks off at the GUT scale $\gtrsim 10^{14}$ GeV.)

The processes (5.3) conserve the following numbers:

$$\begin{aligned} N_1 &= \tfrac{3}{2}N_g + N_q - \tfrac{3}{2}N_L - \tfrac{1}{2}N_Q - \tfrac{1}{2}N_b\,, \\ N_2 &= -\tfrac{3}{2}N_g + N_l + \tfrac{1}{2}N_L - \tfrac{1}{2}N_Q + \tfrac{1}{2}N_b\,. \end{aligned} \tag{5.4}$$

These conservation laws forbid the decay of any ditechnilepton ($N_1 = -3$) into two ordinary fermions ($N_1 \geq -2$).

The four members of the axionic family Π_f^{TF} ($f = 0, 1, 2, 3$) decay as follows: Π_0^{TF} likes to decay into two gluons, $\Pi_0^{TF} \to 2$ gluons, with a width of very rough order $\sim \alpha_s^2 M(\Pi_0^{TF})$. Another possible important decay mode is $\Pi_0^{TF} \to \bar{q}_H + q_H$ where the subscript H denotes the heaviest quark for which this decay can go. Similarly Π_3^{TF} decays into $\Pi_3^{TF} \to \bar{q}_H + q_H$, also $\Pi_3^{T\pi} \to \gamma + \gamma$ and $\Pi_+^{TF} \to \bar{d}_H + u_H$. If Π_3^{TF} and Π_+^{TF} are sufficiently heavy then the modes $\Pi_3^{TF} \to \gamma + Z$ and $\Pi_+^{TF} \to W^+ + \gamma$ will be important.

Careful estimates of the decays of the various pseudos will be presented in a future paper.

6. Production of pseudos

5.1. Pseudos can be singly produced

One of the interesting and useful features of pseudos is that they can be *singly* produced. This may lead to their discovery in the nearer future. However, only a limited number of pseudos have significant couplings to ordinary light quarks and leptons and have a good chance of being singly produced. These include the coloured technipions, the coloured technieta, and the colour singlets Π_f^{TF} ($f = 0, 1, 2, 3$) which

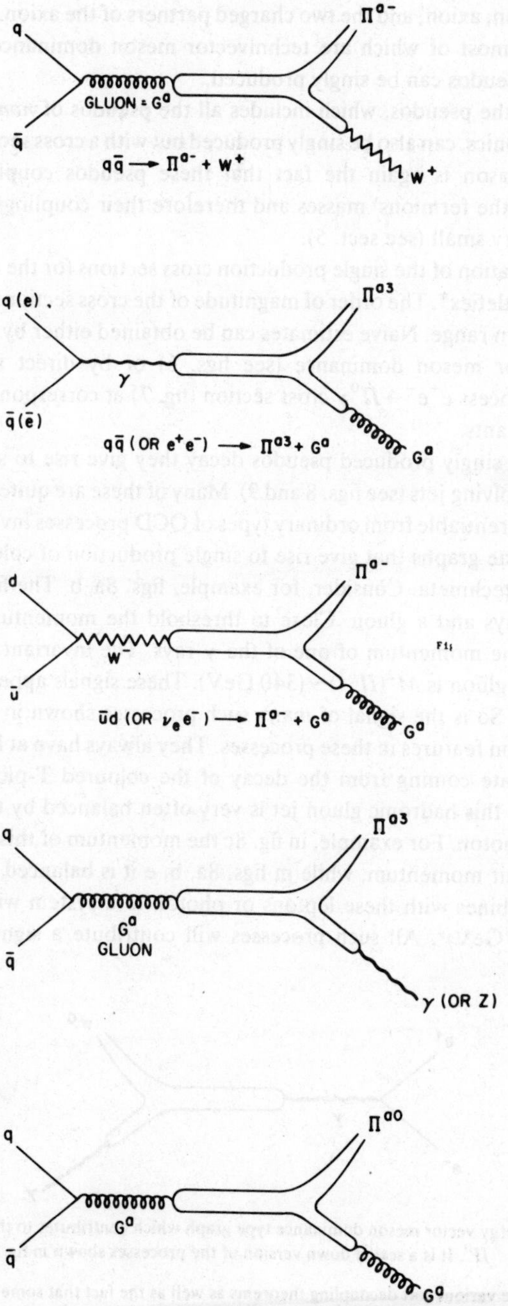

Fig. 6. Possible graphs that may contribute to the single production of coloured T-pions and T-eta.

are the paraxion, axion, and the two charged partners of the axion. In figs. 6 we show some graphs, most of which are technivector meson dominance type graphs, by which these pseudos can be singly produced.

The rest of the pseudos, which includes all the pseudos of *non-vanishing triality* and the dileptonics, can also be singly produced but with a cross section which is quite small. The reason is again the fact that these pseudos couple to fermions in proportion to the fermions' masses and therefore their couplings to ordinary light leptons are very small (see sect. 5).

The computation of the single production cross sections for the various pseudos is not free of subtleties*. The order of magnitude of the cross sections shown in figs. 6 is in the nanobarn range. Naive estimates can be obtained either by direct application of technivector meson dominance (see figs. 6) or by direct scaling up of the isomorphic process $e^+e^- \to \Pi^0\gamma$ cross section (fig. 7) at corresponding energies and coupling constants.

When these singly produced pseudos decay they give rise to several interesting final states involving jets (see figs. 8 and 9). Many of these are quite exotic and should be clearly differentiable from ordinary types of QCD processes involving jets. In figs. 8 we show some graphs that give rise to single production of coloured technipions and coloured technieta. Consider, for example, figs. 8a, b. The final state involves two hard γ-rays and a gluon. Close to threshold the momentum of the gluon is balanced by the momentum of one of the γ-rays. The invariant mass of this hard γ-ray and the gluon is $M^2(\Pi^{a3}) \sim (340\ \mathrm{GeV})$. These signals appear to be clear and unmistakable. So is the signal of many such processes shown in figs. 8. There are several common features in these processes. They always have at least one gluon jet in the final state coming from the decay of the coloured T-pions or T-eta. The momentum of this hadronic gluon jet is very often balanced by the momentum of leptons or a photon. For example, in fig. 8c the momentum of this gluon is balanced by a lepton pair momentum, while in figs. 8a, b, e it is balanced by a photon. The gluon jet combines with these leptons or photon to a system with invariant mass $M^2(\Pi) \sim (340\ \mathrm{GeV})^2$. All such processes will contribute a significant number of

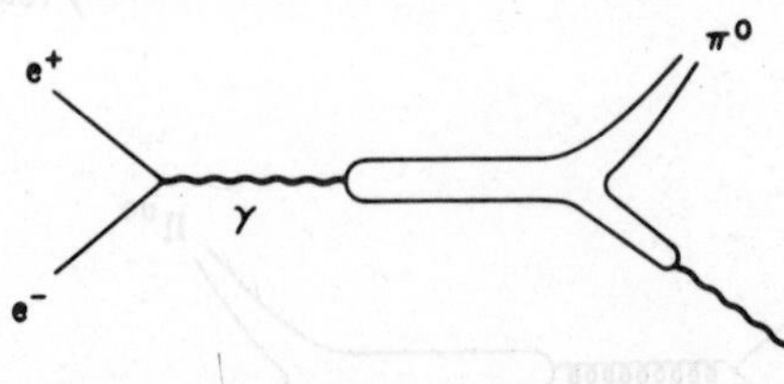

Fig. 7. A low-energy vector meson dominance type graph which contributes to the single production of Π^0. It is a scaled down version of the processes shown in fig. 6.

* We refer to the various soft decoupling theorems as well as the fact that some decay and production graphs can only go through anomalies (à la $\Pi^0 \to \gamma\gamma$).

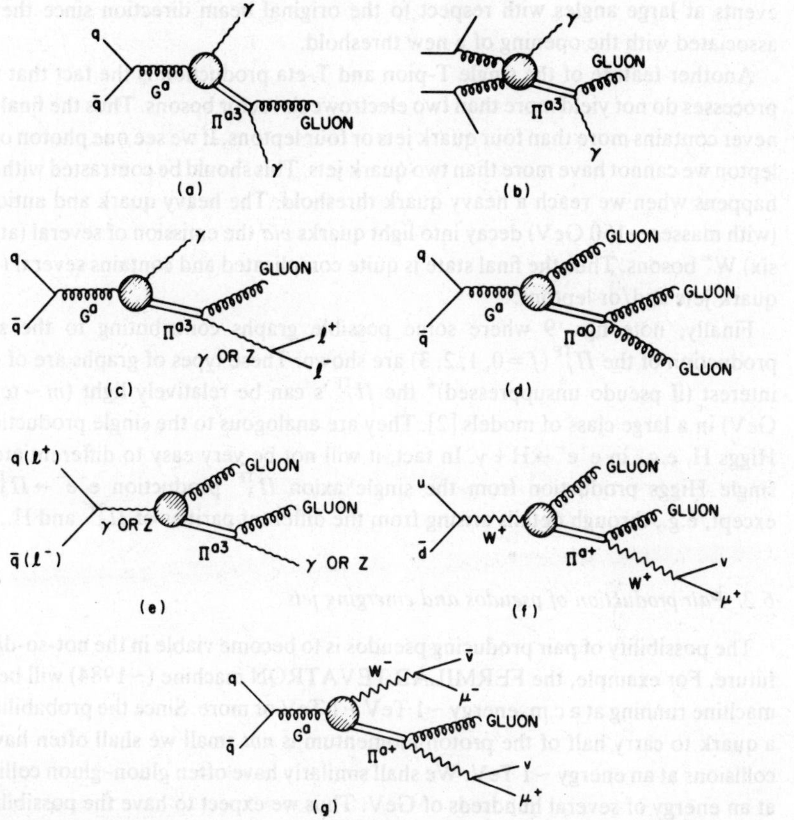

Fig. 8. Processes in which coloured T-pions and T-eta are singly produced result in some interesting events containing jets. Many of them have clear and unmistakable signatures. For example, (a) has a *single* gluon jet and two hard photons.

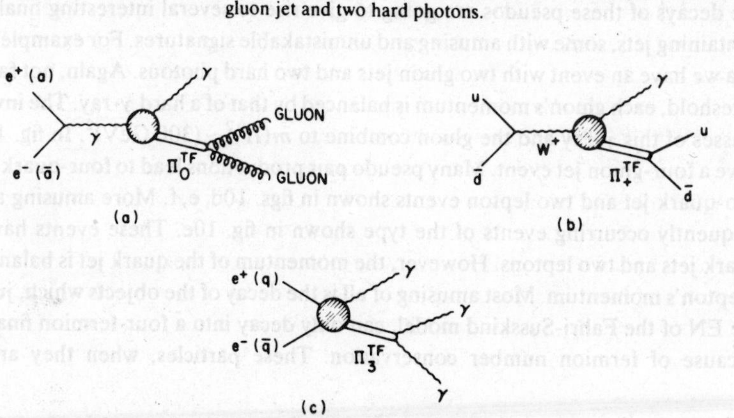

Fig. 9. Possible events arising from the single production of members of the axionic family.

events at large angles with respect to the original beam direction since they are associated with the opening of a new threshold.

Another feature of the single T-pion and T-eta production is the fact that these processes do not yield more than two electroweak vector bosons. Thus the final state never contains more than four quark jets or four leptons. If we see one photon or one lepton we cannot have more than two quark jets. This should be contrasted with what happens when we reach a heavy quark threshold. The heavy quark and antiquark (with masses ~150 GeV) decay into light quarks *via* the emission of several (at least six) $W^{\pm}$ bosons. Thus the final state is quite complicated and contains several (~14) quark jets and/or leptons.

Finally, note figs. 9 where some possible graphs contributing to the single production of the Π_f^{TF} $(f=0,1,2,3)$ are shown. These types of graphs are of great interest (if pseudo unsuppressed)* the Π_f^{TF}'s can be relatively light (m ~ tens of GeV) in a large class of models [2]. They are analogous to the single production of Higgs H, e.g., in $e^+e^- \to H+\gamma$. In fact, it will not be very easy to differentiate this single Higgs production from the single axion Π_3^{TF} production $e^+e^- \to \Pi_3^{TF}+\gamma$ except, e.g., through details arising from the different parities of Π_3^{TF} and H.

6.2. *Pair production of pseudos and emerging jets*

The possibility of pair producing pseudos is to become viable in the not-so-distant future. For example, the FERMILAB TEVATRON machine (~1984) will be a $\bar{p}p$ machine running at a c.m. energy ~1 TeV+1 TeV or more. Since the probability for a quark to carry half of the proton momentum is *not* small we shall often have $q\bar{q}$ collisions at an energy ~1 TeV. We shall similarly have often gluon–gluon collisions at an energy of several hundreds of GeV. Thus we expect to have the possibility of pair producing even the heaviest of the pseudos shown in table 1 (mass ~300 GeV). In figs. 10 we show a few processes that result in pair production of psuedos. Again the decays of these pseudos are going to give rise to several interesting final states containing jets, some with amusing and unmistakable signatures. For example, in fig. 10a we have an event with two gluon jets and two hard photons. Again, not far from threshold, each gluon's momentum is balanced by that of a hard γ-ray. The invariant masses of this γ-ray and the gluon combine to $m(\Pi)^2 \sim (300\ \mathrm{GeV})^2$. In fig. 10b we have a four-gluon jet event. Many pseudo pair productions lead to four-quark jets or two-quark jet and two lepton events shown in figs. 10d, e, f. More amusing are the frequently occurring events of the type shown in fig. 10e. These events have two quark jets and two leptons. However, the momentum of the quark jet is balanced by a lepton's momentum. Most amusing of all is the decay of the objects which, just like the EN of the Fahri-Susskind model, can only decay into a four-fermion final state because of fermion number conservation. These particles, when they are pair

* See previous footnote.

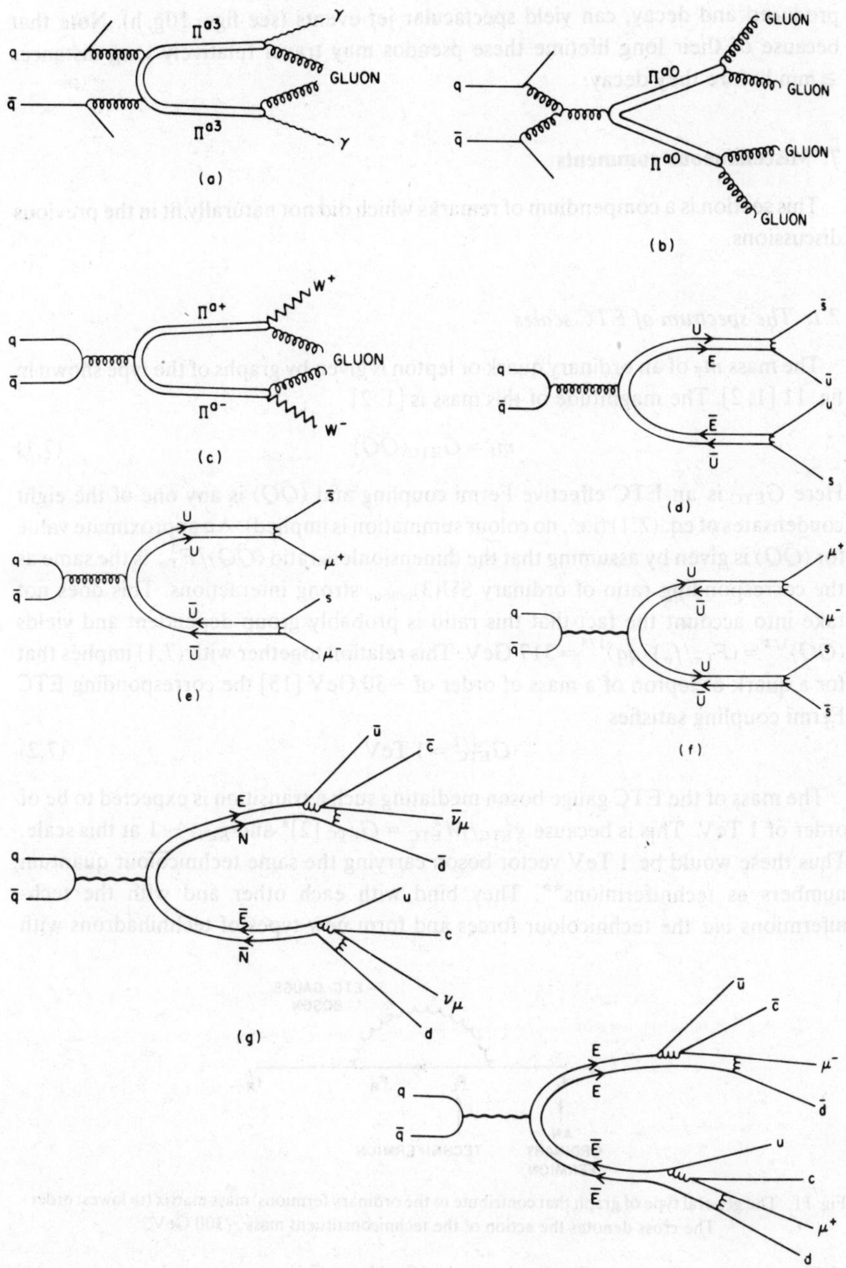

Fig. 10. Some jetful events occurring when various types of pseudos are pair produced. The most spectacular of these are (g) and especially (h).

produced and decay, can yield spectacular jet events (see figs. 10g, h). Note that because of their long lifetime these pseudos may travel relatively long distances $\gtrsim$ mm before they decay.

7. Miscellaneous comments

This section is a compendium of remarks which did not naturally fit in the previous discussions.

7.1. The spectrum of ETC scales

The mass m_F of an ordinary quark or lepton is given by graphs of the type shown in fig. 11 [1, 2]. The magnitude of this mass is [1, 2]

$$m_F \simeq G_{ETC} \langle \bar{Q}Q \rangle \,. \tag{7.1}$$

Here G_{ETC} is an ETC effective Fermi coupling and $\langle \bar{Q}Q \rangle$ is any one of the eight condensates of eq. (2.1) (i.e., no colour summation is implied). An approximate value for $\langle \bar{Q}Q \rangle$ is given by assuming that the dimensionless ratio $\langle \bar{Q}Q \rangle / F^3_{T\pi}$ is the same as the corresponding ratio of ordinary $SU(3)_{colour}$ strong interactions. This does not take into account the fact that this ratio is probably group dependent and yields $\langle \bar{Q}Q \rangle^{1/3} = (F_{T\pi}/f_\pi) \langle \bar{q}q \rangle^{1/3} \simeq 317$ GeV. This relation together with (7.1) implies that for a quark or lepton of a mass of order of ~30 GeV [15] the corresponding ETC Fermi coupling satisfies

$$G_{ETC}^{-1/2} \simeq 1 \text{ TeV} \,. \tag{7.2}$$

The mass of the ETC gauge boson mediating such a transition is expected to be of order of 1 TeV. This is because $g^2_{ETC}/M^2_{ETC} = G_{ETC}$ [2]* and $g_{ETC} \sim 1$ at this scale. Thus these would be 1 TeV vector boson carrying the same technicolour quantum numbers as technifermions**. They bind with each other and with the technifermions *via* the technicolour forces and form new types of technihadrons with

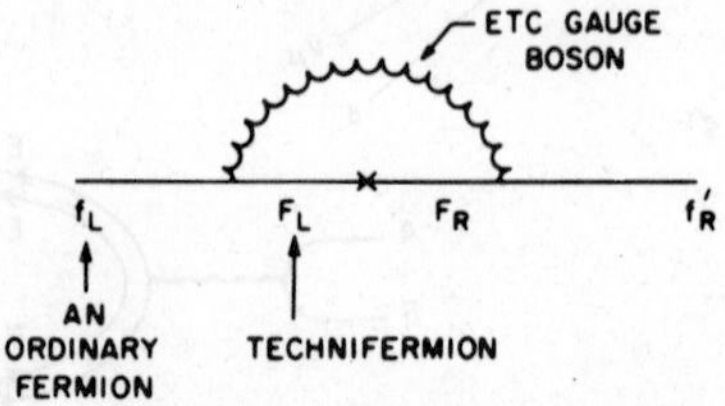

Fig. 11. The general type of graph that contribute to the ordinary fermions' mass matrix (to lowest order). The cross denotes the action of the techniconstituent mass ~300 GeV.

* This can be seen directly by Fierzing the graph of fig. 11 and (7.1).

** The existence of ~1 TeV technicolourless (but possibly colourful and charged) vector bosons is also possible. They can cause transitions among the heaviest ordinary fermions or among technifermions.

masses of order of 1 TeV. Thus, e.g., there will exist ~1 TeV technibaryon made out of such an ETC gauge boson and a technifermion.

On the other extreme in theories where the lightest quarks and leptons receive their mass by the lowest-order graphs of fig. 11 there must exist ETC gauge bosons with masses $G_{\mathrm{ETC}}^{-1/2} \sim 250$ TeV.

7.2. SU(3)$_{\text{colour}}$ β-functions in the presence of several coloured pseudos

The large proliferation of coloured pseudos (see table 1) has some interesting effects on the β-function of ordinary colour forces. These effects mainly arise at an energy region centered around 600 GeV which corresponds to twice the mass of the pseudos that carry large colour (sextets and octets). Consider, for example, the b-function for energies between 600 GeV and a mass of order of the technirho mass ≳1000 GeV. In this region the value of the b-function is expected to be dictated by the ordinary quarks, gluons, and the coloured pseudos. The formula for SU(3)$_{\text{colour}}$ is

$$-b = 11 - \tfrac{2}{3}N_{\mathrm{F}}(3) - \tfrac{1}{6}\sum_{\mathrm{s}} T(R_{\mathrm{s}})$$

$$= 11 - \tfrac{2}{3}N_{\mathrm{F}}(3) - \tfrac{1}{12}N_{\mathrm{s}}(3) - \tfrac{1}{2}N_{\mathrm{s}}(8) - \tfrac{5}{12}N_{\mathrm{s}}(6)\,. \qquad (7.3)$$

Here $N_{\mathrm{F}}(3)$ is the number of 4-component Dirac fermion triplets, and $N_{\mathrm{s}}(3)$, $N_{\mathrm{s}}(8)$, and $N_{\mathrm{s}}(6)$ are the numbers of *real scalar* triplets, octets, and sextets, respectively. Consider now a model with a real technicolour group with an antisymmetric invariant symbol. The pseudo content of such a model is shown in tables 1a, b. If we assume three families of ordinary quarks, then the b becomes equal to $b \sim 2\frac{1}{3}$ instead of the $b \sim 7$ which we would have in the absence of pseudos. For four-quark families $b \sim 1$ instead of $b \sim 5\frac{2}{3}$.

In models where the techniquarks belong to a real representation with a symmetric invariant symbol, the colour contents of pseudos are slightly different (see table 1c). In particular the number of sextets increases from two to six. This changes the value of b from $2\frac{1}{3}$ to 1 for three-quark families. For four-quark families b becomes negative, $b \simeq -\frac{1}{3}$. At energies quite a bit higher than the technirho mass >1 TeV the β-function receives contributions only from the coloured T-quarks, the ordinary quarks and gluons, and also from the lightest coloured ETC vector bosons whose masses are ~1 TeV (see subsect. 7.1).

7.3. Stability of technihadrons

The question of whether there exists a technifermion number which is conserved (except at the superunification [4, 8, 5] scale ≳10^{14} GeV) is an interesting one. The answer to this question is model dependent and no simple criterion has yet been formulated. It does not, for example, have to do with the technicolour group *per se*. To prove this, consider an SU(5)$_{(\mathrm{ETC})}$ which breaks in the pattern SU(5)$_{(\mathrm{ETC})}$ →

$SU(2)_{TC} \times SU(3)_C$. The fermion content, say, consists of SU(5) quintets which break down to T-quark doublets and quark colour triplets. Then the T-baryons consist of two T-quarks and are stable because they carry baryon number equal to $\frac{2}{3}$ and no colour singlet light quark state carries fermion number $= \frac{2}{3}$. In contrast, if we enlarge this model in straight-forward ways [11, 12, 14] to include leptons, always keeping the T-colour group SU(2), then the T-baryon can in general decay into fermion pairs [11, 12, 14]. Thus the stability of T-baryons does not have to do only with the TC group *per se* but also with how the TC group fits with the rest of the symmetries.

In general we expect the theories with stable T-baryon to be problematic in view of the fact that they would naively be expected to predict comparable numbers of technibaryons and baryons in the universe [16]. This is because quarks and T-quarks become unified at energies of order of 10–100 TeV which is much earlier than the energies of $\gtrsim 10^{14}$ GeV [4, 5, 8] at which the fermion numbers of the universe were created [16].

7.4. *Horizontal transitions*

An interesting feature of the ETC models is the necessity of unifying all of the quark and lepton flavours at a scale of $\lesssim 100$ TeV. This has to be done in order that quarks and leptons receive their masses without the appearance of extremely light pseudos.

This early unification has several important consequences. It implies the existence of new horizontal currents which cause transitions between quark flavours, leptons, and also between quarks and leptons. Many of these horizontal currents can be dangerous and screen out several models because they mediate rare transitions at unacceptable levels: these will be discussed in a future paper [17]. A possibly interesting and novel feature of ETC models which we want to just mention is the occurrence of direct quark-lepton transitions. These can happen *via* the emission of an ETC boson (or a coloured pseudo). They will contribute several interesting effects in the future $p\bar{p}$, e^+e^-, and especially in the possible ep machines [17]. These direct $q \leftrightarrow \ell$ transitions at sufficiently large energies are going to be more important than the direct electroweak annihilation graphs which contribute to the Drell-Yan type processes $q\bar{q} \leftrightarrow \ell\bar{\ell}$. This is due to the obvious fact that the direct annihilation cross sections die out as $\sim \alpha^2/s$ while the t-exchange direct $q - \ell$ transitions do not. Also, the angular dependences will be different in the two cases. In the high-energy ep machines the direct $q - \ell$ transition will yield a quark jet in the original e-direction and *vice versa* [17].

7.5. *R*

A quantity of very futuristic interest is $R(\ell^+\ell^- \to \text{anything})$ (ℓ = lepton) at the TeV range. Let us for simplicity discuss $R(\ell^+\ell^- \to \gamma \to \text{any})$. The contribution ΔR_{ps} to this

R coming from the pseudos of tables 1a, b is $\Delta R_{ps} \simeq 9\frac{1}{3}$. This contribution is effective at energies between 600 GeV and ~TeV. At higher energies the contribution ΔR_{TQ} to R comes mainly from techniquarks and also the lighter ETC vector bosons. The contribution of the T-quarks is $\Delta R_{TQ} \simeq \frac{8}{3}N$ where N is the size of the T-colour representation to which the T-fermions belong. Thus for $N \geqslant 4$ the contribution of the T-quarks is bigger than that of the pseudos.

7.6. *The lightest pseudos*

It is important and urgent to study in detail the properties of the colourless pseudos in various types of large classes of models. For they are expected to have masses between few GeV and $\lesssim$100 GeV (see sect. 4) which are smaller than the smallest expected masses of coloured pseudos ~200 GeV. They should be seen in the $\bar{p}p$ CERN machine (1986?) and possibly in the upcoming lepton machines. Unfortunately, their dominant properties are much more model dependent than those of the heavier coloured pseudos (especially the colour octets). This is because their dominant properties strongly depend on the heavy ETC sector which is both model dependent and poorly understood.

The colourless pseudos are the four axionic pseudos Π_f^{TF} ($f = 0, 1, 2, 3$) and, more amusingly, the three possible ditechnileptons EE, EN, and NN. The latter pseudos are likely to be very fascinating. It is, e.g., quite possible that baryon and lepton conservation laws allow them to decay only into final states involving at least four ordinary fermions. In fact, as discussed before (sect. 5), the doubly charged dielectron EE cannot decay into a pair of ordinary fermions if the ETC vertices of fig. 11 are not present in a theory. In this case it would decay into four fermions and would be long enough lived to leave visible tracks. Such a particle would yield spectacular jetful events.

8. Summary

In this paper we made some general observations on some likely features of the scalarless models with extended technicolour. We mainly discussed the spectrum and gross properties of the pseudos that occur in these models if we assume the existence of one family of technifermions. This assumption was motivated by super-unified GUT type models. Obviously none of the features that we mentioned are inescapable. Several of the possible consequences mentioned would constitute a relatively low energy evidence of the ETC ideas. In particular, finding pseudos with the right combinations of strong-electroweak quantum numbers in roughly the right mass range and with the right mass pattern would be a suggestive evidence for ETC. It is unnatural and unlikely to manufacture such a pseudospectrum in models with elementary scalars.

An important feature is the possible existence of several coloured pseudos. In the future we are going to have available high-energy p̄p machines of hopefully decent luminosities. These machines will also be relatively luminous gluon-gluon machines. This will greatly facilitate the production and study of new coloured states. The dominant process at high energies will be 2-gluon collisions like the 2-photon collisions of Brodsky, Takahashi, and Terazawa [18]. At still higher energies (say, the possible 20 + 20 TeV machines) we will have a scaled up replica of ordinary strong interaction physics, e.g., almost constant cross sections, multiperipheral processes, T-pion (coloured or not) proliferation, etc.

It is a pleasure to thank my Stanford friends Marie De Crombrugghe, Stuart Raby, and Leonard Susskind for several very valuable discussions. I also wish to thank James Bjorken, Estia Eichten, Eddie Fahri, Gerald Feinberg, Fokion Hadjioannou, John Iliopoulos and Ken Lane for their comments and interest.

Much of this work was done while I was at Columbia University. I wish to thank the Columbia Theory Group as well as the hospitality of the CERN Theory Group and the Niels Bohr Institute. This work was supported in part by NSF grant PHY 78–26847 at Stanford University and by a Department of Energy grant at Columbia University.

Note added

After completion of this paper I received a paper by Beg, Politzer and Ramond [19]. They discuss some interesting properties of colourless Pseudogoldstones. Some of their conclusions, in particular the mass estimates, agree with those of ref. [20].

References

[1] S. Dimopoulos and L. Susskind, Nucl. Phys. B 155 (1979) 237.
[2] E. Eichten and K. Lane, Dynamical breaking of weak interaction symmetries, Harvard preprint HUTP 79/A002 (1979).
[3] S. Dimopoulos and L. Susskind, A technicolured solution to the strong *CP* problem, Columbia-Stanford preprint (February, 1979);
W. Fishler, The *CP* problem in two dimensions, Los Alamos preprint 79–0263.
[4] H. Georgi and S.L. Glashow, Phys. Rev. Lett. 32 (1974) 438;
J. Pati and A. Salam, Phys. Rev. D8 (1973) 1240; D10 (1974) 275;
H. Fritzch and P. Minkowski, Ann. of Phys. 93 (1975) 193;
F. Hadjioannou, Univ. of Athens preprint (1978);
H. Georgi and D.V. Nanopoulos, Harvard preprints HUTP-78/A039; Nucl. Phys. B155 (1979) 52.
[5] A.J. Buras, J. Ellis, M.K. Gaillard and D.V. Nanopoulos, Nucl. Phys. B 135 (1978) 66.
[6] K. Wilson, unpublished.
[7] L. Susskind, Dynamics of spontaneous symmetry breaking in the Weinberg-Salam theory, SLAC preprint (May, 1978);
S. Weinberg, Phys. Rev. D 19 (1978) 1277
[8] H. Georgi, H.K. Quinn and S. Weinberg, Phys. Rev. Lett. 33 (1974) 451.
[9] S. Weinberg, Phys. Rev. D13 (1976) 974.

[10] S. Weinberg, Phys. Rev. Lett. 19 (1967) 1264;
A. Salam, *in* Elementary particle physics, ed. N. Svartholm (Almqvist and Wiksell, Stockholm, 1968) p. 367.
[11] E. Fahri and L. Susskind, A technicoloured GUT, SLAC Pub. 2361 (1979); unpublished;
M. Gell-Mann, P. Ramond and R. Slansky, in preparation.
[12] S. Raby, unpublished.
[13] R. Peccei and H. Quinn, Phys. Rev. Lett. 38 (1977) 1440; Stanford University Report No. ITP 572 (1977);
S. Weinberg, Phys. Rev. Lett. 40 (1978) 223;
F. Wilczek, Phys. Rev. Lett. 40 (1978) 279.
[14] S. Dimopoulos and L. Susskind, unpublished.
[15] J. Bjorken, Speculations on the pattern of quark and lepton masses, SLAC Pub. 2195 (1978).
[16] M. Yoshimura, Phys. Rev. Lett. 41 (1978) 381;
S. Dimopoulos and L. Susskind, Phys. Rev. D18 (1978) 4500;
D. Toussaint, S. Treiman, F. Wilzcek and A. Zee, Phys. Rev. D19 (1979) 1036;
S. Weinberg, Cosmological production of baryons, Harvard preprint HUTP-78/A040;
S. Dimopoulos and L. Susskind, Baryon asymmetry in the very early universe, Stanford University Report No. ITP 616;
J. Ellis, M.K. Gaillard and D.V. Nanopoulos, CERN preprint (November, 1978).
[17] S. Dimopoulos, S. Raby and L. Susskind, to be published.
[18] H. Terezawa, Rev. Mod. Phys. 45 (1973) 615.
[19] M.A. Beg, H.D. Politzer and P. Ramond, Experimental characteristics of dynamical pseudo-Goldstone bosons, Caltech-Rockefeller preprint COO-22323-189.
[20] S. Dimopoulos, Phys. Lett. 84B (1979) 435.

Nuclear Physics B182 (1981) 529–545

CAN ONE TEST TECHNICOLOUR?

John ELLIS
CERN, Geneva, Switzerland

Mary K. GAILLARD
LAPP, Annecy-le-Vieux, France

D.V. NANOPOULOS and P. SIKIVIE
CERN, Geneva, Switzerland

Received 27 October 1980

We estimate the couplings to ordinary particles of the lightest bound states in technicolour theories and discuss the resulting phenomenology. We compute their couplings to light gauge bosons through axial anomalies and also estimate their non-anomalous couplings at low energies. We estimate their couplings to fermions under the general simplifying assumption that each fermion acquires its mass from a unique technifermion condensate ("monophagy"), in which case they are naturally flavour conserving and relatively well-defined. We find that the classic Higgs search experiments (toponium $\to H^0 + \gamma$, $e^+e^- \to H^+H^-$, $e^+e^- \to Z^0 + H^0$) enable one to make a decisive discrimination between elementary and composite models of spontaneous symmetry breaking. We also emphasize the interest of improving experimental limits on $K_L^0 \to \mu e$ in the context of dynamical symmetry breaking models.

1. Introduction

It is of prime importance to determine the mechanism by which the gauge symmetry of the weak interactions is spontaneously broken. The alternative means available involve explicit Higgs scalar fields [1] or dynamical symmetry breaking [2–4]. There is considerable literature on the possible ways of detecting physical Higgs bosons if they exist [5, 6]. There has recently been a resurgence of interest in dynamical symmetry breaking, both in its theoretical and more phenomenological aspects [7–12]. Phenomenological interest has focused on the relatively light pseudo-Goldstone bosons (PGB's) expected to exist in theories of dynamical symmetry breaking. Calculations have been made [7, 9, 11] of their masses (typically a few to 300 GeV) and estimates made of their couplings to light quarks, leptons and gauge bosons. There have been discussions about the production and detection of these PGB's, dealing for the most part with their appearance in hadron-hadron collisions

[7, 10, 11]. It is not evident that hadron-hadron collisions are the best reactions to search for PGB's, particularly since their couplings to quarks and gluons are expected to be qualitatively similar to those of physical Higgs bosons and it is notorious that Higgs bosons are difficult to see in hadron-hadron collisions [13]. Indeed, it seems likely [6] that a large e^+e^- colliding beam machine such as LEP may be the best device for producing Higgs bosons with masses up to about 100 GeV. Accordingly, we are collaborating [12] with others in a general phenomenological study of the production and detection of PGB's at LEP. In the course of this work we have found it necessary, and possible, to refine previous estimates [7–11] of the couplings of PGB's to light quarks, leptons and gauge bosons. These estimates are presented in this paper, together with some remarks on the phenomenology of PGB's in high energy e^+e^- collisions [12]. It does indeed appear to us that these afford better prospects for the early detection of technicolour than do hadron-hadron collisions.

There are two main aspects of PGB coupling that we develop in this paper. The first is the set of couplings to pairs of light gauge bosons ($g, \gamma, W^{\pm}, Z^0$) induced by anomalous triangle diagrams [14]. We supplement previous [7, 10, 11] results with a complete set of calculations for PGB's in the simplest technicolour model [7], and argue that additional PGB-boson-boson couplings unrelated to anomalies are unlikely to alter quantitatively the decay and production rates we calculate from the anomalies. The second set of couplings that we estimate are those of the PGB's to fermions. It has often been remarked [7–11] that these are as model dependent as those of charged Higgs particles in models with more than one doublet of explicit Higgs fields. However, this model dependence can be reduced by observing that in order to have a chance that flavour-changing neutral interactions may be compatible with the experimental limits, all quarks of the same charge should get their masses from the same condensate of technifermions in the vacuum ("monophagy"). This suppresses PGB exchange contributions to the $K_1 - K_2$ mass difference, and is the analogue of the previous [15] naturalness requirement that all quarks of the same charge should get their masses from the same explicit Higgs vacuum expectation value. This assumption of monophagy relates the fermion couplings of the PGB's directly to the generalized Cabibbo-Kobayashi-Maskawa (KM) angles [16] controlling the $W^{\pm}$ couplings, and enables definite predictions to be made, modulo some ambiguity in the choice of colour representations for extended technicolour gauge bosons. Direct comparisons can then be made with the couplings [5] of the single physical H^0 in the Weinberg-Salam model with just one Higgs doublet. We find readily observable quantitative and qualitative differences in the production of H^0 and neutral PGB's in such processes as $e^+e^- \to \text{toponium} \to X^0 + \gamma$ or $e^+e^- \to Z^0 + X^0$, and make definite predictions for the decay modes of charged PGB's which should be easily testable at PETRA, PEP or LEP.

As an aside, we note that even in monophagic models leptoquark PGB exchange makes a contribution to the $K_L^0 \to \mu e$ decay which approaches the present experimental limit. Further experiments to improve this limit are very interesting for models of dynamical symmetry breaking.

2. The pseudo-Goldstone bosons of extended technicolour

We consider the basic technicolour model [7] with one family of technifermions

$$\begin{pmatrix} U_R & U_W & U_B & N \\ D_R & D_W & D_B & E \end{pmatrix} = \{F_{ia}\} \tag{2.1}$$

($a = 1,2,3,4$ and $i = 1,2$), paralleling the fermion content of the known light fermion generations (u, d, e, ν_e), (c, s, μ, ν_μ), etc. The technicolour group is assumed to be vectorlike with all the technifermions assigned to a complex representation. When there is a need to be specific, we will take $SU(N)_{TC}$ with the technifermions F in the fundamental representation (N) as an illustrative example. The $SU(8)_L \times SU(8)_R$ technicolour flavour symmetry is spontaneously broken down to $SU(8)_{L+R}$ by the technifermion condensates:

$$\langle \bar{F}_{ia} F_{jb} \rangle_0 = \delta^i_j \delta^a_b \mu^3 , \tag{2.2}$$

where μ is of order two to three hundred GeV. 63 (pseudo-) Goldstone bosons $P^{\alpha r}$ [$\alpha = 0, 1, \ldots, 15$ and $r = 0, 1, 2, 3$ $(\alpha, r) \neq 0, 0)$] result which carry the quantum numbers of

$$i\bar{F}\Lambda^\alpha \tau^r \gamma_5 F \sim P^{\alpha r} . \tag{2.3}$$

The axial currents

$$J^{\alpha r}_\mu = \bar{F} \tfrac{1}{2} \Lambda^\alpha \tau^r \gamma_\mu \gamma_5 F \tag{2.4}$$

satisfy the PCAC relations

$$q^\mu J^{\alpha r}_\mu = \sqrt{\tfrac{1}{2}} F_P q^2 P^{\alpha r} , \tag{2.5}$$

where $F_P \simeq 250$ GeV is the VEV of the technisigma field $\Sigma \sim \sqrt{\frac{1}{2}} \bar{F}F$. Among the 63 PGB's, three (P^{01}, P^{02} and P^{03}) are eaten by the $W^\pm$ and Z^0 giving them a mass. The 64th pseudoscalar P^{00} (technieta) acquires a mass of order 1 TeV through the technigluon anomaly. Of the remaining 60 PGB's there are 32 states ($P^{\alpha r}$ for $\alpha = 1, 2, \ldots, 8$ and $r = 0, 1, 2, 3$) which are $SU(3)_c$ octets and either singlets or triplets of isospin, and whose masses are expected [7, 9] to be of order 245 GeV $\times \sqrt{4/N}$. We will call them P^0_8 P^3_8 and $P^\pm_8$ from hereon. Then there are 24 leptoquark PGB's ($P^{\alpha, r}$ for $\alpha = 9, \ldots, 14$ and $r = 0, 1, 2, 3$) which are colour triplets: $P_{\overline{E}U_c}$, $P_{\overline{N}U_c}$, $P_{\overline{E}D_c}$, and $P_{\overline{N}D_c}$ and their antiparticles. They are expected [7, 9] to have masses of order 160 GeV × $\sqrt{4/N}$. Finally, there are four colour singlet PGB's ($P^{\alpha r}$ for $\alpha = 15$, $r = 0, 1, 2, 3$) which we call P^0, P^3 and $P^{\pm}$ and which are expected to be relatively light. The $P^{\pm}$ acquire masses from double-gauge boson exchange which are expected [9, 11] to be of order 7 GeV while all four of these PGB's get contributions [17, 18, 11] to their (mass)^2 of order $\lesssim (2.5\ \text{GeV})^2$ from a necessarily low-energy ($\gtrsim 10^{5.5}$ GeV) quark-lepton unification. We make no attempt to estimate the masses of these light PGB's more precisely, but we take their expected low masses as sufficient motivation to study their couplings to light gauge bosons and fermions.

3. Couplings to gauge bosons

In gauge theories with fermions, the vertex functions of three gauge-invariant currents VVA and AAA are controlled by Adler-Bell-Jackiw anomalies [14] at momenta much smaller than the typical dynamical scale of the gauge theory, $\Lambda \sim$ a few hundred MeV for QCD and $\Lambda_{TC} \sim$ a few hundred GeV for technicolour. These anomalies are then expected to dominate the low momentum couplings to gauge fields of any PGB with mass $\ll \Lambda_{TC}$ by virtue of the coupling F_P of the PGB to the axial current (2.5). Corrections to the low-energy theorems deduced from these axial anomalies are expected to be $O(m_P^2/m_V^2)$ where V is any vector meson with the appropriate quantum numbers to give a pole in any of the external currents [19]. One also expects at low energies that the VVV and VAA vertices will be suppressed by $O(m_P^2/m_V^2)$ relative to the VVA and AAA vertices controlled by anomalies [19]. In the case of QCD the $\pi^0 \to 2\gamma$ decay rate calculated using the anomaly agrees with the experiment to within 10%, and even anomaly calculations of the $\eta \to 2\gamma$ and $\eta' \to 2\gamma$ decay rates agree with experiment to a factor of 2 or better, despite the fact that $m_{\eta'}$ is not expected to go to zero when the quark masses are switched off, and is apparently of order Λ. The P-B_1-B_2 coupling is

$$\frac{1}{2}\left(\frac{\sqrt{2}\,S_{PB_1B_2}}{4\pi^2 F_P}\right)\varepsilon_{\mu\alpha\beta\gamma}\mathcal{E}_1^{\mu}\mathcal{E}_2^{\alpha}k_1^{\beta}k_2^{\gamma}, \tag{3.1}$$

with

$$S_{PB_1B_2} = g_1 g_2 \,\mathrm{Tr}\left(Q_P\{Q_1, Q_2\}\tfrac{1}{2}\right), \tag{3.2}$$

where the contributions from different helicity states are summed separately in the trace. In this convention the original Adler-Bell-Jackiw anomaly is $S_{\pi^0\gamma\gamma} = e^2$. Table 1 shows the anomaly factors S for all vertices of interest. Where the PGB is heavier than the sum of the masses of the gauge bosons, the decay PGB $\to$ 2 bosons is kinematically accessible and occurs with the rate (for $m_P \gg m_{B_1}, m_{B_2}$)

$$\Gamma(P \to B_1 + B_2) = 2^{\delta_{B_1B_2}}\frac{m_P^3}{128\pi}\left(\frac{S\sqrt{2}}{4\pi^2 F_P}\right)^2. \tag{3.3}$$

TABLE 1

Vertex	Anomaly factor S	$\Gamma(\mathrm{P}\to\mathrm{B}_1+\mathrm{B}_2)$	$\Gamma(\mathrm{B}_1\to\mathrm{P}+\mathrm{B}_2)$
$\mathrm{P}^0_{8,c}\mathrm{G}_a\mathrm{G}_b$	$g_s^2 d_{abc} N$	60 MeV $\times(\frac{1}{4}N)^2\left(\frac{m_\mathrm{P}}{245\ \mathrm{GeV}}\right)^3$	
$\gamma\mathrm{G}_a$	$g_s e\frac{1}{3}\delta_{ac} N$	150 keV	
$\mathrm{Z}^0\mathrm{G}_a$	$-g_s e\frac{1}{3}\tan\theta_\mathrm{W}\delta_{ac} N$	45 keV	
$\mathrm{P}^3_{8,c}\gamma\mathrm{G}_a$	$g_s e\delta_{ac} N$	1.4 MeV $\times(\frac{1}{4}N)^2\left(\frac{m_\mathrm{P}}{245\ \mathrm{GeV}}\right)^3$	
$\mathrm{Z}^0\mathrm{G}_a$	$g_s e\left(\frac{1-2\sin^2\theta_\mathrm{W}}{\sin 2\theta_\mathrm{W}}\right)\delta_{ac} N$	0.6 MeV	
$\mathrm{P}^\pm_{8,c}\mathrm{W}^\mp\mathrm{G}_a$	$g_s e\frac{1}{2\sin\theta_\mathrm{W}}\delta_{ac} N$	1.5 MeV $\times(\frac{1}{4}N)^2\left(\frac{m_\mathrm{P}}{245\ \mathrm{GeV}}\right)^3$	
$\mathrm{P}^0\mathrm{G}_a\mathrm{G}_b$	$g_s^2\sqrt{\frac{1}{6}}\,\delta_{ab} N$	30 eV $\times(\frac{1}{4}N)^2\left(\frac{m_\mathrm{P}}{2\ \mathrm{GeV}}\right)^3$	
$\gamma\gamma$	$-e^2\frac{4}{3}\sqrt{\frac{1}{6}}\,N$	0.03 eV	
$\mathrm{Z}^0\gamma$	$e^2\tan\theta_\mathrm{W}\frac{4}{3}\sqrt{\frac{1}{6}}\,N$		0.15 keV $\times(\frac{1}{4}N)^2$
$\mathrm{Z}^0\mathrm{Z}^0$	$-e^2\tan^2\theta_\mathrm{W}\frac{4}{3}\sqrt{\frac{1}{6}}\,N$		
$\mathrm{W}^+\mathrm{W}^-$	0		
$\mathrm{P}^3\gamma\gamma$	$e^2 4\sqrt{\frac{1}{6}}\,N$	0.30 eV $\times(\frac{1}{4}N)^2\left(\frac{m_\mathrm{P}}{2\ \mathrm{GeV}}\right)^3$	
$\mathrm{Z}^0\mathrm{Z}^0$	$-e^2\frac{2-4\sin^2\theta_\mathrm{W}}{\cos^2\theta_\mathrm{W}}\sqrt{\frac{1}{6}}\,N$		
$\mathrm{Z}^0\gamma$	$e^2\frac{1-4\sin^2\theta_\mathrm{W}}{\sin\theta_\mathrm{W}\cos\theta_\mathrm{W}}\sqrt{\frac{1}{6}}\,N$		10 eV $\times(\frac{1}{4}N)^2$
$\mathrm{W}^-\mathrm{W}^+$	0		
$\mathrm{P}^\pm\gamma\mathrm{W}^\mp$	$+e^2\frac{1}{\sin\theta_\mathrm{W}}\sqrt{\frac{1}{6}}\,N$		0.85 keV $(\frac{1}{4}N)^2$
$\mathrm{Z}^0\mathrm{W}^\mp$	$-e^2\frac{1}{\cos\theta_\mathrm{W}}\sqrt{\frac{1}{6}}\,N$		

The numerical values have been calculated using $\alpha_s = 0.1$ and $\sin^2\theta_\mathrm{W} = 0.23$.

In other cases where $m_{\mathrm{B}_1} > m_\mathrm{P} + m_{\mathrm{B}_2}$, the massive gauge boson ($\mathrm{W}^\pm$ or Z^0) may decay into a PGB and a light gauge boson with the rate (for $m_{\mathrm{B}_1} \gg m_\mathrm{P}, m_{\mathrm{B}_2}$):

$$\Gamma(\mathrm{B}_1\to\mathrm{P}+\mathrm{B}_2) = \frac{m_1^3}{384\pi}\left(\frac{S\sqrt{2}}{4\pi^2 F_\mathrm{P}}\right)^2 \tag{3.4}$$

Table 1 contains the decay rates (3.3) and (3.4) wherever appropriate. In all instances where the PGB has a non-zero anomaly involving a γ or a Z^0 the process

$e^+e^- \underset{B_1}{\rightarrow} P + B_2$ will occur with a calculable [20] rate given at c.m. energies $Q \gg m_{B_2,P}$ by [we denote $\alpha_1 = (g_{B_1 e^+ e^-})^2/4\pi$]

$$R_{PB_2} \equiv \frac{\sigma\left(e^+e^- \underset{B_1}{\rightarrow} P + B_2\right)}{\sigma\left(e^+e^- \underset{\gamma}{\rightarrow} \mu^+\mu^-\right)}$$

$$= \frac{\alpha_1}{\alpha} \frac{4^{\delta_{B_1B_2}}\beta^3}{128\pi}\left(\frac{S\sqrt{2}}{4\pi^2 F_P}\right)^2 \frac{Q^6}{\left(Q^2 - m_1^2\right)^2}, \tag{3.5}$$

where β is the PGB velocity. Note that (3.5) needs some modification if one wishes to take proper account of γ-Z^0 interference in $e^+e^- \rightarrow P + B_2$. Away from the Z^0 resonance so that the photon contribution dominates, we find (G = gluon)

$$R_{(P_8^3, P_8^0)G} = (10^{-3}, 10^{-2})(\tfrac{1}{4}N)^2 \beta^3 \left(\frac{Q}{F_P}\right)^2, \tag{3.6}$$

while for P^3 and P^0 production via γ and Z^0 we find

$$R\begin{pmatrix}\gamma \\ Z^0\end{pmatrix}_{(P^3, P^0)} = \begin{pmatrix} 10^{-3} & 10^{-4} \\ 10^{-4.5} & 10^{-5}\end{pmatrix}(\tfrac{1}{4}N)^2 \beta^3 \left(\frac{Q}{F_P}\right)^2. \tag{3.7}$$

We will return to the phenomenological consequences of these results after presenting our considerations on PGB-fermion-antifermion couplings.

4. Couplings to fermions

It has often been remarked that couplings to fermions are of order gm_f/m_W and hence comparable in magnitude to those of conventional Higgs bosons. It has also been observed that since the structure of spontaneous symmetry breaking and mass generation is much richer than in an explicit Higgs model with just one complex doublet, the precise values of the Pff couplings are model-dependent [7, 8, 11] in a manner reminiscent of Higgs models with several complex doublets. Here, however, we note that there is a natural requirement on the structure of the extended technicolour theory which greatly reduces this model dependence. We recall that in explicit Higgs theories the order of magnitude G_H of $(\bar{q}q)^2$ coupling induced by virtual Higgs exchange is

$$G_H \simeq g^2 \left(\frac{m_f}{m_W}\right)^2 \frac{1}{m_H^2}. \tag{4.1a}$$

The $\Delta S = 2$ operator $(\bar{s}d)^2$ of this type should have a coefficient $G_H \lesssim \frac{1}{2} \cdot 10^{-12}$ GeV^{-2} to be compatible with the observed magnitude of the $K_1 - K_2$ mass difference. The exchanges of neutral Higgs bosons with masses $\lesssim 100$ GeV will, in general, violate this bound unless their couplings are diagonal in flavour. This is only natural [15] if all quarks of the same charge get their masses from the vacuum expectation value of the neutral member of the same Higgs doublet, which may, however, be different for the different charges (u,c,t) and (d,s,b) of quarks. The exchange of a PGB has an order of magnitude similar to (4.1a):

$$G_{\mathrm{P}} \simeq g^2 \left(\frac{m_{\mathrm{f}}}{m_{\mathrm{W}}} \right)^2 \frac{1}{m_{\mathrm{P}}^2} . \tag{4.1b}$$

Since m_{P^0} and m_{P^3} are $\lesssim 100$ GeV, we should require that their couplings be diagonal in flavour if we want to avoid some problems with the $\mathrm{K}_1 - \mathrm{K}_2$ mass difference [21]. The naïve analogue of the structure required in explicit Higgs models is that each quark of the same charge gets its mass from the same condensate of technifermions—"monophagy". We find that in this case the $\bar{\mathrm{f}}\mathrm{f}$ coupling of P^0 and P^3, which are potentially dangerous because of their expected low masses, are completely specified in terms of the fermion mass matrices and the KM weak interaction mixing matrix. In simple examples of monophagic models considered below, the couplings of the $\mathrm{P}_8^{3,0,\pm}$ are completely determined as well.

Let us assume that the ETC/TC generators couple each type of ordinary fermion $\mathrm{f} = (\mathrm{u,c,t}\ldots), (\mathrm{d,s,b}\ldots), (\mathrm{e},\mu,\tau\ldots), (\nu_{\mathrm{e}},\nu_\mu,\nu_\tau\ldots)$ to only one (but not necessarily the same) type of technifermion $\mathrm{F} = \mathrm{U,D,E,N}$. Thus we allow for example ETC to couple (u,c,t...) to U, (d,s,b...) to D, $(\mathrm{e},\mu,\tau\ldots)$ to E, and $(\nu_{\mathrm{e}},\nu_\mu,\nu_\tau\ldots)$ to N, but we exclude from consideration cases where ETC couples u to U and c to N for example. Let

$$J_\mu^A = \sum_{\mathrm{f},n} \left(\bar{f}_{\mathrm{L}n} g_{\mathrm{L}n}^A(f) \gamma_\mu F_{\mathrm{L}}(f) + \bar{f}_{\mathrm{R}n} g_{\mathrm{R}n}^A(f) \gamma_\mu F_{\mathrm{R}}(f) + \text{h.c.}, \right. \tag{4.2}$$

be the currents coupled to the ETC/TC vector boson mass eigenstates. The sum is over generations $n = 1,2,3$ and types $\mathrm{f} = (\mathrm{u},\ldots), (\mathrm{d},\ldots), (\mathrm{e},\ldots)\,(\nu,\ldots)$. The coefficients $g_{\mathrm{L}n}^A(f)$ and $g_{\mathrm{R}n}^A(f)$ are constrained by $\mathrm{SU_L(2)} \times \mathrm{U}_Y(1) \times \mathrm{SU(3)_c}$ symmetry. The 4-fermion interactions mediated by the ETC/TC vector bosons are:

$$\sum_{\mathrm{f,f'}} \sum_{n,m} \sum_{A} \left(\frac{g_{\mathrm{ETC}}}{\mu_A} \right)^2 \left[\left(\bar{f}_{\mathrm{L}n} g_{\mathrm{L}n}^A(f) \gamma_\mu F_{\mathrm{L}} \right) \left(\bar{F}'_{\mathrm{R}} g_{\mathrm{R}m}^A(f') \gamma^\mu f'_{\mathrm{R}m} \right) + \text{h.c.} + \cdots \right], \tag{4.3}$$

where the dots represent terms that will not contribute to effective Yukawa couplings between technimesons and ordinary fermions in the case where the technifermion representation is complex. After a Fierz transformation, we obtain

$$-2 \sum_{\mathrm{f,f'}} \left(\bar{F}'_{\mathrm{R}} F_{\mathrm{L}} \right) \left(\sum_{n,m} \bar{f}_{\mathrm{L}n} \Gamma_{nm}^{\mathrm{ff'}} f'_{\mathrm{R}m} \right) + \text{h.c.}, \tag{4.4}$$

where

$$\Gamma_{nm}^{\mathrm{ff}'} = \sum_A \left(\frac{g_{\mathrm{ETC}}}{\mu_A}\right)^2 g_{\mathrm{L}n}^A(f) g_{\mathrm{R}m}^{A*}(f'). \tag{4.5}$$

Eq. (4.4) can be rewritten:

$$-\sum_{\mathrm{f,f'}} \left\{ (\bar{F}'F)\left(\bar{f}\Gamma_1^{\mathrm{ff}'}f' + i\bar{f}\Gamma_2^{\mathrm{ff}'}\gamma_5 f'\right) - (\bar{F}'\gamma_5 F)\left(i\bar{f}\Gamma_2^{\mathrm{ff}'}f' + \bar{f}\Gamma_1^{\mathrm{ff}'}\gamma_5 f'\right)\right\}, \tag{4.6}$$

where the sum over generations is implied and we have decomposed Γ into parts hermitian with respect to the fermion type and generation indices:

$$\Gamma = \Gamma_1 + i\Gamma_2, \qquad \Gamma_1 = \Gamma_1^\dagger, \qquad \Gamma_2 = \Gamma_2^\dagger. \tag{4.7}$$

The TC condensates have the form shown in eq. (2.2) and each type of fermion acquires its mass from a unique technifermion condensate ("monophagy"). The ordinary fermion mass matrices are thus given by

$$m_{\mathrm{f}} = \mu^3\left(\Gamma_1^{\mathrm{ff}} + i\Gamma_2^{\mathrm{ff}}\right), \tag{4.8}$$

and are diagonalized in the usual way:

$$\mathcal{U}_{\mathrm{L}}^{\mathrm{f}} m_{\mathrm{f}} \mathcal{U}_{\mathrm{R}}^{\mathrm{f}\dagger} = \mathrm{diag}\left(m_1^{\mathrm{f}}, m_2^{\mathrm{f}}, m_3^{\mathrm{f}}, \ldots\right). \tag{4.9}$$

The Kobayashi-Maskawa matrix of weak mixing angles is

$$U_{\mathrm{KM}} = \mathcal{U}_{\mathrm{L}}^{\mathrm{u}} \mathcal{U}_{\mathrm{L}}^{\mathrm{d}\dagger}. \tag{4.10}$$

In terms of the ordinary fermion mass eigenstates, eq. (4.6) simplifies since in that basis

$$\Gamma_2^{\mathrm{ff}} = 0,$$

$$\Gamma_1^{\mathrm{ff}} = \frac{1}{\mu^3}\mathrm{diag}\left(m_1^{\mathrm{f}}, m_2^{\mathrm{f}}, m_3^{\mathrm{f}}, \ldots\right),$$

$$\mathrm{f} = (\mathrm{u}, \ldots), (\mathrm{d}, \ldots), (\mathrm{e}, \ldots), (\nu, \ldots). \tag{4.11}$$

Thus the effective Yukawa couplings of the colourless electrically neutral PGB's $P_{\bar{\mathrm{F}}\mathrm{F}} \sim i\bar{F}\gamma_5 F$ are particularly simple in a monophagic model:

$$(\bar{F}\gamma_5 F)\frac{1}{\mu^3}(\bar{f}m\gamma_5 f) = -iP_{\bar{\mathrm{F}}\mathrm{F}}\frac{2\sqrt{2}}{F_{\mathrm{P}}}(\bar{f}m\gamma_5 f), \tag{4.12}$$

where we have used

$$\frac{(i\bar{F}\gamma_5 F)}{\langle \bar{F}F\rangle_0}=\frac{P_{\bar{F}F}}{\langle S_{\bar{F}F}\rangle_0}=\frac{2\sqrt{2}}{F_P}P_{\bar{F}F}, \tag{4.13}$$

with $P_{\bar{F}F}$ and $S_{\bar{F}F}$ the normalized pseudoscalar and scalar $\bar{F}F$ bound states defined previously. Monophagy thus assures that the colourless neutral technimesons do not mediate flavour changing neutral currents. Furthermore, it is clear that this is not in general true in non-monophagic models where each type of fermion couples to more than one technifermion condensate, just as occurred in explicit Higgs models with each type of fermion getting its mass from more than one Higgs doublet. However, we do not exclude the possibility that one may be able to construct models in which the fermions of a given type are, in fact, coupled to more than one type of technifermion, but always in the same combination. Then the contributions to the fermion mass matrix from each type of technifermion would be simultaneously diagonalizable and the $\bar{f}f$ couplings of the neutral physical PGB's would again be a flavour diagonal. The problem is to realize this sort of polyphagic structure in a natural way, and we have not studied this question.

Next we specialize to two particular monophagic models. First, we consider the case where the ETC generators commute with $SU_L(2)\times U_Y(1)\times SU(3)_c$ and couple each ordinary fermion f to the technifermion of the same type, i.e. $(u,c,t\ldots)\leftrightarrow U$, $(d,s,b\ldots)\leftrightarrow D$, $(e,\mu,\tau\ldots)\leftrightarrow E$ and $(\nu_e,\nu_\mu,\nu_\tau\ldots)\leftrightarrow N$. The effective Yukawa interactions of the colour singlet technimesons are

$$\begin{aligned}-2\sum_{c=\text{color}}(\bar{u}_{Lc}\bar{d}_{Lc})&\left[\begin{pmatrix}\bar{U}_{Rc} & U_{Lc}\\ \bar{U}_{Rc} & D_{Lc}\end{pmatrix}\Gamma^{qu}u_{Rc}+\begin{pmatrix}\bar{D}_{Rc} & U_{Lc}\\ \bar{D}_{Rc} & D_{Lc}\end{pmatrix}\Gamma^{qd}d_{Rc}\right]\\ -2(\bar{\nu}_L\bar{e}_L)&\left[\begin{pmatrix}\bar{N}_R & N_L\\ \bar{N}_R & E_L\end{pmatrix}\Gamma^{\ell\nu}\nu_R+\begin{pmatrix}\bar{E}_R & N_L\\ \bar{E}_R & E_L\end{pmatrix}\Gamma^{\ell e}e_R\right]+\text{h.c.},\end{aligned} \tag{4.14}$$

where we have used the constraints $g_L^A(u)=g_L^A(d)$ and $g_L^A(e)=g_L^A(\nu)$ following from $SU(2)_L$ symmetry and where we have renamed [see eq. (4.5)]

$$\Gamma^{uf}=\Gamma^{df}=\Gamma^{qf},\qquad \Gamma^{\nu f}=\Gamma^{ef}=\Gamma^{\ell f},\qquad \text{for}\quad f=u,\,d,\,e\text{ or }\nu. \tag{4.15}$$

Eq. (4.14) has the same form as the Yukawa interactions of the standard model with several Higgs doublets when care has been taken to avoid flavour changing neutral currents by requiring that each fermion type obtains mass from the VEV of only one Higgs doublet. Thus eq. (4.14) is amenable to a well-known [22] set of manipulations which yields for the pseudoscalar technimesons:

$$(-i)\frac{2\sqrt{2}}{F_{\mathrm{P}}}\left\{\sum_{c=\mathrm{color}}\left[P_{\bar{\mathrm{U}}_c\mathrm{U}_c}(\bar{u}_c m^{\mathrm{u}}\gamma_5 u_c)+P_{\bar{\mathrm{D}}_c\mathrm{D}_c}(\bar{d}_c m^{\mathrm{d}}\gamma_5 d_c)\right.\right.$$
$$\left.+P_{\bar{\mathrm{U}}_c\mathrm{D}_c}\bar{d}_c\left(U^{\dagger}_{\mathrm{KM}}m^{\mathrm{u}}\frac{1+\gamma_5}{2}-m^{\mathrm{d}}U^{\dagger}_{\mathrm{KM}}\frac{1-\gamma_5}{2}\right)u_c+\mathrm{h.c.}\right]$$
$$\left.+P_{\bar{\mathrm{E}}\mathrm{E}}(\bar{e}m^{\mathrm{e}}\gamma_5 e)-P_{\bar{\mathrm{N}}\mathrm{E}}\left(\bar{e}m^{\mathrm{e}}\frac{1-\gamma_5}{2}\nu\right)+\mathrm{h.c.}\right\}, \tag{4.16}$$

where the fermions are mass eigenstates and the mass matrices m^{f} are in their diagonal form (4.10). It has been assumed in eq. (4.16) that ν_{R} has a large Majorana mass. This assures its disappearance from the effective low-energy theory even if $\Gamma^{\mathrm{f}\nu}$ is of the same order of magnitude as Γ^{fe}. We can now calculate the effective Yukawa couplings of the four colour singlet PGB's defined earlier. We find

$$\left(\frac{-i}{250\ \mathrm{GeV}}\right)P^0\left[(\bar{u}m^{\mathrm{u}}\gamma_5 u+\bar{d}m^{\mathrm{d}}\gamma_5 d)\sqrt{\tfrac{1}{3}}-\sqrt{3}\,(\bar{e}m^{\mathrm{e}}\gamma_5 e)\right],$$

$$\left(\frac{-i}{250\ \mathrm{GeV}}\right)P^3\left[(\bar{u}m^{\mathrm{u}}\gamma_5 u-\bar{d}m^{\mathrm{d}}\gamma_5 d)\sqrt{\tfrac{1}{3}}+\sqrt{3}\,(\bar{e}m^{\mathrm{e}}\gamma_5 e)\right],$$

$$\left(\frac{-i}{250\ \mathrm{GeV}}\right)P^+\left[\bar{u}\left(U_{\mathrm{KM}}m^{\mathrm{d}}\frac{1+\gamma_5}{2}-m^{\mathrm{u}}U_{\mathrm{KM}}\frac{1-\gamma_5}{2}\right)d\sqrt{\tfrac{2}{3}}\right.$$
$$\left.-\sqrt{6}\left(\bar{\nu}m^{\mathrm{e}}\frac{1+\gamma_5}{2}e\right)\right]+\mathrm{h.c.} \tag{4.17}$$

These differ from the Yukawa couplings of neutral and charged Higgs scalars in the standard model [5, 6] in three important ways. First, the couplings of the neutral PGB's are pseudoscalar. Second, the couplings to the leptons are stronger by a factor of 3 compared to the quarks. Third, the couplings of the $\mathrm{P}^{\pm}$ are directly related to the KM matrix, which is not in general true for physical charged Higgs bosons in models with several Higgs doublets. It should be noted that P^0 and P^3 are not necessarily mass eigenstates; to determine the possible mixing requires a study of PGB mass generation which goes beyond the scope of this paper. The effective Yukawa couplings of the colour octet P_8 and triplet PGB's are calculated in a similar manner. We find

$$\left(\frac{-i}{250\ \mathrm{GeV}}\right)P^0_{8\alpha}[\bar{u}m^{\mathrm{u}}\lambda^{\alpha}\gamma_5 u+\bar{d}m^{\mathrm{d}}\lambda^{\alpha}\gamma_5 d]\sqrt{2}\,,$$

$$\left(\frac{-i}{250\ \mathrm{GeV}}\right)P^3_{8\alpha}[\bar{u}m^{\mathrm{u}}\lambda^{\alpha}\gamma_5 u-\bar{d}m^{\mathrm{d}}\lambda^{\alpha}\gamma_5 d]\sqrt{2}\,,$$

$$\left(\frac{-i}{250\ \mathrm{GeV}}\right)P^+_{8\alpha}\left[\bar{u}\left(U_{\mathrm{KM}}m^{\mathrm{d}}\frac{1+\gamma_5}{2}-m^{\mathrm{u}}U_{\mathrm{KM}}\frac{1-\gamma_5}{2}\right)\lambda^{\alpha}d\right]2+\mathrm{h.c.}, \tag{4.18}$$

$$\left(-i\frac{2\sqrt{2}\,\mu^3}{F_P}\right)P_{\overline{U}_c N}\left[\bar{\nu}\Gamma^{\ell u}\frac{1+\gamma_5}{2}u_c\right],$$

$$\left(-i\frac{2\sqrt{2}\,\mu^3}{F_P}\right)P_{\overline{U}_c E}\left[\bar{e}\left(\Gamma^{\ell u}\frac{1+\gamma_5}{2}-\Gamma^{qe\dagger}\frac{1-\gamma_5}{2}\right)u_c\right],$$

$$\left(-i\frac{2\sqrt{2}\,\mu^3}{F_P}\right)P_{\overline{D}_c E}\left[\bar{e}\left(\Gamma^{\ell d}\frac{1+\gamma_5}{2}-\Gamma^{qe\dagger}\frac{1-\gamma_5}{2}\right)d_c\right],$$

$$\left(-i\frac{2\sqrt{2}\,\mu^3}{F_P}\right)P_{\overline{D}_c N}\left[\bar{\nu}\Gamma^{\ell d}\frac{1+\gamma_5}{2}d_c\right]. \tag{4.19}$$

The effective Yukawa couplings of the colour octet PGB's are very similar to those of the colour singlets in eq. (4.17). The effective Yukawa couplings of the colour triplet PGB's are also related to the masses of the fermions [see eqs. (4.5) and (4.8)] but the relationship is not as simple as for the colour singlet and octet PGB's. From a qualitative viewpoint, however, it is clear that they also couple most strongly to the heaviest fermions.

As an alternative monophagic model, we consider a case where the ETC/TC generators have $SU(2)_L$ isospin $=0$ but couple u, c, t… to $\overline{E}$, d, s, b… e, μ, τ… to $\overline{U}$ and $\nu_e, \nu_\mu, \nu_\tau\ldots$ to $\overline{D}$. This type of coupling, which requires that the right (left)-handed technifermions transform as doublets (singlets) of the electroweak gauge group $SU(2)\times U(1)$, is inspired by a model of Farhi and Susskind [23]. The Farhi-Susskind model, however, has additional couplings which make it non-monophagic. The analysis of our alternative monophagic model is analogous to the previous one. We find for the four colour singlet PGB's:

$$\left(\frac{-i}{250\ \mathrm{GeV}}\right)P^0\left[\left(\bar{u}m^u\gamma_5 u+\bar{d}m^d\gamma_5 d\right)\sqrt{3}-\sqrt{\tfrac{1}{3}}\left(\bar{e}m^e\gamma_5 e\right)\right],$$

$$\left(\frac{-i}{250\ \mathrm{GeV}}\right)P^3\left[\left(\bar{u}m^u\gamma_5 u-\bar{d}m^d\gamma_5 d\right)\sqrt{3}+\sqrt{\tfrac{1}{3}}\left(\bar{e}m^e\gamma_5 e\right)\right]$$

$$\left(\frac{i}{250\ \mathrm{GeV}}\right)P^+\left[\bar{u}\left(U_{\mathrm{KM}}m^d\frac{1+\gamma_5}{2}-m^u U_{\mathrm{KM}}\frac{1-\gamma_5}{2}\right)d\sqrt{6}\right.$$

$$\left.-\sqrt{\tfrac{2}{3}}\left(\bar{\nu}m^e\frac{1-\gamma_5}{2}e\right)\right]+\mathrm{h.c.} \tag{4.20}$$

It is interesting to note that both in this monophagic model and in the previous one, the $\mathrm{Pf\bar{f}}$ couplings differ from those in a standard elementary Higgs model by characteristic factors of $\sqrt{3}$. We will discuss the phenomenological implications of these differences in sect. 5. It is interesting to consider the contribution of colour triplet leptoquark PGB's exchanged in the cross-channel to the decay $K_L^0\to\mu e$. The order of magnitude of effective four-fermion coupling expected from this source is

$$O\left(\frac{m_s m_d}{F_P^2}\frac{1}{m_P^2}\right)(\bar{s}\mu)(\bar{e}d), \tag{4.21}$$

where the relative mixture of pseudoscalar and scalar operators is model dependent and uncertain. To be compatible with the experimental limit of $K_L^0 \to \mu e$ one should [11] have

$$O\left(\frac{m_s m_d}{F_P^2}\frac{1}{m_P^2}\right) \lesssim \left(\frac{1}{300\ \mathrm{TeV}}\right)^2. \tag{4.22}$$

This bound is saturated if one takes the preferred mass of 160 GeV for the colour triplet PGB, and assumes $m_s \simeq 500$ MeV, $m_d \simeq 25$ MeV. We therefore feel (see also ref. [21]) that it would be very interesting to improve the present experimental limit on $K_L^0 \to \mu e$, as it seems likely that in technicolour models this decay should occur at a rate close to the present limit. As a final remark, we note that the contribution of double P^3 exchange to the $K_1^0 - K_2^0$ mass difference is likely to be several orders of magnitude below the experimental value, so that there is no obvious impact of leptoquark PGB's in this quarter.

5. Phenomenological discussion

After the discussion of the couplings of PGB's in extended technicolour theories, we now turn to some of their phenomenological implications, focusing in particular on the prospects for detecting some of the PGB's at high energy e^+e^- colliding beam machines such as LEP. High energy e^+e^- collisions seem to offer the least depressing prospects for detecting the physical Higgs bosons expected in gauge theories with explicit scalar fields [6]. It is natural to ask whether the same is true for the analogous, though composite, PGB's, and how different are the experimental signatures for Higgs bosons and PGB's.

Previous studies [6] have revealed four potentially interesting processes for the production of neutral Higgs bosons in e^+e^- collisions: $e^+e^- \to$ heavy quarkonium $\to H^0 + \gamma$, $e^+e^- \to Z^0 + H^0$, $e^+e^- \to Z^0 \to H^0 + \ell^+\ell^-$ or $H^0 + \gamma$. If they have masses less than that of the heavy quarkonium (toponium), then in the monophagic models discussed earlier the P^0 and P^3 should have branching ratios comparable to that expected for the single neutral boson H^0 in a theory with one explicit complex Higgs doublet:

$$\frac{\Gamma(Q\bar{Q} \to P^0, P^3 + \gamma)}{\Gamma(Q\bar{Q} \to \gamma \to \ell^+\ell^-)} = \left(3 \text{ or } \frac{1}{3}\right)\frac{\Gamma(Q\bar{Q} \to H^0 + \gamma)}{\Gamma(Q\bar{Q} \to \gamma \to \ell^+\ell^-)}, \tag{5.1}$$

where in an explicit Higgs model one would have had

$$\frac{\Gamma(Q\bar{Q} \to H^0 + \gamma)}{\Gamma(Q\bar{Q} \to \gamma \to \ell^+\ell^-)} = \frac{G_F m_V^2}{4\sqrt{2}\,\pi\alpha}\left(1 - \frac{m_H^2}{m_V^2}\right). \tag{5.2}$$

The QCD radiative corrections to the ratio (5.2) have been calculated [24] to be $\lesssim 30\%$ for quarkonia heavier than the Υ. Since the ratio (5.2) would be $\gtrsim \frac{1}{6}$ for any quarkonium in the LEP energy range, it should be relatively easy to discriminate reliably between decays into a P^0 or P^3 and into a H^0 on the basis of quarkonium branching ratios alone. Another discriminator is provided by the different expected decay branching ratios of the neutral bosons into quarks and leptons:

$$\frac{\Gamma(P^0, P^3 \to q\bar{q})}{\Gamma(P^0, P^3 \to \ell^+\ell^-)} = \left(27 \text{ or } \frac{1}{3}\right)\left(\frac{m_q}{m_\ell}\right)^2 \tag{5.3}$$

in the two monophagic models of sect. 4, compared with

$$\frac{\Gamma(H^0 \to q\bar{q})}{\Gamma(H^0 \to \ell^+\ell^-)} = 3\left(\frac{m_q}{m_\ell}\right)^2 \tag{5.4}$$

in the standard elementary Higgs model.

It is instructive to make some comparisons between the $P \to B_1 + B_2$ and $f\bar{f}$ decay rates. Using the results of sect. 3 and $1/F_P^2 = \sqrt{2}\,G_F$, we can write

$$\Gamma(P^0 \to 2 \text{ gluons}) = \frac{G_F m_{P^0}^3}{12\sqrt{2}\,\pi}\left(\frac{\alpha_s}{\pi}\right)^2 N^2, \tag{5.5}$$

$$\Gamma(P_8^0 \to 2 \text{ gluons}) = \frac{5 G_F m_{P_8^0}^3}{48\sqrt{2}\,\pi}\left(\frac{\alpha_s}{\pi}\right)^2 N^2, \tag{5.6}$$

whereas in the first monophagic model of sect. 4

$$\Gamma(P^0 \to q\bar{q}) = \frac{G_F m_q^2 m_{P^0}}{4\sqrt{2}\,\pi}, \tag{5.7}$$

$$\Gamma(P_8^0 \to q\bar{q}) = \frac{G_F m_q^2 m_{P_8^0}}{\sqrt{2}\,\pi}. \tag{5.8}$$

Comparing the rival decay modes (5.5) and (5.7) of the P^0 and (5.6) and (5.8) of the

P_8^0 we find:

$$\frac{\Gamma(P^0 \to 2\text{ gluons})}{\Gamma(P^0 \to q\bar{q})} = \frac{16}{3}\left(\frac{\alpha_s}{\pi}\right)^2 \left(\frac{m_{P^0}}{m_q}\right)^2 \left(\tfrac{1}{4}N\right)^2, \tag{5.9}$$

$$\frac{\Gamma(P_8^0 \to 2\text{ gluons})}{\Gamma(P_8^0 \to q\bar{q})} = \frac{5}{3}\left(\frac{\alpha_s}{\pi}\right)^2 \left(\frac{m_{P_8^0}}{m_q}\right)^2 \left(\tfrac{1}{4}N\right)^2. \tag{5.10}$$

Eq. (5.9) suggests that the dominant decay of P^0 will be some kinematically accessible $q\bar{q}$ or $\ell^+\ell^-$ mode. With $\alpha_s = 0.1$ and $N = 4$, we find that the ratio (5.10) is $\sim 25\%$ if $m_{P_8^0} = 245$ GeV and $m_{\text{top}} = 20$ GeV.

There is a marked difference between the production rates of H^0 and P^0 or P^3 bosons in association with the Z^0. The reason is that the $H^0Z^0Z^0$ coupling is directly related to the Z^0 mass and hence relatively large:

$$g_{H^0Z^0Z^0} = 2^{5/4} m_Z^2 \sqrt{G_F} \tag{5.11}$$

whereas the $P^{0,3}Z^0Z^0$ and $P^{0,3}Z^0\gamma$ couplings are induced by fermion loops and are hence relatively small. We can compare the production rates (3.5), (3.7) with the production rate [5, 25] for $Z^0 + H^0$ at c.m. energies $Q \gg m_Z + m_H$:

$$R_{Z^0H^0} = \frac{1 - 4\sin^2\theta_W + 8\sin^4\theta_W}{128\sin^4\theta_W \cos^2\theta_W} \simeq 0.12, \tag{5.12}$$

to see that for example

$$\frac{R_{Z^0P^3}}{R_{Z^0H^0}} \simeq \frac{32\alpha}{3\pi^2}\tan^2\theta_W \left(1 - 2\sin^2\theta_W\right)^2 \left(\tfrac{1}{4}N\right)^2 \left(\frac{Q}{F_P}\right)^2$$

$$\simeq 2.2 \cdot 10^{-4} \left(\tfrac{1}{4}N\right)^2 \left(\frac{Q}{F_P}\right)^2. \tag{5.13}$$

While $R_{Z^0P^3}$ has a different energy dependence characteristic of the intrinsically non-renormalizable anomaly-dominated (γ, Z^0) Z^0P^3 couplings, we see from eq. (5.13) that a typical ratio of production rates in the LEP energy range is of order 10^{-4}. The decay $Z^0 \to H^0 + \ell^+\ell^-$ would be dominated by the relatively large $Z^0Z^0H^0$ coupling (5.11). Since $B(Z^0 \to H^0 + \ell^+\ell^-)$ is expected [26] to be small ($\lesssim 10^{-4}$ for $m_{H^0} \gtrsim 10$ GeV), the analogous $B(Z^0 \to P^{0,3} + \ell^+\ell^-)$ are presumably too small to be observable even with a Z^0 factory. Indeed, the process $Z^0 \to P^{0,3} + \ell^+\ell^-$ is best regarded as a weak radiative correction to the decay $Z^0 \to P^{0,3} + \gamma$ which has a much larger branching ratio. However, we deduce from the decay widths

$\Gamma(Z^0 \to P^0, P^3 + \gamma)$ in table 1 and the expected total Z^0 decay width of order 3 GeV that

$$B(Z^0 \to P^0, P^3 + \gamma) \simeq 5 \times 10^{-8}, 3 \times 10^{-9}, \tag{5.14}$$

which is even smaller than the anticipated [27]

$$B(Z^0 \to H^0 + \gamma) \simeq 10^{-6}, \tag{5.15}$$

which was already thought [6] to be almost unobservably small. We conclude that a high-energy e^+e^- colliding beam machine such as LEP should be able to produce neutral PGB's in quarkonium decays, and discriminate them from a neutral Higgs on the basis of their production rates and decay branching ratios and the absence of their copious production in the reactions $e^+e^- \to (\gamma \text{ or } Z^0) + (P^0 \text{ or } P^3)$ and $Z^0 \to (P^0 \text{ or } P^3) + \ell^+\ell^-$. The prospects for detecting colour octet PGB's $P_8^{0,3}$ at LEP are relatively remote, unless their masses are substantially less than has been anticipated [7, 9, 11]. The cross-section ratios R for their production in association with a gluon approach the limits

$$R_{(P_8^3, P_8^0)} \simeq (10^{-2}, 10^{-3})\left(\tfrac{1}{4}N\right)^2 \beta^3 \left(\frac{Q}{F_P}\right)^2 . \tag{5.16}$$

The problems of detecting such $e^+e^- \to (P_8^3, P_8^0) + G$ events are discussed elsewhere [10, 11].

The best prospects for detecting charged Higgs bosons $H^\pm$ in e^+e^- collisions were found [6] to be reactions $e^+e^- \to H^+H^-$ and $W^\pm H^\mp$, and heavy quark decays $Q \to H^\pm + q$. At LEP energies the $e^+e^- \to P^+P^-$ cross section would be similar to that for $e^+e^- \to H^+H^-$, with the expected deviations from a point-like cross section only appearing at considerably higher energies. However, the $P^\pm$ decay branching ratios are quite characteristic and predictable: for example in the two monophagic models of sect. 4

$$\frac{\Gamma(P^+ \to t + \bar{b})}{\Gamma(P^+ \to \bar{\tau} + \nu_\tau)} = \left(27 \text{ or } \frac{1}{3}\right)\left(\frac{m_t}{m_\tau}\right)^2 \cos^2\theta_3 \cos^2\theta_2, \tag{5.17}$$

because of the KM matrix and colour factors appearing in the couplings (4.17). By way of contrast, while the decay branching ratios of charged Higgs $H^\pm$ are expected to be correlated with the masses of the final-state fermions, definite predictions are impossible without extra assumptions, though there would be a colour factor of 3 for quarks as for the H^0 decays in eq. (5.4). As in the case of $e^+e^- \to Z^0 + P^0$ or P^3, we expect that the cross section for $e^+e^- \to W^\pm + P^\pm$ shown would be invisibly small, except possibly if the $P^\pm$ are sufficiently light for the direct channel Z^0 pole to provide a large enhancement close to threshold. The cross section for $e^+e^- \to W^\pm + H^\mp$ may be much larger, though it has been pointed out [28] that this would require considerable gymnastics with explicit Higgs representations. Finally, we note that the decay $Q \to P^\pm + q$ would almost certainly be the predominant decay of a heavy quark Q if it were kinematically accessible, and might even lead to competitive decays of heavy quarkonia $Q\overline{Q} \to Q + P^\pm + q \to P^\mp + P^\pm + q + \overline{q}$. Thus heavy quark decays as well as the reaction $e^+e^- \to P^+P^-$ provide good prospects for $P^\pm$ searches in e^+e^- collisions, and the decay modes of any $P^\pm$ produced, such as eq. (5.17), should be a distinctive signature.

In view of the relatively low rates and signal-to-background ratios for PGB production in hadron-hadron collisions, we feel that high energy e^+e^- collisions probably offer the best foreseeable prospects for producing and detecting the lightest PGB's expected in ETC theories, and for discriminating between them and the physical Higgs bosons expected in theories with explicit scalar fields.

We would like to thank Savas Dimopoulos and members of the ECFA/LEP study group on exotic particles for useful discussions.

Note added in proof

After completing this paper we received two papers on related subjects [29]. We thank Dr. A. Ali for drawing this work to our attention. We would also like to point out that small flavour-changing induced Yukawa couplings for P^0 and P^3 exist even in monophagic theories: see [30]. However, they do not alter the quantitative conclusions of this paper.

References

[1] P.W. Higgs, Phys. Rev. Lett. 13 (1964) 508;
F. Englert and R. Brout, Phys. Rev. Lett. 13 (1964) 321;
G.S. Guralnik, C.R. Hagen and T.W. Kibble, Phys. Rev. Lett. 13 (1964) 585;
P.W. Higgs, Phys. Rev. 145 (1966) 1156;
T.W.B. Kibble, Phys. Rev. 155 (1967) 1554

[2] J. Schwinger, Phys. Rev. 125 (1962) 397; 128 (1969) 2425;
R. Jackiw and K. Johnson, Phys. Rev. D8 (1973) 2386;
J.M. Cornwall and R.E. Norton, Phys. Rev. D8 (1973) 3338;
M.A.B. Bég and A. Sirlin, Ann. Rev. Nucl. Sci. 24 (1974) 379;
S. Weinberg, Phys. Rev. D13 (1976) 947; D19 (1979) 1277;
L. Susskind, Phys. Rev. D20 (1979) 2619

[3] S. Dimopoulos and L. Susskind, Nucl. Phys. B155 (1979) 237;
E. Eichten and K.D. Lane, Phys. Lett. 90B (1980) 125

[4] K.D. Lane and M.E. Peskin, Moriond Lectures, 1980, NORDITA preprint 80/33
P. Sikivie, Varenna Lectures, 1980, CERN preprint TH-2951

[5] J. Ellis, M.K. Gaillard and D.V. Nanopoulos, Nucl. Phys. B106 (1976) 292

[6] G. Barbiellini et al. DESY preprint 79/27 (1979)

[7] S. Dimopoulos, Nucl. Phys. B168 (1980) 69

[8] M.A.B. Bég, H.D. Politzer and P. Ramond, Phys. Rev. Lett. 43 (1979) 1701
[9] M.E. Peskin, Nucl. Phys. B175 (1980) 197
[10] F. Hayot and O. Napoly, Z. Phys. C7 (1981) 229
[11] S. Dimopoulos, S. Raby and G.L. Kane, Nucl. Phys. B182 (1981) 77
[12] G. Barbiellini et al., DESY preprint, in preparation (1980)
[13] H. Georgi, S. Glashow, M. Machacek and D.V. Nanopoulos, Phys. Rev. Lett. 40 (1978) 692
[14] J.S. Bell and R. Jackiw, Nuovo Cim. 60A (1969) 47;
S.L. Adler, Phys. Rev. 117 (1969) 2426
[15] S.L. Glashow and S. Weinberg, Phys. Rev. D15 (1977) 1958;
E.A. Paschos, Phys. Rev. D15 (1977) 1966
[16] M. Kobayashi and K. Maskawa, Prog. Theor. Phys. 49 (1973) 652
[17] E. Eichten and K. Lane, ref. [3]
[18] S. Dimopoulos, S. Raby and P. Sikivie, Nucl Phys. B176 (1980) 449
[19] H.J. Schnitzer and S. Weinberg, Phys. Rev. 164 (1967) 1828;
Riazuddin and Fayyazuddin, Phys. Rev. 171 (1968) 1428
[20] N. Cabibbo and R. Gatto, Phys. Rev. 124 (1961) 1577
[21] S. Dimopoulos and J. Ellis, Nucl. Phys. B182 (1981) 505
[22] J. Ellis, M.K. Gaillard and D.V. Nanopoulos, Nucl. Phys. B109 (1976) 213
[23] E. Farhi and L. Susskind, Phys. Rev. D20 (1979) 3404
[24] M.I. Vysotsky, ITEP-preprint-38 (1980)
[25] B.W. Lee, C. Quigg and H. Thacker, Phys. Rev. D16 (1977) 1519;
S.L. Glashow, D.V. Nanopoulos and A. Yildiz, Phys. Rev. D18 (1978) 1724
[26] J.D. Bjorken, Proc. 1976 SLAC Summer Inst. on Particle physics (SLAC-198, 1977)
[27] R.N. Cahn, M.S. Chanowitz and N. Fleishon, Phys. Lett. 82B (1979) 113
[28] J.A. Grifols and A. Méndez, Universitat Antonoma de Barcelona preprint UAB-FT 58 (1980)
[29] A. Ali and M.A.B. Bèg, DESY preprint 80/98 (1980);
A. Ali, H.B. Newman and R.Y. Zhu, DESY preprint 80/110 (1980)
[30] J. Ellis, D.V. Nanopoulos and P. Sikivie, CERN preprint TH-3030 (1981)

Volume 103B, number 4,5 PHYSICS LETTERS 30 July 1981

PRODUCTION OF HIGGS BOSONS AND HYPERPIONS IN e^+e^- ANNIHILATION

A. ALI
Theory Group, DESY, 2 Hamburg 52, Fed. Rep. Germany

and

M.A.B. BÉG [1]
The Rockefeller University, New York, NY 10021, USA

Received 6 October 1980
Revised manuscript received 30 April 1981

The production of hyperpions – low lying pseudo-Goldstone modes of the hypercolor scenario – in e^+e^- annihilation is studied and contrasted with the production of elementary Higgs bosons of the canonical methodology.

1. Introduction. The possibility of implementing the Higgs mechanism in a dynamical way, without the use of elementary spin-0 fields, has lately been the subject of much discussion [1]. In several recent communications [2–6] it has been pointed out that the most popular dynamical scheme, the so-called hypercolor scenario, may be experimentally distinguishable from the canonical Weinberg–Salam scheme via experiments at relatively low (10–100 GeV) energies. [There is, of course, an a priori expectation that differences between the two schemes will be manifest in experiments at TeV energies; this energy regime, however is not likely to be accessible in the near future.] The considerations of refs. [2–4] hinge in a crucial way on one's ability to (a) produce spin-0 particles, of mass ≈10 GeV, coupled semi-weakly to *ordinary quarks* and leptons and (b) distinguish Higgs particles, ϕ's, from the low-lying bosons of the hypercolor scenario, generically called π's, exploiting mainly the difference in parity.

Our purpose in this note is to point out that the pseudoscalar nature of π''s introduces a fundamental difference in the production mechanism of ϕ's and π''s in a large class of reactions. In particular, we calculate the production rates of ϕ's and π''s in e^+e^- annihilation and in the decays of the spin-1 bosons Z and toponium, J_T.

We begin with a discussion of charged hyperpions, $\pi'^{\pm}$, which – if the mass estimates of refs. [2–6] are not too low – can be observed at energies currently available at PETRA and PEP.

2. Charged hyperpions. The cross section for the process [2] $e^+e^- \to \pi'^+\pi'^-$ may be expressed in terms of its contribution to the parameter R

$$\Delta R = \sigma(e^+e^- \to \pi'^+\pi'^-)/\sigma(e^+e^- \to \mu^+\mu^-)$$
$$= \tfrac{1}{4}(1 - 4m^2_{\pi'^{\pm}}/s)^{3/2} , \qquad (1)$$

where $\sqrt{s}$ is the c.m. energy and the factor $\frac{1}{4}$ is due to the spin-0 nature of $\pi'^{\pm}$'s.

What are the signatures of $\pi'^{\pm}$'s in e^+e^- annihilation experiments? The decays of $\pi'^{\pm}$'s are in general model dependent; they may decay via a $W^{\pm}$ exchange or they may require heavier exotic bosons [5,6]. (In either case their couplings to the leptons and quarks involve unknown Cabibbo-like angles.) The $\pi'^{\pm}$ may even be stable. Fortunately, this last possibility has already been ruled out experimentally for $m_{\pi'^{\pm}} < 12$ GeV [‡1] [7]. The details of the decay modes may differ

[1] Work supported in part by the US Department of Energy under Contract Grant Number DE-AC02-76ER02232B.

[‡1] For footnote see next page.

from model to model, but we nevertheless expect the helicity suppression pattern, well known from the decays of ordinary $\pi^{\pm}$. If $m_{\pi'^{\pm}} < m_t + m_b$, the principal decay channels for π'^{+} are $c\bar{s}$ and $\tau^{+}\nu_{\tau}$ with

$$\Gamma(\pi'^{+} \to c\bar{s})/\Gamma(\pi'^{+} \to \tau^{+}\nu_{\tau}) \approx 3\chi(m_c/m_{\tau})^2 , \qquad (2)$$

where $\chi \approx 1$ if the $\pi'^{\pm}$ decay via $W^{\pm}$ boson exchange. (In the simplest models, the W-channel is closed in the limit in which the ultraviolet masses of the hyperquarks vanish.) One expects one (or more) of the following three signatures:

(a) $\chi \approx 1$: about 5–10% of the $\pi'^{\pm}$ will manifest themselves as events with energetic leptons (e, μ) recoiling against a hadronic jet of large multiplicity, sphericity and invariant mass.

(b) $\chi \ll 1$: about 10–15% of the $\pi'^{\pm}$ will manifest themselves as "anomalous eμ events" – very much reminiscent of the $\tau^{\pm}$ albeit with a large acollinearity (from $\pi'^{\pm}$ decay) and a $\sin^2\theta$ angular dependence (from $\pi'^{\pm}$ production).

(c) $\chi \gg 1$: almost all the events will be comprised of two hadronic jets with large sphericity and acoplanarity. This circumstance would have a formidable background from ordinary quark and gluon jets. The detailed distributions due to the $\pi'^{\pm}$ decays and background calculations show that a $\pi'^{\pm}$ induced signal can be observed at PETRA and PEP upto a mass $m_{\pi'^{\pm}} < 15$ GeV [8].

There is one dramatic effect which would stand out in the decays of toponium if $m_{\pi'^{\pm}} < m_t - m_b$. In that case the semi-weak decay, $t \to b + \pi'^{+}$ will dominate the $J_T, J'_T, \ldots$ decays, resulting in the final states [‡2]

$$(J_T, J'_T, \ldots) \to \pi'^{+} + b\bar{t}$$
$$\to \pi'^{+}\pi'^{-} + b\bar{b} , \qquad (3)$$

suppressing the canonical decays of J_T:

$$J_T \to \ell\bar{\ell},\ q\bar{q},\ ggg,\ gg\gamma \quad (\ell \equiv e, \mu \text{ or } \tau) . \qquad (4)$$

The rates for the processes (3) can be calculated in the approximation of free t-quark decay inside J_T, giving

$$\Gamma(t \to b + \pi'^{+}) = (G_F^2 f_{\pi'}^2 / 8\pi m_t^3)\, \lambda^{1/2}(m_t^2, m_b^2, m_{\pi'}^2)$$
$$\times [(m_t^2 - m_b^2)^2 - m_{\pi'}^2(m_t^2 + m_b^2)] , \qquad (5)$$

where $\lambda \equiv [m_t^2 - (m_b + m_{\pi'})^2][m_t^2 - (m_b - m_{\pi'})^2]$, G_F is the Fermi coupling constant and $f_{\pi'}$ is the hyperpion decay coupling constant, normalized such that $\sqrt{2} f_{\pi'}^2 G_F = n_{F'}^{-1}$, $n_{F'}$ being the number of hyperquark doublets [‡3]. In fig. 1 we plot the rate for the decay $J_T \to \pi'^{+} + b\bar{t}$ for various values of m_t and $m_{\pi'^{\pm}}$. We estimate the inclusive rate for the standard decays to be 40–60 keV. It is clear that the $\pi'^{\pm}$ modes (3) dominate the J_T decays upto almost the threshold $m_{\pi'^{\pm}} \approx m_t - m_b$. The signatures of (3) would be (almost) isotropic mixed lepton–hadron events – very different from the pure leptonic and hadronic 2- and 3-jet de-

‡1 More precisely, a stable or quasi-stable ($\tau > 10^{-9}$ s) unit charged particle produced exclusively with $\Delta R \geq 0.02$ is ruled out for the mass ranges 4.5 GeV $\leq m_{\pi'^{\pm}} \leq$ 5.4 GeV and 10 GeV $\leq m_{\pi'^{\pm}} \leq$ 14.8 GeV at 95% c.l. [7,8].

‡2 This decay is very similar to the one involving a charged Higgs scalar which has been discussed in the literature. See, for example, Ellis [10].

‡3 To be definite, we have assumed that hyperquarks transform in the same way as ordinary quarks under the weak gauge group $SU(2)_L \times U(1)$. Moreover, as indicated below, they are taken to transform according to the fundamental representation of the hypercolor group $SU(N_{C'})$. In models in which the hyperquarks belong to a non-trivial, say triplet, representation of $SU(3)_C$, the *ordinary* color group, $n_{F'}$ = 3 × (number of hyperflavor doublets). (See, for example, ref. [4]. The consistency of such models – as indeed of the entire hypercolor scenario – is a question outside the scope of the present note.) Note that the relationship with the notation of ref. [3] is $n_{F'} = N_{F'}/2$.

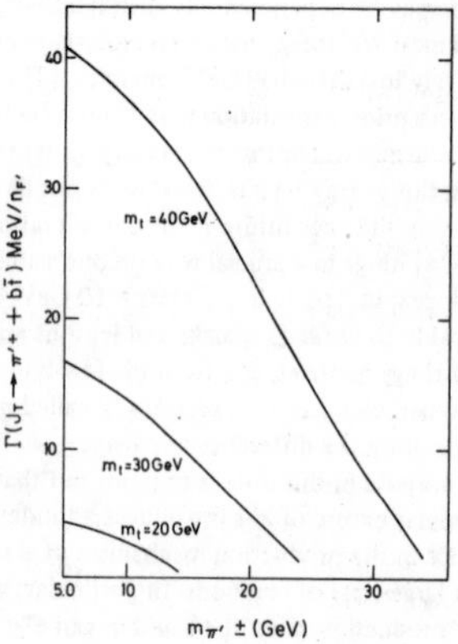

Fig. 1. The rate for the semiweak decay of toponium, J_T, as a function of the charged hyperpion mass. $n_{F'}$ is the number of hyperflavor doublets.

Volume 103B, number 4,5 PHYSICS LETTERS 30 July 1981

cays (4). The $\pi'^{\pm}$ induced J_T decay width (for $m_{J_T} < 80$ GeV) would still be smaller than a typical beam energy resolution ≈ 20 MeV, but the branching ratios for $J_T \to e^+e^-, \mu^+\mu^-$ would become very small.

3. Production of neutral hyperpions. It is generally recognized that if the mass of the Higgs particle ϕ^0 is not very large ($m_{\phi^0} <$ few tens of GeV) then it could be produced in the decays of toponium [9], as well as in the decays of the Z [10]. In particular the following processes are potentially good sources of the Higgs:

$$J_T \to \phi^0 + \gamma\,, \qquad Z \to \phi^0 + \mu^+\mu^-\,, \tag{6,7}$$

$$Z \to \phi^0 + \gamma\,, \qquad e^+e^- \to \phi^0 + Z\,. \tag{8,9}$$

In this section we shall show that whereas toponium still remains an equally good source of the hyperpions, π'^0, the transitions analogous to (7)–(9) do not produce hyperpions at comparable rates. A statistically significant rate for the processes (7)–(9) would amount to an unambiguous confirmation of the canonical Higgs picture as opposed to the hypercolor scenario. Implicit, of course, in the reasoning of this paper is the simplifying assumption that nature uses either elementary Higgs fields or the hypercolor scheme but *not* both.

The radiative transition (6) and the similar reaction $J_T \to \pi'^0 + \gamma$ can be calculated in the approximation of treating the t and $\bar{t}$ quarks at rest [9]. We quote the relevant ratio for equal hyperpion and Higgs mass:

$$\Gamma(J_T \to \pi'^0\gamma)/\Gamma(J_T \to \phi^0\gamma) = \sqrt{2}G_F f_{\pi'}^2 \approx (n_{F'})^{-1}\,. \tag{10}$$

Note that

$$\Gamma(J_T \to \phi^0\gamma)/\Gamma(J_T \to \mu^+\mu^-)$$
$$\approx (G_F m_{J_T}^2/4\sqrt{2}\,\pi\alpha)(1 - m_{\phi^0}^2/m_{J_T}^2)\,, \tag{11}$$

where α is the fine structure constant. The formula (11) has to be corrected to incorporate the characteristic k^3 dipole behaviour if m_{ϕ^0} is close to m_{J_T} [11]. While the discovery of toponium is awaited, we may orient ourselves as to orders of magnitude by taking $m_{J_T} \approx 40$ GeV, $m_{\phi^0} \approx 15$ GeV; these parameters imply that the $\phi^0\gamma$ mode is about 12% of the $\mu^+\mu^-$ mode – a healthy branching ratio indeed, if the charged hyperpions, $\pi'^{\pm}$, are heavier than $m_t - m_b$! Eq. (10) substantiates our assertion that π'^0 production in toponium decay occurs at a rate comparable to that for ϕ^0 production. To clinch the identification [2,3] of the scalar being produced in J_T decays it may be necessary to test whether it decays into $D\bar{D}$(ϕ^0: yes, π'^0: no) or $\pi D\bar{D}$(ϕ^0: no, π'^0: yes). (Note that *CP* invariance is actually sufficient to insure the validity of these tests if the light uncharged pseudo-Goldstone bosons are hyperflavor-neutral – as assumed throughout this paper. Furthermore, the flavor-diagonal couplings of such bosons to ordinary fermions are pure pseudoscalar. These observations are, of course, in no way tied to any specific model for "explicit" breakage of chiral symmetry to generate current fermion masses.)

Next, we turn to processes (7)–(9) and their analogues where the ϕ^0 is replaced by π'^0. Note that the sizeable rates calculated for these reactions [10] stem from the SU(2) × U(1) tree level $ZZ\phi^0$ and $W^+W^-\phi^0$ couplings, which are large in the standard Weinberg–Salam theory. The crucial *difference* between the hypercolor scenario and the canonical theory lies in the observation [2] that there is *no* $\pi'^0 ZZ$ or $\pi'^0 W^+W^-$ coupling at the no-QFD-loop level. At the one-loop level the triangle graph does lead to non-vanishing couplings, the order of magnitude of these couplings compared to the Higgs couplings is thus $\approx(\alpha/\pi)^2$. Our detailed calculations show that these expectations are indeed true.

In what follows we shall calculate the rates for reactions (7)–(9) and the corresponding reactions involving hyperpions. To that end we need to calculate the amplitude for the process

$$V^i \to V^j + \pi'^0\,, \tag{12}$$

where $V^i \equiv Z$ or γ. We specify the process further by taking π'^0 to be the third component of a hyperflavor isotriplet [‡4]; it is then easy to check that only the weak *vector* current comes into play and, in consequence, the amplitudes may be expressed as

$$A(V^i(k_1) \to V^j(k_2) + \pi'^0(q))$$
$$= f(k_1^2/M^2, k_2^2/M^2, q^2/M^2)\, \epsilon_{\mu\nu\sigma\rho} k_1^\mu k_2^\nu \epsilon_1^\sigma \epsilon_2^\rho\,, \tag{13}$$

where $q \equiv k_1 - k_2$, ϵ_1 and ϵ_2 are the polarization vectors of the V's with momenta k_1 and k_2, respectively. M is the constituent mass [3] of the hyperquark ≈ 1 TeV.

‡4 It is possible, of course, to construct models with η-like, K-like, etc. hyperpions. See, for example, ref. [4].

So long as k_1^2, k_2^2, q^2 are all $\ll M^2$, we may approximate f by $f(0, 0, 0)$ which can be calculated using the hyperquark-triangle diagram. Indeed, because of a known extension [12] of the Adler–Bardeen theorem, the value so determined for f has the added and all important virtue of being exact to all orders in QC′D – the QCD-like theory of hypercolor interaction. Without further ado we quote the explicit results:

$$f_{ZZ\pi'^0}(0,0,0) = -\frac{2\alpha}{3\pi f_{\pi'}} g'_A \frac{\sin^2\xi(1-2\sin^2\xi)}{(\sin 2\xi)^2} N_{C'} n_{F'} , \quad (14)$$

$$f_{Z\gamma\pi'^0}(0,0,0) = \frac{\alpha}{6\pi f_{\pi'}} g'_A \frac{1-4\sin^2\xi}{\sin 2\xi} N_{C'} n_{F'} , \quad (15)$$

$$f_{\gamma\gamma\pi'^0}(0,0,0) = \frac{\alpha}{3\pi f_{\pi'}} g'_A N_{C'} n_{F'} . \quad (16)$$

Here ξ is the Glashow–Weinberg–Salam angle, $N_{C'}$ is the number of hypercolors and g'_A is the hyperquark analogue of g_A $[\equiv (G_A/G_V)_{\rm quark}]$; we shall assume that $g'_A \approx 1$. The effective hyperpion coupling, eq. (14), is to be contrasted with the Higgs coupling:

$$g_{ZZ\phi^0} = 2(G_F\sqrt{2})^{1/2} m_Z^2 . \quad (17)$$

Eqs. (14)–(16) permit us to calculate the rates for the processes:

$$Z \to \pi'^0 + \mu^+\mu^- , \quad (7')$$

$$Z \to \pi'^0 + \gamma , \quad (8')$$

$$e^+e^- \to \pi'^0 + Z , \quad (9')$$

as well as the reaction

$$e^+e^- \to \pi'^0 + \gamma , \quad (18)$$

which, though small in absolute rate, may be potentially important in comparison to the Higgs production process $e^+e^- \to \phi^0 + \gamma$ [13].

We present below the rates for the reactions (7) and (7′). The normalized scaled dimuon invariant mass distribution is given by

$$[1/\Gamma(Z\to\mu^+\mu^-)](d/d\kappa^2)\,\Gamma(Z\to\pi'^0+\mu^+\mu^-)$$
$$= \tfrac{1}{6}(\alpha/\pi)^3 \{n_{F'}^3 (N_{C'}/3)^2/[1+(1-4\sin^2\xi)^2]\}\lambda_1^3$$
$$\times [F_1\kappa^2/(1-\kappa^2)^2 + F_2/(1-\kappa^2) + F_3/\kappa^2] , \quad (19)$$

where $\kappa^2 \equiv m_{\mu\mu}^2/m_Z^2$ and

$$\lambda_1 = [(1-\kappa^2+m_{\pi'}^2/m_Z^2)^2 - 4m_{\pi'}^2/m_Z^2]^{1/2} , \quad (20)$$

$$F_1 = [1+(1-4\sin^2\xi)^2]$$
$$\times [\sin^2\xi(1-2\sin^2\xi)/(\sin 2\xi)^3]^2 , \quad (21)$$

$$F_2 = (1-4\sin^2\xi)^2 \sin^2\xi(1-2\sin^2\xi)/(\sin 2\xi)^4 , \quad (22)$$

$$F_3 = \tfrac{1}{4}(1-4\sin^2\xi)^2/(\sin 2\xi)^2 . \quad (23)$$

The corresponding distribution for (7) is [14]

$$[1/\Gamma(Z\to\mu^+\mu^-)](d/d\kappa^2)\,\Gamma(Z\to\phi^0+\mu^+\mu^-)$$
$$= (\alpha/12\pi)(\sin 2\xi)^{-2}\lambda_1(1-\kappa^2)^{-2}(12\kappa^2+\lambda_1^2) , \quad (24)$$

where λ_1 is defined as in eq. (20) with the replacement $m_{\pi'} \to m_\phi$.

The relative rate $\Gamma(Z\to\pi'^0+\mu^+\mu^-)/\Gamma(Z\to\phi^0+\mu^+\mu^-)$ is plotted in fig. 2 as a function of $m_{\pi'^0}$ $(=m_\phi)$ for $\sin^2\xi = 0.20$ (corresponding to $m_Z = 94$ GeV). Typically this ratio is $\approx 10^{-7} n_{F'}^3 N_{C'}^2$ and for reasonable values of $n_{F'}$ and $N_{C'}$ is expected to be $< 10^{-4}$.

We remark that not only are the rates for (7) and (7′) very different, but so are the dimuon invariant mass distributions (19) and (24). The distribution (19) peaks for low values of κ^2 due to the virtual photon contribution in $Z\to\pi'^0+\gamma_v\to\pi'^0+\mu^+\mu^-$, thereby providing a very clean separation even if $n_{F'}$ and $N_{C'}$ are ridiculously large! We plot the normalized distributions (19) and (24) in fig. 3.

The radiative decay (8) can also be an important

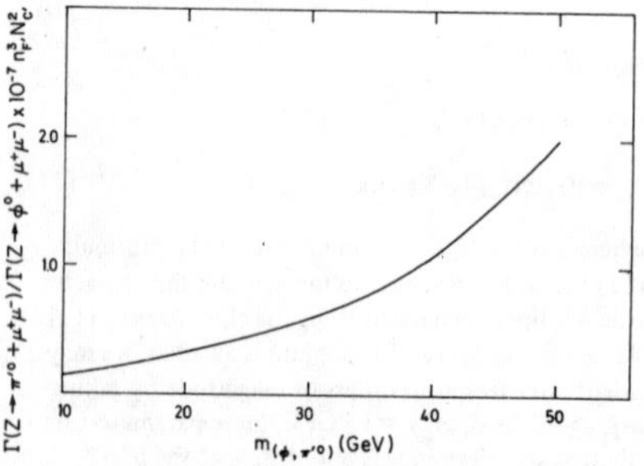

Fig. 2. The ratio $\Gamma(Z\to\pi'^0+\mu^+\mu^-)/\Gamma(Z\to\phi^0+\mu^+\mu^-)$ for equal values of m_{ϕ^0} and $m_{\pi'^0}$. $n_{F'}$ is the number of hyperflavor doublets and $N_{C'}$ the number of hypercolors. We have used $\sin^2\xi = 0.20$ corresponding to $m_Z = 94$ GeV.

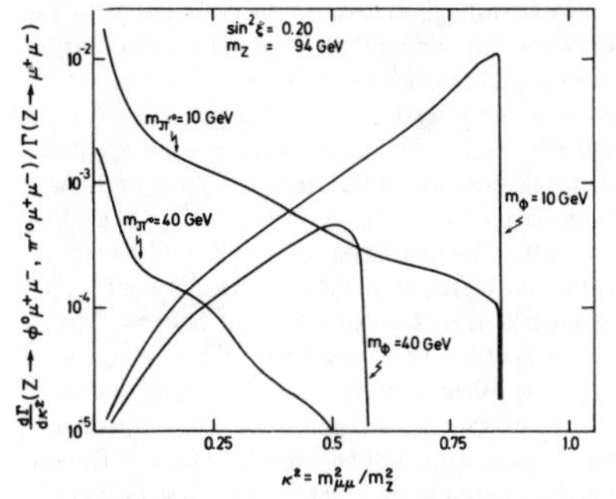

Fig. 3. The dimuon invariant-(mass)^2 distribution from the decays $Z \to \phi^0 + \mu^+\mu^-$ and $Z \to \pi'^0 + \mu^+\mu^-$. We have normalized both the distributions to the same area $[= \Gamma(Z \to \phi^0 + \mu^+\mu^-)]$ for equal m_{ϕ^0} and $m_{\pi'^0}$. The relative scales can be read off from fig. 2.

source of the Higgs boson [15]. This process goes via triangle diagrams involving both fermions and $W^\pm$ bosons. We now have for the relative rates of (8) and (8′):

$$\frac{\Gamma(Z \to \pi'^0\gamma)}{\Gamma(Z \to \phi^0\gamma)} = \frac{N^2_{C'} n^3_{F'}}{9|A|^2} \frac{\sin^2\xi(1 - 4\sin^2\xi)^2}{(\sin 2\xi)^2}, \tag{25}$$

where $|A|^2$ is given by [15]

$$|A|^2 = |A_{\text{fermion}} + A_{W^\pm}|^2, \tag{26}$$

with

$$A_{\text{heavy fermion generation}} = (4/3\cos\xi)(1 - \tfrac{8}{3}\sin^2\xi) \tag{27}$$

and

$$A_{W^\pm} = -(4.9 + 0.3\, m^2_{\phi^0}/m^2_W). \tag{28}$$

Numerically we find for $\sin^2\xi \approx 0.20$:

$$\frac{\Gamma(Z \to \pi'^0\gamma)}{\Gamma(Z \to \phi^0\gamma)} \approx c N^2_{C'} n^3_{F'} \times 10^{-5}, \tag{29}$$

where $c = 8.6$ for $m_{\phi^0} = 10$ GeV and varies slowly with m_{ϕ^0} ($c \approx 8.1$ for $m_{\phi^0} = 60$ GeV). Note that this ratio is small essentially due to the $(1 - 4\sin^2\xi)^2$ factor which enters because only the vector current contributes in $Z \to \pi'^0 + \gamma$.

Next we consider the production of ϕ^0 and π'^0 in e^+e^- annihilation in the processes (9), (9′) and (18). These processes are more relevant for the LEP energies. The cross section for the process (9′) can be written down as

$$\sigma(e^+e^- \to Z\pi'^0) = \frac{\pi}{54}\left(\frac{\alpha}{\pi}\right)^3 \frac{N^2_{C'} n^2_{F'}}{f^2_{\pi'}} \frac{k_{\pi'}}{\sqrt{s}}$$
$$\times [F_1/(s - m^2_Z)^2 - F_2/s(s - m^2_Z) + F_3/s^2]$$
$$\times [k^2_{\pi'} s - \tfrac{3}{4}(s - m^2_Z)^2 + \tfrac{3}{2} m^2_{\pi'}(s + m^2_Z) - \tfrac{3}{4} m^4_{\pi'}]\,. \tag{30}$$

Here s is (c.m. energy)^2, $k_{\pi'}$ is the c.m. momentum of the hyperpion and F_i are the functions already defined in eqs. (21)–(23). Eq. (30) is to be contrasted with the corresponding expression for the production of ϕ^0 [16] ‡5.

$$\sigma(e^+e^- \to \phi^0 Z)$$
$$= \tfrac{2}{3}\pi\alpha^2(2k_\phi/\sqrt{s})[(3m^2_Z + k^2_\phi)/(s - m^2_Z)^2]$$
$$\times (1 - 4\sin^2\xi + 8\sin^4\xi)/(\sin 2\xi)^4, \tag{31}$$

where we have used the same notation as in (30).

In fig. 4 we plot the ratio $\sigma(e^+e^- \to \pi'^0 Z)/\sigma(e^+e^- \to \phi^0 Z)$ as a function of $m_{\pi'^0}$ ($= m_{\phi^0}$) for $\sqrt{s} = 140$, 170 and 200 GeV. The ratio increases with $\sqrt{s}$ but for energies at LEP and a reasonable value of $n_{F'}$ and $N_{C'}$, we still expect it to be $< 10^{-5}$.

Finally, we would like to present the result for the

‡5 The correct expression for $\sigma(e^+e^- \to Z\phi)$, however, is given in ref. [17].

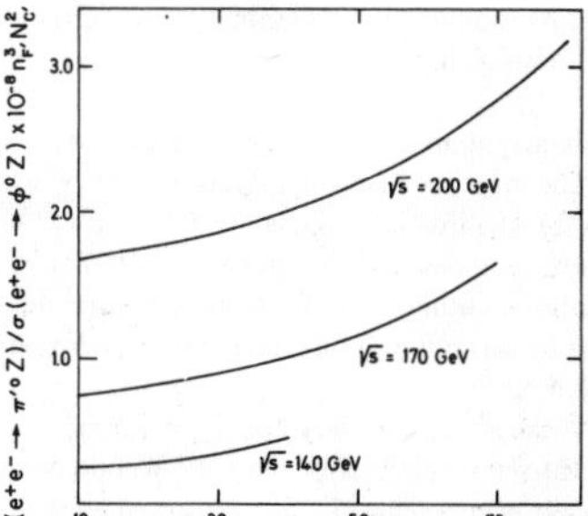

Fig. 4. The ratio $\sigma(e^+e^- \to \pi'^0 Z)/\sigma(e^+e^- \to \phi^0 Z)$ for $\sqrt{s} = 140$ GeV, 170 GeV and 200 GeV and equal values of m_{ϕ^0} and $m_{\pi'^0}$. We have used $\sin^2\xi = 0.20$ and $m_Z = 94$ GeV.

radiative process (18) and the corresponding process involving ϕ^0 [13]. We normalize the cross section for (18) with respect to the point cross section $\sigma(e^+e^- \to \mu^+\mu^-)$:

$$\frac{\sigma(e^+e^- \to \pi'^0\gamma)}{\sigma(e^+e^- \to \mu^+\mu^-)} = \frac{G_F N_{C'}^2 n_{F'}^3}{288\sqrt{2}\pi^2}\left(\frac{\alpha}{\pi}\right)\frac{E_\gamma^3}{\sqrt{s}}$$

$$\left(\frac{(1-4\sin^2\xi)^2}{(\sin 2\xi)^4}[1+(1-4\sin^2\xi)^2]\frac{s^2}{(s-m_Z^2)^2} + \frac{8(1-4\sin^2\xi)^2}{(\sin 2\xi)^2}\frac{s}{(s-m_Z^2)} + 16\right), \tag{32}$$

which is to be compared with the analogous processes involving ϕ^0 [13]:

$$\frac{\sigma(e^+e^- \to \phi^0\gamma)}{\sigma(e^+e^- \to \mu^+\mu^-)} = \frac{\sqrt{2}\,G_F}{\pi^2}\left(\frac{\alpha}{\pi}\right)\frac{E_\gamma^3}{\sqrt{s}}|I'_W - \Sigma_F I'_F|^2, \tag{33}$$

I'_W and I'_F being the W and fermion loop contributions, respectively. To get an estimate we set $\sin^2\xi = 0.25$, so that only the photon contribution survives; we have then:

$$\frac{\sigma(e^+e^- \to \pi'^0\gamma)}{\sigma(e^+e^- \to \phi^0\gamma)} \approx \frac{1}{36}\frac{N_{C'}^2 n_{F'}^3}{|I'_W - \Sigma_F I'_F|^2} \approx \tfrac{1}{36}N_{C'}^2 n_{F'}^3. \tag{34}$$

Thus this *ratio* is potentially large. However, the cross section for $\phi^0\gamma$ production [eq. (33)] is so small that even an enhancement ≈ 200 for the hyperpion case will not help the process $e^+e^- \to \pi'^0\gamma$ very much; it will still be swamped by the normal 1γ background in e^+e^- annihilation.

4. We may summarize our results as follows:

(a) The most promising signals for $\pi'^\pm$ are: $e^+e^- \to (\mu \text{ or } e)$ + hadron jet and $e^+e^- \to \ell^+\ell^- + \nu$'s ($\ell = e, \mu$) with large acollinearity angle and a characteristic $\sin^2\theta$ distribution in production. Furthermore R should not increase by one full unit on crossing the hyperpion threshold.

(b) If the $\pi'^\pm$ are not very heavy, i.e. for $m_{\pi'^\pm} < m_t - m_b$, the semiweak decay $t \to b + \pi'^+$ will dominate the decays of toponium, J_T, J'_T, ..., as well as of the open top mesons T_u, T_d, etc. The width of J_T, J'_T is still expected to be smaller than the beam resolution; however, the topology of J_T, J'_T decays would be very different, as would be the branching ratio for $J_T \to \mu^+\mu^-$, e^+e^-, etc.

(c) To distinguish between the canonical picture of symmetry breaking and the hypercolor scenario it is necessary to produce and detect neutrals. The low rates for the processes $Z \to \pi'^0 + \mu^+\mu^-$, $Z \to \pi'^0 + \gamma$ and $e^+e^- \to Z + \pi'^0$ indicate that a positive signal in these reactions would be evidence in *favor* of the Weinberg–Salam theory. The process $Z \to \pi'^0 + \mu^+\mu^-$ has the further discrimination, vis à vis $Z \to \phi^0 + \mu^+\mu^-$, in that the lepton invariant mass distribution for the two reactions is expected to be very different. The smallness of the $ZZ\pi'^0$ and $W^+W^-\pi'^0$ couplings in hypercolor theories ensures that in all the processes where a ϕ^0 can bremsstrahl off a $W^\pm$ or Z the rates for π'^0 production will be down by $\approx(\alpha/\pi)^2$. Thus, a scalar produced in an ep, νp or γp process with the tell-tale signatures of a semi-weakly coupled particle would indeed be a canonical Higgs.

(d) In the intermediate e^+e^- energy range, toponium may be an excellent laboratory for investigating the properties of hyperpions and thus the nature of the Higgs mechanism.

This work was begun while one of us (A.A.) was a visitor at Rockefeller University during February 1980; it was prepared for publication during the sojourn of M.A.B.B. at the Aspen Center for Physics. The hospitality of the Center is gratefully acknowledged. A.A. would like to acknowledge useful discussions with experimental colleagues at DESY, in particular R. Eichler, R. Felst, D. Haidt, H. Newman, S.C.C. Ting and M. White.

References

[1] For a recent review, see: M.A.B. Bég, RU report No. DOE/EY/2232B-211 (1980), to be published in: Proc. XXth Intern. Conf. on High energy physics;
the original papers on the concept of hypercolor are:
S. Weinberg, Phys. Rev. D13 (1976) 974;
L. Susskind, Phys. Rev. D20 (1979) 2619.

[2] M.A.B. Bég, H.D. Politzer and P. Ramond, Phys. Rev. Lett. 43 (1979) 1701;
the model-independence of the parity tests in this paper is reviewed in: M.A.B. Bég, RU report No. DOE/EY/2232B-220 (1980), to be published in: Proc. VPI Workshop on Weak interactions.

[3] M.A.B. Bég, Proc. Orbis Scientiae (1980, Coral Gables) (Plenum, New York, 1980).

[4] S. Dimopoulos, S. Raby and G.L. Kane, Univ. of Michigan preprint (1980).

[5] E. Eichten and K.D. Lane, Phys. Lett. 90B (1980) 125.

[6] M.E. Peskin, Saclay report DPh-T/80/46 (1980).
[7] JADE Collab., DESY report 80/71 (1980); and private communication.
[8] A. Ali, H. Newman and R. Zhu, Searching for the charged hyperpions in e^+e^- annihilation, DESY report, in preparation.
[9] F. Wilczek, Phys. Rev. Lett. 39 (1977) 1304.
[10] For a review of Higgs production at LEP, see: G. Barbiellini et al., DESY report 79/27 (1979);
see also: J. Ellis, Proc. 1978 Summer Institute, SLAC-215 (1978).
[11] J. Ellis, M.K. Gaillard, D.V. Nanopoulos and C. Sachrajda, CERN report TH-2634 (1979).
[12] S.Y. Pi and S.-S. Shei, Phys. Rev. D11 (1975) 2946.
[13] J.P. Leveille, Wisconsin report COO-881-86 (1979).
[14] J.D. Bjorken, SLAC report No. PUB-1866 (1977).
[15] R.N. Cahn, M.S. Chanowitz and N. Fleishon, LBL report No. LBL-849 (1978).
[16] J. Ellis, M.K. Gaillard and D.V. Nanopoulos, Nucl. Phys. B106 (1976) 292.
[17] B.W. Lee, C. Quigg and H.B. Thacker, Phys. Rev. D16 (1977) 1519;
see also: B.L. Ioffe and V.A. Khoze, Fiz. Elem. Chastits At Yadra 9 (1978) 118 [Sov. J. Part. Nucl. Phys. 9 (1978) 50].

VII. VACUUM ALIGNMENT

Nuclear Physics B175 (1980) 197–233

THE ALIGNMENT OF THE VACUUM IN THEORIES OF TECHNICOLOR

Michael E. PESKIN

Service de Physique Théorique, CEN Saclay, B.P.No. 2, 91190 – Gif-sur-Yvette, France
and
Lyman Laboratory of Physics, Harvard University, Cambridge, Mass. 02138, USA*

Received 28 April 1980

We study a class of models of dynamical weak-interaction symmetry breaking suggested by the work of Weinberg and Susskind. We demonstrate in these models some simple relations which determine the alignment of weakly gauged subgroups and the masses of pseudo-Goldstone bosons. We illustrate these methods by computing the spectrum of pseudo-Goldstone bosons in three models recently examined by Dimopoulos. These models are seen to contain a very light charged Higgs boson, whose mass we estimate by the use of chiral perturbation theory. In one of these models, we observe that the energetically preferred vacuum is a phenomenologically unpleasant one.

1. Introduction

The Weinberg-Salam model [1], in which the weak and electromagnetic interactions derive from a spontaneously broken gauge theory of SU(2) × U(1), has had triumphant success in describing the weak interactions. This success, however, has made troublesome and pressing the question of what it is that breaks the SU(2) × U(1), invariance. The standard answer, that the symmetry breaking is caused by a set of fundamental scalar mesons with negative mass-squared, leads to a consistent but frustrating theory in which the masses and couplings of these scalars — and the masses of the quarks and leptons — are essentially incalculable. It is thus important to search for alternative answers. Among the most attractive of these is the idea that SU(2) × U(1) is broken by a system of strongly interacting fermions and gauge mesons; in this case, one can imagine that the symmetry breaking and the fermion mass generation of the Weinberg-Salam model might be accomplished by a sector which introduces only a single mass scale as its sole adjustable parameter.

The promise offered by such a dynamical scheme for breaking the Weinberg-Salam symmetry is, however, far from being realized in practice through the construction of realistic models. A solution has been found for part of the problem: Weinberg [2] and Susskind [3] have independently shown that the phenomenologically correct pattern of vector boson masses may be rather naturally achieved. But

* Junior Fellow, Harvard Society of Fellows.

the problem of providing fermion masses has proved considerably more intransigent. Eichten and Lane [4] and Dimopoulos and Susskind [5] have felt compelled to introduce a new set of strongly coupled gauge fields, beyond those needed to break SU(2) × U(1), in order to allow this symmetry breaking to produce quark and lepton masses of reasonable magnitude. The resulting models are quite complex but nevertheless still plagued with difficulties; we know of no completely satisfactory model of even a single family of quarks and leptons.

What is the reason for this difficulty? Certainly it is possible that it is intrinsic to the idea that the symmetry breaking is dynamical, but we prefer to believe that it is the result of our limited knowledge of the physics of dynamical symmetry breaking. That physics is, in any event, surprisingly complex: one must worry about the pattern of symmetry breaking produced by the new strongly coupled gauge theory, but one must also understand the interaction of this broken symmetry state with the weakly coupled gauge bosons of SU(2) × U(1). We feel that it is of interest to try to separate and clarify the various aspects of this physics, and, in particular, to point out effects which might be useful or troublesome in the construction of models. Weinberg [6], several years ago, made a first step in this direction, developing a formalism for those aspects of the theory to which current algebra might be applied. Still, though, we have much to learn.

This is the first of a series of papers which we devote to the physics of dynamical symmetry breaking. In this paper, we will examine the simplest class of models: those in which the new strongly coupled gauge group remains unbroken while it breaks the Weinberg-Salam SU(2) × U(1). In this case, the special problems which arise from the interaction between the symmetry-breaking and weakly coupled gauge fields may be resolved in a most transparent way; it is that resolution that we wish to describe. The succeeding papers of this series [7] will deal with more complex situations in which the new strongly coupled gauge group is itself spontaneously broken. We will follow Dimopoulos and Susskind [5] in referring to the new strong interaction group as technicolor; Eichten and Lane [4] have given it the name hypercolor.

To begin our discussion, let us explain more precisely what it is about the interaction of weakly coupled gauge fields with the broken symmetry state that requires our careful attention. The problem is one already raised in the work of Weinberg [6]. Consider a set of fermions coupled strongly to a gauge group G_s (technicolor). To avoid proliferation of parameters, we will, throughout this series of papers, take all fermions to have zero bare mass. Then the system of fermions has, in addition to its gauge symmetry G_s, a large group of additional global symmetries G. We will refer to G as the chiral group, since for the case of quarks in this usual strong interactions, G is just the group chiral $SU(N) \times SU(N)$. Since the gauge coupling of G_s is strong, one might expect to see this symmetry spontaneously broken to a subgroup H which allows fermions coupled to G_s to acquire effective masses. In QCD, this is the familiar spontaneous breaking of chiral

SU(N). A particular broken-symmetry vacuum $|\Omega\rangle$ corresponds to a particular orientation of the subgroup H inside the group G. Now let us couple the fermions weakly to the group G_w, containing SU(2) × U(1). Since G_w is a group of symmetries, it must be a subgroup of G; therefore, it determines a second orientation inside this group. One might picture an angle between the orientations of H and G_w; the energy of a given state $|\Omega\rangle$ will depend on this angle. This energy breaks the degeneracy of the states $|\Omega\rangle$ and forces us to choose a particular one as the true vacuum. The problem, then, is to find this correct vacuum orientation. The fact that the correct vacuum is determined energetically has two major physical consequences. First, since the pattern of symmetry breaking of G_w is determined by the overlap of G_w with H, this pattern cannot be chosen *a priori*; rather it is dynamically determined. Second, since small rotations of the orientation of $|\Omega\rangle$ in the theory decoupled from G_w correspond to the zero-momentum components of Goldstone boson fields, the coupling to G_w gives mass to some of these bosons. Mesons acquiring small masses in this way have been given by Weinberg [8] the generic name of pseudo-Goldstone bosons. In theories of technicolor, such bosons take on a special importance, since they are the lightest and most accessible particles of the new sector. We will refer to such bosons in this context as technions.

In ref. [6], Weinberg gave a formal prescription for resolving the problem we have stated, valid for the most general patterns of dynamical symmetry breaking; unfortunately, however, the very generality of this analysis conceals some intriguing points of physics. In this paper, we will re-examine this problem within a specific and rather limited class of models in which the forces which determine the vacuum orientation may be clearly seen and easily analyzed. These models have the additional advantage of being almost realistic: though none contains enough structure to allow quarks and leptons to acquire masses, their completion by addition of the necessary structure in the manner suggested in refs. [4, 5] would entail only a small perturbation on most of the physics we describe.

The formalism for our analysis will be developed in sect. 2. Here we will review the patterns of symmetry breaking which allow a set of fermions coupled to G_s to acquire effective masses without breaking G_s. These patterns have in common that the symmetry breaking preserves a discrete parity invariance. We will then analyze the energetics of the vacuum orientation, using this parity invariance to simplify the results of Weinberg. We will find in this case that the favored vacuum state is simply the $|\Omega\rangle$ which minimizes the masses acquired by the G_w vector bosons through the Higgs mechanism. We will find, further, a compact set of rules for computing the masses of pseudo-Goldstone bosons. [The casual reader might wish merely to skim this section, noting the results (2.6), (2.23), and (2.37).]

In sect. 3, we will give these rules a first illustration in a very simple model containing two doublets of fermions coupling to G_w = SU(2) × U(1). Here we will review the Weinberg-Susskind scenario for obtaining the correct pattern of SU(2)

× U(1) breaking; we will demonstrate also that this scenario can fail because the energetically preferred vacuum orientation may not be the desired one.

The remainder of the paper will be devoted to the analysis of the spectrum of scalar mesons in three models of SU(2) × U(1) breaking which have been considered recently by Dimopoulos [9]. These models have chiral groups G large enough for the G_s fermions to transform under color SU(3) as well as SU(2) × U(1); this yields a rich spectrum of technions. Dimopoulos has offered estimates of the masses of these bosons. In sect. 4, we will use our rules to refine his estimates in two of these models — in fact, to calculate their spectra, to leading order in α and α_s, up to an overall scale.

Certain of the bosons in these models acquire no mass to leading order in α; this accords with more general observations of Eichten and Lane [4]. These authors note, however, that a pair of electrically charged scalars from this class acquire masses of higher order in α. In sect. 5, we show that these masses may be evaluated by the use of chiral perturbation theory [10]. In the first Dimopoulos model, and in a larger group of theories in which it lies, we find a value of 5 to 8 GeV.

Finally, in sect. 6, we show that the third of Dimopoulos' models is not phenomenologically acceptable because the wrong vacuum is preferred energetically. We clarify the reason for this preference, which provides an amusing illustration of the forces determining vacuum alignment.

2. Formalism

In this section, we will review Weinberg's formalism for the orientation of subgroups [6] in the context of some special scenarios for dynamical symmetry breaking. Our goal will be to find a simple formula for the masses of pseudo-Goldstone bosons; the remainder of the paper will be devoted to exploring the physics of this formula.

We should begin by describing three patterns of breaking of the chiral group G which might be associated with dynamical mass generation for a flavor multiplet of strongly interacting fermions in the case in which their strong interaction symmetry G_s remains unbroken. These three possibilities have also been recognized by Dimopoulos [9] and are applied in the models of his which we will discuss later. In the following, and throughout this paper, we consider only models whose lagrangian contains no fermion mass terms.

In the familiar case of the ordinary strong interactions, we observe the spontaneous breaking of a global symmetry, chiral SU(2) × SU(2), through the generation of dynamical masses for the quarks and the appearance of a non-zero vacuum expectation value of the operator $\bar{\psi}\psi$. We would like to ask how far this schema might be generalized. Our basic assumption will be that the maximum flavor symmetry consistent with dynamical mass generation is preserved; this assumption accords with our limited experience, but, further, it may be checked at least in simple approximation schemes for dynamical symmetry breaking such as those offered in ref. [11]*.

To see clearly the consequences of a non-zero vacuum expectation value for $\bar{\psi}\psi$, it is useful to adopt the following device: rewrite all fermion fields as left-handed fermions by considering the charge conjugates of the right-handed fermion fields as the more fundamental. For the case of QCD, this entails replacing each four-component quark field ψ by two (left-handed) two-component spinor fields, one transforming as a 3 and the other as a $\bar{3}$ under color SU(3). This rewriting sets

$$\bar{\psi}\psi = i\left(\varepsilon^{\alpha\beta}\psi_\alpha^{(3)}\psi_\beta^{(\bar{3})}\right) + \text{h.c.}, \tag{2.1}$$

where α, β are two-component spinor indices and ε is the antisymmetric tenor. In general, spontaneous symmetry breaking must be characterized by a vacuum expectation value of some Lorentz-scalar operator. The only such operators quadratic in fermion fields are those of the form

$$\varepsilon^{\alpha\beta}\psi_\alpha^A\psi_\beta^B I^{AB} \tag{2.2}$$

and their conjugates. Here A, B represent color and flavor indices and I is a c-number tensor. Note that, since the ψ fields anticommute, I must be symmetric under interchange of its indices.

In the case where the gauge symmetry G_s remains unbroken, (2.2) must be a G_s invariant. There are three basic types of models in which such a gauge-invariant, chiral symmetry-breaking form can be constructed:

(i) One may have $2N$ multiplets of (two-component) fermions, N of which belong to a complex representation r of G_s and the other N to the conjugate representation $\bar{r}$. [By the comment above (2.1), QCD with N flavors is a theory of this class.] Call the former set of fermions $\psi_\alpha^{(r)ai}$, where a is the G_s index and $i = 1, \ldots, N$; call the latter set $\psi_{\alpha ai}^{(\bar{r})}$. This theory is symmetric under general independent unitary transformations of the two sets of multiplets, except for an overall phase rotation, which suffers an Adler-Bell-Jackiw anomaly [12]. Thus, the flavor symmetry $G = SU(N) \times SU(N) \times U(1)$. The G_s invariant form (2.2) with the maximal flavor symmetry is

$$\varepsilon^{\alpha\beta}\psi_\alpha^{(r)ai}\psi_{\beta ai}^{(\bar{r})}. \tag{2.3}$$

A vacuum expectation value for the quantity (2.3) signals the spontaneous breakdown of G to $H = SU(N) \times U(1)$.

(ii) One may have $2N$ multiplets of fermions**, each belonging to a representation r of G which is real in the strict sense, that is, such that the symmetric combination of two r's contains an invariant d_{ab}. This theory has a flavor

* This question and the question of when G_s might itself be broken will be discussed at length in [7].

** An odd number is also possible; the notation $2N$ is for later convenience.

symmetry G = SU(2N), if we omit the pure phase transformation which suffers an anomaly. The G_s invariant form (2.2) of maximal flavor symmetry is

$$\varepsilon^{\alpha\beta}\psi_\alpha^{(r)ai}\psi_\beta^{(r)bi}\, d_{ab}\,, \tag{2.4}$$

where now $i = 1, \ldots, 2N$. If (2.4) acquires a vacuum expectation value, G is broken to the set of transformations which preserve the dot product in the flavor space, H = O(2N).

(iii) One may have $2N$ multiplets of fermions, each belonging to a representation r of G_s which is pseudo-real, that is, such that the antisymmetric combination of two r's contains an invariant e_{ab}. This theory again has G = SU(2N). The G_s invariant form of maximal flavor symmetry is

$$\varepsilon^{\alpha\beta}\psi_\alpha^{(r)ai}\psi_\beta^{(r)bj} e_{ab} E_{ij}\,, \qquad E = \left(\begin{array}{c|c} 0 & 1 \\ \hline -1 & 0 \end{array}\right). \tag{2.5}$$

If (2.5) acquires a vacuum expectation value, G is broken to H = Sp(2N), the unitary symplectic group of $2N \times 2N$ matrices.

In the remainder of this paper, we will refer to these scenarios as the SU, O, and Sp cases, respectively. More complex schemes in which G_s is unbroken are all built using several sets of fermions, with each set obeying one of these scenarios. These patterns can be seen even in scenarios in which G_s is broken, as we will show in [7].

It is worth recognizing that the three scenarios can be discussed in a unified way by making manifest the fact that G and H of the O and Sp cases contain as subgroups the G and H of the SU case, as shown in fig. 1. In appendix A, we use this observation to find a useful representation of the generators of G: divide these generators into two classes, calling the generators of H $\{T_i\}$ and the broken generators $\{X_a\}$. Then one may represent them by $2N \times 2N$ hermitian matrices:

$$T_i = \left(\begin{array}{c|c} A & B \\ \hline B^\dagger & -A^{\mathrm{T}} \end{array}\right), \qquad X_a = \left(\begin{array}{c|c} C & D \\ \hline D^\dagger & C^{\mathrm{T}} \end{array}\right), \tag{2.6}$$

where A is hermitian, C is hermitian and traceless, and

$$\begin{array}{lll} B = 0\,, & D = 0\,, & \text{(SU case)}\,, \\ B = -B^{\mathrm{T}}\,, & D = +D^{\mathrm{T}}\,, & \text{(O case)}\,, \\ B = +B^{\mathrm{T}}\,, & D = -D^{\mathrm{T}}\,, & \text{(Sp case)}\,. \end{array} \tag{2.7}$$

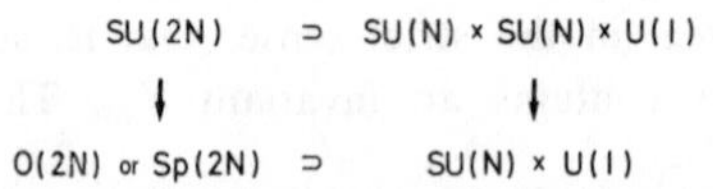

Fig. 1. Relation of three schemes of chiral symmetry breaking.

We normalize these generators to

$$\operatorname{Tr} T_a T_b = \operatorname{Tr} X_a X_b = \delta_{ab}, \qquad \operatorname{Tr} X_a T_i = 0. \tag{2.8}$$

One remarkable feature of our three scenarios is that, in each case, the quotient space G/H is a symmetric space*. By this one means that there exists a parity operation P which preserves the Lie algebra of G, such that

$$PT_iP^{-1} = T_i, \qquad PX_aP^{-1} = -X_a. \tag{2.9}$$

An equivalent criterion is that, if X and T represent arbitrary linear combinations of the broken and unbroken generators, their commutators satisfy the restriction

$$[T,T] = iT, \qquad [X,T] = iX,$$

$$[X,X] = iT. \tag{2.10}$$

The first two restrictions of (2.10) are trivial; they simply express the fact that the Lie algebra of H closes and that the X_a form a representation of H. The third restriction, however, is non-trivial and equivalent to (2.9). The pattern (2.10) may be easily verified from (2.6).

We now proceed to develop the theory of vacuum orientation under the assumption that G/H is a symmetric space. We will see that this hypothesis is a most natural one for the problem, in that it induces considerable simplification.

Consider, then, a theory of fermions with a chiral group G, some of whose generators are coupled weakly to gauge fields of the group G_w. Let us describe this coupling by a lagrangian

$$\delta\mathcal{L} = \sum_{\Lambda} A_\mu^\Lambda J_\mu^\Lambda, \tag{2.11}$$

where $J_\mu^\Lambda = \bar\psi\gamma^\mu\Lambda\psi$ and ψ, following the prescription given above (2.1), contains only left-handed spin components. The various Λ are linear combinations of the generators of G and are defined to contain their coupling constants. Now let G be spontaneously broken to H. This produces a manifold of vacuum states $|\Omega\rangle$, one for each point in G/H, degenerate in energy before we include the effects of (2.11). The exchange of gauge bosons A_μ^Λ will shift the energy density of each $|\Omega\rangle$ by an amount

$$\Delta E = -\tfrac{1}{2}\sum_{\Lambda}\int d^4x\, \Delta^{\mu\nu}(x)\langle\Omega|\mathrm{T}J_\mu^\Lambda(x)J_\nu^\Lambda(0)|\Omega\rangle, \tag{2.12}$$

* See, for example, [3]. Symmetry-breaking schemes in which G/H is a symmetric space have recently excited interest in connection with 2-dimensional sigma models; see [14].

where $\Delta^{\mu\nu}(x)$ is a massless vector boson propagator. This energy will, in general, depend on the orientation of $|\Omega\rangle$ and choose a preferred orientation.

To analyze this effect, it is useful to make a choice of coordinates; this may be done as follows: choose a representation of the broken and unbroken generators, X_a and T_i, of G. Let $|0\rangle$ be the one state $|\Omega\rangle$ invariant to transformations generated by the T_i in this representation. We can now use this vacuum state as a reference point; any other $|\Omega\rangle$ is related to $|0\rangle$ by a unitary transformation $\underline{u}$. This allows us to rewrite the matrix element in (2.12):

$$\begin{aligned}\langle\Omega|\mathrm{T}J_\mu^\Lambda(x)J_\nu^\Lambda(0)|\Omega\rangle &= \langle 0|\underline{u}^\dagger \mathrm{T}J_\mu^\Lambda(x)J_\nu^\Lambda(0)\underline{u}|0\rangle \\ &= \langle 0|\mathrm{T}J_\mu^{U^\dagger\Lambda U}(x)J_\nu^{U^\dagger\Lambda U}(0)|0\rangle\,, \end{aligned} \tag{2.13}$$

where U is a unitary matrix representing an element of G.

We now need to use the fact that G/H is a symmetric space. That condition implies, in particular, that the only H-invariant term in the product X^aX^b is proportional to δ^{ab}. Essentially, this is the sense in which a symmetric space is symmetric; in any event, we offer a simple proof of this restriction in appendix B. It allows us, first, to simplify the two-current correlation function

$$\begin{aligned}\langle 0|\mathrm{T}J_\mu^{X_a}J_\nu^{X_b}|0\rangle &= \delta^{ab}\langle 0|\mathrm{T}J_\mu^X J_\nu^X|0\rangle \\ &= \mathrm{Tr}[X_aX_b]\langle 0|\mathrm{T}J_\mu^X J_\nu^X|0\rangle\,. \end{aligned} \tag{2.14}$$

In this equation, the notation X denotes any single generator X_a (no sum over these generators is implied), and we have used (2.8). Since the T_i span a single real representation of H (the adjoint representation), a similar equation holds for the J_μ^T expectation values. The joint expectation value of one J_μ^T and one J_μ^X is zero by parity. A second application of this restriction will be useful to us later: to every spontaneously broken generator X_a, there corresponds a Goldstone boson π^a which is created from the vacuum by the current $J_\mu^{X_a}$. Our restriction enforces the equality of the corresponding pion decay constants*

$$\langle 0|J_\mu^{X_a}|\pi^b\rangle = ip_\mu f_\pi\delta^{ab} = ip_\mu f_\pi \mathrm{Tr}[X^aX^b]\,. \tag{2.15}$$

Note that both (2.14) and (2.15) require the vacuum $|0\rangle$.

The matrices Λ are linear combinations of the X_a and T_i. Let us denote the decomposition of such a matrix

$$\Lambda = \Lambda_T + \Lambda_X\,. \tag{2.16}$$

* The normalization of f_π used in (2.15) corresponds in the usual strong interactions to $f_\pi = 90$ MeV.

With this notation, we may use (2.14) and its counterpart for the T_i to cast (2.13) into the form

$$
\begin{aligned}
&\langle 0|\mathrm{T}J_\mu^{U^\dagger\Lambda U}J_\nu^{U^\dagger\Lambda U}|0\rangle \\
&\quad= \mathrm{Tr}\big[(U^\dagger\Lambda U)_T(U^\dagger\Lambda U)_T\big]\langle 0|\mathrm{T}J_\mu^T J_\nu^T|0\rangle \\
&\qquad+ \mathrm{Tr}\big[(U^\dagger\Lambda U)_X(U^\dagger\Lambda U)_X\big]\langle 0|\mathrm{T}J_\mu^X J_\nu^X|0\rangle \\
&\quad= \mathrm{Tr}\big[(U^\dagger\Lambda U)_T U^\dagger\Lambda U\big]\langle 0|\mathrm{T}J_\mu^T J_\nu^T|0\rangle + \mathrm{Tr}\big[(U^\dagger\Lambda U)_X U^\dagger\Lambda U\big]\langle 0|\mathrm{T}J_\mu^X J_\nu^X|0\rangle \\
&\quad= \mathrm{Tr}\big[U^\dagger\Lambda U U^\dagger\Lambda U\big]\langle 0|\mathrm{T}J_\mu^T J_\nu^T|0\rangle \\
&\qquad+ \mathrm{Tr}\big[(U^\dagger\lambda U)_X(U^\dagger\Lambda U)_X\big]\langle 0|\mathrm{T}\big(J_\mu^X J_\nu^X - J_\mu^T J_\nu^T\big)|0\rangle\,. \qquad (2.17)
\end{aligned}
$$

We have freely used the orthogonality of X_a and T_i, eq. (2.8). The vacuum energy (2.12) now takes the form

$$
\begin{aligned}
\Delta E = \Bigg(&\Big\{\int \mathrm{d}^4x\,\Delta^{\mu\nu}\langle 0|\mathrm{T}\big(J_\mu^T J_\nu^T - J_\mu^X J_\nu^X\big)|0\rangle\Big\} \\
&\times \tfrac{1}{2}\sum_\Lambda \mathrm{Tr}\big[(U^\dagger\Lambda U)_X(U^\dagger\Lambda U)_X\big]\Bigg) + (\text{independent of } U)\,, \qquad (2.18)
\end{aligned}
$$

in which the dependence on the vacuum orientation U is factored away from the current matrix element. At the end of this section we will argue, though not rigorously, that the quantity in brackets is positive. Thus, one finds the preferred vacuum simply by minimizing the quantity

$$\sum_\Lambda \mathrm{Tr}(\Lambda_X)^2 \qquad (2.19)$$

over possible equivalent representations of the X_a and T_i.

To see more clearly the physical significance of this criterion, we need to make a short digression. If some of the generators of $\mathrm{G_w}$ are indeed spontaneously broken, the corresponding gauge bosons will acquire masses through the Higgs mechanism. Let us compute those masses. It was noted long ago by Schwinger [15] that these masses may be found as the residues of poles in the vacuum polarization tensor of the gauge fields; the utility of this method in the context of dynamically broken theories has been emphasized by Brout and Englert, among others [16]. These authors have noticed that, if the vacuum polarization, computed in the unperturbed theory of $\mathrm{G_s}$ interactions, behaves at zero momentum as

$$\langle \mathrm{T}J_\mu^\Lambda(k)J_\nu^{\Lambda'}(-k)\rangle \underset{k\to 0}{\rightarrow} -i\left(g_{\mu\nu} - \frac{k_\mu k_\nu}{k^2}\right)\mu^2_{\Lambda\Lambda'}, \qquad (2.20)$$

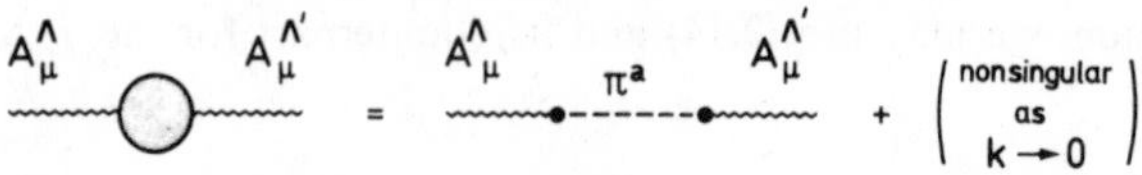

Fig. 2. Origin of poles in the vacuum polarization tensor which signal generation of vector boson masses.

where the μ^2 are a set of constants, the vector bosons acquire, to order k^2, precisely the mass matrix $\mu^2_{\Lambda\Lambda'}$. If one evaluates the left-hand side of (2.20) by inserting a complete set of intermediate states, only the contribution of massless particles, i.e., of Goldstone bosons, yields such a $1/k^2$ singularity (see fig. 2). If the vacuum state were the vacuum $|0\rangle$ defined above, we could use (2.15) to evaluate the contribution of these bosons and find

$$\mu^2_{\Lambda\Lambda'} = \sum_a f_\pi^2 \operatorname{Tr}[X^a\Lambda]\operatorname{Tr}[X^a\Lambda'] = f_\pi^2 \operatorname{Tr}[\Lambda_X \Lambda'_X]. \tag{2.21}$$

Using the fact that a general vacuum $|\Omega\rangle$ and the pion states built upon it are related to the corresponding states of $|0\rangle$ by a unitary transformation, we find that on the vacuum $|\Omega\rangle$,

$$\mu^2_{\Lambda\Lambda'} = f_\pi^2 \operatorname{Tr}\left[(U^\dagger\Lambda U)_X (U^\dagger\Lambda' U)_X\right], \tag{2.22}$$

where U is the unitary matrix introduced in (2.13). Comparing (2.22) with (2.18), we see that

$$\Delta E = (\text{constant}) \times \operatorname{Tr}[\mu^2]. \tag{2.23}$$

This equation makes precise our intuitive notation that the preferred orientation of G_w relative to H is that which breaks the gauge symmetry least; more specifically, it is that orientation which minimizes the trace of the vector boson mass matrix.

We should note parenthetically a simple criterion which guarantees that a given orientation is at least a stationary point of ΔE. Write $U = \exp[i\alpha^a X_a]$; then

$$\frac{\partial}{\partial \alpha_a} \operatorname{Tr}((U^\dagger\Lambda U)_X)^2|_{U=1} = i\operatorname{Tr}([X^a, \Lambda])_X \Lambda_X. \tag{2.24}$$

Using the parity invariance (2.9) or the restrictions on commutators (2.10), this can be written in the form

$$\frac{\partial}{\partial \alpha_a} \operatorname{Tr}((U^\dagger\Lambda U)_X)^2|_{U=1} = i\operatorname{Tr}(X^a[\Lambda_T, \Lambda_X]). \tag{2.25}$$

This if, in a given representation of the generators, $[\Lambda_T, \Lambda_X] = 0$ for all gauge generators Λ, the corresponding vacuum orientation is at least a stationary point of ΔE. This will in fact be the case for all the models we consider in this paper, written in their most natural representations. The stability of this point is, of course, not guaranteed.

Once we have identified the state $|\Omega\rangle$ which minimizes ΔE, we might choose a new representation of the T_i and X_a such that this state becomes the vacuum $|0\rangle$. We will use this new basis in the remainder of the section to study the curvature of the functional ΔE about the preferred vacuum.

It is obvious that every direction in G/H away from the true vacuum along which ΔE has a non-zero curvature corresponds to a massive particle. To see the precise relation, recall the result of Dashen [17] for the mass matrix given to Goldstone bosons by a symmetry-breaking perturbation. If δH is the hamiltonian density associated with the perturbation and Q^a is the charge corresponding to the generator X^a, the mass matrix m^2_{ab} given to the pion fields is [using (2.15)]

$$m^2_{ab} = \frac{1}{f_\pi^2}\langle 0|[Q^a,[Q^b,\delta H]]|0\rangle . \tag{2.26}$$

In our case, δH is the hamiltonian used in (2.12). In the notation of (2.24), (2.26) may be written

$$m^2_{ab} = \frac{1}{f_\pi^2}\frac{\partial^2}{\partial\alpha_a\partial\alpha_b}\Delta E[U]|_{U=1} . \tag{2.27}$$

It is most convenient to differentiate ΔE in the form (2.18).

$$\frac{\partial^2}{\partial\alpha_a\partial\alpha_b}\{\tfrac{1}{2}\operatorname{Tr}(U^\dagger\Lambda U)_X(U^\dagger\Lambda U)_X\}|_{U=1}$$

$$= -\operatorname{Tr}\{([X^a,[X^b,\Lambda]])_X\Lambda_X\} - \operatorname{Tr}\{([X^a,\Lambda])_X([X^b,\Lambda])_X\} . \tag{2.28}$$

Using again (2.9) or (2.10), this becomes

$$(2.28) = -\operatorname{Tr}\{[X^a,[X^b,\Lambda_X]]\Lambda_X\} - \operatorname{Tr}\{[X^a,\Lambda_T][X^b,\Lambda_T]\} . \tag{2.29}$$

After some rearrangement of the traces, (2.27) takes the form

$$m^2_{ab} = \frac{1}{4\pi}M^2\Big(\sum_\Lambda \operatorname{Tr}\{[\Lambda_T,[\Lambda_T,X^a]]X^b - [\Lambda_X,[\Lambda_X,X^a]]X^b\}\Big), \tag{2.30}$$

where

$$M^2 = \frac{4\pi}{f_\pi^2} \int \mathrm{d}^4x \Delta^{\mu\nu}_{(x)} \langle 0 | \mathrm{T}\big(J^T_\mu(x) J^T_\nu(0) - J^X_\mu(x) J^X_\nu(0) \big) | 0 \rangle \,. \tag{2.31}$$

It is not hard to see in this formula the physics we discussed below (2.23): the matrices

$$D^{ab} = \mathrm{Tr}\{ [\Lambda, [\Lambda, X^a]] X^b \} \tag{2.32}$$

are positive, as one can see by writing their diagonal elements as squares of structure constants f^{abc} of the group G. Therefore, unbroken gauged generators Λ_T give a contribution to the mass matrix m^2_{ab} which is strictly positive; the broken gauged generators Λ_X give a contribution which is strictly negative and tends to destabilize the vacuum $|0\rangle$.

The expression (2.30) is not only easy to understand but also easy to compute explicitly. Since the X^a belong to the adjoint representation of G, $[\Lambda, X^a]$ is just the action on X^a of the generator Λ of G. This implies, in particular, that if t^α are the generators of some simple subgroup γ of G and X^a transforms as a member of the irreducible representation r_a of γ,

$$\sum_\alpha \mathrm{Tr}\{ [t^\alpha, [t^\alpha, X^a]] X^b \} = \delta^{ab} T^2_\gamma(\mathrm{r}_a) \,, \tag{2.33}$$

where $T^2_\gamma(\mathrm{r}_a)$ denotes the quadratic Casimir operator of γ in the representation r_a. The unbroken generators Λ_T of $\mathrm{G_w}$ will always generate a subgroup of G; to evaluate their contribution to (2.30) one need only express this group as a product of simple groups and use (2.33).

We have not found a simple rule for the general computation of the contribution of the Λ_X to (2.30), so we will restrict this problem to one closer to our immediate interest. We return, then, to the specific form of the X_a given by our three scenarios [eq. (2.6)] and add one further assumption, that the weakly gauged group $\mathrm{G_w}$ is included in the $\mathrm{SU}(N) \times \mathrm{SU}(N) \times \mathrm{U}(1)$ subgroup of G. Then the Λ_X will be of the form of coupling constants times matrices

$$\hat{t}^\alpha = \left(\begin{array}{c|c} \tau^\alpha & 0 \\ \hline 0 & \tau^{\mathrm{T}\alpha} \end{array} \right), \tag{2.34}$$

where τ^α is a generator of $\mathrm{SU}(N)$. In general, the τ^α derived from the Λ_X will generate a subgroup of this (broken) $\mathrm{SU}(N)$, again, a product of simple groups.

Let us separate the X_a into two classes: in the notation of (2.6),

$$\text{diagonal } X_a \text{ have } C \neq 0, D = 0\,;$$

$$\text{off-diagonal } X_a \text{ have } C = 0, D \neq 0\,. \tag{2.35}$$

If γ is some subgroup of the *unbroken* SU(N), we can classify each X_a as belonging to some irreducible representation r_a of this γ; for off-diagonal X_a, this representation is contained in the product of a representation $\mathrm{r}_{a\mathrm{L}}$, expressing the action on D generated by $D \to \tau^\alpha D$, and the corresponding $\mathrm{r}_{a\mathrm{R}}$ expressing the action generated by $D \to D(\tau^a)^{\mathrm{T}}$.

Now consider the corresponding subgroup γ of the SU(N) group defined by (2.34). The contribution to (2.30) of the Λ_X associated with this subgroup is easily seen to be given in terms of the Casimir operators of γ in the representations $\mathrm{r}_a, \mathrm{r}_{a\mathrm{L}}, \mathrm{r}_{a\mathrm{R}}$ defined in the preceding paragraph:

$$\mathrm{Tr}\left[\hat{t}^\alpha,\left[\hat{t}^\alpha, X^a\right]\right]X^b = \begin{cases} \delta^{ab}T_\gamma^2(\mathrm{r}_a), & (\text{diagonal } X_a), \\ \delta^{ab}\left(2T_\gamma^2(\mathrm{r}_{a\mathrm{L}}) + 2T_\gamma^2(\mathrm{r}_{a\mathrm{R}}) - T_\gamma^2(\mathrm{r}_a)\right), & (\text{off-diagonal } X_a). \end{cases} \tag{2.36}$$

We have now succeeded in making explicit, for the case of our three scenarios, the prescription for computing the masses of pseudo-Goldstone bosons which is contained implicitly in (2.30). We may summarize this prescription as follows: separate the generators Λ of $\mathrm{G_w}$ into unbroken and broken components. Each set of components generates a subgroup of its own SU(N). Write these subgroups as products of simple groups γ and γ'. Then the pseudo-Goldstone boson corresponding to the generator X_a is given a mass:

$$m_a^2 = M^2\left(\sum_\gamma \frac{g_\gamma^2}{4\pi}Q_\gamma - \sum_{\gamma'} \frac{g_{\gamma'}^2}{4\pi}Q_{\gamma'}\right), \tag{2.37}$$

where M^2 is given by (2.31), g_γ and $g_{\gamma'}$ are coupling constants, and

$$Q_\gamma = T_\gamma^2(\mathrm{r}_a),$$
$$Q_{\gamma'} = \begin{cases} T_\gamma^2(\mathrm{r}_a), & (\text{diagonal } X_a), \\ 2T_\gamma^2(\mathrm{r}_{a\mathrm{L}}) + 2T_\gamma^2(\mathrm{r}_{a\mathrm{R}}) - T_\gamma^2(\mathrm{r}_a), & (\text{off-diagonal } X_a). \end{cases} \tag{2.38}$$

$\mathrm{r}_a, \mathrm{r}_{a\mathrm{L}}, \mathrm{r}_{a\mathrm{R}}$ are representations of γ or γ' defined below (2.35). It is worth rewriting this prescription for the special case in which γ is a U(1) subgroup of SU(N) generated by a matrix τ such that, if C is a shorthand for C or D in (2.6),

$$\tau C = q_{\mathrm{L}} C, \qquad C\tau = q_{\mathrm{R}} C, \tag{2.39}$$

where $q_{\mathrm{L}}, q_{\mathrm{R}}$ are c-numbers. Then

$$Q_\gamma = \begin{cases} (q_{\mathrm{L}} - q_{\mathrm{R}})^2, & (\text{diagonal } X_a), \\ (q_{\mathrm{L}} + q_{\mathrm{R}})^2, & (\text{off-diagonal } X_a), \end{cases}$$
$$Q_{\gamma'} = (q_{\mathrm{L}} - q_{\mathrm{R}})^2. \tag{2.40}$$

It is possible that the requirements for constructing irreducible representations of γ and γ' might conflict; in that case, m^2 will be a matrix whose matrix elements are given by (2.37). However, this case will not arise in the applications we will consider here.

We conclude this section by discussing one question which had been deferred in our analysis, that of the sign of M^2 given by (2.31). Let us first note an application of (2.37): return to the usual strong interactions by setting $G = SU(2) \times SU(2)$ and letting γ be the U(1) subgroup of G to which the photon couples. Then for the ordinary $\pi^{\pm}$ mesons, $|q_L - q_R| = 1$, so that the 1-photon exchange contribution to the $\pi^{\pm}$ mass is just

$$m_{\pi^+}^2 - m_{\pi^0}^2 = \alpha M^2 . \tag{2.41}$$

Rewriting the currents in (2.31) as vector and axial vector currents,

$$M^2 = \frac{4\pi}{f_\pi^2} \int d^4x \Delta^{\mu\nu}(x) \langle 0|T\big(V_\mu^3(x)V_\nu^3(0) - A_\mu^3(x)A_\nu^3(0)\big)|0\rangle . \tag{2.42}$$

(2.41) is a very familiar formula, first derived more than a decade ago by Das, Guralnik, Mathur, Low, and Young [18]. Using (2.41) we may evaluate the ratio of M to the ρ-meson mass

$$M^2/m_\rho^2 = 0.29 . \tag{2.43}$$

A first observation is that, if we replace the strong-interaction color SU(3) by SU(N) and take the $N \to \infty$ limit, the ratio computed in (2.43) is independent both of N and of the number of fermion flavors in this limit; further being a dimensionless ratio, it is independent of the quantum scale Λ at which the color coupling constant becomes strong. The phenomenological success of selection rules of the $N \to \infty$ limit lead one to believe that (2.43) provides a good estimate for M^2 in terms of the mass of the ρ meson of technicolored fermions if the technicolor group G_s is any SU(N) group, $N \geqslant 3$. This observation, in particular, fixes the sign of M^2. For our future reference, we note that the ratio

$$\frac{\sqrt{N}\, M}{f_\pi} = 8 \tag{2.44}$$

has this same universality in the $N \to \infty$ limit.

A second line of reasoning allows one to conclude that M^2 is positive in a slightly more general context: it is straightforward to show, for any asymptotically free G_s, that the spectral functions of the currents J_μ^T and J_μ^X satisfy the Weinberg sum rules [19]*

$$\int_0^\infty d\mu^2(\rho_T(\mu^2)-\rho_X(\mu^2))\mu^{-2}=f_\pi^2,$$

$$\int_0^\infty d\mu^2(\rho_T(\mu^2)-\rho_X(\mu^2))=0. \tag{2.45}$$

If we assume that $\rho_T(\mu^2)$ and $\rho_X(\mu^2)$ are each dominated by the contribution of the lowest-lying vector meson of the correct quantum numbers ($\rho(\mu^2)\approx g^2\delta(\mu^2-m^2)$) and use these forms, constrained by (2.45), to evaluate (2.31), we find $M^2>0$. This pole dominance assumption is not an unreasonable one as long as the integral in (2.31) converges rapidly in the ultraviolet; that convergence is itself guaranteed by (2.45). The authors of ref. [18] showed that this assumption is, in fact, quantitatively correct for the evaluation of (2.42).

3. A simple example

In sect. 2, we developed at some length a formalism for discussing vacuum alignment and pseudo-Goldstone boson masses. In this section, we will present a first application of this formalism. This example will be built from a straightforward extension of the model of dynamical breaking of SU(2) × U(1) constructed by Weinberg and Susskind [2, 3].

Let us first review the features of that model. Weinberg and Susskind imagine two multiplets U, D of (1-component) fermions, each transforming under a complex representation of the technicolor group G_s. This theory has a chiral symmetry G = SU(2) × SU(2) × U(1) which we expect to be broken of H = SU(2) × U(1). Now couple this model to the Weinberg-Salam gauge theory in the manner standard for a single quark or lepton doublet. Setting $\psi=(U,D)$ and denoting the gauge bosons of the Weinberg-Salam model by A^a_μ, B_μ, we may write this coupling

$$\delta\mathcal{L}=gA^a_\mu\bar\psi_L\gamma^\mu\tau^a\psi_L+g'B_\mu\left(\bar\psi_R\gamma^\mu\tau^3\psi_R+\tfrac{1}{2}q\bar\psi\gamma^\mu\psi\right), \tag{3.1}$$

where τ^a is an SU(2) generator** and q is the mean electric charge of the doublet. In the notation of sect. 2, this interaction is of the form (2.11) with

$$\Lambda\in\left\{g\begin{pmatrix}\tau^a & \\ & 0\end{pmatrix},\quad g'\begin{bmatrix}\tfrac{1}{2}q & \\ & -\tfrac{1}{2}q-\tau^3\end{bmatrix}\right\}. \tag{3.2}$$

* Modern methods for deriving the Weinberg sum rules are given, for example, in [20].

** Throughout this paper, SU(2) generators τ^a and SU(3) generators λ^a will be normalized to $\mathrm{Tr}\,\tau^a\tau^b=\mathrm{Tr}\,\lambda^a\lambda^b=\frac{1}{2}\delta^{ab}$.

Projecting onto the broken generators of G

$$\Lambda_X \in \left\{ \tfrac{1}{2} g \begin{pmatrix} \tau^a & \\ & \tau^{\mathrm{T}a} \end{pmatrix}, \quad -\tfrac{1}{2} g' \begin{pmatrix} \tau^3 & \\ & \tau^3 \end{pmatrix} \right\}. \tag{3.3}$$

Inserting these matrices into (2.21) yields the familiar Weinberg-Salam vector boson mass matrix and gives, in particular,

$$m_{\mathrm{W}} = \tfrac{1}{2} g f_\pi, \qquad m_{\mathrm{Z}} = \frac{m_{\mathrm{W}}}{\cos\theta_{\mathrm{W}}}, \qquad m_\gamma = 0. \tag{3.4}$$

In this theory, there is no problem of vacuum alignment, since all vacua are equivalent under the gauge symmetry of the Weinberg-Salam model. The three Goldstone bosons of broken chiral SU(2) in the original strong interaction theory remain massless in the presence of the Weinberg-Salam perturbation, so that they might combine with the $W^\pm$ and Z through the Higgs mechanism.

Let us generalize this model by adding a second doublet of technifermions (C, S). The strong-interaction theory is now one of broken chiral SU(4) and produces 15 Goldstone bosons. Of these, three correspond to broken gauge symmetries of the Weinberg-Salam model; these remain massless and act in the Higgs mechanism as before. The remaining 12 bosons appear as physical pseudoscalar particles — technions — and may receive masses from the perturbations which determine subgroup alignment. The resulting mass spectrum may be calculated from (2.37); let us now set up and carry out that calculation.

We need first to write the coupling of U, D, C, S to the Weinberg-Salam bosons. Our choice for this coupling will be the standard one, with one qualification: we need to distinguish the two doublets, in order to avoid SU(2) symmetries linking the doublets which will force some Goldstone bosons to remain massless. Normally, this is accomplished by giving the doublets different bare masses or Yukawa couplings to fundamental scalars, but our model has neither fermion bare masses nor fundamental scalars. We choose the simplest alternative of giving the doublets electric charges differing by an amount Δ. Though the parameter Δ contributes to technion masses, the mean charge q does not; we will therefore set $q = 0$ in the following discussion. To use the analysis of sect. 2, it is convenient to write the fermion fields as a left-handed multiplet:

$$\psi = (U_{\mathrm{L}}, D_{\mathrm{L}}, C_{\mathrm{L}}, S_{\mathrm{L}}, (U_{\mathrm{R}})^{\mathrm{c}}, (D_{\mathrm{R}})^{\mathrm{c}}, (C_{\mathrm{R}})^{\mathrm{c}}, (S_{\mathrm{R}})^{\mathrm{c}}),$$

where c denotes charge conjugation. Then the generators of G = SU(4) × SU(4) × U(1) are 8 × 8 matrices, which, written as in (2.6), divide into 4 × 4 blocks. Each block acts on a vector (U, D, C, S) and is conveniently written as a direct product of a 2 × 2 weak isospin matrix and a 2 × 2 matrix on the space of the two doublets.

Then the Weinberg-Salam coupling to ψ is

$$\delta\mathcal{L} = gA_\mu^a\bar{\psi}\gamma^\mu\left(\begin{array}{c|c}\tau^a\otimes 1 & \\ \hline & 0\end{array}\right)\psi$$

$$+g'B_\mu\bar{\psi}\gamma^\mu\left[\left(\begin{array}{c|c}0 & \\ \hline & -\tau^3\otimes 1\end{array}\right)+\Delta\left(\begin{array}{c|c}1\otimes B & \\ \hline & -1\otimes B\end{array}\right)\right]\psi. \tag{3.5}$$

The matrix $B=\tau^3$ on the space of the doublets, it appears in a pure vector current which splits the charges of (U,D) from (C,S) by an amount Δ.

Let us now assume that the basis we have chosen for the matrices of (3.5) represents the correct vacuum, that is, that U_L pairs with U_R, D_L with D_R, etc. In this vacuum, one can repeat the computation of (3.3), (3.4) to find again the Weinberg-Salam mass relations with $m_w = \sqrt{\frac{1}{2}}\, gf_\pi$. In general, for a theory with L doublets, this choice of vacuum will produce

$$m_W = \sqrt{L}\cdot\left(\tfrac{1}{2}gf_\pi\right) = m_Z\cos\theta_W \tag{3.6}$$

To find the technion spectrum, divide the Λ of (3.5) into unbroken and broken pieces: $\Lambda = \Lambda_T + \Lambda_X$. For the SU(2) currents

$$\Lambda^a = \tfrac{1}{2}g\left(\begin{array}{c|c}\tau^a\otimes 1 & \\ \hline & -\tau^{Ta}\otimes 1\end{array}\right)+\tfrac{1}{2}g\left(\begin{array}{c|c}\tau^a\otimes 1 & \\ \hline & \tau^{Ta}\otimes 1\end{array}\right). \tag{3.7}$$

For the U(1) current

$$\Lambda = \tfrac{1}{2}g'\left(\begin{array}{c|c}A & \\ \hline & -A^T\end{array}\right)-\tfrac{1}{2}g'\left(\begin{array}{c|c}\tau^3\otimes 1 & \\ \hline & \tau^3\otimes 1\end{array}\right), \tag{3.8}$$

where

$$A = \begin{bmatrix}\frac{1}{2}+\Delta & & & 0\\ & -\frac{1}{2}+\Delta & & \\ & & \frac{1}{2}-\Delta & \\ 0 & & & -\frac{1}{2}-\Delta\end{bmatrix}. \tag{3.9}$$

Eqs. (3.7) and (3.8) put the Λ into the form required for the direct application of (2.37).

The 15 Goldstone bosons of broken SU(4) × SU(4) correspond to the generators

$$X_a = \left(\begin{array}{c|c}C & \\ \hline & C^T\end{array}\right),\qquad C^\dagger = C,\qquad \mathrm{Tr}\,C = 0. \tag{3.10}$$

It is convenient to divide the possible matrices C into two classes: class 1 contains the 7 matrices

$$C = \tau^a \otimes 1,\ \tau^a \otimes \tau^3,\ \text{or}\ 1 \otimes \tau^3\,; \tag{3.11}$$

class 2 contains the 8 matrices of the form

$$C = \begin{pmatrix} & 1 \\ 1 & \end{pmatrix} \text{ or } \begin{pmatrix} & -i \\ i & \end{pmatrix}, \tag{3.12}$$

linking one of (U, D) with one of (C, S). We will denote the technions of this class by the flavor quantum numbers of the corresponding current (for example, $\overline{U}C$).

Now we are ready to evaluate (2.37). The evaluation contains three surprises. First, the contributions of the SU(2) generators (3.7) to the two terms of (2.37) cancel precisely. The exchange of the vector bosons A^a_μ does not give mass to any technion in this model. Secondly, for all generators of class 1, the two contributions of the U(1) generator (3.8) also cancel. The technions of this class receive no mass at all from the exchange of one vector boson. Using (2.40), we find the contribution of B_μ exchange to the masses of the technions of the second class to be

$$m^2_{\overline{\rm U}{\rm C}} = m^2_{\overline{\rm D}{\rm S}} = \frac{g'^2}{4\pi}\Delta^2 M^2\,,$$

$$m^2_{\overline{\rm U}{\rm S}} = \frac{g'^2}{4\pi}\Delta(\Delta+1)M^2\,,$$

$$m^2_{\overline{\rm D}{\rm C}} = \frac{g'^2}{4\pi}\Delta(\Delta-1)M^2\,. \tag{3.13}$$

Eqs. (3.13) contain the third surprise: for $|\Delta| < 1$ the vacuum preferred by Weinberg and Susskind is unstable, since the curvature of the energy surface at that point is negative.

The first two of these phenomena are reminiscent of Dashen's well-known result [17] that, in the ordinary strong interactions, the π^0 and K^0 receive no contribution to their masses from one-photon exchange. In fact, they have an equally simple explanation derived from the symmetries of the technicolor theory. This explanation, and a consequence of it, will be the subject of sect. 5. The third phenomenon is of a different, more mechanistic character; let us now complete this section by investigating its origins.

The instability of the Weinberg-Susskind vacuum for $|\Delta| < 1$ results entirely from exchange of the U(1) gauge boson B_μ. Therefore, we need to inquire of the $B_{\bar\mu}$ exchange perturbation in (3.5) what vacuum alignment it should prefer. Let us rewrite the matrix in the B_μ current in (3.5) as

$$\Lambda = \tfrac{1}{2}g' \left[\begin{array}{c|c} \begin{bmatrix} \Delta & & & \\ & \Delta & & \\ & & -\Delta & \\ & & & -\Delta \end{bmatrix} & \\ \hline & (-1)\begin{bmatrix} \Delta+1 & & & \\ & \Delta-1 & & \\ & & -\Delta+1 & \\ & & & -\Delta-1 \end{bmatrix} \end{array} \right] . \tag{3.14}$$

As was demonstrated in sect. 2, the preferred vacuum is that which minimizes the breaking of the symmetry generated by Λ. More specifically, the preferred vacuum is denoted by the unitary transformation U which minimzes $\mathrm{Tr}((U^+ \Lambda U)_X)^2$ or maximizes $\mathrm{Tr}((U^+ \Lambda U)_T)^2$. This is done by transforming Λ so that the lower right-hand block is as close as possible to being equal and opposite to the upper left-hand block. For Δ large, the preferred form is precisely that shown in (3.14); this corresponds to the Weinberg-Susskind vacuum. But for $0 < \Delta < 1$, a lower value of the vacuum energy is obtained by transforming Λ so that its lower right-hand block reads:

$$(-1)\begin{bmatrix} \Delta+1 & & & \\ & -\Delta+1 & & \\ & & \Delta-1 & \\ & & & -\Delta-1 \end{bmatrix} \tag{3.15}$$

This is equivalent to the pairing of D_L with C_R and C_L with D_R. In this vacuum, the electrically charged composite field $\overline{C}D$ acquires a vacuum expectation value; hence, the photon must acquire a mass. In fact, one can readily compute the vector boson mass matrix in this new basis, to find (we give only the 2×2 block mixing A^3_μ and B_μ)

$$\mu^2 \begin{pmatrix} A^3 \\ B \end{pmatrix} = \left(\tfrac{1}{2}f_\pi\right)^2 \begin{pmatrix} 2g^2 & 2\Delta g g' \\ 2\Delta g g' & \left[1 + (1-2\Delta)^2\right] g'^2 \end{pmatrix} \begin{pmatrix} A^3 \\ B \end{pmatrix} . \tag{3.16}$$

The determinant of this matrix vanishes when $\Delta = 1$ only; thus, its two eigenvalues m_γ, m_Z must both be non-zero if $0 \leqslant \Delta < 1$. This theory is therefore phenomenologically disastrous.

We spoke above of three surprises appearing in the technion mass formulae. But the real surprise is the result these phenomena imply: a model of chiral symmetry breaking coupled to the Weinberg-Salam theory is, for dynamical reasons, a sensitive object. Simple modifications to the theory may easily provoke catastrophes by disturbing the delicate energetics of the vacuum alignment.

4. Models with colored technifermions

In sect. 3, we computed the spectrum of technions in the simplest generalization of the Weinberg-Susskind model. In this section, we consider some more complex, but more interesting, generalizations proposed by Dimopoulos [19]. These models are motivated by the idea of Eichten and Lane [4] and Dimopoulos and Susskind [5] that one can produce the bare masses of ordinary quarks and leptons from the dynamical masses of technifermions by introducing a new set of gauge interactions which allow transitions linking the fermions of these two types. If these new interactions conserve color SU(3), quarks can receive mass only from color triplet technifermions and leptons only from color singlet technifermions. One is thus led to consider a technicolor theory with four doublets of technifermions, three of which form a color triplet and the fourth of which is a color singlet. Models of this type contain huge chiral groups, and the spontaneous breaking of these groups gives rise to a rich spectrum of technions. The qualitative features of these particles, and some aspects of their phenomenology, have already been described by Dimopoulos in ref. [9]. We will concern ourselves here with the calculation of the masses of these particles, a calculation easily accomplished using the rules derived at the end of sect. 2.

We might begin our discussion by defining more precisely the models which we will study. Following Dimopoulos, we assume that the color triplet and color singlet technifermions have the same coupling to color SU(3) × weak SU(2) × U(1) as ordinary quarks and leptons. Let us refer to the color triplet weak doublet of technifermions as (U, D) and the color singlet weak doublet as (N, E). For convenience in applying the results of sect. 2, we group these 8 four-component fermions into a multiplet ψ of 16 two-component fields

$$\psi = (U_L, D_L, N_L, E_L, (U_R)^c, (D_R)^c, (N_R)^c, (E_R)^c) \tag{4.1}$$

Again, c denotes charge conjugation. Matrices on this space of 16 fields fall naturally into 8×8 blocks; within each block, there is a further division into a set of 6 quark and 2 lepton indices. As in (3.5), it is convenient to represent these block matrices as direct products. We will write τ^a to denote SU(2) generators, λ^α to denote SU(3) generators, and $\underline{1}_n$ to denote the $n \times n$ unit matrix. In this notation, the coupling of our technifermions to color gluons A_μ^α and the vector bosons A_μ^a, B_μ

of the Weinberg-Salam model is

$$\delta\mathfrak{L} = g_s A^\alpha_\mu \bar\psi \gamma^\mu \left[\begin{array}{c|c} \begin{array}{c|c} \underline{1}_2 \otimes \lambda^\alpha & \\ \hline & 0 \end{array} & \\ \hline & \begin{array}{c|c} -\underline{1}_2 \otimes \lambda^{T\alpha} & \\ \hline & 0 \end{array} \end{array}\right] \psi$$

$$+ g A^a_\mu \bar\psi \gamma^\mu \left(\begin{array}{c|c} \tau^a \otimes \underline{1}_4 & \\ \hline & 0 \end{array}\right) \psi$$

$$+ g' B_\mu \bar\psi \gamma^\mu \left[\left(\begin{array}{c|c} 0 & \\ \hline & -\tau^3 \otimes \underline{1}_4 \end{array}\right) + \tfrac{1}{2}\left(\begin{array}{c|c} \underline{1}_2 \otimes B & \\ \hline & -\underline{1}_2 \otimes B \end{array}\right)\right] \psi, \qquad (4.2)$$

where B is the 4×4 matrix

$$B = \left(\begin{array}{c|c} \frac{1}{3}\underline{1}_3 & \\ \hline & -1 \end{array}\right). \qquad (4.3)$$

We will not need to specify the technicolor group G_s or the specific representation of G_s to which the fields U, D, N, E are assigned. We will insist only that these fields all belong to the same representation r_s of G_s, so that the technifermions may all acquire dynamical masses by one of the scenarios discussed at the beginning of sect. 2. If r_s is a complex representation of G_s, the chiral group G will be $SU(8) \times SU(8) \times U(1)$, broken to $SU(8) \times U(1)$; if r_s is real or pseudo-real, G will be the larger group SU(16), broken to O(16) or Sp(16). In any of these cases, the Goldstone bosons are just those created from the vacuum by currents involving the matrices X_a catalogued in eqs. (2.6).

The first step in studying any one of these cases must be to find the preferred vacuum states, about which the technions are seen as small perturbations. However, it will be clear from our subsequent analysis (and is easily checked explicitly) that the Weinberg-Susskind vacuum is the preferred one in the case of the SU and O scenarios. The Sp case will be discussed in sect. 6. As we have already noted, the choice of the Weinberg-Susskind vacuum insures that the pattern of vector boson masses is the correct one to reproduce the standard phenomenology of charged and neutral weak currents. We have anticipated this choice of vacuum by writing the matrices of (4.2) in the basis appropriate to it.

Given the choices of vacuum state and the catalogue of broken generators, all that remains to compute the technion masses is to apply directly the formula (2.37). Before we begin this exercise, however, let us estimate the magnitude of the parameters which appear in (2.37). The mass scale of the technicolor theory may be determined through its value of f_π; this, in turn, may be found from the relation (3.6), which reads, in this model,

$$m_W = g f_\pi . \tag{4.4}$$

The value of the Fermi constant, or, equivalently, the relation $m_W = 37\ \text{GeV}/\sin\theta_W$, yields

$$f_\pi = 120\ \text{GeV}. \tag{4.5}$$

If G_s is an SU(N) group and the technifermions belong to the fundamental representation, we may use the relations (2.43), (2.44) to find the mass parameter M and the technicolor ρ-meson mass. For technicolor SU(4),

$$M = 480\ \text{GeV}, \qquad m_\rho = 900\ \text{GeV}. \tag{4.6}$$

If G_s is a larger SU(N) group, these values decrease in proportion to $1/\sqrt{N}$. For other groups, nearby SU(N) groups provide a rough estimate of M and m_ρ.

The coupling constants appearing in (4.2) are readily determined from the known values of the strong and electromagnetic coupling constants and the Weinberg angle. The appropriate value of g_s is that measured at $Q^2 \sim m_\rho^2$, which we might estimate by choosing $\Lambda = 500$ MeV and using (4.6) for m_ρ. The constants g and g' may be determined from $\sin^2\theta_W = 0.23$ and $\alpha = (137)^{-1}$. With this input, we find

$$\alpha_s = \frac{g_s^2}{4\pi} = [11]^{-1},$$

$$\alpha_w = \frac{g^2}{4\pi} = [32]^{-1}$$

$$\alpha'_w = \frac{g'^2}{4\pi} = [105]^{-1}. \tag{4.7}$$

The relation $\alpha_s \gg \alpha_w \gg \alpha'_w$ will be useful to us later. Combining (4.6) and (4.7) gives the order of magnitude of technion masses: $m \sim 100$ GeV.

The precise coupling constants $g_\gamma, g_{\gamma'}$ needed for (2.37) are simply related to the couplings of (4.2); the relations may be read directly from the form of the matrices in that lagrangian. We see that the g_γ, associated with the unbroken generators, and the $g_{\gamma'}$, associated with the broken generators, are given by

$$\text{SU(3)}: g_\gamma = g_s, \qquad g_{\gamma'} = 0,$$

$$\text{SU(2)}: g_\gamma = \tfrac{1}{2}g, \qquad g_{\gamma'} = \tfrac{1}{2}g,$$

$$\text{U(1)}: g_\gamma = \tfrac{1}{2}g', \qquad g_{\gamma'} = -\tfrac{1}{2}g'. \tag{4.8}$$

To apply (2.40) for the U(1) generators, it will be useful to tabulate the U(1) charges:

$$q_{(Y)} = \left[\begin{array}{c|c} \begin{bmatrix} \frac{5}{6} & \\ & \frac{1}{6} \end{bmatrix} \otimes \underline{1}_3 & \\ \hline & \begin{array}{cc} -\frac{1}{2} & \\ & -\frac{3}{2} \end{array} \end{array}\right], \qquad q_{(Y')} = \left[\begin{array}{c|c} \begin{bmatrix} \frac{1}{2} & \\ & -\frac{1}{2} \end{bmatrix} \otimes \underline{1}_3 & \\ \hline & \begin{array}{cc} \frac{1}{2} & \\ & -\frac{1}{2} \end{array} \end{array}\right]. \tag{4.9}$$

Now all that remains is to compute, for each X_a, the coefficient $Q_Y, Q_{Y'}$ appearing in (2.37). Let us begin with the SU case. Here there are 63 Goldstone bosons, forming a basis of the adjoint representation of SU(8). From (2.6) and (2.7); we see that these correspond to the generators

$$X^a = \left(\begin{array}{c|c} C & \\ \hline & C^{\mathrm{T}} \end{array}\right), \tag{4.10}$$

where C is an 8×8 hermitian matrix. Such matrices form three classes:

class 1 (color singlets):

$$C \in \left\{ \left(\begin{array}{c|c} \tau^a \otimes \underline{1}_3 & \\ \hline & \tau^a \end{array}\right), \left(\begin{array}{c|c} \frac{1}{3}\tau^a \otimes \underline{1}_3 & \\ \hline & -\tau^a \end{array}\right), \left[\begin{array}{c|c} \frac{1}{3}\underline{1}_6 & \\ \hline & -\underline{1}_2 \end{array}\right] \right\};$$

class 2 (color octets):

$$C \in \left\{ \left(\begin{array}{c|c} \underline{1}_2 \otimes \lambda^\alpha & \\ \hline & 0 \end{array}\right), \left(\begin{array}{c|c} \tau^a \otimes \lambda^\alpha & \\ \hline & 0 \end{array}\right) \right\}, \tag{4.11}$$

class 3 (color triplets):

$$C \in \left\{ \left(\begin{array}{c|c} & \Sigma \otimes \xi \\ \hline \Sigma^\dagger \otimes \xi^\dagger & \end{array}\right) \right\},$$

where, in the last line, ξ is a 3-vector of color and Σ is a 2×2 matrix on the isospin space. It is easiest to compute the Q_Y for generators of class 3 by choosing the basis for Σ:

$$\Sigma \in \{1, \tau^a, i, i\tau^a\}. \tag{4.12}$$

However, one can alternatively choose a basis of 2×2 matrices each of which has only one non-zero element. This makes clear that the bosons of class 3 have the quantum numbers of one technilepton and one antitechniquark, or vice versa. We will label them below by these quantum numbers. Classes 1, 2, and 3 contain 7, 32, and 24 generators, respectively.

The computation of technion masses from the forms (4.1) and the rules (2.37), (2.38), (2.40) is a straightforward exercise; we will simply quote the results. For technions of classes 1 and 2, the weak and electromagnetic interactions make no contribution to the mass, the same phenomenon which we saw for the bosons of class 1 (eq. (3.11)) in section 3. Thus, for the color singlet and octet technions,

$$m^2_{\text{class 1}} = 0\,,$$

$$m^2_{\text{class 2}} = 3\alpha_s M^2\,, \tag{4.13}$$

to leading order in the α's. For technions of class 3,

$$m^2_{\bar{\text{U}}\text{E}} = \left(\tfrac{4}{3}\alpha_s + \tfrac{10}{9}\alpha'_w\right) M^2\,,$$

$$m^2_{\bar{\text{U}}\text{N}} = m^2_{\bar{\text{D}}\text{E}} = \left(\tfrac{4}{3}\alpha_s + \tfrac{4}{9}\alpha'_w\right) M^2\,,$$

$$m^2_{\bar{\text{D}}\text{N}} = \left(\tfrac{4}{3}\alpha_s - \tfrac{2}{9}\alpha'_w\right) M^2\,. \tag{4.14}$$

The cancellation of the A^a_μ exchange contributions and the minus sign in the last line occur just as in (3.13). Using the values of parameters given in (4.6), (4.7), one can locate the various masses numerically for the case G = SU(4). The resulting spectrum is shown in fig. 3.

The Dimopoulos model utilizing the O scenario contains all of these bosons, with precisely the same masses as we have given in (4.13), (4.14) for the SU case. However, it contains also 72 additional technions which fill out the symmetric tensor multiplet of O(16). From (2.6), we may see that these additional particles

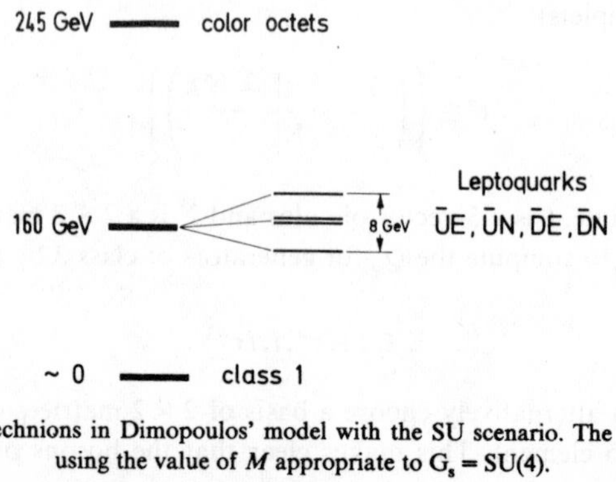

Fig. 3. Spectrum of technions in Dimopoulos' model with the SU scenario. The masses are evaluated using the value of M appropriate to $G_s = SU(4)$.

correspond to generators of the form

$$X_a = \left(\begin{array}{c|c} 0 & D \\ \hline D^\dagger & 0 \end{array}\right), \tag{4.15}$$

where D is a complex 8×8 matrix. These additional bosons have diquark, dilepton, or quark-lepton quantum numbers, and we will use these quantum numbers to refer to them below.

To catalogue these additional bosons, it is useful to introduce a notation for the basis vectors of the required color and isospin representations. Let $\underline{s}$ represent a 3×3 symmetric tensor, $\underline{a}$ a 3×3 antisymmetric tensor, and ξ, as above, a 3-component vector. Let $\underline{\sigma}$ represent a 2×2 symmetric tensor and $\underline{\varepsilon}$ the 2×2 antisymmetric tensor. From these objects, one can easily construct the 36 real 8×8 matrices from which one builds the real and imaginary parts of D in (4.15), or, equivalently, the 36 particle-antiparticle pairs that fill out the multiplet of technions. These 36 matrices divide into three classes, which we add to the classes of (4.11):

class 4 (color sextets):

$$D \in \left\{\left(\begin{array}{c|c} \underline{\sigma}\otimes\underline{s} & \\ \hline & 0 \end{array}\right)\right\}; \tag{4.16}$$

class 5 (color triplets):

$$D \in \left\{\left(\begin{array}{c|c} & \underline{\sigma}\otimes\xi \\ \hline \underline{\sigma}\otimes\xi^\dagger & \end{array}\right), \left(\begin{array}{c|c} & i\underline{\varepsilon}\otimes\xi \\ \hline -i\underline{\varepsilon}\otimes\xi^\dagger & \end{array}\right), \left(\begin{array}{c|c} \underline{\varepsilon}\otimes\underline{a} & \\ \hline & 0 \end{array}\right)\right\};$$

class 6 (color singlets):

$$D \in \left\{\left(\begin{array}{c|c} 0 & \\ \hline & \underline{\sigma} \end{array}\right)\right\}.$$

These classes contain 18, 15, and 3 particle-antiparticle pairs, respectively. The masses of these particules are, again, straightforwardly determined by application of (2.37), (2.38), (2.40):

class 4:

$$m^2_{\mathrm{UU}} = \left(\tfrac{10}{3}\alpha_s + \tfrac{1}{4}\alpha_w + \tfrac{25}{36}\alpha'_w\right)M^2,$$

$$m^2_{\mathrm{DD}} = \left(\tfrac{10}{3}\alpha_s + \tfrac{1}{4}\alpha_w + \tfrac{1}{36}\alpha'_w\right)M^2,$$

$$m^2_{\mathrm{UD}} = \left(\tfrac{10}{3}\alpha_s + \tfrac{1}{4}\alpha_w - \tfrac{5}{36}\alpha'_w\right)M^2;$$

class 5:

$$m^2_{\mathrm{DE}} = \left(\tfrac{4}{3}\alpha_s + \tfrac{1}{4}\alpha_w + \tfrac{25}{36}\alpha'_w\right)M^2,$$

$$m^2_{\mathrm{UN}} = \left(\tfrac{4}{3}\alpha_s + \tfrac{1}{4}\alpha_w + \tfrac{1}{36}\alpha'_w\right)M^2,$$

$$m^2_{(\mathrm{UE+DN})} = \left(\tfrac{4}{3}\alpha_s + \tfrac{1}{4}\alpha_w - \tfrac{5}{36}\alpha'_w\right)M^2,$$

$$m^2_{(\mathrm{UE-DN})} = \left(\tfrac{4}{3}\alpha_s - \tfrac{3}{4}\alpha_w - \tfrac{5}{36}\alpha'_w\right)M^2,$$

$$m^2_{\mathrm{UD}} = \left(\tfrac{4}{3}\alpha_s - \tfrac{3}{4}\alpha_w - \tfrac{5}{36}\alpha'_w\right)M^2;$$

class 6:

$$m^2_{\mathrm{EE}} = \left(\tfrac{1}{4}\alpha_w + \tfrac{9}{4}\alpha'_w\right)M^2,$$

$$m^2_{\mathrm{EN}} = \left(\tfrac{1}{4}\alpha_w + \tfrac{3}{4}\alpha'_w\right)M^2,$$

$$m^2_{\mathrm{NN}} = \left(\tfrac{1}{4}\alpha_w + \tfrac{1}{4}\alpha'_w\right)M^2. \tag{4.17}$$

The full spectrum of the Dimopoulos model based on the O scenario is displayed in fig. 4. The numerical values of the masses have been computed using the value (4.6)

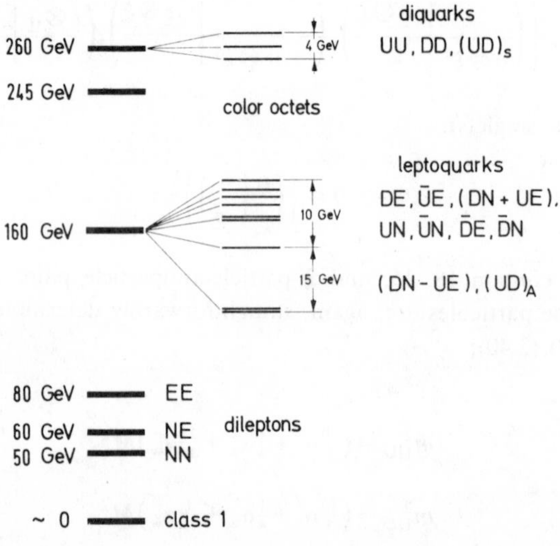

Fig. 4. Spectrum of technions in Dimopoulos' model with the O scenario. The masses are evaluated using the same value of M as that used in fig. 3.

for M, to facilitate comparison with fig. 3. The remarkable feature of this spectrum is the number of thresholds which open near the (unfortunately rather elevated) center of mass energy of 300 GeV. Just above this energy, the e^+e^- annihilation cross section almost doubles, with the entire rise of the cross section resulting from the production of bizarre pseudoscalar mesons with quark-lepton quantum numbers.

5. The lightest technions

The models considered in the previous two sections share the property that several of their technions remain massless at the level of the one gauge boson exchange effects we have involved ourselves in computing. This conclusion was anticipated in the work of Eichten and Lane [4], who showed that a set of technions must be left massless to this order in a general class of models which includes those we have discussed. In this section, we will expand upon the analysis of Eichten and Lane in explaining the effects of contributions of higher order in gauge boson exchanges to the masses of this set of particles. We wish, in particular, to show that two of these particles, a pair of electrically charged mesons, receive masses in order α^2, and that, quite generally in models following the SU scenario, this contribution is exceedingly easy to estimate.

The analysis of this section, like that in the rest of this paper, will be carried out in an incomplete technicolor theory, omitting the couplings of techni and ordinary fermions which generate for the latter their bare masses. We make this restriction to continue our pursuit of theoretical issues which are our main concern. We should warn the reader, then, that the numerical estimates for the masses of light technions which we will derive are sensitive to the manner in which the theory is completed. Eichten and Lane have argued that the interactions which generate the masses of ordinary fermions according to their schema also contribute to the masses of technions. The contribution, though model dependent and difficult to compute precisely, is of order $\Delta m^2 \sim (20\ \mathrm{GeV})^2$. This contribution will not significantly alter the results of sect. 4, but it could well swamp the smaller effects which we will discuss here.

To begin, let us recall the properties of those technions which were prohibited from receiving masses in order α. In each of our models, seven bosons in all were left massless; these formed two isospin triplets and one isosinglet. One of these isotriplets is readily identified as the multiplet of pions which combine with the $W^\pm$ and Z bosons. These are the (exactly massless) Goldstone bosons corresponding to the spontaneously broken generators of the gauged group $SU(2) \times U(1)$. The action of the Higgs mechanism eliminates them as independent physical states. The remaining four particles are physical pseudoscalar mesons and are, in principle observable as such. Following the notation of Eichten and Lane, we will call these physical pseudoscalars $P^0, P'^0, P^{\pm}$. We will see in a moment that the two electrically neutral mesons will remain massless to all orders in vector boson exchanges, since they are the Goldstone bosons corresponding to exact symmetries of the theories we have considered. The masses of the electrically charged pair $P^{\pm}$, on the other hand, will be seen to be not exactly zero, but merely suppressed.

To understand these claims, let us catalogue the symmetries of various segments of the lagrangians of our technicolor theories. The unperturbed lagrangians are, of course, invariant under the full chiral group G. The various terms of the perturbation $\delta\mathcal{L}$ which couples the technifermions to gauge mesons will not, in general, respect the full group G, but each term will respect a certain subgroup of G. In the specific $\delta\mathcal{L}$ given by (4.2), the coupling to gluons A^{α}_{μ} is obvious preserved by the set of chiral symmetries with color singlet generators. But for the Weinberg-Salam bosons, because of their handed coupling, the situation is more interesting. The coupling of the SU(2) bosons A^{α}_{μ}, in both (3.5) and (4.2), is preserved by any generator taking the block form

$$X^a = \left(\begin{array}{c|c} 0 & 0 \\ \hline 0 & H \end{array}\right), \tag{5.1}$$

where H is hermitian and must also be traceless if the symmetry is to be free of anomalies. There is one such symmetry for each broken generator which is diagonal in the sense of (2.35). In the absence of gluon and B_{μ} exchanges, this symmetry would guarantee the absolute masslessness of all technions corresponding to these generators. A corollary of this statement is that technions of this type receive no mass in order $\alpha_{\rm w}$ from A^{α}_{μ} exchange; we have noted this phenomenon in our results (3.13), (4.13), (4.14). The coupling of B_{μ} is, in the same way, preserved by generators of the form

$$X^a = \left(\begin{array}{c|c} J & 0 \\ \hline 0 & 0 \end{array}\right), \tag{5.2}$$

where J commutes with the charge difference matrix B of (3.5) or (4.2). The bosons protected by these symmetries from obtaining masses of order $\alpha'_{\rm w}$ due to B_{μ} exchange are just those of class 1 in (3.11) and of classes 1 and 2 in (4.11). The color singlet bosons of class 1, in both cases, receive no mass at all. The coupling of B_{μ} is also preserved by generators of the form (5.1) provided that the matrix H in (5.1) commutes with both B and τ^3. In our models, the (color singlet) matrices H with this property form precisely the set $\{\tau^3 \otimes \underline{1}, \underline{1} \otimes B, \tau^3 \otimes B\}$. These three generators thus leave invariant the whole of $\delta\mathcal{L}$; since they are also spontaneously broken symmetries, they insure the exact masslessness of three Goldstone bosons. The first of these bosons is obviously the neutral pion (which we already knew was exactly massless). The other two are just the P^0 and P'^0. A similar argument can be made in any technicolor model with more than one weak doublet in which the chiral symmetry is coupled in the standard way to the Weinberg-Salam gauge theory and is broken in accordance with one of the scenarios of sect. 2; one finds, in each case a pair of charged technions left massless to leading order in α and two neutral technions left massless to all orders.

In the above argument, the possible order α contributions to the $P^{\pm}$ masses were prohibited by two quite distinct symmetries of pieces of $\delta\mathcal{L}$, neither of which is a symmetry of the full lagrangian. We must expect, then, that the $P^{\pm}$ will be given small masses m_P by effects of higher order in α. Let us attempt to compute the contribution to m_P of order α^2. The methods of sect. 2 would relate this contribution to a vacuum expectation value of four currents; this makes it clear that we should not be able to compute the effect exactly. However, numerous particles in the theory have been given masses m^2 of order αM^2, where M is the quantity (2.31). We might find graphs with infrared singularities which are cut off by these masses; such graphs would yield an enhanced contribution of magnitude

$$m_P^2 \sim \left(\alpha^2 \log\frac{1}{\alpha}\right) M^2. \tag{5.3}$$

Further, the terms yielding the enhancement would involve integrals dominated by momenta small compared to the scale of momentum set by M or m_ρ. At such small momenta, the couplings of Goldstone bosons ought to be determined by current algebra. Hence, we might expect that terms with the enhancement (5.3) would also be determined as implications of chiral symmetry. The strategy we have just described is precisely that of the chiral perturbation theory of Langacker and Pagels [10]; these authors have applied this strategy to compute corrections to the predictions of current algebra in the ordinary strong interactions. Let us now see how it works in confronting this technicolor application.

To carry through this calculation in all detail, one would proceed as follows: the first step would be to write down the effective lagrangian corresponding to the chiral symmetry G, broken to H*. This lagrangian has as its elementary fields the Goldstone bosons of the broken chiral symmetries; it includes non-linear couplings among these bosons—a four-point vertex with coupling constant f_π^{-2} and higher-point vertices appearing with higher powers of f_π^{-2} — of just the magnitude required by current algebra. Next, one must gauge an SU(3) × SU(2) × U(1) subgroup of G with the strong, weak, and electromagnetic gauge bosons. At the same time, one must add a counterterm, invariant under SU(3) × SU(2) × U(1) but not under the full group G, for the technion masses. This term includes in its coefficient the parameter M, multiplied by the coupling constants of (4.7). Finally, one must search out and compute, using the vertices of this lagrangian, all terms of order $\alpha^2 \log \alpha$ which contribute to m_P. To develop all of its formalism explicitly would

* For a review of the effective lagrangian formalism, see [21].

take us far afield but, fortunately, one needs only this qualitative appreciation of it to understand how to estimate the masses of the $P^{\pm}$. Let us, then, turn immediately to that calculation. We will defer a detailed description of the chiral lagrangian formalism a separate paper devoted specifically to this methodology and its applications to technicolor [22].

We have drawn, in fig. 5, three sets of graphs, to be computed in chiral perturbation theory, which might contribute to m_p. The particles exchanged are the vector bosons W, Z, γ and technions. The first graph of fig. 4c involves the four-point vertex of the original effective lagrangian, the second that of the mass counterterm; both vertices appear with the factor f_π^{-2}. Since the contribution of each graph must have the dimensions of $(\text{mass})^2$, the various contributions may easily be estimated. One should include in this estimate the infrared logarithm associated with each loop. Then

$$\begin{aligned}(\text{fig. 5a}) &\sim \alpha m_W^2 \log m_W^2 ,\\ (\text{fig. 5b}) &\sim \alpha m_Z^2 \log m_Z^2 ,\\ (\text{fig. 5c}) &\sim \frac{m_t^4}{f_\pi^2} \log m_t^2 ,\end{aligned} \tag{5.4}$$

where we have written m_t to denote a generic technion mass. Since $m_W, m_Z \sim \alpha f_\pi^2$ and $m_t^2 \sim M^2$, all these contributions are of order $\alpha^2 \log \alpha$. They comprise, in fact, the only contributions of this order.

Having now simplified the problem to that of calculating the graphs of fig. 5, we may use the symmetries (5.1), (5.2) to simplify it further. The $W^{\pm}$ bosons are built entirely of the SU(2) gauge bosons A_μ^a; the graphs of fig. 5a would be unchanged if the coupling α'_w of technifermions to the U(1) gauge boson were set to zero. But we have seen that, in this limit, the technicolor theory has a symmetry of the type (5.2)

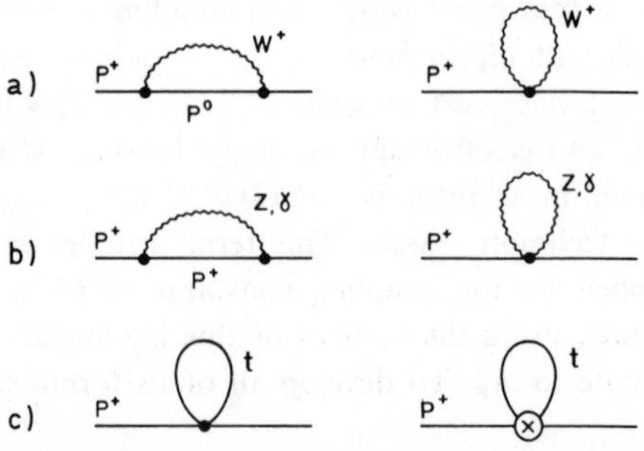

Fig. 5. Contributions from chiral perturbation theory to the masses of the $P^{\pm}$.

Fig. 6. Feynman rules for the evaluation of fig. 5b.

which prohibits the generation of masses for the $P^\pm$. We will see in a moment that the graphs of figs. 5b, c vanish if $\alpha'_w = 0$. Then the graphs of fig. 5a must give zero contribution to m_P, whatever the value of α_w. For the specific case of models where the chiral symmetry breakdown follows the SU scenario, we can make a similar argument about the graphs of fig. 5c: in the SU scenario, as we argued below (5.1), the A^a_μ boson exchange makes no contribution to any technion mass to leading order in α. But then the value of the graphs of fig. 5c would be uncharged if the coupling α_w of the SU(2) gauge bosons were set to zero*. This implies the property we noted above, that the contribution of fig. 5c vanishes if $\alpha'_w = 0$. But it also encourages us to realize that, in the limit $\alpha_w \to 0$, the theory has a symmetry of the type (5.1) which prohibits the generation of $P^\pm$ masses. We have already seen that the contribution of fig. 5a vanishes if $\alpha_w = 0$; we will see shortly that the same is true for fig. 5b. Hence, the contribution of fig. 5c must be zero for all values of α'_w. The computation is further simplified by working in Landau gauge; there the first graph of fig. 5b is proportional, for small external momentum p, to $p^2 \log p^2$, and hence gives no contribution to m_P.

All that remains is to compute the second graph of fig. 5b. This may be done using the Feynman rules shown in fig. 6. These Feynman rules may, after some labor, be read from the gauged effective lagrangian; however, they also follow from a much more elementary argument: the coupling of the P^+ to two photons is precisely that required for a spinless boson of electric charge e. This coupling yields for the photon contribution to m_P an ultraviolet-divergent result $m_P^2 \sim \alpha\Lambda^2$, where Λ is an ultraviolet cutoff, which should be taken to be of order M. However, we have shown that current algebra forbids the appearance of masses for the $P^\pm$ of order α. This divergent term must, then, be cancelled by the Z contribution to m_P. That requirement dictates the coupling of the P^+ to two Z's.

Using the rules of fig. 6, we can evaluate 5b in Landau gauge:

$$m_P^2 = -ie^2 g_{\mu\nu} \int \frac{d^4k}{(2\pi)^4} \left(g^{\mu\nu} - \frac{k^\mu k^\nu}{k^2} \right) \left(\frac{1}{k^2 - m_Z^2} - \frac{1}{k^2} \right), \tag{5.5}$$

* In the O scenario, A^a_μ exchange does contribute to technion masses, and so this argument is not valid. The computation of m_P in this case is more complex and will be discussed in [22].

or

$$m_{\rm P}^2 = \frac{3\alpha}{4\pi} m_{\rm Z}^2 \log \frac{\Lambda^2}{m_{\rm Z}^2}\,, \tag{5.6}$$

where Λ is an ultraviolet cutoff, a mass scale of the technicolor theory at which the effective lagrangian description breaks down. Contributions to $m_{\rm P}^2$ without the logarithmic enhancement may be absorbed as a change in the value of Λ. The expression (5.6) vanishes, because α does, if either $\alpha_{\rm w}$ or $\alpha'_{\rm w}$ is set to zero. A remarkable property of our estimate (5.6) is that all of its parameters except the cutoff Λ are known from the phenomenology of weak interactions. To evaluate (5.6), we set $\sin^2\theta = 0.23$ to determine $m_{\rm Z}$ and vary Λ from 250 GeV to 1 TeV. If the technicolor group were SU(4), such a range of values would correspond, through the scaling argument given below (2.43), to a range of values from 200 to 800 MeV in the ordinary strong interactions. This gives for the $\rm P^{\pm}$ masses:

$$m_{\rm P} = 5 - 8 \text{ GeV}\,. \tag{5.7}$$

We remind the reader that the arguments which led to (5.6) are valid not only in the SU case, but also that they are valid in any technicolor theory following the SU scenario. However, we also recall once more that the interactions which generate masses for the ordinary quarks and leptons may give additional contributions to $m_{\rm P}$. Certainly, in any realistic model, these interactions must give masses to the $\rm P^0$ and $\rm P'^0$, which have been left completely massless in the model we have analyzed.

6. Trouble in Sp(16)

In sect. 4, we analyzed in detail only two of Dimopoulos' models; those based on the SU and O scenarios of sect. 2. In this section we will explore the third Dimopoulos model, the one based on the Sp scenario. It will not be hard for us to discover that this model is troubled. We will profit, however, from exploring the nature and the cause of this trouble.

The difficulty of the model becomes apparent if one attempts to repeat the analysis following (4.15) to compute the technion masses in this scheme. The residual chiral symmetry of the model is Sp(16), and so, just as we did for the O scenario, we must add Goldstone bosons to those of (4.10) to fill out an Sp(16) multiplet. Eq. (2.6) tells us that these bosons correspond to generators of the form (4.15), but with D an antisymmetric matrix. One might catalogue the possible forms for D as we did for the O case in (4.16), but it is clear from the outset that this list must include [in the notation of (4.16)]

$$D = \left(\begin{array}{c|c} 0 & \\ \hline & \underset{=}{\varepsilon} \end{array}\right). \tag{6.1}$$

This generator is a color singlet, so it receives no mass from gluon exchange, and it is an isospin singlet, so, from the rules (2.38), it receives a *negative* contribution to its mass from exchange of the A^a_μ. Indeed, the full expression for this technion mass, derived using (2.37), is

$$m^2 = \left(-\tfrac{3}{4}\alpha_w + \tfrac{3}{4}\alpha'_w\right)M^2. \tag{6.2}$$

Numerically, for technicolor SU(4), this gives $m^2 = -(60\ \text{GeV})^2$; thus, (6.2) signals a strong instability of the Weinberg-Susskind vacuum. This turns out to be the only direction of instability.

This instability corresponds to a purely leptonic generator and would be present in the theory with technileptons only. Let us, then, neglect the techniquarks for a moment and consider this restricted model. We have assumed, in choosing to work in the Weinberg-Susskind vacuum, that the quantity whose vacuum expectation value signals chiral symmetry breaking is

$$\varepsilon^{\alpha\beta}\left(N_{L\alpha}(N_R)^c_\beta + E_{L\alpha}(E_R)^c_\beta\right). \tag{6.3}$$

But, if one acts on this object with the generator (6.1), or, better, with the unitary transformation

$$\begin{pmatrix} N_L \\ E_L \\ (N_R)^c \\ (E_R)^c \end{pmatrix} \to \sqrt{\tfrac{1}{2}} \left(\begin{array}{cc|cc} 1 & 0 & 0 & 1 \\ 0 & 1 & -1 & 0 \\ \hline 0 & 1 & 1 & 0 \\ -1 & 0 & 0 & 1 \end{array}\right) \begin{pmatrix} N_L \\ E_L \\ (N_R)^c \\ (E_R)^c \end{pmatrix}, \tag{6.4}$$

one arrives at another form, equivalent under the symmetry Sp(16),

$$\varepsilon^{\alpha\beta}\left(N_{L\alpha}E_{L\beta} - (N_R)^c_\alpha(E_R)^c_\beta\right). \tag{6.5}$$

This object is invariant under weak isospin. In the vacuum in which (6.5) acquires a vacuum expectation value, weak SU(2) is not broken and the W bosons do not acquire masses; similarly m_Z^2 has no term proportional to α_w. Because $\alpha_w \gg \alpha'_w$, our result (2.23) indicates that this vacuum is clearly preferred. Since the quark and lepton contributions to the vector boson mass matrix in the complete Dimopoulos model are additive (within the class of vacua we consider), the preferred vacuum of this complete model will be the Weinberg-Susskind vacuum for the techniquarks, but the vacuum signalled by an expectation value of (6.5) for the technileptons. In this vacuum, the $W^\pm$ and Z will receive mass as before from the techniquarks; however, since the vacuum contains an electrically charged condensate of technileptons, the photon must also acquire a mass. This mass is easy to compute using

(2.22), (4.2), and the explicit form of the unitary transformation to the new vacuum given in (6.4). We find

$$m_\gamma^2 \approx 2(g')^2 \cos^2\theta_W \left(\tfrac{1}{2}f_\pi\right)^2 \approx (25\ \text{GeV})^2 . \tag{6.6}$$

This is a small mass from the viewpoint of technicolor mass scales, but it is certainly enough to rule out this model on phenomenological grounds.

It is worth asking why we did not find a similar instability in the techniquark sector of this model. The quantity in this sector which receives an expectation value in the Weinberg-Susskind vacuum,

$$\varepsilon^{\alpha\beta}\left(U_{L\alpha}(U_R)^c_\beta + D_{L\alpha}(D_R)^c_\beta\right), \tag{6.7}$$

certainly breaks weak SU(2), but it is also a color singlet. The quantity analogous to (6.5), related by a motion in Sp(16), is

$$\varepsilon^{\alpha\beta}\left(U_{L\alpha}D_{L\beta} - (U_R)^c_\alpha(D_R)^c_\beta\right). \tag{6.8}$$

This object, though an SU(2) invariant, is a color triplet. Its acquisition of a vacuum expectation value would signal the breaking of color SU(3). In the corresponding vacuum, most of the color gluons would acquire masses with m^2 of order α_s. Since $\alpha_s \gg \alpha_w$, our result (2.23) prohibits such a choice of vacuum.

The final result of this argument is an amusing one. In the techniquark sector of this model with Sp symmetry, we have the alternative of breaking either color or weak isospin. We have found that the energetics of the vacuum orientation prefer that color remain unbroken and that weak isospin be broken precisely because color is the more strongly coupled. That is, the decision of whether to break color or weak isospin can be the result, rather than the assumption, of a dynamical model, and the fact that the strong interactions are strong can insist upon the choice.

We have now completed our investigation of the alignment of weakly gauged subgroups with respect to a pattern of dynamical symmetry breaking. We have considered here a special class of theories in which the technicolor symmetry remains unbroken, but even in these relatively simple theories, we have found a host of interesting phenomena. In the succeeding papers of this series, we will turn our attention to models in which the technicolor symmetry is broken through its own strong interactions, producing phenomena yet more strange and more intriguing.

I am grateful to Kenneth Lane and Savas Dimopoulos for sharing with me a bit of their insight into this problem. I am also most indebted to my colleagues at Saclay, particularly, Edouard Brézin, Paul Ginsparg, Sudhir Chadha and Jean Zinn-Justin, for their encouragement and advice.

Appendix A

BREAKING OF SU(2N) TO O(2N) OR Sp(2N)

In this appendix, we will discuss a useful representation of the generators of SU($2N$), O($2N$), and Sp($2N$) which allows one to discuss clearly and in a unified way the patterns of symmetry breaking shown in fig. 1. This analysis is based on the discussion of the symplectic group given by Gilmore [23].

Given a $2N$ vector ϕ belonging to the fundamental representation of SU($2N$), we may consider it written as two N-vectors $\phi = (\phi_1, \phi_2)$. Then we may identity the O($2N$) subgroup of SU($2N$) as the set of transformations preserving the quadratic form on two vectors ϕ, χ:

$$(\phi_1 \cdot \chi_1 + \phi_2 \cdot \chi_2), \tag{A.1}$$

and the Sp($2N$) subgroup as the set of transformations preserving

$$(\phi_1 \cdot \chi_2 - \phi_2 \cdot \chi_1). \tag{A.2}$$

Now define two new combinations of these N-vectors

$$\phi^{\pm} = \sqrt{\tfrac{1}{2}}\,(\phi_1 \pm i\phi_2), \tag{A.3}$$

and note that

$$\begin{aligned} \phi^{+} \cdot \chi^{-} &= \tfrac{1}{2}(\phi_1 \cdot \chi_1 + \phi_2 \cdot \chi_2) - \tfrac{1}{2}i(\phi_1 \cdot \chi_2 - \phi_2 \cdot \chi_1), \\ \phi^{-} \cdot \chi^{+} &= \tfrac{1}{2}(\phi_1 \cdot \chi_1 + \phi_2 \cdot \chi_2) + \tfrac{1}{2}i(\phi_1 \cdot \chi_2 - \phi_2 \cdot \chi_1). \end{aligned} \tag{A.4}$$

We can readily identify a U(N) group of transformation which preserves both lines of (A.4)

$$\phi^{+} \to U\phi^{+}, \qquad \phi^{-} \to U^{*}\phi^{-}. \tag{A.5}$$

This group must be subgroup of both O($2N$) and Sp($2N$). The transformations (A.5), plus the transformations

$$\phi^{+} \to U\phi^{+}, \qquad \phi^{-} \to U^{\mathrm{T}}\phi^{-} \tag{A.6}$$

make up an SU(N) $\times$ SU(N) $\times$ U(1) subgroup of the original SU($2N$). Thus, the basis vectors (A.3) are well-adapted to elucidating the symmetry-breaking patterns of fig. 1.

In the basis (A.3), the invariant forms (A.1), (A.2) become

$$E^{ab}(\phi^{+}, \phi^{-})_a (\chi^{+}, \chi^{-})_b, \tag{A.7}$$

where

$$E=\left(\begin{array}{c|c}0&1\\\hline 1&0\end{array}\right),(\mathrm{O}(2N))\,;\qquad E=\left(\begin{array}{c|c}0&i\\\hline -i&0\end{array}\right),(\mathrm{Sp}(2N))\,. \tag{A.8}$$

In order to preserve (A.7), a generator T of SU(N) must satisfy

$$T^{\mathrm{T}}E+ET=0\,. \tag{A.9}$$

The solutions to this equation are displayed as the matrices T_i in (2.6); the remaining $2N\times 2N$ traceless hermitian matrices comprise the X_a of (2.6). The block diagonal matrices

$$T_i=\left(\begin{array}{c|c}A&0\\\hline 0&-A^{\mathrm{T}}\end{array}\right),\qquad X_a=\left(\begin{array}{c|c}C&0\\\hline 0&C^{\mathrm{T}}\end{array}\right), \tag{A.10}$$

are the generators of the U(N) group (A.5) and the SU(N) group (A.6), respectively.

Appendix B

INVARIANTS OF X^aX^b

We wish to show that, if G/H is a symmetric space, the product X^aX^b contains no H invariant other than δ^{ab}, except in the trivial case that G/H is a direct product of symmetric spaces. This follows from the claim, demonstrated below, that the set of hermitian matrices X^a cannot be decomposed into two distinct representations of H, except in this trivial case.

We will argue first that, if the X^a divide into two distinct representations of H,$\{X_1^a\}$ and $\{X_2^a\}$, then each X_1^a must commute with each X_2^a. The commutation relations (2.10) insist that

$$\left[X_1^a,X_2^b\right]=if_{abj}T^j\,, \tag{B.1}$$

where the f_{abj} are structure constants of G. But then, by the antisymmetry of these structure constants,

$$\left[T^j,X_1^a\right]=if_{abj}X_2^b+(\text{terms with }X_1\text{ only})\,. \tag{B.2}$$

The commutation relation (B.2) expresses the action of a generator of H on X_1^a; thus the X_1 and X_2 belong to the same representation unless the f_{abj} of (B.1) all vanish.

If it is so that the X^a break up into two commuting subsets, the set of matrices $\{X_1,T\}$ form a closed algebra, the Lie algebra of a group G′ such that G ⊃ G′ ⊃ H.

The X_2^a must form a representation of G′. But some generators of G′, the X_1^a, annihilate every X_2^a, so this representation must be rather trivial. The most general possibility is that G′ divides into a direct product $G' = G_1 \otimes G_2$, such that the X_2^a are invariant under G_1 and transform non-trivially under G_2; then the generators of G_2 must be linear combinations of the T_i only. But in this case, the T_i also break up into two mutually commuting subsets $\{T_1^i\}$, $\{T_2^i\}$, such that the T_1^i commute with the X_2^a and the T_2^i commute with the X_1^a. The assumption that the X^a form two disjoint representations of H has thus led us to conclude that the whole symmetric space breaks up into the direct product of two such spaces, the trivial case excluded by assumption above.

Note added in proof:

The analysis reported in this paper has been performed independently by J.P. Preskill, Nucl. Phys. B177 (1981) 21.

References

[1] S. Weinberg, Phys Rev. Lett. 19 (1967) 1264;
A. Salam, *in* Elementary particle theory, ed. N. Svartholm (Almqvist and Wiksell, Stockholm, 1968)
[2] S. Weinberg, Phys. Rev D19 (1979) 1277
[3] L. Susskind Phys. Rev. D20 (1979) 2619
[4] E. Eichten and K.D. Lane, Phys. Lett 90B (1980) 125
[5] S. Dimopoulos and L. Susskind, Nucl. Phys. B155 (1979) 237
[6] S. Weinberg, Phys. Rev. D13 (1976) 974
[7] M.E. Peskin, in preparation
[8] S. Weinberg, Phys. Rev. Lett. 29 (1972) 1698; Phys. Rev D7 (1973) 2887
[9] S. Dimopoulos, Nucl. Phys. B168 (1980) 69
[10] P. Langacker and H. Pagels, Phys. Rev D8 (1973) 4595; 4620
[11] J.M. Cornwall, R. Jackiw and E. Tomboulis, Phys. Rev D10 (1974) 2428;
D. Caldi, Phys. Rev. Lett 39 (1977) 121
[12] J.S. Bell and R. Jackiw, Nuovo Cim. 60A (1969) 47;
S.L. Adler, Phys. Rev. 117 (1969) 2426
[13] S. Helgason, Differential geometry and symmetric spaces (Academic Press, New York, 1962)
[14] H. Eichenherr and M. Forger, Nucl. Phys. B155 (1979) 381;
E. Brézin, S. Hikami and J. Zinn-Justin, Nucl. Phys. B165 (1980) 528
[15] J. Schwinger, Phys. Rev. 125 (1962) 397; 128 (1962) 2425
[16] F. Englert and R. Brout, Phys. Rev. Lett. 13 (1964) 321;
F. Englert, R. Brout and M.F. Thiry, Nuov. Cim. 43 (1966) 244;
R. Jackiw and K. Johnson, Phys. Rev. D8 (1973) 2386;
J.M. Cornwall and R.E. Norton Phys. Rev. D8 (1973) 3338
[17] R. Dashen, Phys. Rev 183 (1969) 1245
[18] T. Das, G.S. Guralnik, V.S. Mathur, F.E. Low and J.E. Young, Phys. Rev. Lett. 18 (1967) 759
[19] S. Weinberg, Phys. Rev. Lett. 18 (1967) 507
[20] C. Bernard, A. Duncan, J. Losecco and S. Weinberg, Phys. Rev D12 (1975) 792
[21] S. Gasiorowicz and D.A. Geffen, Rev. Mod. Phys. 41 (1969) 531
[22] S. Chadha and M.E. Peskin, Nucl. Phys. B185 (1981) 61;
Nucl. Phys. B187 (1981) 541
[23] R. Gilmore, Lie groups, Lie algebras and some of their applications (Wiley, New York, 1974)

Nuclear Physics B177 (1981) 21–59

SUBGROUP ALIGNMENT IN HYPERCOLOR THEORIES

John PRESKILL[1]

Lyman Laboratory of Physics, Harvard University, Cambridge, Massachusetts 02138, USA

Received 7 July 1980

To analyze the physical consequences of a dynamically broken theory of the weak interactions, we must know how the weak gauge group is aligned in an approximate flavor-symmetry group. For a large class of models, spectral-function sum rules enable us to determine this alignment explicitly. We work out the pattern of the electroweak symmetry breakdown for several sample models. Critical values of weak mixing angles are found at which the breakdown pattern changes discontinuously. We compute pseudo-Goldstone boson masses, and find that some models contain unusually light charged or colored pseudo-Goldstone bosons.

1. Introduction

The extremely successful Weinberg-Salam [1] model of the electroweak interactions is marred by one glaring imperfection—elementary scalar fields with negative mass squared are required to drive the breakdown of the electroweak gauge group. Such elementary scalars are distasteful for at least two reasons. First, we must adjust the bare scalar masses very delicately [2–4] to ensure that the mass scale of the electroweak breakdown (300 GeV) is many orders of magnitude smaller than the grand unification [5] mass (10^{15} GeV) and the Planck mass (10^{19} GeV). Second, so many arbitrary parameters are needed to characterize the couplings of the scalars that we are reluctant to accept them as fundamental ingredients in the theory.

Dissatisfaction with the standard Weinberg-Salam model has spawned recent efforts to construct gauge theories of the electroweak interactions without elementary scalar fields. Weinberg [3] and Susskind [4] have proposed that the electroweak gauge group is actually broken dynamically by a postulated new strong interaction, rather than by elementary scalars. This new gauge interaction, which will be called "hypercolor" here, binds the Goldstone bosons which are eaten by the weak W and Z bosons. Dimopoulos and Susskind [6], Eichten and Lane [7], and Weinberg [8] have noted that yet another gauge interaction, called "sideways", is required to generate the masses of quarks and leptons. The sideways gauge group must also be dynamically broken.

[1] Research supported in part by the National Science Foundation under grant number PHY77-22864 and the Harvard Society of Fellows.

No one has yet constructed a fully realistic model based on these ideas. This failure is probably due to the tightly constrained nature of dynamically broken theories. These theories have few free parameters which can be adjusted to fit experiment. Lack of adjustability makes dynamically broken theories very appealing, but also makes it hard to find a theory which works.

To conduct a well-organized search for a realistic model, we must improve our still meager understanding of dynamical symmetry breaking. For example, consider an asymptotically free theory with gauge group G and massless fermions in a reducible representation of G. Because the fermion representation is reducible, this theory respects a global flavor (chiral) symmetry group G_f, which commutes with G. We would like to be able to answer the following questions:

(1) Is the gauge group G dynamically broken? If so, to what subgroup?

(2) Is the flavor-symmetry group G_f dynamically broken? If so, to what subgroup?

But both (1) and (2) are difficult dynamical questions which we do not know how to answer in general.

In this paper, our attention will be focused instead on a somewhat more tractable question. If the flavor-symmetry group G_f is dynamically broken to the subgroup S_f, then the theory has many degenerate vacua. The vacua can be parametrized by the coset space G_f/S_f; each vacuum corresponds to a particular orientation of S_f in G_f. But if the G_f symmetry is explicitly broken by a small perturbation, the degeneracy is lifted. The true vacuum of the theory then corresponds to the orientation of S_f which minimizes the vacuum energy [3, 9]. This orientation of S_f is called the "orientation of the vacuum", or the "alignment of the vacuum". Thus, a third question which arises in dynamically broken theories is:

(3) If G_f is dynamically broken to S_f, and explicitly broken by a small perturbation, what is the orientation of the vacuum?

The orientation of the vacuum, and its consequences in dynamically broken theories of the electroweak interactions, are the topics of this paper.

In electroweak gauge theories without elementary scalar fields, the G_f flavor symmetry of the hypercolored fermions is dynamically broken to S_f by the strong hypercolor interaction [3, 4]. G_f is also explicitly, but weakly, broken by the sideways interaction [6, 7] and by a weak gauge interaction with gauge group $G_W \subset G_f$. [G_W contains the Weinberg-Salam $SU(2) \times U(1)$.] These flavor-symmetry-breaking perturbations determine the orientation of the vacuum.

The influence of the sideways interaction on the vacuum orientation has been considered elsewhere [10–12]. The sideways interaction can cause the vacuum to be oriented in such a way that *CP* is spontaneously broken. It has been proposed [10–12] that the observed *CP* violation is generated by this mechanism.

In this paper we consider the influence of the G_W interaction on the vacuum orientation*. The problem of identifying the correct vacuum is equivalent to the

* The sideways interaction is ignored in this paper. Including it would not alter our results in any essential way.

problem of finding the relative orientation of the subgroups G_W and S_f of G_f which minimizes the vacuum energy. We call this the "subgroup alignment problem". The importance of the alignment of G_W and S_f was emphasized by Weinberg [3], who pointed out that this alignment determines the pattern of the G_W breakdown and the spectrum of pseudo-Goldstone bosons. His general treatment of the subgroup alignment problem is reviewed in sects. 2 and 3.

Previous attempts to solve the subgroup alignment problem in particular cases have often been steeped in folklore*. A popular notion is that gauge symmetries resist being broken, and that G_W and S_f will therefore line up so that the largest possible subgroup of G_W survives. This point of view is sometimes expressed in a different way. The embedding of S_f in G_f can be characterized by a fermion "condensate" [6]; according to the conventional wisdom, the condensate will form in the channel in which the lowest-order G_W interaction is most attractive.

To solve the subgroup alignment problem properly, we must minimize an effective potential, the G_W interaction contribution to the vacuum energy. Even though we assume that the weak G_W interaction can be treated perturbatively, the problem is not trivial, because we cannot calculate the effective potential without solving the strong hypercolor interaction. Weinberg [3] observed that the symmetry properties of the hypercolor interaction, in particular the S_f "isospin" symmetry, provide powerful constraints on the form of the effective potential. However, even in relatively simple cases we cannot find the minimum of the potential unless we know the signs of certain strong-interaction parameters which are not determined by S_f symmetry alone.

The main conclusion** of this paper is that we can in many cases find the signs we need to know to minimize the effective potential, and that the results tend to confirm the conventional wisdom. Spectral function sum rules [15] provide the additional information we require. However, we can confirm the most attractive channel folklore only at the expense of introducing another element of folklore—we must assume that the signs of certain spectral integrals can be determined by saturating with low-lying resonances. Spectral function sum rules are reviewed in sect. 4, and their relevance to the effective potential is explained in sect. 5.

The ramifications of subgroup alignment are most easily appreciated in the context of specific models. The subgroup alignment problem is solved, and pseudo-Goldstone boson masses are calculated, for a number of examples in sect. 6. These examples illustrate three important phenomena:

(i) Some hypercolor theories can be ruled out, because the dynamically determined pattern of G_W breakdown is phenomenologically unacceptable.

(ii) If G_W is not simple, the pattern of G_W breakdown may depend on the relative strength of the different G_W gauge couplings. There are critical values of weak mixing angles at which phase transitions occur.

* See, for example, [6] and [13].

** As this work was being completed, I learned that Peskin [14] has also analyzed subgroup alignment in hypercolor models, and has reached very similar conclusions.

(iii) There may be pseudo-Goldstone bosons which receive mass not in lowest order in the G_W interaction, but in higher order. Hence, some charged or colored pseudo-Goldstone bosons may be considerably lighter than we would naively expect [7].

In this paper, we do not address questions (1) and (2) stated above, but the most attractive channel folklore has been applied to these questions also. Several authors [16] have speculated recently on the pattern of breakdown of gauge and global symmetries in a confining gauge theory with non-real fermion representation content. Lacking the basis for a detailed dynamical analysis, they have assumed that a fermion condensate occurs in the channel which is most attractive in lowest order in the gauge coupling. Of course, the gauge interaction must be strong to bind Goldstone bosons, so lowest-order perturbation theory is completely untrustworthy when dynamical symmetry breaking occurs. We would feel less uncomfortable about applying the most attractive channel condition here if there were some justification for it which goes beyond perturbation theory. No such justification is known.

This paper might be regarded as a modest attempt to justify the most attractive channel condition in a relatively simple context. By assumption, it *is* a good approximation to treat the G_W interaction to lowest order, but the strong hypercolor interaction must be summed to all orders. Taking into account the strong corrections, we find, in the cases we can explicitly analyze, that the subgroups G_W and S_f tend to align so that the fermion condensate occurs in the channel in which the lowest-order G_W interaction is most attractive.

2. The subgroup alignment problem

In this section and sect. 3 we review the general formulation of the subgroup alignment problem given by Weinberg [3].

We consider a gauge theory with massless fermions and no elementary scalars. The gauge group is $G_H \times G_W$. G_H is a simple group, the hypercolor group. The associated running coupling constant becomes strong at a mass scale near 1 TeV. G_W is not necessarily simple. All the G_W gauge couplings are weak at 1 TeV.

It is convenient to choose all fermions to be two-component left-handed spinors. The fermions transform as a representation $\mathfrak{D}$ of G_H which in general is reducible; we have

$$\mathfrak{D} = \prod_\rho n_\rho \mathfrak{D}^{(\rho)} . \tag{2.1}$$

That is, the irreducible representation $\mathfrak{D}^{(\rho)}$ is repeated n_ρ times. The fermions may be denoted $\psi_{ri}^{(\rho)}$, where ρ identifies the irreducible representation of G_H according to which $\psi^{(\rho)}$ transforms, i is the index on which G_H acts, and $r = 1, \ldots, n_\rho$ labels the different flavors of fermions which transform as $\mathfrak{D}^{(\rho)}$. The group G_W acts on the index r.

The lagrangian of this theory is

$$\mathcal{L} = -\tfrac{1}{4}F^a_{\mathrm{H}\mu\nu}F^{\mu\nu a}_{\mathrm{H}} - \tfrac{1}{4}F^{\alpha}_{\mathrm{W}\mu\nu}F^{\mu\nu\alpha}_{\mathrm{W}} + \bar{\psi}^{(\rho)}_{ri}(i\gamma^{\mu}\partial_{\mu})\psi^{(\rho)}_{ri}$$

$$+ \bar{\psi}^{(\rho)}_{ri}\gamma^{\mu}\left(g_{\mathrm{H}}t^{(\rho)a}_{ij}A^a_{\mathrm{H}\mu}\right)\psi^{(\rho)}_{rj}$$

$$+ \bar{\psi}^{(\rho)}_{ri}\gamma^{\mu}\left(e_{\alpha}\theta^{(\rho)\alpha}_{rr'}A^{\alpha}_{\mathrm{W}\mu}\right)\psi^{(\rho)}_{r'i} \tag{2.2}$$

(all indices summed). A_{H} and A_{W} are the $\mathrm{G_H}$ and $\mathrm{G_W}$ gauge fields; F_{H} and F_{W} are the corresponding covariant curls. The $t^{(\rho)}$'s are the $\mathrm{G_H}$ generators in the representation $\mathcal{D}^{(\rho)}$, and the $\theta^{(\rho)}$'s are the $\mathrm{G_W}$ generators. The gauge couplings are g_{H} and e_{α}.

In the limit $e \to 0$, this lagrangian is invariant under transformations of the form

$$\psi^{(\rho)}_{ri} \to U^{(\rho)}_{rr'}\psi^{(\rho)}_{r'i}, \tag{2.3}$$

where $U^{(\rho)}$ is a unitary $n_{\rho} \times n_{\rho}$ matrix. Because there is one $\mathrm{U(1)_A}$ transformation which has a $\mathrm{G_H}$ anomaly [17], the global flavor-symmetry group of this theory is

$$\mathrm{G_f} = \prod_{\rho} \mathrm{U}(n_{\rho})/\mathrm{U_A}(1), \tag{2.4}$$

in the $e = 0$ limit. The weak gauge group $\mathrm{G_W}$ is a subgroup of $\mathrm{G_f}$.

When the hypercolor interaction becomes strong, $\mathrm{G_f}$ is spontaneously broken to an "isospin" subgroup $\mathrm{S_f}$. (If e is not zero but small, the $\mathrm{G_W}$ interaction has little influence on the strong hypercolor dynamics, and we do not expect the pattern of $\mathrm{G_f}$ breaking to be altered.) Although it is generally believed that chiral symmetries break in confining theories [18], we have no reliable way of computing $\mathrm{S_f}$. The only theory for which we have experimental information is QCD; there, $\mathrm{SU(3)} \times \mathrm{SU(3)}$ chiral symmetry appears to break to the maximal diagonal isospin subgroup SU(3), but we do not know if this represents a general phenomenon or is merely one of several natural possibilities [19]. For now, we allow $\mathrm{S_f}$ to be an arbitrary subgroup of $\mathrm{G_f}$. Special cases will be discussed in sects. 5 and 6.

In the $e = 0$ limit, $\mathrm{G_f}$ symmetry is exact, and the vacuum is highly degenerate; the vacua are parametrized by the coset space $\mathrm{G_f}/\mathrm{S_f}$. If e is non-zero but small, $\mathrm{G_W}$-boson exchange generates a weak $\mathrm{G_f}$-breaking perturbation $\mathcal{H}'$. This perturbation lifts the degeneracy and picks out the true vacuum. As Dashen [9] was the first to observe, a perturbation expansion in $\mathcal{H}'$ must be performed about the correct vacuum to avoid paradoxial results.

The correct vacuum can be identified by minimizing an effective potential (the vacuum energy), given to lowest order in $\mathcal{H}'$ by

$$V(g) = \langle 0|U(g)^{-1}\mathcal{H}'U(g)|0\rangle = \langle 0,g|\mathcal{H}'|0,g\rangle, \tag{2.5}$$

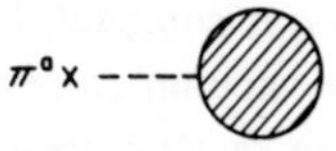

Fig. 1. A Goldstone boson tadpole.

where $U(g)$ represents G_f in Hilbert space. If the vacuum $|0\rangle$ is left invariant by $S_f \subset G_f$, then $U(g)|0\rangle = |0, g\rangle$ is left invariant by the equivalent chirally rotated subgroup $gS_f g^{-1}$.

We can also interpret eq. (2.5) by regarding the vacuum $|0\rangle$ as fixed, so that the chiral rotation is applied to the perturbation $\mathcal{H}'$. Then we minimize $V(g)$ by finding the chiral perturbation $\mathcal{H}'(g) = U(g)^{-1}\mathcal{H}'U(g)$ which gives the lowest contribution to the energy of the S_f-invariant vacuum. If we express $U(gg')$ in terms of the chiral charges Q_a, $U(gg') = U(g)\exp[i\omega_a(g')Q_a]$, the condition satisfied by $\mathcal{H}'(g)$ at the minimum is

$$\left.\frac{\mathrm{d}V}{\mathrm{d}\omega_a}\right|_{\omega=0} = i\langle 0|[\mathcal{H}'(g), Q_a]|0\rangle = 0\,, \tag{2.6}$$

$$\left.\frac{\mathrm{d}^2 V}{\mathrm{d}\omega_a\,\mathrm{d}\omega_b}\right|_{\omega=0} = -\langle 0|[[\mathcal{H}'(g), Q_a], Q_b]|0\rangle \geqslant 0\,. \tag{2.7}$$

(Q may not be defined if the associated symmetry is spontaneously broken, but the commutator of Q_a with a local operator can still be defined.)

These equations have a straightforward current algebra interpretation. Charges Q_a which do not annihilate the vacuum couple to Goldstone bosons. Eq. (2.6) says that Goldstone boson tadpoles vanish (fig. 1). If the perturbation $\mathcal{H}'(g)$ has a non-vanishing matrix element to create a Goldstone boson, $\langle \pi^a|\mathcal{H}'|0\rangle \neq 0$, the Goldstone bosons can be produced spontaneously, and the vacuum is unstable. Eq. (2.7) guarantees that the Goldstone boson mass matrix [20],

$$m_{ab}^2 = -\frac{1}{F_a F_b}\langle 0|[[\mathcal{H}'(g), Q_a], Q_b]|0\rangle\,, \tag{2.8}$$

has no negative eigenvalues (fig. 2). A Goldstone boson tachyon also signals an instability.

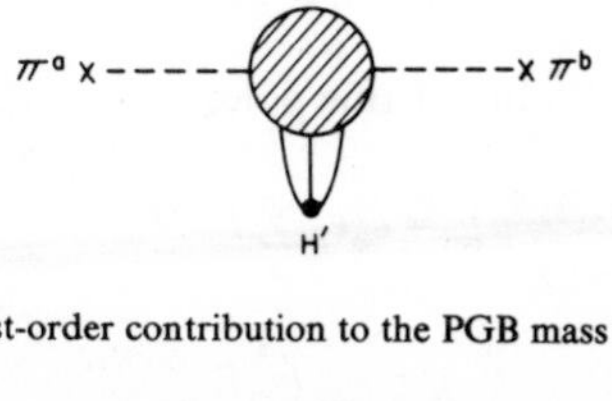

Fig. 2. Lowest-order contribution to the PGB mass matrix m_{ab}^2.

Because the symmetry group G_f is compact, $V(g)$ always has a global minimum and a global maximum. The true vacuum corresponds to the global minimum, but there may also be metastable false vacua corresponding to local minima which are not global minima.

Because $\mathcal{H}'$ is generated by the exchange of G_W gauge bosons, the G_f transformation $U(g)$ has the effect of changing the embedding of G_W in G_f:

$$G_W \to g^{-1} G_W g . \tag{2.9}$$

Minimizing $V(g)$ determines the relative orientation of the subgroups S_f and G_W in G_f. Hence, the problem of finding the minimum of the effective potential when a subgroup of G_f is gauged is called the "subgroup alignment problem" [3].

We must know how the subgroups are aligned to determine how G_W breaks, to classify Goldstone bosons, and to calculate their masses. The general classification of Goldstone bosons in a theory like the one defined by eq. (2.2) was carried out by Weinberg [3]. He observed that the global symmetry group of the G_W interactions may be larger than the local group G_W. Call this global group G_f'. When G_f breaks down to S_f, the resulting Goldstone bosons are in one-to-one correspondence with an orthonormal set of broken currents. The Goldstone bosons fall into three classes:

(i) Fictitious. If the corresponding current is a linear combination of G_W currents and S_f currents, then the Goldstone boson is eaten by a G_W gauge boson.

(ii) Exact. If the corresponding current is a linear combination of G_f' currents which are not G_W currents and S_f currents, then the Goldstone boson is exactly massless.

(iii) Pseudo. If the corresponding current is not a linear combination of G_f' currents and S_f currents, then the Goldstone boson receives mass from the perturbation $\mathcal{H}'$. It is called a pseudo-Goldstone boson (PGB).

We must solve the subgroup alignment problem before we can classify the Goldstone bosons.

The G_W vector bosons acquire masses by eating fictitious Goldstone bosons. If the G_W currents $J_\mu^\alpha = \bar{\psi}_{ri}^{(\rho)} \gamma_\mu \theta_{rr'}^{(\rho)\alpha} \psi_{r'i}^{(\rho)}$ couple to Goldstone bosons,

$$\langle 0 | J_\mu^\alpha | \pi^A \rangle = i k_\mu F^{\alpha A} , \tag{2.10}$$

then the polarization tensor $\pi_{\alpha\beta}$ defined by

$$e_\alpha e_\beta \int d^4x \, e^{ikx} \langle 0 | T\left[J_\mu^\alpha(x) J_\nu^\beta(0) \right] | 0 \rangle = i\left(\eta_{\mu\nu} k^2 - k_\mu k_\nu \right) \pi_{\alpha\beta}(k^2) \tag{2.11}$$

has a pole at $k^2 = 0$. The residue of this pole,

$$\mu_{\alpha\beta}^2 = e_\alpha e_\beta \sum_A F^{\alpha A} F^{\beta A} , \tag{2.12}$$

is the G_W vector boson mass matrix [21]. The $F^{\alpha A}$'s and therefore also μ^2, depend on the relative orientation of G_W and S_f. We must solve the subgroup alignment problem to determine the pattern of G_W breakdown.

Because the Goldstone bosons form S_f isospin multiplets, the $F^{\alpha A}$'s obey relations which are consequences of S_f invariance. Weinberg [3] and Susskind [4] have emphasized that S_f invariance can require the weak vector boson masses given by (2.12) to obey the relation

$$\mu_W^2/\mu_Z^2 = \cos^2\theta_W . \tag{2.13}$$

This relation also holds in the Weinberg-Salam model with an elementary scalar doublet [1], and is known to be well satisfied.

3. The effective potential

Having established the importance of determining the alignment of subgroups, we now turn to the problem of actually constructing and minimizing the effective potential $V(g)$. In this section we show that, by exploiting the S_f invariance of the vacuum, we can write down a compact expression for $V(g)$ in terms of a few unknown matrix elements.

Because the G_W interaction is weak at the mass scale (1 TeV) at which hypercolor becomes strong, we may compute the effective potential in a perturbation series in e. To lowest order in e, the G_f-breaking perturbation $\mathcal{H}'$ is

$$\mathcal{H}' = \left(-\tfrac{1}{2}i\right)\int d^4x\, \Delta^{\mu\nu}(x) e_\alpha^2 \mathrm{T}\left[J_\mu^\alpha(x) J_\nu^\alpha(0)\right], \tag{3.1}$$

where $\Delta^{\mu\nu}$ is the gauge boson propagator and J_μ^α is a G_W current.

In order to clearly exhibit the way the effective potential varies under G_f rotations, it is convenient to express the G_W currents in terms of a standard basis of G_f currents. We denote this basis by

$$J_\mu^A = \bar{\psi}_{ri}^{(\rho)} \gamma_\mu \lambda_{rr'}^{(\rho)A} \psi_{r'i}^{(\rho)}, \tag{3.2}$$

where the λ^A's are a basis for the generators of G_f, normalized so that

$$\mathrm{Tr}\, \lambda^A \lambda^B = \delta^{AB} . \tag{3.3}$$

Under G_f transformations, the currents J^A transform as the adjoint representation of G_f; that is

$$J^A(g) = U^{-1}(g) J^A U(g) = R^{AB}(g) J^B , \tag{3.4}$$

where $R(g)$ is the adjoint representation.

Fig. 3. Lowest-order G_W contribution to the vacuum energy.

The G_W currents are linear combinations of G_f currents,

$$e_\alpha J_\mu^\alpha = e_{\alpha A} J_\mu^A, \qquad (\alpha \text{ not summed}) \tag{3.5}$$

and we can combine (2.5) with (3.1), (3.4) and (3.5) to obtain (fig. 3)

$$V(g) = e_{\alpha A} e_{\alpha B} R^{AC}(g) R^{BD}(g) I^{CD}, \tag{3.6}$$

where

$$I^{CD} = \left(-\tfrac{1}{2}i\right) \int d^4x\, \Delta^{\mu\nu}(x) \langle 0 | T\left[J_\mu^C(x) J_\nu^D(0) \right] | 0 \rangle . \tag{3.7}$$

This expression for $V(g)$ was derived by Weinberg [3].

In (3.6), we have expressed the effective potential in terms of group-theoretic factors, G_W gauge couplings, and the G_f tensor I^{AB}. Because the vacuum $|0\rangle$ is S_f invariant, we see from (3.7) that I^{AB} is also S_f invariant. We can decompose the adjoint representation $\mathrm{Ad}(G_f)$ into irreducible representations of S_f. The tensor I^{AB} can then be expressed in terms of as many unknown constants as there are S_f singlets in $\mathrm{Ad}(G_f) \times \mathrm{Ad}(G_f)$*.

For example, if the currents $J^A, A = 1, 2, \ldots, n_\rho$, are in the (real) irreducible representation $\mathfrak{D}^{(\rho)}$ of S_f, then, by Schur's lemma,

$$I^{AB} = \Delta_\rho \delta^{AB}, \qquad A, B = 1, 2, \ldots, n_\rho . \tag{3.8}$$

However, we cannot determine the constant Δ_ρ without solving the strong hypercolor interaction.

4. Spectral function sum rules

To minimize the effective potential given by (3.6), we must know more about the tensor I^{AB}. In sect. 5, we will argue that the additional information we seek can often be extracted from spectral function sum rules, which are therefore the subject of the present section.

* The number of independent invariants can often be further reduced by invoking additional symmetries, such as CPT and parity.

Spectral function sum rules (SFSR's) were first derived by Weinberg [15] for the case of spontaneously broken $SU(2) \times SU(2)$ chiral symmetry. Wilson [22] and Bernard, Duncan, Lo Secco, and Weinberg [23] put the derivation on a more secure footing by invoking the operator product expansion*. Here we will review the derivation of Bernard et al., which was carried out in the context of spontaneously broken $SU(N) \times SU(N)$ chiral symmetry, and indicate how it can be generalized to arbitrary G_f breakdown patterns.

The starting point of all derivations of the SFSR's is the Lehmann-Källen spectral representation for the (time-ordered) product of two currents. If we define the spin-one and spin-zero spectral functions associated with a pair of currents J_μ^A and J_ν^B by

$$(2\pi)^3 \sum_{\substack{n \\ (\text{spin } 1)}} \langle 0| J_\mu^A(0)|n\rangle\langle n| J_\nu^B(0)|0\rangle \delta^4(k-k_n) = -\left(\eta_{\mu\nu} - \frac{k_\mu k_\nu}{k^2}\right)\rho_{AB}^{(1)}(k^2), \tag{4.1}$$

$$(2\pi)^3 \sum_{\substack{n \\ (\text{spin } 0)}} \langle 0| J_\mu^A(0)|n\rangle\langle n| J_\nu^B(0)|0\rangle \delta^4(k-k_n) = k_\mu k_\nu \rho_{AB}^{(0)}(k^2), \tag{4.2}$$

then we find by inserting a complete set of intermediate states that

$$\langle 0|\mathrm{T}\left[J_\mu^A(x) J_\nu^B(0)\right]|0\rangle = \int_0^\infty \mathrm{d}\mu^2 \int \frac{\mathrm{d}^4 k}{(2\pi)^4}\, \mathrm{e}^{-ikx} \frac{-i}{k^2-\mu^2+i\varepsilon}$$
$$\times\left[\rho_{AB}^{(1)}(\mu^2)\left(\eta_{\mu\nu} - \frac{k_\mu k_\nu}{\mu^2}\right) - \rho_{AB}^{(0)}(\mu^2) k_\mu k_\nu\right]. \tag{4.3}$$

SFSR's are derived by considering the behavior of both sides of eq. (4.3) in the short-distance limit. It is most convenient to expand the Fourier transform in powers of $1/k^2$. We have

$$\int \mathrm{d}^4 x\, \mathrm{e}^{ikx} \langle 0|\mathrm{T}\left[J_\mu^A(x) J_\nu^B(0)\right]|0\rangle = i\frac{k_\mu k_\nu}{k^2}\int_0^\infty \mathrm{d}\mu^2 \left[\frac{\rho_{AB}^{(1)}}{\mu^2} + \rho_{AB}^{(0)}\right]$$
$$- i\frac{\eta_{\mu\nu}}{k^2}\int_0^\infty \mathrm{d}\mu^2 \left[\rho_{AB}^{(1)}\right]$$
$$+ i\frac{k_\mu k_\nu}{(k^2)^2}\int_0^\infty \mathrm{d}\mu^2 \left[\rho_{AB}^{(1)} + \mu^2 \rho_{AB}^{(0)}\right] + \cdots . \tag{4.4}$$

* See also Hagiwara and Mohapatra [29].

Ordinarily the left-hand side of eq. (4.4) is expected to behave like $O(k^2)$ for large k^2. Then the expansion on the right-hand side cannot be valid; presumably it fails because the coefficients are divergent. However, if we construct a linear combination of current products which is softer than $1/k^2$ for large k^2, then the first few terms in the expansion in eq. (4.4) must vanish, and the corresponding linear combinations of spectral functions satisfy the relations

$$\int_0^\infty \mathrm{d}s\left[\rho^{(1)}/s+\rho^{(0)}\right]=0\,,$$
$$\int_0^\infty \mathrm{d}s\,\rho^{(1)}=0\,,$$
$$\int_0^\infty \mathrm{d}s\,s\,\rho^{(0)}=0\,. \tag{4.5}$$

These are the SFSR's.

To find linear combinations of current products with soft high-momentum behavior, we use Wilson's operator product expansion [22]. Bernard et al. [23] have shown that, if G_f chiral symmetry is a symmetry of the lagrangian, then it is respected by the Wilson coefficient functions, whether or not G_f is spontaneously broken. The G_f symmetry of the coefficient functions enables us to find spectral functions which obey SFSR's.

As an example, consider the familiar case in which $G_f = SU(N)_L \times SU(N)_R$ and $S_f = SU(N)_V$, the case realized by QCD. A SFSR can be derived by studying the high-momentum limit of

$$M_{\mu\nu}^{AB}(k)=\int \mathrm{d}^4x\,\mathrm{e}^{ikx}\langle 0|\mathrm{T}\left[J_{L\mu}^A(x)J_{R\nu}^B(0)\right]|0\rangle\,, \tag{4.6}$$

which transforms as (Ad, Ad) under $SU(N)_L \times SU(N)_R$, where Ad is the adjoint representation of $SU(N)$. The asymptotic behavior of $M_{\mu\nu}^{AB}(k)$ is determined by the lowest-dimension operator in the Wilson expansion of $J_{L\mu}^A J_{R\nu}^B$ which has a vacuum expectation value. This operator must be Lorentz invariant, gauge invariant, and S_f invariant, and, because the Wilson coefficient functions respect the G_f symmetry, it must transform as (Ad, Ad) under G_f. The lowest-dimension operator meeting these criteria is a four-fermion operator of dimension (mass)^6. Therefore, $M^{AB} \sim (k^2)^{-2}(\log k^2)^P$, and the SFSR's (4.5) hold for the spectral function ρ_{LR}.

It is customary to express these SFSR's in terms of the vector and axial vector spectral functions ρ_V and ρ_A. Invariance under parity and $SU(N)$ isospin implies

$$\langle 0|V_\mu^A(x)V_\nu^B(0)|0\rangle \propto \delta^{AB}\,,$$
$$\langle 0|A_\mu^A(x)A_\nu^B(0)|0\rangle \propto \delta^{AB}\,,$$
$$\langle 0|V_\mu^A(x)A_\nu^B(0)|0\rangle = 0\,, \tag{4.7}$$

so that

$$M_{\mu\nu}^{AB}(k)=\tfrac{1}{4}\delta^{AB}\int \mathrm{d}^4x\,\mathrm{e}^{ikx}\langle 0|\mathrm{T}\big[V_\mu^A(x)V_\mu^A(0)-A_\mu^A(x)A_\nu^A(0)\big]|0\rangle\,. \tag{4.8}$$

The axial vector currents couple to the Goldstone bosons,

$$\langle 0|A_\mu^B|\pi^A\rangle = ik_\mu F\delta^{AB}\,, \tag{4.9}$$

and therefore,

$$\rho_{\mathrm{A}}^{(0)}(\mu^2)=F^2\delta(\mu^2)\,. \tag{4.10}$$

Now, combining (4.5), (4.8) and (4.10), we find

$$\int\frac{\mathrm{d}s}{s}\left[\rho_{\mathrm{V}}(s)-\rho_{\mathrm{A}}(s)\right]=F^2\,,$$

$$\int \mathrm{d}s\left[\rho_{\mathrm{V}}(s)-\rho_{\mathrm{A}}(s)\right]=0\,. \tag{4.11}$$

These are Weinberg's SFSR's [15], which are exact relations in the chiral limit.

Next, consider the case of N flavors of (left-handed) fermions. $\mathrm{G_f}$ is $\mathrm{SU}(N)$ and $\mathrm{S_f}$ is a subgroup of $\mathrm{G_f}$. The $\mathrm{G_f}$ currents transform as the adjoint representation $\mathrm{Ad(G_f)}$, and the product of two $\mathrm{G_f}$ currents transforms as $\mathrm{Ad(G_f)}\times\mathrm{Ad(G_f)}=1+\mathrm{Ad(G_f)}+\mathrm{Ad(G_f)}+\cdots$. Here $\cdots$ represents other non-trivial representations, all of which, of course, have N-ality zero. Unless $N=2$, the only operators which are gauge invariant, Lorentz invariant, have N-ality zero and dimensionality less than six are $\mathrm{G_f}$-singlet operators which contain no fermion fields. ($\bar\psi\gamma^\mu D_\mu\psi$ can be eliminated by the equations of motion.) Therefore, there is an SFSR associated with each $\mathrm{S_f}$ singlet linear combination of current products which does not contain the $\mathrm{G_f}$ singlet. The number of independent SFSR's is one less than the number of $\mathrm{S_f}$ singlets in $\mathrm{Ad(G_f)}\times\mathrm{Ad(G_f)}$. (Although the above argument breaks down if $N=2$, the SFSR's hold in that case also.)

For each multiplet of broken currents in the representation D of $\mathrm{S_f}$, the corresponding combination $(\rho_{\mathrm{Ad}}-\rho_{\mathrm{D}})$ obeys SFSR's, where ρ_{Ad} is the spectral function associated with an irreducible representation of unbroken currents. We have the relations

$$\int\frac{\mathrm{d}s}{s}\left[\rho_{\mathrm{Ad}}(s)-\rho_{\mathrm{D}}(s)\right]=F_{\mathrm{D}}^2>0\,,$$

$$\int \mathrm{d}s\left[\rho_{\mathrm{Ad}}(s)-\rho_{\mathrm{D}}(s)\right]=0\,. \tag{4.12}$$

If S_f is trivial, SFSR's are satisfied by $\rho_{D_1}-\rho_{D_2}$, with $F^2_{D_2}-F^2_{D_1}$ on the right-hand side of the first sum rule.

If the equivalent but distinct representations D and D′ of S_f occur in $Ad(G_f)$, then the corresponding spectral function $\rho_{DD'}$ obeys the SFSR's

$$\int \frac{ds}{s}\rho_{DD'}(s)=F^2_{DD'},$$

$$\int ds\,\rho_{DD'}(s)=0. \tag{4.13}$$

In general, the D and D′ Goldstone boson multiplets can mix, and $F^2_{DD'}$ need not vanish. However, D and D′ can always be chosen such that there is no mixing, and $F^2_{DD'}=0$.

If S_f is not simple, then we also have

$$\int \frac{ds}{s}(\rho_{Ad}-\rho_{Ad'})=0,$$

$$\int ds(\rho_{Ad}-\rho_{Ad'})=0, \tag{4.14}$$

where ρ_{Ad} and $\rho_{Ad'}$ are spectral functions associated with unbroken currents in different factors of S_f.

It is now clear that SFSR's can be derived for arbitrary G_f and $S_f\subset G_f$. The total number of independent SFSR's is the number of S_f singlets less the number of G_f singlets contained in $Ad(G_f)\times Ad(G_f)$.

The SFSR's (4.12)–(4.14) hold in the limit of exact G_f symmetry. In the theories we are interested in, G_f is explicitly broken by the G_W interaction (and by the sideways interaction, too, in a realistic model). Nevertheless, we are justified in using (4.12)–(4.14) when calculating to lowest order in the G_f-breaking perturbation, as when we calculate the lowest-order G_W contribution to the effective potential.

5. Spectral functions and the effective potential

In sect. 3, the lowest-order G_W-boson exchange contribution to the effective potential was expressed in terms of an S_f-invariant tensor I^{AB}. The tensor I^{AB} depends on several strong-interaction parameters. These parameters are spectral integrals.

If we invoke the spectral representation (4.3), I^{AB} becomes

$$I^{AB} = -\tfrac{1}{2}i\int d^4x\Delta^{\mu\nu}(x)\langle 0|T[J_\mu^A(x)J_\nu^B(0)]|0\rangle$$

$$= \tfrac{1}{2}i\int d\mu^2\int\frac{d^4k}{(2\pi)^4}\frac{3\rho_{AB}^{(1)}(\mu^2)}{k^2(k^2-\mu^2)}$$

$$= -\frac{3}{32\pi^2}\int d\mu^2 \ln\left(\frac{\Lambda^2}{\mu^2}\right)\rho_{AB}^{(1)}(\mu^2), \tag{5.1}$$

where an ultraviolet cutoff Λ has been introduced to regulate the momentum integral. A subtraction must be performed to define the vacuum expectation value of the time-ordered product of two currents. This subtraction can be construed as a redefinition of the zero of the effective potential $V(g)$ given by (3.6). After we subtract the (infinite) G_f-invariant part of I^{AB}, we can express I^{AB} in terms of spectral integrals of the form

$$\int d\mu^2 \ln\left(\frac{\Lambda^2}{\mu^2}\right)\rho(\mu^2), \tag{5.2}$$

where the spectral function ρ, according to the reasoning of sect. 4, obeys either the SFSR's

$$\int\frac{ds}{s}\rho(s) = F^2, \qquad \int ds\,\rho(s) = 0, \tag{5.3}$$

or the SFSR's

$$\int\frac{ds}{s}\rho(s) = 0, \qquad \int ds\,\rho(s) = 0. \tag{5.4}$$

The second SFSR ensures that the coefficient of the logarithmic ultraviolet divergence in (5.2) vanishes.

In general, disregarding an irrelevant G_f-singlet part, we can express $V(g)$ in terms of F's, group-theoretic factors, and quantities of the form

$$\Delta^2 = \frac{1}{F^2}\int ds \ln\left(\frac{s_0}{s}\right)\rho(s), \tag{5.5}$$

where ρ satisfies either (5.3) or (5.4). Δ^2 is finite and independent of s_0.

In special cases to be considered later in this section, it is essential to determine the sign of Δ^2 in order to distinguish minima of $V(g)$ from maxima. Unfortunately,

(5.3) does not unambiguously fix the sign of Δ^2. Both broken and unbroken currents couple to many vector resonances, so ρ oscillates, taking both positive and negative values. However, we can determine the sign if it is a reasonable approximation to saturate the integral (5.5) with two low-lying narrow resonances. This approximation is known to be quite good in QCD, where it leads to successful predictions for the A_1 mass [15] and the electromagnetic mass difference of the pion [24]. We assume that spectral integrals are always rapidly convergent, so that saturating with two resonances makes sense.

If we take

$$\rho(s) = g_1^2\delta(s - M_1^2) - g_2^2\delta(s - M_2^2), \tag{5.6}$$

then eq. (5.3) implies that $g_1^2 = g_2^2$ and that

$$\frac{M_1^2}{M_2^2} = 1 - \frac{F^2 M_1^2}{g_1^2}; \tag{5.7}$$

therefore, eq. (5.5) becomes

$$\Delta^2 = \frac{g_1^2}{F^2}\ln\left(\frac{M_2^2}{M_1^2}\right) > 0. \tag{5.8}$$

Eq. (5.7) suggests that $F^2M_1^2/g_1^2 \sim \frac{1}{2}$, and this relation can actually be "derived" by comparing soft Goldstone boson theorems with the results of a vector dominance approximation [25]. Therefore, when we need a numerical value for Δ^2, we will use

$$\Delta^2 = 2(\ln 2)M_1^2; \tag{5.9}$$

M_1 is the mass of the lightest vector resonance which couples to the currents associated with ρ.

If ρ satisfies (5.4), then in the two-resonance approximation (5.6), we find $\rho = 0$ and $\Delta^2 = 0$.

In the rest of this section, we will consider several special cases in which we can find explicit expressions for $V(g)$, and can make simple observations about the properties of the minimum.

5.1. $G_f = SU(N)_L \times SU(N)_R$, $S_f = SU(N)_V$

We first consider the case in which $G_f = SU(N)_L \times SU(N)_R$ chiral symmetry breaks down to $S_f = SU(N)_V$ isospin. This case is especially simple in that there is only one spectral integral on which the effective potential depends. We will see that $V(g)$ is proportional to a group-theoretic expression which has a simple interpretation.

It is convenient to introduce notation in which all fermions transform as the same (complex) representation of G_H, so we have N left-handed fermions and N right-handed fermions. The G_f currents are

$$J^A_{L\mu} = \bar{\psi}_L \gamma_\mu \lambda^A \psi_L ,$$

$$J^A_{R\mu} = \bar{\psi}_R \gamma_\mu \lambda^A \psi_R , \tag{5.10}$$

where the λ^A's are hermitian traceless $N \times N$ matrices with flavor indices, normalized so that

$$\operatorname{Tr} \lambda^A \lambda^B = \tfrac{1}{2} . \tag{5.11}$$

Both the left-handed and right-handed currents transform as the adjoint representation of S_f. The G_W currents are

$$J^\alpha_{L\mu} = \bar{\psi}_L \gamma_\mu \theta^\alpha_L \psi_L ,$$

$$J^\alpha_{R\mu} = \bar{\psi}_R \gamma_\mu \theta^\alpha_R \psi_R ; \tag{5.12}$$

under G_f transformations they become

$$J'^\alpha_{L\mu} = \bar{\psi}_L \gamma_\mu U^\dagger_L \theta^\alpha_L U_L \psi_L ,$$

$$J'^\alpha_{R\mu} = \bar{\psi}_R \gamma_\mu U^\dagger_R \theta^\alpha_R U_R \psi_R , \tag{5.13}$$

where $U_{L,R}$ are $SU(N)$ matrices.

Now S_f invariance can be exploited to find an explicit form for the effective potential. The effective potential is given by

$$V(U) = -\tfrac{1}{2} i \int d^4x \Delta^{\mu\nu}(x) (e_\alpha)^2 \langle 0 | T\left[J'^\alpha_\mu(x) J'^\alpha_\nu(0) \right] | 0 \rangle , \tag{5.14}$$

with $J'^\alpha_\mu = J'^\alpha_{\mu L} + J'^\alpha_{\mu R}$. To calculate it, we extract the S_f invariants from the time-ordered product of G_W currents. It is clear that S_f invariance implies that

$$\langle 0 | J'^\alpha_L J'^\alpha_L | 0 \rangle \propto \operatorname{Tr}\left(U^\dagger_L \theta^\alpha_L U_L U^\dagger_L \theta^\alpha_L U_L \right) = \operatorname{Tr} \theta^\alpha_L \theta^\alpha_L ,$$

$$\langle 0 | J'^\alpha_R J'^\alpha_R | 0 \rangle \propto \operatorname{Tr}\left(U^\dagger_R \theta^\alpha_R U_R U^\dagger_R \theta^\alpha_R U_R \right) = \operatorname{Tr} \theta^\alpha_R \theta^\alpha_R ,$$

$$\langle 0 | J'^\alpha_L J'^\alpha_R | 0 \rangle \propto \operatorname{Tr}\left(U^\dagger_L \theta^\alpha_L U_L U^\dagger_R \theta^\alpha_R U_R \right) = \operatorname{Tr}\left(\theta^\alpha_L U \theta^\alpha_R U^\dagger \right) , \tag{5.15}$$

where $U = U_L U^\dagger_R$ is an $SU(N)$ matrix. The product of two left-handed or two right-handed currents contributes only an (infinite) G_f-invariant constant to $V(U)$. Subtracting this constant, we are left with

$$V(U) = 4(e_\alpha)^2 \operatorname{Tr}(\theta_L^\alpha U \theta_R^\alpha U^\dagger)(-\tfrac{1}{2}i)\int d^4x \Delta^{\mu\nu}(x)\langle 0|T[J_{\mu L}^B(x) J_{\nu R}^B(0)]|0\rangle$$

(not summed over B)

$$= -\frac{3}{32\pi^2}(e_\alpha)^2 \operatorname{Tr}(\theta_L^\alpha U \theta_R^\alpha U^\dagger)\int ds \ln\left(\frac{s_0}{s}\right)[\rho_V(s) - \rho_A(s)]. \tag{5.16}$$

To obtain the second equality we have used eq. (5.1). The current $A_\mu^B = J_{L\mu}^B - J_{R\mu}^B$ couples to the Goldstone bosons with strength F, and the combination $\rho_V - \rho_A$ satisfies eq. (5.3). In terms of the positive quantity Δ^2 defined by eq. (5.5), we have

$$V(U) = -\frac{3}{32\pi^2}(F^2\Delta^2)(e_\alpha)^2 \operatorname{Tr}\theta_L^\alpha U \theta_R^\alpha U^\dagger. \tag{5.17}$$

One can interpret eq. (5.17) by saying that the alignment of the subgroups S_f and G_W is determined by the condition that the breakdown of G_f to S_f occurs in the most attractive channel of the G_W interactions. To see this, it is best to return to the notation in which all fermions are left-handed. The embedding of S_f in G_f can be characterized by a fermion condensate defined by

$$\sigma\Sigma_{rs} = \langle 0|\varepsilon_{\alpha\beta}\lambda^{ij}\psi_{ri}^\alpha\psi_{sj}^\beta|0\rangle, \tag{5.18}$$

where λ^{ij} is a G_H-invariant tensor, and the Lorentz indices are contracted by $\varepsilon_{\alpha\beta}$ to construct a Lorentz singlet; σ is an overall scale factor. S_f is the subgroup of G_f which leaves Σ invariant. If $G_f = SU(N)\times SU(N)$ and $S_f = SU(N)$, Σ is a $2N\times 2N$ matrix of the form

$$\Sigma = \begin{pmatrix} 0 & U \\ U^T & 0 \end{pmatrix}, \tag{5.19}$$

and the G_W generators can be written

$$\theta^\alpha = \begin{pmatrix} \theta_L^\alpha & 0 \\ 0 & -\theta_R^{\alpha *} \end{pmatrix}. \tag{5.20}$$

Expressed in terms of Σ and θ^α, eq. (5.17) becomes

$$V(\Sigma) = \frac{3}{64\pi^2}F^2\Delta^2(e_\alpha)^2\Sigma_{rs}^*\theta_{rr'}^\alpha\theta_{ss'}^\alpha\Sigma_{r's'}. \tag{5.21}$$

Because $F^2\Delta^2$ is positive, the alignment of the subgroups G_W and S_f is determined by the requirement that the value of the G_W Casimir operator $(e_\alpha)^2\theta^\alpha\theta^\alpha$ acting on

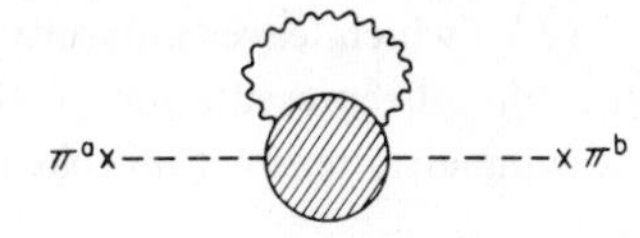

Fig. 4. Lowest-order G_W contribution to PGB masses.

the condensate Σ is as small as possible. In particular, if it is possible for the condensate to be oriented so that it is G_W invariant, this orientation is always the minimum; the G_W gauge group does not break unless it must.

By applying eq. (2.8) we find that the PGB mass matrix is (fig. 4)

$$m_{AB}^2 = -\frac{3}{8\pi^2}\Delta^2(e_\alpha)^2 \operatorname{Tr}\left[U^\dagger\theta_{\rm L}^\alpha U, \lambda^A\right]\left[\theta_{\rm R}^\alpha, \lambda^B\right]. \qquad (5.22)$$

Of course, the eigenvalues of this matrix are non-negative when U is chosen to minimize $V(U)$.

5.2. MAXIMAL ISOSPIN

This case is a generalization of that of subsect. 5.1. We say that S_f is maximal if S_f is simple, all broken G_f currents are in a single irreducible representation of S_f, and the commutator of two broken currents is an unbroken current. If S_f is maximal, there is only one pair of SFSR's, and only one spectral integral on which the effective potential can depend. $V(g)$ can be expressed as a purely group-theoretic factor times a constant of known sign, as in eq. (5.17).

Aside from the breaking of $SU(N)\times SU(N)$ down to $SU(N)$, there are two other notable examples of maximal S_f, both of which can occur only if the fermions are in a real representation of G_H [6, 26]. Suppose there are N flavors of fermions in a representation R of G_H such that the symmetric product of R with itself contains the singlet. Then the tensor λ^{ij} in eq. (5.18) is symmetric, and Fermi statistics requires Σ_{rs} to be symmetric. (The fermions anticommute.) The maximal subgroup of $G_f = SU(N)$ which leaves a symmetric tensor invariant is $S_f = O(N)$. Under $O(N)$, the adjoint representation of $SU(N)$ splits up into two irreducible representations, an antisymmetric tensor (unbroken) and a traceless symmetric tensor (broken),

$$N^2 - 1 \to \tfrac{1}{2}N(N-1) + \left[\tfrac{1}{2}N(N+1) - 1\right],$$

and the commutator of two broken currents is an unbroken current. [The case $N=4$ is exceptional; the broken currents are in the reducible representation $(1,3)+(3,1)$ of $O(4) = SU(2)\times SU(2)$.]

If the fermions are in a representation R of G_H such that the antisymmetric product of R with itself contains the singlet, then Σ_{rs} is antisymmetric. The maximal subgroup of $G_f = SU(2N)$ which leaves an antisymmetric tensor invariant is $Sp(2N)$. Under $Sp(2N)$, the adjoint representation of $SU(2N)$ splits up into two irreducible representations, a symmetric tensor (unbroken) and a "traceless" antisymmetric tensor (broken),

$$4N^2 - 1 \to N(2N+1) + [N(2N-1) - 1],$$

and the commutator of two broken currents is an unbroken current.

Assuming that S_f is maximal, we can apply eqs. (2.6)–(2.8) and (3.1) to derive the conditions satisfied at a local minimum of the effective potential. We first define an orthonormal basis for the G_f generators such that

$$\mathrm{Tr}\, X^A X^B = \mathrm{Tr}\, T^A T^B = \delta^{AB}, \qquad \mathrm{Tr}\, X^A T^B = 0, \tag{5.23}$$

where T^A's are S_f generators and the X^A's are broken generators. The broken currents $J^A_{X\mu} = \bar{\psi}\gamma_\mu X^A \psi$ couple to Goldstone bosons with strength F; that is

$$\langle 0| J^A_{X\mu} |\pi^B\rangle = ik_\mu F \delta^{AB}. \tag{5.24}$$

Given any embedding of G_W in G_f, we can decompose the G_W generators θ^α into broken and unbroken generators, so that

$$J^\alpha_\mu = \bar{\psi}\gamma_\mu \theta^\alpha \psi = \bar{\psi}\gamma_\mu T^\alpha \psi + \bar{\psi}\gamma_\mu X^\alpha \psi, \tag{5.25}$$

where

$$T^\alpha = \sum_A T^A\, \mathrm{Tr}(T^A \theta^\alpha),$$

$$X^\alpha = \sum_A X^A\, \mathrm{Tr}(X^A \theta^\alpha). \tag{5.26}$$

T^α and X^α vary in a complicated way under G_f transformations.

The effective potential is stationary when

$$\begin{aligned} 0 &= \langle 0|[\mathcal{H}', Q^A]|0\rangle \\ &= -\tfrac{1}{2} i \int d^4x \Delta^{\mu\nu}(x)(e_\alpha)^2 \langle 0|[T(J^\alpha_\mu(x) J^\alpha_\nu(0)), Q^A]|0\rangle, \end{aligned} \tag{5.27}$$

where

$$Q^A = \int d^3x J^A_{X_0}. \tag{5.28}$$

In an obvious shorthand, the commutator in eq. (5.27) is

$$\left[(X^\alpha+T^\alpha)(X^\alpha+T^\alpha),X^A\right]=\left\{X^\alpha,\left[X^\alpha,X^A\right]\right\}+\left\{X^\alpha,\left[T^\alpha,X^A\right]\right\}$$
$$=+\left\{T^\alpha,\left[X^\alpha,X^A\right]\right\}+\left\{T^\alpha,\left[T^\alpha,X^A\right]\right\}. \quad (5.29)$$

Because the commutator of two broken generators is unbroken, and, of course, the commutator of a broken generator and an unbroken generator is broken, the first and last terms in eq. (5.29) contain no S_f invariant, and have no vacuum expectation value. Noting that

$$\operatorname{Tr} X^\alpha\left[T^\alpha,X^A\right]=-\operatorname{Tr} T^\alpha\left[X^\alpha,X^A\right], \quad (5.30)$$

and invoking (5.1), eq. (5.27) becomes

$$0=-\frac{3(e_\alpha)^2}{16\pi^2}\operatorname{Tr}\left(T^\alpha\left[X^\alpha,X^A\right]\right)\int \mathrm{d}s\ln\left(\frac{s_0}{s}\right)(\rho_T-\rho_X), \quad (5.31)$$

where ρ_T and ρ_X are the spectral functions associated with the unbroken and broken currents. Hence, the potential is stationary when a purely group-theoretic relation,

$$(e_\alpha)^2\operatorname{Tr} T^\alpha\left[X^\alpha,X^A\right]=0, \quad (5.32)$$

is satisfied.

Similarly, we can compute the double commutator of eq. (2.8), retain only S_f invariants, and obtain an expression for the PGB mass matrix. We find

$$m^2_{AB}=\frac{3}{16\pi^2}\Delta^2(e_\alpha)^2\left(-\operatorname{Tr}\left[T^\alpha,X^A\right]\left[T^\alpha,X^B\right]+\operatorname{Tr}\left[X^\alpha,X^A\right]\left[X^\alpha,X^B\right]\right), \quad (5.33)$$

where

$$\Delta^2=\frac{1}{F^2}\int \mathrm{d}s\ln\left(\frac{s_0}{s}\right)(\rho_T-\rho_X). \quad (5.34)$$

In eq. (5.33), m^2 has been expressed as the difference of two positive matrices. The unbroken G_W generators give a positive contribution to m^2 and the broken G_W generators give a negative contribution. At a local minimum of the potential, m^2 must itself be positive. Roughly speaking, the minimum occurs when as large a subgroup of G_W as possible is unbroken.

Only eqs. (5.32) and (5.33) will be needed to minimize the potential in the examples considered in sect. 6, but the precise sense in which G_W resists being broken can be clarified if we write down an explicit expression for the potential. If S_f is maximal, then the effective potential is given by

$$V = -\tfrac{1}{2}i\int d^4x\Delta^{\mu\nu}(x)(e_\alpha)^2\langle 0|T[J^\alpha_\mu(x)J^\alpha_\nu(0)]|0\rangle$$

$$= -\frac{3}{32\pi^2}(e_\alpha)^2[C_T \operatorname{Tr}(T^\alpha T^\alpha) + C_X \operatorname{Tr}(X^\alpha X^\alpha)], \qquad (5.35)$$

where

$$C_{T,X} = \int ds \ln(\Lambda^2/s)\rho_{T,X}(s). \qquad (5.36)$$

To derive eq. (5.35), we have extracted the S_f invariants from the product of two G_W currents, and have invoked eq. (5.1). T^α and X^α depend implicitly on $g \in G_f$. Because

$$\operatorname{Tr}(\theta^\alpha\theta^\alpha) = \operatorname{Tr}(T^\alpha T^\alpha) + \operatorname{Tr}(X^\alpha X^\alpha) \qquad (5.37)$$

is G_f invariant, the potential V can be written

$$V = \frac{3}{32\pi^2}(F^2\Delta^2)(e_\alpha)^2 \operatorname{Tr} X^\alpha X^\alpha + \text{constant}, \qquad (5.38)$$

where $F^2\Delta^2 = C_T - C_X > 0$.

This expression for V is non-negative. If it is possible to orient G_W so that $X^\alpha = 0$, that orientation minimizes V. G_W will not break unless it must.

Recalling eq. (2.12), we see that the G_W vector boson mass matrix is

$$\mu^2_{\alpha\beta} = e_\alpha e_\beta F^2 \operatorname{Tr}(X^\alpha X^\beta) \qquad (5.39)$$

(not summed). Therefore, the potential is

$$V = \frac{3}{32\pi^2}\Delta^2(\operatorname{Tr}\mu^2). \qquad (5.40)$$

The alignment of G_W and S_f is chosen to minimize the trace of the weak vector boson mass matrix. This condition defines the precise sense in which G_W resists being broken, if S_f is maximal.

Incidentally, one should not conclude from eq. (5.33) that it is impossible for G_W to be completely broken when S_f is maximal. Eq. (5.33) only implies that $T^\alpha \neq 0$ at the minimum; if all G_W currents couple to Goldstone bosons, the currents must also have unbroken components which give a positive contribution to m^2.

5.3. GENERALIZATIONS

In the general case, the potential depends on many independent spectral integrals, and more detailed dynamical information is needed to determine the minimum. However, there are several generalizations of the case of maximal isospin for which we can still make some useful observations.

One generalization is trivial. If $G_f = \Pi_i G_i$ breaks down to $S_f = \Pi_i S_i$, where each S_i is a maximal isospin subgroup of G_i, then all the S_i's align with G_W independently, and the PGB masses are still given by (5.33), but with a different Δ_i^2 for each G_i.

We also observe that, if S_f is simple and it is possible for S_f and G_W to align so that G_W is unbroken, this alignment minimizes the potential. By a suitable orthogonal transformation, we can always define the G_f currents so that the only S_f invariants contributing to the potential are traces as in (5.35)*. We may then decompose the broken G_W generators into generators in the different irreducible representations of S_f **,

$$X^{\alpha} = \sum_{D} X_D^{\alpha}, \tag{5.41}$$

and, if S_f is simple, we find that the potential is given by

$$V = \frac{3}{32\pi^2}(e_\alpha)^2 \left[\sum_{D} (F_D^2 \Delta_D^2) \operatorname{Tr}(X_D^{\alpha} X_D^{\alpha}) \right], \tag{5.42}$$

where

$$F_D^2 \Delta_D^2 = \int ds \ln\left(\frac{s_0}{s}\right)(\rho_T - \rho_D). \tag{5.43}$$

V is non-negative, and is minimized by $X^{\alpha} = 0$.

If S_f is not simple, there are additional terms in V of the form

$$\frac{3}{32\pi^2}(e_\alpha)^2 (C_T - C_{T'}) \operatorname{Tr}(T'^{\alpha} T'^{\alpha}), \tag{5.44}$$

where T^{α} and T'^{α} are components of the unbroken G_W generators in two different factors of S_f, and $C_{T,T'}$ are defined as in (5.36). However, these terms are absent if we assume that spectral integrals can be well-approximated by saturating with two resonances. $C_T - C_{T'}$ is a spectral integral of the form (5.5) where $\rho = \rho_T - \rho_{T'}$ obeys the SFSR's (5.4). This spectral integral vanishes in the two-resonance approximation.

* We assume that this transformation does not mix unbroken currents and broken currents.

** A non-real irreducible representation of S_f plus its complex conjugate are regarded as a single representation in this decomposition.

6. Examples

We now apply the general results of sect. 5 to eight specific models, with various choices of the groups G_f, S_f, and G_W. All of these examples are of genuine interest, because the chosen flavor groups are likely to be subgroups of the flavor group in a realistic hypercolor theory. The weak gauge groups we consider are also realistic. In examples (a)–(f), $G_W = SU(2) \times U(1)$ is the standard electroweak group. In example (g), G_W is $SU(3) \times SU(2) \times U(1)$, where SU(3) is the color group. In example (h), $G_W = SU(3) \times SU(3)$ is a "chiral color" group.

We compute the dynamically determined pattern of G_W breakdown for each sample model. Some models [examples (a), (e), (f)] are found to be phenomenologically unacceptable, because electric charge is broken.

The examples also illustrate an interesting general phenomenon. If the weak gauge group G_W is not simple, the gauge interactions associated with the different factors of G_W may be attractive in different channels, corresponding to different relative orientations of G_W and S_f. Then these interactions compete, and which one wins depends on the relative strength of the gauge couplings. Hence, there are critical values of weak mixing angles at which the pattern of G_W breaking changes discontinuously. This phenomenon occurs in examples (a) and (f), for which critical mixing angles are calculated.

We compute the electroweak and color contributions to the PGB masses for some of the examples. One important feature of the PGB masses was previously emphasized by Eichten and Lane [7]. If a model has a spontaneously broken approximate $SU(N)_L \times SU(N)_R$ chiral symmetry, and G_W commutes with one of the chiral factors, then there will be PGB's which remain massless to lowest order in the G_W interaction. Such a model might contain charged [example (d)] or colored [example (h)] PGB's which are much lighter than we would naively expect.

None of our sample models includes a sideways interaction [6–8]. We should bear in mind that this interaction is also expected to contribute to PGB masses in a truly realistic model. Eichten and Lane have estimated the sideways contribution to be $\Delta m^2 \simeq (10\text{–}30\ \mathrm{GeV})^2$.

(a) Sp(4) ISOSPIN

Suppose that four flavors of fermions are in the real irreducible representation R of the hypercolor group G_H, where the antisymmetric product of R with itself contains the G_H singlet*. The flavor group is $G_f = SU(4)$. The fermion condensate Σ defined by eq. (5.18) is antisymmetric in flavor indices, and we assume that the maximal isospin group $S_f = Sp(4)$ is left unbroken when hypercolor becomes strong. The 15 G_f currents transform as the 10 (unbroken) + 5 (broken) representation of $S_f = Sp(4)$.

* For example, R could be the defining representation of $G_H = Sp(2n)$, or the $(2n+1)$-index antisymmetric tensor representation of $G_H = SU(4n+2)$.

Let the weak group be $G_W = SU(2) \times U(1)$, with the fermions in the G_W representation

$$\begin{array}{lccc} & \begin{pmatrix} N \\ E \end{pmatrix} & N^c & E^c . \\ \tfrac{1}{2}Y = & -\tfrac{1}{2} & 0 & +1 \end{array} \tag{6.1}$$

(All the fermions are left-handed.) These $U(1)_W$ charge assignments are those we might expect for color-singlet hyperleptons. We will not worry about the fact that this model has a G_W anomaly. The anomaly can be cancelled by additional fermions.

The effective potential is a function of the condensate Σ, which specifies the embedding of S_f in G_f. Given Σ, we can express the G_W generators as linear combinations of broken and unbroken G_f generators. It is easy to verify using (5.32) that, if $S_f = Sp(4)$, the effective potential for this model has only two extrema. These extrema are

$$\begin{array}{cc} & \begin{array}{cccc} N & E & N^c & E^c \end{array} \\ \Sigma_1 = & \begin{bmatrix} 0 & 0 & 1 & 0 \\ 0 & 0 & 0 & 1 \\ -1 & 0 & 0 & 0 \\ 0 & -1 & 0 & 0 \end{bmatrix} \begin{array}{c} N \\ E \\ N^c \\ E^c \end{array} , \end{array} \qquad \Sigma_2 = \begin{bmatrix} 0 & 1 & 0 & 0 \\ -1 & 0 & 0 & 0 \\ 0 & 0 & 0 & 1 \\ 0 & 0 & -1 & 0 \end{bmatrix} . \tag{6.2}$$

Σ_1 breaks G_W down to electric charge, but Σ_2 breaks G_W down to $SU(2)_W$. One of these is the minimum of the effective potential; the other is the maximum. But which is which?

To answer this question, we determine whether the PGB masses are positive or negative. If Sp(4) is defined by Σ_1, three of the five Goldstone bosons are eaten. Using (5.33), it is straightforward to calculate the mass of the uneaten PGB (and its *CPT* conjugate). We find

$$m_1^2 = -\frac{9}{64\pi^2}(g^2 - g'^2)\Delta^2 , \tag{6.3}$$

where

$$\Delta^2 = \frac{1}{F^2}\int_0^\infty \mathrm{d}s \ln\left(\frac{s_0}{s}\right)(\rho_{10} - \rho_5) , \tag{6.4}$$

g is the $SU(2)_W$ gauge coupling, and g' is the $U(1)_W$ gauge coupling. If the condensate is Σ_2, $SU(2)_W$ remains unbroken, and there are two uneaten PGB's.

These have equal masses, given by

$$m_2^2 = -m_1^2 . \qquad (6.5)$$

We have discovered something interesting. The $SU(2)_W$ interaction is most attractive in the channel in which $SU(2)_W$ is unbroken, but the $U(1)_W$ interaction prefers to break G_W to $U(1)_{EM}$. Thus, the two interactions compete, and which one wins depends on the relative strengths of the gauge couplings. At a critical value of the Weinberg mixing angle, the unbroken subgroup changes discontinuously from SU(2) to $U(1)_{EM}$; there is a first-order phase transition.

From eq. (6.3), we see that the critical Weinberg angle is

$$\cot^2 \theta_W = g^2/g'^2 = 1 . \qquad (6.6)$$

We argued in sect. 5 that Δ^2 is positive. Therefore, if $\cot^2 \theta_W < 1$, then $m_1^2 > 0$, and $U(1)_{EM}$ is unbroken. If $\cot^2 \theta_W > 1$, then $m_2^2 > 0$, and $SU(2)_W$ is unbroken.

We can repeat the PGB mass calculation for different $U(1)_W$ charge assignments. If the charges are

$$\begin{array}{ccc} \begin{pmatrix} N \\ E \end{pmatrix} & N^c & E^c \\ \tfrac{1}{2} Y = q & -(q+\tfrac{1}{2}) & -(q-\tfrac{1}{2}) , \end{array} \qquad (6.7)$$

then the critical Weinberg angle is

$$\cot^2 \theta_W = g^2/g'^2 = \tfrac{1}{3}(16q^2 - 1) . \qquad (6.8)$$

If $|q| < \frac{1}{4}$, the right-hand side of eq. (6.8) is negative, and no phase transition occurs at all; $SU(2)_W$ is always unbroken. The quark-like charge $q = \frac{1}{6}$ falls within this range.

There *is* a phase transition if $|q| > \frac{1}{4}$. In order to ensure that G_W breaks to $U(1)_{EM}$ when the Weinberg angle has its observed value $\cot^2 \theta_W = 3.5$, we must have

$$|q| > 0.85 . \qquad (6.9)$$

The moral is that the breaking of an SU(4) chiral symmetry to Sp(4), with $G_W \subset SU(4)$ embedded as in eq. (6.7), leads to the wrong kind of G_W breaking unless the fermions have unconventional charges. This restriction should be kept in mind when model building is attempted.

(b) O(4) ISOSPIN

Again, we consider a model with four flavors of a real representation R of G_H, with $G_W = SU(2) \times U(1) \subset SU(4)$, and G_W assignments as in eq. (6.1). But this time we suppose that the *symmetric* product of R with itself contains the singlet, and that the maximal isospin O(4) is unbroken. Then we can show that the minimum of the effective potential is

$$\Sigma = \begin{bmatrix} 0 & 0 & 1 & 0 \\ 0 & 0 & 0 & 1 \\ 1 & 0 & 0 & 0 \\ 0 & 1 & 0 & 0 \end{bmatrix} . \tag{6.10}$$

In this case, it is impossible for $SU(2)_W$ to escape breaking, and there is no phase transition.

Under O(4) = SU(2) × SU(2), the 15 of SU(4) transforms as (1, 3) + (3, 1) + (3, 3); there are nine Goldstone bosons. Because the currents are in three irreducible representations of O(4), this model has two independent pairs of SFSR's, and two spectral integrals on which the effective potential depends.

The unbroken currents are

$$\begin{aligned} J_1^+ &= \sqrt{\tfrac{1}{2}}\,(\bar{E}N - \bar{N}^c E^c), & J_2^+ &= \sqrt{\tfrac{1}{2}}\,(\bar{E}N^c - \bar{N}E^c), \\ J_1^- &= \sqrt{\tfrac{1}{2}}\,(\bar{N}E - \bar{E}^c N^c), & J_2^- &= \sqrt{\tfrac{1}{2}}\,(-\bar{N}^c E + \bar{E}^c N), \\ J_1^3 &= \tfrac{1}{2}(\bar{N}N - \bar{E}E - \bar{N}^c N^c + \bar{E}^c E^c), & J_2^3 &= \tfrac{1}{2}(\bar{N}N + \bar{E}E - \bar{N}^c N^c - \bar{E}^c E^c), \end{aligned} \tag{6.11}$$

and the broken currents are

$$\begin{aligned} K_0^+ &= \bar{N}N^c, & K_1^+ &= \sqrt{\tfrac{1}{2}}\,(\bar{N}E^c + \bar{E}N^c), & K_2^+ &= \bar{E}E^c, \\ K_0^- &= \bar{N}^c N, & K_1^- &= \sqrt{\tfrac{1}{2}}\,(\bar{E}^c N + \bar{N}^c E), & K_2^- &= \bar{E}^c E, \\ L^+ &= \sqrt{\tfrac{1}{2}}\,(\bar{E}N + \bar{N}^c E^c), & L^- &= \sqrt{\tfrac{1}{2}}\,(\bar{N}E + \bar{E}^c N^c), \\ L^0 &= \tfrac{1}{2}(\bar{N}N - \bar{E}E + \bar{N}^c N^c - \bar{E}^c E^c). \end{aligned} \tag{6.12}$$

Here, and throughout this section, we exhibit the flavor structure of the currents, but suppress their Lorentz structure. We have displayed the currents of definite charge, rather than the hermitian currents. The corresponding SU(4) generators are normalized so that $\mathrm{Tr}\,\lambda^A \lambda^{B^+} = \delta^{AB}$. Each broken current couples to a Goldstone boson with strength F.

The G_W currents can be expressed as linear combinations of broken currents and unbroken currents; we have

$$J_W^+ = \sqrt{\tfrac{1}{2}}\ \bar{E}N = \tfrac{1}{2}(L^+ + J_1^+),$$

$$J_W^- = \sqrt{\tfrac{1}{2}}\ \bar{N}E = \tfrac{1}{2}(L^- + J_1^-),$$

$$J_W^3 = \tfrac{1}{2}(\bar{N}N - \bar{E}E) = \tfrac{1}{2}(L^0 + J_1^3),$$

$$J_W^{Y/2} = -\tfrac{1}{2}(\bar{N}N + \bar{E}E) + \bar{E}^c E^c = -\tfrac{1}{2}L^0 + \tfrac{1}{2}J_1^3 - J_2^3. \tag{6.13}$$

It is evident that the electromagnetic current $J_W^3 + J_W^{Y/2}$ is unbroken. Applying eq. (2.12), we see that the weak boson mass matrix is

$$\mu^2 = \begin{bmatrix} g^2 & 0 & 0 & 0 \\ 0 & g^2 & 0 & 0 \\ 0 & 0 & g^2 & gg' \\ 0 & 0 & gg' & g'^2 \end{bmatrix} \tfrac{1}{4}F^2, \tag{6.14}$$

which has the eigenvalues

$$\mu_W^2 = \tfrac{1}{4}g^2F^2, \qquad \mu_Z^2 = \tfrac{1}{4}(g^2 + g'^2)F^2, \qquad \mu_\gamma^2 = 0.$$

The relation $\mu_W^2/\mu_Z^2 = \cos^2\theta_W$ is enforced by the O(4) isospin.

The Goldstone bosons which are not eaten couple to the currents $K_{0,1,2}^{\pm}$. The three PGB masses can be expressed in terms of the quantities

$$\Delta_1^2 = \frac{1}{F^2}\int ds \ln\left(\frac{s_0}{s}\right)\left[\rho_{3(1)} - \rho_9\right],$$

$$\Delta_2^2 = \frac{1}{F^2}\int ds \ln\left(\frac{s_0}{s}\right)\left[\rho_{3(2)} - \rho_9\right], \tag{6.15}$$

where $\rho_{3(1)}$ and $\rho_{3(2)}$ are the spectral functions associated with J_1 and J_2. From (2.8) we find that the masses of the PGB's coupling to the currents $K_{0,1,2}$ are

$$m_0^2 = \frac{3}{64\pi^2}\left[g^2\left(\tfrac{3}{2}\Delta_1^2 - \tfrac{1}{2}\Delta_2^2\right) + g'^2\left(2\Delta_2^2 - \Delta_1^2\right)\right],$$

$$m_1^2 = \frac{3}{64\pi^2}\left[g^2\left(2\Delta_1^2 - \Delta_2^2\right) + 3g'^2\Delta_2^2\right],$$

$$m_2^2 = \frac{3}{64\pi^2}\left[g^2\left(2\Delta_1^2 - \Delta_2^2\right) + g'^2\left(6\Delta_2^2 + 3\Delta_1^2\right)\right]. \tag{6.16}$$

To obtain numerical values for the masses, we use the estimate in eq. (5.9),

$$\Delta_1^2=\Delta_2^2=(2\ln 2)M^2\,, \tag{6.17}$$

where M is the mass of the lightest vector resonance which couples to the unbroken O(4) currents. Roughly, we expect $M\simeq(F/f_\pi)m_\rho$, and $\mu_W^2=\frac{1}{4}g^2F^2$ implies $F\simeq 250$ GeV; therefore, we have $M\simeq 2.0$ TeV and

$$m_0^2=\frac{3(\ln 2)\alpha}{2\pi\sin^2 2\theta_W}M^2\simeq(120\text{ GeV})^2\,,$$

$$m_1^2=m_0^2(1+2\sin^2\theta_W)\simeq(140\text{ GeV})^2\,,$$

$$m_2^2=m_0^2(1+8\sin^2\theta_W)\simeq(200\text{ GeV})^2\,, \tag{6.18}$$

taking $\sin^2\theta_W\simeq 0.23$. In a realistic model, more G_H representations will contribute to μ_W, μ_Z, and F will be smaller than 250 GeV. The PGB masses scale with F.

(c) O(3) ISOSPIN

Now we suppose that there are three real representations of G_H, with G_W representation content given by

$$\begin{array}{lcc} & \begin{pmatrix}N\\E\end{pmatrix} & E^c \\[2ex] \frac{1}{2}Y= & -\frac{1}{2} & 1\,. \end{array} \tag{6.19}$$

This structure might arise if the N^c, which is not protected by G_W quantum numbers, gets a large Majorana mass before the G_H interaction gets strong [27].

Suppose that the condensate is symmetric in flavor indices, and breaks $G_f=SU(3)$ down to $S_f=O(3)$. Then we can verify that the condensate which minimizes the potential is

$$\begin{array}{c} \quad\; N\quad E\quad E^c \\ \Sigma=\begin{bmatrix}1&0&0\\0&0&1\\0&1&0\end{bmatrix}\begin{array}{l}N\\E\\E^c\end{array}, \end{array} \tag{6.20}$$

which breaks G_W to $U(1)_{EM}$.

The Goldstone bosons transform as a 5 under O(3) isospin. Therefore, isospin does not enforce the natural μ_W/μ_Z ratio. Instead we find

$$u_W^2/\mu_Z^2=\tfrac{2}{3}\cos^2\theta_W\,.$$

If this type of flavor symmetry breaking is to be allowed in a realistic model, there must be many other Goldstone boson representations contributing to the W and Z masses which swamp the contribution of the O(3) 5-plet.

(d) SU(2N) ISOSPIN

Next we consider a model with fermions in complex representations of G_H. Suppose we have N left-handed G_W doublets and $2N$ right-handed singlets, with $U(1)_W$ charge assignments given by

$$\begin{array}{lcccc} & \begin{pmatrix} U_i \\ D_i \end{pmatrix}_L & U_{iR} & D_{iR} & \\ \tfrac{1}{2}Y = & q & q+\tfrac{1}{2} & q-\tfrac{1}{2}\,. \end{array} \qquad (6.21)$$

The approximate chiral symmetry group is $G_f = SU(2N)_L \times SU(2N)_R$, which we assume breaks down to $SU(2N)_V$.

If $N = 1$, all three Goldstone bosons are eaten by the $W^{\pm}$ and Z, so there is no subgroup alignment problem to solve; all G_f rotations of the condensate are G_W gauge transformations. If $N > 1$, the global symmetry group respected by the G_W interactions is $G_f' = SU(N) \times SU(N) \times SU(N)$ which breaks down to the exact isospin $S_f' = SU(N)$. Therefore, there are $2(N^2 - 1)$ (electrically neutral) exactly massless Goldstone bosons. In a realistic model, the sideways interaction is needed to provide masses for these Goldstone bosons [7].

The breakdown of the approximate $SU(2N) \times SU(2N)$ chiral symmetry produces $4N^2 - 1$ Goldstone bosons, of which 3 are eaten. Hence there are $2(N^2 - 1)$ remaining charged PGB's which should receive mass from the G_W interactions. However, it is easy to see that the lowest-order G_W interactions give no contribution to these PGB masses [7]. As we observed in sect. 5, the lowest-order G_W contribution to the effective potential has the form

$$V(g) = -i(e_\alpha)^2 \int d^4x \Delta^{\mu\nu}(x) \langle 0| T[J'^{\alpha}_{L\mu}(x) J'^{\alpha}_{R\nu}(0)] |0\rangle$$
$$+ \text{constant}\,. \qquad (6.22)$$

Because the $SU(2)_W$ interactions are purely left-handed, they do not contribute to V. (We can say this another way. The Goldstone bosons couple to axial currents, but we can use right-handed currents as their interpolating currents. The $SU(2)_W$ currents commute with these interpolating currents, so the $SU(2)_W$ interaction cannot give mass to the Goldstone bosons.) The $U(1)_W$ contribution to V also vanishes, because the left-handed $U(1)_W$ current is an $SU(2N)_L$ singlet; this contribution is $SU(2N)_L$ invariant as well as $SU(2N)_V$ invariant, and is therefore invariant under all G_f chiral transformations.

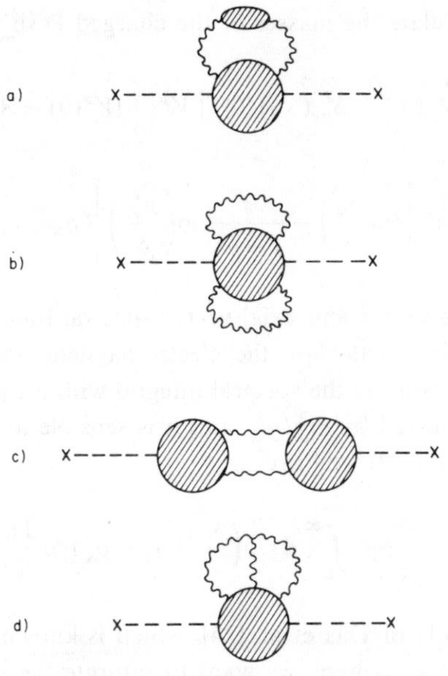

Fig. 5. Higher-order G_W contributions to PGB mass.

Because the masslessness of the charged PGB's is not enforced by any exact symmetry, it will be removed by higher-order G_W corrections. The higher-order corrections can be estimated by taking into account the mass splitting of the γ and Z which is generated in the next order (fig. 5a). Then the minimum of V is found to occur when all fermion generations line up to leave $U(1)_{EM}$ unbroken. To calculate the PGB masses we can use the currents

$$J^{+}_{Rij} = \sqrt{2}\, \bar{D}_{Ri} U_{Rj},$$

$$J^{-}_{Rij} = \sqrt{2}\, \bar{U}_{Rj} D_{Ri}, \tag{6.23}$$

which couple to the Goldstone bosons with strength F. The currents coupling to γ and Z can be written

$$J^{\gamma}_{R} = g' J^{3}_{R} \cos\theta + \text{singlet}, \qquad J^{Z}_{R} = -g' J^{3}_{R} \sin\theta + \text{singlet},$$

$$J^{\gamma}_{L} = g J^{3}_{L} \sin\theta + \text{singlet}, \qquad J^{Z}_{L} = g J^{3}_{L} \cos\theta + \text{singlet}, \tag{6.24}$$

where $J^3 = \Sigma_i \frac{1}{2}(\bar{U}_i U_i - \bar{D}_i D_i)$ and "singlet" represents G_f-invariant terms. Now we can use eq. (2.8) to calculate the masses of the charged PGB's. We find

$$m^2=\frac{e^2}{F_2}i\int \mathrm{d}^4x\left(\Delta^{\gamma}_{\mu\nu}(x)-\Delta^{Z}_{\mu\nu}(x)\right)\langle 0|\mathrm{T}[V^{\mu}(x)V^{\nu}(0)-A^{\mu}(x)A^{\nu}(0)]|0\rangle$$

$$=\frac{3e^2}{16\pi^2F^2}\int_0^{\infty}\mathrm{d}s\left[\ln\left(\frac{s_0}{s}\right)-\frac{s}{s-\mu_Z^2}\ln\left(\frac{\mu_Z^2}{s}\right)\right](\rho_V-\rho_A), \tag{6.25}$$

where ρ_V and ρ_A are the vector and axial vector spectral functions.

We can use eq. (6.25) to calculate the electromagnetic mass difference of the ordinary pion. Then we saturate the spectral integral with the ρ and A_1 resonances; the second term is suppressed by m_{ρ}^2/μ_Z^2, and it is sensible to ignore it. Estimating the spectral integral as in (5.9) we have

$$m_+^2-m_0^2=\frac{3e^2}{16\pi^2F^2}\int_0^{\infty}\mathrm{d}s\ln\left(\frac{s_0}{s}\right)(\rho_V-\rho_A)\simeq\frac{3\ln 2}{2\pi}\alpha m_{\rho}^2. \tag{6.26}$$

This is the classic formula of Das et al. [24], which is known to be well satisfied.

In the case of interest to us here, we want to saturate the spectral integral with hypermeson resonances $M_V, M_A \gg \mu_Z$. Therefore, we have

$$m^2\simeq\frac{3e^2}{16\pi^2F^2}\int\mathrm{d}s\frac{\mu_Z^2}{s}\ln\left(\frac{s}{\mu_Z^2}\right)(\rho_V-\rho_A)$$

$$\simeq\frac{3\alpha}{4\pi}\mu_Z^2\ln\left(\frac{M_V^2}{2\mu_Z^2}\right)\simeq(7\ \mathrm{GeV})^2, \tag{6.27}$$

where we have used eq. (5.6) to obtain the second equality in eq. (6.27), and have taken $M_V\simeq 1$ TeV.

Unlike the PGB mass in example (b), this mass does not scale with F; the dependence on the hypercolor scale is only in the logarithm*. Therefore eq. (6.27) should be a fairly reliable estimate of the electroweak contribution to the mass of these PGB's. As Eichten and Lane [7] have emphasized, this prediction is quite exciting, because relatively light charged PGB's could be produced in e^+e^- colliding beam experiments very soon.

The $2(N^2-1)$ charged PGB's are still degenerate. This degeneracy will not be lifted in higher order. The positively charged PGB's, and their CPT conjugates, are in (N^2-1) multiplets of the exact unbroken SU(N) isospin.

* The graphs in fig. 5b–d have been ignored in our derivation of (6.27). Eichten and Lane [7] conjectured, and Peskin [14] has explicitly verified, that the coefficient of the logarithm in (6.27) is correctly accounted for by fig. 5a alone, at least in the model considered here. The additional graphs only modify the argument of the logarithm.

(e) TWO LEFT-HANDED DOUBLETS

In example (d), all N left-handed $SU(2)_W$ doublets had the same value of the $U(1)_W$ charge $\frac{1}{2}Y$. But if there are left-handed doublets with different values of $\frac{1}{2}Y$, then the $U(1)_W$ interaction contributes to the effective potential in lowest order, and the pattern of G_W breakdown depends on the $U(1)_W$ charge assignments.

Consider a model with four left-handed and four right-handed fermions in a complex representation of G_H, and with G_W representation content given by

$$\begin{array}{lcccccc} & \begin{pmatrix} U_1 \\ D_1 \end{pmatrix}_L & \begin{pmatrix} U_2 \\ D_2 \end{pmatrix}_L & U_{1R} & D_{1R} & U_{2R} & D_{2R} \\ \frac{1}{2}Y = & q_1 & q_2 & q_1+\frac{1}{2} & q_1-\frac{1}{2} & q_2+\frac{1}{2} & q_1-\frac{1}{2}\,. \end{array} \tag{6.28}$$

(G_W anomalies can be eliminated by choosing $q_1 = -q_2$ or by adding more fermions.)

We assume that the $G_f = SU(4)_L \times SU(4)_R$ chiral symmetry breaks down to $S_f = SU(4)_V$. The embedding of S_f in G_f can be specified by a fermion condensate W defined by

$$\langle 0|\bar{\psi}_{Rs}\psi_{Lr}|0\rangle = \sigma W_{rs}\,. \tag{6.29}$$

W is a unimodular unitary matrix. The dependence of the effective potential V on W is given by eq. (5.17). As in example (d), the $SU(2)_W$ interaction does not contribute to $V(W)$ in lowest order.

If $q_1 > q_2$, the minimum of the effective potential occurs at one of the two stationary points

$$\begin{array}{cccc} \bar{U}_{1R} & \bar{D}_{1R} & \bar{U}_{2R} & \bar{D}_{2R} \end{array}$$

$$W_1 = \begin{bmatrix} 1 & 0 & 0 & 0 \\ 0 & 1 & 0 & 0 \\ 0 & 0 & 1 & 0 \\ 0 & 0 & 0 & 1 \end{bmatrix} \begin{matrix} U_{1L} \\ D_{1L} \\ U_{2L} \\ D_{2L} \end{matrix}, \qquad W_2 = \begin{bmatrix} 1 & 0 & 0 & 0 \\ 0 & 0 & 1 & 0 \\ 0 & 1 & 0 & 0 \\ 0 & 0 & 0 & 1 \end{bmatrix}. \tag{6.30}$$

W_1 leaves $U(1)_{EM}$ unbroken, but W_2 breaks G_W completely. From (5.17), we find the value of the effective potential at these points to be

$$V(W_1) = \frac{-3}{32\pi^2} F^2\Delta^2 g'^2 (2q_1^2 + 2q_2^2)\,,$$

$$V(W_2) = \frac{-3}{32\pi^2} F^2\Delta^2 g'^2 \left[(q_1+q_2)^2 + q_1 - q_2\right], \tag{6.31}$$

and therefore we have

$$V(W_1)-V(W_2)=\frac{-3}{32\pi^2}F^2\Delta^2 g'^2[(q_1-q_2)(q_1-q_2-1)]. \tag{6.32}$$

Hence if

$$0<|q_1-q_2|<1, \tag{6.33}$$

W_2 minimizes the effective potential, and G_W is completely broken.

In this model, although there exists a relative orientation of G_W and S_f which leaves $U(1)_{EM}$ unbroken, G_W is completely broken in the dynamically preferred vacuum if (6.33) is satisfied.

(f) RIGHT-HANDED DOUBLETS

If a representation of the sideways group G_S [6–8] contains a complex representation R of $G_H \subset G_S$, it may also contain the conjugate representation $\overline{R}$. Therefore, some hypercolor models contain both left-handed and right-handed G_W doublets. Consider, then, a model with four left-handed and four right-handed fermions in a complex representation of G_H, and with the G_W representation content given by

$$\begin{array}{lcccccc} & \begin{pmatrix} U_1 \\ D_1 \end{pmatrix}_L & U_{1R} & D_{1R} & \begin{pmatrix} U_2 \\ D_2 \end{pmatrix}_R & U_{2L} & D_{2L} \\ \tfrac{1}{2}Y= & q_1 & q_1+\tfrac{1}{2} & q_1-\tfrac{1}{2} & -q_2 & -(q_2+\tfrac{1}{2}) & -(q_2-\tfrac{1}{2}). \end{array} \tag{6.34}$$

This model is similar to example (a) in that the effective potential has two possible minima, at one of which $SU(2)_W$ is left unbroken. The $SU(2)_W$ interaction is most attractive in the $SU(2)_W$ singlet channel, while the $U(1)_W$ interaction is more attractive in the channel in which $U(1)_{EM}$ is unbroken. Therefore, the two interactions compete, and, as in example (a), there is a critical value of the Weinberg angle at which a first-order phase transition occurs.

We assume again that the $G_f = SU(4)_L \times SU(4)_R$ chiral symmetry breaks down to $S_f = SU(4)_V$. The effective potential $V(W)$ given by (5.17) has stationary points at

$$\begin{array}{c} \quad\; \overline{U}_{1R} \;\; \overline{D}_{1R} \;\; \overline{U}_{2R} \;\; \overline{D}_{2R} \\ W_1=\begin{bmatrix} 1 & 0 & 0 & 0 \\ 0 & 1 & 0 & 0 \\ 0 & 0 & 1 & 0 \\ 0 & 0 & 0 & 1 \end{bmatrix} \begin{matrix} U_{1L} \\ D_{1L} \\ U_{2L} \\ D_{2L} \end{matrix}, \end{array} \qquad W_2=\begin{bmatrix} 0 & 0 & 1 & 0 \\ 0 & 0 & 0 & 1 \\ 0 & 1 & 0 & 0 \\ 1 & 0 & 0 & 0 \end{bmatrix}. \tag{6.35}$$

W_1 leaves $U(1)_{EM}$ unbroken and W_2 leaves $SU(2)_W$ unbroken. From (5.17), we find that

$$V(W_1)-V(W_2)=\frac{3}{64\pi^2}F^2\Delta^2\left\{3g^2-\left[4(q_1+q_2)^2-1\right]g'^2\right\}. \tag{6.36}$$

Therefore, W_1 is the minimum of the potential, and G_W breaks down to $U(1)_{EM}$ only if $\cot^2\theta_W$ is smaller than a critical value given by

$$\cot^2\theta_W = g^2/g'^2 = \tfrac{1}{3}\left[4(q_1+q_2)^2 - 1\right]. \tag{6.37}$$

There is no phase transition if $|q_1+q_2| > \frac{1}{2}$; $SU(2)_W$ is always unbroken. For lepton-like charges $q_1 = q_2 = -\frac{1}{2}$, the critical Weinberg angle is $\cot^2\theta_W = 1$. To ensure that G_W breaks down to $U(1)_{EM}$ when $\sin^2\theta = 0.23$, we must have $|q_1+q_2| > 1.7$.

If we add more left-handed G_W doublets and right-handed singlets to the model, they do not help to stabilize the vacuum in which $U(1)_{EM}$ is unbroken. They *do* break $SU(2)_W$, so if we have, for example, a model with two left-handed doublets and one right-handed doublet, plus singlets, G_W is completely broken. This occurs even though there is a relative orientation of G_W and S_f which leaves $U(1)_{EM}$ unbroken.

Therefore, right-handed G_W doublets are very dangerous. Unless they have unconventionally large electric charges, they refuse to allow G_W to break in the observed way.

(g) WEAK COLOR

When hypercolor becomes strong at a renormalization scale near 1 TeV, the color coupling is $\alpha_c \simeq 0.1$, so $SU(3)_c$ can be treated as a chiral-symmetry-breaking perturbation too. The weakly gauged subgroup of G_f is really $G_W = SU(3)\times SU(2)\times U(1)$. As we saw that SU(2) and U(1) may compete with each other to determine the alignment of G_W with S_f, so $SU(3)_c$ may compete with the electroweak group. If so, color will win because α_c is large compared to α.

For instance, suppose that the hyperfermions in example (a) come in three colors; N and E are 3's but N^c and E^c are $\bar{3}$'s. The flavor symmetry is now $G_f = SU(12)$; we assume it breaks down to $S_f = Sp(12)$. As before, the electroweak interactions will prefer that $SU(2)_W$ be unbroken, but color will favor the color-singlet channel, and insist that $SU(2)\times U(1)$ break down to $U(1)_{EM}$.

In any realistic model, there are many PGB's which carry color, and we wish to estimate their masses. Therefore, we consider a model with fermions in a complex representation of G_H, and $G_W = SU(3)\times SU(2)\times U(1)$ representation content given by*

$$\begin{array}{ccccccc} & \begin{pmatrix} U_i \\ D_i \end{pmatrix}_L & U_{iR} & D_{iR} & \begin{pmatrix} N \\ E \end{pmatrix}_L & N_R & E_E\,, \\ \tfrac{1}{2}Y = & \tfrac{1}{6} & \tfrac{2}{3} & -\tfrac{1}{3} & -\tfrac{1}{2} & 0 & -1 \end{array} \tag{6.38}$$

where U, D are color triplets and N, E are color singlets. This model has an approximate $G_f = SU(8)_L \times SU(8)_R$ chiral symmetry which we assume breaks down to $S_f = SU(8)_V$.

* The PGB spectrum of this model has also been considered by Peskin [14] and Dimopoulos [26]. See also Farhi and Susskind [28].

The subgroup of S_f respected by the color interaction is $SU(3)_c \times SU(2)$; the right-handed and left-handed fermions transform as the $(3,2)+(1,2)$ representation of this group. The SU(2) isospin is further broken to electric charge by the electroweak interaction. Under $SU(3) \times SU(2)$, the 63 Goldstone bosons transform as

$$63=(8,3)+(8,1)+(1,3)+(1,3)+(1,1)+(3,3)+(\bar{3},3)+(3,1)+(\bar{3},1). \tag{6.39}$$

A combination of the (1, 3)'s is eaten by the $W^{\pm}$ and Z. The (1, 1) and the neutral member of the remaining (1, 3) are exactly massless Goldstone bosons [7]. The charged members of the uneaten (1, 3) receive masses of order 7 GeV from the second-order electroweak interactions, as in example (d).

The remaining Goldstone bosons are colored, and receive masses from the lowest-order color interaction. Using eq. (5.22), we find that the color-octet and color-triplet PGB masses are

$$m_8^2=\frac{3g_c^2}{16\pi^2}c_2(8)\Delta^2,$$

$$m_3^2=\frac{3g_c^2}{16\pi^2}c_2(3)\Delta^2, \tag{6.40}$$

where

$$\Delta^2=\frac{1}{F^2}\int_0^{\infty} \mathrm{d}s \ln\left(\frac{s_0}{s}\right)(\rho_V-\rho_A);$$

here $c_2(8)=3$ is the quadratic Casimir invariant of the octet and $c_2(3)=\frac{4}{3}$ is the quadratic Casimir invariant of the triplet. Saturating with low-lying resonances as in eq. (5.9) we have $\Delta^2 \simeq 2(\ln 2)M_V^2$, where M_V is the mass of the lightest hypervector resonance. Since four G_H representations contribute to the W, Z mass in this model, we expect $F \simeq 250/\sqrt{4}$ GeV = 125 GeV, and therefore $M_V \simeq (F/f_\pi)m_\rho =$ 1.0 TeV. Hence we estimate

$$m_8^2 \simeq \frac{9(\ln 2)}{2\pi}\alpha_c M_V^2 \simeq (315\ \mathrm{GeV})^2,$$

$$m_3^2 \simeq \tfrac{4}{9} m_8^2 \simeq (210\ \mathrm{GeV})^2, \tag{6.41}$$

taking $\alpha_c \simeq 0.1$ at 1 TeV.

The degeneracy of the SU(2) triplets and singlets found in lowest order is accidental, and will be removed in higher order in α_c. The (8, 1), for example, can annihilate into gluons, while the (8, 3) cannot, and it is presumably pushed up in mass. The splitting is of order $(1+\alpha_c)^{1/2}$, or about 5%.

The electroweak splittings in the (8, 3) are second-order weak, as in example (d). Therefore, these splittings are tiny, being of order

$$\frac{m_8^+ - m_8^0}{m_8} \simeq \frac{1}{2}\frac{\Delta m_8^2}{m_8^2} \simeq \frac{1}{2}\frac{7^2}{300^2} \simeq 3\times 10^{-4}. \tag{6.42}$$

The splitting is about 70–100 MeV.

The electroweak splittings in the (3, 3) are first-order weak. Applying eq. (5.22) we find

$$\Delta m_3^2(\bar{E}U) = \frac{3g'^2}{16\pi^2}\left(\frac{10}{9}\right)\Delta^2 = \frac{5\alpha}{6\alpha_c\cos^2\theta_W} m_3^2,$$

$$\Delta m_3^2\left[\sqrt{\tfrac{1}{2}}\,(\bar{N}U - \bar{E}D)\right] = \frac{3g'^2}{16\pi^2}\left(\frac{4}{9}\right)\Delta^2 = \frac{\alpha}{3\alpha_c\cos^2\theta_W} m_3^2,$$

$$\Delta m_3^2(\bar{N}D) = -\frac{3g'^2}{16\pi^2}\left(\frac{2}{9}\right)\Delta^2 = -\frac{\alpha}{6\alpha_c\cos^2\theta_W} m_3^2. \tag{6.43}$$

Numerically, we have

$$m_3\left(Q=\tfrac{5}{3}\right) - m_3\left(Q=\tfrac{2}{3}\right) \simeq m_3\left(Q=\tfrac{2}{3}\right) - m_3\left(Q=-\tfrac{1}{3}\right) \simeq 5\ \text{GeV}. \tag{6.44}$$

Q is the PGB electric charge.

It is clear that, if there really is a hypercolor interaction, we will have plenty of interesting spectroscopy to study. Of course, all the PGB's will receive mass from the sideways interaction as well, but the sideways contribution is small compared to the color and first-order weak contribution [7]. Electroweak splittings among the color-triplet PGB's of order a few GeV are observable in principle, because these states decay by sideways exchange, and are quite narrow [26].

(h) CHIRAL COLOR

Because we know that the electroweak group $SU(2)\times U(1)$ is broken when hypercolor gets strong, it is tempting to speculate that color SU(3) is embedded in a larger group which also breaks at a few hundred GeV. For example, we have no evidence at present against the possibility that a chiral color group $SU(3)_L \times SU(3)_R$ is broken down to $SU(3)_c$ by the hypercolor interaction. One very interesting consequence of the chiral color is that it allows the colored PGB's to be considerably lighter than we calculated above.

Consider example (g) again, but now suppose that the left-handed hyperquarks are (3, 1)'s and the right-handed hyperquarks are (1, 3)'s under chiral color.

$[SU(3)_L \times SU(3)_R$ anomalies can be cancelled by additional fermions.] The breaking of $SU(8)_L \times SU(8)_R$ chiral symmetry to $SU(8)_V$ forces $SU(3)_L \times SU(3)_R$ to break to the vector subgroup. If the chiral color gauge couplings are $g_L = g_c/\sin\theta$ and $g_R = g_c/\cos\theta$, where θ is a chiral color mixing angle, then the unbroken color currents are

$$J^a = g_c(J_L^a + J_R^a)\,, \tag{6.45}$$

and the massive chiral gluons couple to the currents

$$J'^a = g_c(J_L^a \cot\theta - J_R^a \tan\theta)\,. \tag{6.46}$$

Here $J_{L,R}^a$ are the chiral color currents, normalized as in eq. (5.11).

In lowest order the vector gluons and broken chiral gluons give equal and opposite contributions to PGB masses. (This is just like example (d). Purely left-handed or purely right-handed interactions cannot contribute to the effective potential in lowest order.) But there is a non-vanishing contribution in the next order. Estimating the higher-order corrections as in example (d), we find that the octet and triplet PGB's receive masses

$$m_8^2 = \frac{9}{4\pi}\alpha_c\mu^2 \ln\left(\frac{M_V^2}{2\mu^2}\right),$$

$$m_3^2 = \tfrac{4}{9} m_8^2\,. \tag{6.47}$$

Here μ is the mass of a chiral gluon; since the Goldstone bosons couple with strength F to the axial currents, we see from eq. (6.46) that $\mu^2 = \frac{1}{4} g_c^2 F^2(\cot\theta + \tan\theta)^2 = g_c^2 F^2/\sin^2 2\theta$, and therefore

$$m_8^2 \simeq \frac{9\alpha_c^2 F^2}{\sin^2 2\theta} \ln\left(\frac{M_V^2}{2\mu^2}\right) \simeq \frac{(68\ \text{GeV})^2}{\sin^2 2\theta}\,,$$

$$m_3^2 \simeq (45\ \text{GeV})^2/\sin^2 2\theta\,, \tag{6.48}$$

taking $F = 125$ GeV. For the charge $\frac{5}{3}$ color triplet, the electroweak mass in eq. (6.43), $m_3^2 \simeq (60\ \text{GeV})^2$, may actually be larger than the color contribution.

If chiral color exists, colored PGB's should be experimentally accessible fairly soon. Dimopoulos [26] has pointed out that the color octets can be singly produced, probably most easily in gluon-gluon collisions. The decays of the neutral ones sometimes give rise to exotic events in which a hard photon balances the momentum of a gluon jet. The color triplets can be pair-produced in e^+e^- colliding beam experiments. They decay into distinctive quark-lepton states [26].

We have assumed in this paper that the hypercolor group G_H remains unbroken when hypercolor gets strong, but we cannot exclude the possibility that G_H is actually embedded in a larger group which "self-destructs". In addition to the $W^{\pm}$, Z, and chiral gluons, there may be heavy hypergluons with masses of several hundred GeV.

We have shown that the subgroup alignment problem can be explicitly solved in many hypercolor models, provided the unbroken isospin group S_f is known. The physical consequences of subgroup alignment highlighted by our examples are diverse, and in some cases rather surprising.

We hope that our analysis of subgroup alignment will provide the serious prospective model-builder with a bit of guidance. We can, at least, exclude models in which the electroweak gauge group fails to break down as desired.

We have stressed the impact of the alignment of subgroups on the breakdown of the electroweak gauge group, but we wish to conclude by pointing out another context in which an alignment problem must be solved. In a realistic theory, a sideways interaction [6–8] is required to generate the masses of quarks and leptons. We must account for the dynamical breakdown of the sideways gauge group. One possibility is that the sideways group is broken by another new strong interaction [6], just as the electroweak group is broken by the hypercolor interaction. In that case, the sideways gauge group can be treated as a weakly gauged subgroup of an approximate flavor-symmetry group. Hence, the general formalism developed in this paper can be employed to determine the pattern of the sideways breakdown.

I thank Michael Peskin for informing me about his recent work, and for helpful comments. He has independently obtained many of the results reported here. I am grateful to Kenneth Lane for suggesting that spectral-function sum rules can be used to solve the subgroup alignment problem, and to Estia Eichten for many illuminating conversations while this work was in progress. I also thank Steven Weinberg and Howard Georgi for advice and encouragement.

References

[1] S. Weinberg, Phys. Rev. Lett. 19 (1967) 1264;
A. Salam, *in* Elementary particle theory, ed. N. Svartholm, (Almqvist and Wiksell, Stockholm, 1968)
[2] S. Weinberg, Phys. Lett. 82B (1979) 387
[3] S. Weinberg, Phys. Rev. D13 (1976) 974; D19 (1979) 1277
[4] L. Susskind, Phys. Rev. D20 (1979) 2619
[5] H. Georgi and S.L. Glashow, Phys. Rev. Lett. 32 (1974) 438;
H. Georgi, H.R. Quinn and S. Weinberg, Phys. Rev. Lett. 33 (1974) 451
[6] S. Dimopoulos and L. Susskind, Nucl. Phys. B155 (1979) 237
[7] E. Eichten and K. Lane, Phys. Lett. 90B (1980) 125
[8] S. Weinberg, unpublished
[9] R. Dashen, Phys. Rev. D3 (1971) 1879
[10] E. Eichten, K. Lane and J. Preskill, Phys. Rev. Lett. 45 (1980) 225

[11] E. Eichten, K. Lane and J. Preskill, in preparation
[12] J. Preskill, Harvard Ph.D. thesis (1980), unpublished
[13] S. Dimopoulos and L. Susskind, A technicolored solution to the strong *CP* problem, Columbia preprint 79-0196 (1979)
[14] M.E. Peskin, The alignment of the vacuum in theories of technicolor, Saclay preprint DPh-T/80/46 (1980)
[15] S. Weinberg, Phys. Rev. Lett. 18 (1967) 507
[16] H. Georgi, Nucl. Phys. B156 (1979) 126;
M. Peskin, in preparation;
S. Dimopoulos, S. Raby and L. Susskind, Tumbling gauge theories, Stanford preprint ITP-653 (1979)
[17] J.S. Bell and R. Jackiw, Nuovo Cim. 51 (1969) 47;
S.L. Adler, Phys. Rev. 177 2426 (1969)
[18] Y. Nambu and G. Jona-Lasinio, Phys. Rev. 122 (1961) 345;
G. 't Hooft, Naturalness, chiral symmetry and spontaneous chiral symmetry breaking, Lectures at Cargèse Summer Inst. (1979)
[19] L. Michel and L.A. Radicati, Ann. of Phys. 66 (1971) 758
[20] R. Dashen, Phys. Rev. 183 (1969) 1245
[21] J. Schwinger, Phys. Rev. 125 (1962) 397; 128 (1962) 2425
[22] K. Wilson, Phys. Rev. 179 (1969) 1499
[23] C. Bernard, A. Duncan, J. Lo Secco and S. Weinberg, Phys. Rev. D12 (1975) 792
[24] T. Das, G.S. Guralnik, V.S. Mathur, F.E. Low and J.E. Young, Phys. Rev. Lett. 18 (1967) 759
[25] J.J. Sakurai, Phys. Rev. Lett. 17 (1966) 552
[26] S. Dimopoulos, Nucl. Phys. B168 (1980) 69
[27] B. Holdom, A realistic model with dynamically broken symmetries, Harvard preprint HUTP-80/A010 (1980)
[28] E. Farhi and L. Susskind, Phys. Rev. D20 (1979) 3404
[29] T. Hagiwara and R.N. Mohapatra, Phys. Rev. D11 (1975) 2223

VIII. PROBLEMS WITH EXTENDED HYPERCOLOR

Nuclear Physics B182 (1981) 505–528

CHALLENGES FOR EXTENDED TECHNICOLOUR THEORIES

Savas DIMOPOULOS

Physics Department, Stanford University, Stanford, CA 94305, USA

John ELLIS

CERN, Geneva, Switzerland

Received 20 October 1980

We develop the dialectic between extended technicolour theories and phenomenological constraints on rare processes. The natural suppression of flavour-changing neutral currents and *CP*-violating processes which can occur in models with elementary scalars is not a general feature of naïve extended technicolour (ETC) models. We study the extent to which naïve ETC estimates must in fact be suppressed in order to be compatible with the phenomenology of such rare processes. We emphasize the potential significance of the exchanges of neutral flavour-changing bosons and find that the strongest constraints arise from considering the combination of the D_1^0-D_2^0 and K_1^0-K_2^0 systems. *CP*-violating effects in ETC models must be severely suppressed if they are to be compatible with the observational facts. We point to several rare processes whose further experimental study is of particular concern to ETC theories.

1. Introduction

A classic problem in the construction of weak interaction theories has been to understand the observed strong suppression of flavour-changing neutral interactions. A giant step towards the complete solution of this problem was taken by Glashow, Iliopoulos and Maiani (GIM) [1]. The introduction of charm suppressed to phenomenologically acceptable levels all flavour-changing couplings of neutral vector bosons as well as the effective neutral interactions generated by multiple exchanges of charged vector bosons. It was subsequently shown [2] that the GIM mechanism was essentially the only way of suppressing these neutral interactions naturally, i.e. for arbitrary values of the masses and mixing angles of fermions of given charge and helicity, in the absence of any symmetries or conservation laws restricting the form of the mixing matrix. The naturalness conditions for an arbitrary weak gauge group G can be summarized [3] by the requirements:

(i) all the fermions of a given charge and helicity must occur in equivalent representations of G;

(ii) they must all have the same group-theoretical weights within their representations.

The specific GIM model [1] meets these requirements by assigning all left-handed fermions to SU(2) isodoublets and all right-handed fermions to isosinglets, and giving all helicity states of a given charge identical values of I_3 and Y. In principle, flavour-changing neutral weak interactions could also be mediated by exchanges of spinless bosons such as physical Higgs bosons in theories with fundamental scalar fields. In such theories, the natural suppression of such flavour-changing neutral effects can itself be suppressed naturally [2] by the requirement that

(iii) all fermions of a given charge and helicity acquire their masses from the same Higgs field.

There is much discontentment with gauge theories containing fundamental scalar fields, fuelled by the large number of apparently arbitrary parameters in their couplings, some of which must be adjusted with unnatural precision. It is therefore proposed that the spontaneous symmetry breaking required in gauge theories be generated dynamically [4]. This suggestion first entails the introduction of a new exact non-abelian gauge symmetry called technicolour (TC), whose coupling is strong at a momentum scale of about 300 GeV. It is expected that non-perturbative effects in the strong-coupling regime would cause a breakdown of chiral symmetry accompanied by the appearance of light pseudoscalar bound states analogous to the pion of QCD. Some of these would be massless and "eaten" by the $W^{\pm}$ and Z^0 gauge bosons, becoming their longitudinal helicity states and allowing them to acquire a mass. Others of the pseudoscalars called pseudo-Goldstone bosons (PGBs) acquire masses $\lesssim 300$ GeV from higher-order QCD, weak and electromagnetic interactions as well as so-called extended technicolour (ETC) interactions. These ETC interactions are needed [5] for giving masses to the observed fermions by interconnecting them with technifermions.

ETC theories contain several possible sources of flavour-changing neutral interactions. The most popular scenario [6] for ETC interactions puts all the conventional fermions of a given charge and helicity into a common representation of the ETC group with a multiplet of technifermions irreducible under the unbroken TC subgroup. The full ETC group is then supposed to break down at a sequence of energy scales into the TC subgroup. The ETC group will, therefore, also contain generators coupling different conventional fermions of the same charge and helicity and hence giving rise to flavour-changing neutral weak interactions. Box and other higher order diagrams involving the multiple exchanges of charged or neutral ETC bosons will have similar effects. So, also, will the exchanges of the relatively low-mass PGBs [7,8] whose couplings are similar in nature to those of physical Higgs bosons in theories with fundamental Higgs fields.

By their nature, ETC theories do not obey all the requirements [2,3] necessary for the natural suppression of flavour-changing neutral weak interactions, since all the conventional fermions of a given charge and helicity occupy different weights in a common representation of the ETC group. Also, it is not clear that the PGB couplings will be flavour-diagonal and whether there is an analogue to requirement (iii) above on the form of the Higgs fermion interactions.

It is not clear, however, that the full rigour of the three naturalness requirements above should in fact be applied to ETC theories. For one thing, the masses of the ETC bosons are considerably higher than those of the vector bosons previously encountered in weak interactions, and their single and multiple exchanges are suppressed accordingly. For another, the fermion mass matrices in ETC theories are, in principle, not arbitrary. The conventional fermion masses are believed to be related to the ETC boson masses by

$$m_{\mathrm{f}} \approx (300\ \mathrm{GeV})^3 / m_{\mathrm{ETC}}^2, \tag{1}$$

and all the mixing angles are expected to be calculable in a forthcoming realistic theory. The naturalness requirements above were derived by contemplating arbitrary mass matrices for the fermions of each charge, and it may well be that in the true ETC theory there will be cancellations which now seem to us accidental, but may eventually be seen to follow from presently unsuspected symmetries of the theory.

In this paper we compute the natural orders of magnitude to be expected for various flavour-changing neutral interactions within the present framework of ETC ideas [8]*. Known fermion masses and the mass formula (1) enable us to compute the ETC boson masses and hence estimate the expected effects. Comparing these directly with the experimental values or upper limits on flavour-changing neutral interactions we determine what extra suppression factors will be necessary if ETC theories are to become consistent with experiment. While we briefly indicate some possible mechanisms for achieving these extra suppressions, many of them still remain as challenges for ETC theories.

Our estimates of these further necessary suppression factors are set out in table 1. We have considered four classes of diagrams causing flavour-changing neutral interactions, all of which would be naturally suppressed to levels consistent with experiment in weak gauge theories with explicit spinless fields. These include single gauge boson exchange (SGEX), single PGB exchange (SPEX), and higher order exchanges (HOEX) such as double gauge boson exchange (DGEX). In the cases of SGEX and HOEX we have tabulated the further suppressions necessary given our estimates of the vector boson masses. In the case of SPEX, the considerable uncertainties [6–8] in the expected masses of the PGBs have led us to tabulate the lower limits on their masses suggested by upper limits on the effective neutral interactions they induce in the absence of further suppressions. The most significant challenges seem to arise from purely hadronic $\Delta F = 2$ effects, but we also present results on semi- and purely leptonic processes as well.

* Related computations not specifically in the ETC framework have been made in ref. [9].

TABLE 1
Challenges for ETC theories

Process	Necessary suppression factor		$(\text{Mass})^2$ limit on PGB in absence of suppression of SPEX
	SGEX	HOEX	
Re(K_1-K_2)	10^3	10	$>2\cdot 10^6$ GeV^2
Im(K_1-K_2)	10^6	10^4	$>2\cdot 10^9$ GeV^2
(D_1-D_2)	10^2	OK	$>3\cdot 10^4$ GeV^2
$K_L^0 \to e^- \mu^+$	borderline	OK	$>10^4$ GeV^2
$K^+ \to \pi^+ e^- \mu^+$	OK	OK	less stringent
$K^0 \to \mu^+\mu^-, \gamma\gamma$	OK	OK	less stringent
$\mu N \to eN$	OK	OK	less stringent
$\mu \to e\bar{e}e$	OK	OK	less stringent
$\mu \to e\gamma$		$10^{3/2}$	
d_N		10^2	less stringent

We would like to mention three aspects of our analysis which may be somewhat novel. One is our emphasis on the possible effects of exchanges of the neutral flavour-changing gauge bosons which must acquire masses at each stage of the breakdown of the ETC group to the exact TC subgroup. The second is our emphasis on *CP*-violating effects in the K_1-K_2 mass matrix. Opinions differ as to whether the Kobayashi-Maskawa (KM) mechanism [10] explains naturally the small ratio $O(10^{-3})$ of the imaginary and real $\Delta S=2$ parts of the K_1-K_2 mass matrix, but the smallness of the phenomenon is certainly a challenge to ETC theories [11]. In the absence of some as yet unknown symmetry, the coupling matrices of the ETC bosons have phases of order unity relative to the KM matrix. This means that the ratios of imaginary to real parts of ETC exchange effects are *a priori* of order one, and therefore that the real as well as the imaginary parts of the ETC exchanges should be suppressed 10^3 more than suggested by considerations of the real part alone—unless some new symmetry can be found. The third point we would emphasize concerns the constraints imposed by the D_1-D_2 system. Even the slack present limit [12] on D^0-$\bar{D}^0$ mixing ($\leqslant O(10)\%$) already tells us that the numerical suppression of the D_1-D_2 mass difference must be comparable with the GIM expectation of order G_F^2. This is enough to prevent one from constructing models in which $\Delta F \neq 0$ transitions among the charge $\frac{2}{3}$ quarks are much larger than those among the charge $-\frac{1}{3}$ quarks.

Most of this paper is devoted to developing the challenges for ETC theories set out in table 1. In sect. 2 we review the generic structure of simple ETC models and estimate the masses of the various ETC bosons. In sect. 3 we estimate SGEX diagrams and compare them with the experimental magnitudes of different $\Delta F \neq 0$ effects. In sect. 4 we do the same for SPEX, while sect. 5 is devoted to HOEX such as DGEX. Sect. 6 contains comments on *CP* violation and uncertainties in the ETC estimate [11] for the neutron electric dipole moment, given the problems of table 1 for the imaginary part of the K_1-K_2 mass matrix. Finally, sect. 7 discusses the prospects that ETC theories may meet the challenges raised in earlier parts of this paper. We observe that the leading order exchanges of PGBs are flavour-diagonal in so-called "monophagic" theories [13] where all the fermions of any given charge all get their masses from couplings to the same technifermion condensate. We also mention some of the problems arising in trying to GIM the ETC interactions by xerox copying the technifermions to have one technigeneration per conventional generation. These problems include an immense proliferation in the number of PGBs and the difficulties encountered in giving them masses consistent with the limits of table 1, and the related question of breaking the symmetries between generations. ETC theories are not yet out of the wood. Finally, at the end of this section we remind readers of some specific rare processes: D_1-D_2 mixing, $K_L^0 \to e\mu$ and $K^+ \to \pi^+ e\mu$ decays, $\mu N \to eN$ capture and $\mu \to e\gamma$ decay, where further experimental studies may either see ETC effects or pose more serious challenges for ETC theories.

2. General features of ETC theories

The generic structure of simple ETC theories [5,6] is indicated in fig. 1. To all the classes of conventional fermions, charge $\frac{2}{3}$ quarks, charge $-\frac{1}{3}$ quarks, neutrinos and charged leptons, are associated technifermions (U, D, N and E) which are collected into a technigeneration paralleling the conventional fermion generations. The ETC group is supposed to contain the TC group and to have generators transforming conventional fermions into technifermions and vice versa. It is suggested that the ETC group breaks down to the TC group in stages, perhaps "spitting out" a generation of fermions at each stage of breaking by a dynamical tumbling mechanism. ETC theories give masses to fermions via the diagram shown in fig. 2, from which we abstract the property that ETC bosons must have both left- and right-handed couplings. The ETC boson exchange takes place over a distance short by

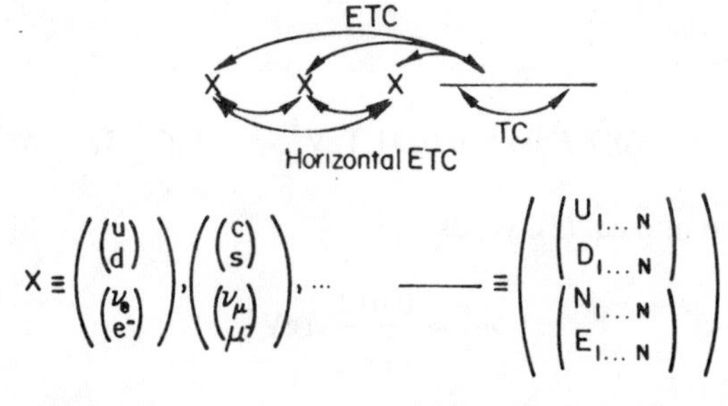

Fig. 1. Generic structure of ETC theories [5,6]. The arrows indicate interactions mediated by gauge bosons. The crosses indicate conventional fermion generations or families with the substructure illustrated beneath. The line indicates a technigeneration whose substructure is also indicated.

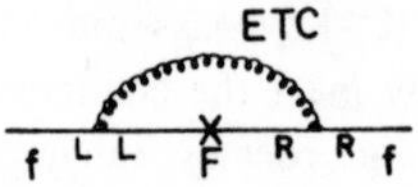

Fig. 2. The mechanism whereby ETC bosons [5] give masses to conventional fermions.

comparison with the rest of the loop, which is cut off by the softness of the technifermion mass X in fig. 2. This arises from the dynamical breakdown of chiral symmetry in the TC subgroup which has a characteristic scale of order 300 GeV.

To calculate the masses of the ETC bosons we first Fierz transform the exchange in fig. 2 to get [5] (assuming a conventional normalization analogous to that of the $W^{\pm}$ for the ETC gauge couplings):

$$m_{\mathrm{f}} = \frac{g_{\mathrm{ETC}}^2}{2m_{\mathrm{ETC_f}}^2} \left\langle 0 \middle| \sum_{\mathrm{TC}} \bar{F}_{\mathrm{TC}} F_{\mathrm{TC}} \middle| 0 \right\rangle . \tag{2}$$

To estimate $\langle 0|\bar{F}F|0\rangle \equiv \Sigma_{\mathrm{TC}}\langle 0|\bar{F}_{\mathrm{TC}}F_{\mathrm{TC}}|0\rangle$ we rescale naïvely from QCD:

$$\langle 0|\bar{F}F|0\rangle = \sum_{\mathrm{C}} \langle 0|\bar{q}_{\mathrm{C}} q_{\mathrm{C}}|0\rangle \times \left(\frac{F_\pi}{f_\pi}\right)^3 \times \left(\frac{3}{N}\right)^{1/2} \tag{3}$$

where N is the dimension of the technirepresentation, f_π is the conventional pion decay constant $\simeq 95$ MeV and F_π is its technianalogue $\simeq 125$ GeV. To estimate $\langle 0|\bar{q}q|0\rangle \equiv \Sigma_{\mathrm{C}}\langle 0|\bar{q}_{\mathrm{C}}q_{\mathrm{C}}|0\rangle$ we use the PCAC relation

$$(m_{\mathrm{d}} + m_{\mathrm{u}})\langle 0|\bar{q}q|0\rangle = f_\pi^2 m_\pi^2 , \tag{4}$$

and the conventional values $m_{\mathrm{d}} = 10$ MeV, $m_{\mathrm{u}} = \frac{1}{1.8} m_{\mathrm{d}}$ to get

$$\langle 0|\bar{q}q|0\rangle = 0.0106\ \mathrm{GeV}^3 . \tag{5}$$

Eq. (3) then gives us

$$\langle 0|\bar{F}F|0\rangle = 0.0241\ \mathrm{TeV}^3, \qquad \text{if } N = 3, \tag{6}$$

which, substituted into eq. (2), yields

$$m_{\mathrm{f}} \approx \frac{0.012}{v_{\mathrm{ETC_f}}^2}\ \mathrm{TeV}^3 , \tag{7}$$

where $v_{\mathrm{ETC}} \equiv m_{\mathrm{ETC}}/g_{\mathrm{ETC}}$. The ETC boson masses and coupling constants generally appear in this combination: of course, $v_{\mathrm{ETC}} \approx m_{\mathrm{ETC}}$ if $g_{\mathrm{ETC}} \approx 1$, which seems a

TABLE 2

q	u	d	s	c	b	t
m_q	6 MeV	10 MeV	200 MeV	1.5 GeV	5 GeV	30 GeV
$v^2_{ETC_q}$	2200 TeV^2	1200 TeV^2	60 TeV^2	8 TeV^2	2.4 TeV^2	0.4 TeV^2

Recall that $v^2_{ETC} \equiv m^2_{ETC}/g^2_{ETC} \approx m^2_{ETC}$ if $g_{ETC} \approx 1$

reasonable estimate. Re-inserting $m_d = 10$ MeV into eq. (7) finally yields for the boson which makes the dominant contribution to the mass of the down quark

$$v^2_{ETC_d} \approx 1200\ \text{TeV}^2. \tag{8}$$

This estimate is, of course, subject to considerable uncertainties. We have neglected various subtleties such as scaling the number of technicolours in eq. (3) and the logarithmic momentum scale variation of the fermion masses and operator matrix elements, but these are unlikely to change the estimate (8) by more than a factor of 2. The principal uncertainty is in the value of $m_d = 10$ MeV we used: because we also used m_d in making the estimate (5), $m^2_{ETC_d} \propto 1/m^2_d$ and an order of magnitude uncertainty in $m^2_{ETC_d}$ is easily conceivable. We can use the approximate formula (7) to estimate masses for the ETC bosons which should give masses to other quarks: the results are shown in table 2 together with the masses of the quarks we assumed. While these estimates are very tentative, they do suggest that the mass of the ETC_t vector bosons in particular may be not much larger than the dynamical scale of the TC interactions.

Before using the masses of table 2 to estimate various higher order flavour-changing neutral interactions we should emphasize one other feature of ETC theories. When the ETC group breaks down giving masses to the ETC_f bosons which give a mass to the fermion f, other electrically neutral vector bosons coupling f to itself and to the other generations of conventional fermions of the same charge (the horizontal ETC bosons of fig. 1) will also acquire masses of similar magnitude to m_{ETC_f}. For example, one expects to encounter both flavour-diagonal and generation-changing bosons with essentially unit coupling strengths transforming $d \leftrightarrow d$, $d \leftrightarrow s$ and $d \rightarrow b$ with $v^2_{ETC} \approx 1000$ TeV^2. To be more specific, suppose that the ETC group commutes with the electroweak group $SU(2)_L$, a reasonable assumption since it would be difficult (though not impossible) to understand the difference in strengths of the technicolour and $SU(2)_L$ couplings if the ETC and $SU(2)_L$ interactions were unified, and baryon-number conservation at low energies would no longer be automatic. If ETC and $SU(2)_L$ commute, then the eigenstates of ETC are the same Cabibbo-rotated* d^L_θ, s^L_θ eigenstates of $SU(2)_L$. Therefore any diagonal currents that cause the

* Of course, the origin of the Cabibbo angle may lie in the up-quark sector, or in the down-quark sector, or can be shared between them.

transition $s_\theta \to s_\theta$ are going to mediate $\Delta S = 2$ amplitudes of order $1/v^2_{ETC_s}$, and similarly for $d_\theta \to d_\theta$. In general, generation-changing $f \leftrightarrow f$ couplings of *unit strength* will be mediated by bosons with masses associated via table 2 with the *lighter* of the two fermions f, f′. In addition, the existence of non-zero Cabibbo mixing suggests that there will, in general, also be *Cabibbo-suppressed couplings* of $f \leftrightarrow f'$ via a boson whose mass is associated via table 2 with the *heavier* of the two fermions f, f′. Generally, we expect that $\Delta G = 0$ processes may be mediated by ETC bosons with unit coupling strengths, while $|\Delta G| = 1, 2$ processes always require a Cabibbo suppression factor of order θ, θ^2, respectively. We will see in sect. 3 that the exchanges of these massive flavour-non-diagonal ETC bosons will raise interesting challenges for ETC theories.

3. Single gauge boson exchange (SGEX)

3.1. K_1-K_2 AND D_1-D_2

We start off by considering the contributions of the SGEX diagrams of fig. 3 to various purely hadronic $\Delta F = 2$ effects: the real parts of the K_1-K_2 and D_1-D_2 mass matrices. From the arguments at the end of sect. 2 one expects SGEX to generate $|\Delta S| = 2$ and $|\Delta C| = 2$ (and hence $|\Delta G| = 2$) four-fermion effective lagrangians of the general orders of magnitude

$$\mathcal{L}^{\text{eff}}_{(\bar{s}d)^2} \approx \tfrac{1}{2}\left\{(\bar{s}_L\gamma_\mu d_L)^2, (\bar{s}_L\gamma_\mu d_L)(\bar{s}_R\gamma^\mu d_R), (\bar{s}_R\gamma_\mu d_R)^2\right\}\left(\frac{\theta^2}{v^2_{\text{ETC}_d}}, \frac{\theta^2}{v^2_{\text{ETC}_s}}\right), \quad (9)$$

$$\mathcal{L}^{\text{eff}}_{(\bar{c}u)^2} \approx \tfrac{1}{2}\left\{(\bar{c}_L\gamma_\mu u_L)^2, (\bar{c}_L\gamma_\mu u_L)(\bar{c}_R\gamma^\mu u_R), (\bar{c}_R\gamma_\mu u_R)^2\right\}\left(\frac{\theta^2}{v^2_{\text{ETC}_u}}, \frac{\theta^2}{v^2_{\text{ETC}_c}}\right), \quad (10)$$

where the subscripts L, R indicate projection operators $\frac{1}{2}(1 \mp \gamma_5)$, and θ^2 indicates mixing angle factors which we take to be $O(\sin^2\theta_C) = O(\frac{1}{20})$. As a standard of comparison we can contrast the $\mathcal{L}^{\text{eff}}_{(\bar{s}d)^2}$ of eq. (9) with that obtained from the

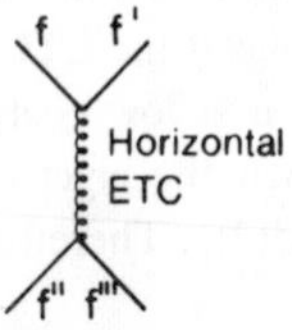

Fig. 3. The single gauge boson exchange (SGEX) contribution to an effective four-fermion lagrangian.

conventional GIM calculation of Gaillard and Lee [14]:

$$\mathcal{L}^{\mathrm{GIM}}_{(\bar{s}d)^2} = \left(\bar{s}_L \gamma_\mu d_L\right)^2 \frac{G_F}{\sqrt{2}} \frac{\alpha}{4\pi} \frac{m_c^2}{m_W^2 \sin^2\theta_W} \sin^2\theta_C \cos^2\theta_C . \tag{11}$$

Requiring that $\mathcal{L}^{\mathrm{eff}}_{(\bar{s}d)^2}$ be no larger than $\mathcal{L}^{\mathrm{GIM}}_{(\bar{s}d)^2}$, we need*

$$\frac{\theta^2}{2v^2_{\mathrm{ETC}_d}}, \frac{\theta^2}{2v^2_{\mathrm{ETC}_s}} \lesssim \tfrac{1}{2}\times 10^{-12}\ \mathrm{GeV}^{-2}, \tag{12}$$

and hence, if $\theta^2 \sim \frac{1}{20}$,

$$v^2_{\mathrm{ETC}_d}, v^2_{\mathrm{ETC}_s} \gtrsim 5\times 10^4\ \mathrm{TeV}^2 . \tag{13}$$

The conflicts between the limits (13) and the previously deduced values of $v^2_{\mathrm{ETC}_d}$ and $v^2_{\mathrm{ETC}_s}$ shown in table 2 imply the need for an additional suppression factor in ETC theories of up to 1000, as tabulated in the challenging table 1. (Having made the point that m_{ETC} generally appears in the combination $v_{\mathrm{ETC}} \equiv m_{\mathrm{ETC}}/g_{\mathrm{ETC}}$, we will henceforth implicitly assume that $g_{\mathrm{ETC}} \approx 1$ unless otherwise stated: this assumption can easily be relaxed.)

We now turn to the D_1-D_2 system. It is known [12] that

$$\frac{\Gamma(D^0 \to K^+\pi^-)}{\Gamma(D^0 \to K^-\pi^+)} \lesssim O(\tfrac{1}{10}), \tag{14}$$

consistent with the absence of D^0-$\bar{D}^0$ mixing. This means that the D_1-D_2 mass difference Δm_D must be less than the D^0 decay rate Γ_D (neutral D's decay before they mix):

$$\tfrac{1}{2}\left(\frac{\Delta m_D}{\Gamma_D}\right)^2 \lesssim \tfrac{1}{10} . \tag{15}$$

The D^0 decay rate is believed theoretically to be of order

$$\Gamma_D \approx O(10)\frac{G_F^2 m_c^5}{192\pi^3} \approx 3.5\times 10^{-12}\ \mathrm{GeV} \tag{16}$$

* We assume that the matrix elements between K^0 and $\bar{K}^0$ of the LL, LR and RR operators of eqs. (9),(10) are all comparable. This seems a reasonable approximation, though it has sometimes been argued [15] that the matrix element of the LR operator might be somewhat larger.

which corresponds to a lifetime of a few $\times 10^{-13}$ seconds and is not grossly different from the present tentative experimental indications. Combining eqs. (15) and (16) we find

$$\frac{\Delta m_{\mathrm{D}}}{m_{\mathrm{D}}} \lesssim 10^{-12}. \tag{17}$$

To see what constraint this imposes on the coefficients C of $(\bar{\mathrm{c}}\mathrm{u})^2$ operators, we need to be able to estimate their matrix elements. Gaillard and Lee [14] estimated the matrix element of the $(\bar{s}_{\mathrm{L}}\gamma_\mu d_{\mathrm{L}})^2$ operator by inserting the vacuum intermediate state to get a value $f_{\mathrm{K}}^2 m_{\mathrm{K}}^2$. This approximation is very crude but more sophisticated approaches [16] give a similar order of magnitude. If we use the same vacuum insertion approximation for the D_1-D_2 system, which is much more questionable, we find

$$10^{-12} \gtrsim \frac{\Delta m_{\mathrm{D}}}{m_{\mathrm{D}}} \approx f_{\mathrm{D}}^2 C, \tag{18}$$

where f_{D} is the D decay constant analogous to f_{K} and f_π. Estimating $F_{\mathrm{D}}^2 \sim 3 \times 10^{-2}$ GeV^2 we conclude from (18) that

$$C \lesssim 3 \times 10^{-11}\ \mathrm{GeV}^{-2}. \tag{19}$$

Comparing this limit on C with the expected form (10) of $\mathcal{L}^{\mathrm{eff}}_{(\bar{\mathrm{c}}\mathrm{u})^2}$ we see that

$$m_{\mathrm{ETC_u}}^2, m_{\mathrm{ETC_c}}^2 \gtrsim 800\ \mathrm{TeV}^2. \tag{20}$$

Comparing the limits (20) and the previous estimates of $m_{\mathrm{ETC_u}}^2$ and $m_{\mathrm{ETC_c}}^2$ in table 2, we see that there is again a discrepancy by two orders of magnitude in the $\mathrm{ETC_c}$ contribution, as tabulated in table 1. However, the numerical estimates in this case are clearly shakier than for the K^0-$\bar{\mathrm{K}}^0$ system, and it could perhaps be argued that the D^0-$\bar{\mathrm{D}}^0$ system does not yet present a serious challenge for ETC theories.

3.2. $\mathrm{K_L^0} \to \mathrm{e}\mu$ AND $\mathrm{K}^+ \to \pi^+ \mathrm{e}\mu$

We now turn to semileptonic flavour-changing neutral interactions. For the sake of argument, we assume that the bosons mediating these interactions as in fig. 3 are the same as those giving masses to the quarks as in table 2, with lepton and quark generations associated in the usual way: $(\mathrm{e}, \nu_{\mathrm{e}} \leftrightarrow \mathrm{u}, \mathrm{d};\ \mu, \nu_\mu \leftrightarrow \mathrm{c}, \mathrm{s};\ \tau, \nu_\tau \leftrightarrow \mathrm{t}, \mathrm{b})$. We therefore expect a $|\Delta S| = 1$ but $|\Delta S| = \Delta L_\mu$ and hence $\Delta G = 0$ interaction of the form

$$\mathcal{L}^{\mathrm{eff}}_{(\bar{\mathrm{s}}\mathrm{d})(\bar{\mathrm{e}}\mu)} = \left(\tfrac{1}{2} g_{\mathrm{ETC}}^2\right)\left\{\bar{s}_{\mathrm{L}}\gamma_\mu d_{\mathrm{L}}, \bar{s}_{\mathrm{R}}\gamma_\mu d_{\mathrm{R}}\right\}\left\{\bar{e}_{\mathrm{L}}\gamma^\mu \mu_{\mathrm{L}}, \bar{e}_{\mathrm{R}}\gamma^\mu \mu_{\mathrm{R}}\right\}\left\{\frac{1}{m_{\mathrm{ETC_d}}^2}, \frac{\theta^2}{m_{\mathrm{ETC_s}}^2}\right\}. \tag{21}$$

Any individual one of these terms would contribute [9] to

$$\frac{\Gamma(K_L^0 \to e^- \mu^+)}{\Gamma(K^+ \to \mu^+ \nu_\mu)} = \frac{\frac{1}{2}\left(\frac{1}{2} g_{ETC}^2\right)^2 \left(1/m_{ETC_d}^2 \text{ or } \theta^2/m_{ETC_s}^2\right)^2}{\sin^2\theta_C \left(g^2/2m_W^2\right)^2}. \tag{22}$$

From the upper limit [17] of 0.63×10^{-9} on this ratio, we would extract

$$m_{ETC_d}^2 \gtrsim 3500 \text{ TeV}^2, \qquad m_{ETC_s}^2 \gtrsim 170 \text{ TeV}^2. \tag{23}$$

These limits are somewhat higher than the ETC boson masses quoted in table 2, but we do not believe that the discrepancy is significant. Furthermore, the limit (23) can easily be avoided by making the ETC quark couplings purely vectorial. A vectorial combination of (21) would, however, show up in $K^+ \to \pi^+ e^- \mu^+$ decay. Any individual one of the terms in (21) would give a ratio

$$\frac{\Gamma(K^+ \to \pi^+ e^- \mu^+)}{\Gamma(K^+ \to \pi^0 \nu_\mu \mu^+)} = \frac{2\left[\frac{1}{2} g_{ETC}^2 \left(1/m_{ETC_d}^2 \text{ or } \theta^2/m_{ETC_s}^2\right)\right]^2}{\sin^2\theta_C \left(g^2/2m_W^2\right)^2}, \tag{24}$$

and this would be a factor of 8 larger if all the terms in (21) were present with equal strength. From the upper limit [18] of 1.5×10^{-7} on the ratio (24), we deduce

$$m_{ETC_d}^2 \gtrsim 200 \text{ TeV}^2, \qquad m_{ETC_s}^2 \gtrsim 10 \text{ TeV}^2, \tag{25}$$

both of which are comfortably consistent with the constraints in table 2. The decays $K_L^0 \to e^- \mu^+$ and $K^+ \to \pi^+ e^- \mu^+$ therefore produce no challenges for ETC theories. It is, however, possible that they may offer the prospect of soon observing a bizarre decay which could be a signature for ETC. If at the quark vertex the ETC interactions are not purely vectorial, as is expected because of the u-d mass splittings, then the decay $K_L^0 \to e^- \mu^+$ should probably show up within an order of magnitude below the present experimental limit. If the ETC interactions are not purely axial at the quark vertex, then the decay $K^+ \to \pi^+ e^- \mu^+$ should probably show up within two orders of magnitude of the present upper limit. It is therefore very interesting to look for these decays as tests of ETC theories, though it is possible that whatever mechanism may suppress ETC contributions to other bizarre amplitudes such as K^0-$\overline{K}^0$ mixing, etc., might also suppress these decay modes.

3.3. $K_L^0 \to \mu^+\mu^-$ AND $K_L^0 \to \gamma\gamma$

We should perhaps note at this point that the classic $|\Delta S| = 1$ semileptonic neutral interactions $K_L^0 \to \mu^+\mu^-$ and $\gamma\gamma$ do not raise any serious problems for SGEX in ETC theories. The rate for the $|\Delta G| = 1$ process $K_L^0 \to \mu^+\mu^-$ is expected to be suppressed by $O(\theta^2)$ relative to the $|\Delta G| = 0$ process $K_L^0 \to e^-\mu^+$, and the experimental limit [17] on the latter is more stringent. The process $K_L^0 \to \gamma\gamma$ is expected to be dominated by the low-mass π^0 and η poles calculated as in the conventional GIM model [14]. Even PGB exchange contributions of the type discussed in sect. 4 would be negligible compared with these conventional contributions because the PGBs couple to quarks in proportion to the quark masses.

3.4. $\mu N \to eN$ CAPTURE

As our last semileptonic process we consider the anomalous muon capture process $\mu N \to eN$ which has $|\Delta G| = 1$. In order to have coherence at the nuclear vertex we need a vector-like quark coupling, so we assume a vector-like effective interaction of the type

$$\mathcal{L}^{\text{eff}}_{(\mu\bar{e})(q\bar{q})} = (\bar{e}\gamma_\mu \mu)\{\bar{d}\gamma^\mu d \text{ or } \bar{u}\gamma^\mu u\}\left(\tfrac{1}{2}g^2_{\text{ETC}}\right)\left\{\frac{\theta}{m^2_{\text{ETC}_d}}\right\}. \tag{26}$$

The calculation of Cahn and Harari [9] yields

$$\frac{\sigma(\mu N \to eN)}{\sigma(\mu N \to \nu N)} = \frac{6}{Z}\cdot 8\cdot\frac{9Z^2\theta^2}{\frac{1}{4}Z(g_V^2 + 3g_A^2)}\left(\frac{m_W^2}{m^2_{\text{ETC}_d}}\right)^2\left(\frac{g^2_{\text{ETC}}}{g^2}\right)^2, \tag{27}$$

and from the upper limit [19] of 1.5×10^{-10} on this process, we find a lower limit

$$m^2_{\text{ETC}_d} \gtrsim 600 \text{ TeV}^2. \tag{28}$$

Comparing this with the value (8) estimated for $m^2_{\text{ETC}_d}$, we see that there is again no incompatibility between ETC and experiment, but another challenging process where ETC theories would expect a signal within an order of magnitude or so of the present experimental upper limit.

3.5. $\mu \to e\bar{e}e$

As a final SGEX contribution to an off-diagonal neutral process we consider the $|\Delta G| = 1$ decay $\mu \to e\bar{e}e$. Here we might expect the relevant effective lagrangian to take the form

$$\mathcal{L}^{\text{eff}}_{(\bar{e}\mu)(\bar{e}e)} = \left(\tfrac{1}{2}g^2_{\text{ETC}}\right)\{\bar{e}_L\gamma_\mu\mu_L, \bar{e}_R\gamma_\mu\mu_R\}\{\bar{e}_L\gamma^\mu e_L, \bar{e}_R\gamma^\mu e_R\}\left(\frac{\theta}{m^2_{\text{ETC}_d}}\right), \tag{29}$$

where we have put in one of the lightest quark ETC although it is possible that a different boson (heavier because $m_e < m_d$?) may contribute to purely leptonic processes. Using the LL interaction in eq. (29) we find

$$\frac{\Gamma(\mu \to e\bar{e}e)}{\Gamma(\mu \to e\bar{\nu}\nu)} = \frac{2\left(g_{\mathrm{ETC}}^2 / m_{\mathrm{ETC}_d}^2\right)^2 \theta^2}{\left(g^2/m_{\mathrm{W}}^2\right)^2}, \tag{30}$$

where we have included a factor of 2 from the identical electrons in the final state. Taking the upper limit [20] of 1.9×10^{-9} on this ratio, we get

$$m_{\mathrm{ETC}_d}^2 \gtrsim 200\ \mathrm{TeV}^2, \tag{31}$$

which is clearly compatible with the estimate in table 2. Again, ETC theories suggest that an effect may be found within two orders of magnitude of the present experimental upper limit.

4. Single pseudo-Goldstone (SPEX)

It is generally agreed [6–8] that there should be many light pseudo-Goldstone bosons (PGBs) in ETC theories. Generically, their couplings to fermions are found to be of order

$$g_{\mathrm{P\bar{f}f}} = \mathrm{O}\left(\frac{m_{\mathrm{f}}}{F_\pi}\right), \tag{32}$$

and we will assume this form, putting in a "Cabibbo-suppression" factor of θ for each change of generation at a vertex. The PGBs include coloured states [6,7] with masses O(300 GeV) which decay into $q\bar{q}$ or $q\bar{\ell}$ combinations according to whether they are octets or triplets. There are also expected to be colour singlet states with masses which are considerably smaller. Just how small is not very clear, but they could [7,8] be as low as a few GeV. The exchanges of these bosons (fig. 4) may give

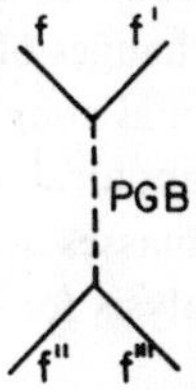

Fig. 4. The single pseudo-Goldstone boson exchange (SPEX) contribution to an effective four-fermion lagrangian.

rise to significant flavour-changing neutral interactions comparable to those induced by Higgs exchange in models with explicit scalar fields.

4.1. K_1-K_2 AND D_1-D_2

We first consider their contributions to $|\Delta S|=2$ and $|\Delta C|=2$ processes, for which we estimate

$$\mathcal{L}^{\text{eff}}_{(\bar{s}d)^2} = \{\bar{s}_L d_R, \bar{s}_R d_L\}\{\bar{s}_L d_R, \bar{s}_R d_L\}\frac{m_s^2}{F_\pi^2}\frac{\theta^2}{m_P^2}, \tag{33}$$

$$\mathcal{L}^{\text{eff}}_{(\bar{c}u)^2} = \{\bar{c}_L u_R, \bar{c}_R u_L\}\{\bar{c}_L u_R, \bar{c}_R u_L\}\frac{m_c^2}{F_\pi^2}\frac{\theta^2}{m_P^2}. \tag{34}$$

If we extend the vacuum insertion approximation of Gaillard and Lee [14] to the effective interactions (33),(34), then we must replace

$$\langle \bar{K}|\bar{s}\gamma_\mu\gamma_5 d|0\rangle\langle 0|\bar{s}\gamma_\mu\gamma_5 d|K\rangle = f_K^2 m_K^2 \tag{35}$$

by

$$\langle \bar{K}|\bar{s}\gamma_5 d|0\rangle\langle 0|\bar{s}\gamma_5 d|K\rangle = \frac{f_K^2 m_K^4}{(m_s+m_d)^2}. \tag{36}$$

We then have from each of the terms in (33)

$$\frac{\Delta m_K}{m_K} = \frac{f_K^2}{F_\pi^2}\frac{m_K^2}{m_P^2}\theta^2. \tag{37}$$

Requiring that this be less than the experimental value of 0.7×10^{-14} means that

$$m_P^2 \gtrsim 4\times 10^7\theta^2\ \text{GeV}^2. \tag{38}$$

Although we have not been careful about colour factors, it may be that the limit (38) poses a challenge in so far as the exchanges of octet PGBs are concerned. However, it is clear that the flavour-non-diagonal couplings of colour singlet PGBs must be much more suppressed than the canonical $\theta^2 \simeq \sin^2\theta_C \approx \frac{1}{20}$, by a factor which depends on their masses. If their masses were O(3) GeV, then the suppression required would be 10^6. A similar analysis for the D_1-D_2 system leads to

$$\frac{\Delta m_D}{m_D} = \frac{f_D^2}{F_\pi^2}\frac{m_D^2}{m_P^2}\theta^2. \tag{39}$$

Taking [17] an experimental upper limit of 10^{-12} on this quantity, we find

$$m_{\mathrm{P}}^2 \gtrsim 7\times 10^6 \theta^2\ \mathrm{GeV}^2, \tag{40}$$

and, hence, another strong need for strong suppressions in the flavour-changing couplings of the PGBs, particularly for the colour singlet ones.

4.2. $\mathrm{K}_{\mathrm{L}}^0 \to e\mu$

As another example of SPEX let us look at the $\mathrm{K}_{\mathrm{L}}^0 \to \mathrm{e}^- \mu^+$ decay. We consider an effective lagrangian of the form

$$\mathcal{L}^{\mathrm{eff}}_{(\bar{s}d)(\bar{\mu}e)} = \frac{m_\mu m_s}{F_\pi^2 m_{\mathrm{P}}^2}\theta^2(\bar{\mu}\gamma_5 e)(\bar{s}\gamma_5 d). \tag{41}$$

An elementary calculation similar to that of appendix B in ref. [8] yields

$$\frac{\Gamma(\mathrm{K}_{\mathrm{L}}^0 \to \mu^+\mathrm{e}^-)}{\Gamma(\mathrm{K}^+ \to \mu^+\nu_\mu)} = 16\left(\frac{\theta^2 m_{\mathrm{W}}^2 m_{\mathrm{K}}^2}{g^2 F_\pi^2 m_{\mathrm{P}}^2 \sin\theta_{\mathrm{C}}}\right)^2. \tag{42}$$

Taking the upper limit [17] of 0.63×10^{-9} on this ratio leads to the limit

$$m_{\mathrm{P}}^2 \gtrsim 2\times 10^5 \theta^2\ \mathrm{GeV}^2. \tag{43}$$

This limit is clearly less stringent that those we found earlier from considering the K^0-$\overline{\mathrm{K}}^0$ and D^0-$\overline{\mathrm{D}}^0$ systems in eqs. (38) and (40). This parallels our finding in sect. 3 that the constraints on semileptonic flavour-changing neutral interactions were not as severe as those from purely non-leptonic processes [cf. eqs. (13) and (23)]. Therefore, any trick that suppresses the off-diagonal neutral PGB couplings sufficiently to be consistent with conditions (38) and (40) will also be automatically consistent with (43). As pointed out elsewhere [13], such a trick is available in the form of the requirement called "monophagy" that all fermions of a given charge acquire their masses from the same technifermion condensate—the analogue of the Higgs naturalness condition (iii) outlined in sect. 1. The crossed channel exchange of a coloured leptoquark PGB is not automatically suppressed in monophagic models, but one may estimate [13] that their exchange gives a rate for $\mathrm{K}_{\mathrm{L}}^0 \to \mathrm{e}^-\mu^+$ just consistent with the experimental limit. We will not continue in this section to reconsider all the other flavour-changing semileptonic and purely leptonic neutral processes ($\mathrm{K}^+ \to \pi^+\mathrm{e}^-\mu^+$, $\mu\mathrm{N} \to \mathrm{eN}$, $\mu \to \mathrm{e\bar{e}e}$) that we considered in the context of SGEX in sect. 3. Based on our experience with the previous calculations of this and the previous section, we believe that no significant new challenges would arise from these processes that are not already met by the trick of monophagy [13].

5. Higher order exchanges (HOEX)

So far we have only been considering second-order single boson exchange diagrams. In this section we go on to consider some higher order diagrams—double gauge boson exchange DGEX (cf. fig. 5) and a vertex correction to single gauge boson emission (cf. fig. 6). We will see that in cases where both SGEX and DGEX are possible then the latter is considerably smaller and poses no serious challenge to ETC theories. We also find that flavour-changing vertex corrections to Z^0 couplings are not troublesome. However, we do find that the purely leptonic $\mu \to e\gamma$ decay challenges ETC theories.

5.1. K_1-K_2 AND D_1-D_2

We compute the diagram of fig. 5 using an ETC-q-Q vertex of the form

$$\mathcal{L}_I = \sqrt{\tfrac{1}{2}}\, g_{ETC} \bar{q}_L \gamma_\mu F_L E^\mu , \tag{44}$$

where F is a technifermion, which yields an expression [14] for the box

$$\mathcal{L}^{eff}_{(\bar{s}d)^2} = \tfrac{1}{8} g^4_{ETC} \theta^2 \int \frac{d^4k}{(2\pi)^4} \frac{\bar{s}_L\{\gamma^\alpha \not{k} \gamma^\beta\} d_L \bar{s}_L \{\gamma_\beta \not{k} \gamma_\alpha\} d_L}{(k^2)^2 (k^2 - m_E^2)^2} . \tag{45}$$

The technifermion propagator is expected to be modified at small momenta $k = O(\Lambda_{TC})$, but the typical momenta in the d^4k integration are $O(m_E)$ so that this modification may be neglected. The expression (45) may be evaluated to give

$$\mathcal{L}^{eff}_{(\bar{s}d)^2} = \frac{g^4_{ETC}\theta^2}{128 m^2_{ETC_s} \pi^2} (\bar{s}_L \gamma_\alpha d_L)^2 , \tag{46}$$

where we have put in the mass of the ETC_s boson, as this is likely to be the lightest which contributes to $\mathcal{L}^{eff}_{(\bar{s}d)^2}$ in order θ^2. We expect similarly to get effective LR and RR interactions. The latter will be of the same magnitude as (46), whereas the LR

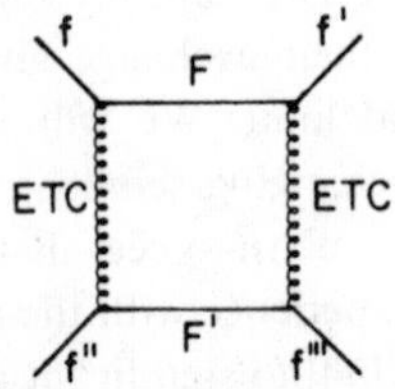

Fig. 5. A double gauge boson exchange (DGEX) box diagram.

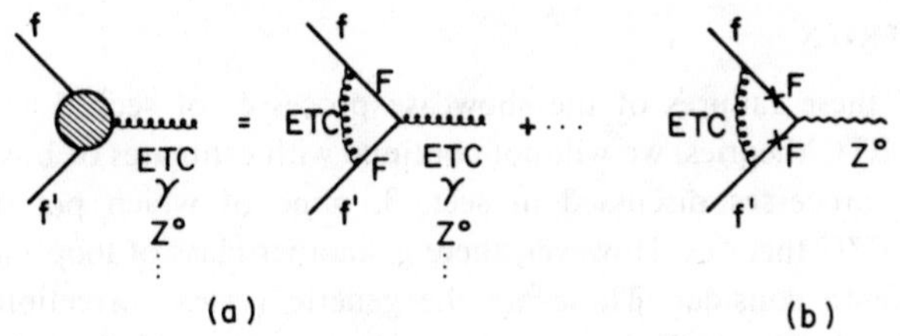

Fig. 6. (a) A typical vertex correction diagram, and (b) a double helicity flip part of it.

diagrams should be a factor of 4 larger. Let us evaluate (46): setting $g_{\mathrm{ETC}} \approx 1$ and using the same upper limit as in eq. (12) we see that from the box diagrams

$$\frac{\theta^2}{128\pi^2 m^2_{\mathrm{ETC}_s}} \lesssim \tfrac{1}{2} \times 10^{-12}\ \mathrm{GeV}^{-2}, \tag{47}$$

which will give a bound on $m^2_{\mathrm{ETC}_s}$ of order 600 times weaker than in eq. (13):

$$m^2_{\mathrm{ETC}_s} \gtrsim 80\ \mathrm{TeV}^2, \tag{48}$$

which is almost compatible with the estimate of $m^2_{\mathrm{ETC}_s}$ in table 2. Adding to the contribution of eq. (46) the corresponding RR and LR contributions, and taking $g_{\mathrm{ETC}} \sim \sqrt{2}$ instead of 1, we would find a one order of magnitude problem for the ETC theories in the real part of the K_1-K_2 mass matrix, which is less than the problem found at the SPEX level. One can go through a similar analysis to deduce a box diagram

$$\mathcal{L}^{\mathrm{eff}}_{(\bar{c}u)^2} = \frac{g^4_{\mathrm{ETC}}\theta^2}{128\pi^2 m^2_{\mathrm{ETC}_c}} (\bar{c}_L \gamma_\alpha u_L)^2, \tag{49}$$

from which we would deduce via the limit of eq. (19) derived from the observed absence of D^0-$\bar{D}^0$ mixing that

$$\frac{\theta^2}{128\pi^2 m^2_{\mathrm{ETC}_c}} \lesssim 3 \times 10^{-11}\ \mathrm{GeV}^{-2}, \tag{50}$$

and hence that

$$m^2_{\mathrm{ETC}_c} \gtrsim 13\ \mathrm{TeV}^2. \tag{51}$$

This is compatible with the estimated ETC boson mass in table 2, and we therefore detect no challenge to ETC theories.

5.2. THE $\bar{s}d\gamma$ VERTEX

In view of these failures of the showcase processes of sect. 3 to pose serious challenges to ETC theories, we will not continue with estimates of box diagrams for all the other processes discussed in sect. 3, none of which posed any serious problems for ETC theories. However, there is another class of loop diagrams which we would like to consider. These are the generic vertex correction diagrams illustrated in fig. 6. They will, in general, give flavour-changing contributions to the couplings of the Weinberg-Salam Z^0 boson [14]. The logarithmic divergences of these diagrams will be absorbed into the definition of the flavour-diagonal lowest-order vertex. As for the finite parts, the leading pieces which have no helicity flip on the internal technifermion line are independent of m_{ETC} and will therefore be subject to a GIM subtraction. The leading contributions to flavour-changing Z^0 vertices involve two helicity flips on the technifermion line as in fig. 4b and give the order of magnitude

$$\left(\frac{g}{2\cos\theta_{\text{W}}}\right)\frac{\alpha_{\text{ETC}}}{2\pi}\theta\left(\bar{s}\gamma_\mu{}_{\text{R}}^{\text{L}}d\right)Z^{0\mu}\left(\frac{\Lambda_{\text{TC}}^2}{m_{\text{ETC}_s}^2}\right), \tag{52}$$

and similarly for other $|\Delta G| = 1$ neutral interactions. In order for these contributions to be no larger than the typical GIM suppressed $|\Delta G| = 1$ neutral currents we need

$$\frac{\Lambda_{\text{TC}}^2}{m_{\text{ETC}_s}^2} \lesssim \frac{m_c^2}{m_{\text{W}}^2}. \tag{53}$$

Taking $\Lambda_{\text{TC}}^2 \approx 0.1\ \text{TeV}^2$, the limit (53) becomes

$$m_{\text{ETC}_s}^2 \gtrsim 200\ \text{TeV}^2. \tag{54}$$

It is difficult to be precise about this number because it depends on the strong coupling régime of the TC interactions. The value of m_{ETC_s} shown in table 2 grazes the limit (54), but there is no acute problem. For example, it is easy to convince oneself that the value of $m_{\text{ETC}_s}^2$ in table 2 gives a Z^0 exchange contribution compatible with the experimental value of $\Gamma(K_L^0 \to \mu^+\mu^-)/\Gamma(K^+ \to \mu^+\nu)$. Also, if we estimate the order of magnitude of $|\Delta S| = 2$ exchange mediated by Z^0 bosons using eq. (52), we find that the limit (53) comfortably assures compatibility with experiment for the K^0-$\bar{K}^0$ system. While we have not attempted a detailed comparison of the ETC vertex correction (52) with all $|\Delta S| \neq 0$ neutral interaction values and limits, we do not expect any of them to pose any serious challenges for ETC theories.

5.3. $\mu \to e\gamma$

As a final one-loop calculation we consider the $\mu \to e\gamma$ decay in ETC theories, as calculated from the diagrams shown in fig. 7. The decay rate for this process can be

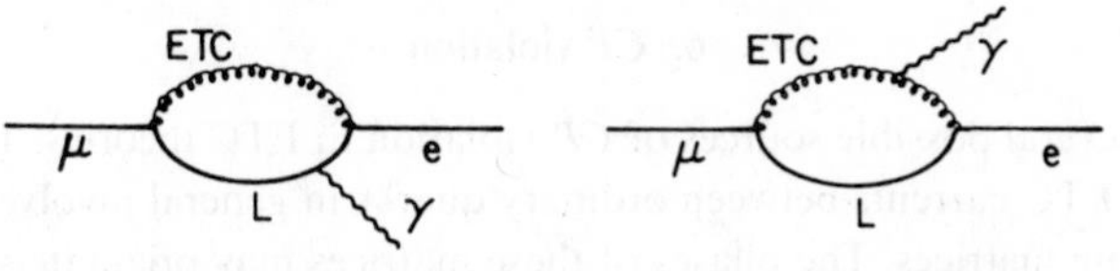

Fig. 7. Diagrams contributing to $\mu \to e\gamma$ in ETC theories.

written as

$$\Gamma(\mu \to e + \gamma) \approx \frac{m_\mu}{8\pi} |F|^2_{\mathrm{LL,LR,RL,RR}}, \tag{55}$$

where we have indicated explicitly the contributions coming from different combinations of L and R helicities at the two ETC $f\bar{f}$ vertices. The distinction is important here because the orders of magnitude of the graphs are very different. For the LL and RR cases it is found [21] that

$$F_{\mathrm{LL,RR}} = \frac{e m_\mu^2 g_{\mathrm{ETC}}^2 \theta}{32\pi^2 m_{\mathrm{ETC}_\mu}^2} \times \mathrm{O}(1), \tag{56}$$

while

$$F_{\mathrm{LR,RL}} = \frac{e m_\mu \Lambda_{\mathrm{TC}} g_{\mathrm{ETC}}^2 \theta}{32\pi^2 m_{\mathrm{ETC}_\mu}^2} \times \mathrm{O}(1). \tag{57}$$

Taking the larger contribution (57) which is only present if the ETC bosons have both L and R couplings, which they must have in order to generate masses, we find

$$\Gamma(\mu \to e\gamma) \approx \frac{m_\mu^3}{64\pi^2} \frac{\alpha^2}{32\pi^2} \frac{g_{\mathrm{ETC}}^4 \Lambda_{\mathrm{TC}}^2}{m_{\mathrm{ETC}_\mu}^4} \theta^2. \tag{58}$$

Taking an experimental upper limit of order (10^{-10}) for the branching ratio of $\mu \to e\gamma$ relative to $\mu \to e\nu\bar{\nu}$ we deduce from eq. (58) using $g_{\mathrm{ETC}} \approx 1$, $\Lambda_{\mathrm{TC}}^2 \approx 0.1$ TeV^2 and $\theta^2 \approx \frac{1}{20}$ as usual that

$$m_{\mathrm{ETC}_\mu}^2 \gtrsim 8 \times 10^4 \ \mathrm{TeV}^2. \tag{59}$$

This compares with a value of $m_{\mathrm{ETC}_\mu}^2 = 2400$ TeV^2 that we would have deduced from the arguments of sect. 2. We therefore conclude that the experimental upper limit on the $\mu \to e\gamma$ decay mode presents a challenge for ETC theories if the assumption of naïve Cabibbo-like ETC couplings with mixing angles $\theta \approx \theta_C$ is also applicable to leptons, as we assumed in order to derive the bound (59) from eq. (58).

6. *CP* violation

There are several possible sources of *CP* violation in ETC theories. The exchanges of horizontal ETC currents between ordinary quarks in general involve *CP*-violating complex mixing matrices. The phases of these matrices may originate spontaneously at some mass scale and cannot be rotated away by chiral transformations of any of the fermion fields. These phases are *a priori* expected to be of order unity. Thus processes mediated by horizontal currents may be expected to exhibit maximal *CP* violation. In view of this and the observed ratio of $O(10^{-3})$ between the imaginary part and real parts of the K^0-$\overline{K}^0$ mass matrix, it seems necessary to suppress all horizontal processes mediating K^0-$\overline{K}^0$ mixing a thousand times more strongly than we deduced in earlier sections on the basis of the real part alone.

CP violation in electroweak interactions may arise in the standard Kobayashi-Maskawa [10] way through phases that appear in the charged weak current coupling matrix $U_d^{L+} U_u^L$. These phases may also arise spontaneously at some mass-scale and cannot be removed by allowed chiral rotations of any fermion fields. This KM *CP* violation predicts [22] a value of $10^{-30\pm1}$ $e\cdot$cm for the neutron's electric dipole moment d_N, and is probably [23] consistent with the small imaginary part of the K_1-K_2 mass matrix which is observed. This source of *CP* violation poses no challenges for ETC theories.

Recently another "U(1)" type of *CP* violation has been suggested [11], and it has been estimated to lead to $d_N \sim 10^{-24}$ to 10^{-25} $e\cdot$cm. This would be very exciting because d_N would then be accessible to the forthcoming generation of experiments. Unfortunately, this ELP mechanism for *CP* violation also suggests that the *CP*-violating part of the K_1-K_2 mass matrix should be comparable with the *CP*-conserving part. It would therefore be desirable to demonstrate that whatever mechanism suppresses *CP* violation in the K^0 system by perhaps six orders of magnitude does not also suppress d_N by a similar amount.

In fact, it may be that in the absence of any suppression the ELP source of *CP* violation may actually give a larger value of d_N than has been suggested. To see this, let us briefly review some of the results of ref. [11]. Consider the chirality breaking part of the effective lagrangian in an ETC theory which describes the interactions of ordinary particles well below ~ 300 GeV. This effective lagrangian is obtained by integrating out all the heavy non-ordinary degrees of freedom such as technifermions, ETC bosons, horizontal bosons, etc. It consists of a series of operators whose lowest dimensional, strongly interacting representatives are

$$\delta\mathcal{L}^{\text{eff}} \ni A_{\ell r}\bar{q}_{\ell_L} q_{r_R} + B_{\ell r}\bar{q}_{\ell_L}\sigma_{\mu\nu}\left(\tfrac{1}{2}\lambda^a\right) q_{r_R} G_a^{\mu\nu} + C_{\ell r,\ell' r'}\bar{q}_{\ell_L} q_{r_R}\bar{q}_{\ell'_L} q_{r'_R} + \text{h.c.} \qquad (60)$$

Here the indices ℓ, r label left- and right-handed helicity quark flavours and $G^a_{\mu\nu}$ is the gluon field strength. The coefficients A, B, C are each expected to have phases of order unity which arise spontaneously in the higher energy sectors. The dominant graphs contributing to A, B and C are shown in fig. 8. Note that the ETC radiative correction to the ETC vertex in fig. 8b is only necessary if the exchange of horizontal ETC bosons does not already produce an interaction of the form $\bar{q}\sigma_{\mu\nu}q\bar{Q}\sigma_{\mu\nu}Q$. If such an interaction is present, then d_N immediately comes out too large, as was noted in ref. [11]. Note also that the gluonic correction to the ETC vertex which was considered in ref. [11] is smaller than the ETC correction that we consider here. From these graphs one easily estimates A, B and C to be

$$A = m_q = \frac{\langle 0|\bar{F}F|0\rangle}{v_{ETC}^2} \equiv \frac{\Lambda_{TC}^3}{v_{ETC}^2},$$

$$B \approx \frac{\alpha_{ETC}(m_{ETC})}{2\pi} g_C \left(\frac{\Lambda_{TC}}{v_{ETC}^2}\right), \tag{61}$$

$$C \approx \frac{1}{v_{ETC}^2}.$$

Here, as before, $v_{ETC} \equiv m_{ETC}/g_{ETC}$ and $\Lambda_{TC} = \langle \bar{F}F \rangle^{1/3} \simeq 300$ GeV is the technicolour scale. A crude estimate of the contributions of A, B and C to the vacuum energy can now be obtained from (60) and (61) by replacing $\bar{q}q \to \Lambda_C^3$ and $G_{\mu\nu}^a \to \Lambda_C^2$, where $\Lambda_C \equiv \langle \bar{q}q \rangle^{1/3} \simeq 240$ MeV is the colour scale. This yields

$$A\Lambda_C^3 = \frac{\Lambda_{TC}^3\Lambda_C^3}{v_{ETC}^2} \equiv a,$$

$$B\Lambda_C^5 = \left(\frac{\Lambda_{TC}^3\Lambda_C^3}{v_{ETC}^2}\right)\left(\frac{\Lambda_C}{\Lambda_{TC}}\right)^2 g_C \frac{\alpha_{ETC}(m_{ETC})}{2\pi} \equiv b, \tag{62}$$

$$C\Lambda_C^6 = \left(\frac{\Lambda_{TC}^3\Lambda_C^3}{v_{ETC}^2}\right)\left(\frac{\Lambda_C}{\Lambda_{TC}}\right)^3 \equiv c.$$

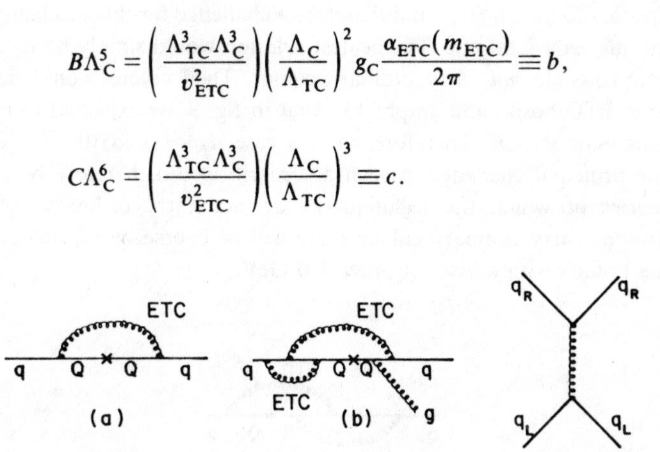

Fig. 8. The lowest order diagrams in ETC theories contributing (a) to A, (b) to B and (c) to C.

In ref. [11] the order of magnitude of d_N is estimated to be

$$d_N \approx \left(\frac{b}{a} + \frac{c}{a}\right) 10^{-15}\ e\cdot\text{cm} \tag{63}$$

which becomes, according to eq. (62),

$$d_N \simeq \left(g_C \frac{\alpha_{ETC}(m_{ETC})}{2\pi} \left(\frac{\Lambda_C}{\Lambda_{TC}}\right)^2 + \left(\frac{\Lambda_C}{\Lambda_{TC}}\right)^3 \right) 10^{-15}\ e\cdot\text{cm}. \tag{64}$$

Since $\Lambda_C/\Lambda_{TC} \approx 10^{-3}$, the c/a term in (63) yields a contribution to d_N of order $\sim 10^{-24}\ e\cdot$cm, whereas the b/a term gives

$$d_N \approx g_C \frac{\alpha_{ETC}(m_{ETC})}{2\pi} 10^{-21}\ e\cdot\text{cm}. \tag{65}$$

We guess that $\alpha_{ETC}(m_{ETC}) \approx 1$ since one has in the leading logarithmic approximation

$$\frac{2\pi}{\alpha_{ETC}(m_{ETC})} \simeq b_{ETC} \ln\left(\frac{m_{ETC}}{\Lambda_{TC}}\right), \qquad b_{ETC} = O(10), \tag{66}$$

so that for $m_{ETC} \sim 1$ TeV corresponding to a top quark of mass 30 GeV one finds $\alpha_{ETC}(m_{ETC})/2\pi \sim 10^{-1}$. Inserting this value into eq. (65) we find $d_N \approx 10^{-22}\ e\cdot$cm from the ELP mechanism.

This appears to be too large and therefore a challenge for this mechanism [11], but it is by no means a fatal one. The neatest solution would simply be to assume that the techniquarks do *not* carry ordinary colour. Then colour would have to flow through the ETC boson and graphs like that in fig. 9 are expected to make a very small contribution to B. Therefore, in this case $d_N \approx (c/a)10^{-15}\ e\cdot\text{cm} \approx 10^{-24}\ e\cdot$cm. The principal challenge in such a scenario would therefore be to construct ETC theories in which the techniquarks do not carry ordinary colour. If no technifermions carry ordinary colour there will of course be no coloured pseudo-Goldstone bosons with masses of order 300 GeV.

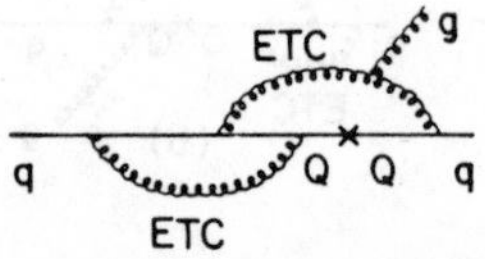

Fig. 9. A smaller contribution to B.

7. Prospects

What are we to conclude from the order of magnitude estimates presented in this paper and summarized in table 1? As optimists, we may hope that the rare processes will once again prove an excellent guide for model building. It is constructive to adopt this point of view and try to see what one might learn from the necessity to suppress rare processes. In the rest of this section we will briefly mention some such possible suppression mechanisms whose developments are still in their pre-embryonic stages.

The first such suppression mechanism is "monophagy" which has already been mentioned [13]. It is an obvious extension of the Glashow-Weinberg mechanism and says that fermions of a given charge get their masses from a single technifermion condensate. Monophagy cures only SPEX but not SGEX.

Another possible mechanism which smells promising is to introduce an $SU(3)_H$ symmetry acting on $\underset{\sim}{d}$, $\underset{\sim}{s}$ and $\underset{\sim}{b}$, or an $SU(2)_H$ symmetry acting on $\underset{\sim}{d}$ and $\underset{\sim}{s}$ (where the tildes indicate the corresponding generations or families), which is unbroken at the ETC scale. Such a symmetry would render $\underset{\sim}{s}$ and $\underset{\sim}{d}$ quarks indistinguishable at ETC energies where the disastrous SGEX mediated mixings normally occur, and thus these problems would be avoided. Another way of seeing the same thing is to note that strangeness conservation is a subsymmetry of $SU(3)_H$ or $SU(2)_H$ and thus valid at ETC energies. This scenario has to be implemented with some low energy (~ 300 GeV ?) mechanism (spontaneous breaking ?) incorporated in order to break the $SU(3)_H$ or $SU(2)_H$ symmetry and thus account for the $\underset{\sim}{d}$-$\underset{\sim}{s}$-$\underset{\sim}{b}$ or $\underset{\sim}{d}$-$\underset{\sim}{s}$ splitting, *without* simultaneously introducing dangerous strangeness-changing light pseudo-Goldstone bosons and/or gauge bosons. No details of how this might be done have been worked out.

There are other ways of introducing strangeness conservation at high energies which involve the existence of strangeness, charm, etc. carrying techniquarks. They would introduce the naturalness conditions (i) and (ii) of sect. 1 into ETC theories and open the way to GIM-like suppressions of gauge boson exchanges [6,2]. These models suffer from the proliferation of possibly non-innocent pseudo-Goldstone bosons associated with the replication of techniquarks, and lack aesthetic appeal.

In conclusion, a solution to all the problems addressed in this paper is not yet in sight. The great wealth of potential problems discussed here suggests that the solution, if any, will have to involve a symmetry valid at ETC energies $E > 300$ GeV, or maybe another GIM-like mechanism. In the meantime, there are ways in which our experimental colleagues can challenge ETC theories further by more searching studies of rare processes, in particular D_1-D_2 mixing, $K_L^0 \to e\mu$ and $K^+ \to \pi^+ e\mu$, $\mu N \to eN$ capture and $\mu \to e\gamma$ decay. Optimists hope to see ETC effects in these processes.

One of us (S.D.) thanks E. Farhi, S. Raby and L. Susskind for discussions, two NSF grants PHY 78-26847 for financial support at Stanford, and the CERN Theory

Division for its hospitality while much of this work was done. The other (J.E.) thanks M.K. Gaillard, D.V. Nanopoulos and P. Sikivie for useful discussions. We both of us thank the Orthodox Academy of Crete, where this work was started, for its kind hospitality under difficult circumstances which we are glad have now been successfully resolved.

References

[1] S.L. Glashow, J. Iliopoulos and L. Maiani, Phys. Rev. D2 (1970) 1285
[2] S.L. Glashow and S. Weinberg, Phys. Rev. D15 (1977) 1958;
E.A. Paschos, Phys. Rev. D15 (1977) 1966
[3] M.S. Chanowitz, J. Ellis and M.K. Gaillard, Nucl. Phys. B128 (1977) 506
[4] J. Schwinger, Phys. Rev. 125 (1962) 397; 128 (1969) 2425;
R. Jackiw and K. Johnson, Phys. Rev. D8 (1973) 2386;
J.M. Cornwall and R.E. Norton, Phys. Rev. D8 (1973) 3338;
M.A.B. Bég and A. Sirlin, Ann. Rev. Nucl. Sci. 24 (1974) 379;
S. Weinberg, Phys. Rev. D13 (1976) 974; D19 (1979) 1277;
L. Susskind, Phys. Rev. D20 (1979) 2619
[5] S. Dimopoulos and L. Susskind, Nucl. Phys. B155 (1979) 237;
E. Eichten and K.D. Lane, Phys. Lett. 90B (1980) 125
[6] S. Dimopoulos, Nucl. Phys. B168 (1980) 69
[7] M.E. Peskin, Nucl. Phys. B175 (1980) 197
[8] S. Dimopoulos, S. Raby and G.L. Kane, Nucl. Phys. B172 (1980) 509
[9] G.L. Kane and R. Thun, Phys. Lett. 94B (1980) 513;
R.N. Cahn and H. Harari, Nucl. Phys. B176 (1980) 135
[10] M. Kobayashi and K. Maskawa, Prog. Theor. Phys. 49 (1973) 652
[11] E. Eichten, K.D. Lane and J.P. Preskill, Phys. Rev. Lett. 45 (1980) 225
[12] S. Wojcicki, Proc. 1978 SLAC Summer Inst. on Particle physics, ed. M.C. Zipf, SLAC-215 (1978), p. 193
[13] J. Ellis, M.K. Gaillard, D.V. Nanopoulos and P. Sikivie, Nucl. Phys. B182 (1981) 529
[14] M.K. Gaillard and B.W. Lee, Phys. Rev. D10 (1974) 897
[15] F. Wilczek, A. Zee, R. Kingsley and S.D. Treiman, Phys. Rev. D12 (1975) 2765
[16] R. Shrock and S.B. Treiman, Phys. Rev. D19 (1979) 2148
[17] A.R. Clark et al., Phys. Rev. Lett. 26 (1970) 1667
[18] A. Diamant-Berger et al., Phys. Lett. 62B (1976) 485
[19] A. Bardetscher et al., Contribution to 19th Int. Conf. on High-energy physics, Tokyo (1978), cited in Proc. by G. Altarelli, p. 412
[20] S. Korenchenko et al., cited by Cahn and Harari, ref. [9]
[21] T.-P. Cheng and L.-F. Li, Phys. Rev. D16 (1977) 1425;
B.W. Lee and R. Shrock, Phys. Rev. D16 (1977) 1444
[22] J. Ellis and M.K. Gaillard, Nucl. Phys. B150 (1979) 141
[23] J. Ellis, M.K. Gaillard and D.V. Nanopoulos, Nucl. Phys. B109 (1979) 213

IX. CRITERIA FOR SYMMETRY BREAKING AND MASSLESS COMPOSITE FERMIONS

NATURALNESS, CHIRAL SYMMETRY, AND SPONTANEOUS

CHIRAL SYMMETRY BREAKING

G. 't Hooft

Institute for Theoretical Fysics

Utrecht, The Netherlands

ABSTRACT

A properly called "naturalness" is imposed on gauge theories. It is an order-of-magnitude restriction that must hold at all energy scales μ. To construct models with complete naturalness for elementary particles one needs more types of confining gauge theories besides quantum chromodynamics. We propose a search program for models with improved naturalness and concentrate on the possibility that presently elementary fermions can be considered as composite. Chiral symmetry must then be responsible for the masslessness of these fermions. Thus we search for QCD-like models where chiral symmetry is not or only partly broken spontaneously. They are restricted by index relations that often cannot be satisfied by other than unphysical fractional indices. This difficulty made the author's own search unsuccessful so far. As a by-product we find yet another reason why in ordinary QCD chiral symmetry must be broken spontaneously.

III1. INTRODUCTION

The concept of causality requires that macroscopic phenomena follow from microscopic equations. Thus the properties of liquids and solids follow from the microscopic properties of molecules and atoms. One may either consider these microscopic properties to have been chosen at random by Nature, or attempt to deduce these from even more fundamental equations at still smaller length and time scales. In either case, it is unlikely that the microscopic equations contain various free parameters that are carefully adjusted by Nature to give cancelling effects such that the macroscopic systems have some special properties. This is a philosophy which we would like to apply to the unified gauge theories: the effective interactions at a large length scale, corresponding to a low energy scale μ_1, should follow from the properties at a much smaller length scale, or higher energy scale μ_2, without the requirement that various different parameters at the energy scale μ_2 match with an accuracy of the order of μ_1/μ_2. That would be unnatural. On the other hand, if at the energy scale μ_2 some parameters would be very small, say

$$\alpha(\mu_2) = \mathcal{O}(\mu_1/\mu_2) \ , \tag{III1}$$

then this may still be natural, provided that this property would not be spoilt by any higher order effects. We now conjecture that the following dogma should be followed:
- at any energy scale μ, a physical parameter or set of physical parameters $\alpha_i(\mu)$ is allowed to be very small only if the replacement $\alpha_i(\mu) = 0$ would increase the symmetry of the system. -
In what follows this is what we mean by naturalness. It is clearly a weaker requirement than that of P. Dirac[1] who insists on having no small numbers at all. It is what one expects if at any mass scale $\mu > \mu_0$ some ununderstood theory with strong interactions determines a spectrum of particles with various good or bad symmetry properties. If at $\mu = \mu_0$ certain parameters come out to be small, say 10^{-5}, then that cannot be an accident; it must be the consequence of a near symmetry.

For instance, at a mass scale

$$\mu = 50 \text{ GeV},$$

the electron mass m_e is 10^{-5}. This is a small parameter. It is acceptable because $m_e = 0$ would imply an additional chiral symmetry corresponding to separate conservation of left handed and right handed electron-like leptons. This guarantees that all renormalizations of m_e are proportional to m_e itself. In sects. III2 and III3 we compare naturalness for quantum electrodynamics and ϕ^4 theory.

Gauge coupling constants and other (sets of) interaction constants may be small because putting them equal to zero would turn the gauge bosons or other particles into free particles so that they are separately conserved.

If within a set of small parameters one is several orders of magnitude smaller than another then the smallest must satisfy our "dogma" separately. As we will see, naturalness will put the severest restriction on the occurrence of scalar particles in renormalizable theories. In fact we conjecture that this is the reason why light, weakly interacting scalar particles are not seen.

It is our aim to use naturalness as a new guideline to construct models of elementary particles (sect. III4). In practice naturalness will be lost beyond a certain mass scale μ_0, to be referred to as "Naturalness Breakdown Mass Scale" (NBMS). This simply means that unknown particles with masses beyond that scale are ignored in our model. The NBMS is only defined as an order of magnitude and can be obtained for each renormalizable field theory. For present "unified theories", including the existing grand unified schemes, it is only about 1000 GeV. In sect. 5 we attempt to construct realistic models with an NBMS some orders of magnitude higher.

One parameter in our world is unnatural, according to our definition, already at a very low mass scale ($\mu_0 \sim 10^{-2}$ eV). This is the cosmological constant. Putting it equal to zero does not seem to increase the symmetry. Apparently gravitational effects do not obey naturalness in our formulation. We have nothing to say about this fundamental problem, accept to suggest that *only* gravitational effects violate naturalness. Quantum gravity is not understood anyhow so we exclude it from our naturalness requirements.

On the other hand it is quite remarkable that all other elementary particle interactions have a high degree of naturalness. No unnatural parameters occur in that energy range where our popular field theories could be checked experimentally. We consider this as important evidence in favor of the general hypothesis of naturalness. Pursuing naturalness beyond 1000 GeV will require theories that are immensely complex compared with some of the grand unified schemes.

A remarkable attempt towards a natural theory was made by Dimopoulos and Susskind [2)]. These authors employ various kinds of confining gauge forces to obtain scalar bound states which may substitute the Higgs fields in the conventional schemes. In their model the observed fermions are still considered to be elementary.

Most likely a complete model of this kind has to be constructed step by step. One starts with the experimentally accessible aspects of the Glashow-Weinberg-Salam-Ward model. This model is natural if one restricts oneself to mass-energy scales below 1000 GeV. Beyond 1000 GeV one has to assume, as Dimopoulos and Susskind do, that the Higgs field is actually a fermion-antifermion composite field. Coupling this field to quarks and leptons in order to produce their mass, requires new scalar fields that cause naturalness to break down at 30 TeV or so. Dimopoulos and Susskind speculate further on how to remedy this. To supplement such ideas, we toyed with the idea that (some of) the presently "elementary" fermions may turn out to be bound states of an odd number of fermions when considered beyond 30 TeV. The binding mechanism would be similar to the one that keeps quarks inside the proton. However, the proton is not particularly light compared with the characteristic mass scale of quantum chromodynamics (QCD). Clearly our idea is only viable if something prevented our "baryons" from obtaining a mass (eventually a small mass may be due to some secondary perturbation).

The proton ows its mass to spontaneous breakdown of chiral symmetry, or so it seems according to a simple, fairly successful model of the mesonic and baryonic states in QCD: the Gell-Mann-Lévy sigma model[3]. Is it possible then that in some variant of QCD chiral symmetry is not spontaneously broken, or only partly, so that at least some chiral symmetry remains in the spectrum of fermionic bound states? In this article we will see that in general in SU(N) binding theories this is not allowed to happen, i.e. chiral symmetry must be broken spontaneously.

III2. NATURALNESS IN QUANTUM ELECTRODYNAMICS

Quantum Electrodynamics as a renormalizable model of electrons (and muons if desired) and photons is an example of a "natural" field theory. The parameters α, m_e (and m_μ) may be small independently. In particular m_e (and m_μ) are very small at large μ. The relevant symmetry here is chiral symmetry, for the electron and the muon separately. We need not be concerned about the Adler-Bell-Jackiw anomaly here because the photon field being Abelian cannot acquire non-trivial topological winding numbers[4].

There is a value of μ where Quantum Electrodynamics ceases to be useful, even as a model. The model is not asymptotically free, so there is an energy scale where all interactions become strong:

$$\mu_o \simeq m_e \exp(6\pi^2/e^2 N_f) \ , \qquad \text{(III2)}$$

where N_f is the number of light fermions. If some world would be described by such a theory at low energies, then a replacement of the theory would be necessary at or below energies of order μ_o.

III3. ϕ^4-THEORY

A renormalizable scalar field theory is described by the Lagrangian

$$\mathcal{L} = -\tfrac{1}{2}(\partial_\mu \phi)^2 - \tfrac{1}{2}m^2\phi^2 - \frac{1}{4!}\lambda\phi^4 \ . \qquad \text{(III3)}$$

the interactions become strong at

$$\mu \simeq m \exp(16\pi^2/3\lambda) \ , \qquad \text{(III4)}$$

but is it still natural there?

There are two parameters, λ and m. Of these, λ may be small because $\lambda = o$ would correspond to a non-interacting theory with total number of ϕ particles conserved. But is small m allowed? If we put m = o in the Lagrangian (III3) then the symmetry is not enhanced*). However we can take both m and λ to be small, because if $\lambda = m = o$ we have invariance under

$$\phi(x) \to \phi(x) + \Lambda . \tag{III5}$$

This would be an approximate symmetry of a new underlying theory at energies of order μ_o. Let the symmetry be broken by effects described by a dimensionless parameter ε. Both the mass term and the interaction term in the effective Lagrangian (III3) result from these symmetry breaking effects. Both are expected to be of order ε. Substituting the correct powers of μ_o to account for the dimensions of these parameters we have

$$\lambda = \mathcal{O}(\varepsilon) , \quad m^2 = \mathcal{O}(\varepsilon\mu_o^2) . \tag{III6}$$

Therefore,

$$\mu_o = \mathcal{O}(m/\sqrt{\lambda}) . \tag{III7}$$

This value is much lower than eq. (III4). We now turn the argument around: if any "natural" underlying theory is to describe a scalar particle whose *effective* Lagrangian at low energies will be eq. (III3), then its energy scale cannot be given by (III4) but at best by (III7). We say that naturalness breaks down beyond $m/\sqrt{\lambda}$. It must be stressed that these are orders of magnitude. For instance one might prefer to consider λ/π^2 rather than λ to be the relevant parameter. μ_o then has to be multiplied by π. Furthermore, λ could be much smaller than ε because $\lambda = o$ separately also enhances the symmetry. Therefore, apart from factors π, eq. (III7) indicates a maximum value for μ_o.

Another way of looking at the problem of naturalness is by comparing field theory with statistical physics. The parameter m/μ would correspond to $(T-T_c)/T$ in a statistical ensemble. Why would the temperature T chosen by Nature to describe the elementary particles be so close to a critical temperature T_c? If $T_c \neq o$ then T may not be close to T_c just by accident.

III4. NATURALNESS IN THE WEINBERG-SALAM-GIM MODEL

The difficulties with the unnatural mass parameters only occur in theories with scalar fields. The only fundamental scalar

*) Conformal symmetry is violated at the quantum level.

field that occurs in the presently fashionable models is the Higgs field in the extended Weinberg-Salam model. The Higgs mass-squared, m_H^2, is up to a coefficient a fundamental parameter in the Lagrangian. It is small at energy scales $\mu \gg m_H$. Is there an approximate symmetry if $m_H \to o$? With some stretch of imagination we might consider a Goldstone-type symmetry:

$$\phi(x) \to \phi(x) + \text{const.} \tag{III8}$$

However we also had the local gauge transformations:

$$\phi(x) \to \Omega(x)\,\phi(x)\ . \tag{III9}$$

The transformations (III8) and (III9) only form a closed group if we also have invariance under

$$\phi(x) \to \phi(x) + C(x)\ . \tag{III10}$$

But then it becomes possible to transform ϕ away completely. The Higgs field would then become an unphysical field and that is not what we want. Alternatively, we could have that (III8) is an approximate symmetry only, and it is broken by all interactions that have to do with the symmetry (III9) which are the weak gauge field interactions. Their strength is $g^2/4\pi = \mathcal{O}(1/137)$. So at best we can have that the symmetry is broken by $\mathcal{O}(1/137)$ effects. Therefore

$$m_H^2/\mu^2 \gtrsim \mathcal{O}(1/137)\ .$$

Also the $\lambda\phi^4$ term in the Higgs field interactions breaks this symmetry. Therefore

$$m_H^2/\mu^2 \gtrsim \mathcal{O}(\lambda) \gtrsim \mathcal{O}(1/137)\ . \tag{III11}$$

Now

$$m_H^2 = \mathcal{O}(\lambda F_H^2)\ , \tag{III12}$$

where F_H is the vacuum expectation value of the Higgs field, known to be*)

$$F_H = (2G\sqrt{2})^{-1/2} = 174\ \text{GeV}\ . \tag{III13}$$

We now read off that

$$\mu \lesssim \mathcal{O}(F_H) = \mathcal{O}(174\ \text{GeV})\ . \tag{III14}$$

*) Some numerical values given during the lecture were incorrect. I here give corrected values.

This means that at energy scales much beyond F_H our model becomes more and more unnatural. Actually, factors of π have been omitted. In practice one factor of 5 or 10 is still not totally unacceptable. Notice that the actual value of m_H dropped out, except that

$$m_H = \mathcal{O}\left(\frac{\sqrt{\lambda}}{g} M_W\right) \gtrsim \mathcal{O}(M_W) \ . \qquad \text{(III15)}$$

Values for m_H of just a few GeV are unnatural.

III5. EXTENDING NATURALNESS

Equation (III14) tells us that at energy scales much beyond 174 GeV the standard model becomes unnatural. As long as the Higgs field H remains a fundamental scalar nothing much can be done about that. We therefore conclude, with Dimopoulos and Susskind[2)] that the "observed" Higgs field must be composite. A non-trivial strongly interacting field theory must be operative at 1000 GeV or so. An obvious and indeed likely possibility is that the Higgs field H can be written as

$$H = Z\bar{\psi}\psi \ , \qquad \text{(III16)}$$

where Z is a renormalization factor and ψ is a new quark-like object, a fermion with a new color-like interaction [2)]. We will refer to the object as meta-quark having meta-color. The theory will have all features of QCD so that we can copy the nomenclature of QCD with the prefix "meta-". The Higgs field is a meta-meson.

It is now tempting to assume that the meta-quarks transform the same way under weak SU(2) x U(1) as ordinary quarks. Take a doublet with left-handed components forming one gauge doublet and right handed components forming two gauge singlets. The meta-quarks are massless. Suppose that the meta-chiral symmetry is broken spontaneously just as in ordinary QCD. What would happen?

What happens is in ordinary QCD well described by the Gell-Mann-Lévy sigma model. The lightest mesons form a quartet of real fields, ϕ_{ij}, transforming as a

$$2^{\text{left}} \otimes 2^{\text{right}}$$

representation of

$$SU(2)^{\text{left}} \otimes SU(2)^{\text{right}} .$$

Since the weak interaction only deals with $SU(2)^{\text{left}}$ this quartet can also be considered as one complex doublet representation of weak SU(2). In ordinary QCD we have

$$\phi_{ij} = \sigma\delta_{ij} + i\tau^a_{ij}\pi^a \ , \tag{III17}$$

and

$$<\sigma>_{vacuum} = \frac{1}{\sqrt{2}} f_\pi = 91 \text{ MeV} \ . \tag{III18}$$

The complex doublet is then

$$\phi_i = \frac{1}{\sqrt{2}} \begin{pmatrix} \sigma + i\pi^3 \\ \pi^2 + i\pi^1 \end{pmatrix} , \tag{III19}$$

and

$$<\phi_i>_{vacuum} = \begin{pmatrix} 1 \\ o \end{pmatrix} \times 64 \text{ MeV} \ . \tag{III20}$$

We conclude that if we transplant this theory to the TeV range then we get a scalar doublet field with a non-vanishing vacuum expectation value for free. All we have to do now is to match the numbers. If we scale all QCD masses by a scaling factor κ then we match

$$F_H = 174 \text{ GeV} = \kappa \ 64 \text{ MeV} \ ;$$
$$\kappa = 2700 \ . \tag{III21}$$

Now the mesonic sector of QCD is usually assumed to be reproduced in the 1/N expansion [5] where N is the number of colors (in QCD we have N = 3). The 4-meson coupling constant goes like 1/N. Then one would expect

$$f_\pi \propto \sqrt{N} \ . \tag{III22}$$

Therefore

$$\kappa = 2700 \sqrt{\frac{3}{N}} \ , \tag{III23}$$

if the metacolor group is SU(N).

Thus we obtain a model that reproduces the W-mass and predicts the Higgs mass. The Higgs is the meta-sigma particle. The ordinary sigma is a wide resonance at about 700 MeV[3], so that we predict

$$m_H = \kappa m_\sigma = 1900 \sqrt{\frac{3}{N}} \text{ GeV} \ , \tag{III24}$$

and it will be extremely difficult to detect among other strongly interacting objects.

III6. WHAT NEXT?

The model of the previous section is to our mind nearly inevitable, but there are problems. These have to do with the observed fermion masses. All leptons and quarks owe their masses to an interaction term of the form

$$g\bar{\psi} H \psi \ , \qquad \text{(III25)}$$

where g is a coupling constant, ψ is the lepton or quark and H is the Higgs field. With (III16) this becomes a four-fermion interaction, a fundamental interaction in the new theory. Because it is non-renormalizable further structure is needed. In ref. 2 the obvious choice is made: a new "meta-weak interaction" gauge theory enters with new super-heavy intermediate vector bosons. But since H is a scalar this boson must be in the crossed channel, a rather awkward situation. (See option a in Figure 1.) A simpler theory is that a new scalar particle is exchanged in the direct channel. (See option b in Figure 1.)

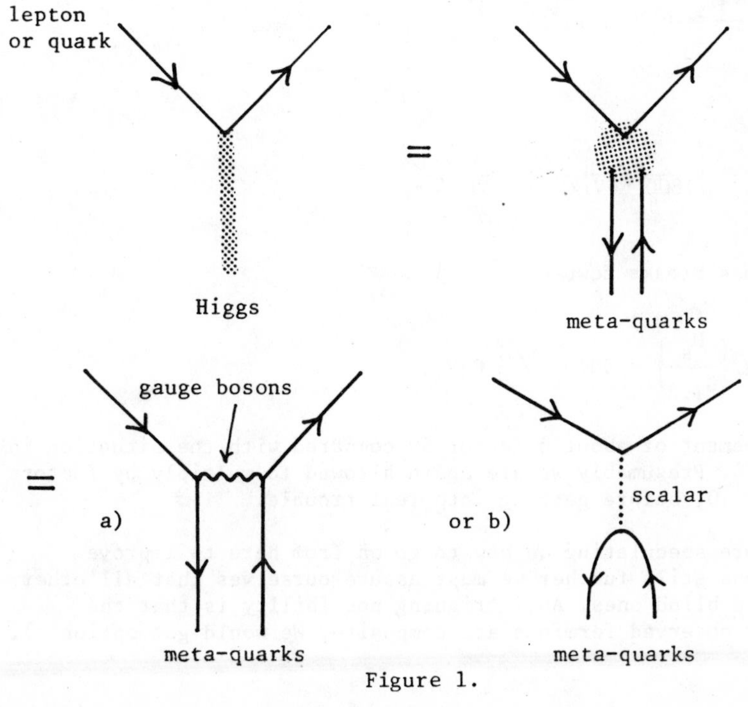

Figure 1.

Notice that in both cases new scalar fields are needed because in case a) something must cause the "spontaneous breakdown" of the new gauge symmetries. Therefore choice b) is simpler.
We removed a Higgs scalar and we get a scalar back. Does naturalness improve? The answer is yes. The coupling constant g in the interaction (III25) satisfies

$$g = g_1 g_2 / M_s^2 Z \ . \tag{III26}$$

Here g_1 and g_2 are the couplings at the new vertices, M_s is the new scalar's mass, and Z is from (III16) and is of order

$$Z \sim \frac{1}{\sqrt{\frac{N}{3}}\,(\kappa\, m_\rho)^2} = \frac{\sqrt{N/3}}{(1800\ \text{GeV})^2} \ . \tag{III27}$$

Suppose that the heaviest lepton or quark is about 10 GeV. For that fermion the coupling constant g is

$$g = \frac{m_f}{F} \simeq 1/20 \ .$$

We get

$$g_1 g_2 \simeq \left(\frac{M_s}{1800\ \text{GeV}}\right)^2 \sqrt{\frac{N}{3}} \cdot \frac{1}{20} \ .$$

Naturalness breaks down at

$$\mu = \mathcal{O}\left(\frac{M_s}{g_{1,2}}\right) = 8000\ \sqrt[4]{\frac{3}{N}}\ \text{GeV} \ ,$$

an improvement of about a factor 50 compared with the situation in sect. III4. Presumably we are again allowed to multiply by factors like 5 or 10, before getting into real trouble.

Before speculating on how to go on from here to improve naturalness still further we must assure ourselves that all other alleys are blind ones. An intriguing possibility is that the presently observed fermions are composite. We would get option c), Figure 2.

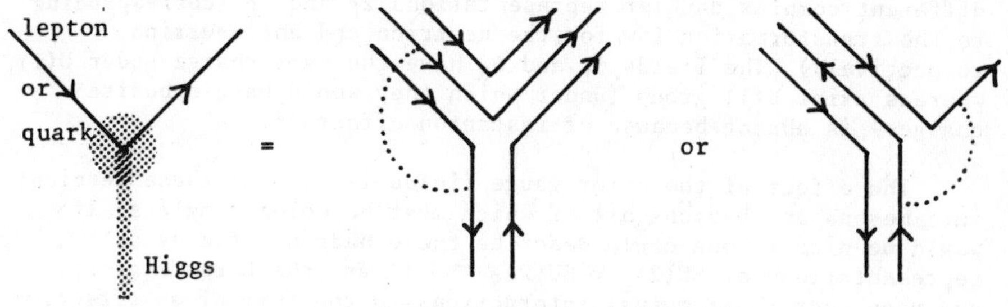

Fig. 2

The dotted line could be an ordinary weak interaction W or photon, that breaks an internal symmetry in the binding force for the new components. The new binding force could either act at the 1 TeV or at the 10-100 TeV range. It could either be an extension of metacolor or be a (color)" or paracolor force. Is such an idea viable?

Clearly, compared with the energy scale on which the binding forces take place, the composite fermions must be nearly massless. Again, this cannot be an accident. The chiral symmetry responsible for this must be present in the underlying theory. Apparently then, the underlying theory will possess a chiral symmetry which is <u>not</u> (or not completely) spontaneously broken, but reflected in the bound state spectrum in the Wigner mode: some massless chiral objects and parity doubled massive fermions. This possibility is most clearly described by the σ-model as a model for the lowest bound states occurring in ordinary quantum chromodynamics.

III7. THE σ MODEL

The fermion system in quantum chromodynamics shows an axial symmetry. To illuminate our problem let us consider the case of two flavors. The local color group is $SU(3)_c$. The subscript c here stands for color. The flavor symmetry group is $SU(2)_L \otimes SU(2)_R \otimes U(1)$ where the subscripts L and R stands for left and right and the group elements must be chosen to be space-time independent. We split the fermion fields ψ into left and right components:

$$\psi = \tfrac{1}{2}(1+\gamma_5)\psi_L + \tfrac{1}{2}(1-\gamma_5)\psi_R \, . \qquad \text{(III28)}$$

ψ_L transforms as a $3_c \otimes 2_L \otimes 1_R \otimes 2_{\mathcal{L}}$ (III29)

and ψ_R transforms as a $3_c \otimes 1_L \otimes 2_R \otimes \bar{2}_{\mathcal{L}}$ (III30)

where the indices refer to the various groups. $\mathcal{L}$ stands for the Lorentzgroup SO(3,1), locally equivalent to SL(2,c) which has two different complex doublet representations $2_\mathcal{L}$ and $\bar{2}_\mathcal{L}$ (corresponding to the transformation law for the neutrino and antineutrino, respectively). The fields ψ_L and ψ_R have the same charge under U(1), whereas axial U(1) group (under which they would have opposite charges) is absent because of instanton effects[4].

The effect of the color gauge fields is to bind these fermions into mesons and baryons all of which must be color singlets. It would be nice if one could describe these hadronic fields as representations of $SU(2)_L \otimes SU(2)_R \otimes U(1)$ and the Lorentz group, and then cast their mutual interactions in the form of an effective Lagrangian, invariant under the flavor symmetry group. In the case at hand this is possible and the resulting construction is a successful and one-time popular model for pions and nucleons: the σ model[3]. We have a nucleon doublet

$$N = \tfrac{1}{2}(1+\gamma_5)N_L + \tfrac{1}{2}(1-\gamma_5)N_R \, , \tag{III31}$$

where

$$N_L \text{ transforms as a } 1_c \otimes 2_L \otimes 1_R \otimes 2 \, , \tag{III32a}$$

$$\text{and } N_R \text{ transforms as a } 1_c \otimes 1_L \otimes 2_R \otimes \bar{2}_\mathcal{L} \, . \tag{III32b}$$

Further we have a quartet of real scalar fields $(\sigma,\vec{\pi})$ which transform as a $1_c \otimes 2_L \otimes 2_\mathcal{L} \otimes 1_\mathcal{L}$. The Lagrangian is

$$\mathcal{L} = -\bar{N}[\gamma\partial+g_o(\sigma+i\vec{\tau}.\vec{\pi}\gamma_5)]N - \tfrac{1}{2}(\partial\pi)^2 - \tfrac{1}{2}(\partial\sigma)^2 - V(\sigma^2+\vec{\pi}^2) \, . \tag{III33}$$

Here V must be a rotationally invariant function.

Usually V is chosen such that its absolute minimum is away from the origin. Let V be minimal at $\sigma = v$ and $\vec{\pi} = o$. Here v is just a c-number. To obtain the physical particle spectrum we write

$$\sigma = v + s \tag{III34}$$

and we find

$$\mathcal{L} = -\tfrac{1}{2}\bar{N}(\gamma\partial+ g_o v)N - \tfrac{1}{2}(\partial\vec{\pi})^2 - \tfrac{1}{2}(\partial s)^2 - 2v^2V''(v^2)s^2 + \text{interaction terms} \, . \tag{III35}$$

Clearly, in this case the nucleons acquire a mass term $m_s = g_o v$ and the s particle has a mass $m_s^2 = 4v^2V''(v^2)$, whereas the pion remains strictly massless. The entire mass of the pion must be due to effects that explicitly break $SU(2)_L \times SU(2)_R$, such as a small mass term $m_q\bar{\psi}\psi$ for the quarks (III28). We say that in this case the flavor group $SU(2)_L \otimes SU(2)_R$ is spontaneously broken into the isospin group SU(2).

Another possibility however, apparently not realised in ordinary quantum chromodynamics, would be that $SU(2)_L \otimes SU(2)_R$ is *not* spontaneously broken. We would read off from the Lagrangian (III33) that the nucleons N would form a massless doublet and that the four fields $(\sigma,\vec{\pi})$ could be heavy. The dynamics of other confining gauge theories could differ sufficiently from ordinary QCD so that, rather than a spontaneous symmetry breakdown, massless "baryons" develop. The principle question we will concentrate on is why do these massless baryons form the representation (III32), and how does this generalize to other systems. We would let future generations worry about the question where exactly the absolute minimum of the effective potential V will appear.

III8. INDICES

We now consider any color group G_c. The fundamental fermions in our system must be non-trivial representation of G_c and we assume "confinement" to occur: all physical particles are bound states that are singlets under G_c. Assume that the fermions are all massless (later mass terms can be considered as a perturbation). We will have automatically some global symmetry which we call the flavor group G_F. (We only consider exact flavor symmetries, not spoilt by instanton effects.) Assume that G_F is not spontaneously broken. Which and how many representations of G_F will occur in the massless fermion spectrum of the baryonic bound states? We must formulate the problem more precisely. The massless nucleons in (III33) being bound states, may have many massive excitations. However, massive Fermion fields cannot transform as a $2_{\mathcal{L}}$ under Lorentz transformations; they must go as a $2_{\mathcal{L}} \oplus \bar{2}_{\mathcal{L}}$. That is because a mass term being a Lorentz invariant product of two fields at one point only links $2_{\mathcal{L}}$ representations with $\bar{2}_{\mathcal{L}}$ representations. Consider a given representation r of G_F. Let p be the number of field multiplets transforming as $r \otimes 2_{\mathcal{L}}$ and q be the number of field multiplets $r \otimes \bar{2}_{\mathcal{L}}$. Mass terms that link the $2_{\mathcal{L}}$ with $\bar{2}_{\mathcal{L}}$ fields are completely invariant and in general to be expected in the effective Lagrangian. But the absolute value of

$$\ell = p - q \qquad \text{(III36)}$$

is the minimal number of surviving massless chiral field multiplets. We will call ℓ the index corresponding to the representation r of G_F. By definition this index must be a (positive or negative) integer. In the sigma model it is postulated that

$$\text{index } (2_L \otimes 1_R) = 1 \tag{III37}$$

$$\text{index } (1_L \otimes 2_R) = -1$$

index $(r) = o$ for all other representations r.

This tells us that if chiral symmetry is not broken spontaneously one massless nucleon doublet emerges. We wish to find out what massless fermionic bound states will come out in more general theories. Our problem is: how does (III37) generalize?

III9. ABSENCE OF MASSLESS BOUND STATES WITH SPIN 3/2 OR HIGHER

In the foregoing we only considered spin o and spin 1/2 bound states. Is it not possible that fundamentally massless bound states develop with higher spin? I believe to have strong arguments that this is indeed not possible. Let us consider the case of spin 3/2. Massive spin 3/2 fermions are described by a Lagrangian of the form

$$\mathcal{L} = \tfrac{1}{2}\bar{\psi}_\mu [\sigma_{\mu\nu}(\gamma\partial+m) + (\gamma\partial+m)\sigma_{\mu\nu}]\psi_\nu \, . \tag{III38}$$

Just like spin-one particles, this has a gauge-invariance if $m \to o$:

$$\psi_\mu \to \psi_\mu + \partial_\mu \eta(x) \, , \tag{III39}$$

where $\eta(x)$ is arbitrary. Indeed, massless spin 3/2 particles only occur in locally supersymmetric field theories. The field $\eta(x)$ is fundamentally unobservable.

Now in our model ψ_μ would be shorthand for some composite field: $\psi_\mu \to \psi\psi\psi$. However, then all components of this, including η, would be observables. If $m = o$ we would be forced to add a gauge fixing term that would turn η into an unacceptable ghost particle*).

We believe, therefore, that unitarity and locality forbid the occurrence of massless bound states with spin 3/2. The case for higher spin will not be any better. And so we concentrate on a bound state spectrum of spin 1/2 particles only.

*) Note added: during the lectures it was suggested by one attendant to consider only gauge-invariant fields as $\Psi_{\mu\nu} = \partial_\mu\psi_\nu - \partial_\nu\psi_\mu$. However, such fields must satisfy constraints: $\partial[\alpha\Psi_{\mu\nu}] = o$. Composite field will never automatically satisfy such constraints.

III10. SPECTATOR GAUGE FIELDS AND -FERMIONS

So far, our model consisted of a strong interaction color gauge theory with gauge group G_c, coupled to chiral fermions in various representations r of G_c but of course in such a way that the anomalies cancel. The fermions are all massless and form multiplets of a global symmetry group, called G_F. For QCD this would be the flavor group. In the metacolor theory G_F would include all other fermion symmetries besides metacolor.

In order to study the mathematical problem raised above we will add another gauge connection field that turns G_F into a local symmetry group. The associated coupling constants may all be arbitrarily small, so that the dynamics of the strong color gauge interactions is not much affected. In particular the massless bound state spectrum should not change. One may either think of this new gauge field as a completely quantized field or simply as an artificial background field with possibly non-trivial topology. We will study the behavior of our system in the presence of this "spectator gauge field". As stated, its gauge group is G_F.

Note however, that some flavor transformations could be associated with anomalies. There are two types of anomalies:

i) those associated with $G_c \times G_F$, only occurring where the color field has a winding number. Only U(1) invariant subgroups of G_F contribute here. They simply correspond to small explicit violations of the G_F symmetry. From now on we will take as G_F only the anomaly-free part. Thus, for QCD with N flavors, G_F is not $U(N) \times U(N)$ but

$$G_F = SU(N) \otimes SU(N) \otimes U(1) .$$

ii) those associated with G_F alone. They only occur if the spectator gauge field is quantized. To remedy these we simply add "spectator fermions" coupled to G_F alone. Again, since these interactions are weak they should not influence the bound state spectrum.

Here, the spectator gauge fields and fermions are introduced as mathematical tools only. It just happens to be that they really do occur in Nature, for instance the weak and electromagnetic $SU(2) \times U(1)$ gauge fields coupled to quarks in QCD. The leptons then play the role of spectator fermions.

III11. ANOMALY CANCELLATION FOR THE BOUND STATE SPECTRUM

Let us now resume the particle content of our theory. At small distances we have a gauge group $G_c \otimes G_F$ with chiral fermions in several representations of this group. Those fermions which are trivial under G_C are only coupled weakly and are called "spectator fermions". All anomalies cancel, by construction.

At low energies, much lower than the mass scale where color binding occurs, we see only the G_F gauge group with its gauge fields. Coupled to these gauge fields are the massless bound states, forming new representations r of G_F, with either left- or right handed chirality. The numbers of left minus right handed fermion fields in the representations r are given by the as yet unknown indices $\ell(r)$. And finally we have the spectator fermions which are unchanged.

We now expect these very light objects to be described by a new local field theory, that is, a theory local with respect to the large distance scale that we now use. The central theme of our reasoning is now that this new theory must again be anomaly free. We simply cannot allow the contradictions that would arise if this were not so. Nature must arrange its new particle spectrum in such a way that unitarity is obeyed, and because of the large distance scale used the effective interactions are either vanishingly small or renormalizable. The requirement of anomaly cancellation in the new particle spectrum gives us equations for the indices $\ell(r)$, as we will see.

The reason why these equations are sometimes difficult or impossible to solve is that the new representations r must be different from the old ones; if $G_C = SU(N)$ then r must also be faithful representations of $G_F/Z(N)$. For instance in QCD we only allow for octet or decuplet representations of $(SU(3))_{flavor}$, whereas the original quarks were triplets.

However, the anomaly cancellation requirement, restrictive as it may be, does not fix the values of $\ell(r)$ completely. We must look for additional limitations.

III12 APPELQUIST-CARAZZONE DECOUPLING AND N-INDEPENDENCE

A further limitation is found by the following argument. Suppose we add a mass term for one of the colored fermions.

$$\Delta\mathcal{L} = m\,\bar{\psi}_{1L}\,\psi_{1R} + \text{h.c.}$$

Clearly this links one of the left handed fermions with one of the right handed ones and thus reduces the flavor group G_F into $G_F' \subset G_F$. Now let us gradually vary m from o to infinity. A famous theorem [5] tells us that in the limit $m \to \infty$ all effects due to this massive quark disappear. All bound states containing this quark should also disappear which they can only do by becoming very heavy. And they can only become heavy if they form representations r' of G_F' with total index $\ell'(r') = o$. Each representation r of G_F forms an array of representations r' of G'_F. Therefore

$$\ell'(r') = \sum_{r \text{ with } r' \subset r} \ell(r) \ . \qquad \text{(III40)}$$

Apparently this expression must vanish.

Thus we found another requirement for the indices $\ell(r)$. The indices will be nearly but not quite uniquely determined now. Calculations show that this second requirement makes our indices $\ell(r)$ practically independent of the dimensions n_i of G_F. For instance, if $G_c = SU(3)$ and if we have left- and righthanded quarks forming triplets and sextets then

$$G_F = SU(n_1)_L \otimes SU(n_2)_R \otimes SU(n_3)_L \otimes SU(n_4)_R \otimes U(1)^3 \qquad \text{(III41)}$$

where $n_{1,2}$ refer to the triplets and $n_{3,4}$ to the sextets. G_c is anomaly-free if

$$n_1 - n_2 + 7(n_3 - n_4) = o \ . \qquad \text{(III42)}$$

Here we have three independent numbers n_i.
If we write the representations r as Young tableaus then $\ell(r)$ could still depend explicitly on n_i.

However, suppose that someone would start as approximation of Bethe-Salpeter type to discover the zero mass bound state spectrum. He would study diagrams such as Fig. 3

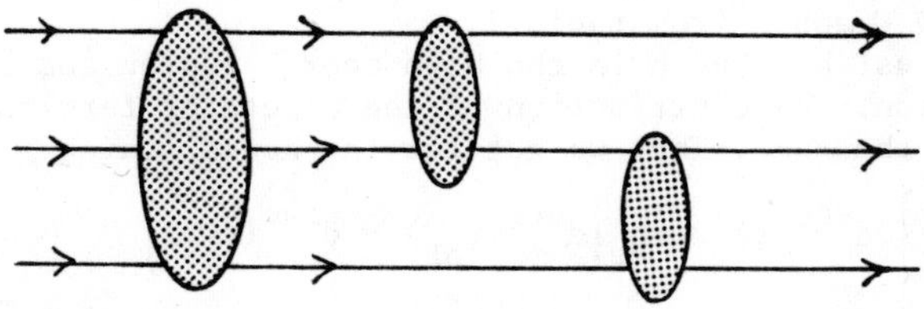

Fig. 3

The resulting indices $\ell(r)$ would follow from topological properties of the interactions represented by the blobs. It is unlikely that this topology would be seriously effected by details such as the contributions of diagrams containing additional closed fermion loops. However, that is the only way in which explicit n-dependence enters. It is therefore natural to assume $\ell(r)$ to be n-independent. This latter assumption fixes $\ell(r)$ completely. What is the result of these calculations?

III13. CALCULATIONS

Let G be any (reducible or irreducible) gauge group. Let chiral fermions in a representation r be coupled to the gauge fields by the covariant derivative

$$D_\mu = \partial_\mu + i\,\lambda^a(r)\,A^a_\mu \ , \qquad \text{(III43)}$$

where A^a_μ are the gauge fields and $\lambda^a(r)$ a set of matrices depending on the representation r. Let the left-handed fermions be in the representations r_L and the right-handed ones in r_R. Then the anomalies cancel if

$$\sum_L \mathrm{Tr}\{\lambda^a(r_L),\ \lambda^b(r_L)\}\ \lambda^c(r_L) =$$

$$\sum_R \mathrm{Tr}\{\lambda^a(r_R),\ {}^b(r_R)\}\ \lambda^c(r_R)\ . \qquad \text{(III44)}$$

The object $d^{abc}(r) = \mathrm{Tr}\{\lambda^a(r),\ \lambda^b(r)\}\ \lambda^c(r)$ can be computed for any r. In table 1 we give some examples. The fundamental representation r_o is represented by a Young tableau: □ . Let it have n components. We take the case that $\mathrm{Tr}\ \lambda(r_o) = o$. Write

$$\mathrm{Tr}\ I(r_o) = n\ , \qquad \mathrm{Tr}\ I(r) = N(r)\ ,$$

$$\mathrm{Tr}\ \lambda(r) = o\ ,$$

$$\mathrm{Tr}\ \lambda^a(r)\ \lambda^b(r) = C(r)\ \mathrm{Tr}\ \lambda^a(r_o)\ \lambda^b(r_o)\ ,$$

$$d^{abc}(r) = K(r)\ d^{abc}(r_o)\ . \qquad \text{(III45)}$$

We read off C and K from table 1.
Now III44 must hold both in the high energy region and in the low energy region. The contribution of the spectator fermions in both regions is the same. Thus we get for the bound states

$$\left(\sum_L - \sum_R\right) d^{abc}(r) = n_c\left(d^{abc}(r_{oL}) - d^{abc}(r_{oR})\right) \qquad \text{(III46)}$$

where a,b,c are indices of G_F and r_o is the fundamental representation of G_F. We have the factor n_c written explicitly, being the number of color components.

Let us now consider the case $G_C = SU(3)$; $G_F = SU_L(n) \otimes SU_R(n) \otimes U(1)$. We have n "quarks" in the fundamental representations. The representations r of the bound states must be in $G_F/Z(3)$. They are assumed to be built from three quarks, but we are free to choose their chirality. The expected representations

Table 1

r	N(r)	C(r)	K(r)
	n	1	1
	n	1	-1
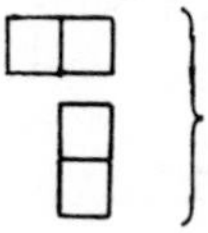	$\frac{n(n\pm 1)}{2}$	$n\pm 2$	$n\pm 4$
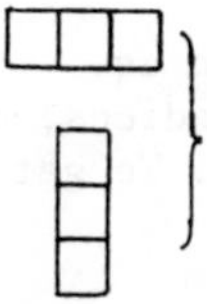	$\frac{n(n\pm 1)(n\pm 2)}{6}$	$\frac{(n\pm 2)(n\pm 3)}{2}$	$\frac{(n\pm 3)(n\pm 6)}{2}$
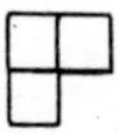	$\frac{n(n^2-1)}{3}$	n^2-3	n^2-9
$A \otimes B$	$N(A)N(B)$	$C(A)N(B) + C(B)N(A)$	$K(A)N(B) + K(B)N(A)$

are given in table 2, where also their indices are defined. Because of left-right symmetry these numbers change sign under interchange of left ↔ right.

Table 2

representation	index	representation	index
	$\ell_{1\pm}$	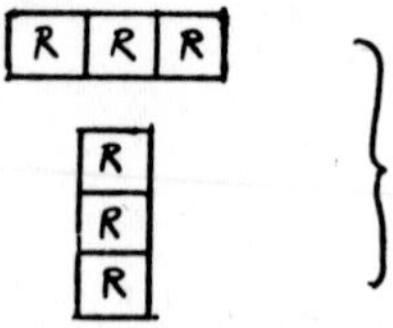	$-\ell_{1\pm}$
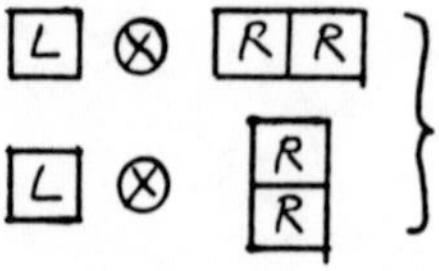	$\ell_{2\pm}$	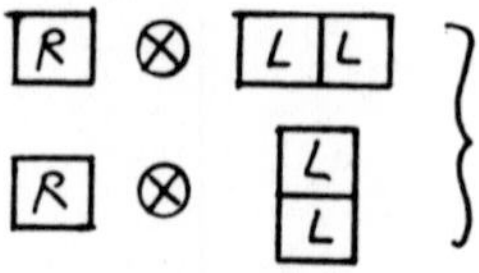	$-\ell_{2\pm}$
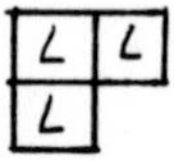	ℓ_3	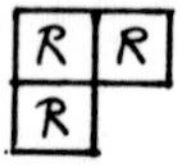	$-\ell_3$

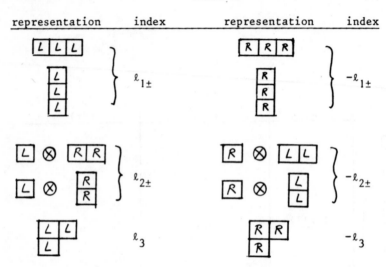

For the time being we assume no other representations. In eq. III46 we may either choose a, b and c all to be $SU(n)_L$ indices, or choose a and b to be $SU(n)_L$ indices and c the U(1) index. We get two independent equations:

$$\sum_{\pm} \tfrac{1}{2}(n\pm3)(n\pm6)\ell_{1\pm} - \sum_{\pm} \tfrac{1}{2}n(n\pm7)\ell_{2\pm} + (n^2-9)\ell_3 = 3, \text{ if } n > 2 ,$$

and

$$\sum_{\pm} \tfrac{1}{2}(n\pm2)(n\pm3)\ell_{1\pm} - \sum_{\pm} \tfrac{1}{2}n(n\pm3)\ell_{2\pm} + (n^2-3)\ell_3 = 1, \text{ if } n > 1 . \quad \text{(III47)}$$

The Appelquist-Carazzone decoupling requirement, eq. (III40), gives us in addition two other equations:

$$\ell_{1+} - \ell_{2+} + \ell_3 = 0 ,$$

$$\ell_{1-} - \ell_{2-} + \ell_3 = 0 , \text{ both if } n > 2 . \quad \text{(III48)}$$

For n > 2 the general solution is

$$\ell_{1+} = \ell_{1-} = \ell \,,$$
$$\ell_{2+} = \ell_{2-} = 3\ell - \frac{1}{3} \,,$$
$$\ell_{3} = 2\ell - \frac{1}{3} \,. \qquad \text{(III49)}$$

Here ℓ is still arbitrary. Clearly this result is unacceptable. We cannot allow any of the indices ℓ to be non-integer. Only for the case n = 2 (QCD with just two flavors) there is another solution. In that case ℓ_{2-} and ℓ_3 describe the same representation, and ℓ_{1-} an empty representation. We get

$$\ell_{2-} + \ell_3 = k = 1 - 10\,\ell_{1+} + 5\,\ell_{2+} \,. \qquad \text{(III50)}$$

According to the σ-model, $\ell_{1+} = \ell_{2+} = 0$; k = 1. The σ-model is therefore a correct solution to our equations.

In the previous section we promised to determine the indices completely. This is done by imposing n-independence for the more general case including also other color representations such as sextets besides triplets. The resulting equations are not very illuminating, with rather ugly coefficients. One finds that in general no solution exists except when one assumes that all mixed representations have vanishing indices. With mixed representations we mean a product of two or more non-trivial representations of two or more non-Abelian invariant subgroups of G_F. If now we assume n-independence this must also hold if the number of sextets is zero. So ℓ_{2+} and ℓ_{2-} must vanish. We get

$$\ell_{1+} = \ell_{1-} = 1/9 \,,$$
$$\ell_3 = -1/9 \,. \qquad \text{(III51)}$$

If all quarks were sextets, not triplets, we would get

$$\ell_{1+} = \ell_{1-} = 2/9 \,,$$
$$\ell_3 = -2/9 \,. \qquad \text{(III52)}$$

In the case G_C = SU(5) the indices were also found. See table 3.

Table 3
indices for G_c = SU(5)

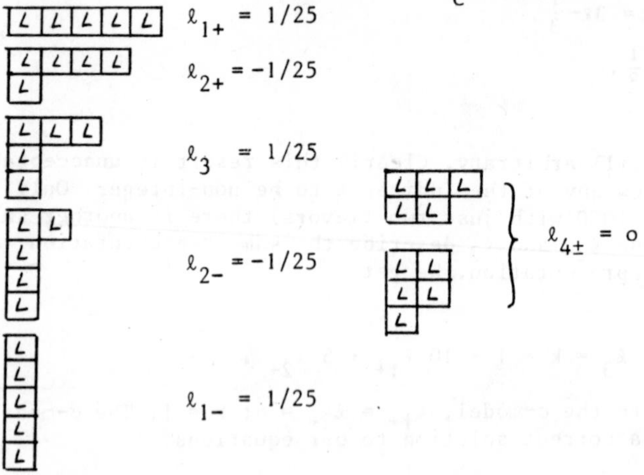

This clearly suggests a general tendency for SU(N) color groups to produce indices $\pm 1/N^2$ or o.

III14. CONCLUSIONS

Our result that the indices we searched for are fractional is clearly absurd. We nevertheless pursued this calculation in order to exhibit the general philosophy of this approach and to find out what a possible cure might be. Our starting point was that chiral symmetry is not broken spontaneously. Most likely this is untenable, as several authors have argued[6]. We find that explicit chiral symmetry in QCD leads to trouble in particular if the number of flavors is more than two. A daring conjecture is then that in QCD the strange quark, being rather light, is responsible for the spontaneous breakdown of chiral symmetry.
An interesting possibility is that in some generalized versions of QCD chiral symmetry is broken only partly, leaving a few massless chiral bound states. Indeed there are examples of models where our philosophy would then give integer indices, but since we must drop the requirement of n-dependence our result was not unique and it was always ugly. No such model seems to reproduce anything resembling the observed quark-lepton spectrum.

Finally there is the remote possibility that the paradoxes associated with higher spin massless bound states can be resolved. Perhaps the $\Delta(1236)$ plays a more subtle role in the σ-model than assumed so far (we took it to be a parity doublet).

We conclude that we are unable to construct a bound state theory for the presently fundamental fermions along the lines suggested above.

We thank R. van Damme for a calculation yielding the indices in the case $G_c = SU(5)$.

REFERENCES

1. P.A.M. Dirac, Nature 139 (1937) 323, Proc. Roy. Soc. A165 (1938) 199, and in: Current Trends in the Theory of Fields, (Tallahassee 1978) AIP Conf. Proc. No 48, Particles and Fields Subseries No 15, ed. by Lannuti and Williams, p. 169.
2. S. Dimopoulos and L. Susskind, Nucl. Phys. B155 (1979) 237.
3. M. Gell-Mann and M. Lévy, Nuovo Cim. 16 (1960) 705.
   B.W. Lee, Chiral Dynamics, Gordon and Breach, New York, London, Paris 1972.
4. G. 't Hooft, Phys. Rev. Lett. 37 (1976) 8; Phys. Rev. D14 (1976) 3432.
   S. Coleman, "The Uses of Instantons", Erice Lectures 1977.
   R. Jackiw and C. Rebbi, Phys. Rev. Lett. 37 (1976) 172.
   C. Callan, R. Dashen and D. Gross, Phys. Lett. 63B (1976) 334.
5. G. 't Hooft, Nucl. Phys. B72 (1974) 461.
6. T. Appelquist and J. Carazzone, Phys. Rev. D11 (1975) 2856.
7. A. Casher, Chiral Symmetry Breaking in Quark Confining Theories, Tel Aviv preprint TAUP 734/79 (1979).

Nuclear Physics B173 (1980) 208–228

LIGHT COMPOSITE FERMIONS

S. DIMOPOULOS[1], S. RABY[2] and L. SUSSKIND[1]

Institute of Theoretical Physics, Department of Physics, Stanford University, Stanford, California 94305, USA

Received 17 March 1980

In this paper we consider the possibility that QCD-like theories can lead to massless or near-massless composite fermions. The method of analysis relies on a conjectured equivalence between the confined and Higgs phases of certain non-abelian gauge theories. This "complementarity" principle allows us to analyze a theory as if the Higgs phenomenon occurred and then reinterpret the results in the language of composite gauge singlets. Those fermions which remain massless in the Higgs picture may then be interpreted as massless fermionic composites.

The principle of complementarity, when applied to a class of extended technicolor models, implies that quarks and leptons are composites bound at a scale of order 1–100 TeV.

1. Introduction

The proliferation of apparently fundamental particles naturally leads to the speculation that some of these degrees of freedom may be composite. Indeed, particularly good reasons exist for thinking that the scalar fields in the standard electroweak theory are composite [1]. In this paper we will argue, with 't Hooft [2], that quarks and leptons may also exhibit composite structure, perhaps in the TeV range.

There is a strong objection to this view. If leptons are composite, then the success of high precision QED [3] requires the range of the forces and wave function to be less than 10^{-3} $(\mathrm{GeV})^{-1}$. It is difficult to understand how forces which operate on the TeV scale can conspire to produce such light particles. The natural expectation is that, like the nucleon, the intrinsic size and mass should define comparable scales.

In a remarkable recent paper 't Hooft [2] has begun a systematic study of these questions by asking under what conditions exactly massless composite fermions can naturally occur in a QCD-like theory. In this paper we will introduce some concepts which, when added to 't Hooft's work, provide a reasonable basis for answering this question.

[1] Supported in part by NSF grant PHY 78-26847.
[2] Supported in part by NSF grant PHY 79-18046.

In general, massless particles, composite or otherwise, do not occur in a non-trivial field theory without some special symmetry insuring their masslessness. In the case of fermions this usually means either discrete or continuous chiral symmetry. Thus the question becomes: Under what conditions will enough chiral symmetry survive spontaneous breakdown in a confining theory to insure the massless character of some set of fermions?

Unfortunately, at the present time the tools do not exist to answer this kind of question. Even in QCD the only clear and persuasive argument that chiral symmetry is spontaneously broken is that it is required to explain the empirical facts*. 't Hooft's work provides only a consistency test and not a dynamical framework. In this paper we take some first steps in filling this gap.

The main method employed in this paper takes advantage of a result due to Fradkin and Shenker [4] in lattice gauge theory. The result states that in gauge theories with scalars in the fundamental representation the Higgs and confined phase are the same. This leads us to conjecture a principle of complementarity which states that a confining theory of fermions and gauge bosons can be analyzed as if a dynamical Higgs phenomenon takes place (Higgs picture) or as if the gauge symmetry is unbroken (symmetric picture). The observable results are the same.

The second technique [5] employed is to use a crude estimate to determine the pattern of dynamical symmetry breaking that takes place in the Higgs picture. The technique has been previously described [5]. The combination of these two ideas leads to a reasonable way to guess the properties of confining theories. In particular it leads to answers to the question of massless fermionic bound states which always agree with 't Hooft's criterion.

2. Complementarity

There is rigorous evidence that in lattice gauge theories [4] with scalars in the fundamental representations no sharp phase boundary separates the confined and Higgs phases. Indeed, Fradkin and Shenker have analyzed the phase diagram of a model with two parameters g and v. These parameters are the lattice versions of the gauge coupling and the classical position of the minimum of the Higgs potential. For small g and large v the theory is forced into conventional Higgs behavior, whereas for large g and small v the dynamics is conventional confinement. The theorem of Fradkin and Shenker asserts that no sharp phase boundary (non-analyticity) separates these two regions. They are one and the same phase.

The theorem of Fradkin and Shenker assures us that anything which is exactly true in the Higgs (confinement) region must extrapolate into the confinement (Higgs) region. Furthermore, the spectra of the two regimes must map into each other, although some states may become unstable as we vary the parameters from

* For recent progress, see also ref. [7].

confinement to Higgs. We shall now illustrate the complementary nature of Higgs and confinement by describing some examples.

EXAMPLE 1

Consider the gauge group SU(2) with gauge bosons W^+, W^-, W^0, a doublet of fermions $\psi = (\psi_1, \psi_2)$ and a scalar Higgs doublet (ϕ_1, ϕ_2). The potential $V(\phi)$ is chosen so as to have a minimum at some non-zero value of $|\phi|$ called v. In the standard Higgs picture we arbitrarily choose the vacuum expectation value of ϕ to satisfy

$$\langle\phi_1\rangle = 0, \qquad \langle\phi_2\rangle = v. \tag{2.1}$$

The spectrum of physical states may be analyzed in the *unitary gauge* defined by

$$\phi_1 = 0, \qquad \phi_2 = \tilde{\phi} = \text{real}. \tag{2.2}$$

In the unitary gauge all fields are physical and the state spectrum contains no gauge artifacts. We will denote fields in unitary gauge by a tilde.

The physical spectrum includes the following particles (labeled by the unitary gauge fields which create them):

(a) a degenerate fermion pair $\tilde{\psi}_1, \tilde{\psi}_2$;

(b) a scalar Higgs boson $(\tilde{\phi} - v)$;

(c) three massive gauge fields, $\tilde{W}^+, \tilde{W}^-, \tilde{W}^0$.

If the coupling g is not too small there may also be some bound states of this essentially perturbative spectrum.

The complementarity principle suggested by the Fradkin-Shenker analysis [4] says that the above spectrum can be interpreted as SU(2) singlet bound states. Moreover it indicates that it evolves in a continuous fashion from the spectrum of a strongly coupled confined theory. For example, the strongly coupled theory will possess composite fermions formed from ϕ's and ψ's. Thus we expect systems with the following quantum numbers*:

(a) two fermionic bound states composed of ψ's and ϕ's:

$$\phi_i^\dagger \psi_i, \qquad \phi_i \psi_j \varepsilon_{ij};$$

(b) a scalar

$$\phi_i^\dagger \phi_i;$$

* The height of the indices has no significance throughout this paper.

(c) three vector states

$$\phi_i \overleftrightarrow{D}_\mu \phi_j \varepsilon_{ij}, \qquad \phi_i^\dagger \overleftrightarrow{D}_\mu \phi_j^\dagger \varepsilon_{ij}, \qquad \phi_i^\dagger \overleftrightarrow{D}_\mu \phi_j .$$

These states evidently correspond to the states enumerated in the Higgs picture.

The bridge between the symmetric and Higgs picture is the *unitary gauge* formulation in which the gauge freedom is used to set $\phi_1 = 0$. This condition is not just on the vacuum expectation value but on the operators. In this gauge the field operators are all physical (gauge invariant). Indeed, we may write them in manifestly gauge-invariant form. Thus

$$\tilde{\psi}_2 = \phi_i^\dagger \psi_i / |\phi| ,$$

$$\tilde{\psi}_1 = \phi_i \psi_j \varepsilon_{ij} / |\phi| . \tag{2.3}$$

Apart from the unimportant factor $|\phi|^{-1}$, these are just the composite fields describing the confined phase.

Similarly

$$\tilde{W}^+ = \frac{\phi_i^\dagger}{|\phi|} \overleftrightarrow{D}_\mu \frac{\phi_j^\dagger}{|\phi|} \varepsilon_{ij} ,$$

$$\tilde{W}^0 = \frac{\phi_i^\dagger}{|\phi|} \overleftrightarrow{D}_\mu \frac{\phi_i}{|\phi|} . \tag{2.4}$$

From this discussion, the physical meaning of the result of Fradkin and Shenker becomes evident.

Next let us add a very weak interaction which couples an abelian vector potential A_μ to the current J_μ

$$J_\mu = \tfrac{1}{2} e \left[\phi^\dagger \overleftrightarrow{D}_\mu \phi + \psi^\dagger \gamma_\mu \psi \right] . \tag{2.5}$$

We shall think of the SU(2) interactions as a strong color-like interaction and $A \cdot J$ as a weak or electromagnetic perturbation. The weak charge Q of the fundamental particles is

$$Q(\psi) = \tfrac{1}{2} e , \qquad Q(\phi) = \tfrac{1}{2} e , \qquad Q(W) = 0 . \tag{2.6}$$

Now in the confined picture all observable particles have integer charges $\pm ne$. One may wonder whether this fact distinguishes the two phases. Thus consider again the Higgs phase in which $\langle \phi_2 \rangle \neq 0$. Since ϕ_2 is not electrically neutral the expectation value spontaneously breaks the U(1) symmetry. Surely this sounds observably different than the symmetric phase in which Q is unbroken and A is massless.

However, in the Higgs picture there is an unbroken U(1) current given by

$$Q' = Q + T_3\,,$$

where T_3 is the SU(2) generator

$$\frac{1}{2}\begin{pmatrix} 1 & 0 \\ 0 & -1 \end{pmatrix}. \tag{2.7}$$

The charge Q' couples to a massless gauge boson $\mathbf{A'}$ and has eigenvalues

$$\begin{aligned} Q'(\psi_1) &= 1\,, & Q'(\mathrm{W}^+) &= 1\,, \\ Q'(\psi_2) &= 0\,, & Q'(\mathrm{W}^-) &= -1\,, \\ Q'(\mathrm{Higgs}) &= 0\,, & Q'(\mathrm{W}^0) &= 0\,. \end{aligned} \tag{2.8}$$

Accordingly the massless gauge boson A′ couples to a set of particles with *integer* electric charge.

Let us return to the symmetric picture in which no Higgs phenomenon occurs. In this case the original U(1) charge Q is unbroken and couples to the massless boson A. The physical states are the composites in a, b, and c. The charge is additive for composites and has values

$$\begin{aligned} Q(\phi\psi) &= 1\,, & Q(\phi^\dagger D_\mu \phi^\dagger \varepsilon) &= 1\,, \\ Q(\phi^\dagger\psi) &= 0\,, & Q(\phi D_\mu \phi) &= -1\,, \\ Q(\phi^\dagger\phi) &= 0\,, & Q(\phi^\dagger D_\mu \phi) &= 0\,. \end{aligned} \tag{2.9}$$

Thus once again a massless "photon" couples to integer charges which have identical values in the Higgs and symmetric pictures.

EXAMPLE 2

The next example illustrates the failure of complementarity for Higgs fields in non-fundamental representations. Consider an SU(2) gauge theory with a triplet Higgs field ϕ_α ($\alpha = 1, 2, 3$) and a fermion doublet ψ_i. In the Higgs phase the spectrum includes massive gauge bosons $\mathrm{W}^\pm$, a massless gauge boson W^0, a massive Higgs boson, and fermions ψ_i. Since the unbroken gauge group is abelian, it does not confine the fermions.

On the other hand, in the confining phase no fermions can exist since no number of scalars or vectors can combine with ψ to make an SU(2) singlet. Accordingly a sharp transition must separate the two regimes. This is in agreement with the analysis of Fradkin and Shenker [4].

EXAMPLE 3

The complementarity principle is not restricted to SU(2) theories. For example, consider the gauge group SU(3) with a Higgs field in the triplet. In the Higgs picture the gauge symmetry breaks down to SU(2). The perturbative spectrum is given by an SU(2) triplet of massless gauge bosons A^α, a pair of massive SU(2) doublet gauge bosons A_i and $A_i^\dagger$, a massive SU(2) singlet gauge boson A_8, and finally an SU(2) singlet Higgs boson ϕ. However, the SU(2) gauge theory, being non-abelian, is expected to confine the SU(2) non-singlet states. Accordingly the physical spectrum contains only the Higgs scalar and the SU(2) singlet vector boson A_8. The vector couples to the "hypercharge" generator of SU(3). Of course there also can exist SU(2) singlet bound states of the confined bosons A_i, $A_i^\dagger$, and A_α.

In the complimentary symmetric picture the same spectrum is described as SU(3) singlets. The scalar is the composite

$$\phi_i^\dagger \phi_i , \tag{2.10}$$

and the massive gauge boson is

$$\phi_i^\dagger D_\mu \phi_i . \tag{2.11}$$

In addition, a variety of other bound states can exist. For example, a bound state of two Higgs fields and two gauge bosons

$$\phi^\dagger [F, F] \phi . \tag{2.12}$$

In the Higgs picture this state corresponds to a bound state of the SU(2) doublet gauge bosons.

3. Dynamical symmetry breaking

In the conventional situations of very weak or very strong coupling the complementarity principle is mainly of academic interest. However, in ref. [5] we studied a class of gauge theories of gluons and fermions without scalars. In these theories the interesting action is at intermediate values of the coupling. In such cases it is extremely valuable to understand what results are common to both weak and strong coupling regimes.

In ref. [5] we developed a simple dynamical framework in which to study such theories in the Higgs picture. It is very illuminating to apply the complementarity principle and in so doing obtain a symmetric confining picture. In this section we will review a simple example of the dynamical Higgs phenomenon.

Consider an SU(5) gauge theory coupled to two multiplets of left-handed fermions [6]. The fermions are in the 5 and 10 representation. We denote these fields ψ_i and χ_{ijk}. The field χ is totally antisymmetric. In the notation of ref. [5], the fermion content is $[1]_5 + [3]_5$ where $[m]_N$ denotes the representation of SU(N) with m antisymmetrized indices.

At some mass scale M_1 the running coupling becomes large enough to cause a scalar chiral-symmetry-breaking fermion condensate. The condensate is assumed to have the quantum numbers of the *most attractive* (scalar) *channel* (MAC). In the present case the MAC is the 5 contained in $\overline{10}\times\overline{10}$. We define*

$$\phi_n = \chi_{ijk}\chi_{lmn}\varepsilon^{ijklm}\,. \tag{3.1}$$

Then we take

$$\langle\phi_5\rangle \neq 0\,. \tag{3.2}$$

When ϕ develops an expectation value at scale M_1 several things happen:

(1) the gauge symmetry breaks down from SU(5) to SU(4);

(2) the 9 gauge bosons in SU(5)/SU(4) gain mass $\sim M_1$;

(3) those fermions participating in the condensate gain mass $\sim M_1$;

(4) at energies much lower than M_1 an effective SU(4) theory describes the physics; the effective theory includes the SU(4) gauge bosons and the uncondensed fermions.

The SU(4) content of the fermion multiplets is

$$\begin{aligned} &[0]_4 = \psi_5\,, \\ &[1]_4 = \psi_i\,, \qquad (i,j,k = 1\ldots 4)\,, \\ &[2]_4 = \chi_{5ij}\,, \\ &[3]_4 = \chi_{ijk}\,. \end{aligned} \tag{3.3}$$

The condensate ϕ_5 is proportional to $\chi_{5ij}\chi_{5hl}\varepsilon^{ijhl}$. This means that the fermion multiplet $[2]_4$ gains a mass $\sim M_1$. The effective low-energy theory involves the remaining multiplets.

As the energy decreases the running coupling increases and the MAC of the effective SU(4) theory condenses at scale M_2. This time the MAC is the SU(4) singlet

$$\psi_i\chi_{jkl}\varepsilon^{ijkl}\,, \qquad (i,j,k,l = 1\ldots 4)\,. \tag{3.4}$$

Accordingly the 2-component fields ψ_i and χ_{jkl} combine to form a Dirac fermion with mass $\sim M_2$. No additional gauge bosons gain mass and the SU(4) theory evolves into the infrared as a confined system.

* See previous footnote.

The physical spectrum consists of various massive SU(4) singlet composites and the massless SU(4) singlet left-handed fermion ψ_5. There is also one massive SU(4) gauge boson which couples to the SU(5) generator

$$\frac{1}{\sqrt{40}}\begin{bmatrix} 1 & 0 & 0 & 0 & 0 \\ 0 & 1 & 0 & 0 & 0 \\ 0 & 0 & 1 & 0 & 0 \\ 0 & 0 & 0 & 1 & 0 \\ 0 & 0 & 0 & 0 & -4 \end{bmatrix} \tag{3.5}$$

Now let us consider the complementary symmetric picture. In particular, consider the spectrum of SU(5)-singlet fermion composites. There are four distinct channels which appear capable of binding into fermions. They are

$$\begin{aligned}
&(\mathrm{a}) \quad \psi_i \chi^{\dagger}_{jkl} \chi^{\dagger}_{mnp} \delta_{ip} \varepsilon_{jklmn}, \\
&(\mathrm{b}) \quad \psi_i \psi_j \chi_{lmn} \varepsilon^{ijlmn}, \\
&(\mathrm{c}) \quad \chi^{\dagger}_{ijk} \chi^{\dagger}_{lmn} \chi^{\dagger}_{pqr} \chi^{\dagger}_{stu} \chi^{\dagger}_{xyz} \varepsilon^{ilpsx} \varepsilon^{jmqty} \varepsilon^{knruz}, \\
&(\mathrm{d}) \quad \psi_i \psi_j \psi_k \psi_l \psi_m \varepsilon_{ijklm}.
\end{aligned} \tag{3.6}$$

Similarly, several boson combinations can be formed:

$$\begin{aligned}
&(\mathrm{e}) \quad \psi^{\dagger} \psi, \\
&(\mathrm{f}) \quad \chi^{\dagger} \chi, \\
&(\mathrm{g}) \quad \psi_i \psi_j \psi_k \chi^{\dagger}_{ijk}, \\
&(\mathrm{h}) \quad \psi_i \chi_{jkl} \chi_{pqr} \chi_{stu} \varepsilon_{ijklp} \varepsilon_{qrstu}.
\end{aligned} \tag{3.7}$$

All other singlets can be formed from products of (a)–(h).

Without detailed further dynamical input it is impossible to prove which if any of these channels has massless particles. However, the Higgs picture suggests that only one linear combination of a, b and c remains massless and corresponds to the massless ψ_5. The state d will be seen to have the wrong quantum numbers to correspond to ψ_5.

Indeed, it appears that two separate global conservation laws follow from the lagrangian. They are associated with phase rotations of the fields ψ and χ:

$$\psi \to e^{i\theta_1}\psi, \qquad \chi \to e^{i\theta_2}\chi. \tag{3.8}$$

The conserved charges are the numbers of 5 and $\overline{10}$ fundamental fermions. For the fermions a, b, c and d we have

$$N_5(\mathrm{a}) = 1\,, \qquad N_{\overline{10}}(\mathrm{a}) = -2\,,$$

$$N_5(\mathrm{b}) = 2\,, \qquad N_{\overline{10}}(\mathrm{b}) = 1\,,$$

$$N_5(\mathrm{c}) = 0\,, \qquad N_{\overline{10}}(\mathrm{c}) = -5\,,$$

$$N_5(\mathrm{d}) = 5\,, \qquad N_{\overline{10}}(\mathrm{d}) = 0\,. \tag{3.9}$$

Apparently the conservation of N_5 and $N_{\overline{10}}$ would forbid the mixing of the states.

The above conclusion is not correct. The currents associated with N_5 and $N_{\overline{10}}$ are anomalous and are violated by SU(5) instantons which cause the transition

$$\psi^\dagger \to \chi\chi\chi\,. \tag{3.10}$$

One linear combination, namely

$$Q = 3N_5 - N_{\overline{10}} \tag{3.11}$$

is conserved. However, the fermions a, b, c all have $Q = 5$. Therefore, instanton-mediated transitions can occur between them.

Evidently there are two distinct fermionic channels with $Q = 5$ and $Q = 15$. To determine which one corresponds to the massless ψ_5 we note that in the Higgs picture

$$\psi_5 = \langle \phi_i^\dagger \rangle \psi_i = \psi_i \langle \chi_{ijk}^\dagger \chi_{lmn}^\dagger \rangle \varepsilon_{jklmn}\,.$$

This corresponds in the symmetric picture to a $Q = 5$ state. Complementarity suggests that this is the only channel containing massless spectrum. In sect. 4 we will see that this conclusion is supported by 't Hooft's analysis which is based on the symmetric picture.

4. 't Hooft's consistency condition

't Hooft has recently given a very beautiful analysis of the consistency of massless bound fermions in confining theories [2]. In this section we will show how 't Hooft's consistency condition is satisfied in the model of sect. 3. We work in the symmetric picture.

The consistency condition is obtained by coupling the gauge system under consideration to additional weakly interacting "spectator" gauge fields via the global or flavor symmetries of the system. In the SU(5) model of sect. 3 the only non-anomalous conserved current is associated with the charge in (3.10). Define the corresponding current to be j_μ:

$$\int j_0 \, d^3x = Q = 3N_5 - N_{\overline{10}} . \tag{4.1}$$

The additional spectator gauge field is an abelian vector potential A_μ coupled to j_μ

$$L_{\text{int}} = e j_\mu A_\mu . \tag{4.2}$$

The theory defined in this way is not quite consistent because of triangle anomalies involving the abelian field A at all three vertices. The anomaly induces a divergence of the current j_μ proportional to

$$F\tilde{F} \sum_i Q(i)^3 \equiv K F\tilde{F} , \tag{4.3}$$

where Σ_i means sum over all left-handed fermions

$$F = \partial_\mu A_\nu - \partial_\nu A_\mu ,$$

and $Q(i)$ is the abelian charge of the ith fermions. Since there are 5 fermions with charge 3, and 10 with charge -1, we find

$$K = 5(3^3) - 10 = 125 . \tag{4.4}$$

To make the model consistent a set of "spectator" fermions must be introduced to cancel this anomaly. It is not important exactly how the spectator fermions are chosen, except that they must be SU(5) singlets and their total U(1) anomaly must be -125.

The condition of 't Hooft asserts that the massless sector of the theory at low energy must be describable by a consistent renormalizable lagrangian. The massless sector consists of

(a) any massless composites of the confined strongly interacting system;

(b) gauge bosons of the unbroken flavor symmetry : in the SU(5) model this means A_μ (in the symmetric picture the current j is not spontaneously violated);

(c) spectator fermions.

Since the low-energy effective theory must be consistent, the triangle anomaly of the U(1) currents must cancel if we replace the original strongly interacting fermions by the massless bound states. This means that the triangle anomalies in the unbroken spectator sector must be the same for fundamental fermions and for massless composite fermions.

As we have seen, the fundamental fermions contribute an anomaly,

$$K = 125 .$$

Now recall that the bound massless fermion in sect. 3 had abelian charge $3N_5 - N_{\overline{10}}$ equal to 5. If the massless bound state replaces the fundamental fermions its anomaly is $5^3 = 125$. Thus we see that 't Hooft's condition is satisfied.

It is interesting to see how the theory including spectator gauge bosons would look in the Higgs picture. The condensate

$$\langle \phi_i \rangle = \langle \chi^\dagger_{ijk} \chi^\dagger_{lmn} \varepsilon^{jklmn} \rangle$$

violates the abelian flavor symmetry so that A is not massless. However, a linear combination

$$Q' = Q + \tfrac{1}{2}\sqrt{40}\,\lambda \tag{4.5}$$

is conserved. Here λ is the diagonal SU(5) generator represented by the matrix

$$\lambda = \begin{bmatrix} 1 & & & & \\ & 1 & & & \\ & & 1 & & \\ & & & 1 & \\ & & & & -4 \end{bmatrix} \sqrt{\frac{1}{40}}\,. \tag{4.6}$$

Thus there is still a linear combination of A_μ and an SU(5) gauge boson which remains massless. It is easy to see that the charges Q' associated with the physical spectrum [SU(4) singlets] in the Higgs picture are the same as the charges Q for the SU(5) singlets in the symmetric picture. For example, the charge Q' of the state ψ_5 is once again 5.

5. Models with SU(n) × U(1) flavor symmetry

5.1. HIGGS PICTURE

In these models the "color" gauge group will be $SU(n+4)$. The fermion content will be $n[1]$'s (denoted by ${}_f[1]$, $f = 1, 2, \ldots, n$) and one $[\bar{2}]$*. The one-instanton process is shown in fig. 1. The instanton emits each one of the [1]'s once and the $[\bar{2}]$ $n+2$ times. The non-anomalous U(1) current is

$$Q = (n+2)\sum_{f=1}^{n} N({}_f[1]) - nN([\bar{2}])\,, \tag{5.1}$$

where $N({}_f[1])$ is the number of ${}_f[1]$'s, etc.

When the gauge forces become strong they are going to bind condensates in the MAC [5]. The theory is going to tumble down to a state of lower symmetry [5].

* $[\bar{2}]$ transforms as the antisymmetrized product of two $[\bar{1}]$'s.

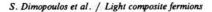

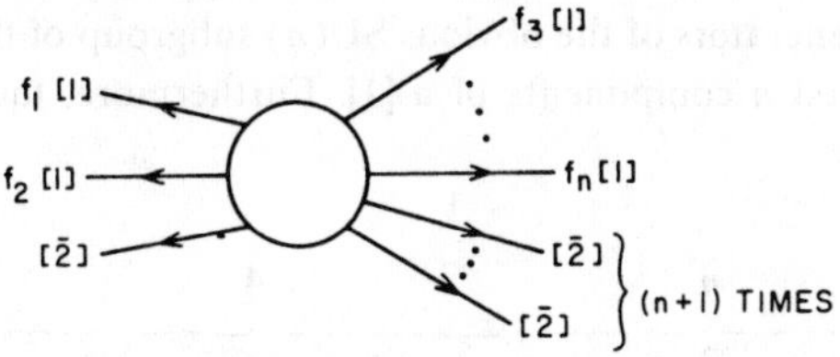

Fig. 1. The one-instanton process. The amplitude conserves the SU(n) but violates the U(n) of flavor. Thus, it is completely antisymmetric in the n flavor indices $f_1, f_2 \ldots f_n$. It follows that this graph mixes the 3-quark baryon $\{_{f_1}[1]\ [\bar{2}]\ _{f_2}[1] - (f_1 \leftrightarrow f_2)\}$ with the $(2n-1)$ quark baryon $\epsilon^{f_1 f_2 f_3 \ldots f_n}\ _{f_3}[1] \ldots {}_{f_n}[1]\ [\bar{2}] \ldots [\bar{2}]$.

The MAC in these models for $n > 2$ binds $_f[1]$ and $[\bar{2}]$ into a condensate in $[\bar{1}]$, i.e.,

$$\langle {}_f[1]_\beta [\bar{2}]^{\alpha\beta} \rangle \equiv {}_f\phi^\alpha \neq 0 , \tag{5.2}$$

where $f = 1, 2, \ldots n$ and $\alpha, \beta = 1, 2, \ldots n+4$.

To understand the possible patterns of symmetry breaking we introduce the hermitian matrix,

$$H(f, f') = \sum_\alpha {}_{f'}\phi^{\alpha\dagger}\ {}_{f'}\phi^\alpha . \tag{5.3}$$

H can always be diagonalized by a flavor rotation. The final unbroken subgroup of color is determined by the number m of non-zero eigenvalues. It is given by SU($n - m + 4$). Furthermore, the condensates give mass to m independent fermions.

Two assumptions about the magnitudes of $_f\phi^\alpha$'s are very natural. One is that only one of the $_f\phi^\alpha$'s is non-vanishing. The second one is that all the $_f\phi^\alpha$'s are non-vanishing and equal, i.e.,

$$_1\phi^1 = {}_2\phi^2 = \cdots = {}_n\phi^n = M \neq 0 . \tag{5.4}$$

Instantons favor this second assumption over the first. One gains instanton mediated potential energy by having more condensates form. This is the case (for instantons) even if the color symmetry has to be broken at the same time.

From now on we shall simply assume this second pattern of symmetry breaking. The condensates of eq. (5.4) break the original SU($n + 4$) color symmetry down to SU(4).

The condensates of eq. (5.4) break the original flavor SU(n) × U(1) symmetries. However, it is easy to define a new flavor group SU′(n)× U′(1) which remains unbroken. The generators of the new SU′(n) will be the sum of the old SU(n)

generators and the generators of the obvious $SU(n)$ subgroup of the color $SU(n + 4)$ which acts on the first n components of a [1]. Furthermore, the charge Q' of the new $U'(1)$ is given by

$$
Q' = Q +
\begin{bmatrix}
2 & & & 0 & & & & & \\
& 2 & & & & & & 0 & \\
& & \ddots & & & & & & \\
0 & & & 2 & & & & & \\
\hline
& & & & -\tfrac{1}{2}n & & & & \\
& & & & & -\tfrac{1}{2}n & & 0 & \\
& 0 & & & & & -\tfrac{1}{2}n & & \\
& & & & 0 & & & -\tfrac{1}{2}n &
\end{bmatrix}
\begin{array}{l} \\ \\ \left.\rule{0pt}{2.5em}\right\} n \\ \\ \\ \left.\rule{0pt}{2.5em}\right\} 4 \\ \end{array}
\tag{5.5}
$$

(acting on $[1]_\alpha$).

The chiral flavor group $SU'(n) \times U'(1)$ is realized in the Wigner mode via the appearance of massless fermions. It is easy to identify them directly from the condensate (5.4). To do this, notice that the components of $_f[1]_a$, where $a = n + 1$, $n + 2$, $n + 3$, $n + 4$, are confined by the $SU(4)$ forces. The remaining $_f[1]_i$, $i = 1, 2, \ldots n$, condense with $[\bar{2}]^{if}$. But since $[\bar{2}]^{if}$ is antisymmetric under interchange of i and f, it follows that the linear combinations of $_f[1]_i$ ($f, i = 1, 2, \ldots n$) that are *symmetric* under interchange of i and f do not condense. Thus,

$$
\psi_{fi} = \sqrt{\tfrac{1}{2}} \left(_f[1]_i + _i[1]_f \right)
\tag{5.6}
$$

forms a massless fermion in the symmetric $\tfrac{1}{2} n(n + 1)$ dimensional representation of $SU'(n)$. The $U'(1)$ charge Q' of ψ_{fi} is

$$
Q'(\psi_{fi}) = n + 4 .
\tag{5.7}
$$

The remaining particles in the spectrum of this theory are the n^2 massive gauge bosons which do *not* carry $SU(4)$ quantum numbers and the mesons and baryons bound via $SU(4)$ forces.

5.2. SYMMETRIC PICTURE

According to this picture, none of the flavor or color symmetries are spontaneously broken. The spectrum of the theory consists of mesons and baryons bound by the strong $SU(n + 4)$ forces. The $\tfrac{1}{2} n(n + 1)$ massless fermion states that we found in the Higgs picture will be replaced by $\tfrac{1}{2} n(n + 1)$ massless composite baryonic bound states in the confining picture. These baryonic bound states obviously have the form [see (5.2)]

$$_f[1]_\alpha[\bar{2}]^{\alpha\beta}{}_{f'}[1]_\beta + (f \leftrightarrow f') \equiv \psi_{ff'} . \tag{5.8}$$

Notice that in the confining picture SU(n) × U(1) is not broken and corresponds to SU′(n) × U′(1). This fact together with eq. (5.6) shows that the $\psi_{ff'}$ indeed have to be the $\frac{1}{2}n(n+1)$ massless composite baryonic states.

What happened to the following $\frac{1}{2}n(n-1)$ antisymmetric combinations?

$$_f[1]_\alpha[\bar{2}]^{\alpha\beta}{}_{f'}[1]_\beta - (f \leftrightarrow f') \equiv \omega_{ff'} . \tag{5.9}$$

These fermions correspond to the $({}_f[1]_i - {}_f[1]_i)\sqrt{\frac{1}{2}}$ that condensed with $[\bar{2}]^{if}$ in the Higgs picture. In the confining picture, condensates are not available to play the role of chirality flippers. However, this role can be successfully played by the instantons. In fact, the instanton amplitude of fig. 1 shows exactly how $\omega_{ff'}$ can form a massive baryon by mixing with $\frac{1}{2}n(n-1)$ baryons consisting of $2n-1$ fermions (i.e., $(n-2)$ ${}_f[1]$'s and $n+1$ $[\bar{2}]$'s).

The n^2 massive gauge bosons of the Higgs picture are now viewed as massive flavored vector mesons containing components of the form

$$_f[1]^*_\alpha(D_\mu\, {}_{f'}[1])_\alpha \equiv \rho^f_{\mu f'} .$$

Finally, we wish to show that 't Hooft's anomaly consistency conditions are satisfied. These conditions [2] require that the flavor anomalies of quarks have to equal the corresponding flavor anomalies of the massless baryons.

In the unbroken flavor group SU(n) × U(1) there exist three types of anomalies (see fig. 2):

(a) the SU(n) anomaly;

(b) the U(1) anomaly;

(c) the anomaly in the U(1) current caused by SU(n) gauge bosons.

The contribution of quarks (A_q) and massless baryons (A_B) to each of these anomalies is as follows.

(a) There are $n+4$ quarks in the fundamental representation (n) of SU(n). Their contribution is $A_q = (n+4)d^{abc}(n)$. There is one baryonic SU(n) multiplet belonging to the symmetric $\frac{1}{2}n(n+1)$ representation. The contribution is

$$A_B = d^{abc}\left(\tfrac{1}{2}n(n+1)\right) = (n+4)d^{abc}(n) . \tag{5.10}$$

Thus $A_B = A_q$ in case (a).

(b) The quarks' contribution is

$$A_q = (n+4)n(n+2)^3 - \tfrac{1}{2}(n+4)(n+3)n^3 \tag{5.11}$$

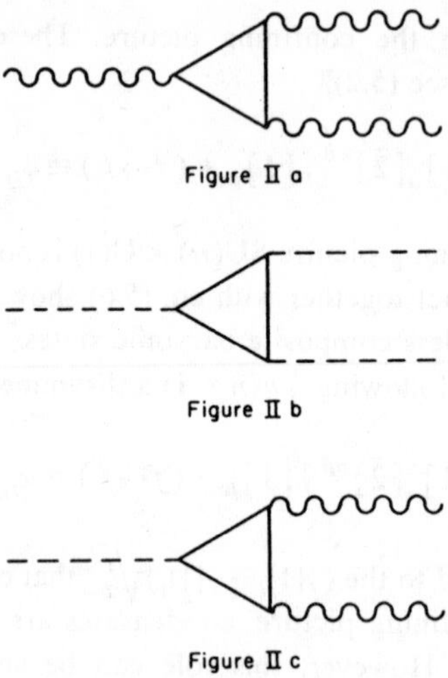

Fig. 2. The 3 types of flavor anomalies of the models of sect. 5. Broken line denotes the U(1) flavor current. Curly lines denote SU(n) flavor currents.

The baryons' contribution is

$$A_{\mathrm{B}} = \tfrac{1}{2}n(n+1)(n+4)^3 . \tag{5.12}$$

It is easily seen that $A_{\mathrm{q}} = A_{\mathrm{B}}$.

(c) The quarks' contribution is

$$A_{\mathrm{q}} = (n+4)(n+2)\tfrac{1}{2} . \tag{5.13}$$

Here $n+4$ is the number of SU(n) n-plets, $(n+2)$ is their U(1) charge, and $\frac{1}{2}$ is the trace $\mathrm{Tr}(\frac{1}{2}\lambda^a \frac{1}{2}\lambda^a)$ for the n of SU(n). The baryons' contribution is also

$$A_{\mathrm{B}} = (n+4)(n+2)\tfrac{1}{2} . \tag{5.14}$$

Here $n+4$ is the U(1) charge of the baryons and $\frac{1}{2}(n+2)$ is the $\mathrm{Tr}\, G^a G^b$ for the symmetric $\frac{1}{2}n(n+1)$ representation of SU(n).

6. Partially broken flavor

't Hooft assumes in his analysis that the entire chiral flavor group remains unbroken. In the previous examples this was indeed the case. However, as 't Hooft himself remarks, only a subgroup of the flavor symmetry may remain unbroken, but this may still be sufficient to guarantee massless fermions. In this event only that subset of the consistency conditions corresponding to the unbroken symmetry will be satisfied. In the following we present an example in which the flavor symmetry breaks down to U(1). We shall see that only the U(1) consistency conditions are satisfied.

Consider an $SU(7)_s$ strong group. The fermion spectrum is

$$3[3]_7 + 2[5]_7 . \qquad (6.1)$$

The flavor symmetry is $SU(3)_{F_1} \times SU(2)_{F_2} \times U(1)_Z$. The charge Z which generates $U(1)_Z$ is given by

$$Z = 3N([5]) - N([3]) . \qquad (6.2)$$

The fermion content under $SU(7)_s \times SU(3)_{F_1} \times SU(2)_{F_2} \times U(1)_Z$ is*

$$([3]_7, \bar{3}, 1, -1) + ([5]_7, 1, 2, 3) .$$

The fermions can also be represented by fields

$$\psi^{\alpha}_{ijk} , \qquad \chi^{ij}_a = \varepsilon^{ijklmnp} \chi_{klmnpa} ,$$

where $(i, j, k, l, m, n, p = 1, \ldots, 7)$, $a = 1, 2$, $\alpha = 1, 2, 3$.

The first condensate (MAC) to occur when the interactions become strong at scale M_1 is in the product

$$\langle \psi^{\alpha}_{ijk} \psi^{\beta}_{lmn} \varepsilon^{ijklmnp} \varepsilon_{\alpha\beta\gamma} \rangle \equiv \phi^p_\gamma . \qquad (6.3)$$

There are two inequivalent symmetry-breaking patterns which are likely. In the first case ϕ may be rotated to the form in which only

$$\phi^1_1 \neq 0 . \qquad (6.4)$$

The symmetry breaks to $SU(6)_{s'} \times SU(2)_{F_1'} \times SU(2)_{F_2} \times U(1)'$. This line eventually tumbles to an unbroken $SU(4)_{s'} \times U(1)$ with three massless fermions in the spectrum. The consistency conditions for the U(1) current are satisfied.

We will, however, follow a different line for which

$$\phi^1_1 = \phi^2_2 = \phi^3_3 \neq 0 . \qquad (6.5)$$

This immediately breaks the symmetry to

$$SU(4)_{s'} \times SU(3)_{F_1'} \times SU(2)_{F_2} \times U(1)_Y . \qquad (6.6)$$

* Note that the strong content is represented by the number of antisymmetrized fundamental indices, while non-abelian flavor representations are labeled by multiplicity. The U(1) content is specified by the value of Z.

The as yet unbroken $SU(3)_{F_1'}$ is the diagonal subgroup of $SU(3)_{F_1}$ and an SU(3) subgroup of $SU(7)_s$. Similarly, $U(1)_Y$ is generated by the combination

$$Y = Z + X, \tag{6.7}$$

where

$$X = \begin{bmatrix} -2 & & & & & & \\ & -2 & & & & & \\ & & -2 & & & & \\ & & & \frac{3}{2} & & & \\ & & & & \frac{3}{2} & & \\ & & & & & \frac{3}{2} & \\ & & & & & & \frac{3}{2} \end{bmatrix} \tag{6.8}$$

is a diagonal generator of $SU(7)_s$.

Thus far, the uncondensed fermions are

$$\left([0]_4, \bar{3}, 1, -7\right) + \left([1]_4, \bar{6}, 1, -\tfrac{7}{2}\right) + \left([4]_4, 3, 2, 7\right) + \left([3]_4, \bar{3}, 2, \tfrac{7}{2}\right) + \left([2]_4, 1, 2, 0\right). \tag{6.9}$$

The process of symmetry breaking continues due to the increasing strength of the $SU(4)_{s'}$ forces. However, because the $SU(4)_{s'}$ representation is real, $SU(4)_{s'}$ is not further broken. We find a second condensate at scale M_2:

$$\begin{aligned} \zeta &= \langle([2]_4, 1, 2, 0) \times ([2]_4, 1, 2, 0)\rangle \\ &= \langle \chi_a^{ij} \chi_b^{kl} \varepsilon_{ijkl123} \rangle . \end{aligned} \tag{6.10}$$

This condensate respects $U(1)_Y$, $SU(3)_{F_1'}$ and $SU(4)_{s'}$ but breaks $SU(2)_{F_2}$ to $U(1)_{F_2}$.

Finally, at scale M_3 the symmetry is broken to $SU(4)_{s'} \times U(1)_Y$ by the condensate

$$\begin{aligned} \left([1]_4, \bar{6}, 1, -\tfrac{7}{2}\right) \times \left([3]_4, \bar{3}, 2, \tfrac{7}{2}\right) &= \sum_{\substack{i=1-7 \\ j=1-7}} \sum_{l=1-4} \left\{ \psi_{lij}^{\alpha} \chi_a^{lp} \varepsilon^{ijk4567} + \psi_{lij}^{k} \chi_a^{lp} \varepsilon^{ij\alpha 4567} \right\} \\ &\equiv \xi_a^{(\alpha k)p}, \qquad (\alpha, p, k = 1, 2, 3), \qquad (a = 1, 2). \end{aligned} \tag{6.11}$$

This condensate combines the $(\bar{6}, 1)_F$ with $(\bar{3}, 2)_F$ to form six massive Dirac fermions. In the process all the symmetry is broken except for $SU(4)_{s'}$ and $U(1)_Y$.

There are 9 uncondensed fermions which transform under $SU(4)_{s'} \times U(1)_Y$ as

$$3([0]_4, -7) + 6([4]_4, 7). \tag{6.12}$$

However, not all of these remain completely massless. The three states $([0]_4, -7)$ combine with three of the $([4]_4, 7)$'s to form three massive Dirac fermions. The process involves a *secondary* coupling of these fermions to the condensate ξ through the broken SU(7) gauge generators. The relevant Feynman graph is shown in fig. 3. It is interesting that this mass is lighter than any of the other scales $M_{1,2,3}$. The diagram in fig. 4 suggests an order of magnitude

$$m \sim M_3^3 / M_1^2 . \tag{6.13}$$

Three fermions remain exactly massless and will be seen to satisfy 't Hooft's conditions.

We shall now re-analyze the model in the symmetric picture. There are several SU(7) singlet fermion composites which can be formed. We have

$$f^\alpha = \psi^\alpha_{ijk} \phi^i_\beta \phi^j_\gamma \phi^k_\delta \varepsilon^{\beta\gamma\delta}, \tag{6.14}$$

$$g_{a\gamma} = \chi^{ij}_a \phi^{*\alpha}_i \phi^{*\beta}_j \varepsilon_{\alpha\beta\gamma}, \tag{6.15}$$

where ϕ is the bilinear given by eq. (6.3). The states f and g have $Z = -7$ and $Z = 7$. They may be identified with the 9 states in eq. (6.12). A variety of other fermionic states with $Z = \pm 7$ can be formed which, as in the SU(5) example, pair by instanton induced interactions into massive fermions.

It is amusing to speculate how the mass matrix among f and g can arise. A mass term obviously requires a spontaneous breakdown of the flavor $SU(3)_{F_1} \times SU(2)_{F_2}$ symmetry. One might speculate that this symmetry does not spontaneously break down although the Higgs picture suggests that it does. However, in this event 't Hooft's consistency conditions would have to be satisfied for the entire flavor group. We have checked and found this to be impossible.

A likely possibility is that a 4-fermion condensate develops and breaks the symmetry. Thus, consider the operator

$$\theta^\alpha_{\beta a} = \psi^\alpha_{ijk} \chi^{jk}_a \phi^i_\beta . \tag{6.16}$$

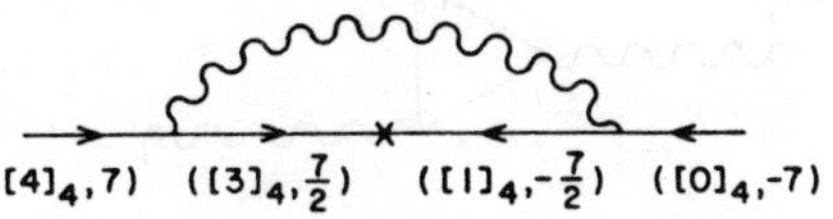

Fig. 3. The secondary mass generation graph resulting in light fermions.

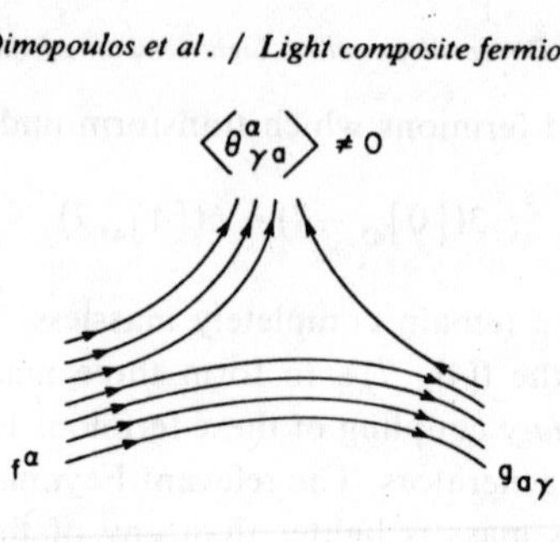

Fig. 4. Secondary mass generation as viewed in the symmetric picture.

A non-vanishing expectation value for θ would have the following effects:

(a) $SU(3)_{F_1} \times SU(2)_{F_2}$ would be completely broken;

(b) $U(1)_Z$ is left unbroken;

(c) The states f and g would be coupled by fig. 4 to form three massive fermions. This leaves three massless states as in the Higgs picture.

Finally, one can check the consistency conditions of 't Hooft for the $U(1)_Z$ current. The entire set of triangle graphs for flavor currents is shown in fig. 5.

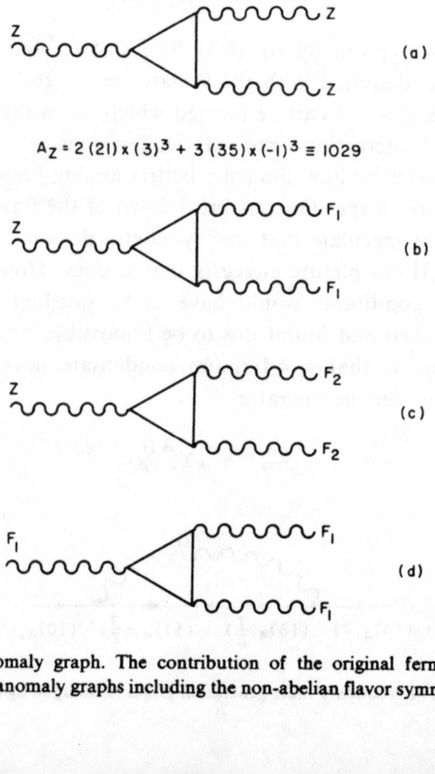

Fig. 5. (a) the $U(1)_Z$ anomaly graph. The contribution of the original fermions (A_Z) is evaluated explicitly. (b), (c), (d) The anomaly graphs including the non-abelian flavor symmetry $SU(3)_{F_1} \times SU(2)_{F_2}$.

Consistency requires the massless composite states to produce the same anomalies as the fundamental fermions for the unbroken subgroup. As 't Hooft has noted, these equations often have no simultaneous solutions and this proves to be the case if we take the entire group to be unbroken.

If only the $U(1)_Z$ subgroup is unbroken, then the only triangle diagram to consider is fig. 5a. We find both the elementary fermion contribution and the massless composites with $Z = 7$ give a total anomaly of 1029.

7. Conclusions

Gauge theories with non-real fermion content may be able to undergo a form of dynamical Higgs phenomenon [5]. In this paper we showed that in many cases the results can be reinterpreted in terms of a symmetric or confined picture in which all observable states are composite. Of particular interest are the massless fermion states which remain uncondensed in the Higgs picture. These states are generally protected from secondary mass generation (fig. 3) by unbroken symmetries. In the confined picture these states must be understood as massless fermionic composites. In fact, we have found that the massless fermions found in this way always satisfy 't Hooft's very nontrivial conditions. This adds to our confidence that the principles of ref. [5] are not unreasonable.

The question of the real significance of this work is not entirely obvious. It is our view that it should be applied to the so-called extended technicolor theory of mass generation. In this scenario quarks, leptons, and technifermions participate in a strong interaction which tumbles to color $\times$ technicolor. Complementarity would then imply that quarks and leptons are composite at a scale $\sim$TeV.

Do we actually want quarks and leptons to appear as massless states satisfying 't Hooft-like equations? There can be two views on this. The first says yes, quarks and leptons are, to some approximation, massless composites. Small effects, either of the electroweak or other interactions, would upset the delicate symmetry relations which keep them massless. The other answer is no, they are merely very light. Indeed, we have argued both here and in other places that secondary mass generation (fig. 3) leads to hierarchies of light fermion masses. An example occurred in sect. 6 in which the fermions underwent secondary mass generation. It is interesting that in the confined picture these small masses are induced by a 4-fermion condensate. It is not unreasonable that the condensate of 4 fermions requires larger coupling strength than the usual bilinear condensates. Accordingly it should occur at lower mass scale.

Whatever the ultimate application of these ideas, it is evident that gauge theories with non-real fermion content are extraordinarily rich and worthy of study for their own sake.

References

[1] S. Dimopoulos and L. Susskind, Nucl. Phys. B155 (1979) 237
[2] G. 't Hooft, Cargèse Summer Lectures (1979)
[3] S. Brodsky and S. Drell, to be published
[4] E. Fradkin and S. Shenker, Phys. Rev. D19 (1979) 3682
G. 't Hooft, Cargèse Summer Lectures (1979)
[5] S. Raby, S. Dimopoulos and L. Susskind, Tumbling gauge theories, Stanford preprint ITP-653 (1979)
[6] H. Georgi, Harvard preprint (1979)
[7] A. Casher, Tel Aviv preprint TAUP 734/79;
B. Svetitsky, S. Drell, H. Quinn and M. Weinstein, SLAC preprint (1980)

Volume 96B, number 1,2 PHYSICS LETTERS 20 October 1980

ANOMALY CONSTRAINTS IN CHIRAL GAUGE THEORIES

T. BANKS [1] and S. YANKIELOWICZ [1]
Department of Physics and Astronomy, Tel Aviv University, Tel Aviv, Israel

and

A. SCHWIMMER [1]
Department of Nuclear Physics, Weizmann Institute, Rehovot, Israel

Received 11 June 1980

A general procedure for constructing solutions to 't Hooft's anomaly equations in left-right asymmetric theories is discussed. All the known solutions and a few new ones are obtained from a unified treatment. The characteristic feature is the appearance of a global symmetry which can be an ordinary group or a superalgebra.

Recently 't Hooft [1] proposed the use of the axial anomaly equations as constraints on the zero mass fermion spectrum of gauge theories.

A gauge theory with a set of zero mass elementary fermions, possesses besides the gauged local symmetry (which in the following we will call colour) also a global flavour symmetry determined by the fermion spectrum. The generators of the flavour symmetry are integrals of gauge invariant currents. The naive conservation of some U(1) flavour currents is spoiled by the triangle anomaly and their divergence becomes proportional to $F^a_{\mu\nu}\widetilde{F}^{\mu\nu}_a$ where $F^a_{\mu\nu}$ is the colour gauge field strength.

The set of flavour currents j^a_μ which continue to be conserved even in the presence of the gauge field generates the flavour group G_F which is usually a direct product of SU(n) and U(1) groups. Three point functions of these currents will have generally a c-number anomaly proportional to Σd_{abc} where the d-symbol is defined as

$$d_{abc} = \mathrm{Tr}(\lambda_a\{\lambda_b, \lambda_c\}) , \tag{1}$$

λ_a being the generators of G_F in a given representation and the summation is over the fermion representations.

If the flavour group G_F is not spontaneously broken the anomaly equations must be saturated by the zero mass fermionic bound-states of the theory [1,2]. These equations put very strong constraints on the theory. In particular 't Hooft has shown that in left right symmetric theories the flavour group cannot be larger than SU(2) and therefore in QCD spontaneous breaking of chiral symmetry is a necessity.

Dimopoulos et al. [3] studied 't Hooft's equations in left handed theories and found a few interesting solutions. Moreover they showed that the massless spectrum obtained from the equations is consistent with the one suggested by dynamical considerations. In the present note we discuss a general method for obtaining solutions of the anomaly equations in left-right asymmetric theories. The previously known solutions are recovered and a few new sets of solutions are obtained. A puzzling feature is the appearance, whenever there is a solution, of a global symmetry algebra or superalgebra which contains the colour and flavour groups as subgroups. This global symmetry is realized in a very strange way the elementary and bound zero mass fermions appearing in the same irreducible representation.

The requirements on the zero mass fermion spectrum are the following:

(a) The elementary fermions should not produce

[1] Supported in part by the Israel Commission for Basic Research and the United States-Israel Binational Science Foundation.

Volume 96B, number 1,2 PHYSICS LETTERS 20 October 1980

anomalies in the colour (gauged) group G_c.

(b) The U(1) currents included in the flavour group should be conserved even in the presence of the gauge fields, i.e. the elementary fermions should not produce anomalies when we consider a U(1) current with two coloured currents.

(c) For three currents in the flavour group the anomaly produced by the elementary fermions should equal the anomaly produced by the bound ones.

A solution to these requirements can be obtained by the following simple procedure:

Consider a group (or superalgebra) G which contains as subgroups G_c and G_F. We choose a complex representation r of G with vanishing d-symbol. Under $G_c \times G_F$, r is generally reducible, i.e.:

$$r = \sum_{c,f} (c,f) .$$

We single out the representations which are colour singlets $(1,f)$ their conjugate being the bound fermion representations while the elementary fermions will be in the other representations:

$$\text{elementary} = \sum_{c \neq 1,f} (c,f) , \quad \text{bound} = \sum_{f} (1,\bar{f}) . \quad (2)$$

The three requirements mentioned above are fulfilled by using repeatedly that the r representation has a vanishing d-symbol and therefore for any combination a, b, e of generators in G:

$$\sum_{c \neq 1,f} d_{abe}(c,f) + \sum_{f} d_{abe}(1,f) = 0 , \quad (3)$$

where $d_{abe}(c,f)$ is the d-symbol corresponding to the (c,f) representation of $G_c \times G_F$. For a, b, e being coloured currents the second term does not contribute and therefore the elementary fermions do not produce an anomaly. For a, the index of a U(1) current in the flavour group and b, e two coloured currents the second term (colour singlets) does not contribute and therefore the U(1) currents in G_F are anomaly free. Finally, for three currents in the flavour group, when the second term is moved to the r.h.s., eq. (3) expresses the equality between the anomalies produced by the elementary and bound fermions.

We emphasize that not every complex anomaly free representation will provide a solution of 't Hooft's constraints. The embedding group G must contain

$$G_c \times G_F = G_c \times \prod_{i=1}^{k} U(n_i)/U(1) ,$$

where G_c is an arbitrary colour group. The n_i are precisely the multiplicities of the nontrivial representations f_i of G_c in the decomposition of the anomaly free irreducible representation of G. Each fundamental fermion must have the following transformation property under $G_c \times G_F$: $(f_i, 0, ..., n_i, ..., 0)$. This is our reason for insisting on a complex representation of G. A real or pseudo real representation would in particular be real under G_F and the physical interpretation of G_F as a flavour group would be impossible. An exception to this is the case G_F = SU(2). This leads to a "doubling" of solutions which we discuss briefly at the end of the paper. In left-right symmetric theories the requirements (a), (b) and part of (c) (the equality of anomalies for three U(1) currents) are automatically satisfied and therefore we do not believe that a bigger group structure appears.

Our strategy then is to find decompositions of anomaly free complex representations (of simple groups or super algebras) which satisfy the above conditions. For simplicity of exposition we consider only left handed fermions; right handed fermions are replaced by their left handed antiparticles.

SO(4m + 2) groups. As is well known all the SO(n) groups (except SO(6)) have vanishing d-symbols. However only the SO(n) groups with $n = 4m + 2$ have complex representations and are therefore able to give non trivial solutions.

The spinor representations (which are 4^m dimensional) are complex. One way to decompose it is through

$$SO(4m+2) \supset SU(2m+1) \times U(1) .$$

In particular for $m = 2$ the 16 dimensional spinor representation of SO(10) is decomposed under SU(5) into

$$16 = 5 + \overline{10} + 1 = [1] + [3] + [5] ,$$

where $[k]$ is the number of antisymmetric indices in SU(5). The U(1) charge Q is given by

$$Q = \sum_{j=1}^{5} \sigma_{2j-1,2j} , \quad (4)$$

$\sigma_{2j-1,2j}$ being SO(10) generators. Using the fermion formalism of ref. [4] it is easy to show that $Q = 5 - 2k$. Therefore the elementary fermions are in the [1] representation with $Q = 3$ and the [3] representation with $Q = -1$ while there is one bound state SU(5) singlet with $Q = 5$. This solution is one of the solutions of ref. [3].

One can break SO(10) to SU(4) × U(1) × U(1) where the U(1) currents are the Q defined above and $Q' = \sigma_{9,10}$ [‡1]. There are 14 elementary fermions in ([1], 3, 1), ([2], −1, −1) and ([3], −1, +1) and two composite ones with charges (1, 5, 1) and (1, −3, 1). In terms of the elementary fields ψ_i, X_{ij} and ϕ_{ijk} the composites are $\psi_k X^*_{ij}\phi^*_{ijk}$ and $X_{im}\phi_{ijk}\phi_{mjk}$.

Though, as we mentioned above, left-right symmetric theories are outside the framework of this analysis, it is amusing to note that 't Hooft's "σ-model solution" for SU(3) colour can be recovered through the decomposition of SO(10) into SU(3) × SU(2) × SU(2) × U(1).

The decomposition of the spinor is:

$$16 = (3, 2, 1, 1) + (\bar{3}, 1, 2, -1) + (1, 2, 1, 3) + (1, 1, 2, -3) ;$$

the U(1) charge is this time given by:

$$Q = \sum_{j=3}^{5} \sigma_{2j-1,2j} = 3 - 2k , \tag{5}$$

k being the number of antisymmetric indices in SU(3) and the SU(2) × SU(2) represents rotations in the first four coordinates of SO(10). In terms of 4-component spinors one has an elementary colour triplet, flavour doublet and a composite flavour doublet, i.e. the "σ-model solution". The other complex representations of SO(10) (higher rank spinors and the self-dual five index antisymmetric tensor) and higher SO(4m + 2) groups do not produce acceptable solutions. Either nonfundamental flavour representations appear or the same colour representation appears a few times leading to an unwanted flavour symmetry.

E6. Among the exceptional groups only E6 has complex representations [5]. The fundamental complex 27-dimensional representation as all the E6 representations has a vanishing [5] d-symbol. The embedding E6 ⊃ G2 × SU(3) under which 27 = (7,3) + (1, 6) gives a solution with three elementary fermions in the 7 dimensional representation of G2 (the colour group) and the bound fermions in the 6 representation of flavour SU(3). The bound fermions are therefore built out of at least five elementary ones. It is interesting to remark that a dynamical analysis analogous to the one used in ref. [3] does not produce these bound massless fermions.

The other embeddings and higher representations like e.g. the 351 dimensional one do not give valid solutions.

Since the symplectic groups have only real representations and only rather exotic complex representations of SU(n) have a vanishing d-symbol we turn now our attention to superalgebras.

SU(n/m). Since this set of superalgebras [6] contains in its commuting sector SU(n) × SU(m) × U(1) it is a natural candidate for embedding solutions of the anomaly equations.

We discuss first a few properties of SU(n/m) which we will need in the following.

The superalgebra SU(n/m) is generated by the bilinears in n bosonic a_i, and m fermionic b_α operators which conserve the sum of the fermion and boson number:

$$X_i^j \equiv a_i^+ a^j , \quad Y_\alpha^\beta \equiv b_\alpha^+ b^\beta , \quad Z_i^\alpha \equiv a_i^+ b^\alpha , \quad W_\alpha^i \equiv b_\alpha^+ a^i . \tag{6}$$

In order to make SU(n/m) simple the diagonal operators in (6) should be replaced by:

$$\widetilde{X}_i^i \equiv X_i^i - \frac{1}{n-m}\Big(\sum_j X_j^j + \sum_\beta Y_\beta^\beta\Big) ,$$
$$\widetilde{Y}_\alpha^\alpha \equiv Y_\alpha^\alpha + \frac{1}{n-m}\Big(\sum_j X_j^j + \sum_\beta Y_\beta^\beta\Big) , \tag{7}$$

such that a U(1) is taken out, i.e.

$$\sum_i \widetilde{X}_i^i + \sum_\alpha \widetilde{Y}_\alpha^\alpha = 0 ,$$

and each generator is supertraceless.

The third order Casimir operator C_3 is defined as [7]:

‡1 In the fermion formalism of ref. [4] Q' corresponds to $1 - 2a_5^+ a_5$.

Volume 96B, number 1,2 PHYSICS LETTERS 20 October 1980

$$C_3 = X^j_i\{X^k_j, X^i_k\} + Z^\alpha_i\{Y^\beta_\alpha, W^i_\beta\} - Z^\alpha_i\{W^k_\alpha, X^i_k\}$$
$$- X^j_i[Z^\alpha_j, W^i_\alpha] + W^i_\alpha\{X^j_i, Z^\alpha_j\} - W^i_\alpha\{Z^\beta_i, Y^\alpha_\beta\}$$
$$- Y^\beta_\alpha[W^i_\beta, Z^\alpha_i] + Y^\beta_\alpha\{Y^\gamma_\beta, Y^\alpha_\gamma\}, \qquad (8)$$

where we used the summation convention and [ ] and { } represent commutators and anticommutators, respectively.

We will be interested in all the representations of SU(n/m) which are obtained when the superalgebra is realized in the space of boson and fermion occupation numbers according to (6). Such a representation is characterized by the total fermion and boson number N and when decomposed according to SU(n) $\times$ SU(m) $\times$ U(1) gives all the symmetric representations of SU(n) with index N_B together with antisymmetric representations of SU(m) with index N_F such that $N_B + N_F = N$.

The supertrace of the representation $\widetilde{\dim}(N, n, m)$, which plays the role of the dimension of the representation for an ordinary group is:

$$\widetilde{\dim}(N, n, m) = \sum_{N_B=0}^{N} \binom{m}{N - N_B}\binom{n + N_B - 1}{N_B}(-1)^{N-N_B}$$
$$= (-1)^N \binom{m-n}{N} = \binom{n - m + N - 1}{N}. \qquad (9)$$

The U(1) charge Q, which must be supertraceless in the representation is:

$$Q = 2\left(\frac{nN}{m-n} + N_B\right). \qquad (10)$$

The third order Casimir can be calculated from (8) and we obtain:

$$C_3 = N(n - m - 1)(n - m - 2)(n - m + N)$$
$$\times (n - m + 2N)(n - m)^{-2}, \qquad (11)$$

and therefore the d-symbol of the representation will be proportional [‡2] to

$$d = C_3 \cdot \widetilde{\dim} = (n - m - 1)(n - m - 2)N(n - m + N)$$
$$\times (n - m + 2N)(n - m)^{-2}\binom{n - m + N - 1}{N}. \qquad (12)$$

[‡2] One should take out a factor $d_{abe}\,d_{abe}$ which we will do by normalizing to the representation $N = 1$.

We remark that since (9), (11), (12) depend on n, m only through their difference SU(n/m) provides a definition for the bosonic representations of SU(n) for $n \leqslant 1$ [‡3].

For our purposes we use the fact that according to (12) d has zeroes for any given N if the difference $m - n$ is chosen appropriately:

$m - n = 4$ for $N = 2$,

$m - n = 3, 6$ for $N = 3$,

$m - n = 3, 4, ..., N, 2N$ for $N > 3$.

For $N = 2$ the decomposition of SU($n/n + 4$) into SU(n) $\times$ SU($n + 4$) $\times$ U(1) is:

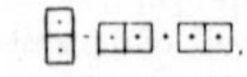 (13)

where □ represents SU($n + 4$) and ⊡ represents SU(n). The U(1) charges, according to (10) are: $n, n + 2, n + 4$. When the d-symbol is decomposed according to SU(n) $\times$ SU($n + 4$) $\times$ U(1) we have alternating signs due to the supertrace. The negative signs are absorbed by going to the conjugate representations. Then from (13) we obtain two solutions:

(i) The colour group is SU(n), the flavour group is SU($n + 4$) $\times$ U(1). There are $n + 4$ elementary fermions in the fundamental representation of SU(n) with $Q = n + 2$ and one elementary fermion in the conjugate of the symmetric two index representation of SU(n) with $Q = -n - 4$. The bound fermions are in the [2] of SU($n + 4$) with $Q = n$.

(ii) The colour group is SU($n + 4$), the flavour group is SU(n) $\times$ U(1). There are n elementary fermions in the fundamental representation of SU($n + 4$) with $Q = n + 2$ and one elementary fermion in [2] of SU($n + 4$) with $Q = -n$. The bound fermions belong to the two index symmetric representation of SU(n) with $Q = n + 4$.

It is amusing to note that for $n = 1$ one recovers the first SU(5) $\times$ U(1) solution (SU(5/1) and SO(10) are rather similar).

Solution (ii) is another one found in ref. [3].

The representation $N = 3$ which has a vanishing d

[‡3] This allows the reduction of bosonic statistical mechanics models for negative ranks to fermionic models with positive ranks and therefore provides an easy interpretation of their peculiar properties [8].

Volume 96B, number 1,2 PHYSICS LETTERS 20 October 1980

for $m = n + 3$ or $m = n + 6$ gives the decomposition.

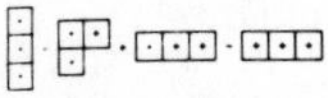 (14)

It is obvious from (14) that nontrivial flavour representations will appear unless $n = 1$ in which case a solution based on SU(4) × U(1) appears. Similar behavior is shown by higher N's.

The solutions obtained can be extended in a trivial way by making direct products with anomaly free groups (e.g. SU(2) which leads to a doubling) and by adding pairs of colour singlets which are in conjugate representations of the flavour group.

We have discussed a general procedure for finding solutions of 't Hooft's constraints for the case in which no spontaneous breakdown of the flavour group occurs. All known solutions were recovered and some new ones found. We suspect that our procedure in fact exhausts all possible solutions of the consistency conditions. The required relations between d symbols hint at the existence of an underlying irreducible representation (note that we have the option of adding real colour singlet flavour representations to complete the irreducible representation of G). We have not however studied solutions with partially broken flavour (e.g. the last scenario of ref. [3]).

The hidden global symmetry group G which "explains" the existence of solutions to 't Hooft's constraints does not seem to admit of any reasonable interpretation as a real symmetry of the underlying Lagrangian. In particular, the particles which make up the irreducible G multiplet are the elementary fermions of the underlying Lagrangian plus the "shadows" of the composite massless fermions. These latter have the same helicity as the physical composite particles but lie in conjugate representations of G_F. It seems then that the introduction of G must be regarded as an algebraic trick rather than the uncovering of a hidden symmetry. The alternative point of view is however so attractive that it merits further study.

Very useful discussions with E. Domany and Y. Frishman are gratefully acknowledged.

References

[1] G. 't Hooft, Cargese Summer Institute Lectures (1979).
[2] T. Banks, S. Yankielowicz, Y. Frishman and A. Schwimmer, Weizmann Institute preprint (1980).
[3] S. Dimopoulos, S. Raby and L. Susskind, Stanford preprint ITP-662 (1980).
[4] R. Casalbuoni and R. Gatto, Phys. Lett. 90B (1980) 81; R.N. Mohapatra and B. Sakita, Phys. Rev. D21 (1980) 1062.
[5] F. Gursey, P. Ramond and P. Sikivie, Phys. Lett. 60B (1976) 177.
[6] P.G.O. Freund and I. Kaplansky, J. Math. Phys. 17 (1976) 228.
[7] N.B. Backhouse, J. Math. Phys. 18 (1977) 239.
[8] R. Balian and G. Toulouse, Phys. Rev. Lett. 30 (1973) 544; M.E. Fisher, Phys. Rev. Lett. 30 (1973) 679.

Nuclear Physics B189 (1981) 547–556

A CONFINING MODEL OF THE WEAK INTERACTIONS

L.F. ABBOTT[1] and E. FARHI
CERN, Geneva, Switzerland

Received 21 April 1981

The weak interactions are analyzed as the low-energy effective interactions of a strong coupling gauge theory whose scale is $G_F^{-1/2}$. The light fermions are shown to be bound states obeying 't Hooft's consistency conditions. The symmetries of the theory are used to analyze the low-energy interactions. Instanton mediated baryon violations are discussed and experimental signatures at high energy are presented.

1. Introduction

At low energies the weak interactions are well described by the usual charged and neutral current four-Fermi lagrangian with coupling strength $G_F(G_F^{-1/2} \approx 300\ \text{GeV})$. In the standard $SU(2)_L \times U(1)$ model [1] these interactions arise from the exchange of weakly coupled massive vector bosons and the mass scale for G_F is set by the vacuum expectation value of a scalar field. However, only the low-energy behaviour of the weak interactions has at present been determined, so there exists another possibility. Perhaps the four-Fermi interactions we see are instead the low-energy remnants of a strong interaction between fermions which are bound states of confined constituents. In this case, G_F is determined by the mass scale of this underlying strong interaction, i.e., the Λ parameter of the theory. Thus, we imagine that the weak interactions are a result of a confining gauge theory [2] which binds constituents into the observed fermions at a scale of about $G_F^{-1/2}$.

In ref. [2], a model based on this idea was presented. This model is very similar to the standard $SU(2)_L \times U(1)$ theory, except that the $SU(2)_L$ gauge theory is taken to be confining rather than spontaneously broken. It was shown that this model has light bound states corresponding to the observed fermions and produces the correct four-Fermi weak interactions. This model may be a viable alternative to the standard spontaneously broken theory. Like the standard theory, the confining model is based on a renormalizable lagrangian and is thus well defined.

In this paper we will address certain questions which were not answered in our previous work. We begin by reviewing the basic assumptions of the model and by discussing its symmetry properties. Using these symmetry properties, we will show that the massless bound states appearing in the theory satisfy the 't Hooft consistency

[1] Permanent address: Brandeis University, Waltham, MA 02254, USA

conditions [3] for the existence of massless bound states in confining theories. Again, using the symmetries we will show that the low-energy effective lagrangian, which correctly describes the weak interactions, is included in the four-Fermi interactions of this theory and that only one other possible interaction is allowed. We will discuss why this extra term might be small. Then we address the queston of whether a strong coupling $SU(2)_L$ theory might have unacceptably high levels of baryon number violation through instantons. The answer is no and this is shown in sect. 5. Finally, we examine some of the experimental signatures of strong coupling weak interactions.

Although our approach has much success, there are certain difficulties. The basic mass scale associated with the weak interactions is $G_F^{-1/2} \simeq 300$ GeV. In order to account for neutral current interactions, the left-handed fermions must have electromagnetic form factors which at low Q^2 go like Q^2/M_1^2 where $M_1 \approx 100$ GeV. On the other hand, a bound-state structure for the muon will produce an anomolous magnetic moment of order m_μ^2/M_2^2 and from experiment we know that $M_2 \geqslant 600$ GeV. The discrepancy between these mass scales is unexplained.

Our entire analysis of the weak interactions depends very little on the nature of the confining force. There are three basic features which a confining gauge theory must have in order to provide a possible description of the low-energy weak interactions. First, it must have light (on the scale $G_F^{-1/2}$) composite fermions. Second, it must have a *global* SU(2) symmetry under which the bound-state fermions are doublets. Third, the strong interaction must only act on left-handed physical fields. These last two conditions almost guarantee the correct form for the low-energy effective lagrangian.

All of these conditions are met by the $SU(2)_L$ confining model, but perhaps it is not unique in this regard. In particular, since there is no spontaneous symmetry breaking in our approach, there is no fundamental reason for a scalar field to be present. In the $SU(2)_L \times U(1)$ model, a scalar doublet is necessary to bind with fermionic preons to produce the observed fermions. However, if there exists a confining theory based on a different group which has the necessary three features, perhaps a weak interaction model with no elementary scalars can be constructed. We are presently searching for such a model.

2. The model and its symmetries

In this section we will review some of the basic features of the strong coupling weak interaction model introduced in ref. [2]. This model is based on the same lagrangian as the standard $SU(2)_L \times U(1)$ model. For three generations of fermions, there are 12 (one lepton and three coloured quarks for each generation) left-handed $SU(2)_L$ doublet Fermi fields $\psi_{Li}^a (a = 1-12; i = 1-2)$ and a complex $SU(2)_L$ doublet scalar field ϕ_i. In addition, there are 24 right-handed Fermi fields $\mathcal{U}_R^a$, $\mathcal{D}_R^a$ which do not feel the $SU(2)_L$ force. These particles have the same U(1) quantum number assignments as in the standard model. They interact through $SU(2)_L$ and U(1) gauge interactions,

Yukawa couplings and scalar self-couplings. The only changes which are made in the strong coupling model are in the values of certain parameters in the lagrangian. These changes are:

(i) The potential of the scalar field is modified so that the scalar vacuum expectation value $\langle\phi\rangle$ vanishes. Therefore, there is no spontaneous symmetry breaking.

(ii) The U(1) coupling constant is set equal to e, the electromagnetic coupling constant where $e^2/4\pi = 1/137$.

(iii) The $SU(2)_L$ coupling constant is increased in value so that the $SU(2)_L$ interactions become strong at an energy scale of order $G_F^{-1/2}$.

Since the $SU(2)_L$ gauge symmetry is unbroken, and this theory becomes strong coupling, confinement presumably occurs and all physical states must be $SU(2)_L$ singlet bound states. Most of these bound states will have masses of order $G_F^{-1/2}$ since this is the mass scale determined by the running $SU(2)_L$ coupling constant. However, this model contains physical fermions which have masses much lighter than $G_F^{-1/2}$. In ref. [2] this was demonstrated by explicitly identifying the chiral symmetry responsible for keeping the fermions light. It was also shown that the U(1) quantum numbers of these light fermions agree with the usual charge assignments [3, 4]. Since the U(1) is unbroken and it has coupling constant e, the physical states have the correct electromagnetic interactions.

In the next section we will rederive the light fermion spectrum using the consistency conditions [3] of 't Hooft. In sect. 4 we will discuss the effective weak interactions of the light fermions. For these discussions it is useful to identify the global symmetries of the confining $SU(2)_L$ lagrangian.

The $SU(2)_L$ confinement occurs at an energy scale of about $G_F^{-1/2}$. The SU(3) colour interactions, electromagnetic interactions and Yukawa couplings are all weak compared to the $SU(2)_L$ interactions at the scale $G_F^{-1/2}$ so these interactions can be neglected. In this case there is no distinction between quarks and leptons or between particles in different generations, so the 12 left-handed $SU(2)_L$ doublets ψ_{Li}^a have an SU(12) symmetry. (We are not interested in the symmetries of the right-handed fields since they do not feel the strong force.) Naively, this symmetry would be U(12) but the U(1) subgroup has an anomaly in the presence of $SU(2)_L$ interactions. The effects of this anomaly will be discussed in sect. 5.

In addition to the SU(12) symmetry of the fermions, there is a global SU(2) symmetry associated with the scalar doublet. If we write the scalar doublet in a matrix form

$$\Omega = \begin{pmatrix} \phi_1 & -\phi_2^* \\ \phi_2 & \phi_1^* \end{pmatrix}, \tag{2.1}$$

then the $SU(2)_L$ standard lagrangian is invariant under the transformation

$$\Omega \to \Omega \mathscr{U}, \tag{2.2}$$

where $\mathscr{U}$ is an arbitrary SU(2) matrix. (This is the same symmetry discussed in ref. [2]

up to a global gauge transformation.) This symmetry is only exact in the absence of electromagnetism and Yukawa couplings. Up to these small corrections, the global symmetry of the strong $SU(2)_L$ theory is $SU(12)\times SU(2)$.

3. Light composite fermions

't Hooft has given the consistency conditions [3] for the existence of light fermion bound states in a confining theory. We will now show that the light fermions in our $SU(2)_L$ strong theory satisfy these conditions.

't Hooft's analysis begins by identifying the global symmetries of a given model. The elementary, confined fermions then produce certain anomalies in these global symmetries through the usual triangle diagrams. We next postulate the existence of certain massless composite fermions, and calculate the anomalies in the global symmetry group produced by these massless states. In order for the postulated existence of these massless composites to be consistent, 't Hooft requires that the anomalies produced by the fundamental fermions be identical to those of the bound state massless fermions.

As we have seen, the global symmetry of the $SU(2)_L$ model is $SU(12)\times SU(2)$. The only possible anomaly can come from triangle diagrams with three SU(12) generators at the vertices. If we normalize the anomaly so that a left-handed fermion in the fundamental representation of SU(12) produces an anomaly of one unit, then the total anomaly of the underlying theory is two since the ψ^a_{Li} are two $\underline{12}$'s of SU(12). This value must be matched by the anomaly of the bound-state massless composites.

The composite fermions consist of bound states of the fundamental fermion doublets and the scalar field doublet. For each fermion type, we can construct two different $SU(2)_L$ singlets with the scalar field. In fact, these two gauge singlets will be a doublet of the global SU(2). Explicitly,

$$\Psi^a_L = \begin{pmatrix} \phi^*_i \psi^a_{Li} \\ \varepsilon_{ij}\phi_i\psi^a_{Lj} \end{pmatrix} = \Omega^\dagger \psi^a_L\,, \tag{3.1}$$

where the Ψ^a_L are the fields of the physical fermions. Note that Ψ^a_L is a doublet under $SU(2)_{global}$ but a singlet under $SU(2)_L$ while ψ^a_{Li} is a global singlet and a doublet under the gauge group. Therefore, the bound-state fermions Ψ^a_L are two $\underline{12}$'s of SU(12) and the anomaly of the composite equals the anomaly in the underlying theory. Thus we have reason to believe that in the limit where $SU(12)\times SU(2)_{global}$ is an exact symmetry, the composites we have identified should be truly massless.

The colour and electromagnetic interactions violate the $SU(12)\times SU(2)_{global}$ symmetry. The $SU(3)\times U(1)$ group of these interactions is a subgroup of $SU(12)\times SU(2)_{global}$ as far as left-handed fields are concerned. Even in the presence of these interactions, there is enough chiral symmetry [before SU(3) dynamical chiral symmetry breaking] to guarantee the masslessness of the states we considered. However, when Yukawa couplings are included, this chiral symmetry is explicitly broken and the composite fermions become massive. Since the masses must vanish when the Yukawa couplings are turned off, the masses will be of order $\lambda G_F^{-1/2}$, where λ is the appropriate dimensionless Yukawa coupling. Typically, λ is much less than one and the physical fermions have masses much less than the scale $G_F^{-1/2}$.

4. Four-Fermi interactions

In our previous paper [2] we showed that the strong coupling $SU(2)_L$ theory can give rise to an effective four-Fermi interaction which describes low-energy weak interaction physics. We now want to show that this interaction is the only type of interaction consistent with the symmetries of the theory except for one other possible term. We will then discuss why this extra term could be small. In fact, the observation of this extra interaction at low energy could signal the existence of a confined theory of the weak interactions [5].

At energies well below $G_F^{-1/2}$, four-Fermi interactions between the bound state fermions are induced by the strong $SU(2)_L$ force. The scale of the $SU(2)_L$ theory is set by requiring that the coefficient of these four-Fermi interactions is G_F which is why we say that the $SU(2)_L$ theory becomes strong at an energy of order $G_F^{-1/2}$. Up to small corrections due to colour, electromagnetic and Yukawa couplings, these low-energy four -Fermi terms must respect the global $SU(12)\times SU(2)$ symmetry of the underlying $SU(2)_L$ theory. Furthermore, the interactions can only involve left-handed Fermi fields since only these feel the $SU(2)_L$ force. This condition, along with the $SU(12)\times SU(2)$ symmetry greatly restricts the possible four-Fermi terms which can appear.

The interactions involve the physical fermions Ψ_L^a which are a (12, 2) under $SU(12)\times SU(2)_{global}$. The four-Fermi interaction must be a singlet under SU(12) so it must contain two Ψ_L^a's and two $\bar{\Psi}_L^a$'s. Writing the interaction in the order $\bar{\Psi}\Psi\bar{\Psi}\Psi$, we see that the left-handed nature of the Fermi fields forces the interaction to be $(V-A)\times(V-A)$. $\bar{\Psi}\Psi$ decomposes into $(1,1)+(1,3)+(143,1)+(143,3)$ under $SU(12)\times SU(2)$. However, the last two representations can be Fierz transformed into the first two so the most general low-energy lagrangian is [5]

$$\frac{4G_F}{\sqrt{2}}[\boldsymbol{J}_{\mu L}\cdot\boldsymbol{J}_L^{\mu}+\xi J_{\mu L}^0 J_L^{0\mu}], \tag{4.1}$$

where

$$\boldsymbol{J}_{\mu L}=\sum_{a=1}^{12}\bar{\Psi}_L^a\gamma_\mu\boldsymbol{\tau}\Psi_L^a, \tag{4.2}$$

$$J_{\mu L}^0=\sum_{a=1}^{12}\bar{\Psi}_L^a\gamma_\mu\mathbb{1}\Psi_L^a. \tag{4.3}$$

The first term in eq. (4.1) gives the correct charged-current interaction and part of the neutral current, so the coefficient must be $4G_F/\sqrt{2}$. Note that the general SU(2) structure of these interactions was generated by the *global* SU(2) symmetry. The universality of the weak interactions is guaranteed by the global SU(12) symmetry.

On the other hand, the interactions of the second term of eq. (4.1) are not observed and it is known [5] from neutral current experiments that $|\xi| \lesssim 0.05$. Thus, we must discuss why ξ is so small in our model and also identify the complete neutral current.

The first term of eq. (4.1) can be produced by the exchange of a spin-one boson of mass $\approx G_F^{-1/2}$ transforming like a triplet under the SU(2) global symmetry group. This state can be made by binding together two scalars into an $SU(2)_L$ singlet state. The lowest lying scalar–scalar bound states [4] are thus identified with the W^+, W^- and W^0 and give the leading contribution to the first term of eq. (4.1). If we try to construct the scalar-scalar spin-one bound state which is the SU(2) global singlet partner of the $W^+W^-W^0$, we find that it is forbidden by Bose symmetry. Thus, as far as scalar–scalar bound state exchange is concerned, the first term in eq. (4.1) can arise, but the second one cannot.

However, we can also construct a spin-one bound state out of two fundamental fermions. Since the fermions are global SU(2) singlets, the bound state is also a singlet and can contribute to the second term of eq. (4.1), but not to the first. Therefore, we must assume that this state is somewhat decoupled. For example, if it is twice as massive as the W^0 and has one half the corresponding coupling to fermions, we will have $|\xi| < 0.07$.

It will take a more detailed understanding of dynamics to know which states really contribute. However, there is a possibility worth entertaining. If the $SU(2)_L$ interactions lead to a confining force which is spin independent, then it may be impossible to make a massive bound state out of massless fermions without chiral symmetry breakdown [6]. We have been assuming all along that $SU(2)_L$ does not break chiral symmetry, otherwise our entire analysis is spoiled. Therefore, it may be impossible to form the massive fermion–fermion bound state responsible for the second term of eq. (4.1). Of course this is only speculation.

As was shown in ref. [2], the complete neutral current interaction can be understood in this strong coupling model. Because the left-handed fermions are bound states, they will have non-trivial electromagnetic form factors at an energy scale of $G_F^{-1/2}$. These form factors produce the remaining terms in the neutral current interaction [5, 7] not present in eq. (4.1). This can be seen by assuming that, as in QCD, these form factors can be computed through vector–meson dominance [7]. We assume that the photon can turn into a W^0 boson with amplitude XQ^2. Then, if X is proportional to the usual $\sin^2\theta$, the neutral current interaction of this model agrees exactly with that of the standard model and thus with the experimental data [2, 5, 7]. If we put in the expected values of the couplings (see ref. [2]), we discover that X is about ten times bigger than the corresponding number describing $\gamma - \rho^0$ mixing in QCD. This may be understandable since the W^0 is two bosons not two fermions and the binding force is SU(2) not SU(3). Still the largeness of X may indicate that the assumption of vector meson dominance is not warranted in this model. With or without vector meson dominance, the form factors required to produce the observed neutral current have a scale only slightly less than $G_F^{-1/2}$ and since they arise from binding at $G_F^{-1/2}$ this seems perfectly consistent.

5. Baryon-number violation through instantons

Baryon number is not conserved in the $SU(2)_L \times U(1)$ model because of $SU(2)_L$ instantons [8]. This is true both for strong and weak couplings. The calculation of the amplitude for a baryon-number violating process yields a factor of $\exp(-8\pi^2/g_2^2)$ which, for the standard weak coupling theory ($g_2 \approx 0.6$), is a tremendous suppression factor. Baryon-number violation occurs, but at an unobservable rate. However, for the strong coupling theory, g_2 gets large and the suppression factor is not operative*. Baryon-number violation may occur at an appreciable rate. We will now estimate this rate and find that it is still orders of magnitude smaller than the experimental bound.

At the scale $G_F^{-1/2}$ where $SU(2)_L$ becomes strong, we can ignore colour and electromagnetism. We will assume that there are three generations of quarks and leptons, so at the scale $G_F^{-1/2}$ there are twelve indistinguishable $SU(2)_L$ doublets $\psi^a (a = 1–12)$ as in sect. 2. The conservation of fermion number for a fermion doublet of type a, N_a, is violated by $SU(2)_L$ instantons, but the difference of any two, $N_a - N_b$, is always conserved. The lowest dimension operator which can mediate baryon-number violation, conserves the various $N_a - N_b$ and is SU(12) symmetric is [8]

$$O = \varepsilon_{a_1 a_2 \cdots a_{12}} \psi^{a_1} \psi^{a_2} \cdots \psi^{a_{12}} , \tag{5.1}$$

where the SU(2) indices are tied up in all possible ways. The amplitudes for baryon-number violating processes are given by the appropriate matrix elements of an effective hamiltonian involving O. The only relevant dimensional parameter here is $G_F^{-1/2}$ so on dimensional grounds this effective hamiltonian must go like

$$\mathcal{H} \approx G_F^7 O . \tag{5.2}$$

The effective hamiltonian of eq. (5.2) mediates processes with $\Delta B = 3$. One such process is nucleon + nucleon → antinucleon + 3 antileptons. For this process, all the mass scales associated with the matrix element of (5.2) and the integrations over final-state momenta are of order m_{proton} or less. Of the 12 fields appearing in O, only three quark fields belong to the first generation. Three more quark fields belong to the second generation so for each of them there is a suppression of $\sin\theta_C$ for a matrix element involving ordinary nucleons. The last three quark fields belong to the third generation and so are suppressed by Kobayashi–Maskawa angle factors. We will assume that each of them is doubly Cabibbo suppressed by a factor $\sin^2\theta_C$. Putting

* This was suggested to us by R. Barbieri.

all this together, we find that the rate for nucleon + nucleon → antinucleon should be about

$$R \simeq (\sin\theta_C)^{18} G_F^{14} m_{\text{proton}}^{29} \simeq (2\times 10^{50}\ \text{years})^{-1}\,. \tag{5.3}$$

The factor of $(m_{\text{proton}})^{29}$ is an upper limit since nuclear physics effects bring factors even smaller than m_{proton}. Still this rate is many orders of magnitude smaller than present limits on the proton lifetime and is unobservable.

A process in which two nucleons go into a state of three strange quarks like the Ω will be somewhat larger than (5.3) because it involves fewer Cabibbo suppression factors. The rate for this process is roughly

$$R \simeq (\sin\theta_C)^{12} G_F^{14} m_{\text{proton}}^{29} \simeq (2\times 10^{46}\ \text{years})^{-1}\,, \tag{5.4}$$

which is still unobservable. However, there is considerable uncertainty in the estimation of these rates. For example, if the factor G_F in (5.4) is replaced by $(1/100\ \text{GeV})^2$ we find a lifetime of about 2×10^{33} years. This is not too far from present bounds on the proton lifetime. Thus, someday, baryon violation from $SU(2)_L$ interactions, if they are strong, may be observable.

6. High-energy phenomenology

The high-energy properties of the $SU(2)_L$ confining theory of the weak interactions are very different from those of the standard model. At energies of the order of $G_F^{-1/2}$ the strong nature of the weak interactions should clearly reveal itself. In this section we discuss some of the distinctive features of this high-energy phenomenology.

Although the conventional $W^\pm$ and Z^0 will not be seen if the confining version of the weak interactions is correct, bound-state vector bosons with similar properties will appear. If we assume that the low-energy weak interactions are dominated by the exchange of the lowest lying bound-state vector bosons, then we find an upper limit for the mass of the bound state $W^\pm$ which was first derived by Bjorken [5] and by Hung and Sakurai [7]. This bound requires that the $W^\pm$ in the confining theory have a mass less than 166 GeV. On the other hand, we would expect [2] the bound-state $W^\pm$ to be more massive than the $W^\pm$ of the standard model. Thus, the $W^\pm$ mass should be somewhere between 100 and 166 GeV. The W^0 mass should be somewhat heavier [7] than the $W^\pm$ mass as in the standard model. The $W^\pm$ and W^0 are quite analogous to the $\rho^\pm$ and ρ^0 of ordinary hadronic physics. The width of the ρ is about 20% of its mass. By analogy then, we expect the $W^\pm$ and W^0 to be broad resonances around 10–30 GeV in width.

The ordinary $W^\pm$ and Z preferentially decay into single fermion pairs. The $W^\pm$ and W^0 of the strong coupling theory will also have these channels open, but in addition they will strongly decay into many fermion final states. At the scale of the mass of the heavy bosons, quarks and leptons of all generations are essentially equivalent so we expect equal numbers of leptons and quarks of each colour divided uniformly between the generations in these final states. The large numbers of leptons produced per event is a particularly good signal [9] of the strong nature of the weak interactions.

The new strong interaction sector will, of course, have a rich spectrum like the hadronic spectrum of QCD. However, since the mass scale of this strong interaction is of order $G_F^{-1/2}$, most of the excited states will lie about 300 GeV above the $W^{\pm}$ and W^0 and so will not be easily detectable. However, in addition to the $W^{\pm}$ and W^0, there should exist the first excited states of the ordinary fermions, the Higgs-like scalar bound state and perhaps a ω like bound state in the region 100–200 GeV. Like the W's, these states will be broad and can decay into numerous particles and in particular produce large numbers of leptons.

We have based many of our arguments about the strongly coupled theory responsible for the weak interactions on analogous properties of QCD and hadronic physics. However, the strong interactions we are describing differ in two important respects. One is that they are purely left-handed and thus we expect parity violation in strong interaction processes [10]. The other is that there is no chiral symmetry breaking in these strong interactions. The effects of this on our analogies with QCD are not known.

7. Conclusions

In this paper we have continued our analysis of a model of the weak interactions without spontaneous breakdown. We have strengthened our arguments for the existence of light bound-state fermions. Using the global symmetries of the theory we have shown how to derive the correct form for the low-energy weak interaction lagrangian including universality. We also discussed other possible interactions which can arise at low energy, at high energy and through instanton effects.

There are two difficulties remaining. The first is to understand why the electric form factors required to give the neutral current are much larger than the magnetic form factors. The second is to better understand the suppression of the isosinglet channel in the neutral current.

Our theory has many phenomenological successes. It suggests the possibility of constructing other models without fundamental scalars based on confining gauge theories. Proton–antiproton experiments should soon determine whether or not the weak interactions are strong.

Notes added

(i) In Sect. 3 we showed that the light composite fermions satisfy 't Hooft's anomaly conditions. In addition it should be noted that 't Hooft's decoupling condition [3] is also satisfied.

(ii) Strangeness changing neutral currents in our model are of the same order of magnitude as in the standard model. Thus, we have no problems with $K^\circ - K^\circ$ mixing, for example.

(iii) In sect. 5 we discussed the effect of strong $SU(2)_L$ instantons. Associated with these instantons is an $SU(2)_L$ θ parameter. However, by redefining the phases of fermion fields, the θ parameter can be set equal to zero and the Yukawa couplings can be made real. Thus no *CP* violation occurs from $SU(2)_L$ instantons. Although the SU(3) θ problem remains as in the standard model, no new θ problem is introduced in our model.

We are grateful to many of our colleagues at CERN for their helpful comments, in particular A. De Rújula, J. Ellis and R. Barbieri. We also thank A. Schwimmer for useful discussions. Some of the ideas presented here were analyzed independently by T. Banks, Y. Frishman, E. Rabinovici and S. Yankielowicz and we have benefited from discussions with them. This work was completed at the Weizmann Institute. We thank H. Harari and the members of the Theory Group for their hospitality.

References

[1] S. Weinberg, Phys. Rev. Lett. 19 (1967) 1264;
A. Salam, *in* Elementary particle theory: relativistic groups and analyticity, Nobel Symp. no. 8, ed. N. Svartholm (Almqvist and Wiksell, Stockholm, 1968)

[2] L. Abbott and E. Farhi, Phys. Lett. 101B (1981) 69

[3] G. 't Hooft, Cargèse Summer Inst. Lectures (1979) (Plenum Press, New York, 1980).

[4] T. Banks and E. Rabinovici, Nucl. Phys. B160 (1979) 349;
E. Fradkin and S.H. Shenker, Phys. Rev. D19 (1979) 3682;
S. Dimopoulos, S. Raby and L. Susskind, Nucl. Phys. B173 (1980) 208

[5] J.D. Bjorken, Arctic Summer School, Äkäslompolo, Finland (1980); Phys. Rev. D19 (1979) 335

[6] A. Casher, Phys. Lett. 83B (1979) 141

[7] P.Q. Hung and J.J. Sakurai, Nucl. Phys. B143 (1978) 81;
N. Dombey, Neutrino 79, vol. II. ed. A. Haatuft and C. Jarlskog, Bergen (1979)

[8] G. 't Hooft, Phys. Rev. Lett. 37 (1976) 8

[9] A. De Rújula, Phys. Lett. 96B (1980) 279

[10] I.H. Dunbar and N. Dombey, University of Sussex preprint (1981)